Tragedy and Triumph in Orbit
The Eighties and Early Nineties

A History of Human Space Exploration

Other Springer-Praxis books by Ben Evans in this Series

Escaping the Bonds of Earth – The Fifties and the Sixties
2009
ISBN: 978-0-387-79093-0

Foothold in the Heavens – The Seventies
2010
ISBN: 978-1-4419-6341-3

At Home in Space – The Late Seventies into the Eighties
2011
ISBN: 978-1-4419-8809-6

Ben Evans

Tragedy and Triumph in Orbit

The Eighties and Early Nineties

 Springer

Published in association with
Praxis Publishing
Chichester, UK

Ben Evans
Space Writer
Atherstone
Warwickshire
UK

SPRINGER–PRAXIS BOOKS IN SPACE EXPLORATION

ISBN 978-1-4614-3429-0   ISBN 978-1-4614-3430-6 (eBook)
DOI 10.1007/978-1-4614-3430-6
Springer New York Heidelberg Dordrecht London

Library of Congress Control Number: 2012931283

Cover design: Jim Wilkie
Project copy editor: David M. Harland
Typesetting: BookEns, Royston, Herts., UK

Printed on acid-free paper

Springer is part of Springer Science+Business Media (www.springer.com)

Contents

Illustrations

Chapter 5

Author's preface

When I set out to write a five-volume series to commemorate the first 50 years of our adventure in space, it seemed a big project, though relatively straightforward and something that I have always wanted to do. An obsessive space enthusiast for as long as I can remember, I received my first space book at the age of five, was given a toy Space Shuttle for a birthday present soon afterwards and by the time I reached my seventh birthday I had watched in childish astonishment as Enterprise – mounted atop a Boeing 747 carrier aircraft – hurtled over my primary school in Birmingham, during sports day. It caused me to drop the egg from my spoon, unfortunately, but the sight was so spectacular and awe-inspiring that it hooked me for life. The Moon landings excited me – and still do – beyond compare and I began writing articles at the age of 15 for the British Interplanetary Society's *Spaceflight* magazine and, later, for *Countdown* and *Astronomy Now*. As I grew older, it became a goal of mine to someday write a 'meaty' history of the human exploration of space, which I continue to believe firmly is one of the greatest adventures ever undertaken by our species, but everything I read seemed to 'lack' something. Some were overloaded with facts and figures, others were devoted solely to the 'popular' audience, while still more simply lacked the detail and human interest factor. I cannot promise the reader that my series fulfils each of these gaps, but what I *can* say with certainty is that I have spent an enjoyable three years researching the history of our adventure, told through news sources, books, the memoirs of those involved, magazines, press kits and oral histories, and have learned an immense amount. The reader may love or hate my book – they may find it hard to put down or may simply find use for it as an expensive additional castor for their sofa – but I have derived great joy from researching and writing it.

It has been impossible to track entire decades within the pages of each volume. The first, *Escaping the Bonds of Earth*, had to take into account some of the achievements of the 1950s, as a prerequisite to focusing on 'its' decade, the 1960s. In a similar vein, the second volume, *Foothold in the Heavens*, needed the focus to fall in considerable depth on some of the most remarkable achievements of the Space Age – Apollo 11 being the obvious example – at the expense of covering an entire decade. Only with this fourth volume did it become clear that I was punching above my weight. My determination to cover each mission with the level of detail that it deserved, including biographies of each spacefarer, turned this book into something much longer; so long, in fact, that I had barely covered the 1980s and the extent was already rapidly approaching 600 pages. As a result, with the approval of Clive Horwood, the series has expanded from five into *six* volumes, to cover the first half-century in its entirety, whilst maintaining the kind of depth that the reader would expect. That depth has been easier to fulfil in some areas than others. The Soviets, even in the late 1980s, with the advent of *glasnost* (openness) and *perestroika* (restructuring), proved notoriously secretive about their activities and in several cases it has been impossible to offer biographies of a handful of their cosmonauts in more than just a few words or paragraphs. That secrecy extended to the West, too, when the Americans staged a number of military Shuttle missions, many of whose precise objectives remain classified to this day. I have attempted, using the sources and contacts at my disposal, to shine a meagre light on these shadowy flights and I fervently hope that a few years from now they will emerge from the shadows to take their rightful place in history.

I have learned much about the human space programmes of the United States and Russia and, equally importantly, have learned a great deal about the political events which shaped their progress. Starting with Yuri Gagarin's pioneering voyage in April 1961, the journey has carried me through a succession of dramatic decades, punctuated by conflict and reconciliation, meddling and political manoeuvring, and has seen the first men land on the Moon, the first men occupy an Earth-circling space station, the first men pilot a reusable vehicle beyond the atmosphere, the first men from other nations – Czechoslovakia, Poland, Vietnam and Mongolia, to name but a few – and, of course, the first members of *womankind* to carry their dreams and aspirations into the heavens. My intention has always been for something a little more than a basic log of crewed expeditions into space, but as time has rolled on, the project evolved into something much larger and more complex than I had envisaged. The *human* side of the story has always been profoundly important to me. Take Sergei Korolev: a man who endured so much physical and psychological trauma in his youth – dreaming, even as he was transported to a living death in the Kolyma gulag, of one day sending a rocket into space. Deke Slayton is another example: a man chosen in his prime as one of America's finest, the 'Mercury Seven', whose hopes were cruelly dashed by a devastating heart condition, yet who sprang back as the man who guided the astronaut corps to the Moon, chose Neil Armstrong to take the first historic steps on its dusty surface, and eventually overcame every obstacle in his path to fly into orbit. Their remarkable stories and their individual trials and tribulations, from childhood to the grave, carry just as much weight and drama and excitement and adventure as the missions they flew. Yuri Gagarin pissing on the wheel of the bus as he prepared to board his space capsule, Wally Schirra conceiving another legendary 'gotcha', Alexei Leonov and his love of cowboy hats and boots, John Glenn and his competitive yearning to be first, Alexander Serebrov and his penchant for fast cars, Christa McAuliffe and her passion to carry education to the final frontier and Shannon Lucid and Svetlana Savitskaya, who both refused to accept that their gender should tie them to Earthbound pursuits and who both strove for the stars. The story of our adventure in space is not simply about who spent the most time there, who performed the most spacewalks or who flew the most missions. It is a collection of *stories*: the stories of how a few hundred remarkable people, all of whom achieved an uncommon goal, forever changed our perspective on the world in which we live and fixed our eyes and our minds and our imagination on the Universe around us.

Ben Evans
Atherstone, December 2011

Acknowledgements

This book would not have been possible without the support of a number of individuals, to whom I am enormously indebted. I must firstly thank my wife, Michelle, for her constant love, support and encouragement throughout the time it has taken to plan, research and write this manuscript. As always, she has been uncomplaining during the weekends and holidays when I sat up late, typing on the laptop, or poring through piles of books, old newspaper cuttings, magazines, interview transcripts, press kits or websites. It is to her, with all my love, that I would like to dedicate this book. My thanks also go to Clive Horwood of Praxis for his enthusiastic support and to David M. Harland for reviewing the manuscript and offering a wealth of advice and guidance; I deeply appreciate not only their support, but also their patience in what has been an overdue project and one which has proven more difficult to write than I had imagined. Additional thanks go to Ed Hengeveld, who has been enormously gracious with his time in identifying suitable illustrations for this book, including many 'unfamiliar' ones which surely bolster the text. Others to whom I owe a debt of gratitude include Sandie Dearn and Malcolm and Helen Chawner. To those friends who have encouraged my fascination with all things 'space' over the years, many thanks: to Andy Salmon and Andy Rowlands and to Dave Evetts and Mike Bryce and to Rob and Jill Wood. Our two golden retrievers – the ever-hungry Rosie and the attention-seeking Milly – have provided a ready source of light relief and a regular opportunity for me to leave the laptop and either play with them or give them a biscuit.

1

"We deliver"

IMPEDANCE MATCHING

Joseph Percival Allen IV spent the eight and a half minutes of his first climb into space with little to see, save a locker-studded bulkhead. He was alone, encased in a lightweight blue flight suit, steel-toed boots and helmet; quite unlike the bulky ensembles worn by earlier astronauts. In fact, even the four Space Shuttle crews who preceded Allen had flown into orbit in fully-pressurised garments. Now, on this fifth flight of the reusable spacecraft, Allen and his three crewmates – Vance Brand, Bob Overmyer and Bill Lenoir – were dispensing with two decades of tradition: for the Space Transportation System (STS) was now fully operational and on the threshold of fulfilling a decade-old promise to open a new era of routine and regular access to space for commercial and scientific gain. It was hoped that a fleet of four orbiters – Columbia, Challenger, Discovery and Atlantis – would fly several dozen times per year, although the first 18 months of Shuttle test flights had already shown signs that these vehicles were more complex and difficult to turn around than had been anticipated. Still, during their five-day mission, the crew of STS-5 expected to take several key steps toward making the dream a reality. Allen and Lenoir would make a spacewalk and the crew would supervise the deployment of two multi-million-dollar communications satellites into orbit.

As Allen sat in the darkened interior of Columbia's middeck, perhaps he had chance to reflect on his own background or those of his crewmates and how they had been brought together to be launched into space on this morning of 11 November 1982. It had been a long journey for all of them, none more so than Allen and Lenoir. The pair had been chosen as scientist-astronauts by the National Aeronautics and Space Administration (NASA) in August 1967 and although both played important roles on the ground, they had waited for 15 years to actually break the shackles of Earth and experience the novelty of weightlessness and space travel. Over the intercom, Allen could hear the voice of Vance Brand, calling out the milestones and acknowledging calls from Mission Control as the Shuttle roared, rocked and rattled its way into space, under the combined thrust of its three main engines and twin Solid Rocket Boosters (SRBs). Allen instinctively knew each and every milestone, his knowledge and skills sharpened by eight months of rigorous training. Since March, they had spent so much time together that they had become almost like family. Yet they were far from alike. When George Abbey, then-head of the Flight Operations Directorate at NASA's Johnson Space Center (JSC) in Houston, Texas, called them to his office to formally assign them to STS-5, Allen joked that they were a true example of "impedance matching". It was an engineering term for the practice of allowing dissimilar electrical circuits to operate in harmony. Abbey had not found the joke in the least amusing, but to Allen it summed up the crew perfectly. Brand and Overmyer were both test pilots, with operational military backgrounds in the US Marine Corps, whilst Lenoir was a civilian electrical engineer and Allen a civilian physicist.

In many ways, the makeup of this crew marked a fundamental shift from the stereotypical image of 'real' astronauts which had been prevalent in the Mercury, Gemini and Apollo days. No longer was space exploration the exclusive domain of the hardened test pilot – although they continued to play an important role – and NASA had long since begun to recognise that scientists, physicians, engineers and ultimately *ordinary civilians* could be considered for seats on the Shuttle. Even Tom Stafford, a former chief astronaut, who flew four times, including a voyage to the Moon and command of the Apollo-Soyuz docking mission with the Soviet Union, admitted that launching into space did not require supermen ... or even *men*, in fact, for a group of six *women* had been picked as astronauts in January 1978. Still, the risks of sitting atop thousands of kilograms of rocket propellants were self-evident and *no* spacefarer, whether military or civilian, was left in any doubt that the Shuttle had the means and the potential to kill them all.

Ironically, as a youth, Allen was not particularly interested in becoming an astronaut. In fact, he only submitted his application, whilst completing a doctorate in physics at Yale University, because he was keen for adventure and eager to learn about NASA's selection process. In September 1966, the space agency announced its intention to pick a team of scientists to support ambitious plans for up to ten manned lunar landings and a series of research missions, known as 'Apollo Applications', which would eventually morph into the Skylab space station. "We announced that we were accepting applications," wrote Deke Slayton, head of Flight Crew Operations from 1963–72, "then asked the Academy of Sciences to do a rating of the applicants based on their scientific credentials." Nine hundred applications arrived at the National Academy of Sciences by the beginning of January 1967 and after two months of evaluations 69 names were recommended to NASA. This number was eventually whittled down to just 11 men, whose names were announced in August. Allen and Lenoir were among them. It had not been an easy process, with several interviews, followed by a week of highly intrusive physical tests at the School of Aerospace Medicine at Brooks Air Force Base in San Antonio, Texas, and eventually the call from Al Shepard, chief of the astronaut office, to inform them of their selection. (Allen intensely disliked the Brooks tests. "*Anyone* who would knowingly subject themselves to *that* kind of a physical examination," he told

NASA's oral historian, "should automatically be disqualified for *mental reasons!*") The good news of their admission into the world's most elite flying fraternity was quickly tempered with a healthy dose of reality: the war in Vietnam was consuming increasing numbers of lives, as well as federal funds, and the space programme was no longer ranked so highly on the list of national priorities. In July 1967, NASA Administrator Jim Webb had been told by the Senate Committee on Appropriations that the following year's authorisation bill would be severely curtailed and behind closed doors it was becoming obvious that there would *not* been ten manned lunar landings and many of the Apollo Applications flights were at similar risk. Deke Slayton tried to cut the number of scientists down to five, but the selection process was already complete and in September he was forced to break the news that there would probably be no space missions for them to fly. There was another complication, too, for in order to *become* qualified astronauts, they needed to become qualified jet pilots. The demands of Vietnam meant that there were simply no available places at Air Force schools to begin this training until at least March 1968. In his typically straightforward manner, Slayton assured them that if they wished to resign, it would not be held against them, and several of the scientists took his advice. Those who remained could hardly have imagined that they would wait almost two decades before entering space. Surplus to requirements, they wryly nicknamed themselves 'The Excess Eleven'.

Amongst the Eleven, Allen would quickly gain the moniker of 'Little Joe', owing to his 1.68 m stature – only five feet and six inches – although he was also an accomplished wrestler and in later years would be invested into wrestling halls of fame in both Oklahoma and his native Indiana. "My dream all through high school," he told Dave Shayler and Colin Burgess in *NASA's Scientist-Astronauts*, "was to become a state champion, a goal I nearly accomplished in all four years of the state tournament." Allen came from the small farming and industrial town of Crawfordsville in Indiana, where he was born on 27 June 1937. With parents and a paternal grandfather who had all graduated from DePauw University, it is perhaps no great surprise that Allen followed in their footsteps, achieving a degree in mathematics and physics in 1959. (Interestingly, his younger brother, David, also attended DePauw and later became a physician.) He subsequently travelled to West Germany on a Fulbright scholarship and then returned to the United States to pursue his master's degree and PhD at Yale. His doctoral thesis, which Allen defended in 1965, focused on the bombardment of oxygen nuclei with proton, deuteron and helium beams and his work improved understanding of theoretical nuclear models. He took a post as a research physicist at Yale's Nuclear Structure Laboratory, followed by a post-doctoral fellowship at the University of Washington in Seattle, and it was during this period that he read about NASA's call for scientist-astronauts.

After selection in the summer of 1967, Allen's group underwent several months of introductory classes and during the course of the following year they were despatched to various Air Force bases across the United States for 53 weeks of instruction in high-performance aircraft. Allen went to Vance Air Force Base in Enid, Oklahoma, and shone there, scoring first place in all five measures: academics, general flying, instrument flying, aerobatics and formation flying. Years later, he would credit his age (at 30, he was several years older than most of the military pilots with whom he trained) and his advanced education as having given him an edge, "and flying, to a degree, is a mind-discipline game, more than it is a physical athlete's game". Allen returned to Houston with an Outstanding Flying Award and received a note of congratulation from Al Shepard. Compliments of *any* kind from the abrasive chief were few and far between, especially for a scientist, and Allen kept the note for posterity.

"The Air Force has a very practical and effective way of training pilots," Allen explained in Deke Slayton's autobiography, referred to as 'the cookbook method'. "It runs you through a basic set of steps, each one building on the next, and is so efficient that it can take just about anybody, from a jock with an aeronautical engineering degree to an English major from Yale, and turn him or her into a pilot, as long as you follow the cookbook. This method drills you in following certain operational rules as if they were the word of God. In the T-38 jets, which were what I flew at NASA, there was a strict airspeed rule: *don't* let the plane go slower than 270 miles an hour, or it'll fall out of the sky. This rule is relatively easy to follow most of the time you're flying: a jet like the T-38 just sort of *overpowers nature*. Hit the gas and you *go*. Wind and rain *aren't* going to slow you down." The trick when it came to land, he added, was that he needed to bleed off airspeed during the final approach to the runway.

None of the Excess Eleven were under any illusion that their performance would bring them significantly closer to a seat on a space mission, but Allen's activities in the following years proved exciting. One of his earliest roles was as a 'capsule communicator' (or 'capcom'), the astronaut in Mission Control with responsibility for talking directly to a crew in space, and by 1970 he was supporting the Apollo 15 lunar landing mission. A year later, he served as one of the mission's capcoms and communicated directly with astronauts Dave Scott and Jim Irwin as they worked on the Moon's dusty surface. He served in the same capacity on the final landing mission, Apollo 17, in December 1972. Allen found the assignment somewhat ironic. He came from a frugal household in Indiana and had *never* even made a transatlantic telephone call, due to the expense … and *now*, his first long-distance call was to the *Moon*! His work on Apollo brought mixed blessings: participation in the most exciting adventure ever undertaken in human history left him, in the spring of 1973, as very much "an astronaut without portfolio", with no mission assignment on the horizon. For a time, he worked on the Outlook for Space study group, headed by Don Hearth, the deputy director of NASA's Langley Research Center, which charted possible roadmaps for America's future in space. One of its recommendations was for a network of tracking and data relay satellites to be placed into orbit by the Shuttle to permit near-continuous voice and data traffic and eliminate the older generation of tracking ships and ground stations. Then, in August 1975, Allen was approached by NASA Administrator Jim Fletcher to become the agency's Associate Administrator for Legislative Affairs in Washington, DC. Allen pondered the assignment before accepting it. He knew of previous astronauts who had disappeared into management and never returned to active duty and, although assured that he *could* return to the astronaut office at the end of his Washington tenure, it was with a heavy heart that he left Houston.

During his three years in the capital, Allen worked on funding issues surrounding a number of major NASA projects, including the Viking voyages to Mars and the genesis of what later became the Galileo Jupiter mission and the Hubble Space Telescope. He also tackled several thorny problems relating to the atmospheric approach and landing trials of the Shuttle. His dealings with other senators and congressmen were equally productive, even in the case of the notoriously anti-space William Proxmire of Wisconsin. However, as 1978 dawned and NASA picked 35 new astronauts for the Shuttle, it became clear to Allen that his return to Houston could not come quickly enough. He asked to return a day or so before the candidates commenced their training; he certainly did not wish to be considered in any way 'junior' to the new group and be pushed even further down the line when flight opportunities began to open up.

According to JSC news releases from the time, Allen actually returned to Houston a couple of weeks *after* the new group, for their formal training started on 11 July 1978 and the announcement of his return did not come until the 26th. Allen need not have worried about being considered junior to the newcomers, for in short order he was assigned to help to develop Shuttle flight techniques and later served as a support crew member for STS-1 – the maiden voyage of Columbia, which eventually took place in April 1981 – during which time he was one of the capcoms in Mission Control. Within a year, he found himself in Abbey's office, along with Brand, Overmyer and Lenoir, to hear that his long spell as an astronaut-in-waiting was at an end. They would form the crew of STS-5.

Allen and Lenoir would serve as 'mission specialists', a new subset of a Shuttle crew, whose responsibilities differed markedly from those of the pilots and encompassed the deployment of payloads, the execution of spacewalks – properly 'Extravehicular Activity' or 'EVA' – and the operation of experiments. Scientists had previously flown aboard Skylab and although their responsibilities focused on research, they cross-trained with the pilots on many tasks, such that the distinction became blurred. In November 1974, at NASA's request, the Space Program Advisory Council undertook a study to define the tasks of scientists on the Shuttle and its report saw them as providing a 'bridge' between the principal investigators of experiments and research facilities and the astronaut office. More significantly, with plans for a European-built laboratory, called 'Spacelab', to be carried in the Shuttle's payload bay, the scientists would play a key role in mission planning and execution from a research standpoint. According to Louis Haughney of the Airborne Science Office at NASA's Ames Research Center, they would serve as 'mission managers'.

By the middle of 1976, the tasks of the mission specialist had steadily crystallised and they would need to be "professionally knowledgeable in the prime discipline of the mission to which [they were] assigned". Years later, Allen felt that the selection of himself and Lenoir as the first pair of mission specialists to join two Marine Corps pilots underlined their ability to speak each other's language. Allen's scientific credentials were as impressive as his flying skills and he regarded William Benjamin Lenoir as "an extremely smart, very well-disciplined, very tightly-wound individual". In fact, when he was chosen as a member of the Excess Eleven, Lenoir became Florida's first 'native' astronaut, having entered the world in Miami on 14 March

1939. From his earliest experiences in high school, he was drawn to engineering and the sciences and it was the guidance of his mathematics teacher, Mr Newmeyer, which "helped me to navigate the MIT admission and scholarship application process" and secure a place at the famed Massachusetts Institute of Technology as a Sloan Scholar in 1957. He studied electrical engineering, received his degree in 1961, a master's credential the following year and eventually his PhD in 1965. Lenoir's doctoral thesis explored the remote sounding of the upper atmosphere with microwave instruments and he led a team conducting experimental research using unmanned high-altitude balloons.

Lenoir remained with MIT as a post-doctoral fellow and assistant professor of electrical engineering and it was during this period that he participated in the design and definition of several remote sensing instruments for future Apollo Applications missions. From 1965 until his selection as an astronaut, he worked with MIT's Research Laboratory of Electronics, helping to design a 60-gigahertz microwave receiver to sense the profile of the atmosphere at different altitudes; many of these experiments eventually found applications in NASA's Nimbus series of global meteorological satellites. By the end of 1966, Lenoir had spent ten years in Massachusetts and was ready for a new challenge ... when he noticed NASA's call for scientist-astronauts in the magazine *Science* and submitted the small clipping with his name and address. If Joe Allen remembered the invasive physical tests at Brooks Air Force Base with the most intense lucidity, then for Lenoir it was the form-filling. "Like magic," he said, "this small clipping became a *one-inch-thick* stack of paperwork to fill out and return." Yet Lenoir did not profess any genuine desire to fly in space. "People are disappointed to hear I didn't grow up wanting to be an astronaut," he told the oral historian. "I had given it no thought." During the Brooks exams, something to which he *did* have to devote some thought was an old hockey injury. On one occasion, a physician approached him with a rather curious question.

"Were you unconscious when you broke your face?"

"I never broke my face," the puzzled Lenoir replied.

The bemused doctor proceeded to explain that there were three spots on the left side of his face, which presented evidence of breaks.

"Oh yeah," Lenoir recalled, "I was playing intramural hockey as a graduate student and I took a check and I hit the ice real hard." He was suffering from concussion and was told to go home and sleep it off. The incident marked Lenoir out as being lucky in two ways. Firstly, the Brooks physician explained, such facial injuries normally do not heal particularly well, and secondly, had Lenoir mentioned that he was even *momentarily* unconscious, or blacked-out, he would have been instantly eliminated from consideration as an astronaut candidate. He sailed through the other tests – describing X-rays, kidney tests and a rapid-decompression run in the altitude chamber as "fun" – until a minor hernia was found. Lenoir contacted a medical friend, who arranged for him to quietly visit Massachusetts General Hospital. Five days later, just before his formal interview, Lenoir took his first flight in one of NASA's T-38 Talon jets. At the interview, later that day, Al Shepard asked him about the hernia.

"I had it fixed," Lenoir said.

"When?"

"Five days ago."

"Oh, wow. Maybe you shouldn't do the T-38 ride, then."

"Too late. I did it *this morning!*"

"Oh. You should have told us."

Years later, Lenoir realised that, subconsciously, he was already playing along with the mentality of a fighter pilot, namely to resolve medical problems discreetly and *never* admit weakness. After selection, he was assigned to Laughlin Air Force Base in Del Rio, Texas, for flight instruction and quickly realised that he had a natural affinity for aviation. At the end of his training, he received the best flying award, the best academic award *and* the Commander's Cup. Returning to Houston in the summer of 1969, Lenoir's experience with Apollo Applications experiments led to an assignment to Skylab, in which he followed the development of the station's airlock and its multiple docking adaptor. It was with little surprise, therefore, that in January 1972 Lenoir was named as the backup science pilot for both the second *and* *third* Skylab missions. Together with Story Musgrave, the backup science pilot for the first mission, this made Lenoir one of the first of his class to receive an actual flight assignment ... albeit only in a backup capacity. Yet Lenoir, more so than Musgrave, came closest to drawing a prime crew assignment. As late as the spring of 1973, NASA was open to the option of flying a fourth Skylab mission, lasting three weeks, to finish up scientific research on the station and mothball it for possible future use. Had this mission transpired, Lenoir and fellow astronauts Vance Brand and Don Lind would have been aboard. This would have secured him the title of becoming the first of his class to fly into space. As circumstances transpired, on STS-5, jointly with Joe Allen, he achieved it.

In the wake of Skylab, from September 1974 until July 1976, Lenoir led NASA's Satellite Power Team, charged with investigating the potential of large-scale satellite power systems for Earth. The team worked with several major corporations, including Boeing and Raytheon, and Lenoir was called to testify before the House and the Senate, since NASA hoped to press the programme forward as a major initiative. Subsequent work included following the development of two major Shuttle components: the European Spacelab and a Canadian-built robotic arm, known as the Remote Manipulator System (RMS). When the new group of astronauts arrived in Houston in the summer of 1978, Lenoir supervised three of them – John Fabian, Sally Ride and Norm Thagard – on RMS systems ... and *all three* wound up flying together, on the same mission, becoming the first of their class to fly. "I always felt pretty good," Lenoir said. "It can't be *too* bad to have worked for me!"

It was not too bad for Lenoir, either, when he received word in March 1982 that he would be one of the mission specialists on STS-5. Yet, despite the outward distinction between mission specialists and pilots, the roles did blur in places and a certain amount of cross-training was required. During launch and ascent, Lenoir would serve as the 'flight engineer', his seat positioned between and behind Brand and Overmyer on the Shuttle's flight deck, and he would be charged with supporting them with checklists and helping them to monitor the systems. "Clearly, two people can get a Shuttle into orbit and back," Lenoir explained, "because they had proven it and they had done it four times. On the other hand, if you've got *another* set of eyes and ears and a brain, you'd be stupid not to use it." Lenoir and Allen would then swap places for descent, with Allen serving as the flight engineer during re-entry and landing. This was somewhat unusual, because STS-6 onwards would feature one astronaut, specifically trained as the flight engineer, for both ascent *and* re-entry. Moreover, on STS-6, which would be the maiden voyage of the second orbiter, Challenger, *four* crew members could be accommodated on the flight deck. For STS-5, it was only possible for three astronauts to sit on Columbia's flight deck ... and the reason for *that* had much to do with a tricky moral problem which had presented Vance Brand with something of a dilemma.

When Columbia flew the first four Shuttle test missions, she carried two pilots, each clad in a bulky high-altitude pressure suit, which afforded them full-body protection, and supported by a pair of ejection seats. If an emergency occurred during certain phases of ascent or descent, the rocket-propelled seats had the capability to fire the astronauts through overhead hatches ... *but* they could only be used up to an altitude of around 30 km, meaning that they would not realistically work during most of the launch phase and could only perform at selected intervals during re-entry. An on-the-pad emergency would have precluded the use of the seats, because the astronauts would have hit the ground before their parachutes opened, and ejections during the first couple of minutes of launch would have hurled them directly into the plumes of the Solid Rocket Boosters.

The ejection seats and their associated hardware were huge and only left enough room at the back of Columbia's tiny flight deck for one additional seat. Challenger and the two subsequent orbiters under construction, Discovery and Atlantis, had been built with more lightweight seats for the commander and pilot, allowing for the inclusion of *two* additional seats at the rear of their flight decks. Columbia's own ejection seats would ultimately be removed and replaced with similar lightweight seats, but the issue raised other concerns. With the expansion of Shuttle crews to four and, ultimately, plans to carry as many as *seven* astronauts on each mission (with three additional seats in the downstairs 'middeck'), it was not practical for more ejection seats to be installed. Others have acknowledged that offering each crew member an ejection seat would have limited the number of astronauts. Even limiting crews to four would have required a major redesign of the orbiter, including its wiring, which would have taken time, incurred great expense and imposed an undesirable weight penalty. After completing her fourth test flight in July 1982, Columbia retained the ejection seats for her pilots, but Brand and Overmyer could not imagine having an escape system for themselves ... and *nothing* for Allen or Lenoir. In Brand's mind, there was only one option: he wanted the ejection seats 'pinned', so that they could not be used. If a valid escape system could not be provided for *all* of them, Brand reasoned, then *none* of them should have it. "He stated it categorically in a flight techniques meeting," Allen recalled, "and NASA officials did not argue with him."

The two scientists saw it quite differently. "I got into a debate with Vance around the ejection seats," Lenoir said. "My take was, why the hell should you lose *four* people, if you could only lose *two*? If I'm one of the two that you're going to lose, I'm going anyway. Why shouldn't two of you get out? Vance was *never* comfortable with that." From Brand's perspective, it was not a selfless act of personal sacrifice on his crewmates' behalf; rather, he actually saw it as *selfish*, for he knew that he could not live the rest of his life knowing that *he* had survived and the two mission specialists had died. "He had been a test pilot in England for a while," said Allen, "and some of the English bomber airplanes enabled the pilots to get out, but *not* the gunners. There was a small body of data that involved psychological studies done on individuals [who] had escaped, but *by* escaping, had left their shipmates to a certain death. *They* had certainly been tormented, in terms of what the data showed, for the *rest* of their lives. Vance … didn't want to be another bit of statistic. He said it was *selfish*. I didn't think it was selfish at all."

During their eight months of training, procedures were devised, principally by Lenoir himself, for the new role of the flight engineer. "I took the lead in inventing the role from the ascent perspective," he explained, "and a lot of that carried into the descent perspective. Joe was a very excellent scientist; less so was he an engineer, whereas I was a very good engineer … so I did the ascent, along with Vance and Bob. In the event of off-nominal [events, I would] immediately get out the [malfunction] book and work us through the procedures and help us get through other things. I would try to back them up, which meant that I needed to learn their systems, so I knew how they worked." Since both Allen and Lenoir were mission specialists, it became necessary to give them numbers. "Officially," Lenoir said, "I was MS2 and Joe was MS1. However, we reversed roles for re-entry, with Joe performing the MS2 orbiter duties. We jokingly called me the MS2/1 and Joe the MS1/2."

The tasks and responsibilities of each mission specialist had themselves been devised by John Fabian, in his early capacity as head of the astronaut office's operations group. "One of the things we did," he told the NASA oral historian, "was to lay out the roles … generically, so you don't have to reinvent a training programme every time you name a crew. We came up with a scheme in which MS1 has overall responsibility for payloads and experiments in orbit and MS2 has primary responsibility for flight engineering and helping the pilot and commander during ascent and re-entry and backup for payloads." Plans were already afoot to fly a third mission specialist (MS3) on later flights and *that* role, continued Fabian, "would be typically the most junior and the lowest training requirement, but heavy on EVA". This was not a hard or fast rule and requirements changed over the years, but MS2 remained a constant, requiring the most ascent and re-entry simulator time in order to support the duties of the flight engineer. The decision brought out the natural comedian in Joe Allen. At one of their pre-flight press conferences, a journalist picked up on the fact that they would be the world's first-ever four-man astronaut crew. "Commander and pilot, we know what *you're* going to do," he began. "Lenoir, what are *you* going to be doing?" Lenoir replied that he would be the flight engineer and promptly explained his duties. Then it was Allen's turn. He would be seated on the middeck, with little to see or do, at least during ascent. "Well," replied Allen, "I think I'm just in charge of *religious activities*." It was a flippant remark, a specimen of gallows humour, perhaps, on the first mission without a viable escape system, but it caused some distress and consternation amongst NASA's public relations staff. It highlighted the need for astronauts to take great care in choosing their words. "They completely mistook what I'd said," Allen recalled later, "and actually thought I *was* going to be in charge of religious activities!"

If the responsibilities of Allen and Lenoir were hard for the media to get their heads around, then the roles of the commander and pilot were fairly straightforward: the former was charged with overall mission success – and with *landing* the Shuttle – whilst the latter handled a myriad of systems, including the main engines, Auxiliary Power Units (APUs) and the deployment of the landing gear a few seconds before touchdown. Commander Vance DeVoe Brand had earned himself a reputation as one of NASA's most gifted fliers – "a hard-nosed aeronautical engineer and an experienced test pilot", according to Deke Slayton – and served as a backup crew member on no fewer than *three* missions. Born in Longmont, Colorado, on 9 May 1931, Brand had been an active Boy Scout during his formative years and achieved the title of Life Scout. After high school, he entered the University of Colorado at Boulder, receiving his business degree in 1953 and accepting a commission into the Marine Corps. For four years, he served as a naval aviator, which included assignments as a fighter pilot in Japan, then resigned from active duty to return to his *alma mater* for a degree in aeronautical engineering. A master's degree in business administration from the University of California at Los Angeles followed in 1964, by which time Brand was working for Lockheed as a flight test engineer for the Navy's P-3 Orion aircraft. His ties with the military remained intact, however: he continued to serve in the Marine Forces Reserve and with Air National Guard jet squadrons until 1966. Along the way, he graduated from the Naval Test Pilot School in 1963 and worked for a time at West Germany's F-104G Flight Test Centre in Istres, France, as an experimental test flier.

After selection by NASA as a member of its fifth group of astronauts in April 1966, Brand focused on thermal vacuum chamber tests of the Apollo spacecraft and later served as a member of the support crews for the Apollo 8 and 13 lunar flights. During the latter, in April 1970, he was a capcom in Mission Control, providing a direct voice link with the astronauts during one of the most harrowing space missions ever attempted. At around the same time, Brand was named as the backup command module pilot for Apollo 15 and, mindful of a well-established three-flight 'rotation' system, confidently anticipated assignment to Apollo 18 and his own journey to the Moon. After the cancellation of that flight in September 1970, Brand continued his Apollo 15 duties and later transferred to Skylab, serving as backup commander for its second and third missions. (In August 1973, Brand and fellow astronaut Don Lind came within days of flying a rescue mission to the station.) In the midst of his Skylab work, Brand was named as command module pilot for the Apollo-Soyuz Test Project, a joint flight with the Soviet Union. This mission eventually took place in July 1975. Brand returned from his first space flight, wrote up his crew report, debriefed with the NASA managers and physicians ... and promptly lost his *parking space*. Reserved spaces outside Building 4 at JSC have become legendary; an indicator that an astronaut-in-waiting has finally received a flight assignment, but *after* a mission their importance steadily deteriorates. "Up to the time of the mission," Brand told the NASA oral historian, "when you're first up to bat, as we used to say, *everyone* will pay attention to you and if there's something that you think might help the mission, why, there are people eagerly waiting to get your ideas. You're sort of on top of the world. Well, after you come *back* from a mission, whereas *before* you had a parking place, when you come back, you *lose* your parking place. That's the *first* thing that happens. After about two weeks post-mission, their interest starts to wane, because they're thinking of the *next* mission." Unfortunately, Apollo-Soyuz was America's *last* manned mission before the arrival of the Shuttle ... and *that* was not expected to fly for several years. Brand found himself working a lot of simulations, including launch abort scenarios, and tracking guidance issues surrounding the Shuttle's ascent into space and descent back through the atmosphere. For a time, he was lined up to command one of the orbital test flights of Columbia, but ultimately wound up with STS-5, the first 'operational' mission.

Seated to his right side on the flight deck was Robert Franklyn Overmyer, a decorated colonel in the Marine Corps, who had been with NASA since September 1969. Born on 14 July 1936 in Lorain, Ohio – a city famed for its shipyards and steel mills and until recently a key manufacturing centre for the Ford Motor Company – Overmyer attended school in Westlake, an affluent suburb of Cleveland, and later studied physics at Baldwin-Wallace College. Graduation in 1958 was followed by admission into the Marines. Overmyer underwent flight instruction in Kingsville, Texas, and joined Marine Attack Squadron 214 in November 1959. He completed a master's degree in aeronautics from the Naval Postgraduate School in 1964 and was detailed firstly to Marine Maintenance Squadron 17 in Iwakuni, Japan, and then to test pilot school at Edwards Air Force Base in California. Overmyer was one of a handful of aviators chosen by the Air Force in 1966 for its Manned Orbiting Laboratory (MOL), a military space station project, for which he later received a Meritorious Service Medal. However, when MOL was cancelled by President Richard Nixon in June 1969, NASA was pressured to accept seven of the pilots into its own astronaut corps. Overmyer was one of them. His first two years with the space agency were devoted to Skylab engineering development. From November 1971 until December 1972, he served on the Apollo 17 support crew, acting as a capcom during the final lunar landing mission. He then worked on the Apollo-Soyuz Test Project from January 1973 until July 1975, as a support crew member, and later moved over to the Shuttle. Overmyer was awarded the Marine Corps' Meritorious Service Medal in 1978 for his duties as a chief T-38 chase pilot during the Enterprise approach and landing tests and later acted as NASA's deputy vehicle manager for Columbia. Following her arrival at the Kennedy Space Center (KSC) on Merritt Island, Florida, in March 1979, the first spaceworthy orbiter was in a pitiful state, particularly her fragile thermal protection system, many of whose thousands of heat-resistant tiles had become debonded from the airframe. These tiles, each of which was *individually* designed and *not* interchangeable, had been a major headache during the Shuttle's construction. For two years, Overmyer oversaw the final work on Columbia, until, at the end of December 1980, she was finally transported to the launch pad for her first mission. A little over a year later, he received word of his first flight assignment.

It has often been said over the years that good crews operate like symphonies, with each member harmonising with the rest. None of the STS-5 crew were perfect – Allen's humour was perhaps more uninhibited than others, Lenoir (by his own admission) had a low tolerance threshold for mistakes and Brand and Overmyer spoke the language of hardened military test pilots, not scientists – but their performance during their five days in orbit would attest to one thing: they *were* a clear example of 'impedance matching', integrating dissimilar circuits and operating with perfection. Their backgrounds and life experiences differed from one another in a thousand ways, but they had been brought together to fulfil a mission, and they would mark themselves out as quite distinct from the spacegoing crews which had come before them and open the door on the truly diverse crews to follow.

WINGS AT THE EDGE OF SPACE

Happiness was a precious commodity during much of the Shuttle's genesis, for the notion of changing the United States' spacegoing philosophy from small capsules, lofted atop the descendants of ballistic missiles, to something the size and shape of a conventional airliner, ferried into orbit attached to a large fuel tank and a pair of rocket boosters, was a vast effort and posed monumental, and frequently maddening, obstacles. Naturally, the idea for the Shuttle did not appear from nowhere; it could trace its development across several decades and many of its systems had been pioneered and proven by a series of aerospace projects organised by NASA and its forerunner, the National Advisory Committee for Aeronautics (NACA). As early as the 1930s, the German aerospace engineer Eugen Sänger developed early concepts for rocket-propelled aircraft, some of which envisaged velocities as high as ten times the speed of sound and altitudes up to 70 km. In his later work, Sänger showed how the addition of wings could enhance a spacegoing rocket: although its initial range would be quite modest, following an arc-like trajectory, much akin to an artillery shell, the *lift* generated by those wings during re-entry would carry it *upward*, allowing it to 'skip' off the atmosphere, thus opening up new possibilities for a craft which could circle the globe and return to its launch site. Within a few years, such dreams rose from the drawing boards and into hardware: in 1947, Chuck Yeager flew the rocket-propelled Bell X-1 aircraft through the sound barrier for the first time and in 1951 the Douglas Skyrocket and its pilot, Scott Crossfield, reached an impressive Mach 2.

At these velocities, issues of aerodynamic overheating was not yet a significant obstacle, but as the US military focused more attention on the prospect of delivering heavy nuclear warheads to Moscow, speed and stability became more important. In January 1952, NACA was urged to pursue a manned research aircraft, capable of exceeding Mach 5, with early reports even suggesting a *commercial* hypersonic vehicle. If such a plan were to be realised, it would need to overcome issues of aerodynamic overheating and stability. The latter problem was resolved by Charles McLellan, an aerodynamicist at NACA's Langley Aeronautical Laboratory in Hampton, Virginia, who proposed that small, wedge-shaped vertical fins and horizontal stabilisers would be much more effective than conventional thin surfaces. Overheating proved a harder nut to crack. If the aircraft re-entered the atmosphere with its nose pointing in the direction of flight, its streamlined shape would subject it to disastrous overheating and destructive aerodynamic stress. A re-entry with the nose positioned at a higher angle, with the flat undersurface presented to the hypersonic airflow, would be more manageable, permitting the craft to lose speed in the upper atmosphere, ease overheating and lower aerodynamic loads.

This realisation that overheating could be reduced did not detract from the reality that such vehicles would be subjected to re-entry temperatures of significantly higher severity than had previously been encountered. Bell Aircraft had already begun to explore temperature-resistant materials, including a chrome-nickel alloy called 'Inconel-X' and stainless steel 'shingles', capable of radiating heat away from the airframe, coupled with techniques for water-cooling areas along the leading edges of the wings. In October 1954, the Aircraft Panel of the Scientific Advisory Board expressed its belief that the next decade's research and development goal should focus on the field of hypersonic flows. "This is one of the fields in which an ingenious and clever application of the existing laws of mechanics is probably not adequate," the panel reported to Air Force Chief of Staff Nathan Twining. "It is one in which much of the necessary physical knowledge still remains unknown at present and must be developed before we arrive at a true understanding and competence." Summing up, the panel considered the time ripe for the construction of a new aircraft to surpass Mach 5 and reach altitudes as high as 150 km. The project received overwhelming support from within the Air Force and NACA and in the spring of 1955 it also received a name: the X-15. In time, it would become the fastest and highest-flying aircraft prior to the arrival of the Shuttle. In August 1963, with NASA test pilot Joe Walker at the controls, the X-15 reached a record altitude of 107 km, and a few years later, the Air Force's William 'Pete' Knight achieved a speed of Mach 6.72 (some 7,270 km/h). The Air Force decided to award its own 'astronaut wings' to pilots who flew higher than 80 km, but the Fédération Aéronautique Internationale (FAI) decreed that only flights which exceeded 100 km officially reached the threshold of space. The FAI's ruling was based on the so-called 'Kármán Line', named after Theodore von Kármán, the Hungarian-American engineer who first described it as an altitude at which the atmosphere is too thin for aeronautical purposes, since a vehicle would need to travel faster than orbital velocity to achieve sufficient lift. Following these two rulings, between July 1962 and August 1968, *thirteen* X-15 missions by eight men exceeded the Air Force's limit, but only *two* of those missions (both flown by Joe Walker) passed the Kármán Line and were considered official 'space flights' by the FAI. Years later, however, NASA honoured the 80 km Air Force limit and in August 2005 awarded 'astronaut wings' to its three civilian X-15 pilots: posthumously in the cases of Walker and Jack McKay and in person to Bill Dana. (The five Air Force X-15 pilots had long since been awarded such wings by their service.) Tim Furniss has referred to this baker's dozen of X-15 sorties as 'astro-flights', thereby differentiating them from orbital or suborbital 'space flights'.

America's space ambitions in the 1960s were driven by the desire to land a man on the Moon, but in the opening months of Richard Nixon's administration a Space Task Group was established with the mandate to chart possible courses for the immediate and long-term future. It was chaired by Vice President Ted Agnew and, sadly, his lack of real clout in the White House meant that only one of his group's four recommendations – the Shuttle – was actually endorsed by Nixon. Plans for a large space station, advanced lunar bases and a manned voyage to Mars fell on deaf ears. The president liked the 'heroic' image of astronauts, but his enthusiasm for space travel itself was low. Unlike John Kennedy and, to a lesser extent, Lyndon Johnson, Nixon saw little political, scientific or technological value in exploring the heavens. The American Institute of Aeronautics and Astronautics (AIAA) also endorsed the Shuttle above the other three options, noting that a low-cost manned spacecraft for delivering medium to large payloads into orbit would enable it to "effectively compete with present expendable boosters" and its members collectively felt that "commitment to an entirely new space station is less urgent than commitment to a new logistics system". The President's Science Advisory Committee (PSAC) went further, suggesting that the Shuttle – "a reusable space transportation system with an early goal of replacing *all* existing launch vehicles" – would not only allow for the deployment and recovery of satellites *and* the orbital assembly of a variety of different structures, but would also lead to a "radical reduction in unit cost of space transportation". In general, the concept of a reusable vehicle was gaining popularity in the opening months of 1969, supported by the AIAA, the PSAC, NASA itself and the Air Force. Since it was expected to be one of the Shuttle's main users, the Air Force became a key player in its final design.

"During the year that followed the landing of Apollo 11 in the Sea of Tranquillity," wrote Thomas Heppenheimer in *The Space Shuttle Decision*, "NASA received a cold bath in the Sea of Reality." A pessimistic remark, perhaps, but appropriate, since 1970 was a time of retreat from lofty goals of permanent lunar bases and expeditions to Mars and saw a refocusing of national priorities: ending the war in Vietnam, beginning the war on poverty, improving the health care system ... and sharply cutting back the space budget. Although Congress *would* ultimately agree that the Shuttle *was* the most important 'next step', and would appropriate $110 million to start the work in Fiscal Year 1971, the chances of this reusable winged spaceplane ever getting off the ground seemed slim. Its initial funding had barely survived in the House and had only won by four votes in the Senate; clearly even this low-cost approach to future space exploration was vulnerable to the slightest change in anti-space sentiment.

The physical form of what would become the Shuttle was also steadily evolving into a delta-winged spacecraft, because this would produce considerably more lift at hypersonic speeds and permit flying substantial distances to the left or right of the nominal track in order to attract the interest of the Air Force. In 1967, Air Force attempts to acquire detailed satellite imagery of troop movements during the Six-Day War in the Middle East and, a year later, during the Soviet invasion of

Czechoslovakia had been thwarted by the limitations of the system then employed by reconnaissance satellites and by the time required for photographic films to be returned and developed. 'Real-time' reconnaissance required not only new technologies, such as electronic imaging devices, but also the ability of the Shuttle to launch into space, perform a single orbit and return to base with film exposed less than an *hour* earlier. Other plans included using the craft to quickly snare Soviet spy satellites from orbit. For the Shuttle to receive Congressional approval, NASA *needed* the Air Force on its side, because such important missions of national security would benefit from the capabilities of a delta-winged, high-cross-range vehicle. "These Air Force leaders knew that they held the upper hand," wrote Heppenheimer. "They were well aware that NASA needed a Shuttle programme and therefore needed both the Air Force's payloads *and* its political support."

Unfortunately, the Air Force did *not* need NASA to launch their satellites; they could rely perfectly well on expendable boosters, such as the Titan III. Their support for the civilian space agency would come at a cost. The Air Force wanted to launch its Shuttles from Vandenberg Air Force Base in California and land there after a single-orbit, 90-minute mission, which would require a 1,700 km cross-range capability – more than twice as large as NASA had envisaged. NASA's Max Faget, who wanted a straight-winged Shuttle, optimised for subsonic performance, was disappointed when the delta wing shape was accepted, but it turned out to be a small price to pay for the Air Force's support. In order to achieve the larger cross-range capability, the Shuttle would glide hypersonically, thereby increasing both the *rate* of aerodynamic heating and the *duration* of that heating ... which, in turn, would require the additional weight (and cost) of a beefed-up thermal protection system.

Linked to the requirements of *both* NASA and the Air Force was the issue of the size of the payload bay. NASA wanted it to measure about 7.5 m in diameter and 10 m long, but, as Heppenheimer explained, "the Air Force needed length, for its reconnaissance satellites amounted to orbiting telescopes and these had to be long to yield the sharpest images. Moreover, such satellites were growing markedly in length. The Corona spacecraft of the 1960s, each with an attached Agena upper stage, had *started* at [5.5 m] and quickly grown to [8 m]." Already, the CIA was preparing for new generations of satellites, known as 'Big Bird' and 'Kennan'. The latter was only in the preliminary design stage, but was expected to have a focal length of around 20 m. Accordingly, the Air Force told NASA that they would require the Shuttle's payload bay to accommodate a 20 m satellite. Any reduction in size would be simply unacceptable: even a 14 m bay would mean that less than *half* of the Air Force's satellites could fly on the Shuttle. "This requirement," they told NASA, "is *still* considered valid. Should you elect to develop the Shuttle with a [smaller] payload compartment, it will preclude our full use of the potential capability and operational flexibility ... Also, if a portion of the present expendable launch vehicle stable must be retained to satisfy some mission requirements, then the potential economic attractiveness and the utility of the Shuttle ... is severely diminished." For the civilian space agency, the bay would enable the Shuttle to carry components for a modular space station, to be launched, piecemeal, in the early 1980s ... but that was a mere side note, compared to the needs of the Air Force and the Department of

From the Shuttle's conception, the Air Force had been a key player in the design and definition of virtually every aspect of the reusable spacecraft ... including the cavernous payload bay. This view from Mission 51J in October 1985 offers a general impression of the size of the bay as a pair of Defense Satellite Communications System (DSCS)-III satellites are deployed atop an Inertial Upper Stage booster.

Defense. In terms of weight, the Shuttle *had* to be capable of delivering an 18,000 kg payload into low polar orbit; Big Bird was known to be around 13,000 kg and future satellites would be even heavier. With few quibbles from NASA, the military got its way. One anecdote from this period concerns an attempt made by Max Faget, one of the world's foremost aerospace engineers, designer of the Mercury spacecraft and NASA's then-head of engineering and development in Houston, to slightly reduce the *diameter* of the payload bay (a dimension in which he thought the military had little interest) from 5 m to 4 m. "It took the Air Force only three days to put him in his place," wrote Heppenheimer, "with a reply that read: *The USAF fully supports and stands firm on the present Level 1 requirement for a payload diameter of 5 m and a length of 20 m.* The reply came from one Patrick Crotty, whose rank was no higher than Major." NASA can have been under no illusions that the Air Force was in control of virtually every detail of the Shuttle. This control was underlined, once and for all, in January 1971, when NASA formally presented the requirements of the Shuttle to all study contractors at a meeting in Williamsburg, Virginia ... and gave the Air Force all it demanded: the cross-range capability, the payload bay length *and* diameter, the delivery capacity to polar orbit; *everything*. "One sometimes hears that when two parties are in a relationship, the one that wants it *more* is the weaker," remarked Heppenheimer. "NASA certainly had been pursuing support for the Shuttle with unmaidenly eagerness and the Williamsburg rules were the result. The agency was now promising to build a bigger and heavier Shuttle than it had *wanted* for its own uses, with considerably more thermal protection. It was also prepared to treat the Shuttle as a *national asset*, which meant the Air Force would not pay for its development or production, yet would receive the equivalent of exclusive use of one or more of these vehicles, entirely *gratis*." The Air Force would be responsible for the development of its own Shuttle launch facility at Vandenberg Air Force Base, but setting that to one side Heppenheimer is not alone in comparing the relationship to a treaty between a superpower and a banana republic ...

As the Shuttle danced to the Air Force's tune, other changes were ongoing in the design of the rocket booster. Potential contractors had considered expendable or only partially-reusable elements, but from August 1969 NASA turned its gaze to a fully-reusable booster. "Partially-reusable designs had represented an effort to meet economic goals by seeking a Shuttle that would cost less to develop than a fully-reusable system," wrote Heppenheimer, "even while imposing higher costs per flight. This approach had held promise prior to the spring of 1969, when the Shuttle had been considered largely as a means of providing space station logistics. Now, its intended uses were broadening to include launches of automated spacecraft, which meant it might fly far more often. The low cost per flight of a fully-reusable [system] now made it attractive and encouraged NASA to accept its higher development cost." Flying more often, it was hoped, would gradually help to reduce overall launch costs and NASA anticipated it might also cut the cost of the *payloads* themselves. As more payloads entered orbit, frequently and reliably, they might stimulate new uses for space, thereby encouraging contractors to build further satellites, with the Shuttle serving much the same role as commercial aviation had done, achieving enormous growth by cutting the prices of its passengers' tickets. Its payload bay offered a vast volume, which eased restrictions on weight and size, and the satellites to be placed into orbit could be fault-checked by the Shuttle crew, *in orbit*, thus lessening the mass of paperwork on the ground. Subsystems aboard these satellites could be standardised and when they started to fail in orbit after a few years of service, the Shuttle would have the capability to retrieve, refurbish and deploy them for a fraction of the cost of a full replacement.

Processing of the Shuttle and its booster was also required to contribute to substantial cost savings. In the early part of 1971, the booster was expected to be longer than a Boeing 747 and considerably fatter in the fuselage and it would be swung into an upright position and mated with the orbiter. An enormous, diesel-powered 'crawler' would then transport it from the assembly building to the pad. Launch would occur under the combined thrust of maybe a dozen main engines – the contract for which went to Rocketdyne, a division of North American Rockwell – and the booster and Shuttle would part company in the high atmosphere, with the former returning to a conventional runway under the power of its own jet engines.

The Shuttle would return to a similar style of landing after completing its mission in space. Processing of the booster and orbiter between flights was expected to require two weeks. Mathematical and economic studies from this period suggest that NASA was confident that each mission could be launched for around $4.6 million – a million dollars less than for an expendable Delta rocket, one of the smallest rockets then in use – and a full order of magnitude lower than the $55 million cost of the Saturn IB booster, used to launch Skylab crews. Early development cost outlays, to be fair, *would* be significant, but within a few years the Shuttle would begin to show massive savings over expendable vehicles; perhaps as high as a billion dollars per annum. This dramatic reduction in costs was expected to spark the new revolution in spacecraft design, but it would not come to pass, for in Thomas Heppenheimer's words, "the *first* statement was a speculation [and] the *second* then amounted to a speculation that *rested* on a speculation".

The final approval of the Shuttle came in April 1972. NASA had drummed up a list of possible names for the new craft – Pegasus, Hermes, Astroplane, Skylark, to name a few – but it was Richard Nixon himself who ultimately made the final choice, by referring to it simply as 'the Space Shuttle'. It would thus break with previous practice of actually *naming* the project, although, as time would tell, individual orbiters *would* receive their own names. Of course, Nixon's acceptance of the Shuttle was rooted securely in down-to-Earth politics and his motivation was to strengthen employment within the aerospace industry as he entered the election year for his second term in office. NASA Administrator Jim Fletcher had told the White House in November 1971 that an early start on the Shuttle "would lead to a direct employment of 8,800 by the end of 1972 and 24,000 by the end of 1973". In terms of employment and the need to pacify America's so-called 'battleground states' – those pivotal areas in control of large blocks of electoral votes – the space programme was recognised to have an importance which was out of proportion to its budget.

On 26 July 1972, out of a pool of proposals which also included Lockheed, McDonnell Douglas and Grumman, the Nixon administration selected North American Rockwell, based in *California*, with its 55-vote monopoly in the Electoral College, to receive the $2.6 billion contract to build the Shuttle. This prompted Jean Westwood, chair of the Democratic National Committee, to harshly criticise Nixon's "calculated use of the American taxpayers' dollars for his own pre-election purposes". Westwood queried the 'help' that five of the corporation's directors provided to Nixon's 1968 election campaign and requested "a full airing" of the background to the contract award. Nevertheless, there *were* solid engineering reasons for not selecting the others. Lockheed's craft, for example, was heavy, "unnecessarily complex", according to Jim Fletcher, and it left a minute-long gap during ascent when there was no emergency abort provision. (In retrospect, this one minute proved a strange criticism, given that the design ultimately adopted imposed a *two-minute* phase, on the Solid Rocket Boosters, during which time there was no viable means of escape.) McDonnell Douglas' proposal was technically deficient and weak, whilst Grumman – which had earlier built the Apollo lunar module for NASA – came second to North American Rockwell. Grumman's presentation was impressive, identifying fundamental problems and offering good solutions, but fell

That the Shuttle was a product of political and technical compromise is amply illustrated in this launch image from STS-5, where its 'butterfly-and-bullet' appearance is evident. The reusable orbiter was attached to a gigantic External Tank, which carried almost two million litres of cryogenic oxygen and hydrogen propellants for the three main engines. Meanwhile, 80 percent of the thrust at liftoff came from a pair of Solid Rocket Boosters. In the programme's 30-year operational history, all three would directly cause or contribute to disaster: the boosters would doom Challenger, the tank would doom Columbia and the orbiter's troublesome main engines and thermal protection system would prove a constant worry for astronauts, engineers and managers alike.

down in terms of costs and overall project management. Even North American Rockwell itself, whose cost estimate was the *lowest* of the four, had a number of weaknesses in its design, not least a crew cabin which would be difficult to build.

High scores and low costs, therefore, were the reason behind the decision to pick North American Rockwell to build the Shuttle. In fact, when NASA Deputy Administrator George Low asked the three losing bidders to comment on the fairness of the competition, all three felt that it was the best and fairest competition that they had ever participated in. The situation was quite the opposite for the leading contenders to build the Shuttle's main engines, with Rocketdyne having been chosen for the $450 million contract over Pratt & Whitney. The latter had confidently *expected* to win, even taking out adverts in major aerospace magazines and declaring themselves ready to get the project started. Pratt & Whitney even lodged a complaint against NASA, alleging that the space agency had performed unfair acts in selecting Rocketdyne, but in March 1972 the Comptroller-General, Elmer Staats, upheld the decision. He slammed Pratt & Whitney, telling their lawyers that it was *also* unfair on NASA, which had helped them "to bring [an] original, *inadequate* proposal up to the level of other, *adequate* proposals, by pointing out those weaknesses which were the result of [Pratt & Whitney's] *own* lack of diligence, competence or inventiveness".

As the contracts rolled out, North American Rockwell (which became 'Rockwell International' in March 1973) proved itself more than honourable, by offering important subcontracts to its rivals. Grumman would take charge of building the orbiter's delta-shaped wings *and* the Gulfstream II Shuttle Training Aircraft, for example, and an unsuccessful attempt had earlier been made to interest Pratt & Whitney in sharing the main engine development programme. McDonnell Douglas would take responsibility for the Shuttle's tail-mounted Orbital Manoeuvring System (OMS) engines. In August 1973 NASA chose Martin Marietta to build a large External Tank which would carry cryogenic propellants for the main engines and in June 1974 picked Morton Thiokol for a pair of reusable, solid-fuelled rocket boosters to provide 80 percent of the thrust needed for the orbiter to reach space. By this time, of course, the shape, design and definition of the Shuttle had evolved into the form that we can recognise today. The orbiter itself was similar in dimensions to the DC-9 airliner, roughly 36 m long with its wings spanning 24 m from tip to tip. The first set of wings, built by Grumman, were among the earliest sections to be completed and were delivered to Rockwell's Palmdale plant in California in April 1975. The habitable area of the Shuttle, meanwhile, consisted of a two-tiered cockpit – a 'flight deck' for operations and a 'middeck' for experimental work, eating and sleeping – which backed onto the 20 m long payload bay and an aft compartment to house the three main engines and two OMS pods and support a vertical stabiliser tail fin. Forty-four tiny Reaction Control System (RCS) thrusters in the Shuttle's nose and tail would provide attitude control and additional manoeuvrability whilst in space. The graphite-epoxy payload bay doors were, at the time, the largest aerospace structures yet built from composite material and *had* to be opened within a couple of hours of reaching orbit, to enable the radiators lining their interior faces to shed excess heat from electrical systems into space. The five-piece doors were hinged at either side of the mid-fuselage, mechanically latched at the forward and aft bulkheads and thermally sealed at the centreline. Ordinarily, they were driven 'open' and 'closed' by electromechanical power, but if they were unable to be opened, it was declared that the vehicle must return to Earth at the earliest opportunity. Conversely, if the doors did not *close* properly at mission's end, two crew members were trained to operate the mechanism manually on a spacewalk. Each orbiter was designed to make a hundred missions before major refurbishment would become necessary, although of the surviving vehicles even the fleet leader, Discovery, had made only 39 voyages when it was finally retired in March 2011.

In the years between the initial contract awards and its maiden voyage, the Shuttle's designers faced a series of setbacks, including frustrating problems with the patchwork of heat-resistant tiles and thermal blankets to shield it during its hypersonic descent through the atmosphere and maddening explosions of the main engines on the test stand. When the crews for the first four orbital missions were named in March 1978, the First Manned Orbital Flight (FMOF) was anticipated in the spring of the following year; however, by September, it had slipped until no earlier than September 1979, with a risk of further delays of nine months or more if additional development funding was not allocated. Other issues surrounded the orbiter's weight. "The weight problem does not present any constraint to early flight tests," explained a JSC news release, dated 25 September 1978, "but does present some problems for both the Galileo mission to Jupiter and certain Air Force missions." Throughout the 1970s, powerful Congressional opponents fought against the allocation of these funds and fiercely questioned the rationale for a reusable manned spacecraft to launch satellites which could be carried aboard expendable boosters. The unusual appearance of the Shuttle 'stack' on the launch pad has been described by astronaut Story Musgrave as "like bolting a butterfly onto a bullet". The *bullet* was the External Tank, which resembled an enormous aluminium zeppelin, standing on end, some 46.6 m tall. It comprised two tanks for liquid oxygen and liquid hydrogen, separated by an 'inter-tank' for instrumentation and umbilicals. The oxygen tank at the top housed up to 542,640 litres of oxidiser and the hydrogen tank held some 1.4 million litres of fuel, both of which were fed through a pair of 43 cm lines into disconnect valves in the Shuttle's aft compartment and thence into the combustion chambers of the three main engines.

At the start of each Shuttle mission, the main engines burned for about eight minutes and were shut down a few seconds before the External Tank was jettisoned, right on the edge of space. Each engine measured 4.2 m in length, weighed 3,400 kg and (similar to the XLR-99 on the X-15 aircraft) was 'throttleable' at one-percent incremental steps from 65 percent to 104 percent rated performance. (In the pre-Challenger period, plans were even advanced to ramp the engines up to 109 percent.) The throttle was controlled by the Shuttle's General Purpose Computers and throttling back reduced stress on the vehicle during periods of maximum aerodynamic turbulence and also served to limit the G-loads in the final phase of ascent. Development of the main engines was undertaken at the National Space Technology Laboratory in Mississippi, a NASA facility previously used for testing the Saturn booster, and got underway in 1974. A full-thrust chamber test of an integrated subsystem demonstrator was conducted in the summer of the following year and in March 1976 the engine was fired successfully for 42 seconds at 65 percent rated performance. These tests uncovered a number of potentially serious problems, including failures of its high-pressure fuel and oxidiser turbopumps, which prompted modifications and the addition of more monitoring instrumentation. Twenty-five further tests got underway in April 1977, with no serious difficulties, although turbopump problems *would* continue to plague the effort throughout its development. Two years later, in July 1979, the main fuel valve experienced a major fracture during a test firing which allowed hydrogen to leak into a mockup aft compartment; the system was commanded to shut down, but not before it suffered serious structural damage. Four months later, the oxidiser turbopump failed again, less than ten seconds into a full-flight-duration, 510-second firing of a three-engine cluster. Triumph followed hard on the heels of failure in the following months: a perfect static test in December 1979, then a premature shutdown in April 1980, then another success ... *and another* shutdown. For astronaut Terry Hart, tasked with following these developments, it was an embarrassing time, for *he* had to stand up in front of his peers, each Monday morning, to explain the situation as each new obstacle arose.

Despite the immense power generated by each engine and the colossal amount of propellant needed to run them, they provided only about 20 percent of the muscle to reach space. The remainder came from the two 45.4 m tall Solid Rocket Boosters, the first solid-fuelled rockets ever to be used in conjunction with a manned spacecraft. Loaded with a powdery aluminium fuel, mixed with an oxidiser of ammonium perchlorate, the boosters, built by Morton Thiokol (now ATK Thiokol) of Utah, were mounted like a pair of big Roman candles on either side of the External Tank. When the decision to use solid-fuelled rockets was made in March 1972, Jim Fletcher explained that they would help to reduce the Shuttle's development costs from $5.5 billion to around $5.15 billion. Typically, during pre-flight preparations, the SRBs were paired in matching sets and filled with propellant ingredients from identical 'batches' to minimise the risk of thrust imbalances during ascent. The SRBs were to be capable of flying 25 times apiece (although they would need to be stripped down and refurbished between missions), but the External Tank was to be discarded and burn up in the atmosphere. It would have been more costly to recover and reuse the tank than to simply build a new one for each mission. The boosters represented relatively new technology for NASA. The Air Force had been using solid-fuelled rockets for some time, but those for the Shuttle were rather larger. The first tests of an empty casing took place in September 1977 at Thiokol's facility in Utah to verify structural integrity and fracture mechanics and to permit analysis of the growth and development of cracks. Four static firings of the motors were conducted between July 1977 and February 1979, followed by three qualification tests from June 1979 to February 1980, which validated the design. Parachute drops were conducted in El Centro, California, and Dennis Jenkins noted that the development and verification of the SRBs over a total of just seven test firings was dwarfed by more than *seven hundred* to qualify the Shuttle's engines.

Preparing for each Shuttle flight required several years, but the actual bringing together of the components began with setting up the boosters on a Mobile Launch

Inside the enormous Vehicle Assembly Building, the orbiters were lifted from a horizontal to a vertical position to be mated to their External Tank and Solid Rocket Boosters. In this image from July 1989, Columbia is prepared for attachment to her 'stack' in readiness for STS-28.

Platform in the gigantic Vehicle Assembly Building (VAB) at KSC. A booster comprised six blocks, called 'segments', each of which was positioned by overhead crane with pinpoint accuracy, one atop the next and joined by a ring of bolts. To prevent a leakage of searing gases whilst operating, a series of rubberised O-rings sealed the joints between the segments. After propelling the Shuttle and External Tank to an altitude of about 45.7 km, about two minutes after launch, pyrotechnics separated the boosters, auxiliary rockets at their nose and tail pushed them away and parachutes were deployed to lower them to a gentle splashdown in the Atlantic Ocean. They were then recovered, stripped down and refurbished for reuse. When the assembly of the SRBs was complete, the External Tank was moved into position between them and connected by a series of spindly, but strong, attachment struts. After checks to verify its mechanical and electrical compatibility, the Shuttle itself was moved from the nearby Orbiter Processing Facility (OPF), tilted by crane onto its tail and mated to the tank. The transfer of the 1.8 million kg stack from the VAB to either Pad 39A or 39B – a distance of 5.6 km – took six hours, with the aptly named 'crawler' inching the precious, $2.2 billion national asset along a track made from specially imported Mississippi river gravel. Once the stack was 'hard down' on the pad surface, the crawler withdrew and a servicing structure rotated into place. Further checks were conducted, payloads installed and the crew participated in a Terminal Countdown Demonstration Test. This was essentially a full dress rehearsal of the final part of the countdown, after which there was a simulated main engine shutdown and emergency evacuation exercise. As for the *return* of the orbiter from space, in October 1974 NASA decided that the first missions would land at Edwards Air Force Base in California; according to Jim Fletcher, this offered "the added safety margins and good weather conditions" needed on those early flights. These margins were particularly evident in the nature of the two primary Edwards landing strips: Runway 17, which crosses a salt flat and measures 13 km long, and the all-concrete Runway 22, whose 4.5 km length was extended by a 500 m asphalt overrun at each end. However, when the vehicle became fully operational, it was expected to routinely use a specially-built 4.5 km runway (the Shuttle Landing Facility) at KSC. Indeed, when Shuttle construction work got underway in Florida in 1974, this runway was one of the first structures to be built.

The name 'Columbia' arose from a NASA decision to give *names*, rather than *numbers*, to each of the orbiters. In May 1978, John Yardley, the agency's Associate Administrator for Space Transportation Systems, sent a list of options to the Public Affairs Office. Each name – 'America', 'Independence', 'Liberty', 'Freedom', to name a few – was required to bear some symbolic relevance to the heritage of the United States. Later that year, Arnold Frutkin chaired a committee which broadened the list of options to include sailing vessels and even Native American tribes. In December, three categories of names were selected: (1) Explorers' Vessels, which included 'Enterprise', 'Endeavour' and 'Discovery', (2) American Tradition and Spirit, which included 'Columbia', 'Constitution' and 'Republic', and (3) Stars and Constellations, which included 'Orion', 'Arcturus' and 'Pegasus'. At length, the committee settled on Explorers' Vessels and even the Star Trek-themed 'Enterprise' had a proud naval heritage, having lent its name to the world's first nuclear-powered aircraft carrier. 'Columbia', meanwhile, owed her name to a privately-owned sailing frigate, built in 1773, which was commanded by Robert Gray in a series of expeditions to the Pacific north-west coast of America in support of the maritime fur trade. Later, in 1790, Columbia (whose full name was 'Columbia Rediviva', Latin for 'Columbia Revived') became the first US vessel to circumnavigate the globe. Significantly, as well as being the female personification of the United States, Columbia also served as the name of Apollo 11's command module.

By the beginning of November 1982, Space Shuttle Columbia had flown four times, logging a total of almost three weeks in orbit, and was promptly declared to be 'operational' by NASA and the media. Preparations for STS-5 had progressed on three separate fronts. Firstly, there was the assembly of the SRBs and External Tank; secondly, there was the processing of Columbia herself; and thirdly, on this first 'operational' mission, there was the preparation of the payload – in this case two commercial telecommunications satellites, called Satellite Business Systems (SBS)-3 and Anik-C3. Both arrived by aircraft at Cape Canaveral Air Force Station in the midsummer of 1982 and were quickly transferred to KSC's Vertical Processing Facility for pre-launch checkout. Columbia, meanwhile, was having unwanted hardware removed and some new hardware installed. Notably, the pyrotechnics for the commander and pilot's ejection seats were disabled and two collapsible seats for the mission specialists – one on the flight deck, the other on the middeck – were fitted. The final removal of the massive ejection seats and their rails was planned during an extensive modification of Columbia, scheduled to take place in 1984, after her STS-9 flight. Further, the middeck floor was strengthened and parts of a Development Flight Instrumentation pallet were removed from the payload bay. On 9 September 1982, the orbiter was rolled into the VAB, tilted on her tail and mated to the boosters and External Tank. Twelve days later, the stack was rolled out to Pad 39A. The two satellites and their attached solid-rocket motors, each encased in a lightweight sunshade with clamshell doors, were moved to the pad on 12 October and loaded aboard Columbia's payload bay shortly thereafter. This procedure, which utilised a device known as the Payload Ground Handling Mechanism, marked the first time in which a piece of cargo had been loaded *vertically* into its launch vehicle. It was a sign of things to come. The Shuttle, as advertised, could carry as many as three communications satellites, together with the motors needed to boost them into geostationary transfer orbits, and their sunshades, and NASA hoped that this alone could generate millions of dollars in revenue each year. The launch of SBS-3 and Anik-C3 by Vance Brand's crew had already netted the agency a first fee of $18 million. Ultimately, it was thought that the Shuttle would out-compete for commercial contracts with the newly-inaugurated European Ariane rocket by also offering 'free' seats into orbit for customers' representatives.

Looking into Columbia's payload bay in the days before STS-5, the observer would be forgiven for thinking that two oversized versions of Pacman were sitting there, for when the protective sunshades opened to release the satellites inside, they looked just like a pair of gaping jaws from the children's game. Fortunately, unlike the real Pacman, these jaws were designed to release something, rather than gobble it up. Each sunshade cradle was composed of a series of machined aluminium frames and chrome-plated steel longeron and keel trunnion fittings, covered with Mylar insulation, and measured 2.4 m long by 4.6 m wide across the width of the payload bay. At the base of the cradle was a turntable with two electrical motors to impart the required spin rate, which could be set in the range 45 to 100 revolutions per minute, depending on the stability needs of the payload, and a spring ejection system to release the satellite and its attached motor. During the Shuttle's ascent, a pair of restraint arms held the precious satellite steady inside its sunshade. Soon after reaching space, the delicate shades were closed to protect the unpowered satellites from the thermal extremes of low-Earth orbit. (When the payloads were released, their spin would even out these thermal stresses.) It was at this point that Lenoir and Allen would respectively supervise the deployment of 'their' individual satellites. The SBS-3 deployment, which would occur eight hours after launch, would be conducted by Lenoir and that of Anik-C3 would take place under Allen's auspices, almost exactly 24 hours later. As each astronaut performed 'his' deployment, the *other* mission specialist would photo-document the procedure, whilst Brand and Overmyer handled Columbia's systems and prepared to manoeuvre her safely out of the way before the satellite ignited its motor to boost itself into a geostationary transfer orbit with its apogee 36,000 km above the equator. Along the way, the spent motor casing would be jettisoned. Later, the satellite would fire an integrated rocket motor to circularise itself into geostationary orbit at its assigned longitude.

The 600 kg SBS-3 satellite was built by the Hughes company at its El Segundo plant in California and, when operational, provided all-digital voice, data, video and email facilities, a high-quality messaging service for business customers and a spare transponder for communications firms, broadcasters and cablecasters. Sitting just behind SBS-3 in Columbia's payload bay was Anik-C3, which was more or less identical in size, shape and mass, but was owned by the Ottawa-based Telesat company. Contracts between Hughes and Telesat for the fabrication of three Anik-C satellites had been signed in April 1978. Anik-C offered, for the first time, rooftop-to-rooftop voice, data and video business communications, as well as Canadian television and other broadcast services. Appropriately, the word 'anik' actually means 'little brother' in Inuit. Although the designation 'C3' also suggests that it was the *third* Anik-C satellite to be placed into orbit, this was actually the *first* of its kind to be launched! Two others, Anik-C1 and Anik-C2, had been built and placed into storage until a suitable date could be established for their launch. By coincidence, Anik-C3's completion occurred at the same time as Telesat's first contracted flight opportunity on the Shuttle and it was decided to simply take it directly from the factory to the launch site. Its siblings were launched in June 1983 and April 1985.

Both SBS-3 and Anik-C3 were of Hughes' HS-376 satellite bus type. These cylindrical, spin-stabilised drums measured 2.7 m tall and 2.1 m wide when stowed in the payload bay, but increased to more than twice that height in their final operating configurations. Both carried two concentric, telescoping solar panels – comprising 14,000 cells in total – which generated 1,100 watts of DC power to support ten-year life spans and also carried their own 100 kg supplies of hydrazine fuel for station-keeping. Each also had an on-board power system with rechargeable nickel-cadmium batteries to run their communications payloads. The 'footprint' of SBS-3 covered the entire contiguous United States and Anik-C3 virtually all of Canada, including its remote northern wilderness, occupied by the Inuit. SBS-3 was equipped with ten transponder channels; Anik-C3 had 16. Attached to the top of each satellite was a 1.7 m shared-aperture grid antenna, with two reflecting surfaces to provide both 'transmission' and 'reception' beams. In the case of Anik-C3, the high transmission power and high-frequency radio bands at 14 and 12 GHz meant that much smaller antennas, just 1.2 m across, could now be situated on rooftops or office blocks. This marked a significant reduction in size from the 3.6 m C-band reception dishes used previously, which had been viable only for hotels and major office buildings. Eventually, however, when joined by its sisters, Anik-C2 and Anik-C1, the anticipated communications traffic did not grow as rapidly as expected. Moreover, Anik-C1, whose main role had been to back up the others against failure, became surplus to requirements, was placed into 'orbital storage' and was eventually sold.

Not only were satellites 'new' to the Shuttle, but so too was the Payload Assist Module (PAM)-D. Together with the cradle in the payload bay, these provided portable launch platforms, fitted with spin motors and deployment springs, and could deliver satellites of up to 1,250 kg into geostationary transfer orbit. Formerly known as the 'spinning solid upper stage', the PAM-D was part of a family of two small boosters destined for use by the Shuttle. The other booster, a PAM-D2, was capable of hauling 1,880 kg and was in the pipeline for subsequent missions. McDonnell Douglas' decision in the 1970s to develop these new boosters arose from dissatisfaction expressed by the commercial communications satellite industry over the Inertial Upper Stage (IUS), which Boeing was developing for Shuttle payloads. Designed for heavy Air Force satellites, the IUS was so large that the payload bay could accommodate only one per flight. Most commercial communications satellites, including Anik-C2 and Palapa-B1, were much smaller and McDonnell Douglas opted for a much more compact turntable capable of spinning and ejecting a satellite mated to a smaller rocket motor.

For Bill Lenoir, this would be his one and only space mission and he remembered the countdown and launch with crystal clarity, even decades later. Fellow astronaut James 'Ox' van Hoften strapped the crew into their seats and departed, leaving them alone in the cabin for perhaps two hours. Lenoir promptly unstrapped, curled himself into as comfortable a position as possible ... and fell asleep. He was awakened by Brand, just half an hour or so before launch, and hurriedly strapped himself back into his seat and began talking his commander through the checklist. Perhaps to gain a better picture of Lenoir's nap, one should consider that the Shuttle was positioned *vertically* on the pad, with the astronauts effectively lying on their backs, with legs elevated. To fall asleep, *at all*, atop a fully-fuelled rocket, would surely have been difficult, let alone only hours before being blasted into orbit. From his first T-38 flight, shortly after having a hernia removed, to this day in November 1982, Lenoir evidently had the *right stuff* in spadefuls. To be fair, though, he instinctively knew how complex the Shuttle was and privately doubted that Columbia would even launch on her first attempt; he was certain that they would be 'waved off' and perhaps try again the next day.

Not so. That is not to say that the countdown was entirely uneventful, for one of the orbiter's computers failed to properly synchronise with its siblings, although it did not delay the launch, which was precisely timed for 7:19 am Eastern Standard Time (EST). Only 40 minutes were available in that day's 'launch window'. This short period of time was partly dictated by the need for daylight conditions at the primary landing site at Edwards Air Force Base and other contingency sites, in case an emergency situation during ascent should force an early return to Earth. However, the main reason for the short window was because the deployment of SBS-3 and Anik-C3 *had* to be timed in order to obtain the correct Sun-angle when they were ready to make the first use of their electricity-generating solar cells.

With five minutes to go before liftoff, Bob Overmyer reached over and switched on the orbiter's Auxiliary Power Units and verified that they were up and running normally. By the time the countdown clock entered its final minute, the excitement of the four men *must* have started to build – they were *really* going to do this, *today*. Launch commentator Hugh Harris ticked off the final milestones, with a technobabble which only served to intensify the excitement: "T-1 minute and counting ... the firing system for the sound suppression system on the pad is armed ... T-55 seconds, the hydrogen igniters under the orbiter's engines have been armed ... T-40 seconds ... we're just seconds away from switching command of the countdown from the ground computers to the on-board computers ... and we've had a Go for auto sequence start." With thirty seconds to go, Columbia's General Purpose Computers assumed primary command of all critical functions. It was now *they*, and they alone, which would make the millisecond-by-millisecond checks and decisions over whether to proceed. "T-21 seconds and counting ... the SRB nozzles are being moved to launch position ... T-15 seconds ... 13, 12, 11, ten ... *we are Go for main engine ignition,*" as a flurry of orange sparks from the hydrogen burn igniters gave way to a sudden, low-pitched rumble, "... six ... we *have* main engine ignition," as a sheet of translucent orange flame and billowing clouds of smoke swept the pad, "... three, two, one and ... solid motor *ignition* ... and *liftoff* ... Liftoff of the first operational Space Shuttle mission, with two satellites on board ... and the Shuttle has *cleared* the tower!"

For the thousands of spectators watching STS-5 take to the skies, the vehicle and the entire launch pad were virtually obliterated for several seconds, so dense was the smoke, but at length Columbia rose majestically beyond the grey murk, atop two dazzling columns of golden flame from her boosters. Vance Brand was heard announcing the onset of the "roll and pitch program" manoeuvre, as the vehicle performed an axial rotation to orient itself onto the proper flight azimuth for a 28.5-degree-inclination orbit. Astronaut Bob Stewart, sitting at the capcom's console in Houston, acknowledged Brand's call with a clipped *"Roger Roll, Columbia!"*

Climbing through the low atmosphere, the wind noise outside gradually intensified into something which could only be likened to a *screaming*. It caused Bob Overmyer some momentary consternation. It was a *rough ride*; none of the four previous crews had mentioned such severe vibrations in their post-mission debriefings. At one point, Overmyer turned to Brand and remarked on it – "This is *too rough*, Vance. I'm afraid we're going to come apart!" – but it was Lenoir, the engineer, the civilian, who settled the worries of the tough, crew-cutted Marine Corps aviator: "Relax, Bob. No use dying all tensed up!"

A minute into the flight, as Columbia approached an altitude of 15 km, she passed through a period of maximum aerodynamic turbulence, which required the computers to throttle the main engines back to just under two-thirds of their rated thrust. The passage through this period, nicknamed 'Max Q', was accompanied by an increase in the noise and vibration of the engines, although their performance remained within structural expectations. Shortly thereafter, the three engines were throttled back up to full power.

"Columbia, this is Houston, you're Go at throttle up," radioed Stewart.

"Roger, go at throttle up," crackled Brand's voice in response.

The sound from the boosters, meanwhile, remained sporadic and decreased to virtually nothing as the time approached, a little more than two minutes into the ascent, for their separation. As the boosters' thrust tailed off, the crew felt as if they had *stopped accelerating*. When the separation motors fired and the SRBs fell away, a bright, orange-yellow 'flash' appeared to stream up in front of the Shuttle's nose and then sweep back above the front windows. Separation was accompanied by a harsh, grating sound, but both boosters performed nominally. At this point Columbia was 35 km high, above much of the sensible atmosphere and travelling at close to four times the speed of sound. With the boosters gone, the astronauts found it much easier to flip switches in the cockpit. It must have been a relief for Overmyer to see the back of the solids, for the rough ride calmed down considerably when they were gone. "Sure enough, it got all quiet and peaceful," recalled Lenoir. "Riding on the main [engines] ... was like an electric engine and you don't feel it." By eight minutes after launch, with three times the force of terrestrial gravity bearing down upon them, each man felt that he had someone heavy sitting on his chest. Breathing became difficult and talking was reduced to a series of guttural grunts. Some eight minutes and 30 seconds since leaving Pad 39A to the cheers and prayers of thousands of spectators, the main engines of Columbia were shut down as planned and the External Tank was jettisoned to follow a ballistic, suborbital re-entry and burn up over a sparsely inhabited stretch of the Indian Ocean. Brand and Overmyer pulsed the RCS thrusters to push themselves away from the discarded tank; there was hardly a murmur of noise associated with the separation. They were in space ... and although still tightly strapped into their seats, the first traces of orbital flight were readily apparent, as washers, filings, screws, wire and other debris began floating freely around the cabin ...

One thing that did *not* seem to be floating, though, was Lenoir himself. As Brand and Overmyer prepared to fire the OMS engines to circularise Columbia's orbit, his first reaction was that they were *not* weightless. Peculiarly, he sat firmly in his seat, when his mind told him that he ought to be *floating*. Then he realised: he was strapped in so tightly that he felt that he still had G forces on his body; only when he lifted his checklist ... and let it go ... and it just *hung there*, comically ... did the evidence of his eyes finally convince his brain of the nature of this new environment. The rise to orbit had been bone-shaking at times, but satisfactory overall, with no major problems. The SRBs on STS-4 had sunk in the Atlantic Ocean owing to parachute failure, but this time improvements worked and both boosters were spotted descending beneath their canopies at a radar-clocked speed of 27 m/sec. They splashed down about 250 km downrange and were recovered for reuse.

In the town of Crawfordsville, Indiana, Joe Allen's mother, Harriet, need not have worried about the coverage her son was receiving as he flew in space. Fifteen years earlier, when Joe had been selected by NASA, her local newspaper, the *Journal Review*, devoted its entire front page to an account of a calf-judging contest at the county fair ... with just a tiny article on the back cover about a young man, born in the area, having recently been chosen as an astronaut. Harriet had been so upset at the time that she phoned the *Review*'s editor ... and was told, quite tersely, that if she wanted an article about Astronaut Joe Allen, she would have to write it *herself*. Now, in November 1982, the *Review*'s front page proudly displayed a picture of her son and the story of his heroic ride into space. She should have been overjoyed ... but for one thing: the *Review* was one of the *few* newspapers to mention STS-5 in more than a couple of sentences. The headlines across the globe had been grabbed by something else; something which had happened, half a world away, the previous day. The head of a superpower was dead.

AN ERA OF MAGNIFICENT FUNERALS

Leonid Ilyich Brezhnev had been the master of the Soviet Union for almost two decades when he finally succumbed to a heart attack on 10 November 1982, aged 75. It was truly the end of an era, for Brezhnev had dominated almost every aspect of life behind the Iron Curtain since toppling his predecessor, the feisty and unpredictable Nikita Khrushchev, in October 1964. He had initially been appointed as First Secretary of the Communist Party, making him *de facto* (though not *de jure*) head of the Soviet Union, a conglomeration of socialist republics and satellite states, scattered across the globe from eastern Europe to the shoreline of the Sea of Okhotsk. Unlike the United States, in which the fundamental positions of power are vested in the President, the Soviets operated a confusing structure of 'collective leadership' under Brezhnev: the General Secretary was administrative head of the Communist Party – the only legal party – whilst the Chairman of the Presidium was the *de jure* head of state and the Premier led the government. For a time, in the mid-1960s, many Western politicians and diplomats, including Lyndon Johnson and Henry Kissinger, viewed Premier Alexei Kosygin as the head of the Soviet Union. He had been sent on key missions to mediate between India and Pakistan in 1966, successfully persuading them to sign the Tashkent Agreement, and he fought vigorously to end an ideological 'split' between the Soviets and the Chinese. Kosygin's eventual fall came from his desire to spearhead economic reforms, in which he sought to transform Russia from a strict (and failing) 'command economy' into something which drew closer parallels to the West, 'guided', rather than 'controlled', by the machinery of the state. Many of his reforms were accepted at first, primarily because the effects of a harsh economic downturn were already being felt. Eventually, Kosygin triggered a strong backlash from hardliners within the

Communist Party, few of whom wanted to relinquish their long-established stranglehold on the economy or agree willingly to 'bourgeois' notions of private ownership and less rigorous oversight by the state. Consequently, by the early 1970s many of Kosygin's former supporters had flocked to Brezhnev's banner, allowing him to consolidate near-absolute power.

For Brezhnev, the years since the overthrow of Khrushchev had brought mixed blessing: unrivalled and unquestioned control of the state, yes, but at the expense of one of the most intensive phases of socioeconomic stagnation and political repression since the days of Joseph Stalin. At Brezhnev's direction, an attempt by Czechoslovakia's communist leadership to institute liberal reforms and improve human rights had been crushed in August 1968 and the new leader would even mention Stalin himself in a positive light for the first time in more than a decade – even adopting the dictator's old title of 'General Secretary'. Relations with China would deteriorate into armed clashes along their borders on the Ussuri River and, under Brezhnev, the feared KGB would attain a similar level of power to that which it had enjoyed under Stalin.

Despite the acute economic stagnation which forced increased attempts to foster détente with the West, the Soviet Union reached an unprecedented level of power and relative internal calm under Brezhnev. A Public Opinion Foundation poll in 2006 found a 61 percent approval rating for his brand of government and many would have preferred to live during the Brezhnev years than any other period of Russian history in the 20th century. Relations with the West – and particularly the United States – improved dramatically, reaching their zenith in the early 1970s, but *declined* with equal speed. The Helsinki Accords, signed in the summer of 1975, attempted to improve relations further, but the involvement of both superpowers in conflicts in the Middle East, Central Asia and the Americas only soured the milk. A new Soviet constitution was implemented in the summer of 1977 and by giving the KGB a free hand to do as it pleased, undermined the recently introduced 'guarantees' of individual liberties.

When Jimmy Carter took the presidency of the United States, his attitude towards the Soviet Union was distinctly unsympathetic and his renewal of diplomatic relations with China in January 1979 enraged Brezhnev. Although the two leaders eventually signed the papers of a second round of Strategic Arms Limitation Talks in Vienna in June of that year, their relationship was tense. Within six months, the Soviets invaded Afghanistan, prompting an "open-mouthed" Carter and the CIA to begin a $40 billion covert effort to equip an insurgency of mujahideen guerrillas, kicking off a vicious, ten-year conflict whose effects reverberate to this day. As the new decade began, fears grew that the Soviets were seeking to expand their sphere into Pakistan and Iran and even that they were positioning themselves for a takeover of oil in the Middle East. For his part, Carter announced that no outside forces would be permitted to gain control of the Persian Gulf, terminated a 'wheat deal' with Russia and made the unpopular move of prohibiting American athletes from participating in the 1980 Summer Olympics in Moscow.

Nor was Carter's successor, Ronald Reagan, who took office in January 1981, any more open to reconciliation. As Brezhnev aged, his paranoia increased, and fears of an imminent nuclear attack led his advisors to implement the 'Operation Ryan' intelligence-gathering exercise, after which the United States executed a series of clandestine naval manoeuvres around Greenland, Iceland and the United Kingdom. During this period, American bombers frequently flew directly towards Soviet airspace, only peeling off at the last minute, in order to test the alertness of their air defence units and the vulnerability of their radar systems. Brezhnev's death was announced on 11 November 1982 and his burial four days later was accorded all of the state honours, culminating in interment within the necropolis of the Kremlin Wall. Thirty-two heads of state, 15 heads of government, 14 foreign ministers and four princes were in attendance; Reagan refused to attend and sent his Vice President, George H.W. Bush, instead. Still, the condolences were surprisingly warm. Reagan penned a two-paragraph note, calling Brezhnev "one of the world's most important figures for nearly two decades", and even the People's Republic of China expressed its "deep condolences".

Brezhnev's funeral has been remarked by some as having ushered in an 'era of magnificent funerals', since several senior Soviet politicians – including his successor, Yuri Andropov, former head of the KGB – would die in rapid succession. In Andropov's mind, 'human rights' were but part of "a wide-ranging imperialist plot to undermine the foundation of the Soviet state". When his name was announced as Brezhnev's successor, there was much apprehension in the West. Andropov's 15 months in office were marked by attempts to improve the economy, clamping down harshly on ill-discipline within the Communist Party and making available the facts of economic stagnation and impaired scientific progress to the public for the first time. He pressed on with the war in Afghanistan, and relations with Reagan deteriorated when the Soviets rejected a compromise plan for the emplacement of nuclear weapons in Europe. In March 1983, Reagan famously labelled the Soviet Union "an evil empire" for the first time. Uncertain and dangerous times lay ahead.

THE 'ACE MOVING COMPANY'

From the astronauts' perspective, the view of Earth from orbit, with its complete absence of lines and borders, seemed totally at peace. "My overall recollection of the flight," explained Joe Allen, "is it was so *extraordinary*, and so short, and I was in kind of a mental saturation for the whole thing. I could hardly believe the beauty of it and the grandeur of it." However, with only five days to complete two satellite deployments *and* a spacewalk, all four men briskly set to work as soon as Columbia was established in orbit. Immediately after Vance Brand and Bob Overmyer had completed the second burn of the OMS engines to circularise the orbit, Allen and Lenoir unstrapped and began to fold away their collapsible seats. The OMS burns were truly a sight to behold ... and to *hear*, too, for their rumble and vibration through the Shuttle's airframe has been compared by more than one astronaut as something akin to a howitzer or a mortar blast. Allen and his camera were at the ready for at least one of these burns. "Since the burn was being done by Vance and Bob," he explained, "Bill and I had only to just look out the back and see the engines

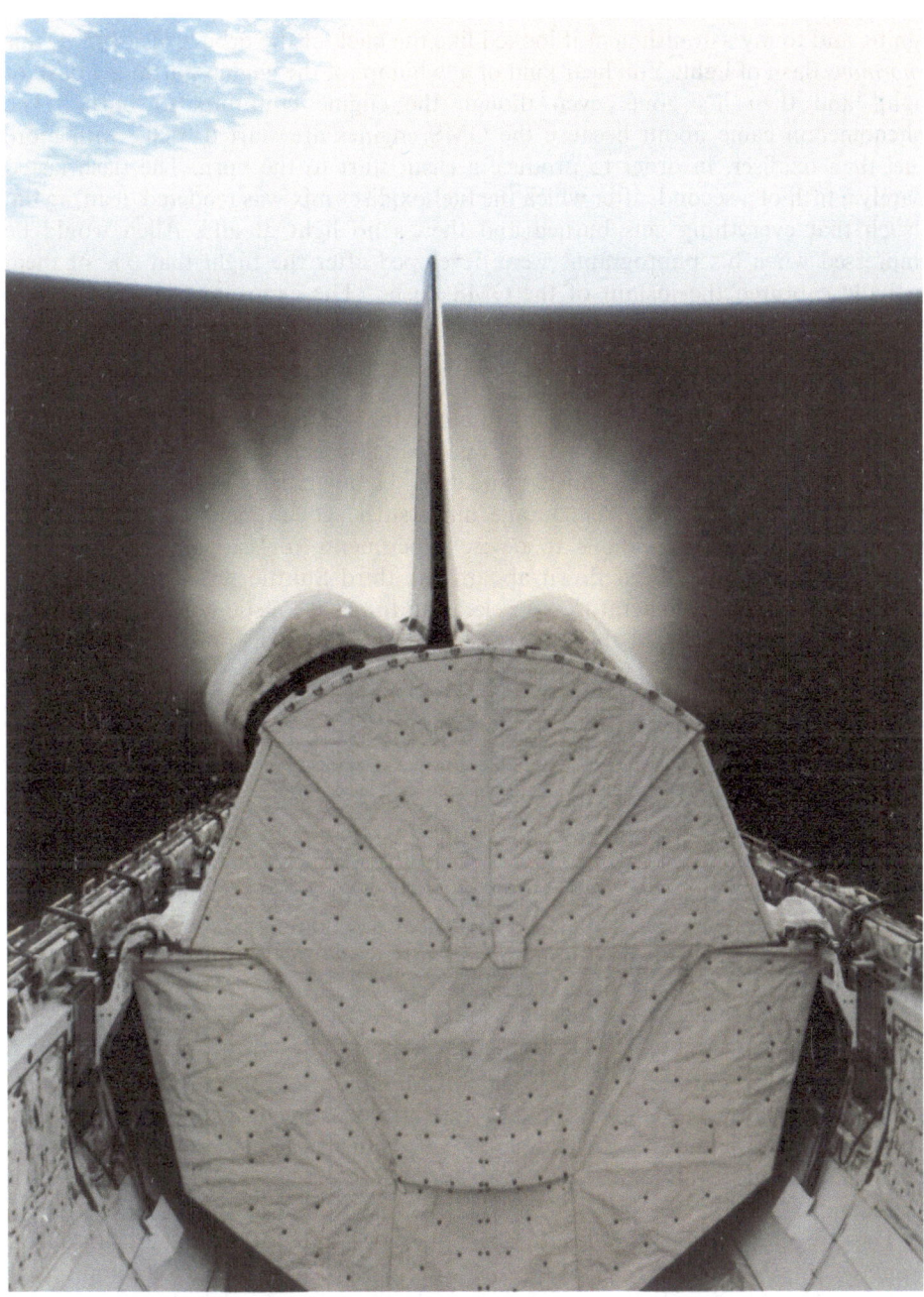

The "enormous flash of light" produced by one of several OMS burns on STS-5 makes Joe Allen's lighthearted remark that the back of the vehicle was about to blow off easier to visualise. The closed cradles for the Anik-C3 and SBS-3 communications satellites are clearly visible.

ignite, and to my astonishment it looked like the back of the orbiter *blew off* ... this *enormous* flash of light. You hear kind of a 'whump' of the engine starting, a flash of light, and then it's gone, even though the engine continues to burn." The phenomenon came about because the OMS engines are started 'rich', with more fuel than oxidiser, in order to produce a clean start to the burn. The flash lasted barely a fifth of a second, after which the fuel-oxidiser mix was rendered 'lean' again, "such that everything gets burned and there's no light at all". Allen would be impressed when his photographs were developed after the flight that *one* of them actually captured the instant of the OMS flash. "The exposure of a camera is a sixtieth of a second," he said, "so you have to put a sixtieth of a second right during that fifth of a second, which is virtually impossible to do, but I got very lucky and was quite pleased by that result!"

During the course of their first day aloft, they activated a 'Getaway Special' (GAS) canister, which was a new type of payload aboard the Shuttle. They took the form of 180 kg dustbin-sized canisters and could be flown in numbers of up to 13 at a time, as part of a drive to encourage universities, government agencies, foreign nationals and even individuals to devise experiments for carriage into orbit. A 'practice' canister had been flown aboard the third Shuttle test flight, STS-3, to evaluate the kind of temperatures, acceleration forces, acoustic noise and pressures imposed on an experiment during launch, ascent, orbital operations and re-entry. Now, on STS-5, the canister was being used for an X-ray study of the solidification of liquid metals. It was funded by the West German Ministry of Research and Technology, but failed completely. It would later become apparent that the canister's battery had leaked its electrolyte and was unable to properly activate the experiment.

Another problem noted by the crew, just a couple of hours into the mission, was the failure of one of Columbia's three cathode ray tube monitors, located on the instrument panel of the forward flight deck. In the early days, before the installation of advanced 'glass cockpits' in the 1990s, there were four such monitors aboard the orbiter: three on the control panel, in front of the commander and pilot's seats, which were needed to display thousands of different functions and a multitude of data, and one on the aft flight deck for payload operations. The aft-mounted monitor would not be required for re-entry, so on 14 November it was decided to switch the cables between it and the failed monitor. This forced the astronauts to switch off the aft monitor, but ensured that Columbia was able to return home with three fully-functional monitors on her forward instrument panel. Several other minor issues cropped up with a troublesome RCS thruster, which showed signs of a possible leak, and Brand and Overmyer kept a close eye on it for the remainder of the mission.

These niggling problems did not impair the first launch of a commercial satellite from the Shuttle. Six hours before the historic deployment, updated computations of Columbia's orbit – including its altitude, velocity and orbital inclination – were radioed to Lenoir from the SBS control centre in Washington, DC. Then, 40 minutes prior to deployment, Brand and Overmyer manoeuvred the orbiter into the correct attitude with the open payload bay facing into the direction of travel. The restraint arms pulled away from the satellite, after which Lenoir flipped a switch on the aft flight deck instrument panel to open the Pacman jaws and impart a 50 rpm spin rate on the payload. This steady rotation helped to stabilise SBS-3 during its deployment. Next, at 3:20:18 pm EST, eight hours after launch and midway through their sixth orbit of Earth, Lenoir issued a command for explosive bolts to fire and release a Marman clamp which held SBS-3 and its PAM-D upper stage in place. Seemingly in slow motion, the spinning satellite departed the bay at just 90 cm/sec, as Allen enthusiastically fired off pictures.

Immediately after leaving the vicinity of the Shuttle, command of the satellite passed to the SBS control centre. Fifteen minutes later, Brand and Overmyer backed Columbia away to a distance of 42 km, aiming the orbiter's belly at the payload in order to protect their delicate topside from the exhaust of the PAM-D's Star-48 solid-propellant rocket. At 4:05 pm, an on-board timer fired the upper stage's perigee motor for about a hundred seconds to boost SBS-3 into a highly elliptical transfer orbit. Overall, the PAM-D's first use aboard the Shuttle was "satisfactory" and it separated from the satellite a few minutes after completing its burn. Shortly thereafter, SBS-3's omni-directional antenna was raised and, over the next couple of days, it employed its own, Thiokol-built Star-30 solid-propellant apogee motor to insert itself into the required near-geostationary orbit, then employed its hydrazine thrusters to manoeuvre onto its 'slot' at 94 degrees West longitude. During this period, SBS-3 unfurled to nearly twice its 'stowed' height and fully deployed its communications payload.

With one of their main objectives completed, it was time for a meal and bed. Sleeping was a more unusual experience than any of them had anticipated. Overmyer used a sleeping bag, whilst Brand preferred to tether himself with clips to a handrail. "One night," remembered Lenoir, "he scared the hell out of himself. When you're relaxed in zero-G, your arms are not down by your side, but [elevated]. Vance woke up in the middle of the night and here's something *right out in front of his face* ... and it scared the *hell* out of him. It was his *own hands!*" Many other astronauts have reported the same strange sensation; unless their hands or arms were properly tucked away, they very quickly worked themselves loose and floated free. Lenoir slept at the rear of the flight deck and Allen was happy to curl up anywhere ... but sometimes ended up *somewhere different*, having floated across the cabin whilst asleep.

A little more than a day into their mission, Brand and Overmyer recircularised Columbia's own orbit with a burst of the OMS engines. This set the stage for the deployment of Anik-C3 a few hours later, under Allen's supervision. Launching the Canadian satellite followed much the same routine. During the afternoon of 12 November, the pilots performed two more OMS burns and would make another (the mission's *sixth* overall) following the deployment to provide a safe separation distance prior to the PAM-D ignition. Repeating Lenoir's procedure of the previous day, Allen opened the sunshade, spun-up the satellite and ejected it into space at 3:24:11 pm, during Columbia's 22nd orbit, as she hurtled high above Hawaii. A perfect firing of the PAM-D, some 45 minutes later, duly inserted Anik-C3 into an elliptical transfer orbit. Like SBS-3, this orbit was a few hundred kilometres 'higher' than its planned 'operational' slot. The satellite then utilised its own motor to place itself into a near-geostationary orbit; by 16 November, it was on-station at 114.9

degrees West longitude, directly in line with Edmonton in Alberta. The astronauts celebrated by turning on music from the movie '2001: A Space Odyssey' and unfurled a sign which would epitomise the growing maturity of the Shuttle as a kind of 'space truck'. This sign has been reproduced countless times in books about the reusable orbiter and its achievements and it is surrounded in a famous photograph by the four astronauts of STS-5. Held by Brand, it read 'Satellite Deployment by The Ace Moving Company. Fast and Courteous Service. We Deliver.' Floating weightlessly around their skipper, head to head, were Overmyer, Allen and Lenoir. The idea for the sign, said Brand, actually came from Allen, "because we *moved stuff* to space".

In those days, unlike today, there was no system of tracking and data-relay satellites to provide near-continuous voice and data coverage; in fact, the crew was only in communication with Mission Control in Houston for perhaps 15-20 percent of each orbit. Significantly, said Lenoir, the SBS-3 and Anik-C3 deployments were to occur whilst crossing the equator and "we never had communication with the ground" at this time. As a result, the astronaut supervising each satellite – Lenoir for SBS-3 and Allen for Anik-C3 – effectively became the satellite's launch director, with the *real* final say over whether to deploy or not. Lenoir coined a new name for the two of them: 'Orbital Launch Directors'.

Months of repetitive training had given Lenoir confidence in his capabilities. "We would go through simulations and I knew, I think, *more* about the system than the trainers did," he explained, "and so I spent as much time training *them* as they spent training *me*!" He was even able to offer guidance to Chuck Shaw, one of the simulation supervisors, who inserted malfunctions to test the reactions of the crew. One potential problem was the temperature of the PAM-D. After deployment, a timer would initiate its ignition sequence, "and you *can't* turn it off, so you want to be *gone*. It's got to be pointed right, and one of the mission rules was the temperature of the rocket engine needed to be ... 55 degrees [Celsius], plus or minus five, or its 'no-go'." In a conversation with Mike Lyons of SBS, Lenoir explained that he *would not launch* if the temperature of the booster was even *a degree* outside of tolerance. *That* would have thrown a spanner in the works, since the Anik deployment was scheduled for the second day of the mission, and would push SBS from the first day to the third day. However, if a similar problem occurred with Anik, then *its* second chance to deploy *also* had to occur on the third day, meaning *two* deployments on *one* day, since the last two days of the mission were devoted to the EVA and preparations for re-entry. Lyons went away and came back with new temperature numbers – this time *65 degrees* – but Lenoir stood firm; he would not launch outside of *those* tolerances, either. At length, Lyons told him not to worry about the temperature reading. "If it's *off-scale high*, just launch!" Lenoir found himself having to ask, with some frustration, what the *real* limits of the PAM-D were, fearful that simply *ignoring* them could lead to a catastrophic failure. He found that at least part of his training was *training others* how to deal properly with their own equipment.

Now, by the late afternoon of 12 November, Columbia's crew could rest assured in the knowledge that the deployments were behind them. It was a triumphant moment for both the astronauts and NASA as an organisation. More than a decade of planning and development and maddening problems with thermal protection tiles and main engines and billions of dollars having been poured into the programme, the Shuttle seemed to be finally proving its commercial value. It still had nowhere to shuttle *to* – certainly no space station was on NASA's agenda yet – but it seemed to be making access to space more routine ... and *this* had always been one of its key goals.

SATELLITES: *TWO*, SPACEWALKS: *NIL*

The triumph was expected to be extended still further by a three and a half hour EVA by Lenoir and Allen on the fourth day of the mission – 14 November – during which they would become the first Americans to leave their spacecraft since February 1974 and give the world its first glimpse of the new Shuttle space suit. Today, it has become increasingly familiar as missions have serviced the Hubble Space Telescope and built the International Space Station. It is a pity, therefore, that what should have been its first outing into the vacuum of space was frustrated by technical difficulties ... and sickness. Originally, in the early days of the Shuttle's genesis, an EVA capability was not considered to be necessary and was not provided. "The NASA perspective of a Shuttle was an *airliner*," explained space suit engineer Jim McBarron in an oral history, "and the people inside it *wouldn't* need suits. It was through prompting and questioning that Aaron Cohen, who was then the Shuttle project manager, finally accepted a contingency capability for closing the payload bay doors – which was an issue they were faced with – to put an EVA capability on the Shuttle."

The plan was for Lenoir, designated 'EV1', with red stripes on the legs of his suit for identification, and Allen ('EV2') to test the snow-white ensembles and practice techniques for a tricky repair of NASA's Solar Max observatory, scheduled for the spring of 1984. They had also been trained to perform an emergency opening and closure of Columbia's payload bay doors, in the event that a malfunction should make it impossible to complete either task automatically. Although the excursion was only expected to last a few hours, its preparation would have occupied virtually the whole day. Assisted by Bob Overmyer, the men would have risen early and begun preparing their suits and the Shuttle's airlock at 2:00 am EST, before spending the next four hours 'pre-breathing' pure oxygen to wash nitrogen from their bloodstreams and thereby avoid an attack of the 'bends'. Their next step would have been to lower the airlock's pressure from the normal 101.3 kPa to just 34.5 kPa to check the integrity of the suits. Finally, a little under 73 hours into the mission, Lenoir and Allen would have opened the outer hatch of the airlock and floated into Columbia's payload bay.

The airlock was a cylindrical structure, about the size of a Volkswagen Beetle, at the rear of the middeck. Its inclusion *within* the pressurised cabin was to preserve the maximum amount of volume in the payload bay and it consisted of two hatches – one for the astronauts to enter from the middeck and another through which they would venture outside. Conditions were cramped. Astronaut Rich Clifford, who made an EVA in March 1996, described floating in the airlock in his bulky suit with

The cramped nature of the Shuttle's airlock (and the bulky space suit) are amply demonstrated in this image from Mission 61B in November 1985, as astronaut Jerry Ross ventures into the payload bay to begin a six-hour EVA. The inward-opening, 'inner' hatch can be seen on the left.

his colleague, barely able to even move his *arms*. Depressurisation and repressurisa-tion could be controlled either from the flight deck or from within the airlock itself. Normally, there were two suits stored in the airlock, although up to four could be accommodated. In fact, many Shuttle missions in the 1990s and into the present century involved as many as four spacewalkers, working in alternating pairs.

The suits worn by Lenoir and Allen were quite distinct from previous ensembles worn by Gemini and Apollo astronauts, yet were designed with the same objective in mind: venturing outside the pressurised confines of the spacecraft. During the EVA, their every move would have been choreographed by Overmyer, whilst Vance Brand would have photo-documented the historic event. The spacewalkers' first task would have been to tether themselves to slidewires running along the sills of the payload bay walls. This safety procedure would have prevented them from floating away from the orbiter. Then, with Lenoir tethered to the starboard sill and Allen on the port side, both men would have translated down the entire 20 m length of the bay to the aft bulkhead. During this time, they would have evaluated the comfort, dexterity, ease of movement and communications and cooling performance of the suits, as well as the floodlights mounted in the bay. They would have also assessed locations at which future spacewalkers could work to repair Solar Max. Practicing 'repair' techniques would have dominated Lenoir and Allen's time: after these initial evaluations, they would have returned to the forward end of the bay to begin work on a dummy set of Solar Max equipment. For more than an hour, they were to test and make comments on a series of fixed and torsion-adjustable bolts, using a special wrench. Lenoir would then have moved to the Solar Max equipment – including a dummy main electronics box from the satellite's coronagraph instrument – to begin a procedure which was intended to be equally as complicated as it would be on the *real* repair mission. Assisted by Allen, handing him parts, Lenoir would have removed thermal blankets, taken off mounting bolts and connectors, cut a grounding strap and painstakingly reattached the connectors; encased in a bulky pressure suit, it would have been no mean feat. Yet this kind of work was absolutely critical, not only for the future of Solar Max, but also for servicing the Hubble Space Telescope, whose own launch was scheduled in the mid-1980s. Although the Solar Max tests would have occupied a large portion of their time, the spacewalkers would also have evaluated a manual system for closing the payload bay doors. This involved a winch system, attached to the forward bulkhead, and they would have used it both with and without the benefit of foot restraints. They would also have moved a bag of tools and evaluated a small black-and-white television camera, mounted atop Allen's helmet; its postage-stamp-sized lens was expected to yield some impressive images from the EVA.

Unfortunately, none of this happened; at least, not on STS-5. The excursion was initially postponed by 24 hours until 15 November – the day before Columbia's scheduled landing – after both Lenoir and Overmyer suffered a particularly severe dose of space sickness. Properly termed 'Space Adaptation Syndrome' (SAS), this ailment is now known to affect around half of all astronauts and cosmonauts. Research over the last five decades has concluded that it is a nauseous malaise akin to motion sickness, which typically lasts no more than two or three days. It was first noted by Soviet cosmonaut Gherman Titov in 1961 and usually manifests itself in sensations of disorientation and discomfort, coupled with dizziness and recurrent headaches. Even today, explanations and countermeasures for it remain imprecise. It appears to be aggravated by a subject's ability to move around freely in weightlessness, and seems to be more prevalent in 'larger' spacecraft, as indicated by the fact that 60 percent of Shuttle astronauts report the complaint. Modern thinking postulates that the influence of weightlessness on the human vestibular system – the workings of the inner ear, which control balance – could offer a possible root cause. This disorientation arises when sensations from the eyes and other sensory organs conflict with those from the vestibular apparatus and with information in the brain derived from a lifetime of 'normality' in terrestrial gravity. A 'repatterning' of the central memory network occurs over the first few days, such that unfamiliar sensations from eyes and ears begin to be correctly interpreted. Today, motion sickness medicines have been shown to help, but are rarely used, because most space fliers prefer to adapt naturally in orbit, rather than risk starting their missions in a drowsy state.

During the 1960s, the sickness was virtually unknown and most of the military test pilots in the astronaut corps, imbued as they were with seemingly limitless stores of testosterone, tended not to report experiencing it, lest their susceptibility impair their chances of being assigned further missions.

Nor was it possible to accurately predict *who* might be susceptible; for on STS-5, it was hardened Marine test pilot Overmyer and T-38 jock Lenoir who fell victim, whilst Brand and Allen were unaffected. Lenoir later recalled that Overmyer's suffering began as early as the second day of the mission. "He filled a couple of bags," Lenoir told the NASA oral historian, "but I never stopped giving Bob credit. We had a bunch of engineering tests to do. Sitting up in the pilot's seat, taking data, doing this, that and the other, he *never* missed a step. He'd puke his guts out and he'd get back to work ... and he felt crappy for *two days*!" Shortly after Overmyer began to feel bad, it was Lenoir's turn. In subsequent conversations with physician Bill Thornton, Lenoir felt that he had psyched himself up for the two satellite deployments, then allowed himself to relax; a wrong move. It felt, said Lenoir, very much like a low-grade hangover. He could *work*, but did not want to exert himself physically or mentally, other than to curl up and sleep it off. As circumstances transpired, that is exactly what happened. The EVA with Allen was delayed, allowing Lenoir some time to "sack out in the middeck".

Sacking out on the middeck also created another problem, as his three crewmates were granted the opportunity to raid Columbia's pantry ... which included green jalapeños. Lenoir had a particular penchant for the hot chili peppers and grew them in his Houston backyard, frequently bringing bagfuls into JSC to munch whilst training. He would offer them to other astronauts, wrote Dave Shayler and Colin Burgess, "and take a little impish pleasure at their reaction when they bit down and found out the true nature of what they were eating". On STS-5, Lenoir hoped to be able to playfully chew on them, particularly during televised broadcasts, but his brush with space sickness meant that most of his beloved jalapeños rapidly vanished instead into the mouths of his three crewmates.

More trouble was afoot, however, when they tried again to perform the EVA on 15 November. When the men finally donned their suits and ran through the standard checks, a problem was noticed with Allen's ventilation fan; it sounded, said the crew, "like a motorboat". In effect, it was starting itself up, running unexpectedly slowly, surging, struggling and finally shutting down. Nor was Allen's suit the only one giving trouble. Lenoir's primary oxygen regulator – which would have been used during his pre-breathing exercises and throughout the EVA – failed to generate enough pressure, regulating to 26.2 kPa, instead of the required 29.6 kPa. Some of the astronauts' helmet-mounted floodlights also refused to work properly. After fruitless efforts to troubleshoot the problems, the EVA was deferred to STS-6. This proved bitterly disappointing to Lenoir. "I guess I was the bad guy," Vance Brand told the NASA oral historian. "I recommended to the ground that we [cancel] the EVA, because we had a unit in each space suit fail in the *same* way. It looked like we had a generic failure there. It was the first time out of the ship. We didn't want to get two guys – or even *one* guy – outside and then have [another failure]. We could have taken a chance and done it, but we didn't."

As the chief spacewalker, Lenoir was upset at losing the chance to put months of EVA training to the test. He wondered if it was possible for just *one* of them to go outside. "We tried to talk them into it," he recalled, "but it was made more difficult by Joe being inop[erable], because now I'm trying to talk them into: not only are we going to go EVA with a suit that you don't really understand, but there's only *one* person going to be out there, so you don't have a buddy system." To be fair, as Allen noted, they did not really *have* to do the EVA; the satellite deployments and a safe landing were their top-priority objectives. "On the day before the spacewalk," he said, "we commented to ourselves that we really had just *two* important things [remaining] to do: one was the spacewalk and the second was the re-entry and a safe landing." Allen made the observation to Brand: *he* would prefer the safe landing! In Allen's mind, the *bad* news was that the suits had failed ... but the *good* news was that they had *not* failed whilst the men were outside in the payload bay. Had the worst happened, Allen was confident that it would not have been a fatal situation, but would have required him to scramble back into the airlock and get himself out of the suit before further failures occurred.

Immediately after STS-5 landed, on 16 November, a task force, headed by Richard Colonna, manager of the Program Operations Office, was established to investigate the problem. The fault in Lenoir's suit was ultimately traced to two missing 'locking' devices – each the size of a grain of rice – in his primary oxygen regulator. These effectively allowed a locking ring in the suit to open, thereby triggering a pressure leak. According to paperwork, provided by the suit's manufacturer, Hamilton Standard of Hartford, Connecticut, the locking devices *were* fitted and signed-off by a supervisor in August 1982, but actually had not been fitted *at all* and were not checked. The responsible employee, and the supervisor, were "barred from further NASA work", according to *Time* magazine's summing-up of the debacle on 20 December. The absence of the locking devices allowed the pressure in Lenoir's suit to creep back from 29.6 kPa to 26.2 kPa. The problem in Allen's case was a faulty magnetic sensor in the fan electronics. Colonna's final report pointed out that "even with no improvements, if the regulator were fabricated properly, the PLSS [the Portable Life Support System, the suit's backpack] would function properly". It also listed ways to test and inspect future regulators and motors and recommended testing inside the Shuttle's airlock on the day before launch. Additional plans were set in motion, though not in time for STS-6, to provide sensors with better moisture resistance for future motors and new tests to allow defects to be identified earlier.

With the disappointment of the cancelled EVA behind them, the astronauts spent the last few days completing experiments and preparing for the voyage home. Five days of work at breakneck speed had exhausted them. "Joe kind of ran himself out of gas the last day," Lenoir told the NASA oral historian, "because he wasn't sleeping real well and some combination of that. He needed to get some sleep." At one stage, Allen asked Lenoir to take over a student experiment he was working on. From Lenoir's perspective, having suffered space sickness, it was all part and parcel of *adapting oneself* to the new environment ... and *that* invariably took time. On a five-day mission, the men had barely enough time to acclimatise to the new conditions around them, before it was time to come home. Brains attuned to several decades of acquiring data from the eyes and ears – which way is 'up' and reorienting oneself, for example – were thrown temporarily into disarray. "Even today," Lenoir said, years later, "I have days where I realise *this* is a day where my brain won't listen to my ears ... and at least I've figured that out and I *don't* close my eyes a lot in the shower, when I'm standing on one foot. Humans are humans."

It had been hoped, initially, to land Columbia in a crosswind at Edwards Air Force Base in order to evaluate the Shuttle's handling characteristics under duress, but instead the spacecraft would touch down in California under calm weather conditions. The re-entry was particularly dynamic from the astronauts' perspective and photographs taken by Joe Allen from the flight engineer's seat, looking over the shoulders of Brand and Overmyer, recorded a hellish, pinkish-orange glow outside the front windows. As Columbia slowed from an orbital velocity of Mach 25 down to around Mach 20, decelerating from 28,300 km/h to around 22,600 km/h – it was possible to see sweeping white 'beams', or shockwaves, rippling across the Shuttle's nose. At Mach 18, still travelling at more than 20,000 km/h, Brand took over manual control from Columbia's computers to perform a flight test manoeuvre. He pushed down from a 40-degree angle of attack to 35 degrees, then up to 45 degrees, and back to 40 degrees. This so-called Push-Over-Pull-Up (POPU) exercise had been demonstrated during earlier Shuttle test flights. "When I did that," explained Brand, "Joe Allen said a shockwave came from the nose and *attached* to the windows right in front of us! *That* was a little worrisome, because he knew it was *hot*, but then about as soon as it got there I was on my way to a higher angle of attack, so it 'walked' back to the nose."

Thanks to Allen and his camera, many of these astonishing images – "interesting things", according to Brand – were captured for posterity. Several images found their way into Allen's 1986 book about his experiences, *Entering Space*. Some were shot through the overhead flight deck windows and reveal astonishing flickers of white-hot plasma as the orbiter knifed her way hypersonically towards a landing

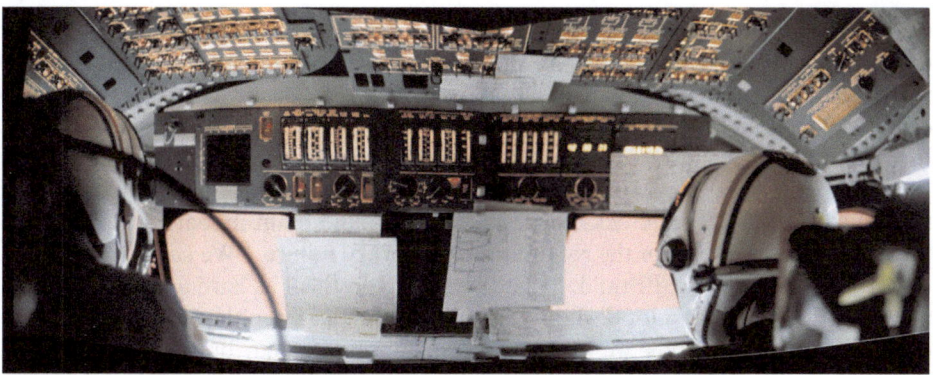

In a remarkable panorama of Columbia's forward flight deck, Vance Brand (left) and Bob Overmyer monitor their displays and instruments as the Shuttle plummets hypersonically back to Earth. Outside, re-entry temperatures climbed to several thousand degrees Celsius. This image was taken by flight engineer Joe Allen, seated behind and between the pilots.

strip, halfway across the planet. "The engineers were very interested in that," Allen wrote. "They'd *never* seen it before." The first real indication of entrance into the atmosphere, Allen wrote, came when the needle of the G-meter on the instrument panel, which had been fixed at zero for five days, suddenly began to quiver, shook itself to life and began steadily climbing. In his book, he vividly described the descent through the steadily thickening atmosphere, as Columbia's computers rolled the vehicle to the left and to the right, gradually bleeding off velocity through a series of wide, sweeping turns. "As the orbiter banks into a left turn," he wrote, "the crew on the flight deck watches in the forward windows the distant, dim horizon *turn* until it is almost *vertical*. Then, as the autopilot commands a bank reversal from left to right, the crew watches the line of the horizon roll until it is momentarily level again, then continue to roll until the right limb of the horizon disappears out the top of the window and the left limb falls away and is hidden by the orbiter's nose." The small RCS thrusters continued to be pulsed for corrections along the way, but were gradually phased out as Columbia descended further into the thicker air and the role of its aerodynamic surfaces – the elevons on the wings for pitch and roll and the rudder on the vertical stabiliser tail fin for yaw and speed braking – became effective. The onset of gravity, Allen wrote, pushed him *hard* against his seat; he could *feel* increasing pressures on his thighs and his back as the punishing deceleration continued. After almost a week in microgravity, he now felt more than double his normal Earthly weight. It was not extremely uncomfortable, but definitely *not* a pleasant experience.

"When you get down to Mach 2," Brand said, as the orbiter fell, brick-like, at a velocity of 1,400 km/h, "you're getting into thick atmosphere and it's *rumbling* outside ... you can *hear* it rumble and you're decelerating such that it's pushing you into your straps." On reaching the west coast of the United States and crossing over Monterey Bay, Brand, Overmyer and Allen beheld the first glimmers of a pale desert sunrise over the eastern horizon. Passing over California's Tehachapi Mountains, the orbiter went subsonic with a pair of trademark sonic booms, sounding, for all the world, like the roar of 12-gauge shotguns. "We went through a very thin cloud deck," Brand said. "I was on instruments flying and circled down and landed at Edwards." The touchdown, which came at 6:33:26 am Pacific Standard Time (PST) – or 9:33:26 am in Florida – was immediately followed by a test of Columbia's maximum braking capabilities. However, the left-hand main inboard wheel 'locked up' during the final part of the rollout, due to a brake failure. "We *completely* ruined the brakes," remembered Brand. "I had to stomp on them as hard as I could, which points out that we had a lot of flight test on the mission. Even though it was the first commercial flight, I think we had 50-50 test objectives. That braking test was just one of them. We *ruined* the brakes – completely ruined them – but it was a *test* to see how well they would hold together if you did that."

RISE OF THE CHALLENGER

It had long been hoped that Columbia's predecessor, the Shuttle Enterprise, would be extensively upgraded to make her spaceworthy. Designated 'OV-101', she had undertaken a series of captive and free flights over the Californian desert in the summer of 1977 and lessons learned during her construction were incorporated into the design of Columbia, but NASA quickly recognised that Enterprise simply weighed too much to transport a full payload into orbit. She would need a new set of plans to render her capable of travelling into space. Moreover, she had no propulsion systems, plumbing, fuel lines or propellant tanks. Her three main engines were all dummies, her payload bay had no mounting hardware for cargo, its doors had no opening mechanisms or radiators and her thermal protection system was little more than black and white polyurethane and fibreglass. Modifying her for space missions was envisaged to be a long, complex and costly process. Transportation and modification funds were simply unavailable and, as 1978 wore on, NASA was already looking to a high-fidelity structural test article – known as 'STA-99' – as a cheaper alternative to upgrade for orbital service. Since the signing of the original Shuttle contracts in July 1972, the reusable spacecraft's design had evolved under such weight-saving pressures that virtually all airframe components were required to handle significant structural stress. Furthermore, in view of the difficulties involved in accurately predicting mechanical and thermal loads on the vehicle using the limited computers of the 1970s, NASA opted to build STA-99 specifically as an engineering tool. As a result, after its completion in February 1978, the structural test article underwent a year of intensive evaluation in a steel rig at Lockheed's Plant 42 facility in Palmdale, California. Originally intended to test the Lockheed TriStar aircraft, the rig's 256 jacks subjected 836 load application points to pressures equivalent to those of launch, orbital flight, re-entry and landing. Even the tremendous jolt of main engine ignition was simulated by three hydraulic cylinders, each imparting a force of 450,000 kg. Additionally, cold nitrogen gas and thermal blankets were employed to recreate the frigid conditions of orbital flight and the

Enshrouded in a gloomy midwinter fog, Challenger rolls towards Pad 39A.

intense heat of re-entry. The decision to modify STA-99 as a 'true' orbiter came about because, unlike Enterprise, it was a bare, incomplete airframe and could be more economically upgraded.

Traditionally, manned spacecraft had been tested to 140 percent of their design strength, but NASA engineers recognised that this might cause so much damage to STA-99 as to make it inadvisable to do so. Consequently, Tom Moser and his team at JSC developed an analytical computer model to simulate over 3,000 measurement points on the airframe. Their results confirmed that it could easily withstand 140 percent loads, with actual stress distributions in critical areas comparing favourably with pre-test model data. On 29 January 1979, it was official. Under a $1.9 billion contract between NASA and Rockwell International, STA-99 would follow Columbia as the second spaceworthy vehicle and two more orbiters would also be built. Four days later, on 2 February, the structural test article was renamed 'Challenger'. Like Columbia, she was named for a seafaring vessel which had made an outstanding contribution to exploration. The nautical Challenger was a steam-assisted Royal Navy corvette, which undertook a prolonged cruise – the 'Challenger Expedition' – from December 1872 until May 1876. In support of this expedition, all but two of Challenger's guns were removed and her spars reduced to increase the availability of space for laboratories, additional cabins and a special dredging platform. Equipped with specimen jars, alcohol to preserve samples, microscopes and chemical apparatus, thermometers, sounding leads and sediment collectors, the 127,580 km Challenger Expedition gathered the equivalent of 50 volumes of information about the Atlantic and Pacific Oceans. Sadly, she was broken up for her copper bottom in 1921 and, today, nothing, save her figurehead remains; this is enshrined in the National Oceanography Centre in Southampton.

For her spacegoing namesake, ground evaluations, practice landings and structural test articles were no substitute for actually operating in space. Before she could be declared ready for flight, Challenger required substantial disassembly and rework at Rockwell's Palmdale plant in November 1979. Her payload bay doors, aft body flap and elevons were removed and returned to their vendors for refurbishment, followed by her vertical stabiliser tail fin in January 1980. She had been built with a simulated crew cabin, which required the two halves of her forward fuselage to be 'cracked open' to remove it for modifications. In July 1981, after its own series of improvements, the aft fuselage returned to Palmdale. In physical appearance, the rebuilt Challenger looked similar to Columbia. External appearances, though, proved deceptive. "Challenger would end up some 2,889 lb [1,300 kg] lighter," wrote Dennis Jenkins, "in spite of having additional operational equipment installed and the more robust structure." This saving was also achieved through the absence of ejection seats, the replacement of several hundred thermal protection tiles with new insulating blankets, the removal of a number of tube-supporting frames, the use of lightweight 'honeycomb' for her landing gear doors and vertical stabiliser and the incorporation of less weighty main engine heat shields.

By December 1982, most of the preparations in support of Challenger's maiden voyage, STS-6, had already been completed; in fact, attached to her External Tank and boosters, and partially enshrouded in a gloomy midwinter fog, she crept out to

Pad 39A on 30 November. She had been ferried to Florida almost five months earlier and, despite the ostensibly 'operational' status conferred upon her, she was given a battery of test equipment to monitor her performance during launch, ascent, orbital operations, re-entry and landing.

Following Challenger's rollout to the launch pad, several milestones had remained to be achieved before final preparations for STS-6 could commence. One of the most critical exercises was a Wet Countdown Demonstration Test, which was scheduled to culminate on 18 December in an approximately 15-second firing of her three main engines. This Flight Readiness Firing (FRF) was needed to demonstrate the engines' ability to throttle between 94 and 100 percent rated thrust and 'gimbal' under hydraulic command, just as they would be required to do during launch. Similar 'wet' – or fully-fuelled – tests had been performed before each Saturn V launch, although on those occasions the rocket's engines were not fired. Preparations for the FRF proceeded in a manner not dissimilar to a real countdown: launch controllers started the clock by powering up the SRBs and ground support equipment and activating Challenger's flight systems. With a minute to go before engine ignition, the NASA commentator picked up the coverage: "T-1 minute and counting ... the firing system that releases the sound suppression water onto the pad has been armed ... T-50 seconds and counting ... T-45 seconds and counting ... T-40 seconds and counting; SRB development flight recorders are being turned on ... T-37 ... gaseous oxygen vent arm will *not* be retracted on this particular test ... T-31 seconds, we have a Go from LPS [Launch Processing System] for auto-sequence start ... [Challenger's] four primary flight computers taking over control of the terminal count ... final LPS command for engine start will occur at approximately ten seconds ... T-15 seconds and counting ..." At this stage, the relative silence and serenity on the pad began to change markedly. Firstly, the sound suppression water, thousands of litres of it, began to gush across the launch pad from four gigantic 'rain birds'. "T-10 ... *Go for main engine start...we have* main engine start ..." as the now-familiar sheet of orange flame gave way to a trio of shock diamonds from the three main engines, combined with a thunderous roar and vast cloud of smoke. The engines ignited in a ripple-like sequence, starting up at 120-millisecond intervals, reaching 90 percent of rated performance within three seconds and hitting the 100-percent mark precisely as the countdown clock touched zero: "T-0, engines throttled at 100 percent, all engines up and burning ... T+5 seconds, engines continuing to burn ... T+10 seconds ... *twelve* ... first [engine] cutoff at T+15 seconds ... [Number One] engine cutoff ... and engines Two and Three also cutoff at 16.8 seconds ... T+25 seconds; GLS [Ground Launch Sequencer] safing now in progress ..."

As well as helping to validate Challenger's integrity, the test also evaluated her External Tank, which was of a lighter design, weighing 4,500 kg less than earlier models. This had been accomplished by eliminating portions of longitudinal structural stiffeners (known as 'stringers') and milling the tank with a thinner aluminium skin. Additional weight savings included replacing heavy SRB attachment points with lighter, yet stronger and cheaper, titanium alloy ones and removing an 'anti-geyser' line previously used to circulate liquid oxygen during the lengthy tank-filling process. Moreover, the SRBs themselves were lighter, with walls 0.08 to 0.12 mm thinner than previous boosters. This saved 1,800 kg in weight, although in the wake of Challenger's second mission it was feared that *too much* material had been removed and NASA reverted, for a time, to the original thickness. Finally, Challenger's main engines were capable of achieving 104 percent thrust – a four percent increase over the capabilities of Columbia – which enabled her to transport heavier payloads aloft. In fact, for each one-percent of performance increase over rated thrust, the new Shuttle gained 450 kg of additional payload-to-orbit capability. Challenger's higher thrust was accomplished by incorporating redesigned components into each engine. Such changes became necessary to deal with the higher temperatures, pressures and pump speeds that they would encounter at the greater thrust levels. A series of test firings, lasting over 62,000 seconds, had been performed to validate the engines in readiness for their first orbital mission; additionally, the main injectors employed stronger liquid oxygen posts and modified fuel preburners to overcome turbine blade erosion and thicker tubes and coolant supply lines to handle higher aerodynamic loads at liftoff.

However, their performance during the FRF was not entirely successful.

As they blazed at full power on 18 December, engineers detected levels of gaseous hydrogen in Challenger's aft compartment which significantly exceeded allowable limits. When it became impossible to pinpoint the cause or location of a leak, the decision was taken to perform a second FRF. New instrumentation was installed both inside and outside the aft compartment to determine whether the hydrogen was leaking from an internal or external source. Suspicion focused initially on the latter possibility, because vibration and current had found their way into the aft compartment, behind the engines' heat shields. Extra sensors and a higher than ambient pressurisation level were duly installed to prohibit any penetration by 'external' hydrogen sources. By the beginning of the new year, launch was not anticipated before 1 February at the earliest. However, the second test firing on 25 January, during which the engines were run at 100 percent for 23 seconds, *again* revealed the presence of leaking hydrogen gas. Several more days of troubleshooting eventually identified a cracked weld in tubing leading to the uppermost (Number One) engine, which was promptly removed on 4 February. A replacement arrived from the National Space Technology Laboratory in Mississippi, but initial inspections in the VAB uncovered a leak in an inlet line to its liquid oxygen heat exchanger. Before it could even be installed onto Challenger, the 'replacement' was *itself* replaced by a third engine. After more checks in Mississippi, including a 500-second, full-flight-duration test firing, it was despatched to Florida on 3 March and fitted a week later.

Unfortunately, while this work was ongoing, painstaking efforts were underway to ensure that the other two original engines did not exhibit any leaks – and the bad news seemed to be that they *did*! Towards the end of February, hairline cracks were found in one of the fuel lines to the left-hand (Number Two) engine and borescope observations of the right-hand (Number Three) engine revealed a similar problem. Both were removed, returned to the VAB and repairs were conducted. With the arrival of the replacement engine from Mississippi, all three were installed by mid-

March and verified as being ready to support a launch. The leaks from the Number Two and Three engines were apparently caused by a generic 'seepage' in a 45 cm inconel-625 tube in their ignition systems. It apparently occurred underneath a protective sleeve brazed onto a small hydrogen line which sent fuel to the engine's augmented spark igniter. The sleeve was designed to counter possible chafing. After practicing cutting off the sleeve on the Mississippi plant's test stand, Rocketdyne technicians proceeded to Florida and replaced it with a non-sleeved inconel-625 tube on each of Challenger's engines.

By the time their orbiter was finally declared 'flight ready', the four-man crew of STS-6 had already performed their Terminal Countdown Demonstration Test and, on 5 February, the payload for their five-day mission – the first in a series of large Tracking and Data Relay Satellites (TDRS), attached to an Air Force-built Inertial Upper Stage (IUS) booster – was transferred to the launch pad and inserted into the payload bay. The impressive TDRS/IUS combination, when accommodated in a doughnut-shaped 'tilt table', consumed three quarters of the bay and was by far the largest single payload yet carried by the Shuttle. After deployment, ten hours into the mission, the two-stage, solid-fuelled IUS would boost the TDRS into a geostationary transfer orbit and finally circularise that orbit at an altitude of 35,600 km above Earth. Before launch, the satellite was alphabetically designated 'TDRS-A', but when fully operational it would be numerically renamed 'TDRS-1'. In spite of the engine leaks, the payload had originally been moved to the pad a couple of days after Christmas, but when it became clear that Challenger would not be flying in January 1983 and another FRF would be needed, it was returned to the Vertical Processing Facility for temporary storage. By the end of the first week in February, it was back at Pad 39A and ensconced in the Payload Changeout Room of the Rotating Service Structure for installation aboard the Shuttle. Then, on 28 February, strong winds whipped across the launch area and breached a weather seal between the changeout room and Challenger's payload bay, depositing a fine layer of particulate material on the satellite's solar array deployment springs. This resulted in an additional delay from 26 March until 4 April. After thorough inspections, TDRS-A was removed and carefully cleaned, before being replaced aboard the Shuttle on 19 March. The plan was to follow up with the second TDRS in the summer of 1983. They would provide near-continuous voice and data relay traffic between Mission Control and future Shuttle crews, as well as support an ambitious series of future scientific missions, including the Hubble Space Telescope. Unfortunately, a long and difficult road would have to be travelled before TDRS-A could enter operational service.

FLIGHT OF THE F-TROOP

For the four men assigned to fly STS-6, as with their predecessors, the wait for space had been a long one. Commander Paul Weitz was making his second flight and yet even he had waited almost a decade since last savouring weightlessness; his colleagues Karol Bobko in the pilot's seat and mission specialists Don Peterson and Story Musgrave had waited more than a decade apiece for their first voyage into orbit. Publicly, Weitz' crew had nicknamed themselves 'The F-Troop' – from a television series about an aging cavalry unit. This reflected the fact that they were the sixth ('F') Shuttle crew and also their military backgrounds. The idea for the nickname came from Weitz himself and they even had tongue-in-cheek 'official' photographs and memorabilia produced, including a crew portrait, with the quartet clad in Civil War attire, with cavalry hats, braces and red and white neckties: Weitz seated, very stern-looking, sword in hand, Peterson with a Winchester lever-action rifle, Musgrave with a bugle and Bobko tending a cavalry flag, emblazoned with the legend 'F-Troop'. The sword, according to Peterson in a NASA oral history, had once belonged to a lieutenant in Napoleon's army. Another image, made available for television transmission, involved the crew donning vintage spectacles.

Certainly, the aged cowboy image was apt, because when Challenger launched on 4 April 1983 to begin STS-6, her first orbital voyage, Weitz, Bobko, Peterson and Musgrave may have inspired the movie 'Space Cowboys', since they were the oldest astronaut crew to date. Behind their backs, fellow astronauts dubbed them 'The Geritol Bunch', after the dietary supplement famously associated with aging. Peterson, for one, did not recall *that* nickname with quite as much fondness. "Maybe that was something everybody said about us when we *weren't* around," he told the oral historian.

Donald Herod Peterson was born in Winona, Mississippi, on 22 October 1933. He grew up with an interest in adventure stories and science fiction and his first exposure to aviation came from a close neighbour, Joe Glenn, who had served as a fighter pilot in the Second World War, flying the Lockheed P-38 Lightning. "He was kind of the town hero," Peterson explained. "Since he lived next door, I could go visit with him and he used to talk about flying." At Winona City High School, Peterson gravitated towards mathematics and physics and entered the Military Academy at West Point in New York to study for a science degree. "I'm *so old*," he explained, "that when I graduated, there was no Air Force Academy and the rule then was that one-third of the West Point graduates and one third of the Annapolis [Naval Academy] graduates went into the Air Force; so it wasn't like a big individual decision. All assignments out of the Military Academy were based on class standing, but you didn't have to stand very high in the class in order to get your choice. Since *a third* of the whole class went into the Air Force, there were *a lot* of us in those days that did."

Having gained his degree in 1955, Peterson spent his first four years of active duty in the Air Force as an instructor pilot in the Air Training Command; it was an assignment he enjoyed, since it offered formation and instrument flying and aerobatics. He would credit his experience in these few years as having aided his selection to enter test pilot school. Peterson was then detailed to follow a nuclear-powered aircraft programme, which included the opportunity to pursue a master's degree in nuclear engineering at the Air Force's Institute of Technology, based at Wright-Patterson Air Force Base in Ohio. Ironically, a mere *six months* before he graduated, in early 1962, the nuclear-powered aircraft was cancelled! He moved instead to work in technical intelligence at Wright-Patterson, focusing on nuclear reactor systems, and later served as a fighter pilot with Tactical Air Command,

The crew of the F-Troop disembarks from Challenger at Edwards Air Force Base, concluding their five-day mission to deploy TDRS-A and perform the first Shuttle EVA. Paul Weitz leads Story Musgrave, Don Peterson and Karol 'Bo' Bobko. As well as forming the oldest spacegoing crew ever assembled at that time, these four men had participated in the design, development and testing of the Shuttle from its conception.

before entering the Aerospace Research Pilots School at Edwards Air Force Base in California.

Peterson's introduction to space exploration came in June 1967 when he was selected as one of four pilots in the third group for the Air Force's ill-fated Manned Orbiting Laboratory (MOL), a military station for orbital surveillance and reconnaissance. Years later, he would credit not only his flying experience, but also his background in technical intelligence, as having helped with his selection. Training included the kind of water and jungle survival skills undertaken by NASA's astronauts, but due to the classified nature of MOL, Peterson had to travel around the United States on false identification papers. Unfortunately, as the war in Vietnam steadily drained military coffers, MOL was finally cancelled by President Richard Nixon in June 1969. It had swallowed close to a billion dollars and was still several years away from its first launch. George Mueller, NASA's Associate Administrator for Manned Space Flight, pressed Deke Slayton to accept the ex-MOL pilots into the astronaut office. Mueller knew that NASA *needed* the Air Force on its side as it pushed the Shuttle towards Congressional approval. "It won't hurt," he told Slayton, an ex-Air Force officer, "to make them happy, just this once."

Mindful of his awkward conversation with the Excess Eleven, less than two years before, Slayton considered it unfair to pick a new crop of astronauts and then let them down by offering them little chance of ever flying into space. Eventually, he and Mueller reached agreement: they would pick the ex-MOL pilots who were 36 years of age or younger. Seven were ultimately selected and, of those, Peterson only narrowly made the cut: at the time of the NASA announcement in September 1969, he was just a few weeks shy of his 36th birthday! He was 'in', but with very little idea of when – or *if* – he might ever gain a seat on a space mission. Peterson and fellow selectee Hank Hartsfield opted to study for their PhDs, but when NASA changed its criteria for when it wanted them to report to Houston for training, the idea came to nothing. Hartsfield went on to pursue a master's credential, but since Peterson already held one, "it really wasn't worth doing". In his oral history, Peterson recalled that he needed special permission from the Air Force to be what he referred to as "a PhD dropout". Years later, he supposed that if he had been allowed to pursue it, his doctoral research would have focused on aerodynamics and thermodynamics, and been based at the Arnold Engineering Development Center at Arnold Air Force Base, near Nashville in Tennessee.

Although no space missions were on the horizon, Peterson found himself working on the support crew of Apollo 16, the penultimate lunar landing, in April 1972, and was later assigned to the development of the Shuttle. "In fact," he said, "most of my career was spent working Shuttle." Over the next few years, he was involved in the testing of the new vehicle's avionics, computers, guidance equipment, navigational controls and air data systems. Working alongside Hartsfield during this period, Peterson helped to address flaws in the flight controls using the Shuttle Avionics Integration Laboratory. Twenty-four years of active service in the Air Force came to an end in 1979, when Peterson retired from the service, having reached the rank of colonel. However, he elected to remain with NASA as a civilian.

When he was assigned to STS-6 in March 1982, it seemed inconceivable that Peterson and fellow mission specialist Story Musgrave would prove to be the first men to perform an EVA from the Shuttle. The problems encountered by Bill Lenoir and Joe Allen changed all that and, as early as 27 November, *Flight International* was reporting that it was "likely" that the EVA would be rescheduled to either STS-6 or STS-7. NASA was well aware that the new space suits needed to be thoroughly tested in readiness for the ambitious Solar Max repair – a mission which had long been on the cards, but which had only been officially approved by the agency in August 1982 – and within days of the completion of Richard Colonna's report it was formally decided to reassign the EVA to STS-6. This was confirmed by *Flight International* on 25 December. According to NASA, the addition of the spacewalk caused the mission to be extended from an original two days to five days. "It didn't give us much time to train," Peterson recalled. "I didn't have much experience in the suit, but the advantage we had was that Story was the astronaut office's point of contact for the suit development, so he knew *everything* there was to know. He'd spent 400 hours in the water tank, so he didn't really have to be trained!" This water tank was the Weightless Environment Training Facility (WET-F) in Building 29 at JSC and employed 'neutral buoyancy' – placing the astronauts, fully-suited and laden with lead weights, underwater – as a technique to simulate the microgravity conditions of low-Earth orbit. Accompanied by scuba divers to ensure their safety, Musgrave and Peterson were able to rehearse procedures for their spacewalk in a water tank which measured 7.6 m deep, 23.8 m long and 10.1 m wide, large enough to contain a full-sized model of the Shuttle's crew cabin, its airlock and its payload bay.

There were, however, disadvantages to the WET-F. Astronaut Bruce McCandless, whose expertise and role in the suit development process closely paralleled Musgrave's own, explained that the optical distortion, caused by looking through a concave helmet in the optically-dense medium of water, caused everything to appear 'smaller'. Of course, during training, neither Musgrave or Peterson were truly 'weightless' and two major differences between operating in the tank and working in space was that they could still 'feel' the weight of their 125 kg suits and the viscous drag of the water, which tended to make some tasks easier to perform on the ground and others easier to perform in space. Still, even today, neutral buoyancy remains the closest terrestrial analogue to the real thing.

During typical training runs, the scuba divers guided Musgrave and Peterson into the WET-F and fitted the weights onto their suits, to enable them to 'hang' in the water, neither rising or sinking. Encased inside the suits, conditions were often uncomfortable and sometimes painful. Since their bodies were supported by the weight of the suit, said McCandless, "the blood ran to your head when upside down and your *weight* was supported on your collar bone!" This made a precise fit essential: both astronauts' heels had to be firmly planted against the backs of their boots, their shoulders secured against their harnesses and their heads crisply against the crowns of their helmets. "It used to be a form of medieval torture to hurt people's fingernails," said Peterson, "but the *gloves*, if they're wrong, can be *really* bad." Hand fatigue over several hours in the suit created effects similar to squeezing a balloon and, according to veteran spacewalker George 'Pinky' Nelson, any irritation or pressure points quickly led to extreme soreness in the fingers.

By his own admission, Peterson's EVA training for STS-6 "was pretty rushed". He recalled being underwater in the WET-F on 15-20 occasions, "and that's not really enough to know everything you need to know". However, the tasks for their excursion were relatively straightforward evaluations of the performance of the suit and of the airlock; if any of the equipment tests went awry, he and Musgrave could quickly curtail their spacewalk and return to the safety of Challenger's cabin. "Story was a fun guy to work with," Peterson told the NASA historian. "On the job, he was extremely dedicated and would do anything; he'd work 20 hours a day. He didn't argue about anything, but just did whatever needed to be done." Musgrave was actually the ideal choice to lead the first Shuttle EVA, having closely followed, virtually from conception, the development of the suit. More than that, Musgrave's experience, education and delightfully unique persona became the stuff of legend and an inspiration to several generations of astronauts. Tom Jones, who flew with him, wrote that Musgrave was "the office's Renaissance Man" and that his 16-year wait for a space mission "put me in awe of him".

Scientist, physician, engineer, pilot, mechanic, poet and literary critic, Franklin

Story Musgrave had amassed extensive flying time in the Marine Corps and by 1987 would have secured no fewer than *six* academic degrees. He was never called by his first name – not even by his parents – and his middle name, 'Story', actually honoured an old family surname from several generations back. "I got into this business to be on the intellectual and physical frontier," he explained of his decision to pursue a career with NASA. "I wanted a *transcendental* experience – an existential reaction to the environment. I'm not talking about an illusion, or seeing something that wasn't there, but a magical, emotional reaction to the environment. *That's* what I've been after all my life: to experience and feel new sensations." That life, certainly, began badly for Musgrave, and in *NASA's Scientist-Astronauts*, Dave Shayler and Colin Burgess would characterise his formative years as "a childhood filled with despair". Born in Stockbridge, Massachusetts, on 19 August 1935, his parents were alcoholics – his mother meek and acquiescent, his father malicious and brutal – and the family's isolation on a farm meant that visitors were rare. Two things 'saved' him: one was that his ancestry on his father's side had boasted *nine* straight generations of doctors and the other was the ability to escape into the natural environment. "The unhappy situation," wrote Shayler and Burgess, "would often cause young Story to flee his home by night, making his way into the embrace of a nearby forest, where he would lie on his back, look up and marvel at the stars. He recalls doing this when aged *only three*, but the darkened forest held no fears for him." Even at this age, Musgrave considered nature to be his home – "my solace," he said, "a place where there was beauty, in which there was order" – and he was soon building rafts and becoming more self-reliant.

His salvation came in 1945, when his abused mother finally decided that she could no longer bear a violent existence with her husband and fled, taking Story with her. Yet tragedy was never far away: both of his parents *and* his younger brother would eventually commit suicide, whilst his older brother died in an aircraft accident. Psychologically, these calamities helped to improve Musgrave's self-reliance and mould him into the man that he would become. Surprisingly, in view of his later academic accomplishments, he did not shine at high school. "He hated school and all that was associated with it," wrote his biographer, Anne Lenehan, in *A Passionate Life* on Musgrave's website, www.storymusgrave.com. "He was constantly in trouble with the school authorities and was subjected to almost continual disciplinary action." A motorcycle accident caused him to miss out on his final exams, but Musgrave's perspective was that home life and school life offered a "fantastically narrow" window on the world. "I felt the urge to expand my horizons," he told Lenehan, "and to see other worlds." He joined the Marines in 1953 and trained as an aviation and instrument technician, completing active-duty assignments in Korea, Japan and Hawaii and serving aboard the aircraft carrier USS *Wasp* in the Far East. These years also rekindled an earlier interest in aviation and he resumed his studies to gain his pilot's licence.

Clutching a National Defense Service Medal and an Outstanding Unit Citation from his Marine Corps squadron, Musgrave left the military and enrolled at Syracuse University in New York to study mathematics and statistics. Shortly before gaining his bachelor's degree, he was employed as a mathematician and operations analyst by the Eastman Kodak Company in Rochester, New York. He then followed up with a master's in business administration from the University of California in 1959, a bachelor's credential in chemistry from Marietta College in 1960 and a doctorate in medicine from Columbia University in 1964. With his medical degree under his belt, Musgrave began his own research into the human nervous system, with a one-year surgical residency at the University of Kentucky Medical Center in Lexington. His achievements enabled him to win post-doctoral fellowships from both the Air Force and the National Heart Institute. During this period, Musgrave's interests broadened to encompass aerospace physiology, temperature regulation and clinical surgery ... and *another* master's degree (this time in physiology and biophysics) in 1966. When NASA announced its intention to select a group of scientist-astronauts, Musgrave was convinced that his experiences had led him to the door of space. At first, NASA considered him to be *over-qualified* – five degrees so far, an active laboratory, a surgical practice and a licenced commercial pilot, flight instructor and accomplished parachutist (he would ultimately make more than 500 jumps) – but certainly his potential was noticed and he was selected in August 1967.

Musgrave's advantage over the other members of the Excess Eleven, aside from his sheer number of academic degrees, was that he was alone among them in being a qualified pilot ... although even he had never flown in high-performance jets. Jungle survival training and flight instruction were therefore mandatory for all group members. Musgrave went to Reece Air Force Base in Lubbock, Texas, completing his 53 weeks of training with the highest scores ever recorded at the base and a commendation. In April 1969, he was detailed to the Apollo Applications Program, subsequently renamed 'Skylab'. Less than two years later, his name was announced as the backup science pilot for the first Skylab mission. At the time, plans existed for a second space station – Skylab B – and Shayler and Burgess have speculated that had this actually flown, in around 1975-77, Musgrave would have been a primary candidate for a seat on one of its missions. Sadly, budget cuts and the emphasis on getting the Shuttle ready to fly ultimately sounded the death knell for Skylab B. In 1974, he was assigned to the life sciences branch of the astronaut office and in October of that year participated in a medical development test of Spacelab with Dennis Morrison of JSC's Bioscience Payloads Office. The two men spent seven days inside a Spacelab mockup and conducted a series of biomedical demonstrations in order to perfect operational procedures for real missions. Musgrave participated in a second such test for five days in January 1976, together with nuclear chemist Robert Clark and cardiopulmonary physiologist Charles Sawin; on this occasion, they lived and worked in a full-scale mockup of not only the Spacelab facility, but also the middeck and flight deck of the orbiter itself.

During this period, Musgrave continued clinical and scientific training as a part-time surgeon at Denver General Hospital and as a part-time professor of physiology and biophysics at the University of Kentucky Medical Center. In his NASA capacity, he participated in the development of the Shuttle space suit – together with the airlock, life-support systems and a new Manned Manoeuvring Unit – as well as working in the Shuttle Avionics Integration Laboratory. Moreover, for an astronaut who was selected with *no* jet experience, Musgrave would eventually amass 17,700 hours of experience; more than a third of this time was in NASA's T-38 Talon jets, but in all he flew over 160 different types of aircraft. By comparison, no other astronaut has come close to this total, with most former chiefs of the office averaging 7,000 hours. In fact, even the most flight-experienced chief astronaut, John Young, logged 15,200 hours in his career. "From the beginning of his NASA days," wrote Anne Lenehan, "Story flew just about every day, sometimes *twice* a day, and around a hundred hours a month ... an extraordinary amount, given his other commitments." Nor was Musgrave a cautious aviator. Fellow astronaut Mike Mullane recalled one stomach-churning episode aboard a T-38, shortly before STS-6. In his autobiography, *Riding Rockets*, Mullane wrote that Musgrave "asked [Air Traffic Control] for a block of altitude and then went into a series of spiralling rolls and violent manoeuvres that alternately had me slammed into my seat at 4 Gs and lifted from it in negative Gs. My head snapped back and forth like a palm tree in a hurricane. Within a minute I was ready to blow my last meal ... and had to plead with him to stop." Musgrave's astonishing array of talents would also encompass *another* degree (a bachelor's qualification in literature from the University of Houston in 1987) and his remarkable attention to detail became legendary. He was nicknamed 'Dr Details' by his fellow astronauts. He exuded self-confidence both in the missions he flew and in the tasks he fulfilled. "It was sheer play for me," said the man whose life had begun under such a cloud of menace, "to be able to so completely interact with my environment."

For the spacewalk on STS-6, Musgrave would take the lead as 'EV1', wearing red stripes on the legs of his suit, and although it was scheduled to last for barely four hours, the *preparation* and *conduct* of the excursion consumed the crew's entire working day on 6 April 1983. Aided by Bobko, who would choreograph the spacewalk, Musgrave and Peterson rose early that morning to begin readying their suits and equipment in Challenger's airlock. They spent three and a half hours 'pre-breathing' to avoid an attack of the bends. Otherwise known as 'caisson disease', the bends are triggered by the formation and expansion of nitrogen gas bubbles in the blood when subjected to a rapid decrease in external pressure. The consequences can be dire: ranging from severe pain in the joints to paralysis and eventually death. Indeed, the name 'bends' comes from the fact that sufferers instinctively bend into a foetal position. To sidestep this danger, in a procedure similar to that commonly followed by deep-sea divers, Musgrave and Peterson spent time in Challenger's 101.3 kPal middeck, pre-breathing pure oxygen from face masks to prepare themselves for operating inside the suits at 29.6 kPal pure oxygen. Shortly before the onset of pre-breathing, the entire cabin had been reduced from its normal pressure to around 70.3 kPal, whilst the percentage of oxygen in the atmosphere was slightly increased. As Musgrave and Peterson breathed using their masks, they were still able to attend to their other tasks on the middeck. At the end of pre-breathing, with sufficient dissolved nitrogen now cleared from their blood, they were at last ready to begin donning their space suits.

Since reaching orbit, two days earlier, Weitz, Bobko, Peterson and Musgrave had repeatedly checked their equipment for the long-awaited EVA: testing a third, 'spare' upper torso in accordance with flight rules, verifying that oxygen regulators and fans operated normally, inspecting for leaks and confirming that communications were satisfactory. In fact, the only problem raised was a need to replace some flat floodlight batteries. With everything in place, space suit donning began and, in true F-Troop fashion, it ran as crisply as a military campaign. "We were instrumented with little stick-on patches to measure our heart rates," Peterson recalled. "Then we put on what looked like long underwear – a cooling garment – which had water tubes that ran through it and hooked through a connector to the suit." This long underwear, officially known as the 'liquid cooling and ventilation garment', was a one-piece, zip-up suit, based on one developed for moonwalkers, composed of stretchable spandex fabric and laced with 91.5 m of plastic tubing. During the course of their spacewalk, cooling water would be pumped through this tubing to control Musgrave and Peterson's body heat, exhaled gases and perspiration. Next, anti-fog compound was rubbed onto the insides of their helmets. "I wore glasses," said Peterson, "and we rubbed this on the lenses so they wouldn't fog up, because I was inside a helmet and couldn't get my hands inside the suit." To provide an additional measure of safety and prevent them from falling off, Peterson's glasses were tied to an elasticated strap around the back of his head.

Before actually clambering into the two-piece space suits, electrical harnesses were attached to their 'hard' upper torsos to provide biomedical and communications links through the backpack. A wrist mirror and spiral-bound, 27-page checklist were placed on each suit's left arm, followed by the insertion of a small fruit and nut food bar and water-filled drink bag. The next step was the connection of a black-and-white communications hat – famously nicknamed the 'Snoopy cap' since Apollo days – to the top of the torso. Physically, the so-called Extravehicular Mobility Unit was a $2.5 million miniature spacecraft in its own right, consisting of 'upper' (above-waist) and 'lower' (below-waist) segments, together with helmet, gloves and backpack, known as the Portable Life Support System. The suits had been developed under a series of contracts with Hamilton Standard of Connecticut, signed between July 1976 and March 1977. Musgrave and Peterson firstly pulled themselves into the lower torso, which featured joints at its hips, knees and ankles and a metal body seal closure for connecting to a ring on the upper torso. It also included a large bearing at its waist, which offered greater mobility and allowed the astronauts to twist whilst their feet were held firmly in restraints.

After donning the trousers of the suit, their next step was to plug the airlock's service and cooling umbilical into a display and control panel on the front of the upper torso. This would provide cooling water, oxygen and electrical power from the Shuttle until shortly before they were scheduled to go outside, thereby conserving the limited consumables available in their backpacks. The two men finally entered the airlock, where the upper torsos 'hung' on opposing walls and, through a half-diving, half-squirming motion, manoeuvred themselves into the top halves of their suits. With arms outstretched, and Bobko nearby to assist, they slipped themselves into the upper torsos and their waist rings were brought together, connecting the cooling water tubing and ventilation ducting of the long underwear and the biomedical sensors to their backpacks. Bobko helped them to lock the body seal closure rings at their waists. The hard upper torso was essentially a fibreglass shell under several

Musgrave works in the payload bay during the first EVA of the Shuttle era. The doughnut-like ASE tilt table, used to support and deploy TDRS-A, earlier in the mission, is clearly visible. So too is the Earth – a vista which Musgrave never tired of gazing upon.

fabric layers of a thermal and micrometeoroid garment. On its back, it held the life-support system and on its chest the display and control unit by which the spacewalker would manage oxygen, coolant and other consumables; in fact, due to the difficulties in seeing 'down' to read labels on the unit, the mirrors on the suits' left wrists would help immeasurably. For additional ease, the labels were written backwards!

Next they pulled on their Snoopy caps, equipped with microphone and headphones to provide two-way communications with his crewmates and Mission Control. Then came the gloves. Snapped into place on the wrist rings of the upper torso, these had silicone rubber fingertips to provide a measure of tactile sensitivity when handling tools in Challenger's payload bay. Finally, the enormous polycarbonate bubble helmets were lifted over the astronauts' heads and clicked into place on the neck rings of their upper torsos. Over the top of each helmet was an assembly containing manually adjustable visors to shield their eyes from solar glare, together with two EVA lamps to illuminate work areas out of range of the Sun or the Shuttle's own payload bay floodlights. Mobility in the neck rings was unnecessary, because the helmets were easily big enough to allow the astronauts to move their heads around.

Unlike previous Apollo space suits, the modularised Shuttle ensemble, with its waist closure ring, eliminated the need for pressure-sealing zips and therefore had a much longer shelf life. Additionally, the use of newer, stronger and more durable fabrics enabled space suit engineers to design joints with better mobility, resulting in lower weight and a reduction in overall cost. Story Musgrave and Don Peterson, by now floating motionless in Challenger's tiny airlock, were, in effect, small spacecraft in their own right. However, they were not yet 'self-contained', as their oxygen, electricity and cooling water were still being provided by the Shuttle's systems; not until shortly before the two men ventured outside would they transfer to their suits' life-support consumables. Before they could do that, they had to lower the airlock's pressure to 29.6 kPal to check their integrity, necessitating a final 45 minutes of pre-breathing. At length, at 4:05 pm EST on 6 April 1983, Musgrave initiated the final depressurisation of the airlock and, 16 minutes later, pushed open the outer hatch into the payload bay. The plan called for three hours of activities, but in order to accommodate delays he and Peterson had up to six hours' worth of consumables. Entering the overwhelmingly floodlit bay for the first time, one of his first comments, somewhat understatedly, was that "it's so *bright* out here!"

Watching through the aft flight deck windows, Paul Joseph Weitz, commander of STS-6, knew exactly what Musgrave and Peterson were experiencing, because he had performed an EVA himself, a decade earlier. On seeing Musgrave poke his head outside into the floodlit bay, Weitz would later quip that "Story seemed like a butterfly coming out of a chrysalis – only *he's* not as pretty!" A retired captain in the US Navy, Weitz, known almost universally by his nickname, 'PJ', was born in Erie, Pennsylvania, on 25 July 1932 and like many pilot-astronauts from the Apollo era he had a natural affinity for aviation. With a chief petty officer father, who fought at the battles of Midway and the Coral Sea, Weitz has described himself as "an impressionable young lad during World War II" and the Navy drew him like a magnet. Graduation from high school led to a scholarship from the Navy to study aeronautical engineering at Pennsylvania State University. Whilst there, his instructor advised him to go to sea aboard a destroyer before entering flight training. Weitz would later regret doing this, "because it put me a year and a half behind my contemporaries". When he finally entered flight school in Jacksonville, Florida, he found himself a classmate of Al Bean, who would also become an astronaut and command Skylab's second crew.

Weitz' first application to attend the coveted Naval Test Pilot School at Patuxent River, Maryland, was rejected and he moved instead to an air development squadron at China Lake in California's Mojave Desert. His second attempt to enter 'Pax River' was accepted – and then *rejected*, because "I had just been moved from the East Coast to the West Coast and they weren't going to move me back to the East Coast again!" Nonetheless, Weitz enjoyed his time at China Lake, serving as a project pilot for five different types of aircraft and helping to develop tactics for delivering air-to-ground weapons. With a degree under his belt, his next (unsolicited) assignment was to the Naval Postgraduate School in Monterey, California, to study for a master's credential in aeronautical engineering. "I did not apply for it," he told NASA's oral historian. "I didn't want to go, because I had what I thought was a good job." At the time, Weitz had no desire to return to full-time education and believed it to be the wrong choice; however, years later, he would admit that "since I did not have a test pilot background directly ... the reason I got selected finally for the astronaut programme was because I did have a master's degree". By the time of his graduation in 1964, he had met three other astronauts-to-be: Gene Cernan, Ron Evans and Jack Lousma.

In fact, by completing 'extra work', thanks to the input of a kindly professor, both Weitz and Lousma were able to complete their degrees ahead of schedule, in two years, rather than the standard three. Weitz then moved to Whidbey Island in Washington State's Puget Sound to fly the A-3 Skywarrior heavy bomber, undertaking his first combat mission in Vietnam in 1965. It was during this period that "another one of those strange forks in the road" came up and he received a message from the Bureau of Naval Personnel, informing him that NASA was looking for astronaut candidates. Lacking test pilot credentials, Weitz assumed that his chances of selection were slim and so was surprised when he made the cut as one of 19 new astronauts in April 1966. His association with Apollo Applications began early – in his oral history, he recalled undergoing EVA training with Jack Lousma to make an S-IVB hydrogen tank habitable as a 'wet workshop' – and certainly his performance on the support crew for Pete Conrad's Apollo 12 mission in 1969 caught the attention of his superiors and was a contributory factor in his selection as pilot of the first Skylab mission. In the early summer of 1973, Weitz, Conrad and science pilot Joe Kerwin undertook one of the most challenging missions in space history: Skylab had been partially crippled by a launch accident and during their record-breaking 28-day flight, the astronauts not only saved it from the brink of total failure, but also kicked off an ambitious programme of scientific research.

Returning from space, though, brought with it the grim reality that the *next* opportunity to fly would not come for many years and Weitz was not alone in considering whether to return to active military duty. "The impression I had," he said, "was they'd give me a good job to see how I'd perform for a couple of years, but by then ... I'd done nothing directly for the Navy [since 1966]. All these other people had done their time aboard ship and I just felt that my best opportunities were probably within NASA, rather than back in the Navy." Weitz retired from naval duty, with the rank of a captain, in June 1976, but remained with the space agency as a civilian. The lean astronaut corps was working almost exclusively on the Shuttle and Weitz' duties included the cathode ray tube displays for the flight deck instruments and he was named to what he called a "pseudo crew" for one of the early missions. "These crews, with *no* assurances of ever flying, were assigned, just to exercise the whole system," he explained. The crews were alphabetically labelled 'A' through to 'F', with Weitz in command of the F mission.

Teamed with Weitz during this period as pilot of the F crew was Air Force Colonel Karol Joseph Bobko, nicknamed 'Bo', whose military background mirrored that of Don Peterson in many ways: he too had worked in the MOL project and had been selected by NASA in September 1969. Of Polish descent, Bobko came from New York, where he was born on 23 December 1937. He attended Brooklyn Technical High School and joined the new Air Force Academy, graduating as a member of its first class in 1959. "I had thought about going to West Point," Bobko remembered, "and there was a lieutenant-colonel that lived in our neighbourhood and told me about the new Air Force Academy that was being developed." Many of the academy's instructors – one of whom had been the range safety officer at the Eniwetok Atoll atomic tests in the Marshall Islands of the South Pacific – were already discussing the possibilities of a future human space programme and the awareness dawned on Bobko that becoming a pilot would be an essential prerequisite. After graduation from the academy, he underwent flight instruction at Barstow Air Force Base in Florida and Vance Air Force Base in Oklahoma and became a fighter pilot, serving two tours of duty. From 1961 until 1965, he flew F-100 Super Sabre and F-105 Thunderchief aircraft, before being selected to attend the Aerospace Research Pilots School at Edwards Air Force Base in California. Shortly thereafter, in June 1966, Bobko was chosen as a member of the second group of MOL pilots, alongside future Shuttle astronauts Bob Crippen, Gordon Fullerton, Hank Hartsfield and Bob Overmyer.

Cancellation of this military space station in June 1969 brought with it the decision to return to his studies and Bobko enrolled on a master's course to study aerospace engineering at the University of Southern California. Selected by NASA in September of that year, he received his degree in 1970. Those first few months as an 'astronaut', though, were filled with frustration. Bobko knew that a space mission would not come his way in the near future, but his group was surprised to be told *twice* that year that they were *fired*. In later years, astronaut John Fabian would even recall tongue-in-cheek horror stories that Bobko's group was treated like a leper colony. "They [NASA] said, 'We don't need you folks and so you're going to go back to the Air Force,'" Bobko said, "but that was *rescinded* and we stayed and got involved in the programmes that were going on here." Among those programmes was the early definition of a modular space station and by 1975 Bobko was working as a support crew member for Apollo-Soyuz. He had also participated as a test subject in the 56-day Skylab Medical Experiments Altitude Test (SMEAT) in the summer of 1972. As the Shuttle effort gathered momentum, Bobko found himself serving as a capcom during the approach and landing tests of Enterprise in 1977, as well as validating the orbiter's performance as a *landing aircraft*. His tasks included helping to design the Shuttle's approach and landing profile and even using *radar* to study the structure of the lakebed runway surface at Edwards "to see if there were any voids the Shuttle might fall through". Six years later, floating in Challenger's flight deck, alongside Weitz, Bobko's long-held dream of flying into space was now a reality.

For Story Musgrave, whom the two pilots now watched, mesmerised, as he ventured outside, it was a moment of intense pride. "This [EVA] was to be only three hours of experience on top of 48 years," he said later, "but it's like a surgeon who's been training 16 years to operate. Sooner or later, the surgeon *has* to operate. Sooner or later, I *knew* I was meant to walk in space." The poetic justice of being first to venture into space wearing the fruit of so many of his labours was clearly not lost on the intensely philosophical Musgrave.

Although somewhat different from the ensembles worn on previous Gemini and Apollo missions, the suits were designed with the same objective in mind: to leave the pressurised confinement of a spacecraft. However, the near-flawless procedures followed by Musgrave, Peterson and Bobko to don the suits masked a complex, tumultuous and near-tragic developmental history. A multitude of glitches also characterised the space suit's early days: problems with CO_2-scrubbing lithium hydroxide cartridges, clogged sublimators, cycling water pressure regulators and battery failures.

At length, with these and other technical woes long since resolved, Musgrave and Peterson read status reports showing that all their suit systems were running near-perfectly. During the excursion, Bobko directed their every move from Challenger's aft flight deck, while Weitz photo-documented the historic event. The spacewalkers' first task was to tether themselves to slide wires running along the sills of the payload bay walls (one on either side, to prevent mutual interference) and move towards the aft bulkhead, in the process evaluating their ability to reach, pick up and handle tools. During the spacewalk, the 14 m long slide wires were used as part of a safety procedure to prevent Musgrave or Peterson from inadvertently floating away from the Shuttle. Meanwhile, the two men began conducting their first 'real' evaluations of the new suits: their comfort, dexterity, ease of movement and the performance of their communications and cooling systems and the payload bay floodlights. One of very few concerns expressed by Peterson after the mission was that "the gloves are hard to work in – extremely stiff – and I had to get my hands strengthened with a little hand exerciser". Despite this, both astronauts reported that the suits' mobility enabled them to satisfactorily accomplish each of their tasks. Most of their work focused on identifying suitable locations from which future spacewalkers could best work on the malfunctioning Solar Max, and on practicing some of the intended repair techniques.

This kind of work was essential, not only for the successful reactivation of the $240 million solar observatory, but also for future servicing missions to the still-to-be-launched Hubble Space Telescope, at that time scheduled for the mid-1980s. Musgrave and Peterson finally evaluated the manual system for closing Challenger's payload bay doors in the event of a failure. This involved using a hand operated winch attached to the forward bulkhead, and was performed both with and without foot restraints. It was during tests of the 7.3 m winch line that they encountered difficulties. "Story got the rope hung over something," Peterson recalled, "and couldn't release the winch. It was under a lot of tension. There was some talk about how we could get this thing loose so we could get it restored. We couldn't just leave it where it was because it was on the rollers that were used to latch the doors down." After the crew's suggestion to cut the Kevlar line was rejected by Mission Control, Musgrave eventually pried it free with his gloved hands. At another point during the spacewalk, in what NASA labelled a "high metabolic period", Peterson received a 'high O_2 usage' warning on his chest display. Although the message cleared quickly and did not recur, it was attributed to flexure within the suit and his high work rate. "I stopped and said 'I've got an alarm'," Peterson recalled. "Story stopped what he was doing and came over. We were trying to check what was going on and the seal popped back in place and the leak stopped. Now, in those days, we didn't have constant contact with the ground. They weren't watching at the time that happened. By the time we dumped the data from the computers to the ground that showed the leak, we were back inside the orbiter."

It seemed that Peterson's alarm was caused by overworking and breathing excessively rapidly; this depleted his oxygen, forced a higher feed level and triggered the warning. Biomedical data confirmed that his heart rate was around 192 beats per minute whilst cranking the wrench, but Peterson doubted he could have worked so hard as to breathe enough oxygen to set off the alarm. In spite of the problems, both Musgrave and Peterson clearly savoured the opportunity to venture outside the vehicle in orbit. They were even able to look 'outwards' into the Universe. "The Shuttle flies with the payload bay towards Earth all the time," recalled Peterson, "but we thought it would be neat when we got on the dark side if we could look out at the night sky and see all the stars. When we were on the daylight side, we went into 'the Ferris wheel mode'. Just like a Ferris wheel seat goes around and never changes attitude, we went around the world holding one attitude, so when we got on the dark side we faced exactly away."

Musgrave asked that the floodlights be turned off, but Weitz would have none of it. He was concerned that, on Challenger's maiden mission, problems switching them back on again would have left the two spacewalkers with the difficult task of finding their way back to the airlock in pitch darkness. "We didn't get much of a view," continued Peterson, "but what we could see was pretty interesting." Musgrave, for his part, was hoping for some sort of 'religious' experience, but kept an open mind. In the event, the EVA ran so smoothly that he was actually *disappointed* by what he felt. "I never got that transcendental jolt," he said. "I never experienced a separation phenomenon. I had no sense of Earth being 'down'. In fact, I had no 'down' reference at all. My frame of reference was always the payload bay of the orbiter."

A SWITCHBOARD IN THE SKY

After returning to the airlock, which was repressurised at 8:15 pm EST, the data on Peterson's alarm was pored over by flight controllers with dismay. "They were upset about it," he said later and, this being the suit's first outing in space, the astronauts would almost certainly have been directed back to the airlock had the problem occurred whilst in communication with Mission Control. At the time of STS-6, however, the Shuttle still relied heavily on a network of ground stations to relay its communications and data traffic during part of each orbit, typically only 20 percent. It was therefore ironic that on STS-6, which, in Peterson's mind, *benefitted* from having gaps in communications with the ground, the first in a series of huge Tracking and Data Relay Satellites (TDRS) had been deployed to improve communications with future Shuttle crews. It was optimistically hoped that, after the launch of a second, identical satellite on STS-8 later in 1983, it would be possible to talk to crews not only during the majority of their orbital time, but also throughout re-entry, eliminating the radio blackout normally experienced during this phase of the mission. Moreover, the existing network of 20-year-old ground stations, which were capable of supporting barely one or two spacecraft at a time, could be retired to save money. Once in place, the TDRS network was expected to be capable of supporting the Shuttle and up to 26 other satellites simultaneously.

In the eyes of the world, Musgrave and Peterson's EVA was the defining moment of STS-6, but when Paul Weitz' crew was announced by NASA in March 1982 their key tasks were to evaluate Challenger's spaceworthiness and insert the first TDRS into orbit. No EVA was planned and the flight was to last barely two days. Originally scheduled to fly on 24 January of the following year, the hydrogen leaks in the main engines led to more than two months of delays. After 13 years with NASA, a slightly longer wait proved of little consequence to Peterson. "It wasn't a huge big deal," he said. "I just figured, sooner or later, I'd get a chance to fly." The delays gave the astronauts and their control teams an opportunity to sharpen their skills. "This was about the most challenging job you could ask for," said STS-6 Flight Director Jay Greene. "The simulations were mental simulations that were as challenging as anything NASA has to offer. There were two things going on: one was the goal to train the crew to work with the control centre and, at the same time, train maybe a dozen different operators to the max extent possible. Instead of having one failure – which is about the most you'd expect during a launch – they'd try and give everybody something to play with and the flight director would have to co-ordinate everybody's problems and come out with a solution that got the crew safely to orbit or resulted in a successful abort and recovery. During the course of a day, we'd run maybe eight launch abort sims and every sim had maybe ten different faults that the [simulation supervisors] would put in. By the end of the day, you had somewhere between 80-100 problems that you dealt with." Training, Peterson admitted, was tough, but he did not recollect a single cross word between himself and his crewmates during their 13 months together. "And that's unusual," he pointed out, "because the training's really intense and very demanding and you're working long hours and things go wrong and there are many delays." However, he remembered a characteristically unflappable Musgrave happily going off to fetch sandwiches for the entire crew after one gruelling session in the simulator. "It changes your outlook a lot," said Bobko of the training grind. "You've got to do it, so it just changes the way you look at things. It's not some thing out there in the mist; it's up close and personal. There's a lot to be learned for a space flight and it doesn't seem like there's ever enough time. You feel you want to learn it all and it gives you a lot of incentive to work hard and try to learn as much as you can and get things as squared away as you can."

Finally, on 4 April 1983, after a decade and a half waiting for his first mission, which felt like being "a cosine wave in a sine-wave world", Bobko accompanied Weitz, Peterson and Musgrave out to Pad 39A for the real thing. Challenger's flight deck, despite being roomier than that of Columbia, was still cramped, with Weitz in the left seat and Bobko to his right. Sitting just between and behind them was Musgrave, to whom fell the job of flight engineer during the Shuttle's ascent and re-entry. Shoulder to shoulder with Musgrave, and directly behind Bobko, was Peterson. Six seconds before 1:30 pm EST, the three main engines roared to life. Then, as the countdown clock hit T-zero, came the crackle of the SRBs. The milestones clicked off perfectly: the SRBs burned out and separated a couple of minutes into the ascent and by 1:40 pm the Shuttle's main engines had been shut down and the External Tank jettisoned.

"The *value* of our simulators *ends* when those engines light and you lift off," Weitz recalled. "They try to fake you out a little bit by tipping the Shuttle simulator, but it doesn't compare with ... three main engines and two solids going. You *know* you're on your way and you hope they keep pointed in the *right* direction, because it's an *awesome* feeling." Challenger's maiden voyage had gotten off to a fine start. After firing the twin OMS engines on two occasions – one to achieve an elliptical orbit with a specific apogee, and the second time, half a revolution later, to circularise it – and opening the payload bay doors an hour and 45 minutes into the mission, the first order of business was to prepare TDRS-A for deployment. However, with three rookie spacefarers aboard, the experience of absorbing where they were proved over-whelming. "Every hour and a half," Musgrave said, "we made a complete orbit and it was like getting a crash course in world geography. Seeing entire continents with the naked eye is something special. We saw oil slicks off India, oil tankers in the Persian Gulf, the swirls in the Earth's crust where Iran, Pakistan and India collided millions of years ago and the mountains thrust upward by the force. We saw the White Nile and the Blue Nile converge in the Sudan, the dust storms in Mexico, thunderstorms over Africa and the tranquil beauty of the Bahamian islands."

The astronauts' first fleeting glimpses of their home planet from orbit were, however, eclipsed by a hectic schedule, which was already ticking down towards a first deployment opportunity for TDRS-A a little over ten hours into the mission. Even two decades later, the network – which now boasts six 'first generation' satellites and three updated 'second generation' ones – has proven instrumental in providing near-continuous voice and data contact between Mission Control and orbiting astronauts. Unusually for a Shuttle mission at the time, no fewer than three Payload Operations Control Centers (POCCs) would follow TDRS-A during its manoeuvres to geostationary orbit. Throughout the pre-deployment checks, right up until the satellite drifted away from Challenger, Harold Draughton's team at JSC in Houston had responsibility. Command would then pass to Pete Frank at the Air Force's Satellite Control Facility in Sunnyvale, California, until TDRS-A reached its operating position. Finally, the Spacecom concern, based at White Sands in New Mexico, would take over the day-to-day operation of the satellite on NASA's behalf.

Under an initial ten-year contract, signed with NASA in December 1976, Spacecom had agreed to lease TDRS communications, tracking and data-relay services to the space agency at a cost of some $250 million per annum. As such, it is the world's largest and most powerful privately owned tracking, communications and data relay system currently in orbit. Housed inside Challenger's payload bay, TDRS-A was huge, even though still in its 'launch' configuration with its communications hardware hidden from view, its umbrella-like antennas closed and its two electricity-generating solar arrays each folded into three parts. The satellite comprised three main segments: an equipment module, a communications payload and a battery of relay antennas. At this stage, however, it bore little resemblance to the enormous 'windmill' into which it would transform itself in geostationary orbit.

The TDRS concept was initiated in the early 1970s – one of the recommendations of the Hearth study group, of which Joe Allen was a member – as a means of not only supporting Shuttle crews, but other 'users', including the Hubble Space Telescope. It was recognised that a system of relay platforms operated from a single ground terminal in New Mexico would provide more adequate and near-constant support than the worldwide network of tracking stations previously employed. In fact, as well as supporting low-orbiting missions, it could relay data from satellites up to 5,000 km above Earth. Since the dawn of human space flight, astronauts had been out of contact with Mission Control for up to 80 percent of every orbit; furthermore, satellites had to tape record data and transmit it when they came within range of a tracking ship or ground station. As the Shuttle effort gained momentum in the mid-1970s, it was envisaged that two geostationary TDRS relays – one over the equator, just off the north-eastern coast of Brazil, near the city of Fortaleza, and a second over the central Pacific Ocean, near the Phoenix Islands – would provide astronauts with space-to-ground voice and data links for between 85-98 percent of each orbit. Yet, TDRS was no miracle worker. It could not process or adjust communications traffic in either direction. Rather, it operated as a 'bent pipe' repeater, relaying signals and data between its Earth-circling users and the highly automated ground terminal. Signals processing, therefore, was done on the ground and the satellite's sophistication was devoted to its very high throughput. Located in the inhospitable New Mexico desert, White Sands provided a clear line of sight with both satellites and its very small amount of annual rainfall meant that weather conditions would not interfere with their Ku-band uplink or downlink channels.

Responsibility for deploying TDRS-A was shared by the entire STS-6 crew, although the two mission specialists led the effort. There had already been some confusion in the weeks preceding liftoff. "We were in quarantine in the crew quarters at the Cape," remembered Peterson, "and a couple of nights before launch, two guys showed up from Boeing. It turned out that the software we'd trained on in the simulator was not exactly the same as the software that was flying and a lot of the codes were different. Story and I copied a bunch of stuff down with pen and ink and used that on orbit and that's really scary because we were taking these [Boeing] guys' words for it! We'd never seen some of this stuff in the simulator. Suppose what they told was not right and we messed up the payload? We'd never find those two guys again! Story called somebody in Houston to confirm the codes, but it was pretty

For almost three decades, Tracking and Data Relay Satellites (TDRS) have provided near-continuous voice and data communications coverage for Shuttle and International Space Station crews. One of the first-generation TDRS relays, its solar panels folded around the central bus and stowed golden antennas, is here pictured being raised to its deployment position on STS-26. The Air Force's Inertial Upper Stage booster is visible at the base of the satellite.

vague. That bothered us, because TDRS-A was extremely expensive and important to get working properly. When there were last-minute changes, we wondered if it had really been tested and thrashed out. The way we did commands was by dialling in a set of three numbers and then hitting a switch to execute them. The command was determined by what three numbers were set in there and they gave us numbers we'd never used before. We had no way of knowing whether they were right or not right." Fortunately, the commands proved accurate and some eight and a half hours after launch the crew was in position to raise TDRS-A and its IUS booster to a pre-deployment angle of 29 degrees above the payload bay. This was followed by radio frequency checks and, finally, at 10:51 pm EST on 4 April, a final 'Go' was given for deployment. Less than 20 minutes later, the IUS was switched onto its internal power source.

Originally intended as a temporary substitute for a reusable 'space tug' when it was designed in the 1970s, the booster was dubbed the 'Interim' Upper Stage and later became 'Inertial' in recognition of its sophisticated internal guidance system. Losing its 'interim' status also reflected a growing awareness, when the space tug was cancelled in late 1977, that the IUS' services would be needed throughout the 1980s. In fact, not until the early years of the present century did it fly for the last time as a 'standalone' booster. Prime contractor for the IUS was Boeing, which began developing the two-stage vehicle in August 1976 and supported its first launch aboard a Titan 34D rocket six years later. Measuring 5 m long and a little under 3 m in diameter and weighing some 14,740 kg, the cylindrical booster – made from Kevlar-wound aluminium – was capable of delivering 2,270 kg payloads to geostationary altitudes. Its first stage carried 9,700 kg of solid propellant and a large motor, capable of firing for up to 145 seconds; this made it the longest burning solid-fuelled engine ever used in space applications. Meanwhile, the second stage carried 2,720 kg of propellant. Both the first and second stage nozzles, commanded by redundant electromechanical actuators, could steer the former by up to four degrees and the latter up to seven degrees. Although solid rockets were known to generate a harsh impulse, the separation mechanism between the first and second stages employed a low-shock ordnance device in order to avoid damaging its payload. Moreover, solid propellant was chosen over a liquid-fuelled booster because of its simplicity, safety, high reliability and low cost. Hydrazine-fed reaction control thrusters provided the IUS with additional stability during the 'coasting' phase between the first and second stage firings, as well as ensuring accurate roll control and assisting with the satellite's insertion into geostationary orbit. Situated between the two stages was an equipment section with avionics systems to provide guidance, navigation, control, telemetry and data management services to TDRS-A. Importantly, most critical components, except the bellows for the gimbal actuator, were fully redundant to provide a reliability of more than 98 percent. In the early days of the IUS' development, Boeing even proposed adding a smaller third stage to propel planetary missions out of Earth orbit, although, with a payload attached, it would have been a tight squeeze in the payload bay.

Mounted on the base of TDRS-A, the booster was held securely in Challenger's payload bay by the doughnut-shaped tilt table known as the Airborne Support

Equipment (ASE). As well as providing the crew with an ability to hoist the entire 14 m long stack from a horizontal position to the deployment angle of 59 degrees, it incorporated electronics, batteries and cabling to enable Peterson and Musgrave to issue commands during the lengthy checkout of both the satellite and booster. The ASE included a low-response spreader beam and torsion bar mechanism to reduce spacecraft dynamic loads to less than a third of what might otherwise have been imposed. It was secured into Challenger's payload bay by means of six standard, non-deployable attachment fittings, which mated to the ASE's forward and aft frames, and two payload-retention latch actuators.

Upon arrival in orbit, the Shuttle was oriented with her payload bay facing Earthward, in order that the IUS and its precious satellite did not experience excessive thermal loads from the varying solar illumination during orbital flight. The entire payload, during this time, was supported by Challenger's on-board electricity-generating fuel cells, although the ASE included its own batteries to take over in the event of a power interruption. After releasing the forward retention latch actuators and raising the stack to 29 degrees above the payload bay, telemetry checks were performed by the ground to ensure that both were ready for deployment and the IUS was transferred to its internal batteries. By 11:18 pm EST, payload-to-orbiter umbilical cables were released and TDRS-A and the IUS were positioned at their deployment angle of 59 degrees and, precisely on time, Musgrave flipped the switches to release the combination from the tilt table. At that point, the booster's ordnance separation device, fitted with compressed springs, physically ejected the payload from Challenger at a rate of just 12 cm per second. Deployment occurred at 11:31:58 pm, during the eighth orbit, and the stack swept silently into the inky blackness, passing directly over the flight deck, surprising Bobko as it did so. "I learned how *big* it was when it came out," he said later. "I was up in the front, in the pilot's seat, and [Musgrave and Peterson] both said something like 'Oh, my God!' when this big satellite came out over the cockpit."

The astronauts' surprise was short lived. Nineteen minutes after the combo left Challenger's vicinity, Weitz and Bobko fired both OMS engines to create a safe separation distance before the ignition of the IUS' first stage. When this two and a half minute firing got underway at 12:26 am on 5 April, a complex ballet of celestial mechanics was set in motion to insert TDRS-A into geostationary transfer orbit. First stage separation was then timed to occur at 5:44 am and second stage ignition two minutes after that. By 6:01 am, it was expected that the satellite should have reached its final operational location. Following the deployment of its solar arrays and communications payload, by the late evening of 6 April, TDRS-A should have been fully functional and ready to begin several weeks of testing.

Unfortunately, problems with the booster set the satellite on a rocky path that would not see it rendered wholly operational until the winter of the following year. During the IUS' development, problems had been encountered with its propulsion system, including burst body casings, tacky liners, soft and cracked propellants and nozzle delaminations, in addition to problems with its on-board software and avionics. The first operational use of the booster was supposed to be on the Shuttle with TDRS-A, although delays and scheduling conflicts caused it to be leapfrogged by a pair of military communications satellites launched together on an Air Force Titan 34D in October 1982. In spite of telemetry dropouts, that IUS mission proved successful. TDRS-A was not quite so lucky.

As the firing of the second stage was underway, a critical seal (a manifold in the baffle of the gimbal actuator) failed and the control system lost its ability to accurately point the motor nozzle, which 'canted' and caused both the IUS and attached satellite to tumble wildly through space; furthermore, unusually high levels of cosmic radiation disrupted the normal automatic sequencing activities. Instead of reaching a circular orbit, TDRS-A was left in an elliptical one with a perigee of 22,000 km and an apogee of 34,000 km. Worse, communication with the tumbling payload were also lost. Throughout the night, flight controllers tried to separate TDRS-A from the IUS. "We started to transmit separation commands 'in the blind', not having any communications with the vehicle," recalled Glynn Lunney, who was the Shuttle payload integration manager, "and we jacked up the power on the network to as high as we could get and we radiated commands to jettison the IUS. We never got any signal. The deployment happened at night and by the next morning, around breakfast time, we were getting ready to give up when the network picked up a signal from the satellite." The first task was to regain control of TDRS-A, which was spinning at 30 revolutions per minute. Fortunately, it was able to be stabilised.

On 2 May, NASA began an ambitious series of experimental firings of the satellite's hydrazine-fuelled attitude-control thrusters and, despite frustrating instances of overheating and 'choking', steadily extended their duration to as long as two and even three hours. Robert Aller, head of NASA's TDRS Division, told *Flight International* on 28 May that he anticipated the satellite would achieve a circular geostationary orbit by mid-June, at the expense of consuming 380 kg of the attitude propellant which would otherwise have been used for stationkeeping during a ten-year operational life span. This represented more than two-thirds of the available supply, but there was no other way to save the mission. The procedure had been devised by NASA, in conjunction with White Sands terminal operator, Contel and the satellite's manufacturer, TRW Defense and Space Systems Group of Redondo Beach, California. In total, some three dozen thruster firings were executed. On 6 July, TDRS-A was finally activated to begin tests of its communications payload. After three months, one of its Ku-band single access diplexers failed; followed, later, by the loss of a Ku-band travelling wave tube amplifier. Despite these failures, which meant that TDRS-A could only provide one Ku-band single access forward link – thus postponing its ability to support the Shuttle's new middeck text and graphics machine – the satellite was nonetheless declared operational for communications purposes in December 1984.

By this time, however, it *had* already supported some data traffic on the first Spacelab mission, launched in November 1983. When fully deployed in orbit, the newly-renumbered TDRS-1 resembled a colossal 'windmill', measuring 17.4 m across its fully unfurled solar panels, which extended from a hexagonal 'bus'. The dual panels generated 1,800 watts of electrical power, supplemented by on-board nickel-cadmium batteries when in Earth's shadow. Inside the bus, the communications payload was capable of transmitting in a single second the entire contents of a 20-volume encyclopaedia. Mounted atop the bus were the satellite's antennas, capable of receiving transmissions from White Sands, amplifying them and retransmitting them to the 'user' spacecraft and vice versa. The main space-to-ground link was a circular, 2 m antenna, which operated across the Ku-band frequency, while data from other spacecraft was routed through one of two umbrella-shaped, 4.9 m diameter dual feed S-band/Ku-band single access parabolic dishes. Constructed from gold-clad molybdenum wire mesh, both had transmission rates in the order of 300 megabits per second, capable of handling heavy traffic from Hubble, the Shuttle and, specifically, the 1982-launched Landsat-4 Earth resources platform. For multiple access service, an S-band 'phased' array of 30 helical antennas was mounted directly onto the body. This incorporated a forward link, which transmitted command data to the 'user' spacecraft, and a return link to relay signal outputs directly to White Sands. Upon receipt, the ground terminal 'demultiplexed' the signal and distributed it to 20 sets of beam-forming equipment, which discriminated among the 30 signals to select the unique signatures of individual users. However, the success story that TDRS has since become could not have been further from NASA's collective mind during those first few weeks of April 1983 and, before Challenger had even landed, an investigation board was put in place to establish the cause of the IUS failure. Co-chaired by Don Henderson of the embarrassed Air Force and Thomas 'Jack' Lee of NASA's Marshall Space Flight Center, the board's initial work obliged the agency to delete TDRS-B from the eighth Shuttle mission in August 1983. (Additionally, it had recently been announced that STS-12 would deploy TDRS-C in March 1984; its payload was also suspended.) Not until the booster's performance had been satisfactorily demonstrated would another TDRS ride the Shuttle into orbit.

A number of modifications were implemented, including improvements to prevent hot gases impinging on the oil-filled Techroll seal in the region at which the booster's nozzle was connected to its solid rocket motor. It was the failure of this seal, 83 seconds into a planned 107-second firing, which prevented the insertion of TDRS-A into the target orbit. Additionally, a new nose cap to protect the forward part of the seal was thickened in the IUS' second stage and higher density carbon was employed in a Grafoil seal between the Techroll and nozzles in both the first and second stages. Tests of the repaired booster were conducted from March 1984 onwards and, in January of the following year, it successfully launched a top-secret Department of Defense payload. Meanwhile, it was soon recognised that with TDRS-B removed from the STS-8 payload manifest, the high-profile Spacelab-1 science mission on STS-9 was under threat. "Spacelab-1 needs *two* TDRS in orbit to ensure a complete record of the results of its experiments," *Flight International* explained on 21 May. "One estimate suggests that a single TDRS will allow 60-70 percent of Spacelab-1 data to be achieved."

Despite these problems, the STS-6 crew could have done nothing to prevent the fault in the IUS' second stage and, indeed, the remainder of their five-day mission proceeded extremely well. "Technically, the mission was extraordinarily exciting," Musgrave told the post-flight press conference, "because we accomplished everything we set out to do – launching TDRS, performing the EVA, conducting medical experiments and bringing the Shuttle home in great shape. It was also personally fulfilling, because I've been waiting for this for a long time. For some reason, I immediately oriented to weightlessness. I was totally at home in zero gravity and felt extraordinarily comfortable in a 'no down' environment. I trained myself not to expect to see a 'down'. I was prepared to tell myself that the floor of the spaceship was 'down' and to keep myself oriented that way, but I found that I didn't need a 'down'! To me, the Earth was neither 'down' nor 'up'. It was just … there."

In fact, as Musgrave dozed off each night on Challenger's flight deck, alongside Paul Weitz, climbing into his sleeping bag was the only time at which he actually missed the presence of full gravity. "I tied it horizontally and slept horizontally, up near the commander, just to keep him company, since Bo and Don were sleeping on the middeck. I'm a side sleeper and like to change to different positions throughout the night, but since there's no 'up' or 'down' in space, I really couldn't sleep on my side. No matter what position I tried to take, the zero gravity would keep me locked in a neutral position; neither 'up', 'down' or 'sideways'. I couldn't twist and turn or hold a new position. I was tempted to take a strap and lock my knees in a crouched position, just to get some variety, but I never did. It's amazing that man – a creature genetically coded to live in gravity – can survive in zero gravity. When the space programme began, there were people who said that man wouldn't be able to breathe or swallow in zero gravity!" Over the years, Musgrave would share his own ideas and tips for preparing for the unusual sensation of weightlessness. When a new group of astronauts was selected in January 1990, he advised them to set the foot of their beds on cinder blocks and sleep with their heads 'down' for several nights before launch. "Story claimed that the resulting congestion and headaches were a *good* thing," wrote Tom Jones in his memoir, *Skywalking*, "because they gave your body a several-day jump on coping with the same symptoms in flight. [He] did admit that his strategy had its faults: it was *tough* getting a good night's sleep in quarantine when your body kept sliding into the *headboard* …"

One of Musgrave's responsibilities during STS-6 was a machine known as the Continuous Flow Electrophoresis System (CFES), assigned to him by Weitz in new of his expertise as a medical doctor. The 'electrophoresis' technique worked by passing an electric field through a fluid as it moved from one end of a processing chamber to the other. Akin to a prism splitting white light into its constituent colours, the 1.8 m tall CFES device – situated on Challenger's middeck – had the ability to separate cells and proteins, but its effectiveness on Earth was limited by the gravity-induced effects of convection and sedimentation. In low orbit, where these influences were a million times weaker, it was anticipated that such separation processes could achieve higher levels of perfection and purity. For example, components of some biological substances could become hotter or colder than on Earth, enabling greater control over the solidification process. Additionally, in weightlessness it was possible for particles to flow freely without risk of contamination caused by contact with the container. On STS-6, during a pair of seven-hour-long experiment runs, CFES processed 700 times more biological material than was achievable in operations on Earth. Moreover, their purity was some four times higher.

"We saw the prospect of purifying medical materials like hormones and enzymes, which make up the basic components of treatments for a variety of diseases," explained Charlie Walker, formerly an engineer for CFES' sponsor, the McDonnell Douglas company. "In space, you can do that purification to a degree that's impossible here on Earth. In others, we did the mathematics of the purification process using fluid dynamic equations and theorised that purities of four or five times what could be done in the best processes on Earth could be achieved by taking this process into space." First flown aboard Columbia on STS-4 in the summer of 1982, McDonnell Douglas and NASA intended to jointly carry the 250 kg electrophoresis machine six times, culminating in the development of a much larger processing facility for carriage in the payload bay. Unlike the middeck-borne CFES unit, which had only one processing chamber, the payload bay version – known as Electrophoresis Operations in Space (EOS) and weighing 2,270 kg – had two dozen. By the beginning of 1986, it was slated to fly for the first time aboard Challenger in July of that year.

The promise of significant breakthroughs in pharmaceutical research seemed just around the corner in January 1986, when Challenger lifted off on what was expected to be a routine mission. Her destruction 73 seconds later, and the deaths of her crew, put McDonnell Douglas' plans on indefinite hold. Subsequent plans implemented by NASA to limit the commercial utilisation of the Shuttle meant that neither the CFES, nor the payload bay mounted EOS facility, ever flew again. Moreover, the development of new gene-splicing techniques towards the end of the 1980s rendered electrophoresis effectively redundant. Attitudes were different in 1983, with the promise of regular, fortnightly missions. Significant improvements, including software changes, better cooling and a greater separation capability, had been implemented during the interval between STS-4 and STS-6.

Samples of rat and egg albumin and cell culture fluid were successfully separated during STS-4 and, then on STS-6, Musgrave evaluated the flow profile of high concentrations of haemoglobin and monitored a mixture of haemoglobin and polysaccharide to investigate the separation of different molecular configurations. Each sample was satisfactorily processed, although post-flight removal indicated that the refrigerator had been inadvertently turned off. The condition of the samples, nonetheless, was considered acceptable. "It showed great potential," explained scientist astronaut Don Lind, who worked closely with McDonnell Douglas and the experiment before flying aboard Challenger in the spring of 1985. "If we ever get to the point where we can have a guaranteed schedule – so companies can know if they send their samples up on Tuesday, they'll get them back three weeks later – there are a number of companies that have very productive manufacturing experiments in space. Electrophoresis just happened to be one of the first ones."

After what Musgrave called "a five-day high", with the investigation board just getting its teeth into resolving the IUS failure and TDRS-A shortly to begin limping to geostationary orbit, Challenger's crew prepared for their return to Earth on 9 April. Although this was Weitz' second re-entry, it was his first in the Shuttle. He and Bobko also considered the inclusion of a new Heads-Up Display (HUD) in Challenger's cockpit as "a great aid, in my mind, of performing a landing". Akin to the displays used in military and civilian aircraft at the time, the HUD projected instantaneous data on velocity, descent rate, altitude and other critical parameters onto a transparent viewing glass in line of sight to the pilots' cockpit windows. This provided them with the ability to assimilate data from both the 'heads down' world of instrument flying and the 'heads up' domain of looking directly through the windows at the approaching runway. Astronaut John Blaha worked on developing the HUD at Kaiser Electronics in San Jose, California, and remembered it being cluttered with information and disliked by many pilots in the corps. "The biggest challenge," he said, "was the older, established, astronauts had not flown military aircraft with a HUD. The younger guys had all flown aircraft with a HUD, so there was some resistance." Ultimately, Blaha concluded, it was declared fully operational, in a less cluttered form, on STS-8.

During Challenger's re-entry, quite contrary to standard operating procedures, Musgrave unstrapped and stood up on the flight deck – and would be reprimanded by chief astronaut John Young after landing. Weitz admitted to Musgrave's indiscretion at the post-flight press conference. For his part, Musgrave would explain his reasoning as a desire to show that an astronaut could indeed stand during the transition from weightlessness to terrestrial gravity. "I had my Hasselblad camera and was taking some photos," he said. "Also, I wanted to prove that I could do it. That's important if an astronaut ever has to leave the flight deck and go below to throw a switch or circuit breaker. I wanted to show that the cardiovascular system doesn't have any problem going back into gravity and you don't have to be strapped down. My standing was smooth and steady and it shows that the Shuttle is maturing. Standing up throughout re-entry, instead of being strapped down, was the perfect end to a perfect trip." The onset of gravity, though, was evident. In *NASA's Scientist-Astronauts*, Dave Shayler and Colin Burgess noted that Musgrave's camera was "difficult to hold" during descent and as he disembarked after touchdown, he had to hold onto the handrail "for dear life" to prevent himself toppling over.

From the perspective of the four STS-6 astronauts, the second half of the hour-long hypersonic dive through the atmosphere was akin to hurtling through a blast furnace. "During the dark time of your approach," Bobko recalled years later, "the plasma sheath around the Shuttle recombines over the top and there's a big tongue of flame following you down." Swooping into Edwards Air Force Base, deep in California's inhospitable Mojave Desert, Challenger alighted on concrete Runway 22 at 10:53:42 am PST (1:53:42 pm in Florida), a few minutes over five days since leaving Earth. "We landed on the solid surface, rather than the lakebed," Bobko recalled. "If the lakebed is dry, it gives you a little more latitude. Edwards has large runways but, luckily, on this flight, nothing went wrong, so we didn't take that extra margin in any way. Landing on the concrete was just fine." Since the beginning of the Shuttle era, it had ordinarily been the pilot's job to both arm and deploy the landing gear, some 15 seconds before the commander performed the actual touchdown. Like a conventional airliner, the 'business end' of the orbiter's landing hardware is arranged in a tricycle fashion, with two fixed main gears in the belly and a nose gear slightly aft of the nose cap. Normally, the gear is deployed at an altitude of around 75 m above the runway, whilst travelling at a ground speed no higher than 550 km/h. Brake, axle and wheel damage suffered by Columbia at the end of STS-5 had already led to the incorporation of successful 'saddle' modifications. However, Challenger's landing was not as perfect as believed. During post-landing disassembly, six cracks were detected on three stators in her right hand inboard brake. Subsequent investigation revealed an undersized machining template had caused expansion slots in the stator disks to be produced 'undersized'; it was possible, NASA's report said, that similar problems had arisen on STS-5, although on that mission the stators were in such poor condition that it was difficult to prove.

Inspections also highlighted damage to the Advanced Flexible Reusable Surface Insulation (AFRSI) thermal blankets on Challenger's OMS pods. These consisted of silica tile material sandwiched between sewn composite quilted fabric which were much lighter than the Low Temperature Reusable Surface Insulation (LRSI) tiles used on other sections of the orbiter's airframe, including the vertical stabiliser and parts of the main fuselage. In total, the new blankets took the place of more than 600 LRSI tiles. The AFRSI damage ranged from missing outermost sheets and insulation to broken stitches. The worst cases were attributed to "some type of undetermined flow phenomena" during re-entry. To further examine the problem, four AFRSI 'test' blankets would be carried on the wings, upper surface and side fuselage of Challenger during STS-8 in the summer of 1983. Capable of shielding sections of the airframe from temperatures of up to 650 degrees Celsius, AFRSI blankets were incorporated after the completion of Columbia as one of the features that helped shave weight from later orbiters, including Challenger herself. They were more durable, cheaper and faster to produce and fit than LRSI tiles and varied in thickness – depending upon their expected heating load – between 1.1 and 2.4 cm. Elsewhere, other thermal protection included white tiles, black tiles capable of withstanding up to 1,260 degrees Celsius and grey reinforced carbon-carbon panels for the nose cap and leading edges of Challenger's wings. The orbiter's shielding experienced varying levels of degradation and discolouration, but, in general, NASA's second orbiter had returned in good condition from her maiden voyage.

Efforts to manoeuvre TDRS-1 to its correct orbital position ultimately allowed it to provide communications and data relay support for the first Spacelab mission. In view of the dramatic reduction of its station-keeping hydrazine supply by two thirds and problems with its own Ku-band system, it is remarkable that it worked solo to provide near-continuous communications coverage of each Shuttle mission until the second TDRS was launched at the end of 1988. In early 1992, with the network expanded to four satellites, TDRS-1, by then in a state of semi-retirement, was called upon at short notice to support the Compton Gamma Ray Observatory, whose data recorders had failed. A ground station was quickly assembled at Tidbinbilla, near Canberra in Australia, and TDRS-1 was repositioned with line of sight of the new terminal. The result was that the observatory was granted a downlink capability over previously inaccessible portions of its orbit. Other firsts achieved by a doddery, yet venerable, old satellite included the first live web cast from the North Pole and the first pole to pole phone call. In 1998, NASA abandoned plans to retire it and instead allowed scientists at the Amundsen-Scott base to employ it as a relay for transmitting research data to the continental United States. Additionally, it supported a medical emergency at McMurdo station, allowing scientists to conduct a telemedicine conference with doctors in the United States; this enabled a welder to be guided through a real operation on a woman diagnosed with breast cancer. Two years later, it aided an extended scientific expedition, jointly funded by the National Science Foundation and the Coast Guard, to the Gakal Ridge, just below the North Pole. In the early years of the present century, it continued to support polar communications, before its last travelling wave tube amplifier failed and rendered it unable to support its users' Ku-band needs. Retiring TDRS-1 in October 2009, Roger Flaherty, the space network project manager at NASA's Goddard Space Flight Center in Greenbelt, Maryland, glowingly described the old satellite – which had operated *four times* longer than expected, despite having started off short of propellant – as "the queen of our fleet". Eight months later, it entered a geostationary storage 'slot', its remaining volatiles were drained and it was finally shut down.

Although the arrival of TDRS-1 undoubtedly made the world a more interconnected place, it was to Story Musgrave's intense regret that his fellow humans seemed far more interested in building barriers than bridges. "I'm an optimist," he said quietly at the post-flight press conference. "I like to think positive, but man is *not* a social animal. One of my biggest disappointments is the absolute failure of the human being as a social animal. You get back here on Earth and open the newspapers and, every week, there are ten or 12 new wars breaking out all over the world. When I was in space, I never thought about war. I never had one negative thought. It was an incredibly positive experience – there was no time or inclination to think of war or problems, disease or death. I had absolute confidence that this mission would go as smoothly as it did," he continued. "This is my career and though I'm not scheduled for another flight as yet, I hope I don't have to wait another 16 years."

He wouldn't.

THE RIDE OF SALLY RIDE

A strange paradox occurred in the summer of 1976. For nearly a year, no Americans had ventured into orbit; nor would they do so for at least another four years. The space ambitions of the United States were by no means directionless, but its 30-strong astronaut corps faced a crisis: no missions were available in the foreseeable future, yet more astronauts were urgently required. By the time the Shuttle entered operational service sometime in 1980, NASA optimistically hoped that missions would be launching as often as once every fortnight. In other words, more crews would rocket into the heavens during its first couple of years than had previously ridden every American spacecraft since May 1961. A corps of less than three dozen could not support such an ambitious flight rate, obliging NASA, in an 8 July 1976 press release, to announce plans to hire "at least 15 pilot candidates and 15 mission specialist candidates" for the Shuttle effort. Crucially, and totally at odds with previous astronaut candidates, this group would specifically include both ethnic minorities and women, as part of an "affirmative action programme". Yet this was no lip-service payment to the equal opportunities brigade. In October 1976, a team led by cardiologist Robert Johnson of JSC's Cardiopulmonary Laboratory, began an extensive effort to gather baseline data on female physiological performance and tolerance levels and women volunteers were invited to undertake tests on the treadmill and a lower-body negative pressure machine. Pilots, declared the agency, would need to possess a bachelor's in engineering, biological or physical science or mathematics, with advanced qualifications desirable. Moreover, they needed to have accrued at least a thousand hours of pilot-in-command time in high-performance aircraft, with flight test experience preferable. For the mission specialists, similar academic credentials, plus three years of related professional expertise or advanced degrees, were demanded, although flight experience was not mandatory.

More than 8,000 applications reached NASA by the closing date of 30 June 1977. Between August and November, NASA summoned 208 candidates – pilots and mission specialists, military and civilian, male and female – in ten groups to JSC for intensive interviews, physical exams and psychiatric and psychological screening. One of the earliest to arrive in Houston, in mid-September, was a 36-year-old naval officer named Frederick 'Rick' Hauck. "I was a project test pilot for the Navy's carrier acceptance trials on the F-14 [Tomcat fighter]," he recalled, "and at the end of that tour of duty, I went to Air Wing 14 on the USS *Enterprise*, which was the first time the F-14 was deployed overseas. It was a concentration of some of the up and coming pilots in the Navy. During my second cruise on *Enterprise* in 1977, there was a [leaflet] from NASA, saying they were looking for applicants for the astronaut programme to fly the Shuttle and, in fact, four of us on *Enterprise* wound in my astronaut class: myself, Robert 'Hoot' Gibson, Dale Gardner and John Creighton. Three of the 15 pilots were from that air wing! Dale was a mission specialist, which is interesting. Twenty percent of the pilots came from that ship."

At length, on 16 January 1978, NASA announced the names of 35 new astronaut candidates and, in the words of Mike Mullane, "the diversity of America was represented" within their ranks, for they included six women, three blacks, an Asian-American of the Buddhist faith, a mother of three and a pair of Jews. Ironically, with seven previous groups having been exclusively white males, the white males in *this* selection – including Mullane and Hauck – were virtually invisible to the gathered media. Yet Frederick Hamilton Hauck cemented his credentials almost immediately upon arriving in Houston to begin astronaut training. Holding the rank of a commander in the Navy, he was the senior military officer and would become one of two informal leaders of the group, who eventually dubbed themselves 'The Thirty-Five New Guys', or 'TFNGs'. Mullane described Hauck's actions at one of the astronaut office's Monday morning meetings. "A large table dominated the room," he wrote. "On it sat some conference phones and an overhead projector ... Chairs ringed the table, but I gave no thought to taking one. This was the sacred table of Apollo. Alan Shepard and Jim Lovell and Neil Armstrong had sat here. At the moment, moonwalker John Young was sitting at its head ... Several rows of chairs had been placed at the back of the room and I aimed for these cheap sets. Most of my [fellows] did likewise. Most, but not all. Rick ... took a seat at *the* table. *You don't get more alpha male than this*, I thought."

Alpha male, perhaps, but Hauck came from a proud tradition of military duty and personal sacrifice. Born in Long Beach, California, on 11 April 1941, his father had risen to the rank of captain in the Navy, whilst his maternal grandfather and paternal uncle had both done naval service. In his childhood, Hauck had particular admiration for the military exploits of his paternal uncle, although when he happened to mention the possibility of joining the *Air Force* to his aunt, she threatened – only half-jokingly – to wash the teenage boy's mouth out with *soap!* Adventure, though, was in his blood: rollercoasters and Flexible Flyers both terrified and excited him and it was a next-door neighbour, who happened to be a naval aviator and a civilian flight instructor, who first offered Hauck a ride in a small Piper Cub aircraft. *That*, even more so than his aunt's stern admonitions, convinced him that naval aviation was the route to take.

Graduation from St Albans High School in Washington, DC, in 1958, was followed by admission to Tufts University, on a scholarship from the Navy's Reserve Officer Training Corps, to study physics. Receiving his degree in 1962, Hauck was commissioned as an ensign and spent two years as a communications officer aboard the destroyer USS *Warrington*, based at Newport, Rhode Island. While giving thought to resigning from the service when his time was up, he was unexpectedly offered the opportunity to enroll at the Naval Postgraduate School in Monterey, California, for a master's degree in physics. It seemed the perfect option – getting paid to study – and Hauck, by now married with a young child, began his studies in physics and mathematics. "The Navy selects a small number of their students to go on at a civilian university for doctoral studies," he told the NASA oral historian in November 2003, "and I was selected for what was called their 'advanced science programme'. They would send me to any university in the United States that would accept me, pay all tuition, to pursue an advanced degree in physics." It seemed too good to be true, said Hauck, and instead of simply writing a short thesis, he opted for the Army's Defense Language Institute for a ten-week course in scientific Russian, then moved to the Massachusetts Institute of Technology to study thermonuclear physics. "It was a *very* challenging education," he remembered. "I'd never had an engineering course before in my *life*; it had always been mathematics or physics, which is a *scientific* discipline, and it was a rude awakening." Still, Hauck received his master's degree in nuclear physics in 1966.

Doctoral research held no great appeal and he began flight instruction at the Naval Air Station in Pensacola, Florida, received his Navy wings of gold in 1968 and focused on qualifying to fly the A-6 Intruder attack aircraft. Hauck subsequently deployed to the Tonkin Gulf and flew 114 combat and support missions from the USS *Coral Sea* over Vietnam. He was selected for test pilot training and attended Pax River in 1971, before entering a three-year tour in the Naval Air Test Center's Carrier Suitability Branch of the Flight Test Division. In this role, Hauck served as the project test pilot for automatic carrier-landing systems in a range of aircraft. In fact, in the case of the F-14, he was team leader of the Navy's board of inspection and survey for carrier trials. Assigned as an operations officer to the USS *Enterprise* in 1974, he flew the A-6, A-7 Corsair and F-14 during daytime and nighttime carrier trials, before reporting to Attack Squadron 145 in the spring of 1977 as the executive officer, effectively the second in command of the unit. It was during this tour aboard the *Enterprise* that Hauck – and several of his shipmates – first saw the NASA call for Shuttle astronauts.

Physical exams, psychiatric evaluations and psychological probing followed – "everything they could measure," Hauck recalled of his week of tests in September 1977 – and placed them under varying levels of stress to examine their responses and how easily they became flustered or antagonistic. They had to count backwards from a certain number by a certain amount of numbers, decide what kind of animal they might like to see on the front of a T-shirt and curl themselves into a foetal position inside a fabric sphere to test their ability to withstand claustrophobia. "I found it most comfortable to sort of lie on my back with my knees up," Hauck laughed, "and I almost fell asleep!" After the tests, the candidates returned to their respective jobs and Hauck remembered being called one day by John Young, the chief of the astronaut office … and asked for his opinions about *another* candidate, a naval aviator called Dan Brandenstein. Hauck knew Brandenstein well, since they had been in the same class at test pilot school in 1971, and wondered if Brandenstein had received a similar call about *him*. (In fact, Brandenstein had been a member of the very first group of candidates, summoned to Houston in early August 1977.) One interesting anecdote centred on astronaut Dick Truly's wife, Cody, who approached Hauck and Brandenstein at around this time to make a bet with them: a *fifth* of Chivas Regal scotch if they were accepted into the astronaut corps. Years later, Hauck was convinced that Cody "had a little bit of insight" that both men had already been selected …

When the call came from George Abbey, Hauck was overjoyed … but anxious to approach his close friend; he did not want to break his 'good' news, lest Brandenstein had received 'bad' news. "And *he* didn't want to call *me* and say *he* was selected," Hauck said. "Within a few hours, the word was out and we heard that each other had been selected, so that had a good ending." As circumstances transpired, Hauck and Brandenstein would become the first two pilots of the TFNGs to actually fly into space, underlining in many minds the suspicion that Abbey was offering preferential treatment to the Navy astronauts. Training officially began on 11 July 1978, with initial instruction on the T-38 jet, followed by a week-long water survival course at Homestead Air Force Base in Florida in August, then a series of lectures in space history, technical assignment methods and procedures, spacecraft engineering, the Shuttle programme, aerodynamics and flight operations and various scientific disciplines. The training had been expected to last a full two years, but the class was graduated at the end of August 1979. In making the announcement, George Abbey noted that NASA had been impressed with their performance and that the initial training of future groups would also be reduced. "Based on the experience with this group," explained a JSC news release, "future candidates … will also undergo a one-year training and evaluation period." Training had been both long and arduous, though, and the existing astronaut corps desperately needed an influx of new blood to support preparatory work on the Shuttle. One of Rick Hauck's early duties was to work for Dick Truly on cue cards and pocket checklists for emergency ascent procedures. Alongside Joe Allen, Hauck later sat at the capcom's console in Houston during the first Shuttle mission and, a year later, on 19 April 1982, received his own flight assignment: as one of the first TFNGs to fly – *and* the first pilot – on the second voyage of Challenger, STS-7.

In view of Hauck's military seniority, his achievements and his leadership, it is perhaps hardly surprising that he wound up as one of the first to fly. Yet in some minds, not least Mike Mullane, it was the *other* TFNG names announced that day in April 1982 which would *really* go down in the history books. The crews were named for both STS-7, which included the first American woman astronaut, and STS-8, which included the first African-American spacefarer. For Sally Kristen Ride, a civilian physics PhD from Stanford University, it was recognised that from that day onwards she would become an icon. "As the first American woman in space," wrote Mullane, "she could look forward to book deals, speech honorariums, corporate board seats and consulting fees that could earn her millions." Perhaps Mullane's words of financial gain and lasting fame are a little unfair, for the *real* honour, surely, was a seat on a space mission; a seat which, even today, few have had the chance to experience. Of Norwegian ancestry, Ride came from Los Angeles, born on 26 May 1951. In her youth, she aspired to become a professional tennis player and, for a time at Westlake High School, was captain of the team. After graduation from Westlake, she entered Stanford to study physics and English. Whilst there, Billie Jean King watched her play and advised Ride to leave college and turn professional. She rejected King's advice and continued her studies; it is interesting that, since her astronaut days, Ride has become an outspoken advocate for getting more women involved in science and engineering. She received her degree in 1973, a master's credential in 1975 and her doctorate in astrophysics and free electron laser physics in 1978, only days – *hours*, even – before she drove to Houston to commence astronaut training.

"I saw an ad in the Stanford University student newspaper ... that NASA was accepting applications," Ride told the agency's oral historian. "They *wanted* applications from women, which is presumably the reason the Center for Research on Women [at Stanford] was contacted and the reason they offered to place the ad in the newspaper." Two weeks after Rick Hauck's screening, early in October 1977, the 26-year-old Ride was called to Houston as part of another group of 20 candidates. "It was a group I'd never met before," she said, "and I didn't meet any of the other 180 who were interviewed. The only ones I met were the ones in my little group of 20. We spent a week going from briefing to briefing, from dinner to medical evaluations, psychological exams and individual interviews with the astronaut selection committee."

A month later, 23 other candidates – including Air Force test engineer John Fabian and physician Norm Thagard – came to Houston for their own screening. Little could Hauck, Ride, Fabian and Thagard know that they would fly together aboard Challenger a little more than five years later. For Ride, though, the media attention at becoming one of six female candidates was especially intense. "The impact started before I left for Houston," she remembered. "There was a lot of attention surrounding the announcement, because not only was it the first astronaut selection in nearly ten years, it was the first time that women were part of a class. There was a lot of press attention surrounding all six of us. Stanford arranged a press conference for me on the day of the announcement! I was a PhD physics student. Press conferences were *not* a normal part of my day! A lot of newspaper and magazine articles were written, primarily about the women in the group, even before we arrived. The media attention settled down quite a bit once we got to Houston. There were still the occasional stories and we definitely found ourselves being sent on plenty of public appearances." The pressure on NASA to select female astronauts was strong and, in Deke Slayton's words, "there was some last-minute political bullshit". This appeared to centre on the fact that only *one* woman originally made the space agency's final cut and five pilots had to be dropped in favour of five female mission specialists. "They got selected a couple of years later," Slayton said of the pilots and, indeed, *six* pilots who reached the semi-final stage (John Blaha, Roy Bridges, Guy Gardner, Ron Grabe, Bryan O'Connor and Dick Richards) *were* chosen in 1980. The identity of the 'one woman' has never been divulged, but whatever the truth the incident underlines the importance that NASA placed in its public image and its need to hire an astronaut class which truly represented the depth and breadth of America.

The selection committee, co-chaired by John Young, was looking not just for academic and technical talent, but also for the ability of men and women to work effectively together. "And they succeeded," added Ride. "It was a congenial class and we really didn't have any issues at all within our group. They were very respectful and incorporated us as part of the group from the beginning. We all walked in as rookies; as neophytes in the astronaut corps. None of us knew anything about what was going to happen to us and so, as you can imagine, we were a pretty close-knit group. None of the astronauts who applied did it for publicity. Everybody applied because this is what they wanted to do, so the males in the group didn't really want to be spending their time with reporters – they wanted to be spending their time training and learning things. Frankly, the women would have preferred less attention." The new astronauts were desperately needed, said Hauck, due to natural attrition from the corps since the end of the Apollo missions. Over a period of just a year, by mid-1977, four veteran astronauts had retired from NASA to pursue other interests. The remainder, Hauck explained, "wanted to get us as smart about the Shuttle's systems as soon as they could." In spite of the teamwork, however, each member of the TFNGs knew that one day their performance in the simulators and through their technical assignments would drive the decision as to which of them would fly first. Then, in April 1982, a few weeks after Space Shuttle Columbia landed from her third test flight, Sally Ride was called into George Abbey's office at JSC. Abbey had chaired the selection committee and it was he who gave final approval on the choice of astronaut crews. His power in determining the fate of many a spacefarer has become legendary and the subject of considerable praise and criticism over the years, but Abbey's influence in crew selections is indisputable. For Ride, being summoned to his office, alone, that spring day, was unusual. "The commander is the first to know about a flight assignment," she remembered. "Bob Crippen, who would be the commander of my crew, had already been told, but then usually the rest of the crew is told together; at least, that was the way it was done then. In this case, Mr Abbey told me first, before he called over the other members of the crew. He took me up to JSC Director Chris Kraft's office, who talked about the implications of being the first American woman astronaut. He reminded me that I would get a lot of press attention and asked if I was ready for that. His message was 'Let us know if you need help. We're here to help you in any way and can offer whatever help you need'. It was a very reassuring message, coming from the head of the space centre."

Ride's colleagues on the STS-7 mission would be Crippen, a veteran of the first Shuttle flight, joined by Rick Hauck in the pilot's seat and fellow mission specialist John Fabian. They were destined to train for a year, with a tentatively scheduled launch date sometime in April 1983 aboard Challenger to deploy two communications satellites and release and later retrieve a free-flying platform (the Shuttle Pallet Satellite, or 'SPAS') using the Canadian-built mechanical arm. In the same way that Ride found out about her assignment, alone, John Fabian had a similar recollection. "I didn't know right off the bat that Sally was going to fly with me," he explained, "and that Rick Hauck was going to fly with me. I'm sure that the decision had been made, but maybe because *they* hadn't been told, *I* wasn't told. It wasn't a gathering. I don't know why." Certainly, being told *as individuals* was unusual, for most Shuttle crews were informed as a group ... and even the crew of Apollo 11 was gathered together to be told of their impending assignment. Little did they know at the time that their crew would ultimately expand to five members with the inclusion of a third mission specialist, Norm Thagard.

The $100 million mechanical arm, officially known as the Remote Manipulator System (RMS), was Canada's contribution to the Shuttle – a contribution that dated back to 1974, when Spar Space Robotics Corporation was contracted by the country's National Research Council to build a device for deploying and retrieving satellites from orbit and, ultimately, assembling the components of a space station. In May 1979, the first contracts were signed with NASA for the production of three RMS units and their supporting hardware and software. The challenges involved in building an arm of such complexity and dexterity were enormous: it needed to operate both autonomously and under manual control and meet strict weight and safety requirements. Moreover, nothing quite like it had ever been built or used in space before, which made Spar's task yet more difficult. Although a functional floor rig was built to test its joints, the first real demonstration did not come until it was actually uncradled on the STS-2 mission in November 1981. Measuring 15.2 m long in order to be able to reach the far end of the payload bay, it consisted – just like a human arm – of shoulder, elbow and wrist joints, linked by two graphite epoxy booms. Other components were constructed from titanium and stainless steel. To protect it from thermal extremes in space, the arm was covered in white insulation and fitted with heaters to maintain its temperature within required limits. Without a payload attached, it could move at up to 60 cm/min, but this was reduced to a tenth of that speed when fully loaded.

Ingeniously, the means by which the arm 'picked up' and 'put down' objects was achieved by the so-called 'end effector' – essentially a hand that employed a kind of three-tie wire snare to capture a prong-like grapple fixture attached to deployable or retrievable payloads. Already, the Hubble Space Telescope, at that time scheduled for launch in the mid-1980s, had an in-built grapple fixture that would enable it to be deployed, retrieved and serviced by future Shuttle crews. During 'operational' missions, like STS-7, astronauts would use two television cameras on the arm's wrist and elbow to guide the end effector over a target's grapple fixture, before commanding the three metal ties of the snare to close around it at precisely the right instant. When this was done, it would impart a force of 500 kg onto the grapple fixture, establishing a grip sufficient for the RMS to move the target. Although the arm was controlled by the Shuttle's General Purpose Computers, its movements were directed by an astronaut using a joystick on the aft flight deck. As the astronauts issued their instructions, the computers examined them and determined which joints needed moving, their direction and their speed and angle. Meanwhile, the computers also looked at each joint at 80-millisecond intervals and, in the event of a failure, automatically applied a series of brakes and notified the crew.

"One of my first assignments was on the RMS," Ride said. "I was one of a couple of astronauts that became heavily involved in the work to verify that the simulators accurately modelled the arm: to develop procedures for using the arm in orbit, to develop the malfunction procedures, so astronauts would know what to do if something went wrong. There weren't any checklists when we started; we developed them all! We also helped with the testing of the hardware itself at the contractor's facility in Canada. Until you actually start using something, it's very difficult to make predictions on how well it's going to work, what it's used for and how to accomplish the tasks that it's designed to accomplish. Many of the recommendations came in the form of the procedures that we developed. It was rewarding work, because it was at a time when the system was just being developed and nobody had paid attention to those things yet."

Colonel John McCreary Fabian of the US Air Force had also amassed considerable expertise in developing the techniques for operating the RMS. It was entirely appropriate, therefore, that on STS-7 – which involved the first ever deployment and recovery of a payload by the Canadian arm – its movements would be under the watchful gaze of both himself and Ride. Fabian came from Goosecreek, Texas, where he was born on 28 January 1939. He received his degree in mechanical engineering from Washington State University in 1962 and entered the Air Force, with the expectation of admission to pilot training. However, "the Air Force initiated a new programme to get young lieutenants into graduate school ... and so I was sent to Wright-Patterson Air Force Base, directly into graduate school at the Air Force Institute of Technology". He duly received his master's in aerospace engineering in 1964 and worked on flight tests of the F-106 Delta Dart aircraft in the service engineering division of Kelly Air Force Base in San Antonio, Texas. It was at this point in his career that Fabian, almost, crossed paths with NASA for the first time ... by *losing* his place at flight school. "It turns out that one of the people that was selected into the Apollo programme as a scientist was at pilot training with me at Williams Air Force Base [in Arizona]," Fabian told the NASA oral historian. "He was Owen Garriott, who was exactly one class ahead of me. In fact, Owen took my pilot training slot. I had to wait an additional six months to go into pilot training! Owen and I laughed about that over the years." At length, Fabian was accepted for flight training at Williams and later joined his brother's squadron, flying the KC-135 Stratotanker in Michigan, and saw action in Vietnam, participating in two deployments and flying 96 combat missions. Upon his return to the United States, he was sent to the University of Washington at Seattle to study for a PhD in aeronautics and astronautics, which he received in 1974, and served four years on the faculty of the Air Force Academy in Colorado.

Whilst there, in the summer of 1977, Fabian heard the first mutterings that NASA was selecting new astronaut candidates. He had not given much consideration to an application, since he had always considered himself – at 183 cm, or six feet and one inch – too tall; but *now*, with the arrival of the Shuttle, candidates could be as tall as 193 cm, or six feet and *four* inches. "So, with a relatively fresh PhD and operational flying experience and the six-foot-one height, I applied for the programme ... and got lucky," he recalled. *Lucky* would seem to be a modest choice of word, for Fabian soon became aware that the military services *only* screened the *top percentage* of their applicants, "so getting through was a great thrill to me". Among Fabian's earliest assignments in the astronaut office was the RMS and he would later calculate that he spent around 400 days at Spar's facility in Toronto, simulating and developing procedures for the mechanical arm's various failure modes. He spent so much time in Canada that it became a running joke to spend no more than a hundred days in the *same* motel. "It really is quite easy to use," he said of the arm, "but it's also a little bit intimidating, because you've got this thing which is 50 feet long, out there in the cargo bay, and if you're not careful, you could punch a hole in the wing or do something *really* stupid with it ... so learning *how* to operate it and learning *what* constraints need to be applied to that operation was ... part of the job that we set out to do."

Before STS-7 even left Earth, however, the most famous aspect of the mission was Sally Ride herself. In some of the more cynical areas of the media, it was speculated that she had been added to the crew purely as a public relations ploy, in response to the Soviet Union launching its second female cosmonaut, Svetlana Savitskaya, in August 1982. "NASA's crew allocation procedure is a closely-guarded secret, though it is known to involve seniority and an attempt to match education and experience to the mission," *Flight International* told its readers in April 1982, "but since NASA is financed by the US taxpayer, its public image is also important. So it is likely that NASA is capitalising on the publicity of having a woman fly early ..." Whilst it may be a little more than pure coincidence that a female astronaut happened to be one of the earliest to fly, Bob Crippen vehemently disagreed with the notion that Ride was simply a politically-driven 'token' on the mission. "She is flying with us because she is the very best person for the job," he told the press. "There is no man I would rather have in her place." Still, the importance of her presence was evident. President Ronald Reagan invited the entire crew to the White House before launch ... and *again*, for a state dinner, after the flight ... and at various functions the white-male astronauts – including Steve Hawley, whom she had married in July 1982 – were largely ignored or unrecognised by the press; the journalists were interested *only* in Ride. Years later, Rick Hauck felt that, despite a few "awkward" occasions, training and execution of the first American mixed-sex space flight went without many problems. "There were situations," he acquiesced, "where, maybe in the potty training, I'd never been involved in professional discussions with women about those! It was uncomfortable in a few situations, but the discomfort disappeared easily. Sally was great and Crip set the right tone in terms of what his expectations were of the crew. We just did it." Awkwardness was also a problem faced by NASA's male-dominated engineering community, who decided that the female astronauts were bound to require a makeup kit! "So they came to me," laughed Ride, "figuring that I could give them advice. It was about the last thing in the world that I wanted to be spending my time training on, so I didn't spend much time on it at all. There were a couple of other female astronauts who were given the job of determining what should go in the makeup kit and how many tampons should fly as part of a flight kit. I remember the engineers trying to decide how many tampons should fly on a one-week flight and there were probably other issues, just because they had never thought about what kind of personal equipment a female astronaut would take. They knew that a man might want a shaving kit, but they didn't know what a woman would carry."

Confining four people to a volume the size of a camper van for six days made for cramped accommodation. Then, eight months into their training, the quartet became a quintet. When Vance Brand's STS-5 crew rocketed into orbit on November 1982, one of their objectives had been to perform the first-ever Shuttle EVA. Unfortunately, the excursion was cancelled because of equipment failures in Bill Lenoir and Joe Allen's space suits. However, the day before these problems materialised, another area of concern – space sickness – reared its ugly head. According to a NASA news release of June 1982, the ailment was akin to a nauseous motion sickness and could not be resolved entirely through ground-based medical research. As a result, a series of Detailed Supplementary Objectives (DSOs) were timetabled into several Shuttle missions, comparing in-flight observations and crew-completed questionnaires with medical data acquired before launch and after touchdown. It was hoped that this might finally enable doctors to identify unique parameters and predict which individuals would be especially susceptible to the condition. During training, these crews did everything from filling in the questionnaires to having motion sickness artificially induced in the unforgiving rotating chair of NASA's neurophysiology laboratory. This allowed doctors to provide each astronaut with a 'data point' against which their predicted in-flight susceptibility could be compared. Ultimately, medication was provided in the form of Dexedrine and Scopolamine tablets, taken minutes after arrival in orbit, and bags of salted water were consumed before re-entry to 'fill out' the blood stream, ward off orthostatic intolerance and lessen the punishing effects of terrestrial gravity. "The symptoms were not only occasional nausea, but also what the docs called 'episodic vomiting'," explained sufferer Charlie Walker. "The only other symptoms were a malaise and slightly sweaty palms, like symptoms that others have with a cold or the flu. You feel low in energy, a little stuffy in the sinuses and I felt, with weightlessness, that the blood was rushing to my head, which is exactly the case! I still had these symptoms for about 72 hours and then they went away, just like that. Three days into the flight and I felt fine."

NASA considered the impact of space sickness to be minimal, highlighting the fact that only four previous crews had been directed affected. None of these episodes proved detrimental to the mission, but they did reflect a problem that had affected ten percent of all American manned flights to date. Another key issue was that, although 'cross-trained' Shuttle astronauts could accommodate a sick colleague for a limited time, serious obstacles could arise if the syndrome affected the *entire crew*. Consequently, following STS-5, NASA decided to add a pair of medical doctors to STS-7 and STS-8. Norm Thagard, the physician joining Crippen's crew, was already well known to Rick Hauck. "He and I had first met when we were both on the USS *Lake Champlain*, learning to land airplanes on aircraft carriers," in the mid-1960s, recalled Hauck. "In order to try to learn more about space sickness, NASA generated a bunch of tests and I was one of the guinea pigs! As soon as we got on orbit, Norm had these visual, spinning things that I had to watch and, boy, I felt *miserable*. They sure accomplished the purpose! At one point, I said 'Hey guys, I've *had it*. I'm going to go into the airlock', which was a nice place to hide. I said 'I'm going to close my eyes and please don't bother me until I come out'. I didn't know whether I was going to throw up. It was after about four hours that I started to come out of it and that resolved itself." A similar perspective was offered by Fabian – "I told people that *if you had one*, Norm Thagard *measured it*!" – who was also brought uncomfortably close to sickness whilst in orbit, though only through the good doctor's battery of tests.

The "blip", as Fabian called it, of having Thagard join the crew was not a negative one, but came at a time when two-thirds of their mission-specific training was done. "We tried really hard to integrate Norm into the rest of the crew activities," he recalled. "Sally, for example, was our lead on our electrophoresis experiment ... and I was the lead on doing the deployment and the retrieval, using the [RMS] ... but remember, Norm Thagard was the third member working with Bill Lenoir on the RMS, so he was no stranger to this system. We worked hard to try to find a way to work Norm into that and we ended up having him do one of the retrievals, without it being written in the checklist that way." Thagard's arrival also benefited the crew from an EVA standpoint. Although none were planned on STS-7, two crew members *had* to be trained for contingencies: Sally Ride would not fit the space suit, according to Fabian, so it was initially himself and Rick Hauck who were given EVA duties. "They [NASA] don't like using a pilot to do that," said Fabian, "so when Norm came along, Rick became the support person, helping us get into the suits, and Norm and I would have gone out if we'd had an emergency."

At the time of Thagard's assignment – just four days before Christmas 1982 – the STS-7 launch was still officially scheduled for April of the following year, which also provided NASA with invaluable data about the length of time needed by astronauts to prepare themselves for missions. Eventually, due to hydrogen leaks which pushed Challenger's maiden voyage from late January into early April, Bob Crippen's team found themselves rescheduled for mid-June. Despite the late addition of Thagard, Sally Ride recalled that he blended into the crew seamlessly. "We didn't spend every waking hour together," she said, "but we did spend almost all our time together, either as an entire crew or in groups of two or three. I was spending almost all my time with Crip and Rick in launch and re-entry simulations or with John and Norm in orbit or RMS training. Also, because we had things that required the whole crew, we did a lot of training together. We got to know each other very well. We never had any issues at all and got to be very good friends through the training."

As the last member of the STS-7 crew to be assigned, it is ironic that Norman Earl Thagard would go on to fly five missions and log 140 days in space – more than *double* the cumulative total of Crippen, Hauck, Fabian *and* Ride. Born on 3 July 1943, his birthplace is given as Marianna in Florida and in a NASA oral history he admitted that his father was a Greyhound bus driver and Marianna was the bus changeover point, "so I was born at a time when he was doing that". Thagard attended high school in Jacksonville and studied engineering science at Florida State University, gaining bachelor's and master's qualifications in 1965 and 1966. (As a high school senior, Thagard told his classmates that his aspiration was to become an electrical engineer, a jet pilot, a medical doctor and an astronaut, in *that* order . . . which, incidentally, is precisely the path that his career ultimately took.) Shortly after receiving his master's degree, he entered active duty as a reservist with the Marine Corps, quickly attained the rank of captain and was designated a naval aviator in 1968. Under normal circumstances, as a designate for the Platoon Leaders Class-Aviation (PLC-A), Thagard would have been put through Marine basic school, but "the Vietnam War was on by that time and they needed pilots desperately, so they basically eliminated the requirement . . . for PLC-A designates". Interestingly, Captain Thagard won his wings in the very same squadron as one Lieutenant Rick Hauck. After initial flight instruction at Naval Air Station Pensacola in Florida, he flew the F-4 Phantom at Marine Air Corps Station Beaufort in South Carolina, and then, in January 1969 went to Vietnam.

Thagard flew 163 combat missions in total and, although he admitted to never flying over North Vietnam, his role was primarily to support the fleet in the Tonkin Gulf or provide fighter cover for B-52 Stratofortress bombers or close aerial support for ground units in the South. In later years, his opinion of the conflict would harden, but at the time, "I thought that we were very altruistic . . . I saw nothing wrong with what we were doing in South-East Asia, but I saw a *lot* wrong with perhaps being killed and having people in the United States think that was a good thing . . . my *just* desserts! I had a *lot* of problems with that!" Returning to civilian life in the early 1970s, he would meet with derogatory comments when he revealed that he was a former fighter pilot, which irritated him greatly, not least because, in Thagard's mind, "anybody that thinks the North Vietnamese were good people has a very poor idea about human character . . . *that* was a pretty bad group we were fighting". To Thagard, the principle was the seemingly endless march of communism, and the fear that so many countries went from their previous political orientation *to* communism, but rarely *went back*. "There was a real concern," he said, "that if *that* continued, ultimately, the end result would have been all the world would be communist . . . so it didn't seem to me wrong or unreasonable for us to try to keep a country from being taken over by a communist regime." Three decades later, in his conversation with NASA's oral historian, clear anger was evident in

Thagard's words: that it was impossible to fight a war without adequate support – moral and otherwise – and that much of the opposition, back home in the United States, including the refusal to accept the military draft, was little more than a "self-serving" desire "to avoid risk and responsibility".

Returning to the United States, Thagard served as an aviation weapons division officer back at Beaufort and resumed his academic studies in March 1971, pursuing additional work in electrical engineering and a medical degree. At first, he attended Florida State University on a PhD in electrical engineering, but it was decided soon afterwards to terminate the whole school – "the Apollo programme was winding to an end," Thagard recalled, "[and] engineering was kind of on a down cycle that year" – and medicine seemed the next alternative. Here, he found severe age discrimination: Baylor College of Medicine in Waco, Texas, warned him that applicants over 25 were rarely given serious consideration and Duke University in North Carolina told him that would take the same view. However, a trump card appeared from the most unlikely of places: Thagard's *mother-in-law*! "She knew the dean of the School of Medicine, University of Texas Health Science Center [in San Antonio, Texas]," he explained, "because she was working there at the time." She asked the dean on Thagard's behalf and was told that if he moved to Texas and established residency there, he would be admitted. More trouble was afoot after the Thagards' arrival in Texas in June 1972. "We got there on a Friday evening at about five o'clock," he continued, "and got up the next morning, Saturday, and the headlines in the *San Antonio Light* [newspaper] read 'Medical School Dean Fired'. *This* was the fellow that said if I'd move there, they'd admit me to his school!"

Thagard worked initially as an engineer, whilst awaiting responses from other medical schools. He eventually accepted a place at the University of Texas Southwestern Medical School in Dallas and received his doctorate in 1977, only to learn that NASA was taking applications for its astronaut corps. His wife actually sent off for the application pack, "before I ever even *knew* about it", and Thagard applied for *both* the pilot and mission specialist categories, although his 804 hours of jet experience fell short of the minimum 1,000 hours needed for a pilot. "Realistically," he recalled, "the pilot [way] wasn't really going to happen. I didn't meet 1,000, but ... it wasn't *just* 1,000: you also needed to be a graduate of a military test pilot school, which I also was not." The mission specialist application, though, bore fruit and a month after starting his medical internship, in the summer of 1977, paperwork for security clearances, background checks and requests to attend his local police station for fingerprint profiles began to arrive at the Thagards' home. Late in October, he received the call from NASA to come to Houston for a week of screening. He heard nothing more until January 1978. "I actually had one of those rare Saturday evenings off," he told the oral historian, "and we had some friends over to the house. The news was on and there was an announcement ... that NASA on Monday was going to announce the first new group of astronauts. I was a little embarrassed, because to me that meant that I *wasn't* taken, because it seemed to me that this is *Saturday* and they're going to announce *Monday*, they've already told the people who are selected."

Two days later, on Monday morning, Thagard was at the Veterans Administration Hospital, about to start work, when he received *the* call from George Abbey which would change the direction of his life ...

Five years later, Thagard felt comfortable with STS-7. "I was already assigned to support the crew," he told this author, "so I had been working with them for months before being added to the crew, so I was familiar with the mission before my assignment. I performed a lot of the photography ... and operated the RMS to capture the SPAS. Except to add my space sickness activities to the mission, there was little change to the pre-existing flight data file. As I was the physician most familiar with STS-7 and my previous technical assignments included rendezvous and proximity operations similar to those involved in releasing and recapturing the SPAS satellite, as well as operations of the Canadian-built robot arm, I was the obvious choice to fly."

By the time Paul Weitz' crew brought Challenger swooping into Edwards Air Force Base on 9 April 1983, the STS-7 launch date had slipped to "no earlier than" 18 June. Even as the new orbiter slowed to a halt, two-thirds of the components for her next mission were already in place on the other side of the United States. In the VAB, the Solid Rocket Boosters had been attached to the External Tank on 2 March. When Challenger returned to Florida on 17 April, atop the modified 747 carrier aircraft, it was a race against the clock to ready her for a mid-June liftoff. In a quite remarkable turnaround that was trumpeted by NASA in its STS-7 press kit, Challenger was overhauled, the payloads from her last mission removed and support hardware for her next set of equipment installed and she was rolled into the VAB on 21 May. At just 34 days, turnaround was accomplished a week faster than the previous record holder, STS-4 in June 1982. The speedy processing flow was, however, a worrying harbinger of future problems and would be one of several issues highlighted by the Rogers Commission inquiry into flawed decision-making processes which contributed to Challenger's loss in January 1986. Many of the time savings were achieved by deleting the need to repeat tests of systems that had operated perfectly throughout STS-6. Other important tasks included repairing damaged areas of the two OMS engine pods with around 170 white tiles; similarly, sections of Challenger's elevons – the flap-like assemblies at the rear of her wings – required replacement with new thermal protection material. Elsewhere, an additional seat for Thagard was installed on the middeck and the RMS (which was not carried on STS-6) was fitted on the port sill of the payload bay. Following attachment to her boosters and tank on 24 May and rollout to Pad 39A two days later, preparations to load Challenger's cargo began in earnest. Although SPAS, with its rendezvous commitment, attracted the most press attention, the commercial focus was a pair of communications satellites: Anik-C2 and Palapa-B1. Both arrived by aircraft at Cape Canaveral Air Force Station on 30 November 1982 and were transferred to the Vertical Processing Facility for checkout. Less than six months later, on 23 May 1983, the satellites and their attached solid rockets – each encased in a lightweight sunshade – were moved to the pad and loaded aboard Challenger.

The reader will recall that Anik-C3 had already flown into orbit aboard STS-5, becoming the first of three such satellites to be launched. The Canadian Telesat

Sally Ride, America's first female astronaut, floats above the pilot's seat on Challenger's flight deck.

organisation reportedly paid NASA an additional $9 million to carry this second satellite, Anik-C2, aloft, with a third scheduled (at the time) for mid-1984. Together with its sister, Anik-C2 would focus four transmission beams to cover virtually all of Canada – including its remote northern wilderness – and when joined by the final satellite, Anik-C1, the trio would operate exclusively at 12 and 14 GHz in the Ku-band, with 16 transponders apiece, each capable of carrying two colour television channels and their associated audio and control circuits.

Physically identical to the Anik-C series, Palapa-B1 was the first in a second-generation series of satellites to offer regional telecommunications across Indonesia's 6,000 inhabited islands, 150 million inhabitants and 250 languages. Unlike the earlier Palapa-A series, launched by NASA in the 1970s on conventional rockets to provide telephone, television and fax services, the newer version was four times more powerful and extended communications to the Philippines, Malaysia, Thailand, Singapore and Papua New Guinea. Appropriately, 'palapa' translates to 'fruits of labour' and, in Indonesian political ideology, has symbolised harmony and unity for centuries. Indeed, in ancient Javanese, the oath "amuktl palapa" literally means "relaxation after exertion".

Displaying a similar theme of unity as a crew, and heralded by signs which screamed *Ride, Sally Ride*, Bob Crippen led his team out of the Operations and

Checkout Building into the glare of media flashbulbs in the early hours of 18 June 1983. It had been a peculiar morning. John Fabian remembered it lucidly: their 'last breakfast', a cake on the table, designed with their crew patch, followed by suiting-up and the long drive to Pad 39A. "You go through these various steps along the way," he recalled of the drive. "At each place, they're checking your ID and less and less people can proceed beyond each one of those." When they finally passed through the last of those checks, Crippen turned to the others and told them that they had just said goodbye to the *last* sane people in the facility, "because we've got to be *crazy* to do what *we're* doing!"

Robert Laurel Crippen was the 'old man' of the crew and STS-7's only veteran spacefarer. Born in Beaumont, Texas, part of the Gulf Coast's Golden Triangle of industry, on 11 September 1937. He completed his education at New Caney High School, before moving to the University of Texas at Austin to study for a degree in aerospace engineering, which he received in 1960. Crippen joined the Navy immediately after graduation and underwent aviator training in Pensacola, Florida, then Whiting Field and finally to Beeville, back in his home state of Texas, to gain his wings. For more than two years, from June 1962 until November 1964, he served as an attack pilot aboard the USS *Independence*, flying the A-4 Skyhawk. This was prior to the outbreak of the Vietnam conflict and Crippen found himself involved in operations to support President Kennedy's firm stance in Berlin and during the Cuban Missile Crisis, as well as a series of NATO exercises in the Mediterranean and Adriatic Seas. His naval career progressed rapidly and he was sent to the Aerospace Research Pilot School at Edwards Air Force Base in California, from which he graduated as a test pilot and was selected for the Air Force's Manned Orbiting Laboratory (MOL) project in October 1966. "The opportunity came to apply *both* to NASA *and* the military," he told the oral historian. "I applied for both. At some point in the process, I had to decide one way or the other and ended up picking MOL, because I thought NASA had more astronauts than they knew what to do with and the [Apollo] programme … was already starting to have some of the flights cancelled. I ended up being selected for MOL and, sure enough, after a couple of years on that programme, it got cancelled!" The Air Force pressured NASA into hiring the ex-MOL pilots as astronauts and, when the new class was announced in September 1969, Crippen was among them. In his early career with the space agency, he commanded the 56-day Skylab Medical Experiments Altitude Test in the summer of 1972 and served on the support crew for Apollo-Soyuz. Following the conclusion of this final Apollo mission, the focus of everyone moved onto the Shuttle and Crippen was assigned to work on the development of its computers, software and cockpit displays. In March 1978, whilst walking around the test vehicle Enterprise on the runway, George Abbey turned to Crippen and offered him the pilot's seat on the first orbital mission. Years later, he would wonder if his computer knowledge played a part in Abbey's decision to offer him this coveted position on a mission which was arguably one of the most dangerous ever attempted in space history.

Having said this, in a sense, STS-7 was even riskier, because at least Crippen and John Young had the benefit of ejection seats and full-pressure suits on Columbia's maiden voyage. Now, two years later, the ejection seats were gone and the crew wore blue flight coveralls, helmets and boots – in South-East Asia, John Fabian remarked, such garments were nicknamed 'party suits'. At the base of the launch pad, Fabian wondered darkly if a sign reading 'Enter If You Dare' should be placed on the gantry door. The five astronauts took the elevator ride to the 60 m level, where they disembarked and headed across a narrow walkway to the 'white room', adjacent to Challenger's access hatch, where technicians awaited them. Whilst Crippen and Hauck were being strapped into their seats, Fabian had a few minutes to look around, "take a last-minute nervous pee" and watch the twinkling of car headlights, to the north, the south and the west, crowding the roadways of the Kennedy Space Center. As MS1, Fabian took his seat directly behind Hauck, with Ride to his left in the flight engineer's position and Thagard downstairs on the middeck. "Once you get in the vehicle," he explained, "and get strapped down and the door's closed and latched and the technicians who are out there have *run like hell*, which is the *right* thing to do, you have just a little bit of time to think about all this, about what you're going to do and about why you're out there and about how you feel doing it." That feeling lasted all the way down to the last built-in hold in the countdown, at T-9 minutes; after *that*, Fabian continued, "you're only set on one thing: and that is you *really* want to fly today!" Unlike some astronauts, he admitted that he was aware of the risks involved, but 'fear' did not factor into his emotional state. "I tell people I've been married to the same woman for 44 years," he reasoned, "so I *don't* scare easily!"

Thankfully, their countdown and liftoff at precisely 11:33 am EST was one of the smoothest ever conducted. Challenger's three main engines shut down on time, eight minutes and 20 seconds into the mission, and by 12:19 pm, Crippen and Hauck had completed the second OMS burn needed to circularise their 28.45 degree orbit at an altitude of 260 km. To the untrained eye, the perfect ascent demonstrated NASA's seemingly effortless ability to fly on time and within the tolerance of very brief 'launch windows'. Only five minutes were available to the STS-7 crew for their first opportunity on 18 June and only two minutes for a second shot, beginning at 12:24 pm. The shorter than normal windows were dictated by three considerations: Earth horizon sensor constraints on Anik-C2 for a deployment during Challenger's eighth orbit and on Palapa-B1 some 11 orbits later, together with a requirement for adequate lighting conditions at Edwards Air Force Base in California, in the event that an emergency landing should become necessary.

For the four rookies on the crew, their years of training had paid off. "Physically, the simulator does a pretty good job," Sally Ride said of its closeness to the real thing. "It shakes about right and the sound level is about right and the sensation of being on your back is right. It can't simulate the g-forces that you feel, but that's not too dramatic on a Shuttle launch. The physical sensations are pretty close and, of course, the details of what you see in the cockpit are very realistic. The simulator is the same as the Shuttle cockpit and what you see on the computer screens is what you'd see in flight." There, however, the similarities ended. "The actual experience of a launch is not even close to the simulators," Ride exclaimed. "The simulators just don't capture the psychological and emotional feelings that come along with the actual launch. Those are fuelled by the realisation that you're not in a simulator –

you're sitting on top of tons of rocket fuel and it's basically exploding underneath you! It's an emotionally and psychologically overwhelming experience; very exhilarating and terrifying, all at the same time."

During ascent and re-entry, Ride served as the flight engineer and helped Crippen and Hauck to keep track of Challenger's systems. "My job was primarily to keep track of where we were in the checklists and be prepared with the malfunction checklists should anything go wrong," she remembered. "I was the one that was expected to be first to find and turn to the procedures should anything go wrong. I was also monitoring systems and status on the computer screens. My main job, though, assuming nothing went wrong, was to read the checklist and tick off the milestones. One of the first things that I was supposed to do – seven seconds after booster ignition – was, once the Shuttle started to roll, to say 'Roll program'. I'll guarantee that those were the hardest words I ever had to get out of my mouth. It's not easy to speak seven seconds after launch!"

Meanwhile, in the right-hand pilot's seat, Rick Hauck recalled seeing the sky outside his cockpit window change colour as Challenger climbed higher. "Seeing the sky turn from blue to black in a fraction of a second was amazing," he said later, "because as you leave the atmosphere, the Sun's rays are no longer being scattered by the air molecules. I remember as I was glancing out the window, startled, Crip said 'Eyes on the cockpit!' Back to work. Watch all the gauges. I guess that's one thing that stands out in my memory. Everything about it was thrilling." From his position, Fabian remembered that there was no chit-chat, no jokes – "it was all taken very professionally," he said, "and very seriously" – although he did get the chance to crane his neck, a few seconds after launch, to look through one of the overhead windows and watch the fire from the SRBs and the launch pad gradually recede, the view broadening to take in the entire launch complex, then the whole of Cape Canaveral. After unstrapping, all five astronauts had little time to contemplate their new surroundings. The main objective of their first day in space was the deployment of Anik-C2, performed by Fabian and Ride. Three hours into the mission, updated computations of Challenger's orbital path – including her altitude, velocity and inclination – were radioed to the mission specialists. Then, about 40 minutes before deployment, Crippen and Hauck manoeuvred the Shuttle into the correct attitude with its long axis 'horizontal', one wing down, and the open payload bay doors facing into the direction of travel. At length, the restraint arms pulled away from the $160 million Anik-C2 and the astronauts flipped a switch on Challenger's aft flight deck to open the Pacman jaws and impart a spin rate of 50 revolutions per minute on the payload. This steady rotation helped to stabilise the satellite during deployment. Next, at 9:01:42 pm EST, nine and a half hours since leaving Florida and flying high over the Pacific, Fabian and Ride fired and released a Marman clamp that held the satellite and its booster in place. Seeming to move in slow motion, the payload left the bay at just 90 cm per second. Fifteen minutes later, Crippen and Hauck backed the Shuttle away to a distance of around 40 km, aiming their spacecraft's belly towards the satellite to protect their delicate topside from the exhaust of the PAM-D. At 9:46 pm, as the combination hurtled over Africa, an on-board timer automatically fired the motor for approximately a hundred seconds to push Anik-

C2 into a highly elliptical geostationary transfer orbit. The performance of the booster and its titanium-skinned Star-48 solid rocket motor were described as "satisfactory" on STS-7, with the only minor concern being a slight hesitation of Anik-C2's Pacman sunshield during closure. Post-flight inspections revealed that a small Teflon rub strip, laced into one of its insulating panels, had inadvertently pulled itself loose.

After insertion into geostationary transfer orbit, Anik-C2 raised its omni-directional antenna and, over the following three days, employed its own apogee motor to position itself into the required slot at 112.5 degrees West longitude, south of central Alberta. Initially used by an American provider, the GTE Satellite Corporation of Stamford, Connecticut, in support of one of the world's first direct-to-home pay television services until December 1984, it supported Anik-C3 and ultimately handled educational broadcasts and the TransCanada Telephone System. American involvement stemmed from an agreement with Telesat to purchase temporarily surplus satellite capacity for US companies.

Launching the Indonesian Palapa-B1 followed much the same routine. Once more under the watchful eyes of Fabian and Ride, it was sent spinning out of the payload bay at 1:36 pm on 19 June. Forty-five minutes later, its PAM-D ignited to insert it perfectly into an accurate transfer orbit. Commanded from an Indonesian ground station at Cibinong, near Jakarta, the satellite was manoeuvred, in a similar manner to Anik-C2, into its operational slot at 108 degrees East longitude. In doing so, when it became operational on 30 July, it eventually replaced the earlier Palapa-A1 satellite. Operating on C-band (4/6 GHz) frequencies with 24 transponders, Palapa-B1 increased the telecommunications coverage of small rural satellite terminals in remote locations over its eight-year operational lifetime. Its advertised capabilities included carrying a thousand two-way voice circuits or a colour television channel on each of its transponders. Initially used by the Indonesian government's Perumtel organisation, it was replaced in April 1990 by the Palapa-B2R satellite and later sold to PT Pasifik Satelit Nusantara (PSN) for its 'inclined satellite' business.

"I didn't really know what to expect, because there isn't a way to train for being weightless," said Sally Ride of her first experience of life off the planet. "It's so far removed from a person's everyday experience that even hearing other astronauts describe it didn't give me a clue how to prepare for it. What I discovered was that, although it took an hour or so to get used to moving around, I adapted to it pretty quickly. I loved it! I really enjoyed being weightless." It was a pity that physician Norm Thagard, with his battery of space sickness tests to operate, could not have applied some of his expertise to the third deployable payload aboard Challenger. For, had the Shuttle Pallet Satellite been a human crew member, its manoeuvres in space during the second half of the STS-7 mission would undoubtedly have rendered it somewhat queasy. The aim of flying the research platform was to demonstrate the Shuttle's ability to conduct close range 'proximity' operations, including rendezvous, station-keeping and retrieval. Such operations on STS-7 would provide critical, real world data in support of one of Challenger's most important assignments planned for the spring of 1984: the recovery and repair of NASA's crippled Solar Max. To further underline the link between these two flights, and exploiting his knowledge of the orbiter's computers, rendezvous and navigation hardware, Bob Crippen would command both missions.

Indeed, the SPAS operations were, admitted Rick Hauck, one of the most challenging aspects of STS-7. "It was going to be the first time that the Shuttle had flown in close proximity to another object," he explained. "We knew that the Shuttle had a lot of capabilities that had been designed into it and one of our major objectives was to flight test the ability to do the last stages of rendezvous and fly very close to another object when you're both going at [28,000 km/h]. The objective was, using the RMS, Sally was to lift it out of the bay and release it. Crip would fly the Shuttle with it just sitting there, because we could always drift relative to each other. We needed to make sure we could fly close to it comfortably, then back away, fire the jets to go back to it, eventually up to [300 m], fly around it and see if we could fly [the Shuttle] without having the reaction jets upset the satellite."

Designed and built by the West German aerospace firm Messerschmitt-Bolkow-Blohm (MBB), under a June 1981 agreement with NASA, it was designed to accommodate scientific and technical experiments provided by fee-paying customers. Roughly triangular in shape, it measured 4.2 m across, 0.7 wide and 1.5 m high and weighed 1,500 kg when fully laden. During missions, it could operate in the payload bay – secured by one keel and two longeron trunnions – or be deployed for up to 40 hours in autonomous free flight. For STS-7, the $13 million platform was laden with several scientific and technical experiments, funded by what was then West Germany, the European Space Agency and NASA. Although crammed with experiments – ranging from studies of metallic alloys to a state-of-the-art remote sensing scanner – it became most famous for its NASA-provided cameras, which yielded the first picture of the full Shuttle in space.

Getting such a historic photograph *was* planned, said Bob Crippen, thanks to the inputs of Bill Green from NASA Headquarters, but what was *not* intended was positioning the RMS in such a way that the mechanical arm's joints created the number '7' to honour their own mission number. As a crew, they had practiced the manoeuvre on the ground – in fact, the design of their mission patch included a similar image – and it was Ride who placed the arm into this configuration in orbit. Some flight controllers were decidedly unhappy about the astronauts' antics. They had not seen the RMS in such a position before and were concerned that to do so, for nothing more than a photo opportunity, might risk stretching the arm to its structural limits. Still, the imagery acquired by SPAS of the Shuttle in orbit, with the glittering blue and white marble of Earth drifting serenely beneath, proved truly stunning. Years later, Ride would admit that she still used the famous photograph as a slide during her lectures. "We worked hard on that," Fabian remembered of the planning for the 'SPAS Photo'. "We worked out the position [with] the [RMS] arm in the shape of a 'seven' for the seventh flight and we didn't tell anybody about this, of course. We had this on a back-of-our-hand-type of procedure – what angles each joint had to be in order for it to look like that – and then we had worked on the timing, so that we could catch the Space Shuttle against the black sky, with the horizon down below. *That* was the picture we most wanted. Now, we got a lot of good pictures, against the cloud background and against the total black sky ... It had just a whole battery of cameras: a still camera, a TV camera, a motion-picture camera, and so we're running these various cameras by remote as we fly the Shuttle around it so that we can get the Shuttle in various types of positions."

Beginning on 20 June, the first of two phases of SPAS activities got underway with initial testing in the payload bay. During this time, seven of its 11 experiments were switched on and allowed to run continuously for 24 hours. Then, next day, with the satellite held securely by the RMS, Crippen and Hauck pulsed Challenger's RCS thrusters to evaluate movements within the arm. Again, Sally Ride found her months of practice on the ground had prepared her amply for operating the real thing in orbit. "The simulators did a good job," she said later. "It was a little easier to use the arm in space than it was in the simulators, because I could look out the window and see a real arm! Although the visuals in the simulator were very good, there's nothing quite like being able to look out of the window and see the real thing. It felt very comfortable and familiar. The simulators had prepared me very well." Early on 22 June, the second phase – actually releasing SPAS into space – got underway. Shortly before 9:00 am EST, under John Fabian's control, it was released from the arm. The crew reported that the satellite's handling characteristics were exactly as expected and the RMS imparted no appreciable motion. For the next nine and a half hours, the astronauts tested the arm, fired off RCS plumes to deliberately disturb the satellite and practiced the rendezvous and proximity operations needed during the Solar Max repair. During deployed proximity operations, Challenger flew 'down' and 'forward' of SPAS to a distance of 300 m, during which time the crew used a newly-fitted Ku-band antenna as a 'rendezvous radar' to track the satellite. Crippen then approached SPAS and Fabian retrieved it, before releasing it again and recapturing it as it rotated. An hour later, it was deployed yet again, this time under Ride's control, and Challenger flew 'forward', 'up' and 'down' to a distance of 60 m. Later, at closer gaps of a few tens of metres, Crippen and Hauck fired the RCS jets at nine different locations to evaluate the effects of plume impingement on the satellite. Subsequent tests included releasing the satellite with the RMS in automatic mode, before finally capturing it, reberthing it in the payload bay and then deactivating it.

"It was a big deal," Crippen reflected on the first deployment and retrieval by the Shuttle, "and we wanted to make sure that we *could* rendezvous with satellites; could come back in and grab them. It turned out that it all went extremely well. It was a little bit different, in that what we called the 'digital autopilot', or the 'DAP' – which is the way the computer fires the various jets – when we got in close to the satellite, I found that when you tried to slow down sometimes, the attitude control thrusters would also start going, and it kept 'walking' you in when you didn't want to go in ... We ended up learning a few things about how the autopilot worked that we corrected subsequently and makes it very nice for rendezvous today, which is extremely important on things like working with the [International Space] Station. It all really worked very well."

As this celestial ballet was ongoing, the remaining experiments aboard SPAS, costing around $10 million, were activated. One of the most important was West Germany's Modular Optoelectronic Multi-spectral Scanner (MOMS), which acquired high-resolution imagery and conducted thematic mapping of ground-based

Seen from the first SPAS free-flying platform, Challenger drifts serenely, high above Earth, on her second mission. The Canadian-built RMS mechanical arm, angled into a figure of seven, is visible to the left, and the other contents of the payload bay – the empty sunshades of Palapa-B1 and Anik-C2, the wall-mounted GAS canisters and the bridge-like OSTA-2 assembly – can also be discerned.

targets, including arid regions, coastal areas, islands and mountains. This proved extraordinarily successful for geological mapping, mineral exploration, hydrology and the monitoring of renewable resources for agriculture, forestry and urban or regional planning. Twenty-six minutes of high-resolution imagery validated the concept and cleared the way for the instrument's inclusion on a West German-dedicated Spacelab mission in the autumn of 1985. Additionally, the University of Bonn provided a double focusing magnetic mass spectrometer to measure the gaseous contaminants in the payload bay. An experimental heat pipe, a yaw sensor package and a variety of developmental solar cells were also affixed to the satellite, together with a number of investigations associated with STS-7's fourth major payload: the NASA-funded Office of Space and Terrestrial Applications (OSTA)-2 payload.

Proximity operations with SPAS were aided immeasurably by the maiden flight of the Shuttle's steerable Ku-band communications antenna. Since the antenna could only be effectively operated in conjunction with an active Tracking and Data Relay Satellite (TDRS) in geostationary orbit, STS-7 marked the first time the high-data-rate device had flown aboard the Shuttle; on STS-6 in April 1983, the slower S-band communications link was used instead. The 91 cm Ku-band dish, mounted on the starboard wall of the forward payload bay, enabled high-data-rate communications to be transmitted to Mission Control. Although the S-band link could operate through TDRS, it could only do so at a lower data rate, since it did not have a high enough signal gain to support high-rate traffic. A drawback of the Ku-band system, however, was its narrow 'pencil' beam, which rendered it more difficult for TDRS antennas to lock onto its signal. Consequently, the wider beam of the S-band was used to 'locate' TDRS-1 and lock the Ku-band dish into position. When this had been achieved, the latter's signal was switched on and the device conducted a three-minute-long spiral conical scan to zero in on the satellite. During proximity operations, the dish served as a rendezvous radar capable of 'skin-tracking' a satellite in order to provide target-angle and range data to Challenger's navigational software. The only problems in this instance were occasional communications 'dropouts' from the payload interrogator, which provided a telemetry link between the Shuttle and SPAS.

Meanwhile, the crew described the retrieval – both in stable and slowly rotating attitudes – as easy to perform, "but the act of going up and capturing it was a little scary," admitted Ride. "What if we couldn't capture this satellite? It was easy in the simulators, but was it going to be easy in orbit? The experience was different because it was real! In the simulator, it wasn't that important and if you missed, it was just a virtual arm going through a virtual payload. In orbit, it really mattered that I captured the satellite." Fortunately, the retrieval went perfectly.

Although the most visible elements of STS-7 were the carriage of three satellites and the presence of Sally Ride, a vast amount of valuable research was being conducted autonomously by OSTA-2. The first OSTA experiments had flown aboard Columbia on STS-2 and included a powerful synthetic-aperture radar. The 1,448 kg OSTA-2 package was the first use of the Mission Peculiar Equipment Support Structure (MPESS), which formed a 'bridge' across the payload bay. Developed jointly by NASA's Marshall Space Flight Center of Huntsville, Alabama, and West Germany's Aerospace Research Establishment, OSTA-2 included the Materials Experiment Assembly (MEA), which included three studies of advanced semiconductor crystal growth, metallurgy and containerless glass technology. Elsewhere was the Materialwissenschaftliche Autonome Experimente unter Schwerelosigkeit (MAUS), consisting of three cylindrical Getaway Special canisters laden with investigations into the melting and solidification of metals, alloys and industrial glasses. It was already known that, on Earth, gravitational effects influenced the formation of materials in ways that yielded undesirable effects such as 'sedimentation' – the 'settling' of melts of composite materials whose constituents had different densities. Other results of terrestrial gravity included hydrostatic pressure and convection currents, both of which were known to cause 'stirring' in fluids. Even the walls of the containers in which such materials were solidified could cause stresses and imperfections. As a result, samples processed on Earth were often flawed in structure and composition and much less suitable for advanced technologies than more 'homogeneous' materials would be. Prior to the flight of OSTA-2, it had proven extremely difficult to observe some aspects of the theoretical properties of specific materials because near-freedom from Earth's gravitational constraints could only be achieved for a few seconds at a time in ground-based laboratories. Skylab experiments in the early 1970s had already hinted at the effectiveness of the microgravity environment for advanced materials processing. During typical OSTA-2 experiment runs, Challenger adopted 'gravity gradient' attitude – a stable orientation with her heavy rear end pointing towards Earth – which achieved the minimum quantity of vehicle-induced g-forces and restricted the required number of thrusters firings. The development of MEA began in 1977, when NASA issued an announcement for proposals of materials investigations for carriage aboard the Shuttle. It was anticipated that the reusability of the system, which flew again aboard Challenger as part of the West German-sponsored Spacelab-D1 payload in late 1985, would provide a cost-effective means of getting experiments into space for longer periods of time than had been possible aboard sub-orbital rockets. Activated by Fabian and Ride from instrument panels on the aft flight deck, the rectangular, boxy MEA began operations on 19 June, barely 24 hours into the mission. During the course of the next five days, it processed samples of germanium selenide – which, it was hoped, would be of benefit to the semiconductor industry – as well as mixing liquid metals and exploring the viability of producing high-temperature, containerless glass-forming substances. Meanwhile, one of the MAUS canisters operated for almost its full programmed duration of 80 hours, while the second prematurely shut down at the end of its first processing run. Two of the MAUS experiments also had components installed aboard SPAS and were conducted during its free flight to minimise the impact of Shuttle-induced mass and stabilisation movements. Sponsored by the West German Federal Ministry for Research and Technology, the SPAS-mounted MAUS experiments explored the processing of a new permanent magnetic alloy using the properties of bismuth and manganese, two metals which have proven difficult to mix uniformly on Earth. A second investigation measured oscillatory convection in fusion processes, while a third determined the effects of gravity on ground-based pneumatic conveyor systems.

Meanwhile, mounted in Challenger's middeck were the Monodisperse Latex Reactor and Continuous Flow Electrophoresis System, both of which had been aboard STS-6. The former, designed to produce large quantities of ten-micron-sized latex beads as part of ongoing studies, operated extremely well, using all four reactor chambers. Already, the National Bureau of Standards had shown interest in using such mass-produced beads for calibration standards in ground-based medical and scientific equipment. On STS-6, the CFES had successfully separated one sample containing haemoglobin and a second containing a mixture of haemoglobin and a complex sugar known as polysaccharide. For its STS-7 flight, polystyrene latex particles were carried to further investigate the concentration limitations of continuous flow electrophoresis in space and better calibrate the machine. Its success on both missions guaranteed manufacturer McDonnell Douglas a seat for one of its employees – engineer Charlie Walker – on a Shuttle flight, specifically to operate the apparatus on a future mission. "As I remember it," said Walker, who had been assigned in June 1983 to the STS-12 crew, "the initial agreement was for six flights of the proof-of-concept CFES. It was very limited in terms of the number of flights available. I think there was wording in the contract of optional additional flights, to be negotiated later, if the concept proved to be of merit to both the industry and NASA and there were future needs to move into." Little did he know that those 'optional additional' flights would not only be added, but would lead to no fewer than *three* missions for himself.

One of the more worrying problems was a 4 mm wide pit in one of the Shuttle's six forward flight deck windows; caused, it turned out, by the impact of 'space debris'. It was first noticed by the astronauts on 20 June, but they did not report it. "Crippen decided not to tell the ground that we'd been hit and it didn't come up until after the flight," John Fabian explained later. "His rationale for that, I assume, was that there wasn't anything that the ground could do to help us. The event had already occurred. We were perfectly safe ... and so he elected not to say anything. I think it was the right decision." The window was subjected to detailed energy-discursive X-ray analysis after landing, and titanium oxide and small quantities of aluminium, carbon and potassium were found in addition to pit glass. The morphology of the impact was suggestive of an impacting particle (most likely a tiny fleck of paint), measuring just 0.2 mm in diameter, but travelling at 6 km *per second*! "The results," Fabian continued, "are *so* much larger than the event itself that it's staggering."

Six windows wrapped, airliner-like, around the orbiter's cockpit, representing the thickest ever manufactured as optical-quality viewing ports. Each window consisted of no fewer than three individual layers. The innermost pane, measuring 15.8 mm thick and made from tempered alumino-silicate glass, helped to maintain the cabin's pressure. Next came a 3.3 cm thick sheet of low-expansion, fused-silica glass to provide high optical quality and excellent thermal shock resistance. Lastly, came the 15.8 mm thick outermost pane, also of fused silica, but containing a high-efficiency, anti-reflection coating and capable of withstanding temperature extremes up to 420 degrees Celsius. It was fortunate that Challenger's windows were thus equipped with these three layers, for the outer pane – primarily employed to provide thermal protection during the later stages of atmospheric re-entry – was the only one affected by the pit. However, post-flight inspections noted that significant structural weaknesses caused to the outermost panes by such minutely sized debris particles could lead to further problems during a particularly harsh re-entry. Moreover, NASA's post-mission report added that more debris damage was experienced generally by Challenger's thermal protection system on STS-7 than any previous Shuttle flight. Much of this damage was close to the left 'chine' – the region between the leading edge of the left wing and the main fuselage – and was caused, apparently, during ascent on 18 June, as breakaway foam and ice tumbled from the External Tank. More discolouration of her insulating blanketing was evident, compared to STS-6, and several tiles were lost, including a fragment from one belonging to the left main landing gear door.

Originally, STS-7 was scheduled to perform the first Shuttle landing at the Kennedy Space Center, a fact highlighted in the mission's press kit, which would have helped to reduce turnaround times significantly. "We were looking forward to that," remembered Sally Ride. "They had a red carpet ready to roll out for us and our families were all waiting for us in Florida." Unfortunately, the touchdown on 24 June, due to occur on Challenger's 96th orbit, was postponed by two further revolutions in the hope that conditions would improve or facilitate a landing attempt in California. It was expected that bringing each Shuttle mission back home to the East Coast launch site would save around $1 million and five days' worth of processing for the next flight. Moreover, KSC landings would remove the necessity to expose the orbiter to the uncertainties and potential dangers of a cross-country ferry flight atop NASA's modified 747. However, as Crippen's crew discovered that June day in 1983, the West Coast landing site exhibited far more stable weather conditions than Florida. The KSC runway – the Shuttle Landing Facility – had been officially opened in 1976 and is located a few kilometres north-west of the VAB. Measuring 4.6 km long and 91 m wide, with a 300 m overrun at each end, it is all concrete and slopes slightly from the centreline to facilitate drainage. Two options were available to Shuttle crews returning to KSC: they could either approach from the south-eastern 'end' of the landing strip, designated 'Runway 33', or the north-western 'end, known as 'Runway 15'. The decision over which runway to use was largely dependent upon wind speed and direction, but in STS-7's press kit, Crippen was aiming for Runway 15. Sadly, not until February 1984 would a Shuttle crew make landfall in Florida, although Crippen, Hauck, Ride and Thagard would all make landings there later in their careers.

After the hopes of an East Coast touchdown came to nothing, Crippen and Hauck duly fired Challenger's OMS engines to begin the hour-long hypersonic glide to Earth. Sally Ride was pleased. "I remember being disappointed that we weren't going to land in Florida," she said later, "but I grew up in California and we'd spent a lot of time at Edwards Air Force Base. The pilots had done a lot of approach and landing practice at Edwards, so it almost felt like a second home. But there weren't many people there waiting for us!" Nonetheless, Challenger's second touchdown in just over two months, occurring at 10:56:59 am PST (1:56:59 pm in Florida), was near perfect. Her systems performed well during re-entry and landing, but during towing operations a chattering noise was heard from one of the wheels on her right hand main gear. The Shuttle had to be jacked up, the wheel removed, its brake assembly disassembled and the wheel remounted before towing could resume.

"I'm not a Shuttle pilot," said Fabian, "but I *am* a pilot and I know a thing or two about kicking rudders and moving ailerons ... and *this* is a *very* difficult machine to fly. I have had an opportunity to fly the [Shuttle] simulator. It's not nearly as easy to fly as a big air transport, like a Boeing 707 or 757, and certainly a *lot* more difficult to fly than a little NASA T-38. You've got to stay on top of it all the time. You've got to be thinking well ahead of the vehicle, so this is not just a flying job for ... the guy who really knows how to manoeuvre the airplane. This is a machine that is flown by people who are of great intellect as well as great skill. But when you come back down and you finally roll out on final and you can see the runway in front of you, even though you've seen this in the simulator before — and those of us who weren't part of the landing crew only saw it a few times, a half a dozen maybe – it's still startling when you look out there and see how rapidly you descend down towards that runway. You're *really* coming down fast, about a twenty degree glide slope, and *that's* really noticeable." *Twenty degrees* represents an angle of attack more than six times steeper than a commercial aircraft – indeed, for the final minutes of each Shuttle flight, the vehicle fell to Earth with all the grace of a brick ...

Detailed inspections of Challenger following her second landing revealed that the right hand inboard brake had actually suffered major structural damage to two of its rotors, including the beryllium heat sink and carbon lining segments. Additionally, the right hand outboard brake had two loose carbon pads with retainer washers missing. Cracked retaining washers were found in all brake assemblies and it was discovered that a similar situation might have occurred on previous Shuttle missions with no adverse effects. None, however, had previously been positively identified. It became clear that the washers had probably cracked during their manufacture or pre-flight assembly, with structural and thermal analyses confirming that neither the flight nor landing could have caused the damage. One of the main 'to-do' tasks on the list for Challenger's processing team at KSC before her next mission, STS-8 in August 1983, would be the replacement of all cracked or suspect brake washers. The NASA convoy responsible for recovering STS-7 after touchdown was somewhat smaller than intended, due to the diverted landing site: instead of the 24 vehicles and 110 personnel normally in attendance, only six trucks and 24 people were at Edwards that day. As with previous flights, they determined that residual hazardous vapours were below significant levels and began attaching purging and coolant equipment to Challenger's aft fuselage. These measures enabled them to remove re-entry heat from the Shuttle and better protect its electronic hardware. Half an hour after wheelstop, the astronauts departed Challenger, using an airliner-like mobile stairway. Ground personnel then climbed into the flight deck to complete safing activities and prepare the vehicle for transfer to the enormous Mate-Demate Device hoisting crane, which would later install her atop the Boeing 747 and attach a tail cone to protect her main engines and OMS pods during the return flight to Florida.

When the crew returned to Houston, the media frenzy was more intense than previous missions, although their opportunities to relax were limited. Hauck and Fabian visited Indonesia and the whole crew was invited to a White House state dinner, hosted by President Ronald Reagan in honour of the Emir of Bahrain. As the first American woman in space, Sally Ride naturally drew the spotlight, to the extent that Crippen and NASA management were obliged to shield her on occasion. At one glitzy function, a group of unrecognised males – the rest of the STS-7 crew, together with Ride's husband, fellow astronaut Steve Hawley – were almost turned away. Everyone knew Ride, but no one recognised *them*. Norm Thagard was pushed up against a wall by a particularly discourteous photographer, such was the urgency with which the latter *needed* to get to Ride and present his lens to her face. Instinctively, each of them knew that it was all part of the post-flight circus ("Your turn in the barrel," as John Fabian put it) and the price to be paid for having flown into space. Still, Ride only half-jokingly told the NASA oral historian that she was relieved to be assigned to her second mission in November 1983, because training kept her safe from the media. At least *there* she could be left to get on with her job. For the male members of the STS-7 crew who also found themselves named to upcoming missions, it was less traumatic. Almost immediately, Crippen was immersed in training to lead the high-profile Solar Max repair – to which he had been assigned several months prior to STS-7's liftoff. Thagard had joined the Spacelab-3 flight. Before 1983 was out, Fabian was attached to a new crew and Hauck had received his first command. Ride, too, would fly again, teamed with Crippen once more. In fact, by the end of 1984, 'Crip' would acquire the new and perhaps more fitting nickname of 'Mr Shuttle'.

LAUGHING ALL THE WAY

Guy Bluford, the first black American spacefarer, laughed with excitement all the way into orbit on STS-8.

It was around midnight EST, local time, on the rainy evening of 30 August 1983 when he and crewmates Dick Truly, Dan Brandenstein, Dale Gardner and Bill Thornton left the Operations and Checkout Building, bound for Pad 39A. Sitting out at the launch complex, resplendent in the dazzling glare of powerful xenon floodlights, Space Shuttle Challenger was ready for her third orbital voyage in less than five months. Admittedly, the reusable spacecraft was far from achieving NASA's vision of a flight every fortnight, but it was certainly beginning to prove its commercial worth. Tucked into the Shuttle's payload bay for the planned five-day flight was an Indian communications satellite called Insat-1B, which had netted the space agency $4 million in fees and which Gardner and Bluford would deploy a few hours into the mission. Unfortunately, another major cargo element – the second Tracking and Data Relay Satellite (TDRS-B) – had already been deleted from STS-8's roster as a result of the embarrassing IUS booster failure in April 1983.

In some minds, not least the media, it was more than coincidental that NASA had chosen to fly two of its astronaut trump cards – a woman and an African-American – on the first two Shuttle missions to feature members of the racially and culturally diverse TFNG class. Certainly, in April 1982, when the names of Bluford and Sally Ride were announced, *Flight International* had hinted strongly that the space agency's decision may have been politically motivated. Having said this, Lieutenant-Colonel Guion Stewart Bluford Jr had accrued an impressive educational record – one of only two military TFNGs with a doctorate – and had pursued an equally impressive Air Force career. Born in Philadelphia, Pennsylvania, on 22 November 1942, the son of a mechanical engineer father and a special education teacher mother, Bluford attended Overbrook High School and Pennsylvania State University, where he studied aerospace engineering on an Air Force officers' training programme. Graduation in September 1964 was followed by a commission as a second lieutenant, flight training at Williams Air Force Base in Phoenix, Arizona, and the award of his pilot's wings in February 1966. With many fellow Air Force pilots going to Vietnam, Bluford was no exception; after completing survival school and several months of radar and intercept training, he ultimately flew 144 combat missions – half of which were directly over the communist North – in F-4C Phantom jets. "These missions included combat air patrol, close air-to-ground support and air superiority flights," he told the NASA oral historian, "throughout North and South Vietnam, as well as Laos." By the summer of 1967, back in the United States, he had been designated as a T-38A Talon instructor pilot at Sheppard Air Force Base, near Wichita Falls in Texas, and later served as a standardisation and evaluation officer and an assistant flight commander. Bluford attended Squadron Officers School in 1971 and returned as executive support officer to the deputy commander of operations.

Whilst working at Sheppard, he began to explore opportunities for becoming an aerospace engineer. His parent service was not particularly enamoured by the idea – "the Air Force was critically short of pilots at that time," Bluford explained, "and thus needed my skills as an instructor pilot, versus as an engineer" – but was prepared to support him on a master's degree course. He completed the course in aerospace engineering at the Air Force Institute of Technology in 1974 and, whilst studying, one of Bluford' professors advised him to continue towards a doctorate. "I applied and got accepted … whilst still completing my master's degree requirements," he explained. "I dovetailed some of the PhD coursework among my master's degree courses, so that I could complete the coursework for *both* programmes in *two and a quarter years!*" In March 1974, after completing his PhD coursework, Bluford was assigned to the Flight Dynamics Laboratory at Wright-Patterson Air Force Base in Dayton, Ohio, as a staff development engineer and, later, as deputy for advanced concepts in the Aeromechanics Division and as branch chief of the Aerodynamics and Airframe Branch. Whilst there, he completed his doctoral thesis. "It was a great opportunity," he said, "for me to use both my technical skills and my flying experience in developing advanced technologies for future aircraft. I led an organisation of 45-50 engineers, who were doing basic aerodynamic research in such areas as forward swept wings, supercritical airfoils, advanced analytical aircraft techniques, inlets, axisymmetric nozzles and computational fluid dynamics."

By 1977, the Air Force was pressuring Bluford to return to active flying, as an instructor pilot for the T-37 Tweet training aircraft, and during this period a notice from NASA, calling for Shuttle astronaut candidates, caught his attention. In

Bluford's mind, it was the perfect opportunity to fulfil his flying requirements for the military, whilst also putting his technical skills to good use and expanding his knowledge. "I could do it *all* as a NASA astronaut," he exclaimed. "What a deal!" With more than the minimum mandated 1,000 hours of pilot-in-command jet time, Bluford was able to apply for both the pilot *and* mission specialist categories – although he was not a test pilot – and was ultimately summoned to Houston in early November, part of a 23-strong group which also included John Fabian, Norm Thagard and a young naval aviator who would later join him as a crewmate on STS-8: Dale Gardner. It was the ninth and second-to-last group to be interviewed for the 1978 selection. "From what I later learned," said Bluford, "there were *more* astronaut candidates selected from *that* group than from any other astronaut finalist group." Indeed, looking at the names, seven were ultimately chosen in 1978 (Bluford, John Fabian, Gardner, Terry Hart, Judy Resnik, Norm Thagard and James van Hoften) and two others were picked two years later (Bill Fisher and Bob Springer). Like Thagard, Bluford heard the news of the astronaut selection one day in January 1978 and assumed that he had been unsuccessful ... only to arrive at work one Monday morning and receive a call from George Abbey informing him of his selection. "I later discovered that NASA had called *all* 200 finalists that morning," Bluford recalled, "and told them of their decision." Triumph, however, was tinged with sadness, for that very same month of January brought the news that his mother was gravely ill, with barely six months to live.

Interestingly, several of the new candidates – Bluford included – were feverishly working to complete their doctoral dissertations at the time of selection. "I had given myself until the end of the year to complete the document," he told the oral historian. "NASA wanted me in Houston in July and thus I had to expedite the writing! I later learned that both Sally Ride and Kathy Sullivan were in the *same* situation with their PhD dissertations. I defended my research and completed my dissertation in June 1978, just before I left for my new assignment as a NASA astronaut." In fact, the Blufords – Guy, his wife and their two children – were in the process of moving house from Dayton to Philadelphia, early that month, and he stayed behind to finish the dissertation. "I eventually completed the document," Bluford concluded, "made six or seven copies of it, dropped it off on my dissertation advisor's desk one Sunday evening and left for Philadelphia to pick up the family." Years later, he would consider getting his PhD from the Air Force Institute of Technology as his crowning achievement.

Bluford's first four years with NASA included work in the Shuttle Avionics Integration Laboratory and Rockwell International's Flight Systems Laboratory and requalification as a T-38 pilot; by the end of his career, he would have accumulated more than four and a half thousand hours of flying time in the jet trainer. When he received notification of his first crew assignment as a mission specialist on STS-8 in April 1982, Bluford felt comfortable serving as MS2 – the flight engineer, seated behind and between Dick Truly and Dan Brandenstein – since his experience in the Shuttle simulators and with the flight data files was extensive.

Had TDRS-B and its IUS remained aboard Challenger, alongside Insat-1B, for this mission, it would have been the heaviest cargo complement yet ferried into orbit at over 29,000 kg. "There was very little weight-growth margin," Bluford said later. "During the training, Dale and I made several trips to Boeing Aircraft Corporation in Seattle, Washington, to learn about the IUS. We were becoming well versed in the operation of the IUS when it malfunctioned on STS-6 and, because of that, NASA decided not to fly the TDRS on our flight until after the mishap was investigated." The presence of two of these communications and data relay platforms in geostationary orbit – one at 171 degrees West longitude, above the central Pacific Ocean to the south of Hawaii, and another just off the Atlantic coast of Brazil, at 41 degrees West – was highly desirable to support the first Spacelab research flight in late 1983. A third orbital 'spare' (TDRS-C) was then to be launched on STS-12 in the late spring of 1984 and placed over the equator at 79 degrees West. However, by the end of May 1983, as investigators got to grips with finding out why the IUS had failed to inject TDRS-A into its 35,600 km orbit, NASA opted not to risk launching another one until the problems were resolved. Efforts were already underway to raise TDRS-A into its correct 'slot', at the expense of using two thirds of its valuable hydrazine station-keeping fuel. As late as the middle of July 1983, *Flight International* noted that NASA was hopeful that TDRS-A could be recovered in time to support the first Spacelab mission and the second satellite, TDRS-B, was provisionally remanifested onto STS-12.

In place of the Tracking and Data Relay Satellite would fly an unusual contraption called the Payload Flight Test Article (PFTA). According to NASA's Shuttle manifest of April 1982, this had been scheduled to fly aboard STS-16 in June 1984, but within a month of the return of STS-6 it had been moved forward to STS-11 and ultimately STS-8. Measuring 4.6 m long by 4.9 m high and weighing 3,900 kg, it was, in effect, a giant dumb-bell structure – a pair of 'wheels', connected by a 6 m central axle – to evaluate the performance and handling characteristics of the RMS arm. The PFTA was constructed from aluminium and stainless steel and equipped with four grapple fixtures; two of which would be used on STS-8. As with the Shuttle Pallet Satellite tests undertaken on the last mission, the aim was to acquire 'real world' data and develop crew expertise on elbow, wrist and shoulder joint reactions before the RMS was committed to the Solar Max repair. The experience gained on STS-8 would thus help to prepare the Solar Max crew not only for the repair procedure, but also to deploy their own payload: a 9,750 kg monster of a satellite called the Long Duration Exposure Facility. As a result, Dale Gardner's performance as lead RMS operator on STS-8 was being carefully scrutinised by NASA and Bob Crippen's next crew to ensure that the mechanical arm could indeed handle and manoeuvre large payloads with dexterity.

Yet it was the deployment and tracking requirements of their other payload – the Indian National Satellite, known as Insat-1B – that brought about one of the most historic features of the mission: the first Shuttle launch in darkness. After returning from California to Florida at the end of STS-7, Challenger spent a little under a month in the Orbiter Processing Facility and the PFTA was installed into her payload bay on 21 July. Following rollout to Pad 39A less than a fortnight later, Insat-1B was also loaded aboard. In doing so, preparation for STS-8 snared a new record for the fastest processing time between missions so far – a mere 62 days –

which was attained primarily by Challenger's personnel working around-the-clock to get her flight-ready. Seventy-six thermal protection tiles were replaced, as were the damaged brakes in her landing gear, the pitted flight deck windowpane and a failed Auxiliary Power Unit. Other experiments, including a record dozen Getaway Special canisters, were also affixed to her port and starboard payload bay walls.

When Dick Truly's crew arrived in Florida in their T-38 jet trainers on 27 August, they included among their number no fewer than *three* TFNGs. Although they were assigned at the same time as Bob Crippen's STS-7 team, they would actually become the second subset of TFNGs to fly. Years later, Dan Brandenstein recalled the excitement of the call to George Abbey's office and reception of the news. "By April 1982, the first six Shuttle flights had been assigned and they were all experienced people that had been around a long time. Nobody from our class had flown. I got called over one day and they said that I was going to fly STS-8. One of the neat things about it was that it was going to be a night launch and a night landing. What drove that was we were launching Insat and, to get it in the proper place, we worked the problem backwards. They wanted the satellite 'here', so then we had to go back down our orbital mechanics and it meant we had to launch at night. The fact we launched at night meant that we would end up landing at night. Dick and I had both done night carrier landings and, judging from the way the Shuttle flies and doing that at night, we both looked at each other and said 'Oooh. This is going to be interesting!' We got very much involved in developing a lighting system to enable us to safely land at night. We didn't have enough time to focus 'just' on that, although we got involved because we were the ones doing it first." This nocturnal launch commitment had been simulated, to an extent, on the ground. "We concentrated on flying night launches and night landings in a darkened simulator," Bluford recalled. "We learned to set our light levels low enough in the cockpit that we could maintain our night vision and I had a special lamp mounted on the back of my seat so that I could read the checklist in the dark. The only thing that wasn't simulated was the lighting associated with the Solid Rocket Booster ignition and the firing of the pyros for SRB and External Tank separation." STS-8's boosters had themselves changed from the set flown aboard Challenger's previous mission, since they contained new, high performance motors, which expanded the initial thrust by four percent. This improvement was achieved by lengthening the exit cones of their nozzles by 25.4 cm and decreasing the diameter of the nozzles' throats by 10.1 cm; the result was an increase in the velocity of solid fuel gases as they departed the booster. Moreover, some of the propellant inhibitor used in previous SRBs was removed, allowing the fuel to burn more rapidly.

Like STS-7, four crew members had become five in December 1982, when physician Bill Thornton, a member of the 'Excess Eleven' class, which also included Joe Allen, Bill Lenoir and Story Musgrave, was assigned to supervise a series of investigations into Space Adaptation Syndrome; the nauseous sickness which had plagued several earlier astronauts and threatened to plague many more. Standing just over six feet tall, William Edgar Thornton was one of NASA's largest astronauts – both in terms of height and physical bulk – and his size had inspired the nickname of 'Moose' in his younger days. Born in Faison, North Carolina, on 14 April 1929, he grew into "a bright and inquisitive boy", according to Dave Shayler and Colin Burgess, with a fascination for aircraft, science, electronics and building or repairing things. In his teens, he read about rockets and the practicalities of space flight for the first time and in high school operated a radio electronics repair shop, whose proceeds he later used to put himself through university. In 1952, Thornton received his bachelor's degree in physics from the University of North Carolina and entered the Air Force, later serving as the officer in charge of the Flight Test Photo Optics Instrumentation Laboratory at the Air Proving Ground of Eglin Air Force Base, near Valparaiso, Florida. His responsibilities involved the in-flight testing of all-weather interceptor aircraft and Thornton developed the first successful airborne target and evaluation missile scoring system; this work led to an Air Force Legion of Merit award in 1956.

At around this time, his engineering work drew Thornton towards medicine and in the summer of 1958 – after marrying an English-born medical student, Jennifer Fowler – he entered medical school at the University of North Carolina and received his doctorate in 1963. By this time, the first humans had begun to journey into space and he was determined the join them; consequently, his interest in aerospace medicine broadened and in 1964 he rejoined the Air Force as a captain and completed a rotating internship and a primary flight surgeon's course. At Brooks Air Force Base in Texas, he focused on medical instrumentation for the Manned Orbiting Laboratory project and helped to develop a non-gravimetric mass measuring system, which allowed 'weight' to be calculated in 'weightlessness'. A descendant of this system ultimately flew aboard Skylab and the Shuttle. Thornton's chances of entering the ranks of the NASA astronaut corps had seemed slim, even in 1965, when his age was several months beyond the rigid upper limit set by the space agency. A year later, he contacted Deke Slayton and was encouraged to hold out for another selection in 1967, in which previously disqualifying factors (notably height and age) would be relaxed "in exceptional cases".

Clearly, Thornton *was* an exceptional candidate, for he made the cut as one of the oldest members of the Excess Eleven and proceeded through his initial classroom instruction and on to flight school. His completion of jet training was slightly delayed in June 1969, due to an eye problem: he needed specially-designed spectacles to allow him to achieve the level of stereoscopic vision required for certain piloting exercises. It made little difference, Thornton told Shayler and Burgess, since he was close to the end of his flight training when the eye problem was uncovered. Indeed, two months *earlier*, in April, Thornton had already been assigned to the Apollo Applications branch of the astronaut office. Had the Skylab-B space station actually flown in the late 1970s, it has been suggested by Shayler and Burgess that Thornton would have been a prime candidate for a science pilot slot. He served on the support crew of all three Skylab missions, as well as a 56-day medical experiments altitude test at JSC in the summer of 1972 *and* on the Spacelab Medical Development (SMD)-III test in May 1977.

When Thornton finally received his first mission assignment in December 1982, it was certainly the highpoint of his professional life; indeed, his wife Jennifer described *that* Christmas as his happiest since joining NASA. As the astronaut office's foremost expert on Space Adaptation Syndrome, he recalled that STS-8 "was the first and probably only flight that an investigator was *ever* allowed to make his *own* selection of experiments ... and *fly* with it." Thornton's assignment had actually led to the creation of an extra, unofficial crew patch. Historically, astronauts avoided doctors like the plague, remarking that there were only two ways a pilot could emerge from a consultation: either 'fine' or 'grounded'. None of the STS-8 astronauts was at risk of being grounded by Thornton, but his experiments – which included a series of blood tests *on himself* – resulted in a tongue-in-cheek mission patch featuring his bespectacled eyes peering at a cluster of four pairs of frightened eyes in Challenger's flight deck. This good-natured 'fear' of the good doctor continued into space, during a telecast in which Thornton explained the purpose of his medical tests to the terrestrial audience. At the end of the telecast, Dick Truly quipped that the rest of the crew were now fed up with this "chamber of horrors" and picked up a hammer, floated across the middeck and revealed Thornton being restrained to a bulkhead with grey tape. "His three colleagues," wrote Shayler and Burgess, "each wielded knives, wrenches, pliers and hammers and, as the screen faded, a muffled *scream* from the good doctor was heard to close the telecast."

Behind the humour, however, there were serious concerns among NASA's senior management that space sickness could detrimentally affect future missions if crew members reacted severely to it. During a lecture in Birmingham, England, in October 1991, attended by this author, Thornton admitted that it remained difficult to predict which individuals were susceptible, although he pointed out that Dale Gardner experienced the nauseous ailment, yet was still able to complete all of his assigned tasks, including the hours-long Insat-1B deployment. "You can't redesign the human body," Thornton said, "but human beings have learned and will continue to learn to adapt and work in zero gravity." During STS-8, his investigations encompassed seven medical disciplines: testing aural sensitivity thresholds ('audiometry'), tracking his crewmates' general health ('biomonitoring'), recording electrical signals generated by their eye movements ('electro-oculography'), studying the effect of repeating physical movements ('kinesymmetry'), examining changes in their limb-volume circumference ('plethysmography'), measuring external tissue pressures ('tonometry') and photographing changes in leg volume throughout the mission. Thornton's main conclusions were twofold: that none of the astronauts were directly 'motor control affected' by the condition and that symptoms had more or less disappeared within 72 hours of launch. Since the earliest reported instance by Soviet cosmonaut Gherman Titov in 1961, around 40 percent of space travellers have experienced the problem, although detailed investigations during the Spacelab-1 mission in late 1983 identified the practice of rolling or pitching the head to be a helpful countermeasure. 'Ambiguous visual cues', on the other hand, such as viewing a crewmate from an unusual orientation, generally exacerbated the sensation of malaise and sickness. However, the near-impossibility of determining which astronauts were most likely to fall prey to space sickness came as a surprise to Challenger's pilot, Dan Brandenstein. "I'd never been seasick, airsick or anything in my life," he said. "I don't understand half of those medical experiments, but during training they put us in a spinning chair and put a blindfold on each of us. They spun the chair and then they had us move our heads down, up, right, left, down, up, right, left. I was convinced I could never get motion sickness but, man, in about 30 seconds, I was sick puppy!"

By the late summer of 1983, the five astronauts had become a close-knit miniature family in their own right. "You spend so much time working together," said Brandenstein, "and that's part of the process of crew selections. You don't put oil and water together. When I ran the astronaut office [from 1987-1992], I was responsible for the crew assignments and you specifically look for people that are compatible. I can't speak for assignments that were made on me before I was doing them, but it was obvious by even looking at it that NASA looked for a good mix. They looked for people with specialities that mesh with the mission requirements. STS-8 was a good crew. Dick Truly had been around a long time and was a good commander; he taught us a lot. Everybody had their strengths and their area of expertise and they focused on those and shared their experience and wisdom with the other folks. We got the job done."

The man responsible for knitting them into a team was Captain Richard Harrison Truly, the commander, who had previously become – on 12 November 1981 – the world's first spacefarer to blast off on his first mission *on his birthday*, for he had been born on 12 November 1937 in Fayette, Mississippi. "I *was* interested in flying" as a child, he told the NASA oral historian, "but I *never* really intended to be a pilot. It just never occurred to me that *that* would be a possibility." He attended local schools and studied aeronautical engineering at Georgia Institute of Technology, where he received his degree in 1959. A military career, by now, was firmly on his radar, since he held a scholarship with the Reserve Officer Training Corps, but even at this late stage becoming a pilot was not really part of the plan. "It wasn't that I *didn't* want to be," he explained, "it just never really occurred to me. I was going to be an *engineer*." A handful of training flights changed his mind and Truly entered flight school and was designated as a naval aviator in October 1960. His first tour was with Fighter Squadron 33, in which he flew F-8 Crusaders from the USS *Intrepid* and (propitiously) the USS *Enterprise*, performing more than 300 carrier landings. Whilst aboard the *Enterprise*, Truly's squadron leader, Commander Larry Ned Smith, advised him to consider test pilot school and for two years, from 1963 until 1965, he worked at Edwards Air Force Base, first as a student and later as an instructor in the Aerospace Research Pilot School. During this period, he was the youngest of nine candidates put forward for the Air Force's Manned Orbiting Laboratory project; additionally, he and another aviator, Bob Crippen, were the only Navy members of the group. When MOL was cancelled in June 1969, Truly and Crippen were amongst a handful of pilots forced down NASA's throat by the Pentagon. "The agreement that was cut was that they would assign the seven youngest MOL crewmen ... to NASA," he recalled. "It turned out that, of the original MOL crew that had been announced in 1965, I was the only one that was young enough [in 1969, by which time others had joined the MOL team] to *still* be in the youngest seven, so I never filled out an application to NASA. I'm the *only* person who has *ever* flown in space that *never* applied!" The unbridled joy at having entered the ranks of the world's most elite flying fraternity had its ups and downs, though, and one of the notable 'downs' was the lack of seats on space missions. Dave Scott asked him to serve on the Apollo 15 support crew, but Truly felt his chances of a flight were better in Skylab and elected to take duties there instead. "I foolishly thought that maybe I would actually get to fly on Skylab," he said, "but I really didn't account for the fact that … you were *in a line*. There were a *lot* of people ahead of me." Capcom duties in Skylab were followed by a support role on the Apollo-Soyuz Test Project and finally, in the spring of 1976, assignment to the Enterprise approach and landing tests and the pilot's seat on the second Shuttle mission. Less than six months after returning from his first space mission, Truly found himself in George Abbey's office, welcoming Dan Brandenstein, Dale Gardner and Guy Bluford as his STS-8 crewmates.

Despite his experience, Truly had never flown a night launch, so it was with an air of excitement and trepidation that Challenger's crew headed into a bewildering glare of flashbulbs in the opening minutes of 30 August 1983. Their liftoff, at 2:32 am EST, came 17 minutes into a half-hour 'window', due to thunderstorms in the area, and lit up the sky of a slumbering Florida. For Daniel Charles Brandenstein, STS-8 was the opportunity to put five years of training to the ultimate test. He came from Watertown, Wisconsin, where he was born on 17 January 1943. After attending high school in his home town, Brandenstein enrolled at the University of Wisconsin in River Falls and received a degree in mathematics and physics in 1965. Aviation had always been at the back of his mind and he considered America's fledgling manned space programme as "the *ultimate* form of aviation" and thus a goal for the future. He read the biographies of the Original Seven Mercury astronauts and identified the main requirements: active-duty military officers and test pilots with degrees in science or engineering. To Brandenstein, mathematics and physics "were always my favourite courses" and "as close to engineering as you could get". The decision over which branch of the armed services to join came in his final year of college: Air Force pilots landed on several kilometres' worth of runway, whereas naval 'aviators' brought their jets screaming onto a couple of hundred metres of *pitching steel* in the middle of the ocean! "Looking at what looked to be the most interesting and challenging," Brandenstein told the NASA oral historian, "the naval aspect of aviation caught my fancy."

He entered active Navy duty in September 1965, was attached to the Naval Air Training Command for flight instruction and received the designation of a naval aviator a little under two years later. Flying A-6 Intruders, Brandenstein participated in two cruises to Vietnam between 1968 and 1970, aboard the USS *Constellation* and the USS *Ranger*, logging 192 combat missions. Subsequent work focused on operational tests of weapons systems and tactics for the A-6 and he was selected (alongside Rick Hauck) for the Navy's test pilot school at Pax River in 1971. After graduation, he conducted electronic warfare systems tests in a variety of fighter aircraft and deployed to the Western Pacific aboard the *Ranger*, again flying A-6 jets, from March 1975 until September 1977. Each step in his education and experience had guided him closer to the astronaut corps, but by this point the Shuttle was on the horizon and in the summer of 1977 Brandenstein was invited to Houston – part of the first group in a total of ten groups which would be interviewed between

August and November of that year – for an extensive series of tests. Less than six months later, his wife pulled him, dripping wet, out of the shower one Monday morning to take a call from someone called George Abbey.

Astronaut Candidate training could only be compared to drinking water from a fire hose, since it kept coming and coming *and coming* and the 1978 class captured as much as they could, knowing that someday it might help to save their lives. The civilians included amongst their ranks medical doctors, astronomers, geologists, engineers, physicists and biochemists, whereas the military officers carried with them many years of experience in the cockpit. Yet all of them were 'rookies' as far as operating the Shuttle was concerned. "We got a full set of briefs on each system, so we knew how the electrical system worked and how the hydraulic system worked and the computers," Brandenstein recalled. Part of his requirement for being on 'active' flight status was also having the ability to maintain proficiency in NASA's fleet of T-38s. These legendary – some observers have called them 'antique' – aircraft continue to be used by today's Shuttle astronauts for flight training and, literally, as personal taxis to reach appointments across the United States. Unfortunately, the ability of this sleek, supersonic dart to precisely mirror the handling characteristics of the stubby, delta-winged orbiter has long proven problematic: the lift-to-drag ratios of the two vehicles are quite dissimilar. In order to best simulate the steep-angled Shuttle approach to the runway, astronauts typically opened the T-38's speed brakes as wide as possible and deployed its landing gear at the very start of their descent. "There was an area, just outside Houston, over the Gulf of Mexico, where we could go out and do what we called 'turn and burn'," Brandenstein explained of his T-38 escapades, "which is do aerobatics and loops and rolls and chase around clouds and stuff like that. All the time, that's a way of maintaining your piloting skills. Obviously, it's a kick for people that had flown thousands of hours, but for somebody who had never flown before or had very little experience, it was a 'real' kick, because you could go supersonic, pulling 7 G. All the pilots had been test pilots before, so we'd go out and run the mission specialists over the wringer, showing them the various things you'd do if you're testing a new airplane. We'd do simulated combat runs and show them what it was like to have a dogfight and all those sorts of things." As mission-specific training got underway, Brandenstein and Truly found themselves practicing Shuttle landing approaches at least once or twice per month, not only in Houston, but also at KSC, Edwards Air Force Base in California and White Sands in New Mexico.

Since the crew would be launching at night, it became necessary in the final week before the mission for them to enter quarantine and shift their sleep patterns into the daytime hours. "It took us about a week to get comfortable with that," recalled Bluford, who ended up 'sleep shifting' in readiness for three of his four space voyages. "Some of us slept at home, while others slept in the crew quarters ... in Houston. We ate food prepared at the centre and practiced in the simulators at night. About three to four days before launch, we flew to the Cape for the final launch countdown. On 29 August, we were awakened at 10:00 pm [EST]. We had breakfast and suited up for the mission, then headed downstairs for the van ride to the launch pad. I noticed it was raining. There was lightning in the area and there

Carrying America's first black spacefarer, Guy Bluford, and the oldest person yet launched into orbit, 54-year-old Bill Thornton, Challenger ascends on a pillar of golden flame in the first night launch of the Shuttle era.

was some concern expressed by the launch control centre about our safety as we proceeded out to the pad. Finally, they left it up to Dick to decide if it was safe for the crew to go to the pad. He made the decision for us to proceed and went out to Challenger. As we climbed into the vehicle and completed our pre-flight checks with the launch control centre, the rain began to subside and the clouds began to clear away. The ride into orbit was really exciting! We had darkened the cockpit to prepare for liftoff; however, when the SRBs ignited, they turned night into day inside! Whatever night vision we hoped to maintain, we lost right away at liftoff. The ride on the SRBs was noisy and bumpy as Challenger rotated to align us to a 28.45 degree inclination. The orbiter pitched down as we headed downrange, upside down. Approximately two minutes into the mission, we jettisoned the boosters. There was a large, momentary flash of light in the windows when the SRB pyros fired. We continued to ride on the three main engines for the next six and a half minutes and then jettisoned the External Tank at eight minutes and 45 seconds into the flight."

As the flight engineer, Bluford checked off each stage of the violent climb to orbit and was prepared to read out procedures to support the pilots in the event of problems. Next to him on the flight deck, literally shoulder to shoulder, and directly behind Brandenstein, was Dale Gardner. Bill Thornton sat alone in the darkened, locker-studded middeck. From his vantage point, the 54-year-old physician had little to see: the only window was a small circular one in the side hatch, although, craning his neck, he could see 'upwards' into the flight deck and through the overhead windows. At the instant of ignition, Thornton recalled years later, the sensation was similar to "taking a fast ride on the London Underground". From his perspective, all was dark during the first two minutes of ascent, but as soon as Challenger shed her twin SRBs, the entire cockpit was eerily lit up. Upstairs, Gardner's main view was through the overhead windows – and what he saw worried him sufficiently to call Brandenstein over the intercom. As the pilot, one of Brandenstein's key roles during ascent was to monitor the performance of Challenger's three main engines.

"Obviously, Dick Truly and I were up front, watching the instruments," recalled Brandenstein, "and Dale was looking back over his head out the [overhead] window and back at the ground. At night, he could see how it lit everything up. During the first stage, it was really bright, because we had the boosters going. In fact, from the front cockpit, looking out, it was like we were inside a fire, because we didn't really see the flame, but we did see the reflection and the light. We weren't very far into the launch and Dale said 'Dan, how do the engines look?' I said 'Yes, look fine'. Thirty seconds later, he said again 'Dan, how do the engines look?' 'Fine'. I don't know how many times this happened going uphill. We didn't have a lot of time to chat about it, so finally we got all settled down on orbit, I said 'What was going on?' He said 'I was looking out the window', and when you watch a Shuttle launch the flame from the [main] engine is solid. It comes out of the nozzle and just 'sits' there. During all those engine tests before STS-1, you'd have an engine running on the test stand and the flame would be solid and then, all of a sudden, the flame would 'flutter' and the engine would blow up! As you get higher in altitude, and from the perspective Dale had, the flames from the engines seemed to be fluttering, so his connection was that when the flames flutter, the engine blows up. You just have a different perspective as you get higher. The air pressure goes way down and you get into a vacuum, so basically what holds your flame real tight is the atmospheric pressure factors in that. When you get outside atmospheric pressure, they expand and flutter and little bit more."

Lieutenant-Commander Dale Allen Gardner was, at just 34, the youngest member of the STS-8 crew. He came from Fairmont, Minnesota, where he was born on 8 November 1948, and began to overachieve from an early age: he graduated as *valedictorian* – the highest-ranking academic member of his class – from Savannah Community High School in Illinois in 1966 and entered the University of Illinois at Urbana-Champaign to study engineering physics. Gardner received his bachelor's degree in 1970 and entered active duty with the US Navy. As an ensign at the Aviation Officer Candidate School in Pensacola, Florida, he was the most promising naval officer in his class and graduated from basic officer training with the highest academic average *ever achieved* in the history of his VT-10 squadron. Next came the Naval Technical Training Center at Glynco, Georgia, from which Gardner emerged as a Distinguished Naval Graduate, receiving his flight officer's wings in May 1971. For the next two years, he was detailed to the Naval Air Test Center at Patuxent River, working in the weapons systems test division, conducting initial development tests of the new F-14 Tomcat fighter. He subsequently participated in two cruises to the Western Pacific and Indian Oceans aboard the USS *Enterprise*, flying the

Tomcat, and in December 1976 he joined Test and Evaluation Squadron Four at Naval Air Station Point Mugu in California. During the next year, Gardner was heavily involved in the operational testing of advanced fighters for the Navy. It was whilst at Point Mugu that he noticed NASA's invitation for astronaut candidates and, in November 1977, found himself among 23 finalists invited to Houston for screening. Selection in January of the following year was followed by assignment as project manager for the Shuttle's flight software and as a support crew member for STS-4; by the time the latter flew, he had already been assigned to his first mission, STS-8. Even amongst the over-achievers of the astronaut corps, Gardner stood out: indeed, fellow astronaut Joe Allen, who flew with him in November 1984, would describe him glowingly as "a premier Navy test engineer".

After STS-8, as the five astronauts listened back on their cockpit intercom tapes from ascent, they were puzzled to hear someone chuckling all the way into orbit. It was Bluford. Years later, he remembered being so excited by the whole event that his only feeling at the time was not fear, but sheer elation. His journey into space had taken much longer than eight and a half minutes and represented an enormous leap for African-Americans, who came to regard him as a new role model for their own aspirations and dreams. However, the astronaut himself has since remarked that it was never his intention to become the first black American in space and, fortunately, the media circus surrounding the achievement of STS-7's Sally Ride had left him largely ignored. "I recognised the importance of it," Bluford admitted, "but I didn't want it to be a distraction for my crew. We were all contributing to history and to our continued exploration of space." Nonetheless, he felt the Tuskegee Airmen – the United States' first all-black flight squadron in the Second World War – helped pave the way for his achievement. When he was chosen in January 1978, Bluford was joined by two other black astronauts: an Air Force helicopter pilot named Fred Gregory and a civilian physicist, Ron McNair. Gregory would later become the first African-American to command a space mission and, ultimately, served as Deputy Administrator of NASA, whilst McNair, tragically, would die aboard Challenger in January 1986. Seeing his ancestral homeland from space was profoundly moving for Bluford. "I still remember seeing the African coast and the Sahara Desert coming up over the horizon," he said later. "It was beautiful. Once we completed our [OMS] burns, I unstrapped from my seat and started floating at the 'top' of the cockpit. Like all the other astronauts before me, I fumbled around in zero-g for quite a while before I got my 'space legs'. However, it was a great feeling and I knew right away that I was going to enjoy this experience."

In a sense, Bluford's desire to blend in with the rest of the crew, and thereby avoid allegations of 'positive' discrimination, helped to shield NASA from accusations of creating a racially-motivated stunt, but no one at the space agency was unaware of the significance of the event. In Mike Mullane's autobiography, *Riding Rockets*, he made insightful reference to the issue of race and the fact that it never created a problem in the astronaut office. Having said that, Mullane related a story from January 1983, when he had to sit in for Bluford one day in the Shuttle Mission Simulator in Houston. During the course of the day, the Sim Sup asked them to invent a medical emergency for the flight surgeon to deal with. In the cockpit, the ideas flowed – Brandenstein had stomach pains, Truly had flu symptoms, Gardner had toothache – until one of them came up with the perfect idea: Bluford had turned *white*! "NASA ... had been in orgasmic ecstasy over the impending flight of America's first black astronaut," Mullane wrote. "Knowing this, the suggestion was outrageously funny."

Outrageous, indeed. When Dick Truly heard the conversation, he warned them that if they made the call, the *closest* they would get to space would be the office of JSC Director Chris Kraft ... to be *fired*.

Upon reaching orbit, Dan Brandenstein discovered that, despite having been sick during the ground tests, he adapted to microgravity exceptionally well. "I'm one of the lucky ones in that I did a back flip out of my seat and never looked back," he said, "and never had a hiccup in any of my missions. It certainly makes your mission more enjoyable if you don't have to deal with that, but NASA was trying to decide what made people sick and how to prevent it and it turned out, after a while, they quit trying and there was no correlation. Some guys could ride the spinning chair until the motor burned up and didn't get sick and then got into orbit and, within ten minutes, they were as sick as could be. Ultimately, they found Phenegren worked on almost everybody. Doctors use it on people that have had chemotherapy. So as soon as somebody would start getting a symptom of space sickness, you'd give them a shot and, in about 15 minutes, they'd be as good as new for the rest of the flight." Bluford, too, did not recall any problems. "We had little sandwiches tied to our seats," he said later, "and when we got on orbit, a couple of crew members weren't feeling well as they adapted to space, so they 'passed' on lunch. I felt fine. I not only ate my lunch, but part of theirs, too!"

Despite concerns about space sickness and the fact that Dale Gardner, as lead crew member for both the Insat-1B deployment and RMS operations, suffered from the ailment, all five astronauts were able to conduct their prescribed tasks without problems. Releasing the $50 million Indian communications satellite and its attached PAM-D booster followed a similar protocol to that of the Anik-C2 and Palapa-B1 deployments on STS-7. As its name implies, Insat-1B was the second in a series of multi-purpose geostationary platforms to provide telecommunications, television broadcasting, meteorology and search and rescue services to most of the Indian subcontinent and Indian Ocean. Its predecessor, Insat-1A, was launched atop a Delta rocket in April 1982. However, despite reaching its 35,600 km geostationary orbit, successfully deploying a jammed C-band antenna and returning valuable meteorological imagery, it failed to deploy its solar sail – which provided a 'counterbalance' for its single solar array – and later lost its 'lock' on Earth, began to tumble and inadvertently exhausted its entire supply of attitude control propellant. The satellite was abandoned that September, far short of its advertised seven-year life span, but India's Department of Space received a $70 million insurance payout from the debacle. Incidentally, only eight weeks before STS-8 flew, India had contracted a Shuttle launch slot for Insat-1C – to replace the lost Insat-1A – on STS-35 in February 1986. "India will buy Insat-1C from prime contractor Ford Aerospace [Corporation]," *Flight International* revealed, "apparently as an extension to the contract covering the first two craft."

In the wake of the Insat-1A loss, Ford Aerospace introduced an automatic switching mechanism for the antenna, to prevent any future loss of 'lock' on Earth, made alterations to the attitude control propellant valves and modified the design to ensure that the solar sail deployed properly. Like its predecessor, Insat-1B was cube-shaped, weighed 1,150 kg and carried a dozen C-band and three S-band transponders for its communications and television services. Its meteorological payload consisted of a Very High Resolution Radiometer (VHRR), capable of acquiring visible and infrared images of Earth every 30 minutes, and a system for taking environmental data from unattended land-based and ocean-based stations. Between 1982 and 1990, four Insat-1s surveyed India's natural resources. Their data provided estimates of major crops, conducted drought monitoring, assessed the condition of vegetation, mapped areas at risk of flooding and identified new underground water supplies. The deployment of Insat-1B was timed to occur during Challenger's 18th orbit, a little over a day into the mission, and, precisely on time at 3:48:54 am EST on 31 August, Gardner and Bluford flipped switches on the aft flight deck instrument panel to send the satellite on its way. Fifteen minutes later, Truly and Brandenstein performed a now-customary separation burn in readiness for the PAM-D ignition. Deployment from the Shuttle was so precise (within a tenth of a degree) that it saved Insat some 230 kg of station-keeping propellant which might otherwise have been needed had it been launched aboard an expendable rocket. At 4:34 am, the PAM-D fired to lift Insat to geostationary transfer orbit with a 35,600 km apogee. Later, ground controllers used the satellite's own hypergolic motor, which mixed nitrogen tetroxide and monomethyl hydrazine, to circularise the orbit.

However, during its first few days of operations, under the direction of controllers at India's Department of Space, it came close to suffering the same fate as Insat-1A. Unconfirmed video recordings from the crew suggested that it may have been hit by debris just 19.5 seconds after leaving the payload bay and, indeed, it was not until mid-September that ground operators at the Master Control Facility in Hassan succeeded in unfurling its single, five-panel planar solar array. Due to the presence of the radiometer on the opposite side, it was not practical to install two solar panels on the Insat-1 design. However, a 12.6 m solar sail had been installed on its VHRR 'side' to provide passive compensation of the solar pressure torque about the satellite's body; effectively, to act as a 'counterbalance' against the effect of the solar wind. By this stage, Insat-1B was on station at 74 degrees East longitude – replacing its failed predecessor – and commenced full operations the following month. The debris, meanwhile, appeared to have originated from the orbiter's payload bay and a detailed, six-hour-long television scan was conducted after touchdown. Nothing on the satellite's sunshade or deployment mechanism appeared to be either missing or damaged and, upon inspecting still and video camera footage, no evidence of a direct strike on the satellite was confirmed. It seemed more likely, NASA's post-flight anomaly report concluded, that a stray particle had been spotted by the astronauts as it drifted between themselves and the satellite. For almost seven years, Insat-1B provided satisfactory services, returning 36,000 images of Earth and providing communications and direct nationwide television services to thousands of remote Indian villages. On the ground, more than 5,000 Indian-built satellite dishes, some just 3 m in diameter, were established to allow the satellite to broadcast social and educational programmes to rural communities. Insat-1B operated until July 1990, after which it served in a 'standby' capacity until it was replaced at 93.5 degrees East by Insat-2B in August 1993.

Despite the astronauts' intense focus on their mission, memories of simply being in space were aplenty. "The first impression," said Brandenstein, "is still the biggest. We were crossing Africa when I saw my first sunrise in orbit and, to this day, that is the 'wow' of my space flight career. Sunrises and sunsets from orbit are just phenomenal and the first one knocked my socks off! It happens relatively quickly because you're going so fast and you get this vivid spectrum forming at the horizon. When the Sun finally pops up, it's so bright; not attenuated by smog or clouds." Throughout STS-8, they received daily updates from Mission Control on terrestrial events. "They kept me abreast of how Penn State was doing in football," said Bluford, "and how the Philadelphia Phillies were doing in baseball. Each morning, we were awakened by a school song. We were informed about the shooting down of a Korean airliner, Dick Truly told me he was leaving the astronaut office to become Commander of the Naval Space Command and my wife sent me a message saying we had termites in our house!"

In addition to Insat-1B, the crew had a range of middeck investigations to tend. One was the Continuous Flow Electrophoresis System, which had ridden on all three of Challenger's missions and, on STS-8, carried live human cells from a pancreas and a kidney, together with a rat pituitary gland. It was the first time that 'living' cells had been carried for electrophoretic separation in orbit. All of the samples were used to separate specific secretory cells with no apparent problems, although post-mission analysis revealed a larger residue of cells inside the spent CFES syringes than was anticipated. Although not considered a problem with the machine itself, it was noted by NASA that it might represent a 'shortcoming' in the design of the CFES equipment to handle and separate living cells. Nonetheless, results from its three previous missions – one aboard Columbia – amply demonstrated its ability to separate 700 times more material in space than was achievable on Earth. It was hoped that the pancreas cells in particular, which had been provided by McDonnell Douglas through an agreement with researchers at Washington University's School of Medicine, could be used in studies of purification techniques and, ultimately, new treatments for diabetes. The kidney cells, meanwhile, were supplied by NASA and the pituitary cells by Pennsylvania State University. In view of their 'living' status, one of Bluford and Gardner's key challenges was to keep them alive both before and after electrophoretic separation had taken place. To accomplish this, the CFES hardware was fitted with a tray on which samples were carried aloft on a surface of micro-carrier beads in a fluid that was compatible with the living cells. Bluford, who tended the machine for several hours on 30 August, and Gardner, who monitored it the following day, transferred the cells to syringes before inserting into the separation chambers. Maintaining the cells and keeping them alive made it necessary to schedule CFES runs as soon as Challenger entered orbit. Hence, it was operated only on the first and second days of the flight.

Elsewhere, in addition to monitoring the astronauts' adaptation to microgravity,

Bill Thornton kept a close watch on the behaviour of six male albino rats in an Animal Enclosure Module (AEM), housed in a middeck locker. One of the aims of the device, which would fly in support of a student experiment on Challenger's next mission in early 1984, was to assess how well the AEM contained micro-organisms and prevented 'leaks'. Apart from two micro-organisms, presumably introduced by the potatoes provided as a food and water source, the device maintained the rats' health satisfactorily during the mission. Moreover, by posing no danger to the astronauts' own well being, it demonstrated the device's ability to maintain biological materials in full isolation. In fact, STS-8 marked the very first occasion on which a cage of animals had flown in the Shuttle's crew cabin. The rats consumed less food than predicted and hence did not gain weight at expected rates, compared to ground-based 'control' animals, although they returned to Earth in a healthy state. Their lower-than-expected food consumption level was attributed to the AEM's delivery system, which differed from ground-based units.

By 1 September, with the Insat-1B deployment and completion of the lengthy electrophoresis experiment runs behind them, Challenger's astronauts set to work on their next major objective: testing the muscle of their ship's mechanical arm with the Payload Flight Test Article. Although it would not be released into space, the device was still the largest payload yet manipulated by the RMS – twice as heavy as the SPAS platform carried by Bob Crippen's crew – and, true to its nickname, was entirely passive, with no power, command or attitude control functions of its own. Yet even PFTA was barely a third of the weight of the enormous Long Duration Exposure Facility, destined to be placed into orbit by another Shuttle crew in the spring of 1984. Nonetheless, its forward and aft screens closely mimicked the visibility and manoeuvrability obstacles that future astronauts deploying large, cylindrical structures might face. In particular, PFTA became the first Shuttle-borne cargo with a 'five point' attachment to the payload bay – a keel and four longeron fittings – all of which were out of the direct view of the crew. Consequently, Gardner and Bluford would be reliant upon cameras fitted to the RMS. With Dale Gardner at the controls, the dumb-bell was first grappled by one of its two 'active' fixtures and subjected to a variety of tests, including evaluations of the arm's performance as Truly pulsed Challenger's RCS thrusters. These tasks helped to satisfy a number of test objectives to verify ground-based simulations, assess visual cues for payload handling and demonstrate both hardware and computer software. During each activity, the RMS was employed in both 'manual' and 'automatic' modes. The two grapple fixtures on the payload provided different geometries and mass properties for the arm and one of them – in the centre of the PFTA's forward screen – offered a larger moment of inertia. The second active fixture was attached to the upper port side 'corner' of the aft screen. Much of the payload's mass was situated at its aft end, thanks to a quantity of lead ballast, and Gardner's evaluations helped to verify that the RMS could position a large structure within 50 mm and one degree of accuracy in respect to the Shuttle's axes.

Although the TDRS-B satellite had long since been deleted from the STS-8 roster, a number of important tests were performed during the mission to ensure that its doddery sibling, recently established in geostationary orbit at 67 degrees West longitude, would be able to support the Spacelab-1 flight, alone, later in 1983. Among these tests were evaluations of TDRS-1's ability to relay voice transmissions, commands and Shuttle housekeeping telemetry through its S-band communications channels, as well as demonstrating its high capacity Ku-band link. This began only minutes after liftoff on 30 August, shortly after Challenger flew over Bermuda, and proved largely successful, although S-band telemetry was lost for a period of three hours at one point. However, crew voice communications were still available through other channels and the crew was asked to switch their data over to the S-band link. In addition, the White Sands Ground Terminal in New Mexico suffered a series of computer failures, which, in most cases, led to the loss of data. In total, TDRS-1 supported Challenger during 65 orbital 'passes' – exactly two dozen fewer than originally planned – and, of these, approximately two thirds were deemed fully successful. In particular, the performance of the TDRS-to-Shuttle S-band link was found to be highly dependent upon antenna 'look' angles, with instances in which the satellite was able to maintain return-link telemetry data, but forward-link lock could not be maintained. Still, TDRS-1's support of the Ku-band communications link proved excellent.

Meanwhile, slightly forward of the PFTA in the payload bay sat the U-shaped Development Flight Instrumentation pallet, equipped with two scientific and engineering experiments, including a heat pipe that investigators hoped could provide a useful means of maintaining systems temperatures on future satellites. Dan Brandenstein activated the pipe's heater power switch and photographed temperature-sensitive tape through Challenger's aft flight deck windows. The investigation, performed early on 31 August, proved highly successful, requiring 15 minutes to warm up and running at stable temperatures, which varied slowly in response to changes in the external environment. Although 36 photographs were taken, fewer than two dozen proved usable, due to a problem with the camera's film-advance mechanism. Still, about one and a half hours' worth of data was recorded and transcripts of the astronauts' visual observations were incorporated into the results. The second experiment, known as the Evaluation of Oxygen Interaction with Materials, consisted of a passive array of various samples – including coatings, composites and polymeric films – exposed to bombardment by molecular and atomic oxygen present in low Earth orbit. Previous tests aboard Columbia had revealed that atomic oxygen was extremely reactive when in contact with solid surfaces; causing chemical changes, altering optical and electrical characteristics and even removing complete layers of material. This could, NASA feared, cause problems. It was expected that the Hubble Space Telescope's relatively high orbit would make atomic oxygen reaction rates fairly low, but the long-duration nature of other missions could give rise to significant erosion of solar arrays, optical coatings, light baffles and thermal control films. Among the materials flown on STS-8 were specimens of the Shuttle's new Advanced Flexible Reusable Surface Insulation blanketing and thermal protection tiles to assess their degradation during orbital flight. As well as being mounted in trays and atop canisters on the DFI pallet, several samples were affixed to the RMS and exposed to space for a total of 40 hours.

Elsewhere in the payload bay were a record number of Getaway Special canisters, four of which held scientific investigations and eight carried more than a quarter of a million first-day philatelic covers, intended to be sold by the US Postal Service after the mission. Each cover bore a recently released $9.35 postage stamp and featured the STS-8 crew's patch and logo to commemorate NASA's 25th anniversary that year. Unfortunately, the covers also bore the mission's originally scheduled launch date of 14 August – which was also the stamp's release date – but this was rectified after landing. Eclipsed by the philatelic covers, but no less important, were the other four GAS canisters, which included a Japanese effort to grow artificial snow crystals, a NASA-funded cosmic ray experiment, a test of the sensitivity of ultraviolet films in space and a contamination monitor to measure the impact of atomic oxygen particles on samples of carbon and osmium. The Japanese study was actually a repeat of an experiment carried on Challenger's maiden mission, albeit with new and improved hardware. Sponsored by the newspaper *Asahi Shimbun*, its principal investigator, Shigeru Kimura, observed the growth of artificial snow particles in microgravity. Post-flight analysis after STS-6 had revealed that the temperature of the experiment's GAS end plate had fallen lower than expected. This prompted a hardware redesign to warm the experiment's water in two tanks and thus provide sufficient vapour to generate crystals. On STS-8, it successfully produced the crystals and acquired high quality video imagery. Meanwhile, the Cosmic Ray Upset Experiment, provided by NASA's Goddard Space Flight Center of Greenbelt, Maryland, helped to resolve long-standing questions about the probability of highly charged particles causing errors in memory-type integrated circuits. In some technologies, Principal Investigator John Adolphson explained, enough energy could be deposited to cause an effect known as 'latch-up', in which electronic devices literally destroyed themselves by drawing too much electrical current. Also from Goddard was the ultraviolet-sensitive photographic emulsion experiment, whose results would pave the way for a major astrophysical instrument – the Naval Research Laboratory's High Resolution Telescope and Spectrograph – scheduled to be flown operationally aboard Spacelab-2 in the winter of 1984. In evaluating the effect of Challenger's gaseous environment, the emulsion experiment, led by Principal Investigator Werner Neupert, provided valuable insights into the impact of orbital hypersonic flight regimes on ultraviolet-sensitive films. In fact, STS-8 was ideal for this kind of study, since the Shuttle's flight path was deliberately adjusted during several thermal tests. As a result, the six ultraviolet-sensitive films were oriented in the direction of travel – the 'velocity vector' – that produced a 'ram' effect, whilst the vehicle was in full sunlight. Laboratory experiments had already shown that charged particles, caused, perhaps, by clouds of ions produced in space by the action of solar ultraviolet radiation on residual gases from the orbiter, could cause chemical reactions and blacken emulsions. During STS-8, films were typically exposed for between a few minutes and almost a full hour. Lastly, the Contamination Monitor Package, previously flown aboard Columbia on STS-3 in March 1982, was actually mounted *outside* of its GAS canister. Led by Principal Investigator Jack Triolo of Goddard, the experiment employed samples of carbon and osmium – two materials known to readily oxidise – to determine the detrimental effects of atomic oxygen flux in low Earth orbit.

At the expense of being dubbed 'dull', Challenger's third mission went exceptionally smoothly, with the only problems being a minor cabin pressure leak, later isolated to the toilet, and the presence of increasing amounts of floating dust. This became especially uncomfortable on 4 September – the night before landing – when the crew unstowed their clothes bags to prepare their flight suits for re-entry, stirring up the dust in the process. Cabin filters appeared to work properly, although Truly and Brandenstein were obliged to wipe dust from their flight deck computer displays before commencing re-entry preparations. STS-8's nocturnal launch and Insat deployment requirement also meant that touchdown, too, would, for the first time, occur in darkness. To provide additional margins of safety, Challenger would return to Edwards Air Force Base in California, rather than attempt to land at KSC. "In other words," Brandenstein recalled, "if we had some problem and ran off the side of the runway, we wouldn't go into the moat!"

Additionally, the decision was taken to land on concrete Runway 22, rather than the dry lakebed, because "if we landed on the lakebed with the lights that we had devised to do the night landing, we'd kick up a cloud of dust, which attenuated the light," said Brandenstein. "We felt it was safer to take the approach to land on the concrete rather than the lakebed." The lights devised to support STS-8's homecoming were called Precision Approach Path Indicators (PAPI) and kept the pilots on their correct outer glide path of 19 degrees with a beam of half-white, half-red light. The PAPI system was situated 2.3 km from the end of the runway and 3 km from the Shuttle's point of touchdown; the correct flight path was determined by Truly and Brandenstein by centring the white light onto the 'band' of red lights. Transition and area lighting, consisting of 800 million candlepower xenon floodlights, illuminated the whole area, with green marker lights indicating the 'end' of the runway.

The mission had already been extended from five to six days to obtain an extra few hours' worth of TDRS-1 tests, by the time Challenger fired its OMS engines for two and a half minutes, beginning at 2:47:30 am EST on 5 September, to start the glide to California. "As we re-entered the Earth's atmosphere," remembered Bluford, "we began to feel the effects of gravity and saw the fiery plasma of hot air outside the front windows of the orbiter. Dale took pictures of the hot plasma as it enveloped us and he would occasionally hand me the camera. I could feel the camera getting heavier and heavier as we got closer to home." For Truly, whose previous Shuttle landing aboard Columbia in November 1981 had been in daylight, STS-8 presented a new series of challenges. "No engines. No moon. No correct dashboard info," he recalled years later. "The stars were blanked out because the window was frosted over. Then, finally, there were the lights of the California coast and Edwards. On the runway were the lines of red and white lights and that's what brought us in."

Touchdown itself came at 00:40:43 am PST (3:40:43 am EST), completing a six-day journey which, although demonstrating that space sickness could not be effectively predicted, had helped immeasurably to further certify the RMS for the Solar Max repair and prepare TDRS-1 for its vital role supporting the Spacelab-1 mission. Before the Solar Max crew could fly, however, there was one more set of RMS handling tasks pencilled in for Challenger's next trip in January 1984.

Moreover, the crew of that flight, the tenth Shuttle mission overall, would be required to conduct two ambitious spacewalks using another piece of equipment that was crucial for the Solar Max repair: a jet propelled backpack, which would turn astronauts Bruce McCandless and Bob Stewart into self contained, free-flying spacecraft in their own right.

A LABORATORY IN SPACE

Despite a steady shift from notions of "evil empire" to *glasnost* ('openness') and *perestroika* ('restructuring') in the latter part of the decade, American distrust of the Soviets remained alive and well and in October 1982 NASA had announced Challenger's crew for STS-10, which would be the first classified flight for the Department of Defense. A crew of four – Ken Mattingly, Loren Shriver, Ellison Onizuka and Jim Buchli – were assigned, although *their* payload (codenamed 'DoD-81-1') would have required Boeing's Inertial Upper Stage to boost it into its operational orbit. Following the TDRS-A debacle and doubts about the IUS' reliability, the STS-10 'slot' in early November 1983 was kept open – "NASA is calling the flight a payload of opportunity," *Flight International* explained in July, whilst admitting that "it seems unlikely that a replacement cargo can be found in the relatively short time" – and, predictably, the mission was cancelled. This left a sizeable gap in the Shuttle manifest, with STS-9 scheduled for late September 1983 and STS-11 unable to move forward from its target 29 January 1984 launch date, "because its cargo is not ready".

Another element of the Shuttle which appeared to be unready were the SRBs, which had returned from STS-8 with evidence of excessive corrosion in the 'throat' of the left-hand booster's nozzle. An 8 cm thick carbon-fibre resin lining, responsible for protecting the nozzle and normally designed to erode about half of its thickness during firing, was found to have burned down to just a few *millimetres* in places. Engineers were shocked and there were concerns that this would have left only about 14 seconds of firing time before the nozzle ruptured ... a catastrophe which would almost certainly have claimed the entire vehicle and the lives of Dick Truly and his crew.

The fault was ultimately traced to a particular batch of resin used on Challenger's boosters and the *next* mission, STS-9, was postponed to allow adequate time for the nozzles to be replaced. "NASA now believes that excessive corrosion of the SRB nozzle throat ... relates to processing of the nozzle during the cure cycle," *Flight International* told its readers in late October. "The resin used to line the throat of the nozzle is available from two manufacturers and material supplied by one of them is apparently more sensitive to the manufacturing process than the other. High pressure applied early in the cure cycle of the 'sensitive' material is said to have prevented proper escape of volatiles, resulting in a weaker lining." The consequence was that 'spalling' occurred in the char-layer of the throat material, causing it to 'flake' away, rather than steadily erode. NASA revealed that *both* resin materials *had* met its stringent specifications and that the phenomenon had been previously unseen. The problem required the STS-9 stack, which had already been rolled out to Pad 39A in early October, to be returned to the VAB for repairs. Since the nozzle resided at the *bottom* of the SRB, the Shuttle and External Tank had to be detached from the stack and the *entire* booster disassembled in order for its removal and replacement. "The substituted aft elements," continued *Flight International*, "incorporating nozzle linings made from less sensitive material, were due to fly on STS-11." In the meantime, Columbia was transferred back to the Orbiter Processing Facility and the External Tank was kept in the VAB transfer aisle whilst the repairs on the boosters were performed. The focus of the world was fixed on the flight for a number of reasons: as well as carrying a record-sized crew of six men, flying for a record-breaking nine days, it would feature the first non-American astronaut to ride a US spacecraft *and* operate Spacelab-1, a unique, European-built research facility. A delay to the end of November threatened to compromise the results from seven of the laboratory's astronomy, plasma physics, atmospheric and Earth observation experiments. "In general," *Flight International* explained in early November, "the scientists associated with these experiments would prefer to delay Spacelab-1 until observing conditions are more favourable." However, a launch in December or January 1984 was unacceptable to the Europeans because of potential losses of scientific data ... *and*, factored into this equation, was the inevitable $300,000 cost for *every month* of additional delay.

There was another problem which, if not addressed, threatened to delay Spacelab-1 even further. Back in the mid-1970s, a requirement had been laid down by NASA and ESA to have at least one, and preferably two, Tracking and Data Relay Satellites in orbit to support the mission's scientific yield for at least 80 percent of each orbit. "A further complication," wrote Douglas Lord, "was the requirement for several months' wait after putting the first satellite in orbit before launching the second *and* to have the second in orbit for several months to assure that the total system was checked out properly, before it supported a mission as complicated and important as Spacelab-1." Clearly, only *one* satellite, TDRS-1, was in orbit, and even it had barely limped into geostationary orbit a few months earlier. A *second* satellite was ready for launch, but its booster – the Air Force's IUS – had proven unreliable and was several months away from another flight. "*Now* the question was, should Spacelab-1 be flown with only a single relay satellite in place," asked Lord, "or should it be delayed until the two-satellite system was ready?" Many scientists on both sides of the Atlantic felt that a meaningful mission could not be accomplished without dual-TDRS support. At length, it was decided to carry out Spacelab-1 with a single relay satellite, but, as Lord concluded, "a number of fingers would be crossed". It did not bode well for STS-9, which, in terms of science and international participation, represented one of the most ambitious Shuttle missions yet attempted.

Towards the end of 1969, not long after Neil Armstrong and Buzz Aldrin became the first humans to set foot on the Moon, NASA outlined a number of key directions for the US space programme after Apollo. President John Kennedy's challenge to land a man on the lunar surface before the end of the decade had been met, but it was not until this time that serious consideration was given to what would come next. The establishment of some kind of permanent, or at least 'frequent', human presence in space was of major importance – hence the Shuttle – and in December 1972, the European Space Research Organisation (ESRO) – which, in 1974, merged with the European Launcher Development Organisation (ELDO) to establish today's European Space Agency (ESA) – agreed at a ministerial conference in Brussels to develop a modular, multi-purpose laboratory to fly in the payload bay of the winged orbiter. Originally described as a 'sortie module', it was later renamed 'Spacelab'. To NASA, the word 'sortie' merely reflected the low-cost, short-duration nature of the research facility … but in Europe, and particularly France, the similarity of the word with the verb 'to leave' – *sortir* – generated some distaste. Indeed, the French word *sortie* is equivalent to the English word *exit*. In the United States, even the name 'Spacelab' was initially accepted with some hesitation, for the agency had only recently concluded its 'Skylab' series of missions and was worried that the two might become confused. "However," wrote Douglas Lord, "despite NASA's objections, once the Europeans had committed to the programme, they unilaterally decided to use the name 'Spacelab', and 'Spacelab' it became." The deal called for Europe to develop the facility in exchange for flying its own astronauts on specific missions. Spacelab, it seemed, would permit NASA to neatly sidestep the biggest obstacle in its financial battles with Congress: how to have both the Shuttle *and* a temporary space station in a decade of decreasing budgets. Unfortunately, the reality proved an unhappy prelude to the project's eventual success.

For Europe, involvement in Spacelab initially proved the consolation prize, after two other proposals were rejected by NASA and the Department of Defense. During 1971, ESRO and ELDO had invested $20 million in a series of studies to develop one of three components for the Shuttle: its payload bay doors, the sortie module or a reusable 'space tug'. Consensus favoured the latter, but it was turned down by NASA in June 1972, apparently because the Pentagon – envisaged to be the tug's main user – was reluctant to have a 'foreign entity' building a booster to place its top-secret payloads into orbit. (In April of that year, the Air Force had cited concerns over "national security issues" and told NASA that it would not guarantee buying a foreign-built tug if that did not meet its required specifications.) Constructing the clamshell-like payload bay doors was also promptly rejected because, said NASA, the Europeans lacked the expertise or organisation to make such a vital contribution to the orbiter's structure. The sortie module, on the other hand, required less sophisticated technology, had a well-defined interface with the Shuttle and schedule slips or budgetary overruns would not hamper the reusable spacecraft's ability to fly. When NASA offered this option to Europe, it received a lukewarm reception. Many ESRO member states hesitated to participate, questioning what they had to gain from an effort that demanded a $250 million investment and yet would be used principally by the United States. These lingering doubts were overruled, said political scientist John Logsdon, by a desire to "pursue co-operation on almost any terms, no matter how one-sided". In truth, it provided western Europe with its best chance of advancing its ambitions in space and sending its own astronauts aloft.

Indecision gave way to approval of involvement with the sortie module in December 1972; moreover, ministers agreed to form a single European space organisation. In January 1973, ESRO's member states voted in favour of building the sortie module and, from May, began working with NASA on the details of a formal Memorandum of Understanding. On 24 September, the contracts between the two organisations were finally signed in Washington, DC, by NASA Administrator Jim Fletcher and ESRO Director-General Alexander Hocker and the wheels of the largest international collaborative venture in space so far were set in motion. The majority of the funding for Spacelab (54.1 percent) came from West Germany, with Italy ranking second at 18 percent. In the terms of the contract, ESRO assumed responsibility for the "definition, design, development, manufacturing, qualification, acceptance testing and delivery" of an engineering model and spaceworthy flight unit to NASA by 1978. For its part, NASA would operate Spacelab, fly European astronauts as 'payload specialists' on selected missions and possibly procure additional hardware. Project management in the United States went to the Marshall Space Flight Center (MSFC) in Huntsville, Alabama, and its growing involvement in selecting and training scientists for payload specialist positions on Spacelab missions led to an expression of concern from JSC. Director Chris Kraft was reluctant to permit MSFC to choose and train payload specialists, arguing that they should be "selected from the present corps residing in Houston". Unfortunately for Kraft, the decision had already been made by NASA Headquarters, but it highlighted the different way in which Spacelab would operate, compared to other flights. Unlike pilots and mission specialists, who were selected formally as 'career' astronauts by the agency, payload specialists would be picked by an Investigators Working Group (IWG), whose panel included principal scientists responsible for Spacelab experiments. For the first time, a centre outside Houston was infringing on JSC's territory by providing mission-specific training. In all other areas, however, the two centres remained separate: JSC having responsibility for Shuttle operations and MSFC for Spacelab payloads.

Difficulties with the European partnership also arose when it became clear that NASA would not purchase more than one additional flight unit; in its 1973 plan for Shuttle utilisation, the agency expected to buy five Spacelabs to support two dozen research missions annually. However, a multitude of technical problems pushed the reusable spacecraft's first flight beyond 1980, only two units had been commissioned and, as ESA's budget declined, it became worryingly clear that the Europeans would have little funding available to even use Spacelab for their own experiments. The $250 million facility, they feared, would inevitably slip under American control. Perhaps the most important defining phase came in March 1978, when NASA and ESA initiated a nine-month critical design review, which provided a final opportunity to incorporate significant technical changes. Concerns remained, however. The first three Spacelab missions would all exceed the Shuttle's cargo-carrying limit as a result of orbiter-supplied equipment. MSFC opted to upgrade the orbiter's landing capability as a possible solution, although NASA Headquarters felt this would set a bad precedent. Nor was reducing Spacelab's payload weight desirable. Ultimately, ways were found to absorb the weight excess without seriously impacting each of the missions. Ongoing budgetary problems caused further woes. The European member states initially pledged up to 120 percent of their individual financial commitments to accommodate cost overruns, but even those had been consumed by September 1979. After protracted deliberations, ESA was obliged to propose increasing its members' funding to 140 percent. Only Italy refused to accept the new cost ceiling and by the end of the decade both the Europeans and Americans had reason for disappointment: the former resented paying more than expected, while the latter was disappointed that ESA was unprepared to take risks or bear full responsibility for Spacelab's early missions.

Inevitably, these worries impacted the flight schedule. When ESA delivered a pair of Spacelab modules to the United States – one under the terms of the original Memorandum of Understanding, the second as a follow-on procurement – NASA's budget was slashed in the first year of Ronald Reagan's presidency. This imposed one-year delays to several Spacelab missions. "Over the past four years," lamented James Harrington, head of the Spacelab effort at NASA Headquarters, "the Spacelab-1 launch has slipped three years! Additionally, the manifest of Spacelab flights has been reduced from four or five flights per year to the current two flights per year through 1986." As the problems of this increasingly unequal partnership were being thrashed out, the miniature space laboratory gradually took shape throughout the mid to late 1970s. Its bus-sized pressurised module comprised two components: a 'core' segment, which housed data-processing equipment, a work-bench and a set of air-conditioned research racks lining its walls, and an 'experiment' segment, providing additional room for scientific operations. Although the core could be flown on its own, this configuration was ultimately never used and all module flights employed both segments joined together, with a pressurised volume of 75 m^3, to form a 'long module'. When one considers the dimensions of the short module, it is clear why NASA opted to use the longer version for a total of 16 missions between November 1983 and May 1998: the long module was over 7 m long and virtually doubled the amount of 'rack' space in which to carry experiments. Racks were refrigerator-sized facilities which could be 'rolled' into the module's cylindrical shell and the facility also offered a central aisle of floor space, onto which experiments could be affixed, and provided a pair of ceiling openings for viewing windows or scientific airlocks. "The racks were pretty much standard," remembered Gene Rice, a former head of space life sciences with NASA. "You either had a drawer in a rack or you had a whole rack or ... a double rack, depending on the magnitude or size of the experiment. We would help [experiment customers] through the process of designing their experiment, integrating it into a Spacelab rack, doing the testing that they needed to do [and] getting it to a [NASA] centre. They would have to show that they met the safety requirements to put it into the Spacelab and to fly it." The racks contained air ducts to cool experiments and power-switching panels. On the first dedicated Spacelab mission, STS-9, the module contained 12 racks, of which the pair nearest the entrance to the module were devoted to control subsystems. The ceiling of the core section provided a 0.3 m wide opening for a high-optical-quality Scientific Window Adaptor Assembly (SWAA), through which Earth observation cameras could be directed, and that of the experiment section provided a Scientific Airlock (SAL), into which samples requiring exposure to space could be inserted and retrieved. As with other payload bay hardware, the exterior of Spacelab was covered with a layer of passive thermal protection material to protect it from the extremes of sunlight or frigid orbital darkness. Closed at each end by a pair of truncated cones, the module was held in place by three longeron fittings on the payload bay walls and one in its floor and was situated at the mid-point of the bay to avoid violating the Shuttle's centre-of-gravity constraints during ascent and re-entry. It was linked to the crew cabin by a 5.8 m long tunnel, built by McDonnell Douglas. Since the Spacelab hatch was 1.5 m 'higher' than that of the middeck airlock, a 'joggle' section was included in the tunnel to provide this vertical offset. Of course, astronauts might still need to perform contingency EVAs – perhaps to close the payload bay doors – and with this in mind a 'mini-airlock' was built into the roof of the tunnel and spacewalkers could close off the whole section without having to depressurise the crew cabin or the Spacelab. (Nevertheless, mission rules prohibited the module from being occupied during an EVA.) Although an EVA was never actually performed during a module flight, towards the end of one mission in June 1991 such action was briefly considered when a problem with some thermal insulation threatened to interfere with the closure of the payload bay doors.

Spacelab would be a multi-purpose scientific research platform, capable of supporting experiments in a pressurised, shirt-sleeve environment and exposing others to the harsh vacuum of space. The segment designed to expose experiments to the space environment had already been flight-tested on STS-2 and STS-3 and was known as a 'pallet'. It took the form of a U-shaped rigid metal frame, measuring about 3 m long by 4 m wide and covered with aluminium panels onto which large telescopes, antennas or sensors requiring unobstructed fields of view could be attached. On STS-2, for example, a single pallet was used to carry a large synthetic aperture radar for mapping Earth's surface. Up to *five* pallets, bolted together into a rigid 'train', could be accommodated in the payload bay, although the greatest number actually flown was three, on the Spacelab-2 voyage in July 1985. Missions which required pallets were also accompanied by a 2 m cylinder, called an 'igloo', which stood vertically at the forward end of the 'front' pallet, and provided a temperature-controlled, pressurised container for subsystems and equipment needed to support the experiments on the pallets.

In October 1974, the first of three simulated 'missions', known as Spacelab Medical Development (SMD), was undertaken by physician-astronaut Story Musgrave and bioscientist Dennis Morrison in a full-sized mockup of the module at JSC. The two men fulfilled the roles of what would become a 'mission specialist' and a 'payload specialist' and spent a week running through operational procedures for a Spacelab flight, including a number of scientific experiments. A little over a year later, in January 1976, Musgrave returned to the facility with nuclear chemist Robert Clark and cardiopulmonary physiologist Charles Sawin for a five-day simulation, which included a full-size mockup of the Shuttle's flight deck and middeck. (In the previous test, Musgrave and Morrison had spent their off-duty hours in a mobile home; now, in SMD-II, daily activities were bearing a closer parallel to an actual space mission.) During this second test, Musgrave, Clark and Sawin performed almost two dozen experiments and participated in a series of tests to evaluate personal hygiene, cleaning and maintenance of the mockup orbiter. The

The Spacelab hardware undergoes processing for flight, with three pallets visible in the foreground and the long module beyond. All are covered with thermal insulation material. The module's end cone can be clearly seen and, above it, the covered window for observing operations on the pallets; it is also possible to discern the roof-mounted Scientific Airlock. Between November 1983 and May 1998, the module flew 16 times, whilst the pallets have continued to be used into the International Space Station era, carrying experiments and logistics to and from the orbital outpost.

third and final SMD test took place in May 1977, led by physician-astronaut Bill Thornton. Its complexity lay in its experiments, which numbered over three dozen, including studies of rats, monkeys, frogs, mice, thousands of flies and the three men themselves. In the months leading up to SMD-III, Thornton had expressed concerns that they would be little more than a "puppet on an electronic string", manipulated by flight controllers, and felt that the role of the mission and payload specialists *should* be an extension of the principal investigator; in other words, *wholly responsible* for overall scientific success. Moreover, the process for choosing 'payload specialists' for SMD-III was fraught with difficulty, particularly with regard to candidates from NASA's Ames Research Center in Moffett Field, California. "A tentative selection of PS was being made between JSC and Ames," wrote Dave Shayler and Colin Burgess, "and unofficial joint 'training' was being accomplished to an acceptable degree ... But news that the PS selection might not be ratified and that Ames was looking to train its own selection ... did not go down well at JSC. There was even a rumour that the JSC PS selected might be replaced by a second person of unknown technical experience." Although SMD-III was a simulation, and not a *real* space mission, Thornton felt that the selection process for its 'payload specialists' would serve to establish guidelines for the eventual selection of 'real' flight crews. He was worried that making late-in-the-day selections of payload specialists would have a detrimental effect on the scientific success of a mission *and* would impose a greater weight on the mission specialist, who would struggle with his responsibility of ensuring scientific success. Thornton discussed the issue with Al Bean, then-deputy chief of the astronaut office, and a number of scientist-astronauts, and it was decided that the payload specialists would share duties as primary or backup on the whole range of experiments, thereby providing dual coverage and 'redundancy' where required. This would enable Thornton, as the mission specialist, to take a 'global' outlook on the experiments and supervise troubleshooting.

At length, late in 1976, four candidates – Carter Alexander of JSC and Patricia Cowing, Richard Grindeland and Bill Williams of Ames Research Center – were chosen for medical tests before the selection of two payload specialists on SMD-III. "The exams are similar to those to be given to persons named to operate scientific experiments to be carried into orbit by Space Shuttle," revealed a JSC news release, dated 1 December. "Among the tests are electrocardiograms and vectorcardiograms, made while the subjects exercise on a treadmill and lie in a horizontal tank called a lower body negative pressure (LBNP) device." The treadmill and LBNP allowed for the monitoring of heart rate, blood pressure and the reactions of the cardiovascular system to pressure changes. Ultimately, biochemist Alexander and biophysicist Williams were chosen to accompany Thornton on SMD-III. When the week-long simulation took place in May 1977, it identified numerous problems – notably communications between the principal investigators and the 'flight crew' – but Thornton was convinced that it was an essential preparation for 'real' Spacelab missions. "I shudder to think what might have been if we had not done it," he told Shayler and Burgess. "There were some first-rate trainers, but the procedures in getting some of the experiments defined for the simulation were hard enough, without trying to evaluate them for space flight conditions."

A year after the completion of SMD-III, the first crew positions were announced for Spacelab-1, the first 'dedicated' flight of the facility, in which a long module and a single pallet would be flown together. Despite carrying a pallet, the mission did not require an igloo, because all of the subsystems needed to run its experiments would be housed inside the pressurised module. Additionally, this payload was not dedicated to a single scientific discipline, but was a 'free-for-all', covering many areas of research; in fact, more than 70 investigations in the life sciences, technology, astronomy, solar physics, Earth observations, plasma physics and materials processing would be conducted. Science activities would be conducted by a pair of mission specialists and a pair of payload specialists; the latter would include a US and a European scientist. Selection of the payload specialists was the responsibility of a committee, known as the Investigators Working Group (IWG), which included the principal investigator for each Spacelab-1 experiment and was chaired by the Mission Scientist from NASA's Marshall Space Flight Center in Huntsville, Alabama. In May 1978, University of Berkeley physicist Mike Lampton and MIT biomedical engineer Byron Lichtenberg were selected by the IWG as candidates for the US payload specialist seat and West German physicist Ulf Merbold of the Max Planck Institute and Dutch physicist Wubbo Ockels of the University of Groningen were selected as candidates for the European payload specialist seat. (In fact, Bill Thornton himself had also been nominated for the US payload specialist position in December 1977.)

On 1 October 1982, a year before launch, Lichtenberg and Merbold were formally announced as the two prime payload specialists by the IWG after a two-day meeting at the Marshall Space Flight Center. "Thus, the researchers who *designed* and *built* the experiments, and who are to *analyse* the results," wrote Walter Froehlich in the *Spacelab: An International Short-Stay Orbiting Laboratory*, "helped select from their peers the two in-flight specialists who are to be in charge of carrying out research in orbit." Byron Kurt Lichtenberg came from Stroudsburg, Pennsylvania, where he was born on 19 February 1948. His yearning to become an astronaut started from an early age; reading science fiction books and seeing the first pioneers set off on their missions of exploration in the early 1960s. Lichtenberg knew that they were military test pilots with advanced engineering or science credentials and he dedicated himself to following a similar career path. He received a bachelor's degree in aerospace engineering from Brown University in 1969 and entered the Air Force, flying the F-4 Phantom and F-100 Super Sabre supersonic fighters and the A-10 Thunderbolt ground attack aircraft. During the course of the Vietnam conflict, Lichtenberg flew 238 combat missions, for which he received two Distinguished Flying Crosses and ten Air Medals. The pull of the space programme remained strong, however, and he left active duty in 1973 to return to academia. However, he remained a military reservist, serving with the Massachusetts Air National Guard and providing close aerial support in the A-10. A master's degree in mechanical engineering from Massachusetts Institute of Technology came in 1975 and a doctorate in biomedical engineering – from the same institution – in 1979. "I realised if I was a fighter pilot with a doctorate," he explained years later, "I would have a better chance of getting into space." Whilst undertaking his doctoral studies, Lichtenberg applied, unsuccessfully, for both the 1978 *and* 1980 NASA astronaut selections. At around the same time, he was selected by the IWG as a candidate for the US payload specialist slot on the Spacelab-1 flight.

The role of the 'payload specialist' had by now crystallised into something which described a non-career astronaut; someone who would be chosen to represent research institutions, space agencies, large aerospace corporations and even foreign governments and assigned to fly a specific mission to operate a specific piece of equipment or experiment. The second payload specialist on STS-9 spoke with a very different accent and came from a very different part of the world; in fact, Ulf Dietrich Merbold, though 'West German' by nationality and political status, had actually been born in Greiz, Thuringia, less than 40 km from the birthplace of Germany's first astronaut, the 'East German' Sigmund Jähn. Both Merbold (born on 20 June 1941) and Jähn (four years his senior) had spent their formative years in the German Democratic Republic, the communist-led East Germany, one of the Soviet Union's satellite states. When he finished high school in 1960, not long before the erection of the Berlin Wall, the young Merbold was one of thousands who defected to the Federal Republic and the democratic West. He attended the University of Stuttgart, receiving a diploma in physics in 1968 and a doctorate in 1976, then joined the Max Planck Institute for Metals Research as a scholar of the Max Planck Society. He rose to become a faculty staff member, working on solid-state and low-temperature physics, with particular research interests in lattice defects in body-centred cubic metals. In December 1977, along with Dutch physicist Wubbo Ockels and Swiss astrophysicist Claude Nicollier, he was selected by ESA from more than 2,000 applicants as a candidate for Spacelab-1 and, in May of the following year, he and Ockels commenced payload specialist training with Lampton and Lichtenberg. (Claude Nicollier, meanwhile, was selected by NASA for mission specialist training and was therefore removed from consideration for Spacelab-1.) Three months later, on the first day of August 1978, NASA assigned the two mission specialist positions to scientist-astronauts Owen Garriott and Bob Parker. Both had extensive experience, particularly in Spacelab development; the former was a Skylab veteran who had spent several years as head of NASA's Science and Applications Directorate, whilst the latter, though unflown in space, had served as the chief of the astronaut office for that same directorate. Parker had also participated in an extensive airborne simulation of Spacelab, ASSESS-I.

Owen Kay Garriott was born in Enid, Oklahoma, on 22 November 1930, his Christian names honouring his father and paying homage to a diminutive of his mother's middle name, Catherine. In *NASA's Scientist-Astronauts*, Garriott told Dave Shayler and Colin Burgess that his ancestors on his father's side were French farmers and his own grandfather had routinely travelled all day to transport goods and supplies by horse and wagon from the closest railway stop to Enid's tiny convenience store. Years later, Garriott – the first of *two* generations of his family who would ultimately break the surly bonds of Earth and venture into the heavens – would express astonishment that, flying aboard the Skylab space station for eight weeks in the summer of 1973, he would routinely cover the same distance as his grandfather ... albeit in barely *four seconds*. Garriott's fascination with physics and engineering emerged at a young age, thanks to his father, a geologist by training who spent his career working as a chemist and as an oil and gasoline distributor. One day in 1944, he came home and invited his teenage son to attend an adult class on radio theory with him. The class included electronics, an understanding of Morse code and the construction of radio and transmitting equipment and they both quickly passed their Federal Communications Commission exams. (Amateur radio, in particular, would later become a fascination for Garriott and during Spacelab-1, he would speak to many terrestrial 'hams', including none other than King Hussein of Jordan.) Engineering was now in the young man's blood and from high school he entered the University of Oklahoma on a Navy scholarship and received a degree in electrical engineering in 1953. As part of his commitment to the Navy, Garriott – by now married to his school sweetheart, Helen – undertook three years of active military service, acting as an electronics officer and participating in several tours aboard destroyers at sea. After the completion of his military duty, Garriott moved to Stanford University in Palo Alto, California, working in the Radio Propagation Laboratory, and completed his master's degree in 1957. Next came his doctorate, which he gained in 1960 and which offered him an inroad into the Space Age for the first time. "My dissertation," he told Shayler and Burgess, "used the radio signals from Sputnik 3 to study the electron content of the ionosphere." Garriott remained at Stanford for five years as a faculty member, specialising in electromagnetic theory and ionospheric physics and rising to the position of Associate Professor of Electrical Engineering.

A casual conversation with a friend over dinner, early in 1965, prompted Garriott to secure for himself a private pilot's licence, with an instrument rating, and to submit an application to NASA for its much-publicised first selection of scientist-astronauts. In his oral history, more than three decades later, he doubted that his doctoral research or even his work as a member of the Stanford faculty was *necessary* for admission into the hallowed ranks of the astronaut corps, but the *credentials* – the degrees and the published papers – served to qualify him. "What NASA wanted then," he explained, "is not a world-class athlete, but somebody who has *nothing wrong with them.* They want average, everything down the middle, in terms of your biological characteristics. You wanted to be as normal as you can possibly be!" One area in which Garriott was 'borderline' was his eyesight: astronaut candidates needed 20/20 uncorrected vision. His left eye was fine, but his right eye was on the edge.

"Maybe 20/25," the optometrist told him.

"Well, let me blink a little bit," Garriott replied. "Let me see if I can't focus a little bit more carefully."

The optometrist retested him and agreed that he had just hit 20/20, although Garriott knew that his right eye at near distances was hazy. Years later, he recognised that if the optometrist had classified him as 20/25, he would never have been accepted by NASA.

Soon after selection, the six new scientist-astronauts were despatched for flight training. Most of them had never flown an aircraft and even Garriott's private licence did not suffice. "Even though we were light airplane pilots," he explained, "three of us were not jet-qualified. The other three of us went to flight school at Williams Air Force Base in Arizona." After a year of instruction, the old heads in the astronaut offices must have been a little surprised to see the scientists coming back to Houston with their jet credentials. Additionally, all of them underwent advanced helicopter training later in their NASA careers. Upon their return to the heat and humidity of Houston, however, it was back to the grinding business of preparing for future space missions. Jungle survival training in Panama and geological expeditions to Alaska and Iceland and Hawaii occupied their time, as did the steadily growing space station project, which was at that time known as Apollo Applications, later renamed 'Skylab'. In January 1972, Garriott was assigned as the science pilot for the second Skylab mission, which took place in the late summer of the following year. With no further flights on the immediate horizon, he was reassigned as deputy head of NASA's Science and Applications Directorate, working extensively on the evolution of Spacelab for two years. Although he remained in Houston, it was a post which left him uncertain as to whether he was *still* an astronaut; but Garriott "preferred it that way, because I wanted to come back to the astronaut office at the appropriate time, when the time was right for another kind of flight". A one-year sabbatical at Stanford University was followed by a return to NASA as deputy head of physical sciences in the agency's new Physical and Life Sciences Directorate, after which, in August 1978, Garriott was assigned to Spacelab-1.

The second mission specialist, Robert Allan Ridley Parker, was a member of the 'Excess Eleven', selected in the summer of 1967, and who were now able to cast a wry smile at the naysayers who doubted that they would ever fly in space. (In fact, of the 11 scientist-astronauts, *eight* would ultimately fly at least one Shuttle mission.) Parker was a New Yorker, born on 14 December 1936, although he grew up and received schooling in Shrewsbury, a suburb of Worcester, Massachusetts. Education and learning was in Parker's blood: his paternal grandfather had been a high school teacher and librarian, his father the chairman of the physics department at Worcester Polytechnic Institute, whilst his twin brother would later teach physics at Yale University and his younger brother became a systems analyst. Parker received a degree in astronomy and physics from Amherst College in 1958 and a doctorate from the California Institute of Technology in 1962, after which he accepted a National Science Foundation fellowship and joined the Badger faculty of the University of Wisconsin at Madison. Whilst there, Parker and one of his colleagues were caught up in the excitement of America's drive to land a man on the Moon – "it was way, way, *way* beyond saving the whales or the rainforest or the ozone layer," he told the NASA oral historian – and a call for applicants for scientist-astronauts in the spring of 1965 captured their attention. Unfortunately, NASA's requirements were strict: six feet (1.8 m) was the maximum height allowable and perfect eyesight was mandatory. "I had an office mate at Wisconsin," Parker recalled. "He was over six feet; I wasn't. He had perfect eyes; but I *didn't*." It did not seem worth the effort to complete the stack of application paperwork and Parker thought no more of becoming an astronaut for two more years, until he heard of another selection in 1967, in which the restrictions on eyesight had softened. He soon found himself among a group of 69 finalists, summoned to Brooks Air Force Base in San Antonio, Texas, for medical screening, interviewed by NASA in Houston and given his first flight in a high-performance T-38 jet. "As a matter of fact, I got *sick*," he recalled. It did not affect his selection in August and, years later, Parker would strongly suspect that the selection committee were more interested in finding out whether he enjoyed flying, rather than simply dismissing him on the basis of airsickness. And Parker certainly *loved* flying.

Flight school, at Williams Air Force Base in Arizona, would not begin until

March 1968, so intense was the need for the US military to prepare its own pilots for Vietnam. A year later, Parker returned from Williams with his jet credentials and worked initially on Skylab issues, before an assignment to the support crew of the Apollo 15 lunar mission. During this time, he undertook trips to Iceland, Hawaii, the Grand Canyon and other exotic destinations in support of the geology training, serving as the capcom for the backup crew – Dick Gordon, Vance Brand and Jack Schmitt – and liaising with the scientific community. Parker extended this work in his next role, as the mission scientist for Apollo 17, the final lunar landing flight. He returned to work on Skylab and in May 1974 was assigned as chief of the astronaut office for NASA's Science and Applications Directorate. One of Parker's primary focuses was upon Spacelab and just six months into his tenure, he was approached by Philip Culbertson, the head of mission and payload integration at NASA Headquarters, and also by Owen Garriott, to participate in the first mission of a project known as the Airborne Science/Spacelab Experiment System Simulation (ASSESS). This was a joint ESA/NASA effort to closely examine the operational requirements for future Spacelab missions and consisted of five flights of a CV-990 Galileo II aircraft over a six-day period in June 1975, in which a four-man crew simulated high-altitude experiments in atmospheric physics and infrared astronomy. Each flight lasted around seven hours and after touchdown they completed scientific debriefings, then bedded down in specially-designed crew quarters, adjacent to the aircraft parking area.

The ASSESS-I mission provided valuable scientific data, as well as raising important issues of crew training and demonstrating the ability of two international space agencies to work together. Certainly, it must have factored into the eventual decision to assign Parker to one of the two mission specialist spots on Spacelab-1. At the time of the NASA press release, in August 1978, the seven-day mission was scheduled "for the early 1980s", although in his oral history Parker anticipated a launch sometime in 1981. Unfortunately, delays to the Shuttle, including problems with the reusable spacecraft's thermal protection system and difficulties with the certification of the main engines, pushed its maiden voyage further to the right ... and caused Spacelab-1 to correspondingly slip. "Training expands to fill a vacuum," Parker said, and, for a time, it seemed that he and Garriott did not participate in a great deal of 'active' training, other than visits to various European countries – Denmark, France, West Germany and others – to observe the development and testing of the Spacelab-1 experiments.

It was in anticipation of this ambitious make-or-break mission to demonstrate the new laboratory's capabilities that Columbia returned to Florida a few weeks before Christmas 1982, following STS-5. Even with five flights to her credit, it was clear that Spacelab-1 would be her most ambitious venture to date. In addition to the normal 'flow' in the Orbiter Processing Facility at KSC, she underwent a series of 'Spacelab-Only' modifications. A seat was fitted for Parker on the flight deck (the reader will recall that Columbia still carried massive ejection seats during this period and could therefore only house three 'upstairs' crew members) and three additional seats on the middeck for Garriott and the two payload specialists. A fax machine was installed, improvements were made to the orbiter's brakes and tyres, structural and electrical provisions were made in readiness for Spacelab-1, the payload bay floor was strengthened and large, phonebox-sized 'sleep stations' were placed in the middeck for the crew members. These stations were provided with a 'sleeping pallet' and restraint, personal storage space, a light, ventilation and overhead light shields and herein lay another fundamental difference between STS-9 and 'ordinary' flights: it would feature around-the-clock science operations, effected by two 12-hour shifts: a 'Red Team' and a 'Blue Team'. Each team would feature one of the pilots to monitor Columbia's systems, together with a mission specialist and payload specialist to tend the Spacelab experiments. Not all of the modifications were in direct support of Spacelab-1, but the focus of the work was upon this mission. The Shuttle spent the first three months after STS-5 in Orbiter Processing Facility Bay One, having her propellants drained and her RCS and OMS units removed for maintenance, before being transferred to Bay Two on 27 February 1983 to begin the Spacelab improvements. In October of the previous year, the mission duration had been extended from seven to nine days, requiring the installation of three 'substack' fuel cells, using five (rather than three) sets of cryogenic reactant tanks underneath the payload bay floor. According to *Flight International*, the mission extension had been effected "to relax crew workload", although ESA's Spacelab Programme Board had been pushing for a mission of longer than seven days since September 1980. Other preparations were ongoing at the National Space Technology Laboratory at Bay St Louis in Missouri, where a set of upgraded main engines, capable of 104 percent rated thrust, were successfully tested in the early summer. On 20 July, they arrived in Florida for installation aboard Columbia.

Elsewhere at KSC, in the Operations and Checkout Building, the Spacelab-1 hardware had been arriving in dribs and drabs since a life science mini-lab in October 1981. The two showpieces of the mission – the long module and the pallet – arrived shortly before Christmas of that year and an official unveiling ceremony occurred the following February, attended by Vice President George Bush. "More than 300 invited guests from Europe and the US gathered in the high bay area of the Operations and Checkout Building," wrote Douglas Lord, "where they could see in the background both the engineering model and flight unit hardware on the Spacelab work stands." In his address, Bush described the facility as "the fruit of a lot of hard work", praised the European involvement in the project and likened the workforce responsible for bringing Spacelab to fruition to a gathering of proud parents. By the end of the summer, all of the experiment facilities for Spacelab-1 had been installed. A mission sequence test in November served to verify their compatibility with each other and with a set of 'dummy' Spacelab systems. "The experiments were brought in by their various scientific teams," recalled Spacelab-1 Mission Manager Harry Craft of NASA's Marshall Space Flight Center. "We would let them check the experiments out initially in an off-line capability and then we'd bring them into a room and just make sure the instrument had met the transportation environment and still worked, [then] they'd turn it over to us." A mission sequence test in the spring of 1983 simulated 79 hours of the planned nine-day mission, with ground support equipment taking the place of the 'orbiter' and demonstrating high-data-rate recording and playback. In May 1983, Spacelab-1 was hooked up to the Cargo

Integration Test Equipment, which duplicated the Shuttle's systems in high fidelity and verified the compatibility of the hardware. On 16 August, the 15,265 kg integrated Spacelab-1 system was moved to the OPF and inserted into Columbia's payload bay. Two weeks later, the tunnel was connected and fit and leak checks were carried out. Early in September, further compatibility testing occurred and the Spacelab-1 experiments were briefly operated, via remote control, from the Payload Operations Control Center (POCC) in Houston.

By this time, problems with the first Tracking and Data Relay Satellite, which NASA intended to use during the Spacelab-1 mission, had caused STS-9 to be postponed from 28 September until late the following month. As a consequence, rollout to Pad 39A was pushed back until the beginning of October. Then, on the 14th, NASA and ESA jointly agreed to delay the mission yet again, due to the SRB nozzle problem discovered in the wake of the STS-8 mission. Columbia returned to the VAB on the 17th and was removed from the stack. Back in the Orbiter Processing Facility, film and batteries were replaced in several Spacelab-1 experiments and a glitch with her Ku-band antenna was fixed. At length, Columbia returned to the VAB on 4 November, was restacked and rolled back to the pad on the 8th, with an expectation that launch could be attempted at the end of the month.

Launch morning, Monday 28 November 1983, dawned fine and reasonably dry, although there were concerns about thunderstorms in the Cape Canaveral area; an emergency landing site in Spain was also iffy in terms of the weather. STS-9, explained Douglas Lord, was an important mission from Europe's perspective, but the general perspective of the American public was that Shuttle missions were now becoming 'routine' and the viewing turnout was correspondingly smaller than earlier launches. Still, the VIPs were in attendance, including many from Japan, which had provided one of the Spacelab-1 experiments. "The causeways were lined with the usual assortment of campers, cars, signs, flags, vendors, public address speakers, portajohns and sunbathers," wrote Douglas Lord. "The VIP stands were filled with enthusiastic supporters of the Spacelab programme and a scattering of dignitaries and luminaries from the entertainment, political and international arenas. A thriving business was underway in Spacelab and STS-9 mission mementoes, first-day covers, hats and T-shirts. The huge countdown clock in front of the viewing area moved ever so slowly and paused at the planned holds for what seemed an eternity. Photographers manoeuvred for the best spots and the telephoto lenses looked like small howitzers aimed at the distant Shuttle launch complex. The public address announcer droned on with a running monologue of the countdown, but most people concentrated on looking around to see who they could recognise. Members of the Spacelab team not needed in the Launch Control Center ... exchanged greetings and wished each other good luck."

The countdown proceeded crisply and at T-31 seconds, the ground launch sequencer transferred control to Columbia's on-board computers; the pivotal event known as 'Auto-Sequence Start'.

"... Coming up on the 30-second point ... and we Go for autosequence start," came the call from the launch commentator, "... the SRB hydraulic power units have started; these move the solid motor nozzles to steer the vehicle ... T-20 seconds

... 18, 17, 16, 15, 14, 13, 12 ... ten ... we have *Go for main engine start* ... eight, seven six ... we *have* main engine start ... three, two, one and ... solid motor ignition and *liftoff* ... Liftoff of Columbia and the first flight of the European Space Agency's Spacelab ... the Shuttle has *cleared the tower* ..."

The launch came precisely at 11:00 am EST, at the start of a short, 14-minute 'window' for that day. "Soon, the reverberations from the Shuttle main engines and its boosters reached the viewing stands," wrote Douglas Lord, "and overwhelmed the cheers from those looking on. The Shuttle quickly rolled around its axis and started to pitch over as it passed through the layers of clouds. Camera shutters clicked rapidly, old friends hugged each other with delight and tears coursed the cheeks of many space-hardened veterans. There is *nothing* quite like those few moments after liftoff, when *everyone* is of a single mind, trying to help *push* the launch vehicle into orbit ..." The ninth Shuttle flight was officially underway and, for her mission specialists and payload specialists, more than five years of training had long since knitted them into a close team. "When we were training the science crew," said Bob Parker, "we had prime and backups for the American payload specialist and for the European payload specialist, and then Owen and myself. The six of us almost always travelled together and did the training together. At the very end, in the last year or so, when it became obvious this is going to be 'your' shift and this is going to be on their shift, we separated a bit in that sense, but still you could do an experiment on both shifts."

Parker also provided something of a cross-over between the science team and the pilots. His role during ascent and re-entry was to serve as the flight engineer, seated behind and between John Young and Brewster Shaw, and this had produced quite a diverse training schedule, particularly during the final few months before launch. "At the same time as we were training on the experiments, I was training with Brewster and John on ascents and entries," he told the NASA oral historian. "For a good six months or so, I'd be training on Mondays with them, fly to Huntsville [the Marshall Space Flight Center in Alabama] for experiment training, fly back and do ascents and entries on Wednesday and back and forth." In this way, Parker was probably the only crew member who maintained regular contact with the entire crew during training.

Demonstrative of the importance of this international venture, the six-man crew of STS-9 was commanded by the chief of NASA's astronaut corps, John Watts Young Jr, who had already flown five times into space. Even today, spacefarers continue to refer to him as 'The Astronauts' Astronaut' and even those who did not get along with him or disliked his management style are unfailing in their praise of his piloting and engineering skills. Over the years, he came to be regarded as something of a mystery; his misleading, "aw, shucks" demeanour and country-boy drawl cleverly concealed a sharp wit and a talented engineering mind. Bob Crippen, who flew with him on STS-1, once commented that if Young was worried about something, "then *I* should be worried about it as well!" Young came from San Francisco, where he was born on 24 September 1930, but when he was three years old his family moved to Cartersville, Georgia, then settled permanently in Florida, in the city of Orlando. It was at around this time that Young began to build model aircraft, a hobby that would remain with him throughout high school, together with rockets, which he chose for a speech to his classmates in the 11th grade.

In 1952, Young earned his degree in aeronautical engineering from Georgia Institute of Technology with highest honours, receiving coveted membership of the institute's prestigious Anak Society. Among his earliest assignments after joining the US Navy in June of that year, he served as fire control officer aboard the destroyer USS *Laws*. During this time, he completed a tour in Korea and a former shipmate would remember his coolness under duress. Joseph LaMantia (quoted on the website www.johnwyoung.com) recalled: "Though only an ensign at the time, he was the most respected officer on the ship. When we sustained counter-battery fire and enemy rounds were striking the ship, it was John Young's leadership which kept us all cool and focused on returning that enemy fire … which won the day." On returning home, he entered flight school at Naval Basic Air Training Command in Pensacola, Florida, where he learned to fly props, jets and helicopters. Later, he undertook a six-month course at the Navy's Advanced Training School in Corpus Christi, Texas. With receipt of his wings came four years of service as a pilot in Fighter Squadron 103, flying F-9 Cougars from the USS *Coral Sea* and F-8 Crusaders from the USS *Forrestal*, one of the new supercarriers. During these years, colleagues would describe him as "the epitome of swashbuckling aviators … he exuded confidence coupled with uncommon ability".

This ability would ultimately guide him into the hallowed ranks of NASA's spacefaring corps. But not yet. The selection process for the Mercury Seven began early in 1959, at which time Young was just starting Naval Test Pilot School at Patuxent River, Maryland, and test-flying credentials were a prerequisite for astronaut selection. After graduation, he served as a project test pilot and programme manager for the F-4H Phantom II weapons system at the Naval Air Test Center in Maryland, evaluating armaments, radar and bombing fire controls for both the Crusader and Phantom fighters. During one air-to-air missile test, he and another pilot approached each other's aircraft at closing speeds of more than three times the speed of sound. "I got a telegram from the chief of naval operations," Young later quipped, "asking me *not* to do this anymore!" In early 1962, he set two time-to-climb world records. By now a lieutenant-commander, Young's experience with the 'Phabulous' Phantom had made him the obvious choice to set the records as part of Project High Jump. The first, on 21 February, saw him climb to 3,000 m above Naval Air Station Brunswick in Maine in 34.5 seconds; and six weeks later he made another attempt from Point Mugu in California and achieved 25,000 m in 230.4 seconds. In September of that year, after leaving active naval duties as a maintenance officer in Phantom Fighter Squadron 143, he received a call from Deke Slayton which marked the start of his astronaut career. The training was arduous. "You had to learn a lot of stuff," he said later. "You probably only needed to know one percent of all the stuff you had to learn … but you *didn't* know which one percent it was!" In the two decades that followed, Young journeyed into space five times: twice aboard Gemini, twice aboard Apollo (becoming the first man to fly solo in lunar orbit and the ninth to leave his footprints on the Moon) and command of the first Space Shuttle mission. He retired from the Navy with the rank of a captain in September 1976. Now fifty-three years old and preparing for his record-breaking *sixth* flight into space, Young was showing no sign of slowing down; he had served as head of NASA's astronaut corps since April 1974 and would hold the post for 13 years, longer than any other incumbent. By the end of his flying career he would have logged over 15,200 hours in many different aircraft ... nearly *double* the amount of any other chief astronaut. Many have paid tribute to his endless supply of dry one-liners, which endured, undiluted, over the decades. As any son-in-law will understand, arguably the best came on the eve of Young's very first mission, way back in March 1965. A journalist asked him if he any qualms about flying into space with his somewhat fiery crewmate, Virgil 'Gus' Grissom. In a split-second response, timed to perfection, Young deadpanned: "Are you *kidding*? I'd have gone with my mother-in-law!"

Seated alongside Young on the flight deck for STS-9 was Major Brewster Hopkinson Shaw Jr, the first Air Force pilot from the 1978 TFNG class to draw a flight assignment. George Abbey's alleged preferential treatment of Navy pilots has already been noted and some astronauts have openly accused him of assigning less challenging missions, in terms of flying skills, to Air Force pilots. *That* certainly appeared to be the case during the first few years of the Shuttle programme, when *every* Spacelab mission featured a pilot from the Air Force ... while complex rendezvous, RMS or EVA missions often had Navy pilots. Even Shaw himself would admit that his role as the 'pilot' on STS-9 was limited in scope, with very few manoeuvres. Born in Cass City, Michigan, on 16 May 1945, Shaw grew up on a farm; his father was a qualified engineer, "although he didn't really practice engineering; he was in construction and farming was his hobby". Shaw graduated from high school in Cass City and whilst flipping through a book from the University of Wisconsin at Madison, his attention was caught by a degree programme in engineering mechanics. The career paths seemed compatible with his natural interests in mathematics and science and he pursued the course, receiving his degree in 1968. Whilst in Wisconsin, he joined a rock 'n' roll band, called 'The Gentlemen', and encountered a fellow student, Steve Schimming, who played drums ... and who also happened to possess a pilot's licence. One day, Schimming took Shaw for a flight ... and got the young man hooked on aviation. He completed his degree in 1968 and a master's the following year *and* also qualified for his private pilot's licence.

By this time, at the height of the Vietnam War, Shaw's military draft had already been deferred to enable him to study for his master's degree and in early 1969 he entered the Air Force, hoping to become a fighter pilot. In his NASA oral history, he remembered talking to the recruiter in Madison. After looking at Shaw's academic background in engineering mechanics, the recruiter told him that he was sorry, but there were no engineering opportunities available ... "the only thing we could offer you would be a pilot slot!" Far from being an unfortunate second-best option, pilot training was precisely what Shaw wanted and he plunged enthusiastically into officers' training school at Lackland Air Force Base in San Antonio, Texas, marrying his wife, Kathy, at around this time. Undergraduate pilot training followed at Craig Air Force Base in Selma, Alabama – a place Kathy *least* wanted to go – but they enjoyed their time in the Deep South and Shaw received his wings in 1970. He was initially assigned to the F-100 Super Sabre training unit at Luke Air Force Base in Arizona and in March 1971 went to Vietnam, during which time he also flew the F-4 Phantom ... and contracted hepatitis, requiring a period of grounding. Back in the United States, Shaw reported to George Air Force Base in California for F-4 duties, then headed to Thailand for a year in March 1972. It was a long year, he reflected, "a bad experience all around ... The *whole damn war* was a bad experience!" Returning to George Air Force Base in April 1973, he was becoming increasingly disillusioned with the war and considered leaving the military entirely and entering law school. This idea was quashed in July 1975, when Shaw was selected for a place at the famed Air Force Test Pilot School at Edwards Air Force Base; after graduation, he became an instructor in August 1977. Next, he applied to NASA and was called to Houston in mid-September for a week of interviews. It is interesting that two other candidates in the 20-strong group, Bryan O'Connor and Dick Richards, though not selected in 1978, would later become astronauts and would fly their first missions with Shaw as their commander.

When Shaw was named as John Young's pilot for STS-9 in April 1982, a lesser man might have felt somewhat intimidated to be flying right-seat to one of America's most experienced astronauts. In reality, the two men had actually flown together sometime before Shaw was even selected by NASA. Veteran astronaut Tom Stafford was commander of Edwards Air Force Base in the late 1970s and once invited Young to give a talk to the members of the test pilot school, which, at the time, included Shaw amongst their number. "I gave him a ride back to Houston in an F-4," Shaw recalled with some glee, "so John and I flew back all the way ... and he wondered why we couldn't do it in *one flight* and I kept telling him that we'd run out of gas before we got here and that we ought to stop! Anyhow, we got to fly together, so John *knew* I could fly airplanes and I figured that wasn't an issue." Or so it seemed. Years later, on one Friday afternoon in Houston, Shaw was flying T-38 approaches over Ellington Field and handed over the controls to his backseater, fellow astronaut Mike Mullane, to land. "He made this one landing," Shaw recalled, "and he didn't hold the nose off the ground [as he should have]. When the nose wheel came down, it came down right on the barrier cable that runs across the runway and it bent the flange of the wheel ... and let *all* of the air out of the tyre!" Shaw knew that backseaters were not supposed to land T-38s and so when the mechanics on the flight line asked what had happened, *he* took the blame. The pair thought nothing more of the incident until they arrived for work on Monday morning and astronaut Dave Walker – then serving as the office's safety representative for flight operations – walked into the conference room with *Shaw's flat tyre* under his arm!

"Brewster, do you want to explain this?" asked Walker, paraphrased in Mullane's autobiography. "The incident report says you forgot to hold the nose up for aerobraking."

Shaw shot him an awkward glance.

In Mullane's recollection, *no one* in the conference room seriously believed that Brewster Shaw – an experienced fighter pilot and exceptional test pilot – had 'forgotten' to hold up the jet's nose and the two astronauts promptly found themselves in John Young's office to admit what had happened and were verbally roasted for it. "We were finally dismissed to return to our offices," Mullane wrote, "to worry about whether we'd ever fly in space." Ironically, Shaw had considered coming clean to Young earlier that morning and even approached deputy chief astronaut Al Bean for his advice. Bean had told him to keep quiet about it. "*That* was a mistake," Shaw told the NASA oral historian. "I should have gone and told John ... This is one of these times when you see your career as a fledgling astronaut pass in front of your face ... but, fortunately, it didn't happen." Neither astronaut's career was ultimately harmed and, to Mullane, one of the greatest ironies was that not only were the two of them *not* fired ... but Shaw wound up as Young's pilot on STS-9!

Another "situation", as Shaw called it, cropped up during their mission training, when the pilots and the mission specialists had spent some time at the Marshall Space Flight Center in Alabama, working on the Spacelab hardware, and were aboard their T-38s, ready to return to Houston. Young and Shaw were in one aircraft, Parker and Garriott in the other. Often, they would do formation takeoffs, but the runway was narrow and the foursome decided against it; they would 'stagger' their departures, with Parker and Garriott going first. On the runway, Shaw was in the front seat of his jet, head down, studying the engine instruments, with Young behind him, when out of the corner of his eye he saw Parker's jet take to the air. Instinctively, Shaw reacted to hundreds of hours of formation takeoffs, released the brakes, stroked the T-38's afterburners ... and realised that he had created precisely the *wrong* spacing between the two aircraft. "I [got] it into the air," he explained, "and we're right in [Parker's] wake, his vortices coming off his airplane; dirty, dirty air. The next thing I know, we're about 20 feet in the air and we're in 90 degrees of bank, because the dirty wind kicked the airplane over, so I'm *standing* on top rudder as hard as I can and I've got full aileron in."

From the back seat came the drawl of his boss, the chief of the astronaut office: a clipped *"Brewster!"*

"This dirty air is *spinning air*," continued Shaw to the oral historian, "and as you move back and forth, it does different things to the airplane ... so the next thing I know, we're in 90 degrees of bank the *other way* and I got the other top rudder and the aileron the other way."

Again, from the backseat, came Young's voice: *"Brewster!"*

Finally, Shaw succeeded in getting away from the dirty air and rejoined Parker and Garriott and the two jets proceeded back to Houston. "Not *another word* is said between John and I the whole way back," Shaw recalled, "until we're letting down in Houston." He *had* to say something. "And all I could think to say was, 'God, John, I'm sorry!'" Privately, Shaw was certain that Young would remove him from STS-9, "but, bless his heart, he didn't do that. So we got to go fly STS-9 together".

Once in orbit, the six astronauts divided themselves into their teams to begin work. The 'Red Team', consisting of Young, Parker and Merbold, worked the 'night' shift from 9:00 pm through to 9:00 am EST, whilst the 'Blue Team' of Shaw, Garriott and Lichtenberg aligned their schedule with that of the Houston 'day'. This required Young's team to begin their first sleep period a few hours after Columbia reached orbit, whilst at the same time giving Shaw's team an extremely long Flight Day One. "The standard definition of 'Flight Day' would be confusing if applied to a two-shift, 24-hour operations flight," admitted the STS-9 Crew Activity Plan. "Launch Day will be FD1 for *both* teams. When the Red Team awakens after their first sleep period, this will be the start of Red Flight Day Two. When the Blue Team awakens ... it will be the start of Blue Flight Day Two. This will continue through the flight for a total of ten Red FDs and ten Blue FDs." Typically, the two groups would meet twice daily for a brief 'handover', to discuss experiments or orbiter operations, and to share meals – even though half of them would be wolfing down *dinner* and the other half would be eating *breakfast* – and the 24-hour operation proved quite unlike anything attempted previously in human space flight. In fact, the making of meals benefitted from the first use of the Shuttle's galley. This modular facility, built by General Electric under a September 1978 contract with NASA, provided hot and cold water dispensers, a pantry, an oven, food serving trays, a personal hygiene station – including a hand-washing area, light and shaving mirror – and a water heater. No refrigerators or freezers for food were carried aboard the Shuttle and foodstuffs were typically dehydrated (water came as a byproduct of the fuel cells), thermostabilised, irradiated or in natural form.

Spacelab-1 represented one of two so-called 'Verification Flight Tests' of the new system and would evaluate the long module and the pallet, whilst Spacelab-2 – at the time scheduled for launch in late March 1985 – would put a 'train' of three pallets, an igloo and a battery of astronomical and solar physics instruments through their paces. Only after these two pathfinding missions had been successfully accomplished would the Spacelab system be declared fully operational. For STS-9, more than 200 sensors were installed throughout the vehicle, providing data on their combined performance during launch, ascent, orbital operations, re-entry and landing. This equipment verified that Spacelab-1's passive thermal control system kept temperatures within the module and outside, on the pallet, within required limits and prevented condensation or heat leaks. The thermal, acoustic and structural response of the entire payload during the most critical and dynamic parts of the mission were carefully monitored and the astronauts checked the Scientific Airlock and the Spacelab tunnel and evaluated their ability to communicate and transmit data through the Tracking and Data Relay Satellite; the latter, by now, had come through the worst of its problems and was close to being declared operational. All six astronauts confirmed that the module itself was habitable and provided a pleasant environment for work. In general, Spacelab-1 performed admirably and the men even demonstrated the optical quality of a small window in the module's rear end cone by shooting television pictures of the experiments on the pallet. The two teams, Red and Blue, worked extremely well, the astronauts having adjusted their sleep cycles a fortnight before launch to support the shift pattern. Although 24-hour operations would not be needed on *all* Spacelab missions, the exercise was a valuable demonstration of how to maximise the use of the crew's time in orbit.

Despite its ultimate success, the mission hit a snag shortly after Columbia reached orbit. "We couldn't get the [Spacelab] hatch open," recalled Shaw. "We couldn't get into the [module] at all!" It was a momentary scare, particularly looking into the faces of Garriott and Parker – "these guys are seeing their *lives* pass in front of their faces, if we can't get in there and do the stuff they've been training to do". However, by the late evening of 28 November, Spacelab-1 was open for business, the lights turned on and the first experiments coming to life. In fact, several experiments were *already* underway, even as the Shuttle speared into orbit, for Lichtenberg and Merbold were outfitted with biomedical headgear to monitor their eye movements throughout ascent. Typically, during shift operations, the pilots remained on the flight deck and spoke to their crewmates in the module by means of an intercom. "Since there was only one of us [himself or Young] awake on the flight deck ... you didn't want to leave the vehicle unattended very much," Shaw told the NASA oral historian, "because this is *still* STS-9, fairly early in the programme. We hadn't worked out all the bugs and everything and neither John or I felt too comfortable leaving the flight deck unattended, so we spent most of our time there." Orbital manoeuvres occurred occasionally, but the majority of their role was spent monitoring Columbia's systems. "After a few days of *that*, boy, it got pretty boring, quite frankly!" Every so often, they would perform a manoeuvre to place Spacelab-1 into a different thermal attitude – exposing it to varying degrees of heat and cold to test its response – but Shaw found himself spending much time gazing out of the window, taking photographs for the Earth observation tasks. With no one to talk to, it became a monotonous time. Earth observations, including photography, were important to Young, too. "One of our first experiments," Owen Garriott told the oral historian, "had to do with a rotating dome, which would be taking a picture of your eye. That camera ... *broke* very early in the flight and ... all of this film and camera was sitting there without any particular use, so John Young helped to find a use for *that*. He took the camera and all of these rolls of film up to the flight deck. I think he spent a major fraction of the flight ... taking pictures out that left-hand window with the nice format camera." In fact, the post-flight mission report would acknowledge that more than 7,000 frames of 'out-of-the-window' photography were acquired in total. Young also accommodated other crew members who wished to venture up to the flight deck to take photographs. Venturing *down* to the Spacelab, though, for the pilots, was something to be feared. Life sciences experiments were conducted by the entire crew and Young and Shaw would be pounced upon by the science team for blood tests whenever they dared to enter the module.

The only real problem of note which occurred with Spacelab-1 was a temperature glitch in a remote acquisition unit, which served all of NASA's experiments on the pallet. Subsequent analysis by ground-based engineers revealed that the temperature of Columbia's freon coolant loop was partly to blame and the issue was quickly resolved by adjusting the unit to 22 degrees Celsius. Unfortunately, as part of the resolution of this problem, a 'patch' inserted into the Spacelab software caused the module's computer to crash and temporarily affect data-collection efforts from the pallet experiments. Otherwise, said Shaw, the mission gathered a great deal of valuable scientific data. Its experiments – more than 70 in total – were divided into five disciplines: astronomy and solar physics, space plasma physics, atmospheric physics and Earth observations, life sciences and materials sciences and technology.

These had been jointly sponsored by NASA and ESA, although they featured extensive co-operation from Canadian and Japanese research teams. The NASA side was run by the agency's Marshall Space Flight Center, whilst the ESA side was controlled by a team called the Spacelab Payload Integration and Co-ordination in Europe (SPICE), based in Cologne.

In recognition of ESA's immense contribution, Garriott, Parker and the payload specialists had spent a great deal of their training time in West Germany, where the European experiments were centralised. The genesis of these experiments dated back to 1976, when 400 proposals were jointly solicited by NASA and ESA. Ultimately, 70 finalists were selected and the principal investigator for each one took responsibility for forming a working group, which, aided by a NASA mission scientist, followed its development from conception through to manufacture and ensuring that it met the strict requirements for integration into Spacelab-1. During the mission itself, science operations were co-ordinated from a Payload Operations Control Center (POCC) at JSC in Houston and it was possible – for the first time – for ground-based scientists to speak directly to the crew, rather than indirectly, via the capcom. In flight, this worked generally well, although on occasion the science crew found themselves overloaded with requests. On one occasion, Bob Parker lost his patience when the POCC asked him to change a medical procedure on which he was busily working ... *and* recharge a battery ... *and* restart one of the Spacelab's materials processing furnaces ... *and* check an experiment on the pallet. "If you guys would recognise that there are *two people* up here trying to get all your stuff done," Parker snapped, in a recollection paraphrased by Douglas Lord, "I think you might be *quiet* until we got one or the other of them done!"

Six of the experiments were devoted to astronomy and solar physics. These included a NASA-provided far ultraviolet telescope, mounted on the pallet, which acquired spectra of high-temperature celestial objects as part of efforts to gain a clearer insight into the life cycles of distant stars and galaxies. Unfortunately, a fogged film ruined many of the images and an investigation after the flight recommended that, when it flew next, the telescope should record photons *electronically*, rather than on film as time exposures, to better pinpoint the cause. It did, however, achieve 95 percent of its scientific objectives and took the first far ultraviolet image of the Cygnus Loop, a relatively close supernova remnant. The roof-mounted Scientific Airlock accommodated ESA's wide field camera, which photographed ten astronomical sources on 3 December, including ultraviolet images of a 'bridge' of hot gas between the Large and Small Magellanic Clouds, indicative of an interaction between them. Sadly, the month-long delay of STS-9 from October until November resulted in shorter-than-expected orbital 'nights', reducing the camera's viewing opportunities by 60 percent. The third astronomical instrument was ESA's highly-successful X-ray spectrometer, mounted on the pallet. Other experiments – the Active Cavity Radiometer (ACR), the Measurement of Solar Constant (SOLCON) and Measurement of Solar Spectrum (SOLSPEC) – tracked with extreme precision the total amount of solar energy received by Earth's atmosphere and its impact upon our planet's environment to further the study of the solar-terrestrial relationship.

Five experiments came under the space plasma physics banner, all of them mounted on the pallet. Of these, the most noticeable was the Space Experiments with Particle Accelerators (SEPAC), designed by Japan's Institute of Space and Astronautical Science, based in Tokyo. It comprised an 'electron gun' to investigate the dynamics of the ionosphere and during the mission was used to fire streams of gas and high-intensity electrons, as part of efforts to gain a clearer understanding of aurorae, the nature of the planet's magnetic and electric fields and the effects of plasma on the Shuttle herself. In spite of the failure of its electron beam assembly to function in a high-power mode, SEPAC was very successful and achieved almost all of its planned tasks. A similar gun, provided by France, was the Phenomena Induced by Charged Particle Beams (PICPAB), which carried an 'active' unit on the Spacelab pallet and a 'passive' recorder inside the module's Scientific Airlock. Like SEPAC, this experiment was capable of producing and examining artificial aurorae. Some of its data was lost when one of its gas bottles failed, but its success rate topped 60 percent. Working in conjunction with the two electron-gun investigations were experiments which sought to measure atmospheric constituents and examine real and artificial aurorae under ultraviolet and visible light. Six atmospheric physics and Earth observations investigations were aboard Spacelab-1, five of them provided by

The first glimmer of orbital daybreak creates a kaleidoscope of colour in the limb of Earth's atmosphere, as viewed from STS-9. A large complement of the Spacelab-1 research was focused upon the interaction between the atmosphere and the Sun. For the crew, experiencing no fewer than *sixteen* sunrises in each 24-hour period, the complication of working 'day' or 'night' shifts in the Spacelab must have paled into insignificance.

ESA and one – the Imaging Spectrometric Observatory (ISO) – by NASA. The latter comprised no fewer than *five* spectrometers, housed in a single unit, and examined the presence and relative abundances of oxygen, nitrogen and sodium in the 'mesosphere', a region of the 'middle atmosphere' located some 80-100 km above the surface. ISO formed part of an extensive project to build a comprehensive database of the atmosphere's vertical structure. Meanwhile, the five European experiments included a metric camera in the Spacelab module's roof-mounted optical window, which yielded a wide variety of high-resolution mapping images using both black and white and infrared films. Three others were mounted on the pallet and examined oxygen and hydrogen emissions from the atmosphere and the Grille Spectrometer was designed to study the dynamic behaviour of the gaseous constituents of the stratosphere, mesosphere and thermosphere. Unfortunately, this device – so-called because it employed a special 'grille' as a window for part of its optical system and as a mirror for the other – was only partially successful, due to the lengthy STS-9 launch delay. Observing conditions in early December, compared to late September, proved unfavourable and only 16 percent of its objectives were accomplished. Nevertheless, the Grille was subsequently reflown on a dedicated Earth sciences mission in 1992 with greater success.

The bulk of the Spacelab-1 experiments, however, were housed inside the cylindrical module and consisted of 16 life sciences investigations and three dozen studies of the behaviour and processing of materials and fluids in the microgravity environment. The predominance of the latter was clearly reflective of the fact that STS-9 was the first mission in which extensive materials research could be conducted; previous studies aboard Skylab had been comparatively primitive and offered little more than a taste of what could be achieved. Scientists had long anticipated that in microgravity the virtual absence of the complications induced by buoyancy and other factors should make it possible to produce lighter and stronger building materials, more reliable and less costly electronics, new alloys, plastics, ceramics, composites and glasses for industrial machinery and perhaps even 'grow' crystals of important proteins to support research into the development of new pharmaceutical drugs to battle cancer and other diseases. Thirty of the materials science experiments were accommodated in ESA's dedicated Materials Science Double Rack, a huge, refrigerator-sized facility which was activated on 29 November, less than a day into the mission. It housed four furnaces and materials processing chambers – a fluid physics module for liquids research, a gradient heating furnace and mirror heating facility for crystal growth investigations and an isothermal heating facility for studies of solidification and casting processes. All four members of the science crew had undergone extensive training with the rack, but on occasion they felt little more than lab technicians; "We put materials sealed in cartridges into furnaces, heated, melted, solidified the materials, pushed buttons and started computer programs," recalled Ulf Merbold. The experiments were successful, although the mirror and isothermal heating facilities experienced partial power failures on 30 November and returned only limited results. (The mirror furnace, however, was later restored through a maintenance procedure by the crew.)

Alongside the materials research, Spacelab-1's life sciences experiments were conducted by the entire crew, *on* the entire crew, much to John Young's chagrin, since the pilots would be pounced upon for blood draws whenever they dared to enter the module. Nine investigations were provided by ESA and seven by NASA and focused principally upon the effects of the strange microgravity environment and high-energy cosmic radiation on the human body. One particular experiment was remembered clearly by Brewster Shaw: "Helen's Balls!" Principal investigator Helen Ross, a psychologist from the University of Stirling in Scotland, had devised a study of mass discrimination in microgravity, which involved a series of yellow balls of differing sizes and masses. "Since there's no weight – only *mass* – in microgravity, we had to differentiate between the mass of these balls," recalled Shaw. "You would take a ball in your hand and shake it and feel the mass of it by the inertia and the momentum of the ball as you would start and stop the motion. Then you'd take another one and try to differentiate between them and eventually you'd try to rank-order the balls. They were numbered as to which was the most massive to the least massive. We did that several times." A number of vestibular experiments, some provided by Canadian physical scientists, were also conducted to investigate the behaviour of the human vestibular system in the inner ear – responsible for controlling balance and orientation – and help to identify a relationship between an astronaut's sense of balance and eye motions. These experiments were also insightful in terms of linking head movements to the onset of Space Adaptation Syndrome; indeed, three of the four scientists suffered varying degrees of the ailment whilst in orbit, although, as Douglas Lord commented, many of them were *designed* to drive them to the *brink* of sickness, "so a high percentage of such problems in this mission would not have seemed unusual". Other studies explored the role of microgravity in the reduction of red blood cell mass and its effect on the crew's immune systems.

Overall, with a few minor problems, the mission proceeded smoothly; so smoothly, in fact, that on 3 December, NASA and ESA agreed to extend the astronauts' time in orbit by 24 hours, when it became clear that on-board consumables were not being used as quickly as expected. Overall, Spacelab-1's power consumption rate averaged about 1.2 kilowatts *less* than engineers had conservatively predicted before launch. Even at STS-9's halfway point, Mission Scientist Rick Chappell described the flight as "a very successful merger of manned space flight and space science". The astronauts were also performing well, although, at least to Shaw, it was obvious who had been here before and which ones were rookies. "John and Owen were the experienced guys," he told the NASA oral historian, "and they were mentors of the rest of us. It was fun to watch Owen back in the module, because you could tell, right from the beginning, he'd been in space before. He knew *exactly* how to handle himself, how to keep himself still, how to move without banging all around the place. The rest of us were *bouncing off the walls* until we learned how to operate." Even though Garriott's most recent mission had taken place a full decade earlier, what amazed Shaw the most was the incredible *adaptability* of the human body and its ability to 'remember' previous states.

Following a spectacular nine days in space, late on 7 December, the science crew began the process of shutting down Spacelab-1's experiments, storing and securing the samples and closing the module and pallet. Meanwhile, Young and Shaw, assisted by flight engineer Parker, set to work on the procedures to prepare Columbia for her hypersonic descent back through the atmosphere to a landing at Edwards Air Force Base in California. Then, barely five hours before the scheduled landing, at 6:10 am EST on the 8th, as they configured the orbiter's five General Purpose Computers for re-entry, the gremlins struck STS-9: one of the computers failed during an RCS manoeuvre. Any such failure generated a bad prognosis, since the computers were responsible for monitoring thousands of separate functions during one of the most dynamic phases of a mission. Worse was to come. Six minutes after the first failure, a *second* GPC abruptly stopped working! Young and Shaw successfully brought the second computer back on-line, but their efforts to restart the first one were fruitless and it was powered down for the remainder of the mission. "A ground review of the [second GPC] memory dump indicated that memory alterations had occurred," noted NASA's post-flight mission report. "However, [it] was reinitiated and placed back in the redundant set." The mood on the flight deck was tense. "My knees started *shaking*," Young told a post-flight press conference of his reaction to the *first* GPC failure. "When the *next* computer failed, I turned to *jelly*. Our eyes opened a lot wider than they were before!"

Eventually, the GPC which had been earlier restored by Young and Shaw, and two others, were used to support Columbia's return to Earth. Yet there was even more trouble in store. Later that same morning, one of the Inertial Measurement Units (IMUs) – a crucial piece of navigational hardware – failed. Re-entry was postponed to allow ground specialists to analyse the problems. No crew activities were scheduled, except that the six men were kept in a state of preparedness for re-entry. Eventually, after nearly eight hours, the de-orbit burn was performed at 5:52 pm EST to begin the irreversible descent to Earth. Overall, Columbia's performance was nominal ... until about four minutes before touchdown, when the temperature of one of the Auxiliary Power Units (APUs) began to rise sharply. "Then we had another lesson," Shaw told the oral historian. "*Never* let them change the software in the flight control system without having adequate opportunity to train with it. There were 'gains' in the flight control system and [these] changed depending upon what phase of flight you're in. When you're flying a final [approach], there are certain gains that make the vehicle respond a certain way to the inputs the pilot makes on the stick. When the main gear touch down, the gains *change* and the gains are set up so you can de-rotate the vehicle and get the nose on the ground in an appropriate way. We had done *all* our training in a simulator with a certain set of gains ... and then they *changed* the flight software and these gains, so that when it came time for John to land the vehicle in real flight, the gains were *different* than he'd done all his training on. Certainly, when John started to de-rotate the vehicle, it responded differently than he had trained on." With Shaw calling out airspeed and altitude data, Young brought Columbia's main landing gear gently onto the runway and began the process of de-rotating the nose gear. Everyone on the middeck started cheering. "We were down to about 150 knots," continued Shaw, "when the nose hits the ground and it goes *smash*!" The changed gains in the flight software and the 'different' response of the vehicle presented them with a hard, but safe, landing.

It subsequently became clear that the first GPC failure was caused by a sliver of

After a hugely successful mission, Columbia sits silently on the runway at Edwards Air Force Base as dusk falls over the Californian desert. The last 24 hours of the longest Shuttle flight to date had been eventful, to say the least: two GPC failures, a waved-off landing, a harder-than-expected touchdown and an APU fire. Yet the triumph of STS-9 and Spacelab-1 would contribute to President Ronald Reagan's decision in early 1984 to authorise NASA to proceed with studies into the establishment of a permanent international space station.

solder, which had dislodged itself during an RCS thruster firing and shorted out the computer. The *second* failed GPC, which Young and Shaw had earlier managed to revive, failed *again*, seconds after Columbia's 3:47:24 PST (6:47:24 pm EST) touchdown. And six and a half minutes after *that*, one of the APUs shut itself down, as did a second just four minutes later. Little did the crew know at the time, but one of the APUs had actually been *on fire*, due to a hydrazine leak, as the Shuttle raced down the runway. "We got called the next day," Brewster Shaw recalled, "because we had an APU fire. The reason the first one shut down was [because] it was on fire and the fuel wasn't getting to the catalyst bed and so it 'undersped' and automatically shut itself down." In response to that situation, Shaw – who, as pilot, was responsible for the APUs – reconfigured systems to prevent the third APU from shutting down. "The next one [APU-3] didn't shut down *until* we shut it down," he said. "We had a generic failure of a little tube of metal where the fuel went through and was injected into the catalyst bed and it *cracked*. When we shut [the APUs] down and shut the ammonia off to them, the fires went out. We had some damage back there, but the fires stopped. We didn't know anything about that 'til the next day!"

Six weeks after Columbia's landing, on 19 January 1984, a report landed on

NASA Administrator Jim Beggs' desk, with a glowing summary of the scientific achievements of the Spacelab-1 mission. Already, Beggs had spoken informally to President Ronald Reagan and had gained his verbal approval to proceed with studies for a permanent, Earth-circling space station ... which would involve co-operation from Europe, Canada and Japan. Nor would Europe be relegated to any kind of 'junior' role: its remarkable achievement of taking on the challenge of Spacelab and seeing the project through to a triumphant completion would henceforth assure it of a place as a full partner in future endeavours. By March, Spacelab-1 had been judged an unqualified success and many scientific journals in the summer of 1984 were overflowing, not only with reports, papers and letters on the results of different research facilities, but also with *praise* for a quite remarkable mission. Ulf Merbold, always outspoken, had been somewhat critical, even provocative: Europe had constructed a world-class research facility, he said, and yet had been granted only *half* of the *first* mission. West Germany, as the dominant partner in the effort, and having contributed more than half of the finances, already had its eye keenly on a series of dedicated Spacelab missions of its own, beginning in 1985. And when 'West' and 'East' united in 1990, they continued to press forward their space ambitions. It is no accident that when Europe's showpiece of the International Space Station – the Columbus laboratory – finally reached orbit in February 2008, one of the spacewalkers who helped install it was German. In many ways, the Shuttle and Spacelab had spurred a new era of international co-operation in human space flight. But as this first chapter of the Shuttle came to a close – with satellite deployments, EVAs, complicated RMS operations, rendezvous and the first flight of a fully-fledged science laboratory, together with women astronauts, black astronauts, West German astronauts and a handful of fiftysomething astronauts – it served as but a prelude for the *next* chapter. The next two years of Shuttle operations, beginning with the tenth mission in February 1984, would reach even greater heights and score even more remarkable successes. However, this apparent success would screen ugly realities, fatal flaws, poor decision-making and a culture of appalling negligence from the eyes of the world. At least two Shuttle crews would feel Death's breath on their faces, missing oblivion by seconds, until finally, a catastrophe of unimaginable proportions would unfold on millions of television screens across the world, one cold morning in January 1986.

2

A final Soviet salute

A TOUCH OF *DÉJÀ VU*?

The ten years which elapsed between 1975 and 1985 were amongst the most decisive, and divisive, in the history of the United States and the Soviet Union. At few other times in the 20th century did the pendulum of change in their relationship swing so sharply between harmony and hostility. At its root, of course, lay their quite distinct political doctrines, their underlying distrust of one another, their respective opinions about freedom of speech and basic human rights and their mutual meddling in international affairs. With the Soviets wary of Chinese aggression and America embroiled in the bloodbath of Vietnam, sincere efforts had been made between the two superpowers to reduce their ballistic missile stocks in the early 1970s, with General Secretary Leonid Brezhnev and President Richard Nixon signing a series of agreements to refrain from hostilities. Efforts to improve the political situation with the Helsinki Accords quickly soured when it became clear that neither side was willing to disengage from involvement in foreign conflicts; perhaps most fundamentally, the Soviet invasion of Afghanistan in 1979. In response to the latter, President Jimmy Carter and his hawkish national security advisor, Zbigniew Brzezinski, started a $40 billion CIA initiative to covertly arm a mujahideen insurgency. Fears grew that the Soviets were positioning themselves for an oil takeover in the Middle East. American athletes were prohibited from participating in the 1980 Summer Olympics in Moscow – a tit-for-tat move to which the Soviets reciprocated with their own boycott of the Los Angeles Games, four years later – and the misinterpretation of mock military operations (notably 'Able Archer') aggravated the paranoia of General Secretary Yuri Andropov.

Andropov was a former head of the feared KGB and had assumed mastery of the Soviet Union in November 1982, upon the death of Brezhnev. His accession brought with it an ominous cloud of uncertainty about his intentions and designs on the West, although, to be fair, the administration of President Ronald Reagan was equally aggressive in its stance. Already, Reagan had implemented the Strategic

Defense Initiative to transport weapons of mass destruction to the final frontier and his provocative words of "evil empire" placed both leaders on an equal footing of distrust, dislike and unwillingness to compromise. It was a worrying sign that the 1980s might be an unhappy decade. Against this backdrop, of course, were renewed noises from within the satellite states of the Eastern Bloc for genuine economic and political change – noises which made themselves heard with more vehemence as many policies of the Soviet-style 'planned economies' began to fail. Economic stagnation in Czechoslovakia, a hard-line dictatorship and martial law in Poland, an East Germany from which thousands tried desperately to escape across the barbed wire and death strips of the Berlin Wall, the increasingly erratic regime of Todor Zhivkov in Bulgaria and the cruelty of Nicolae Ceauşescu towards his own people in Romania added to an international storm of protest. Few could have foreseen how dramatically the fall of communism in Eastern Europe would occur, but under such a cloud of oppression, few could have doubted that at some stage it *would* occur.

As these internal struggles gathered momentum, late on the evening of 19 April 1982, in a desolate and windswept area of central-southern Kazakhstan, the silence of the barren steppe was rudely shaken by the roar of six powerful rocket engines, producing close to a million kilograms of thrust, as the Soviet Union's seventh space station speared for the heavens, atop one of the largest, mightiest and most reliable boosters in operation: the Proton. It had not always been that way, however, and the Proton had suffered an astonishingly high failure rate during its genesis in 1965-70. It originated as an oversized intercontinental ballistic missile, designed to deliver a huge nuclear warhead across distances of 13,000 km to hardened targets in the West. Today, the Proton's descendants – still built at the same Khrunichev plant in Moscow – can lay claim to a proud heritage and their boast of a 96 percent launch success rate is among the highest in the world. Standing 160 m tall, with the Salyut 7 space station – a 'salute' to the world's first cosmonaut, Yuri Gagarin – tucked away inside its bullet-like nose and upper main body, the Proton was a visually impressive creation. Its three stages were each fed by a combination of unsymmetrical dimethyl hydrazine fuel and an oxidiser of nitrogen tetroxide; both are highly toxic and highly corrosive propellants, but with the advantage that they can be kept at ambient temperatures and are hypergolic, meaning that they burn on coming into contact, eliminating the need for an igniter. As a consequence, the Proton could sit on the pad for far longer than a cryogenic booster.

The perfect launch of Salyut 7 marked yet another achievement for a launch site which had long since become synonymous with space exploration ... and yet whose history could hardly have been more removed from science and technology. 'Tyuratam', as it is properly known, is as a railway junction, some 200 km east of the Aral Sea, and in the local tongue its name is roughly translatable as the gravesite of Tyura, beloved son of the great Mongol conqueror Genghis Khan, whose medieval empire spanned much of Asia. According to some sources, it began as an ancient cattle-rearing settlement on the north bank of the Syr Darya River, although at least one Soviet-era journalist has given it a more modern origin, hinting at its foundation in 1901 as an outpost to replenish steam engines passing between Orenburg and Tashkent. Its importance over the last half a century, though, cannot be disputed. It was from this sparsely inhabited expanse of steppe, five decades ago, that the first strides of a journey far more audacious, much longer and considerably harder than any the Great Khan could have foreseen were taken. It was from this place that Yuri Gagarin, the first man in space, began his historic flight in April 1961, changing our perception of the Universe forever. Today, it is still variously called 'Tyuratam', after the tiny railhead, or, more often, 'Baikonur', which covers a broader and different geographical location to the north-east. It was still shrouded in secrecy in June 1982, when the first Western European cosmonaut – a Frenchman named Jean-Loup Chrétien – arrived for his journey into space. After almost three years of training, Chrétien had visited Tyuratam … only *three* times: once for the launch of Salyut 7 in April, again for the launch of the station's first resident crew and, finally, for his own flight.

Yet the new space station, Salyut 7, was not the revolution that was expected. It was the backup vehicle for its predecessor, Salyut 6, and although it featured improved systems and updated instruments, its overall capabilities were the same and it represented no quantum leap in technology. Today, it is generally accepted that the Soviets' next-generation space station, 'Mir', a 'modular' outpost, whose development had been approved in February 1976, was running several years behind schedule and the Salyut 6 backup unit was pressed into service as a stop-gap measure. By 1982, the Soviets had already established themselves as world record holders in terms of space endurance, with crews having occupied the Salyut 6 station for 684 days, with the longest single visit lasting more than six months. With Salyut 7, a new era of long-term space flight, with missions lasting perhaps as much as a full year, was expected and the first international 'Intercosmos' visitor, a test pilot from France, was scheduled to fly in June 1982. For now, there was simply exultation that the new station was proceeding perfectly to orbit. Each of the Proton's three stages burned brightly in the night sky and, within minutes, the payload shroud was jettisoned and Salyut 7 was inserted into orbit.

Physically, the machine was virtually identical to its predecessor, Salyut 6, weighing almost 20,000 kg and measuring 14.4 m long and 4.15 m across its broadest diameter and providing a habitable volume of around 90 m^3 for its crew. Salyut 7 comprised a trio of cylinders – a forward transfer compartment and two working and living areas – and possessed a pair of docking ports, one at each end of the station, to permit the arrival of visiting cosmonaut crews aboard the Soyuz spacecraft and deliveries aboard unmanned Progress resupply freighters. Early Salyuts carried only a single docking port and featured a propulsion module at the rear of the station, although the provision of the second port traced its origins back to December 1973, when Vasili Mishin, then-head of the TsKBEM design bureau, 'borrowed' the concept from a military configuration designed by Vladimir Chelomei. In order to describe the basic layout of the new station, it is perhaps best to imagine oneself aboard a Soyuz ferry which has just docked at Salyut 7's forward port. Upon opening the hatch, a cosmonaut visitor would first enter the cramped forward transfer compartment, a tapering cylinder some 3.5 m in length and 2 m in diameter. Within this compartment was stowage provision for the crew's extravehicular suits, together with bottles of compressed air and the main EVA controls and an inward-opening hatch through which spacewalkers could exit the station. It also housed an astronomical camera, a sextant port and a fixture for an experiment to survey Earth's horizon. Outside was a large, dish-like rendezvous antenna, thermal control panels and EVA handholds. A second axial hatch led into the first work compartment, which measured 3.5 m long and 2.9 m wide and upon which (like the earlier Salyut 4 station) were mounted three steerable solar arrays. In total, these covered an area of 51 m^2 and produced around 4 kW of electrical power. Inside the first work compartment were the main control consoles – together with bicycle-like seats for the cosmonauts – and single drives for the three arrays. Elsewhere were cupboards and other storage facilities, a veloergometer and body mass meter on the 'ceiling', the Rodnik ('Spring') drinking water system, the Stroka ('Line') teleprinter and the KATE-140 wide-angle stereographic camera, with which to create topographic maps of Earth's surface. It could take single or strip photographs under cosmonaut or automatic ground command. At a conical 'frustrum', the first work compartment broadened into the second compartment, which measured 2.7 m in length and 4.15 m across; this represented the maximum diameter supportable by the Proton launch vehicle. Within the frustrum was the MKF-6M Earth resources camera, which enabled photographs to be taken in six different spectral bands and had been heavily modified (hence the 'M' suffix) to support operations over several years in space. Its film cassettes were each capable of taking 1,200 exposures and spares would be frequently delivered to the station aboard Progress freighters. Phillip Clark estimated that the MKF-6M's resolution was probably about 20 m or so. The presence of these two instruments had already led the Soviet Ministry of Agriculture to plant a number of specially selected crops at test sites in the Ukraine and the region around Lake Baikal to examine their capabilities. The main work compartment was dominated by an X-ray detection system, a huge conical structure which included the XT-4M telescope, prepared by the Lebedev Institute of Physics, and the XS-02M spectrometer, developed by the Stenberg Institute. The spectrometer had a sensitive surface of 3,000 cm^2, compared to just 450 cm^2 on the earlier Filin ('Eagle Owl') detector, flown aboard Salyut 4.

At other places in Salyut 7 were creature comforts for the crew: a small shower, a treadmill, electric stoves for heating food, a refrigerator, constant hot water, a trio of sleeping bags, a pair of trash airlocks and food lockers. One of the trash airlocks, interestingly, would later be used to house the Splav ('Alloy') experimental materials furnace. Two of Salyut 7's portholes were modified to allow for the penetration of ultraviolet radiation; part of a design measure to kill infections. These portholes had transparent covers which could be closed when not in use, thereby limiting the effect of micrometeorite impacts. At the far end of the complex was the second docking port, to be occupied by Soyuz or Progress visitors. Although Salyut 7 did not possess a large propulsion unit, it *did* carry manoeuvring thrusters: two primary engines, fed by unsymmetrical dimethyl hydrazine and nitrogen tetroxide, each of which Clark believed to have a thrust of some 300 kg. The provision of a second docking port should be neither underestimated, nor overlooked, for it was perhaps the pivotal feature of the new station which permitted the Soviets to conduct Soyuz changeovers and regular Progress deliveries of supplies and equipment, thereby empowering

Salyut 6 and 7 to operate on a totally new level over their predecessors. Additionally, noted *Flight International* on 29 May, the Salyut 7 ports were "more reliable" and capable of allowing "larger craft to link up". The magazine speculated on a possible Salyut-to-Salyut docking and suggested – correctly – that larger visiting craft, "such as derivatives of ... Cosmos", may also be in the pipeline.

The development of Progress had been in the works since 1973 and the first draft plans had been laid in February of the following year. Early Salyuts had to be launched with almost all of the cosmonauts' supplies aboard and, over time, they would also require periodic reboosts to maintain their altitude against orbital decay ... reboosts which demanded many hundreds of kilograms of propellant expenditure *every year*. "To maintain a 250 km orbit," wrote Phillip Clark, "would require [4,300 kg] of propellant. At 350 km, this requirement dropped to about 600 kg. Of course, the Soviets allowed the Salyuts to decay to lower orbits and then raised them again in large manoeuvres; this was more economical in terms of fuel expenditure. All in all, to maintain a fully-functioning Salyut in orbit for two years with a crew permanently on-board required about [18,000 kg] of consumables. It was ... impractical to launch all these supplies with the station – even if a Soviet booster had the necessary lifting power, which none had at the time." Progress was the obvious solution. It was modelled closely on Soyuz and the descendents of both spacecraft are still in service today, supporting the International Space Station.

'Soyuz' was the brainchild of Sergei Korolev, the famous 'Chief Designer' of early Soviet spacecraft and rockets, with the original intention of undertaking both Earth-circling missions and lunar ventures to rival the United States' Apollo effort. As early as 1964, the design and definition of Soyuz was well underway and technical documentation and a mockup revealed it as a craft capable of lofting two or even three cosmonauts. Even its *name* was no accident: for the Soviet Union's official moniker – *Soyuz Sovietskikh Sotsialisticheskikh Respublik*, the Union of Soviet Socialist Republics – was often popularly known amongst its citizenry as 'Soyuz' (the Union). Therefore, the name of the spacecraft not only reflected its role in supporting rendezvous and orbital stations, but was also a highly symbolic and political statement. In *Challenge to Apollo*, a history of the early Soviet manned space programme, Asif Siddiqi noted that when Korolev first saw the mockup, he proudly declared that Soyuz was "the machine of the future". With the exception of his vague title, it was not until long after his death on a hospital operating table in January 1966 that the world learned anything of substance about Korolev. Yet this man of outstanding engineering genius had masterminded some of the most remarkable triumphs in the exploration of space. It was he who had designed the R-7 intercontinental ballistic missile which launched Sputnik – the world's first artificial satellite – and later Yuri Gagarin. It was his Kaliningrad-based design bureau, OKB-1 (later renamed TsKBEM and, in the early 1970s, Energia), which assembled the first piloted spacecraft, Vostok, and it was he who oversaw the first three-man orbital mission, the first spacewalk and his nation's first (and ultimately fruitless) steps toward the Moon. His brilliance and unwavering devotion to a lifelong dream of exploring space was balanced by an all-or-nothing obstinacy which often manifested itself in a violent temper, capable of exploding without warning. Korolev lived a hard, thankless life of service to the Soviet state and it was this, ultimately, which consumed him. It was Korolev's development of Soyuz which has had the most long-lasting impact on the world. Since its first manned flight in April 1967, Soyuz and a modified version of the original R-7 continue to be used operationally today; a fitting legacy to an enduring talent.

Phillip Clark has traced its history back to a three-part 'Soyuz complex' – a manned craft, a 'dry' rocket block and a propellant-carrying tanker – which Korolev had envisaged in the early 1960s being assembled in low-Earth orbit to fly circumlunar missions. The first part, which Clark identified as 'Soyuz-A', but which the Soviets catalogued as 'Soyuz-7K', was closest in physical appearance to the spacecraft which actually flew and it was to the construction of this that Korolev committed OKB-1 in March 1963. Measuring 7.7 m long, Soyuz-7K had three components: a cylindrical 'orbital module', a bell-shaped 'descent module' for the crew and a drum-like 'instrument module' for manoeuvring engines, propellant and electrical power. According to Korolev's earliest blueprints, it weighed around 6,450 kg, but, unlike the final design, was not equipped with solar panels, relying instead upon chemical batteries. Supporting Soyuz-7K were the 'dry' Soyuz-B rocket block and the Soyuz-V propellant tanker, known to the Soviets by the designations of '9K' and '11K'. Clark hinted that a typical flight profile would have begun with the launch of a 9K, followed, at 24-hour intervals, by as many as four 11Ks, which would dock, transfer their propellant loads and then separate. When the 9K had been fully fuelled, a manned 7K would be despatched to dock with the rocket block. "Mastering rendezvous and docking operations in Earth orbit may have been one of the primary objectives of the Soyuz complex," wrote Asif Siddiqi, "but the incorporation of five consecutive dockings in Earth orbit to carry out a circumlunar mission was purely because of a lack of rocket-lifting power in the Soviet space programme." In fact, it was the sheer 'complexity' of the Soyuz complex which seems to have foreshadowed its restructuring sometime in 1964 and effected a delay of its maiden voyage until at least the spring of 1966.

By the end of the decade, seven manned Soyuz spacecraft had rocketed into orbit. However, a key physical difference between these vehicles and the original 7K was that they employed a pair of large rectangular solar panels, mounted on the instrument module, to generate electrical power. The total surface area of these wing-like appendages was 14 m^2, with each wing measuring 3.6 m long and 1.9 m wide. The remainder of the craft's design was strikingly similar to the 7K: a spheroidal orbital module, 2.65 m long and 2.25 m wide, the bell-shaped descent module, itself 2.2 m long and 2.3 m wide at the base, and the instrument module, a cylinder 2.3 m long and 2.3 m wide. This shape emerged at the end of almost a decade of planning, theoretical work and aerodynamic modelling. As early as 1958, Mikhail Tikhonravov and Konstantin Feoktistov, both engineers at Korolev's bureau, envisaged a multi-purpose craft capable of both Earth-orbiting and circumlunar missions. Space historians Rex Hall and Dave Shayler noted that the shape of the descent module was decided at least partly by a desire to touch down on land, rather than in water, and several designs were sketched out. The first utilised aerodynamic surfaces, facilitating an aircraft-like return to a runway, whilst the second adopted a 'missile principle', entering space in a ballistic manner and descending beneath parachutes. By 1961, concerns about mass and the need for adequate thermal protection during re-entry had eliminated the winged design from consideration. The missile principle, though, needed further work to man-rate it: a ballistic descent would impose significant duress on the vehicle and its occupants and Tikhonravov and Feoktistov moved instead toward the concept of a 'glancing' re-entry to reduce stress. If the new craft was ever to undertake lunar flights, its return trajectory from the Moon would produce correspondingly higher re-entry speeds of perhaps 40,000 km/h, prompting the engineers to design a 'double-dip' profile, which, by reducing the velocity in stages, would lessen the G loads on the cosmonauts. When consensus had been reached on the method of re-entry, OKB-1 engineers and researchers at the NII-1 and NII-88 aerodynamic institutes explored a trio of designs: one nicknamed the 'segmented sphere', another called the 'sphere with a needle' and a third dubbed the 'sliced sphere'. The segmented version emerged as the most promising design, with Vladimir Roshchin's group at OKB-1 promoting a descent module with a displaced centre of mass as a means of generating aerodynamic lift. By 1962, this had evolved into a shape approximating a car's headlamp, which aerodynamic simulations predicted would avoid the high deceleration and thermal loads of a ballistic descent and have sufficient lift to be able to steer towards a given landing site. A plethora of proposals also surrounded the means of landing, with helicopter-like rotors, fan-jet or liquid-propelled engines, controlled parachutes, ejection seats and shock-absorbing inflatable balloons all being considered. By 1963, however, Korolev had approved the design which remains in use today: a combination of braking parachutes and a soft-landing apparatus of solid-fuelled rockets. Even as the descent module was taking shape, the appearance of the spacecraft remained somewhat fluid and early designs for a space station ferry and a lunar-going concept both utilised a descent module for the crew, attached to an instrument module for propulsion and power. Already, the design was expanding further to encompass a habitable orbital module and there was disagreement about where this should be located. In some initial drawings it appeared *between* the instrument module and the descent module and in others it was *above* the descent module. The idea of placing the orbital module below the descent module was soon rejected, since it would require cutting a hatch into the descent module's base, potentially compromising its heat shield. The final layout, with the descent module in the middle, was in place by the end of 1962. By this time, it had also received the name of 'Soyuz'.

In spite of Korolev's assertion that it was the machine of the future, Soyuz had been mired for some years in technical and bureaucratic problems, to such an extent that by 1964 its development was virtually paralysed by the Soviet drive for the Moon. Early plans called for it to carry one or two cosmonauts, but by December 1963 the basic design of the Earth-circling version, known as 'Soyuz-7K-OK' ('Orbitalny Korabl' or 'Orbital Ship'), had grown to accommodate a three-man crew. Its purpose was to support automated rendezvous and docking, spacewalking, manoeuvring and scientific research, thereby fulfilling the key requirements for a space station ferry. During 1964, Korolev directed a small group under Boris

Chertok, one of his deputies at OKB-1, to explore other uses for the basic 7K-OK craft. One proposal called for docking two Soyuz together in orbit to demonstrate their rendezvous capabilities and having a cosmonaut spacewalk from one ship to the other. Not only would this ambitious plan offer valuable engineering experience, but it also supported early ideas for a Soyuz-based Moon mission in which a cosmonaut would transfer from the command ship to the landing craft in lunar orbit by 'extravehicular activity' (EVA). In February 1965, Korolev presented this 'new' version of Soyuz to the Scientific-Technical Council of the State Committee for Defence Technology and was told to proceed.

Beginning at its base, the instrument module, also known as the 'service module', carried chemical batteries and two large solar panels to charge them, together with a thermo-regulation radiator and an integrated propulsion and attitude-control system. The latter, designated 'KTDU-35', comprised a pair of engines, one primary and one backup, sharing the same oxidiser and fuel supply. The primary engine had a thrust of 417 kg and was capable of a change in velocity of some 2,750 m/sec, equivalent to a specific impulse of around 280 seconds. On the basis of early reports, which speculated that this engine could boost Soyuz to an altitude of 1,300 km, Phillip Clark suggested that the spacecraft required a propellant capacity of 755 kg. Propellants took the form of unsymmetrical dimethyl hydrazine and an oxidiser of nitric acid, housed in spherical tanks within the instrument module. Attitude control came from 22 primary and eight backup hydrogen peroxide thrusters. Guidance, rendezvous, communications and environmental gear filled the remainder of the cylindrical compartment. The descent module sat directly above the instrument module and housed the crew during ascent and re-entry. It had a habitable volume of some 2.5 m^3. The commander's seat was located in the centre, flanked by positions for a flight engineer and a research cosmonaut or 'test' engineer. Many of Soyuz' flight regimes were pre-programmed from the ground. Consequently, the main instrument panel presented the crew with readouts and visual displays of the performance of on-board systems, together with a monitor for the external television camera, an optical orientation viewfinder called 'Vzor' ('Visor') for attitude manoeuvres and the Globus ('Globe') device to show the spacecraft's position above Earth. In the event of a failure of the automatic systems, and to facilitate rendezvous and docking, it was expected that the commander could assume manual control. As a result, two hand controllers (one for velocity, the other for attitude) were located directly underneath the instrument panel.

Rendezvous and docking were supported by the Vzor, together with a system of gyroscopes, attitude-control sensors and thrusters and the Igla ('Needle') radar. The latter would automatically navigate the spacecraft to its target and draw to a halt at a range of 200-300 m, after which the crew would take charge and accomplish the final approach and docking. The systems to facilitate physical contact had undergone extensive development since 1962. At first, OKB-1 engineers Viktor Legostayev and Vladimir Syromiatnikov advocated a 'pin-cone' device to allow two vehicles to dock. At this stage, however, there was no provision for an internal transfer of cosmonauts from one craft to the other and, sometime in 1965, Korolev's proposal to change this was rejected by Feoktistov on the basis that a significant amount of work had already been done and additional revisions would put the development further behind schedule. The docking system featured a pin on the active spacecraft, which would be captured by a cone-like funnel on the passive one, essentially cancelling any remaining velocity or angular displacement.

The descent module would be the only component capable of surviving the intense heat of atmospheric re-entry and bringing the cosmonauts back to Earth. At the end of a mission, the instrument and orbital modules would be jettisoned and the descent module would employ half a dozen hydrogen peroxide engines, each producing a thrust of 10 kg, to provide roll, pitch and yaw controllability during the early stages of re-entry. To protect its occupants, it was coated with a heat-resistant ablator, together with a thermal shield at its base that would detach shortly before touchdown to expose the four solid-propellant landing rockets. A 14 m² drogue parachute would deploy 9.5 km above the ground in order to stabilise the craft, prior to deploying the main canopy. If a problem occurred, a secondary canopy could be deployed. Seconds before touchdown, an altimeter would command the landing rockets to fire to cushion the impact. Atop the descent module in space, the spheroidal orbital module held a bunk, a cupboard for food and water, life-support gear, controls for experiments, cameras and a variety of other equipment appropriate to each individual mission.

Difficulties aside, Soyuz promised to be one of the safest manned craft ever built, possessing as it did the Soviets' first 'true' launch escape system. This consisted of a tower atop the R-7's payload shroud and a multiple-nozzle, solid-fuelled rocket engine. In the event of an emergency from 20 minutes before launch until 160 seconds into the ascent, the shroud would split at the base of the descent module and the escape tower's engine would lift the descent and orbital modules to safety. At the top of the arc, the descent module would be released to parachute back to Earth, landing a couple of kilometres from the pad. Early predictions estimated that the crew could be exposed to acceleration loads as high as 10 G during such a scenario. Launching Soyuz, which was considerably heavier and more complex than the earlier Vostok craft, demanded further improvement of Korolev's original R-7. The basic design of the missile, physically, remained the same: a two-stage behemoth, fed by liquid oxygen and a refined form of kerosene known as 'Rocket Propellant-1' (RP-1). Strapped around its lower stage were four tapering boosters, each 19.6 m long. The upper stage had an upgraded engine which enhanced its thrust from 27,210 kg to 27,573 kg. With the escape tower in place, the upgraded R-7 stood 49.3 m tall, and produced 411,650 kg of thrust at liftoff. This 3 percent increase over the earlier Vostok version enabled it to insert a 6,900 kg payload into a 200 × 450 km orbit. Like Vostok before it, the R-7 was rolled to the launch pad horizontally on a railcar; a method still used today. Four cradling arms, known as the 'tulip', supported the booster and a pair of towering gantries provided pre-launch access.

Originally designed for a crew of three, Soyuz had effectively become a two-person vehicle after the deaths of cosmonauts Georgi Dobrovolski, Vladislav Volkov and Viktor Patsayev in June 1971. During the spacecraft's development, Korolev felt that wearing a pressurised space suit would be just as uncomfortable and impractical as wearing a wetsuit inside a submarine and opted to do away with them. It was a fatal decision, for during the re-entry of Soyuz 11 a pressure valve inadvertently

Atop a descendent of Sergei Korolev's R-7, the crew of Soyuz T-10B is launched towards the Salyut 7 station in February 1984. Note the four tapering boosters, clustered around the central core of the rocket.

opened and Dobrovolski, Volkov and Patsayev died when their air leaked out of the cabin. In the wake of the tragedy, the valves were modified and it was decreed that, in future, space suits would be worn for *all phases* of a mission in which depressurisation was a possibility. In response to this requirement, a Sokol-K ('Space Falcon') suit would be tailored for each cosmonaut and was compatible with the seat liners aboard Soyuz. A prototype was completed within weeks of the disaster and by the spring of 1972 had been fully tested and signed off as flight-ready. Since the Soyuz 12 mission in September 1973, the suit and its descendents have been worn by every cosmonaut during launch, docking, undocking, re-entry and landing. "In the event of decompression," wrote Hall and Shayler, "the [Sokol] is automatically isolated from the cabin environment and supplied directly with either pure oxygen or an oxygen-rich mixture from a supply in the cabin or from self-contained systems." It included a soft helmet which could be pushed back over the head when not in use, a removable, white-topped 'skull-cap' for communications headgear and pressure-sealed gloves. The Sokol could also be used in the emergency transfer of cosmonauts from one spacecraft to another, with the aid of small hoses connected to the spacecraft's life-support system or through a portable backpack, although this has never been done. Testimony to its success is that, since 1971, no other cosmonaut has lost his or her life through the decompression of their spacecraft; indeed, the hardware has proven so reliable that there have been no other instances of depressurisation, *at all*, aboard a Soyuz. However, in order for the suits to be properly accommodated in the confines of the spacecraft, the third crew seat – that of the research cosmonaut or test engineer – was eliminated and its place taken by a system which could automatically pump air into the cabin in the event of decompression. Not until November 1980 and the arrival of an upgraded version of Soyuz would another three-man crew venture aloft.

This upgraded vehicle was known as 'Soyuz-T' ('Soyuz-Transport') and, although it looked much the same as its predecessors, its capabilities were much improved. It could survive, independently, for up to 14 days, *without* the full powering-down of its systems, and could be kept in 'orbital storage' for six full months. Its orbital module could also be left accessible during long-duration missions docked to a Salyut to provide a few extra cubic metres of storage space *and* could be jettisoned *prior* to the de-orbit burn, thereby allowing a ten-percent reduction in propellant to around 250 kg at the end of each mission. This allowed for the inclusion of a third (fully-suited) crew member or two cosmonauts and a hundred kilograms of cargo. As for the descent module itself, this component included improved window covers, capable of being jettisoned after re-entry to allow better views of parachute deployment, and the cosmonauts themselves would benefit from an improved Sokol suit, weighing only eight kilograms, which was lighter and considerably more flexible than earlier models. The escape tower atop the R-7 booster was also modified. It could be jettisoned 123 seconds after liftoff, rather than the previous 160 seconds, and the upgraded solid-fuel rocket meant that in an abort situation the orbital and descent modules would be pulled to a higher altitude, thereby enabling the *main* parachute – rather than the less reliable backup canopy – to deploy and bring the craft safely down to the ground. Six soft-landing rockets in the base of the Soyuz-T descent module replaced four in the previous model and, internally, the 'Chaika' ('Seagull') control systems incorporated a digital computer called 'Argon', cathode ray tube displays and lightweight circuitry. The new instrument module had been designed in a similar manner to that of Progress, with smaller attitude-control thrusters incorporated into the main propulsion system, so that both could draw their nitrogen tetroxide and unsymmetrical dimethyl hydrazine propellants from the same supply. Finally, two wing-like solar arrays spanned 10.6 m (slightly smaller than those of the original Soyuz) and generated a little over half a kilowatt of power. Soyuz-T was actually under development for Salyut 7, whose launch slipped because Salyut 6 was still usable and it made its first piloted flight in the summer of 1980. It is reasonable to assume that original plans for this manned test would have seen it fly sometime after the abandonment of one station and before the launch of the next. With Salyut 6 still available, however, it made sense to send the new mission to it.

As we have seen, Progress closely mirrored Soyuz in design, with the orbital and 'descent' modules altered to house up to 1,300 kg of foodstuffs, water, experiments – including biological payloads which had to be operated within a short period of time – and up to 1,000 kg of propellant. It was not reusable and the entire craft was intended to burn up during re-entry. For ease of removal, the lattice-like framework of cargo racks within the Progress could be unfastened with simple half-turn bolts. A Soyuz-T would normally automatically rendezvous and position itself several hundred metres from a station, whereupon the commander would take manual control and complete a docking. The Progress was capable of docking automatically. Its orbital module carried two television cameras to permit flight controllers on Earth to observe the station as the procedure was executed. The descent module was replaced by a framework with four large tanks to carry several hundred kilograms of propellants for the station. These would be fed into the storage tanks in Salyut 7's aft compartment using an ingenious system of pipes which mated at the exterior of the docking collar. "The refuelling can be conducted either by the crew," explained Clark, "or automatically under the control of the ground. Once Progress has docked at the back of Salyut, the propellant lines and connections are checked for integrity. This done, a compressor slowly reduces the nitrogen pressure in the propellant tanks. The nitrogen is pumped back into its storage bottles ready for the refuelling operation itself. The fuel and oxidiser are transferred at different times for safety reasons." After the fuel had been fed through the system of pipes into Salyut, the oxidiser would be transferred and the connecting lines would be purged with high-pressure nitrogen to prevent contamination or spillage. In practical terms, as time would tell, the Soviets would be able to proudly boast that from the first Progress mission in January 1978, the propellant transfer system never once failed them. In their history of Soyuz, Hall and Shayler speculated that the name 'Progress' may have come from the implication that it suggested significant progress in space station operations, although its precise heritage is unclear.

Since Soyuz-T was capable of supporting up to three cosmonauts, it came as something of a surprise to Western observers when the two-man Soyuz T-5 crew was blasted into orbit at 12:58 pm Moscow Time. Commander Anatoli Nikolayevich Berezovoi was making his first flight, having been born in the town of Enem in Krasnodar Krai, deep in the northern foothills of the Caucasus Mountains, on 11 April 1982. Today, Enem forms part of the post-Soviet Republic of Adygea, enclaved within Krasnodar Krai. Berezovoi came from an ethnic Ukrainian family and after schooling entered the Soviet Air Force and was selected for cosmonaut training in April 1970. He had hoped to make a two-week flight to the Salyut 5 military outpost, alongside Mikhail Lisun, in July 1977, but the mission was cancelled due to delays in completing their Soyuz craft and the resultant impact on the station's dwindling propellant supplies. (In addition to commanding the *first* flight to Salyut 7, Berezovoi would also have commanded the *last* flight, in 1986, had not problems with the space station's controllability ruled this out.) Not only was

Berezovoi overjoyed with his first opportunity to travel into orbit, but praised the craft aboard which he would fly. Soyuz T-5 was, he said, "indisputably more perfect than previous Soyuz space ships". In particular, its computer-based control system was better and the equipment supporting rendezvous and docking were much improved over earlier models. Berezovoi's flight engineer, Valentin Vitalyevich Lebedev, came from Moscow, where he was born on 14 April 1942. After high school in Naro-Fominsk, he studied for a year at the Higher Air Force Navigators School, near Orenburg. However, due to the reduction in the number of personnel in the armed forces, he was discharged and continued his studies at the Moscow Aviation Institute, graduating in 1966. He then worked as an aircraft designer, specialising in structures and materials, before joining the cosmonaut team a month shy of his 30th birthday. However, Lebedev's involvement in space-related matters had actually begun a few years earlier, when he served aboard an expedition to the Indian Ocean and Bombay – today's Mumbai – to support rescue operations for two unmanned Zond lunar missions. He first flew into space aboard Soyuz 13 in December 1973, accumulating eight days in orbit, and had been in line to serve aboard the long-duration Soyuz 35 mission to Salyut 6, launched in April 1980 ... but a trampolining accident caused him to be replaced by his backup, Valeri Ryumin. Between his first and second missions, Lebedev also defended his PhD thesis and in 2000 became the first cosmonaut ever elected to the Russian Academy of Sciences.

Many in the West had expected the arrival of Salyut 7's first resident crew to occur within maybe a week or two of the station's launch ... but not *three weeks*. Phillip Clark later suggested that the longer-than-normal interval might have been due to the focus of Soviet controllers on tracking a British military task force as it made its way into the South Atlantic Ocean, toward an obscure collection of islands which would soon make their presence known, exploding dramatically onto the world stage. Even today, three decades later, the events which precipitated and followed on the Falkland Islands and South Georgia remain a significant stumbling block in the relationship between the British and Argentine governments and, at the time of writing, with recent discoveries of massive oil reserves in the area, tensions remain high.

RETURN TO SPACE

The conflict in the Falklands had reached its zenith by the time that Anatoli Berezovoi and Valentin Lebedev completed a flawless rendezvous and docked their Soyuz T-5 spacecraft onto Salyut 7 at 2:36 pm Moscow Time on 14 May 1982. The first 'principal expedition' to the new station – the maiden *expeditsya osnovnoi* (EO) – was underway. Despite Berezovoi's praise of his ship, the rendezvous hardware proved somewhat sluggish: it failed to lock onto the station on its first attempt and did not 'find' its quarry until Soyuz T-5 had closed to a distance of just 30 km, closing at a rate of 45 m/sec. However, the Igla radar quickly acquired the target and the spacecraft's Argon computer was able to control the final approach and docking.

Certainly, Berezovoi had found Soyuz T-5 a very cold environment, requiring him to wear his space suit in order to keep warm, whilst Lebedev considered it uncomfortable: he tried sleeping in his seat, *above* his seat, and even spread-eagled across both seats, with little success. It must have been a welcome relief to leave the cramped Soyuz and enter the larger volume of the station. Among their first tasks aboard the new station were checking the orientation system, loading cameras with film and 'mothballing' their Soyuz craft, which indicated to Western observers that their residency would a lengthy one. Three days later, on the 17th, they scored a minor 'first', by ejecting a small amateur radio satellite – Iskra-2 ('Spark') – into space through Salyut's airlock. Not only did this represent the first satellite ever released from an orbital station, but it also preceded the first satellite deployment from the Shuttle by almost six months. Iskra came from the student design office at the Sergo Ordzhonikidze Aviation Institute, near Moscow, and was destined to be used by youth groups in eleven countries of the 'Comecon' – the Soviet-led economic assistance organisation in the Eastern Bloc – and the similarity between its orbit (357 × 342 km) and that of Salyut (360 × 343 km) was seen as indicative that the 'deployment' mechanism was probably relatively crude, perhaps based on springs. Iskra-2 weighed 28 kg and followed on the heels of an earlier satellite, flown aboard a Meteor spacecraft in July 1981, which failed to deploy properly. The second Iskra was considerably more successful, utilising its radio repeater for amateur communications experiments and data transmission. In claiming that Iskra was the first satellite to be deployed from a manned spacecraft, however, the Soviets failed to note that Apollo 15 and 16 had each left a particles and fields subsatellite in orbit around the Moon, more than a decade earlier.

Berezovoi and Lebedev's workload was aided by the 'Delta' computer, which assumed many of the routine responsibilities of attitude corrections and navigational tasks, and its presence enabled the men to focus upon their scientific objectives, which primarily encompassed Earth and atmospheric studies, astrophysics, medical and biological research and the conduct of technological experiments. The station's MKF-6M multispectral camera was used for observations of the Krasnodar region, then an integral part of the Soviet Union's winter granary, together with cotton fields and pastures in Central Asia, crops along the fertile banks of the mighty Volga River – together with promising indicators of oil deposits, close to Astrakan – and lead reserves in East Yakutia. Among their other tasks was the Oazis ('Oasis') garden, intended to study the growth of plants in the microgravity environment. Ten days into the mission, the Progress 13 spacecraft was launched and successfully docked at Salyut 7 on the 25th, bringing with it supplies of propellant, water, camera film, clothing and hygiene supplies and scientific equipment. Later that same day, Berezevoi and Lebedev reoriented the station, such that Progress' aft end pointed directly toward Earth, thereby establishing 'gravity gradient' stabilisation. In his diary, Lebedev wrote that the attitude control thrusters were "very noisy" and sounded like "hitting a barrel with a sledgehammer". Propellants were transferred automatically to the station's tanks and on 29 May the cosmonauts began organising the supplies and filling the freighter with unneeded materials. Whenever Lebedev entered Progress, it seemed to 'jingle' with a metallic sound, to such an extent that he

Jean-Loup Chrétien (left) and Patrick Baudry, clad in Sokol space suits, are pictured outside the Soyuz-T simulator during training.

wondered if it would sound like a brass *band* when it finally undocked. A few days later, on 2 June, the freighter's engines were used to slightly lower Salyut 7's orbit to around 300 km to receive the next group of human visitors: the Soyuz T-6 crew. In fact, around 250 kg of the scientific equipment carried aboard Progress 13 was in support of the most unlikely of Soviet guest cosmonauts: a *Frenchman*, Colonel Jean-Loup Jacques Marie Chrétien of the Armée de l'Air, the French Air Force.

Born on 20 August 1938 in La Rochelle, Chrétien's middle names honour those of his parents: his father, Jacques, was a sailor in the French Navy, whilst his mother, Marie-Blanche, was a housewife. As a child, he lived in Brittany and grew up in a country still occupied by the forces of Nazi Germany: an airfield lay just a few kilometres from his home and he remembers vividly watching fighter aircraft taking off and landing. "I was living under a part of the theatre of fighter airplanes of World War II," he told a NASA oral historian. "I had a *permanent* 3D movie under my eyes." These sights, sounds and experiences laid the foundations of a desire to become a pilot and, after assembling model aircraft in his youth, finally learned to fly in 1953 and gained his private licence a year later. Chrétien received his education at L'École Communale à Ploujean, the Collège Saint-Charles à Saint-Brieuc and the Lycée de Morlaix, before entering France's prestigious Air Force Academy – the École de l'Air – at Salon-de-Provence in 1959. Two years later, he received a master's degree in aeronautical engineering. He earned his military pilot's wings in 1962 and served in a fighter squadron in Orange, in the south of France, for the next seven years. During this period, Chrétien flew the Dassault Super-Mystère fighter-bomber and the Mirage III interceptor and, in 1970, entered the French test pilot school, the École du Personnel Navigant d'Essais et de Réception, and served at the Istres Flight Test Centre, working on the Mirage F-1. Chrétien was appointed as deputy commander of the South Air Defence Division in Aix en Provence in 1977 and was selected as an Intercosmos candidate by France's national space agency, the Centre National d'Études Spatiales (CNES), alongside fellow pilot Patrick Baudry, in June 1980.

In fact, Chrétien's interest in space long predated his selection as an Intercosmos candidate. He had been fascinated by the idea of venturing beyond the atmosphere since 1961, when two pivotal events occurred: Yuri Gagarin made his pioneering flight and CNES was established by President Charles de Gaulle. When the Soviet Union offered France the opportunity to fly one of its nationals aboard a Soyuz spacecraft in April 1979, it is hardly surprising that Chrétien applied. The selection process for France's first 'spationaut' began in September 1979 and ultimately 400 applicants were shortlisted for physiological and psychological evaluation. "Slightly tougher tests than those for Europe's first Spacelab crew," wrote *Flight International*, were carried out between December 1979 and February 1980, resulting in the selection of Chrétien and Baudry in June. Their training commenced at the top-secret Zvezdny Gorodok ('Star City') cosmonauts' training centre, on the forested outskirts of Moscow, in September of the following year. Entering the closed realm of Star City, it was strange for the two Frenchmen to be working with old Cold War foes. "Just before the selection," said Chrétien, "I was deeply involved in US-French co-operation ... within the south of France and I was responsible for the organisation of exercises over the Mediterranean Sea between the US Sixth Fleet and the air defence system in the south of France. I was *really* involved and spending a lot of my time on the US aircraft carriers and I was feeling that, if, one day, I was going to fly in space, it would be from *Houston* ... so I was very surprised when they asked me to go to *Moscow*!" Star City provided the two Frenchmen with a profound cultural shock: small apartments, no telephones – they had to use the base's telegraph system to contact family members – and Soviet insistence on using the transportation system, rather than cars. If they *did* ride in a car, it was aboard a chauffeur-driven black Volga, which often ferried them to the Marine Club at the US Embassy ... and the poor driver was obliged to *wait*, sometimes until six o'clock in the morning, to take them back to their apartments. Despite the spartan living situation, Chrétien was impressed by the friendliness of their hosts ... but disappointed at the official position of CNES towards its first spacefarers. Initially, he and Baudry were told that they would remain at Star City for two full years, with only short summer breaks, which *had* to be spent in the Soviet Union. Chrétien balked at this and prompted a week of talks, involving CNES President Hubert Curien, which ultimately decided to allow the Frenchmen home, three or four times per year, "for scientific purposes ... to follow the experiments". Curien's visit also guided the Soviets to agree that Chrétien and Baudry's time, after work, would be their own.

Still, the training was arduous. Theoretical preparation in the classroom, with blackboards and notebooks, was followed in their second year by practical experience in the Soyuz-T simulator, survival training in the Black Sea ... and intensive instruction in Russian, one of the world's most complex languages. The training was given in Russian, the instructors spoke *only* Russian and the two Frenchmen were required to not only become functionally adept with the language itself, but also with the technical aspects of their mission, the Salyut station and the Soyuz-T. In Chrétien's mind, it was the best possible route to getting his head around the new language. It was something he tried to convince his superiors in Houston, years later, to do: getting people speaking Russian, as *soon* as possible in the training flow, so that interpreters would not be necessary. "What good are interpreters?" he asked. "Most of the time, they cannot be real technicians. I've been following the courses for NASA many times at Star City, taking a group of astronaut candidates. The very first is a waste of time, because *half* of the time is spent in *translation*, plus *thinking*. The guy who translates has to think, so he takes more than half of the time, by definition. At the end, only 30-35 percent of the 'useful' time is dedicated to the candidate and there are *lots* of mistakes during translation, just because the guy does his best, but he doesn't understand all of it." Learning Russian, as soon as possible, was also one of Chrétien's recommendations for future International Space Station crews; he felt that it doubled, or even tripled, their efficiency. He and Baudry had spent six months before going to Star City, working on the language, twelve hours per day, including Saturdays, although by his own admission it provided them with "the good luggage", but not the fluency. By the end of their first year, in late 1981, they had developed "a good technical package" and they had drawn closer to full fluency in the language, enabling them to integrate effectively with their crews. During their first year, the question of who would fly – Chrétien or Baudry – remained open, which placed two good friends in the decidedly unpleasant situation of being in constant competition. "It's kind of a poker game," said Chrétien, "and one of us will be identified, but we don't think it depends so much on us. It mostly depends on the people and they had probably had already their own choice in mind for different reasons." One of his recommendations was *not* to place the two finalists in such an unfair position of competition, which he felt impaired their ability to train effectively together.

By the end of 1981, two teams were thus training in tandem for the French flight: Chrétien would be joined on the prime crew by commander Yuri Malyshev and flight engineer Alexander Ivanchenkov, both of whom had flown in space before. In fact, Malyshev had commanded the first manned flight of the Soyuz-T spacecraft in June 1980, whilst Ivanchenkov had participated in a record-setting 139-day mission to the Salyut 6 station in the summer of 1978. The backup crew would consist of Baudry, teamed with veteran cosmonaut Leonid Kizim and rookie flight engineer Vladimir Solovyov. Shortly before the mission, however, it would appear that some sort of 'personality clash' occurred between Malyshev and Chrétien and the Soviet commander was dropped and replaced by another experienced cosmonaut, Vladimir Alexandrovich Dzhanibekov. (Rex Hall and Dave Shayler told a different account, however, noting that Malyshev's removal was the result of "failing a medical

Two commanders – Vladimir Dzhanibekov (left) and Anatoli Berezovoi – pictured inside Salyut 7 during the Franco-Soviet Soyuz T-6 mission. Both men would play a pivotal role in the new space station's fortunes: Berezovoi would open it up for business in May 1982 as commander of its first expedition, whilst Dzhanibekov would lead the mission to raise it from the dead in June 1985.

examination".) The replacement commander, Dzhanibekov, was born on 13 May 1942 in the remote area of Iskandar, in the region of Tashkent, today's capital of independent Uzbekistan. His father was a fireman, his mother a nurse. Of his parents, he would later comment with admiration that Alexander Dzhanibekov was "ready to deploy, 24/7, in any weather conditions and emergency situation to help people" and that his mother was "a dear, taking care of my younger brother and me". His birth surname was 'Krysin', but he appears to have changed it later in life to honour his wife's family, who were descendants of Jani Beg (Russianised to 'Dzhanibek'), a 14th-century khan of the Golden Horde. At the age of 18, Dzhanibekov entered Leningrad University to study physics and, whilst there, developed a love of aviation. The following year, he entered a higher military fighter pilot school at Yeisk and on graduating in 1965 he became a Soviet Air Force instructor. In an interview, decades later, he noted that he could never match his friends in sports or gymnastics and feared being considered to be "the weak son of a strong father". One problem for Dzhanibekov in the cosmonaut selection process was his weight; "10 kg too much! I first went to hospital, where therapists helped me reduce. For a week, I consumed only *water*...and was able to down my weight to 75 kg from 85." He was one of nine pilots picked as cosmonauts in April 1970.

In his early years, Dzhanibekov worked on the joint US-Soviet Apollo-Soyuz Test Project, commanding the backup crew for the Soyuz 16 dress rehearsal flight with engineer Boris Andreyev. According to Mark Wade on his website, Dzhanibekov was then assigned to fly with engineer Pyotr Kolodin on the Soyuz 26 visiting mission in November 1977. Their primary objective would have been to perform a spacecraft swap, flying home in Kovalyonok and Ryumin's Soyuz and leaving their own 'fresh' craft behind. This mission was deleted when a mishap with Soyuz 25 led to new Soviet guidelines to include at least one veteran cosmonaut aboard each crew; Kolodin was replaced by Oleg Makarov and he and Dzhanibekov entered space aboard Soyuz 27 in January 1978. Command of a second crew, with a Mongolian cosmonaut, followed in March 1981 and Dzhanibekov might next have aspired to a long-duration flight aboard Salyut 7. In fact, he was in training, alongside fellow cosmonaut Alexander Alexandrov, for such a flight when he learned of his reassignment to command Soyuz T-6. This was certainly done within the last few months before flight, for Dzhanibekov's place on the Salyut 7 long-duration crew was taken by fellow cosmonaut Vladimir Lyakhov in January 1982.

By this time, long-duration flight was in the blood of the Soyuz T-6 flight engineer, Alexander Sergeyevich Ivanchenkov. He came from the town of Ivanteyevka, on the Ucha River, a couple of dozen kilometres north-east of Moscow, where he was born on 28 September 1940. A keen sportsman, traveller and guitarist, academia suited Ivanchenkov well and he departed middle school with a gold medal, then entered the Moscow Aviation Institute and graduated in 1964. He spent some years thereafter working for Sergei Korolev's OKB-1 design bureau, "engaged in designing space technology products", according to Mark Wade, then transferred to the test flight division, under Sergei Anokhin, in 1971. Two years later, he was selected as one of four civilian cosmonaut engineers and in 1974 was paired with Yuri Romanenko on the support crew for the Apollo-Soyuz rehearsal mission, Soyuz 16, and also for the joint flight itself, Soyuz 19. He remained with Romanenko to back up Soyuz 25 in October 1977. By the logic of the assignment process, these two would have gone on to fly a long-duration Salyut mission together, but this was prevented by the decision at the end of 1977 that every crew must include a veteran cosmonaut. The team was broken and Ivanchenkov found himself reassigned, with veteran commander Vladimir Kovalyonok, to back up *both* Soyuz 26 and 27. With so many reserve stints under his belt, it must have been with a sense of euphoria that Ivanchenkov finally set off in June 1978, aboard Soyuz 29, on a mission of his own. He and Kovalyonok set an empirical record of more than four and a half months in orbit.

In his history of the early Soviet manned space programme, Phillip Clark noted that it was possible to determine which crews would consist of two members and which would consist of three members, judging from adjustments to Salyut 7's orbital altitude. "When a three-manned crew was to be launched," he wrote, "Salyut was operating in an orbit some 35-45 km lower than when a two-manned crew was scheduled. Therefore, as a landing window was approaching – implying the impending launch of a 1-2 week visiting mission – the altitude of Salyut 7 revealed the size of the crew to be launched." Clark explained that calculations of minimum Soyuz propellant requirements for the differing orbits, ranging from 110 kg to 175 kg, closely paralleled the mass of a cosmonaut. "Therefore," he concluded, "in order to maintain an approximately consistent Soyuz-T launch mass, the extra weight of a third cosmonaut was compensated for by a reduction in the propellant load, thus reducing the altitude which the Soyuz-T could reach."

As Dzhanibekov, Ivanchenkov and Chrétien prepared for their voyage, the research activity aboard Salyut 7 had moved into a routine, with Berezovoi and Lebedev working on a variety of medical, astrophysical and technical studies. They conducted 'echography' experiments to monitor the dynamics of their cardiovascular systems and examined changes in their physical posture and balance mechanisms in the new environment of low-Earth orbit. Meanwhile, other instruments observed celestial targets, as well as terrestrial and atmospheric ones, and the first materials processing experiments got underway. At 7:29 pm Moscow Time on 24 June 1982, Soyuz T-6 rose perfectly into the darkened Tyuratam sky, the brilliance of its rocket exhaust plume illuminating the barren steppe for many kilometres as Chrétien became only the third 'Westerner' – after the United States and Cuba – to travel into space. Unfortunately for the Frenchman, his watch was not properly set to the correct time, "so I was expecting *every* event at the *wrong* time", with everything seeming to occur a minute *earlier* than it should have done. After the tremendous buffeting and vibration of the first stage, a peculiar sense of calm and ethereal quietness elapsed for a few seconds, before the second stage roared to life and continued the boost into space. At first, *that* boost, and his incorrect watch, convinced him that something had gone wrong ... but as he glanced across the cabin to look at Dzhanibekov and Ivanchenkov, they both seemed unperturbed. "*Then* I understood that it was wrong with my watch," he recalled, "but I could not ... recycle [it during] the ascent, so everything happened in disorder." Around four minutes into the climb, the escape tower was jettisoned, as planned, and the cosmonauts – and France's first spationaut – were granted their first glimpse of the Home Planet, far below.

Pencils and pens, books and checklists, began to float comically before their eyes ... and, through the window, the curvature of Earth and the darkness of space were cut by the ribbon-like phosphorescent line of the horizon. To Chrétien, two years of difficult training, both physically and mentally exhausting, and culturally shocking, was suddenly and instantaneously *worth it*. With a 25-hour rendezvous profile ahead of them to reach Salyut 7, it was imperative to get some rest. This was easier said than done. In the cramped confines of the Soyuz, Chrétien managed perhaps six hours of light sleep on his first night in orbit, but the *strangeness* of their surroundings got unnerving at times: the hum of the fans, the mixed signals entering eyes and ears and brain and the realisation that *everything* was weightless, including their Sokol space suits, which floated like three additional crew members in the orbital module. The beauty of Earth was captivating, though they were all keenly aware of the ever-present influence of the actions of their fellow men: deep within the South Atlantic, Chrétien easily spotted a huge black stain of smoke, presumably from one of the sunken vessels, in the Falklands. It was a disturbing reminder of humanity's often destructive impact on the very cradle which had given the species life. Unlike many space travellers who preceded and followed him, the Frenchman did not suffer any instance of space sickness, although his appetite was noticeably reduced during the first day, and he expressed mild annoyance at his two veteran crewmates, who seemed perfectly adapted and ready to gobble down large meals from Soyuz' pantry. The meal which *really* got Chrétien back to normal was the one which they shared with Berezovoi and Lebedev. Docking provided some additional drama, when Soyuz T-6's automatic rendezvous system failed, forcing Dzhanibekov to assume manual control. The computer correctly performed a braking manoeuvre at a distance of 900 m, then reoriented the craft to position its docking mechanism in the proper position ... but quickly sensed that the gyros were approaching 'gimbal lock' and halted the attempt. Soyuz T-6 was sent into an end-over-end spin and Dzhanibekov took over. From inside the station, Berezovoi and Lebedev saw the craft "just sitting there", 200 m away, with the occasional wedge-shaped spurt from an attitude thruster as Dzhanibekov struggled to determine his position in relation to Salyut. At length, after receiving permission for the manual docking from Mission Control, routed through a tracking ship in the Strait of Gibraltar, he brought the two space vehicles safely into a metallic embrace at 10:46 pm on 25 June. Three hours later, the new arrivals floated through Salyut's hatch into the warm expanse of the space station and the welcome bear hugs of the two resident cosmonauts. For the first time in history, *five* cosmonauts were together in space.

Berezovoi and Lebedev had readied a hearty Russian dinner and Dzhanibekov and Ivanchenkov – whom Chrétien good-naturedly referred to as "the two monsters", in view of their voracious appetites – wolfed down their second large meal of the day. Almost immediately, however, the work began on a complex programme of French experiments, which encompassed life sciences, astronomy and materials research. Studies were undertaken into the blood vessels of the brain, the acuity and depth of vision in space, the effects of cosmic radiation on biological specimens and the threshold of colour sensitivity, whilst observations of the night sky and investigations in infrared astronomy were made. The Frenchman employed much of the equipment aboard Salyut 7, including the Kristall ('Crystal') furnace, and components ferried aloft by Progress 13, to support a series of materials experiments, focusing on temperature variations, the influence of capillary forces on the formation of aluminium and indium alloys and the smelting and cooling of new metals.

Returning to Earth on 2 July, after eight days in space, was almost delayed by Dzhanibekov, the commander, himself. "He was late to get on his seat," in the Soyuz, noted Chrétien, "because he was doing some stuff in the orbital [module] and he was still in his underwear when we were already sitting in our suits on our seats." Every few minutes, the Frenchman would summon him and a flustered Ivanchenkov would alert him that, if the hatch was not closed soon, they would miss their undocking window, miss their re-entry window, be forced to wave off the landing attempt ... "and they *won't* be very happy down there" in Mission Control. At length, Dzhanibekov donned his suit and positioned himself in the central commander's seat. Re-entry felt like being in the midst of a ferocious storm, said Chrétien, and brought with it a sense of helplessness which was palpable. For a veteran test pilot, it was worrisome that he felt he had absolutely no control over the enormous events going on beyond the Soyuz' thin walls. With the worst of the re-entry heating behind them, their spacecraft descended, meteor-like, into Soviet territorial airspace and, at length, Chrétien could actually *see* their landing spot. Through his tiny window, he beheld the desolate steppes of Kazakhstan, across which scurried hundreds of military personnel, dozens of trucks, a handful of helicopters ... and the tractors and horses of intrigued locals, awaiting the latest visitors from the heavens. Touchdown came in the gathering gloom of the late summer's afternoon, at 5:21 pm Moscow Time, bringing to a close the first Intercosmos mission to feature a participant from a Western-aligned, non-communist nation.

Yet the flight of a Frenchman aboard a Soviet spacecraft, whilst cementing a measure of harmony in the heavens, did little to convey goodwill on the ground. More than two years earlier, in December 1979, tanks bearing the menacing crimson and yellow of the Hammer and Sickle had rumbled into Kabul, kicking off a vicious, decade-long war of attrition, which met with enormous international condemnation. Cast against this ugly backdrop of communist aggression, France had the misfortune of becoming the first Intercosmos nation to fly with the Soviets in the wake of the invasion of Afghanistan ... and her close ties with the United States and, more broadly, her intimate involvement with NATO, meant that relations were extremely strained in the early 1980s. In fact, for Chrétien and Baudry, there was no official welcoming ceremony upon their return to French soil. "They had told the press that we were *not coming* that day," Chrétien recalled, "so we arrived like tourists and the only people welcoming us at the airport in Paris ... were the people from the Soviet Embassy!" For the nation's first spationaut, it was fine, because he could go on vacation and he was free from demands for speaking engagements and public relations visits. Nevertheless, he said, it was strange that the very people who had decided to send him into space were now keen to ignore him and his *Premier Vol Habité* ('First Manned Flight'). It must also have been saddening that such a historic episode in France's long history should have been so neglected, on the basis of disagreements over political ideology and foreign policy. By the time Chrétien next flew into space, again with the Soviets, in late 1988, relations had warmed between the two nations. The stance of Mikhail Gorbachev, who became General Secretary of the Communist Party in 1985 and promulgated a new attitude of openness and political restructuring, placed the Soviet Union on a quite different diplomatic footing with the West than had been experienced during the tense days of his predecessors, Yuri Andropov and Konstantin Chernenko.

However, in 1982, the old Soviet propagandist desire to surpass the United States had returned drearily to the fore. Years earlier, under Nikita Khrushchev, the first man, first woman, first spacewalker and first multi-crew mission had illustrated that many of the Soviet Union's policies in the heavens were dictated by one overarching consideration: to be *first*. That had changed somewhat in the late 1970s, when America's Shuttle programme was years away from its first launch and the Soviets could rest safe in the knowledge that their dominance in human space flight was, for a time, unassailable. With the arrival of the Shuttle in April 1981, however, it became evident that the United States was back in the game, with plans to fly joint missions with Western European nations *and* fly their own woman into space. *Six* women, in fact, had been chosen by NASA in January 1978, and far from simply being factory workers, like Valentina Tereshkova, who had flown Vostok 6 in June 1963 as a political stunt, the NASA women were scientists, engineers and doctors. With tensions rising high yet again, something had to be done to restore the sense that the Soviets were ahead in the game. The timing of the agreement to fly a Frenchman was no accident: it was signed as the SALT II talks in Vienna between Jimmy Carter and Leonid Brezhnev faltered, only months before the Soviet invasion of Afghanistan, and it occurred at one of the most difficult junctures in the East-West relationship. The Soviets knew that America planned to fly a Western European astronaut aboard the Shuttle and Chrétien's mission neatly beat them to it. Then, in April 1982, NASA announced its intention to fly its first female astronaut, Sally Ride, aboard Shuttle mission STS-7 in the early summer of the following year. As if by magic, it seemed, a *female* cosmonaut, Svetlana Yevgenyevna Savitskaya, appeared on the crew roster for Soyuz T-7, scheduled for August 1982. As circumstances would transpire, Savitskaya's career would accomplish two space firsts – and come close to securing a third: in July 1984, she would be the first woman to perform a spacewalk *and* the first to undertake two space missions and, but for a quirk of fate, might have commanded the first all-female crew of cosmonauts in the autumn of 1986.

Savitskaya was no Tereshkova. She had not been chosen as a cosmonaut on the whim of a boorish General Secretary. She was not a factory worker, nor was she uneducated or inexperienced. Rather, she was a veteran test pilot, an accomplished parachutist, a former member of the Soviet Union's National Aerobatics Team – for which she had earned the accolade of World Champion in 1970 – and had established records in supersonic and turboprop aircraft. (Doubtless it also did not hurt that she was a staunch, unbending and steely member of the Communist Party and, at the time of writing, in 2011, currently serves in the Russian State Duma.) At the time of her most recent election in 2007, one Western journalist who met Savitskaya referred to her as exhibiting a "deadly serious, cosmetically unretouched image … what one political analyst calls an 'imitation of a monument to the peasant worker'." She was born in Moscow on 8 August 1948, the daughter of Yevgeni Yakovlevich Savitsky, a veteran flying ace from the Great Patriotic War, twice decorated as a Hero of the Soviet Union and one-time commanding general of aviation for the Soviet Union's Air Defence Forces. In her late teens, Savitskaya entered the Moscow Aviation Institute and later attracted international attention at the Sixth FAI World Aerobatic Championship, held in the United Kingdom, as a member of the Soviet National Aerobatics Team. She won first place and was nicknamed 'Miss Sensation' in the British press. After graduation from the aviation institute in 1972, she pursued a career as pilot, flying 20 different types of aircraft, including a record as the first women to attain 2,683 km/h in a MiG-21. Her fascination with aviation also extended to parachuting and, as a girl, she had hidden these activities from her father and was only discovered when he found a parachute

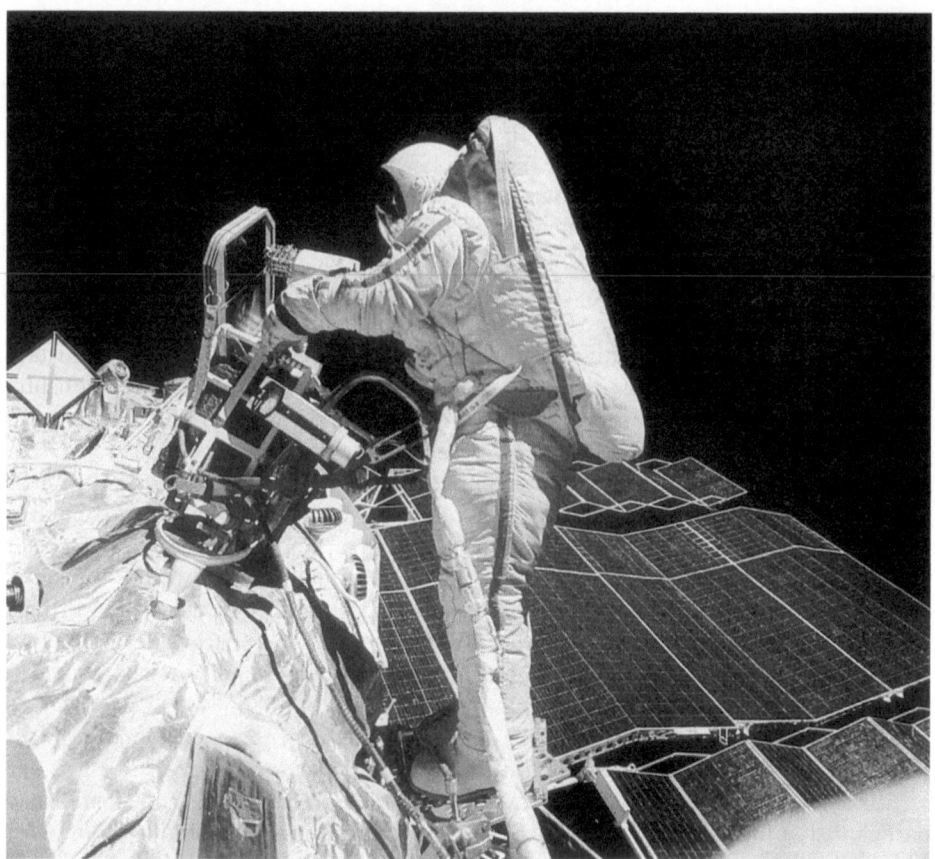

During her career as a cosmonaut, Svetlana Savitskaya would become the first woman to complete two missions and – here, on Soyuz T-12 in July 1984 – would become the first woman to perform an EVA. Had history played out a little differently, she might also have been the first woman to command a space mission, more than a decade before Eileen Collins secured that achievement.

knife in her school bag. By the age of 17, she had already logged 450 jumps and in 1965 she jumped from 14,252 m, opening her chute a mere 500 m above the ground. In 1968, she soloed in a Yak-18 training aircraft. With such impressive credentials, it is perhaps little surprising that Savitskaya made the cut in the July 1980 selection of nine female cosmonauts. She was the only test pilot in the group; her group also included four physicians (Galina Amelkina, Yelena Dobrokvashina, Larisa Pozharskaya and Tamara Zakharova), three engineers (Yekaterina Ivanova, Natalya Kuleshova and Irina Pronina) and a physicist (Irina Latysheva). Seven of the women, including Savitskaya, completed training in February 1982, whilst the others (Dobrokvashina and Latysheva) graduated in July 1984. The announcement that Soviet women cosmonauts were back in training was actually made during Jean-Loup Chrétien's mission, although it was met with great cynicism in the West.

When she was named as a member of the Soyuz T-7 crew, joining commander Leonid Popov and flight engineer Alexander Serebrov, Savitskaya was under no illusion as to the importance of her role ... not just as a *woman*, but as a competent *cosmonaut*. "Americans were preparing to send women into space," she told an interviewer, years later. "I was called to our Ministry. The question was: *What about us?* Would *we* be able to do it? Was there enough time? I said 'Why not?' We had to keep our priority positions, where possible." Still, even within the predominantly male cosmonaut corps, there was sexual discrimination. In early 1984, whilst preparing to conduct a spacewalk on her second mission, Savitskaya was asked *why* women needed to perform welding and other tasks in orbit – surely, they might *burn* each other's suits or even the exterior of the space station, it was argued – but she remained firm and confident in her own capabilities. "After *my* space flight, *everyone* had to shut up!"

Her work ethic clearly originated from her father and mother, as did her staunch support for the Communist Party ... and ingrained distrust of the motives of the West. Even years later, when talking to *Baltimore Sun* journalist Clara Germani, she insisted on being interviewed in a darkened room, in overcoats and boots, and recorded on *two* tape recorders, "so she can be certain that her words are not twisted". Germani described Savitskaya as exuding a strong, patriotic image of the old regime and she certainly spoke for many Russians in her desire to abandon the failed free market economy of the post-Soviet era and return to the predictable stability of the past. She told Germani that her parents (her mother was a Moscow Communist Party leader) would suffer "a second death" if they had lived long enough to see modern Russia. To Savitskaya, the collapse of the Eastern Bloc was perhaps the greatest calamity to befall the state – and the most unforgivable – and she saw it as a betrayal of her parents' generation, who had fought and bled in the Great Patriotic War, endured unimaginable hardship, had been first into work and first into battle and had sidelined their personal lives to take on responsibilities to the party and to the nation. (Left to one side were the violent purges of Stalin, catastrophic economic policies of Khrushchev and Brezhnev and a string of failed Five-Year Plans.) On the other hand, her words are understandable, for the Russians had endured perhaps more than any other nation the brutality and wholesale destruction of their country at the hands of Nazi Germany. "It's difficult for Americans to understand," she told Germani, "because no other country saw such ruin as this one after the war. And Communists managed to build the country to conquer the virgin lands and we were the first to reach outer space."

It was too early for democracy of any kind on 19 August 1982, when Soyuz T-7 speared for the heavens at 8:12 pm Moscow Time, kicking off the second Salyut 7 'visiting expedition', or *expeditsya poseshchenya* (EP). Leonid Brezhnev had three months left to live and two more uncompromising General Secretaries – Andropov and Chernenko – would assume power, before Gorbachev arrived to begin his efforts of reform. In a sense, Soyuz T-7 was one of the last Soviet attempts to secure a propaganda 'first'. Seated alongside Sovitskaya were Leonid Ivanovich Popov and Alexander Alexandrovich Serebrov. The former came from Oleksandria in the central portion of today's Ukraine, where he was born on 31 August 1945. Popov graduated from the Chernigov Higher Air Force School with an electrical engineering degree in 1968 and was selected as one of nine cosmonauts in March 1970. After obtaining further credentials from the Gagarin Military Academy, he served on the support crew for Valeri Bykovsky's Soyuz 22 mission, before being paired with Valentin Lebedev on the backup team for Soyuz 32. Popov would have completed each of his missions, and spent a grand total of some 200 days in space, before his 37th birthday. He made his first flight as commander of the 185-day Soyuz 35 mission to Salyut 6 in 1980 and also led Soyuz 40 into orbit in May of the following year. His flight engineer, the gregarious Alexander Serebrov, was making his first mission on Soyuz T-7. In his book *Dragonfly*, Bryan Burrough described Serebrov as being "garrulous" and "headline-loving", with a penchant for fast cars – quite the opposite, it seems, of Savitskaya. Like his female colleague, Serebrov was born in Moscow, on 15 February 1944, and graduated from the capital's Institute of Physics and Technology. He was chosen as one of seven civilian cosmonaut engineers in December 1978.

For Savitskaya, the launch was one of the highlights of her first mission: the grumbling, groaning and shuddering of the rocket, she later told a journalist, left her in no doubt of the enormity of the events occurring outside ... although the expected colossal overloads did not materialise. Maybe it was her steely feminism talking, but she was convinced that physically healthy and properly trained individuals could withstand the stresses of launch and ascent to orbit, without the slightest inconvenience. Like Jean-Loup Chrétien before her, she anticipated weightlessness and tested it, firstly by releasing a pencil from her logbook, and watched it float in front of her eyes. Unlike the Frenchman, she slept fitfully. It was already late into the evening by the time orbital insertion had been achieved and, in words possibly chosen to make Soyuz appear 'larger' than it actually was, she later commented that "to prevent myself from sailing into another compartment ... I tied myself with a belt". A day after launch, at 9:32 pm Moscow Time on the 20th, Leonid Popov guided his ship to a smooth docking with Salyut 7 ... and Savitskaya was met by bear hugs and a lighthearted, though undeniable, example of Soviet male chauvinism: Valentin Lebedev insisted that she get straight to work on the housekeeping duties! The steely female pilot assured him that she would do no such thing. "Housekeeping chores," she said, firmly, "are the responsibility of the *host* cosmonauts!"

Experiments *were* undertaken over the next few days, including use of the French-supplied equipment to examine Savitskaya's adaptation to the microgravity environment, and the Tavria investigation into the separation of biological mixtures, but the primary objectives of Soyuz T-7 were twofold: to score a triumph over the United States, by flying another woman into space, 'proving', yet again, the superiority of the beautiful socialist state, and to replace Berezovoi and Lebedev's Soyuz T-5 craft with a fresh vehicle. Popov's team undocked on 27 August and were back on the ground, in central Kazakhstan, at 6:04 pm Moscow Time. Two days later, Berezovoi and Lebedev boarded Soyuz T-7, separated from Salyut 7's aft port, manoeuvred around the station and redocked onto the front port. This opened the way for a new Progress resupply freighter in mid-September – the third of their expedition, having already hosted one in May and another in July – which brought much-needed supplies. The two resident cosmonauts would henceforth be alone for the remainder of their 211-day mission, which, in the words of Phillip Clark, posed a few interpersonal difficulties. He noted that the two men had "verbally clashed", early in their mission, and had "hardly spoken to one another for four months". Little more detail ever emerged from behind the Iron Curtain, but it certainly offered a good deal of food for thought for the psychological training of future long-duration crews.

By September, Berezovoi and Lebedev were well into their fourth month aloft, but they had been far from unoccupied. A lengthy EVA had been performed on 30 July, in which the cosmonauts spent two hours and 33 minutes practicing techniques for the assembly of large structures in space. "After we got up at 10 pm, we ate breakfast and did medical tests," Lebedev wrote in his diary. "My blood pressure was 106/86 and my pulse was 100; the result of a sleepless night. Then we reoriented the station, deactivated the gyroscopes ... and switched it over to the Kaskad ('Cascade') automatic stabilisation control system. After that, our station was in a fixed position, pointing toward the stars. We put on our undergarment accessories and prepared [Soyuz T-5] for an emergency departure ... We deactivated the station, closed the hatch between the station and the transport vehicle as well as the hatch between the [transfer] and working compartments. We put on our suits. Ground Control told us to be completely dressed by 3:50 am..." In fact, the EVA began at 2:39 am Moscow Time, when Lebedev exited Salyut's forward transfer compartment and quickly anchored himself close to a package of materials exposure experiments. With Berezovoi assisting him, Lebedev replaced canisters of the Medusa cosmic radiation study, then removed part of a micrometeoroid detector. Berezovoi's first impression as he moved from the transfer compartment into the full glare of orbital sunlight was that it was like "being on a street on a bright, sunny day, with the ground covered in pure white snow".

The main focus, though, was on construction methods, with plans to augment Salyut 7's solar arrays on a subsequent mission. Lebedev worked with the Pamyat ('Memory') investigation into the "thermo-mechanical joining" of pipes and girders, checked the reliability of bolts and other threaded connectors, tested a new wrench and made observations of how materials behaved under duress. His only concern was that the Orlan-D ('Sea Eagle') space suit's cooling system made his feet cold. In addition to the spacewalk, the cosmonauts continued with their activities inside Salyut 7. Plants were grown, the Oasis garden was tended – producing peas, oats and onions – and the effects of cosmic radiation were traced in flax seeds. Astrophysical research was also undertaken with the Yelena gamma ray detector and the Astra-1 spectrometer. The departure of Progress 15 in mid-October prompted some speculation in the West that a *third* visiting crew might arrive, but this proved groundless, particularly since Soyuz T-7 had been in orbit for barely two months and did not require replacement. Visitors *did* arrive in early November, though, in the form of Progress 16, which carried among its cargo the Iskra-3 satellite, virtually identical to its predecessor, launched earlier in the year. Berezovoi and Lebedev ejected it from Salyut's airlock on 18 November. By this time, they had eclipsed the

Soviet space endurance record of 185 days and air sports guidelines, enshrined by the Fédération Aéronautique Internationale (FAI), only recognised a 'new' record if it broke the 'old' record by at least a ten-percent margin; Berezovoi and Lebedev would need to remain in orbit until at least 24 November in order to secure it.

In fact, many observers in the West had already predicted the cosmonauts' return to Earth to occur in late December, due to the predictable cycle of 'landing windows'. As early as September 1982, the Soviets announced that the mission would be extended by two months, producing a landing around Christmas week, although Berezovoi fell ill in early November and there was talk of an early return. Fortunately, he made a good recovery and the marathon flight continued. An understanding of the landing requirements for Soyuz missions had enabled early space historians to make generally accurate predictors for the estimated return of cosmonauts to Earth ... and also generally accurate assessments of which flights had been curtailed, due to mechanical or other difficulties. "A successful mission," wrote Phillip Clark, "is ideally recovered before sunset and not significantly earlier than three hours before sunset." Since the prime recovery zone was in Kazakhstan, Soyuz crews descended back to Earth on a north-easterly track across Africa, preparing for their de-orbit burn in daylight, performing the manoeuvre over the South Atlantic and landing in Soviet Central Asia in the late afternoon. Additionally, the Soviets attempted, wherever possible, to impose a pair of primary criteria for each landing: that it should not occur in the hours of darkness and that the retrofire procedure should happen in sunlight, in order to permit manual orientation of the spacecraft. This latter condition, continued Clark, "means that the lower limit on the landing time is not strictly speaking three hours before sunset, but the approximation ... to define landing opportunities will not be greatly in error." Moreover, the new Soyuz-T spacecraft could typically remain aloft for a maximum of around four months. With these factors in mind, Clark and others had been able to reliably predict, with an error margin of a few days, when missions were typically scheduled to begin and end. It came as something of a surprise when Berezovoi and Lebedev landed at 10:03 pm Moscow Time on 10 December, completing a record-setting mission of 211 days in space. Rex Hall and Dave Shayler, and Phillip Clark, have suggested that seven and a half months – maybe 225 days or more – was the original expectation. Clark suggested that the cosmonauts' taking of "air samples" from the station's atmosphere in their final few days may lend credence to some sort of atmospheric contamination which prompted an early return home, although Hall and Shayler highlighted the continuing problems (and eventual failure) of Salyut 7's Delta navigation computer.

The earlier than scheduled return to Earth meant that Berezovoi and Lebedev touched down in highly undesirable conditions: they landed in the hours of darkness, in the thick of a midwinter blizzard, with weather forecasts predicting 21 km/h winds, temperatures of -9 degrees Celsius and visibility of no more than 10 km. High winds dragged the Soyuz T-7 descent module over a small incline, causing it to roll down a hill, exacerbating the discomfort of the cosmonauts, who had already been unable to properly undertake their normal pre-landing medical and physical conditioning. "Despite being strapped into their seats," wrote Hall and Shayler,

"Lebedev found himself on top of Berezovoi when the capsule finally stopped." One of the rescue helicopter pilots spotted the Soyuz' flashing beacon and attempted to land on a dry river bed – breaking a wheel and support strut in the process – and it was eventually decided to resort to ground vehicles to pick up the cosmonauts. Mission Control advised the men to remain inside the Soyuz and after 40 minutes the first medical teams arrived at the landing site. A target of 225 days had been missed by a mere fortnight, but it mattered little, for as 1982 faded into history the Soviets remained securely in the lead, not only in terms of long-duration flight, but also in terms of a fair amount of socialist propaganda. They had 'deployed' satellites, ahead of the Americans, they had flown a Western European national, ahead of the Americans, and they had flown a woman, ahead of the Americans. All of these achievements were regarded in the West, rightly, with a healthy helping of cynicism, for they portrayed the Soviet Union in a light which reminded many people of the early 1960s. The Cold War was in full swing, with nuclear tensions reinvigorated ... and *that* caused concern to many on both sides of the Iron Curtain.

ONE PLANET, ONE HOME

In time, Vladimir Georgyevich Titov would spend a full year of his life circling Earth on a single mission, yet to him the notion of being a cosmonaut was 'just' his job ... and it was a job that he loved with every fibre of his being. Even as he grew older, commanding a 366-day space station mission and flying the Shuttle twice with American crewmates, he would yearn for more. It was an enthusiasm that he would pass on to his son, although not only a love for space, but a love for the Home Planet, too. "He doesn't understand about border," Titov told a NASA oral historian in the summer of 1998. "For him, it's the same: America or Russia. He doesn't want to understand ... different countries ... because, for him, it's his house here, house there. It's the same planet, same home." In 1983, Titov was granted the uncommon opportunity to view that planet from space, not once, but *twice*. Both opportunities would meet with intense disappointment: on the first occasion, in the late spring, he and fellow cosmonauts Gennadi Strekalov and Alexander Serebrov would fail to dock with the Salyut 7 station and, on the second, in the autumn, again teamed with Strekalov, Titov would come within a hair's breadth of death when their rocket exploded on the launch pad.

Titov came from Sretensk, born on 1 January 1947 in the Zabaykalsky Krai region of southern Siberia, which shares lengthy borders with both China and Mongolia. After completing his secondary education, he entered the Higher Air Force College at Chernihiv in the Ukraine in 1970 and remained for several years at the college, serving as a pilot-instructor, responsible for up to a dozen students. Titov later worked as a flight commander with the air regiment, flew ten different types of aircraft – including the MiG-21 supersonic jet fighter – and accrued 1,400 hours in his logbook and the qualifications of Military Pilot First Class and Test Pilot Third Class. He was selected as a cosmonaut in August 1976 and, five years later, was teamed with veteran flight engineer Gennadi Strekalov for training for a long-duration Salyut residency. The pair backed up Berezovoi and Lebedev on Soyuz T-5, before being recycled, along with Alexander Serebrov, to fly aboard Soyuz T-8. During this period, Titov and Strekalov underwent intensive EVA training to install new solar cells aboard Salyut 7, to augment the station's own arrays, which had suffered rapid degradation in their ability to produce electrical power, soon after launch. Much of this degradation was caused by exposure to ultraviolet radiation and atomic oxygen in low-Earth orbit. Strekalov already had some pedigree when it came to conducting repair and maintenance work in space, having flown his first mission in November 1980 aboard Soyuz T-3, which successfully upgraded Salyut 6's thermal control and electrical systems, in support of future crews. Gennadi Mikhailovich Strekalov was born in Mytishchi, a major industrial hub, situated to the north-east of Moscow, on 26 October 1940, and judging from the nature of his birthplace it is perhaps not surprising that he forged a career for himself in science and engineering. His father was killed in 1945, during the Red Army's liberation of Poland, only weeks before the end of the bloody conflict with Nazi Germany. The young Strekalov completed his schooling and became an apprentice coppersmith, before enrolling at the prestigious Bauman Moscow Higher Technical School. He received his engineer's diploma in 1965 and moved directly to work for the organisation which evolved from OKB-1 into TsKBEM and eventually Energia, helping with the design of Soyuz. Strekalov was chosen as a civilian cosmonaut in March 1973 and within months began formal training. His first crew assignment was as backup flight engineer on the Soyuz 22 mission in September 1976. Two years later, he recommenced flight training and at the end of 1979 was teamed with Vasili Lazarev for Soyuz T-3. Six months later, following changes to the mission, he was reassigned as backup flight engineer to Konstantin Feoktistov. When Feoktistov was grounded in October 1980, Strekalov found himself back on the prime crew.

Together with Alexander Serebrov, the Soyuz T-8 crew were expected to become the first three-man cosmonaut team to attempt a long-duration mission beyond the 84 days achieved by America's final Skylab crew in February 1974. At first glance, Serebrov's name seems a surprising one, since he had returned from space only eight months earlier, following Soyuz T-7. Indeed, it would appear that his place was originally assigned to a *female* cosmonaut, Irina Pronina, who had served as Savitskaya's backup, and whom the Soviets wanted to fly in order to gather data on a woman's adaptation to long-duration flight ... and, doubtless, score another propaganda triumph over the West. Pronina's mission, flown during the late spring and into the summer of 1983, would conveniently overlap – and hopefully overshadow – the flight of America's first female astronaut, Sally Ride, in June. Sadly, Pronina's chance did not come to pass. In their history of Soyuz, Hall and Shayler commented that in March 1983 – only weeks before launch – "the internal politics of the Soviet programme" led to "heavy pressure" to remove Pronina from the Soyuz T-8 crew. She was replaced by Serebrov. "Several Western accounts," Hall and Shayler wrote, "suggest that Pronina was supposed to participate in the EVAs." How many EVAs would have been needed to complete the work ranges between one and two, according to Clark. At length, this was dropped and the EVAs were assigned to Titov and Strekalov ... but *this* created another problem. It would appear that senior management, or even senior political figures within the Central Committee of the Communist Party, were worried that Pronina would be left alone aboard Salyut 7 during the EVAs and, if something went wrong, according to Strekalov, she would not know how to return to Earth aboard the Soyuz capsule. Hall and Shayler inferred that this pointed, perhaps, to a certain lack of training on the part of the female cosmonauts in Soyuz systems and operations. Still others have argued that the predominantly military leadership of the cosmonaut corps strongly opposed the idea of women flying into space and their voices thus reinforced the call for Pronina's removal from Soyuz T-8.

Phillip Clark suggested that a mission of between eight and nine months – somewhere between 240 and 270 days – was anticipated by many Western observers, although when the cosmonauts rose from Earth, at 4:11 pm Moscow Time on 20 April 1983, their launch occurred during a Salyut landing window. To Clark, this implied "that possibly a mission of four, six or even eight months had been planned", but he cautioned that there were "comments made at the time ... that this was *not* expected to be a record-breaking mission ... and therefore a four-month mission seems to be likely". More recently, Hall and Shayler have broadly concluded along the same lines, noting that a landing in the summer of 1983, perhaps July, was likely. Speculation was rife that Salyut 7 would pioneer efforts to permanently man a space station, with the Soyuz T-9 crew of Lyakhov and Alexandrov lifting off in August for their own mission. (Previous long-duration expeditions had taken the form of 'stand-alone' missions, with maybe one being conducted each year. With the next-generation Mir station, the Soviets planned for an era of permanent occupancy, with one crew directly handing over operations to the next, and Salyut 7 was expected to come close to achieving that goal.) Had the Soyuz T-8 and T-9 expeditions flown as planned, Hall and Shayler suggested that a series of all-Soviet visiting expeditions would have been flown to regularly exchange Soyuz craft, in a similar manner to the procedure followed by Popov's crew ... but with a difference. The Soviets were already well advanced in their efforts to develop a reusable manned Space Shuttle, known as 'Buran' ('Snowstorm') to challenge America's own, and it was intended to fly a handful of civilian Buran pilots – Igor Volk, Anatoli Levchenko and Rimantas Stankyavichus – on Soyuz missions for a specific purpose. They would undertake short-duration missions to Salyut 7 and, within hours of landing, they would immediately board a modified Tu-154 aircraft, specifically outfitted with Buran-type controls, and fly from Baikonur to Moscow, *then* a MiG-25 from Moscow back to Baikonur. It was hoped that this would satisfy lingering worries about a cosmonaut's ability to pilot a Shuttle back to a conventional runway after several days in the disorientating environment of weightlessness. During Lyakhov and Alexandrov's expedition, sometime late in 1983, a visiting crew of Leonid Kizim, Vladimir Solovyov and Buran pilot Volk were expected to fly such a mission, aboard Soyuz T-10. Although two of the Buran pilots, Volk and Levchenko, *did* eventually fly visiting Soyuz missions, the chance of any of them piloting the Shuttle itself would ultimately come to nothing, due to budgetary woes in the early post-Soviet era.

Many of these plans for 1983 were shelved when Soyuz T-8 failed to dock with

Salyut 7. Titov, Strekalov and Serebrov ascended into orbit perfectly, unfurled their craft's electricity-producing solar arrays and set to work checking their systems. Problems arose during their second circuit of Earth, when the Igla rendezvous radar's parabolic antenna refused to yield data onto the Soyuz display panels. At first, the cosmonauts suspected that the Igla had not properly deployed and recycled the switches, but with no success. Worrisome telemetry had already alerted ground controllers that the antenna boom *had* opened, but *not* to the proper extent. Strekalov wondered if it had become caught on something and the crew were allowed to use their ship's thrusters in an attempt to jolt it open. Their efforts were fruitless. It was decided to continue the rendezvous profile, whilst mission managers analysed the problem, but the mood aboard Soyuz T-8 was tense. "To tell you honestly," Titov said later, "we did not feel like rejoicing, so we just got down to work, calmly and busily, as we had trained to do." By their sixth orbit, it was time to rest, although none of the crew were psychologically or physically in any position to do so. The spectre of a failed mission loomed large, for without Igla it would be impossible for the cosmonauts to navigate the final distance to Salyut 7, lacking the necessary ranging and closure data for either an automatic or manual docking. On the ground, controllers seemed more concerned about concealing the problem – telling the outside world that the thruster firings were simply 'a test' – but the station drew nearer with every passing hour and *something* would have to be done. Early on the 21st, Titov was told to attempt a manual docking, using visual cues alone, although ground simulations had already shown that the chances of success were slim.

'Zero hour' for Titov would commence on Soyuz T-8's nineteenth orbit, when the craft would be positioned a thousand metres from Salyut 7's rear port. The station itself had grown substantially in size since the departure of Berezovoi and Lebedev. On 2 March 1983, a large module, designated 'Cosmos 1443', had been launched aboard a Proton booster and had docked onto Salyut's front port a few days later. The Soviets compared the new craft to an earlier mission, Cosmos 1267, which had operated in conjunction with the aging Salyut 6 in 1981. Both were descendants of the *Transportniy Korabl Snabzheniya* (TKS), designed by Vladimir Chelomei's OKB-52 bureau as a manned alternative to Soyuz and originally intended to support a series of 'Almaz' ('Diamond') military space stations. It composed two segments: a functional cargo block, the *Funktsionalna Gruzovoy Blok* (FGB), which provided a pressurised laboratory and working module, and a conical launch and re-entry capsule for a three-man crew. This capsule was known as the *Vozvraschaemyi Apparat* (VA, or 'Return Vehicle') and seems to have mirrored the United States' Apollo command module in terms of shape, although its re-entry thrusters, batteries and radio equipment were mounted at its apex. It measured 3.6 m high and 2.8 m wide, across its base. Despite its outward similarity to Apollo, the TKS-VA was perhaps closer in function to the US Air Force's Manned Orbiting Laboratory, an unrealised project in which a piloted Gemini vehicle would have been launched into orbit, attached to a laboratory module. Astronauts would have transferred to the laboratory by means of a hatch, cut into their craft's base, before undocking and returning to Earth aboard the Gemini at the end of their mission. In a similar fashion, after reaching orbit, the cosmonauts would have transferred to the main TKS laboratory and returned home aboard their VA. The 21,600 kg TKS-VA combination measured 17.5 m long and the laboratory itself featured 40 m² of solar arrays, providing 2,400 watts of electrical power. Supporting these ambitious missions, Soviet planners executed a number of unmanned flights, with *two* VA capsules aboard each Proton booster, to maximise the test data. The first pair of VAs flew perfectly in December 1976, followed by an equally successful shakedown of the complete TKS-VA system in July of the following year. In the latter, the VA undocked and returned correctly to Earth in August, whilst the main TKS laboratory stayed in orbit, under full ground control, until February 1978. Unfortunately, a failure of the Proton failure ruined a second VA test, whilst a third, in March 1978, was more successful. A year later, the booster's engines shut down, whilst still on the pad. The abort system worked as intended and the VA was pulled to safety ... before experiencing a parachute failure and crashing to the ground. Another pair of VAs flew atop the same Proton in May 1979, but one of them suffered an electrical failure and returned to Earth after just two orbits. At around this time, Almaz was cancelled and several of its military experiments were incorporated into plans for a next-generation 'modular' space station, called 'Mir' ('Peace'). The TKS-VA project, though, did not die. A group of six cosmonauts had already begun training in December 1978 and late in the following year, crews were assembled for two flights: the first would comprise Yuri Glazkov, Valeri Makrushkin and Eduard Stepanov, whilst the second would be manned by Gennadi Sarafanov, Valeri Romanov and Vladimir Preobrazhensky. They trained for two years, weathering further technical woes with the TKS-VA, but in December 1981 the project was finally cancelled and its modules assigned instead to the growing Mir effort. When Cosmos 1443 rose into orbit, unmanned, atop a Proton booster, in March 1983, it was the very same TKS-VA that Glazkov's crew might have been aboard.

The Soviet intention appeared to be a joint mission between Cosmos 1443 (known as a 'Heavy Cosmos' in the West) and the Salyut 7 station, with Vladimir Titov, Gennadi Strekalov and Alexander Serebrov checking out the new module. Indeed, the Soviet media noted precisely as much in their official announcement: the craft was "a try-out of on-board systems, equipment and structural elements ... in various modes of flight". Aboard Cosmos 1443 was 2,700 kg of payload and 4,000 kg of propellant. Unfortunately, Titov's crew would not benefit from any of it. Not only were they faced with the daunting prospect of attempting a *manual* docking, without Igla support, and reliant on eyeballs alone, but Soyuz T-8's drift indicator was providing unreliable measurements of the craft's distance from the Salyut-Cosmos combination. From the commander's seat, Titov found it acutely difficult to judge his rate of closure – on the small screen, the target did not appear to grow 'larger', but was visible only as a 'dot' – and was told by mission controllers to execute a 50-second thruster firing to bring the two vehicles closer. At length, Salyut-Cosmos began to grow in size on his screen and he was told to watch the station and switch on his craft's searchlights. Shortly thereafter, Soyuz T-8 drifted out of radio communications range and the cosmonauts were on their own. By his own

A Soviet painting of the Salyut 7 and Cosmos 1443 combination, with the conical VA module clearly visible at the top of the image. Approaching the aft port of the station, at the bottom, is a Soyuz-T spacecraft.

admission, Titov had never practiced a manual docking and doubted his depth perception when judging the ship's closing velocity. That closure rate grew quite alarming, to such an extent that he feared that they would collide with the station. "We hurtled on," he recalled, "and the distance to the station was 280 m. We [maintained] the station's position on the screen, but we could *feel* that the rendezvous speed was high, so I fired the de-boosting engine. The distance was 160 m." At this point, Titov felt sure that he could not attempt a successful docking; the rate of closure was still too high and he aborted the attempt, descending and flying past his quarry. The crew were in orbital darkness for the next 35 minutes and by the time they entered their next sunrise, Salyut-Cosmos was 4 km away. With visual aids as his only option, Titov knew that there could be no option for a second attempt. *Another* half hour transpired before communications with the ground were restored and the cosmonauts were told to return to Earth. Propellant was low, rendezvous hardware was unreliable and, as noted by the state-run news agency, Tass, in typically ambiguous fashion, "because of deviations from the planned approach regime ... the docking of the Soyuz T-8 craft with the Salyut 7 orbital station was cancelled".

The descent module, bearing Titov, Strekalov and Serebrov, touched down safely on Soviet soil at 4:29 pm Moscow Time on 22 April, concluding a tension-filled mission of barely two days. It subsequently became clear that the Igla system's parabolic antenna had been torn away during launch – hardly surprising, therefore, that the cosmonauts' efforts had been fruitless – and the chance of a successful docking with Salyut-Cosmos was effectively eliminated before it could even begin. Years later, in *Dragonfly*, Bryan Burrough recalled Serebrov's accusation that former cosmonaut Valeri Ryumin, then-head of flight operations for Salyut 7, deliberately misled an investigative commission in the wake of Soyuz T-8. According to Serebrov, Ryumin tried to divert blame for the docking failure away from his own flight controllers and onto the heads of Titov and his men. In Burrough's words, Titov had been unfairly instructed to attempt a manual docking, for which he was untrained, and the lack of communications coverage had "robbed him of the crucial telemetry data" which he needed from the ground. "Serebrov says the Soyuz nearly collided with the station," wrote Burrough, "missing a collision by barely 10 m." Whatever the reality, in the meantime, the Soyuz T-9 crew of Vladimir Lyakhov and Alexander Alexandrov were moved forward from their previously scheduled launch target of August to late June to begin the laborious process of unpacking Cosmos 1443. (Launch and landing lighting constraints prevented Soyuz T-9 from flying any sooner.) Vladimir Titov and Gennadi Strekalov, having already trained for EVA work on Salyut 7's solar arrays, remained together and were reassigned to the long-duration Soyuz T-10 mission, due to fly in September 1983, whilst Leonid Kizim, Vladimir Solovyov and Igor Volk were disbanded as a visiting crew. As circumstances would transpire, Kizim and Solovyov, teamed with physician Oleg Atkov, would fly a long-duration mission in 1984, whilst Volk would get his own flight on a visiting EP expedition. Before any of those flights could be attempted, in 1983, the cards of success and calamity would fall in equal measure as summer ran its course into autumn. Lyakhov and Alexandrov would triumphantly reoccupy Salyut 7, whilst Titov and Strekalov would endure one of the most dire and harrowing near-disasters ever experienced in the Soviet space programme: a catastrophic booster explosion, a fire on the launch pad and a hair-raising, bone-jarring rollercoaster of a ride on the Soyuz escape system.

With the removal of Vladimir Dzhanibekov from the long-duration training programme in order to fly the Franco-Soviet joint mission, his place in command of Soyuz T-9 was taken by veteran cosmonaut Vladimir Afanasyevich Lyakhov. A colonel in the Soviet Air Force, he was born on 20 July 1941 in Antratsyt, in what is today south-eastern Ukraine. He entered the military pilots' school at Chuguyev in his early twenties and was selected as a cosmonaut candidate in May 1967. Only months into his two-year training programme, he was one of a handful of cosmonauts detailed to the Soviet Union's Spiral orbital spaceplane project and, after this was cancelled, he was transferred to Salyut. At the end of 1977, he was assigned to the EO training group, for principal expeditions, and served as backup commander for Soyuz 29. Despite a number of adjustments, he commanded Soyuz 32 in 1979, during which he and crewmate Valeri Ryumin spent a then-record-breaking 175 days in orbit. Lyakhov next backed-up the Soyuz 39 mission and also Vladimir Titov's flight. His crewmate aboard Soyuz T-9 was Alexander Pavlovich Alexandrov. A born and bred 'Muscovite', Alexandrov entered the world on 20 February 1943 and received virtually his entire education in Moscow, culminating in a doctorate, whose specialism was in electrical engineering, specifically spacecraft navigational and control systems. Despite failing the medical testing for cosmonaut selection in 1967, he finally made the cut in December 1978 and served alongside Lyakhov on the Soyuz T-8 backup crew. The composition of Soviet crews has remained broadly the same to this day: military pilots normally assume command positions, with civilians selected as flight engineers. Yet there were similarities across the selection process. "The requirements are basically the same to both categories of the cosmonauts," Alexandrov told NASA's Shuttle-Mir Oral History Project. "First is health. You have to have the graduate degree, specifically in the technical field, preferably. Of course, there is also an age limit. It's not that rigid, it's not that strict, but of course we prefer to hire younger people." Interestingly, out of 16 candidates in the December 1978 cosmonaut selection – which included two military pilots – Alexandrov was the third-oldest.

Alexandrov's first launch into orbit, at precisely 12 minutes after midday, Moscow Time, on 27 June 1982, was something for which even his training and technical background left him unprepared. "My *first* launch," he exulted. "You cannot compare it with *anything* in the world!" A little more than a day later, Vladimir Lyakhov guided Soyuz T-9 to a smooth docking with Salyut 7's aft port and the two cosmonauts quickly settled down for what Tass had already revealed would be a mission of a somewhat shorter duration than the record-breaking marathon of Berezovoi and Lebedev. What was not disclosed, though, was that an attempt would be made to overlap two long-duration expeditions – creating, for the first time, a semi-permanent occupancy of the station – with Vladimir Titov and Gennadi Strekalov scheduled to launch aboard Soyuz T-10 in late September, shortly before Lyakhov and Alexandrov's return to Earth. According to Hall and Shayler, it would appear that Titov and Strekalov would perform their EVAs whilst their predecessors were still aboard Salyut, providing a measure of technical support. For now, at the end of June 1983, Lyakhov and Alexandrov's role was to unload Cosmos 1443 of its cargo, whose mass was estimated by Phillip Clark as totalling three and a half times the amount that a Progress could carry. By the end of the first week in July, the cosmonauts had emptied the new module, loaded their camera equipment with film, activated the MKF-6M and KATE-140 cameras and were preparing to begin work on growing plants in the station's greenhouse. "It was hoped," wrote Clark, "that their work would bring the time nearer when fresh vegetables would be available from an orbiting garden."

As July wore into August, Lyakhov and Alexandrov concluded the loading of equipment – photographic materials, old pieces of scientific gear, an air regenerator, "which has completed its period of service", and a defunct memory unit from the Delta navigation computer – into Cosmos 1443's VA capsule for the autonomous return to Earth. It came as something of a surprise to many Western observers when the *entire* TKS undocked on 14 August; the VA craft later separated, completed a few days of independent manoeuvres and successfully re-entered and landed, 100 km south-east of Arkalyk, on the 23rd. (Ten years later, in December 1993, the Cosmos 1443 VA was sold to the Perot Foundation at a Sotheby's auction and today, ironically, resides in the National Air & Space Museum in Washington, DC.) The main body of Cosmos 1443, meanwhile, remained in orbit for another month, burning up in the atmosphere on 19 September. *Surprise* was not too strong an emotion, for the sheer habitable *volume* of Cosmos 1443 convinced many that Lyakhov and Alexandrov effectively possessed *two* space stations in which to operate. What was not revealed until later was that the aft end of the TKS was not fitted with a suitable docking mechanism for Soyuz-T or Progress. Some analysts expected the VA to separate and open up a compatible port on the end of the TKS, which the cosmonauts could then use to relocate their Soyuz T-9 and await new visitors at Salyut's 'aft' port. "If part of Cosmos 1443 is still attached," *Flight International* told its readers in late August, "then it *must* have a docking port – to which Soyuz T-9 is now attached. It is likely that the 'free' end of Cosmos 1443 has a rocket engine, which would mean a docking port at right angles to Salyut 7's axis. If *none* of Cosmos 1443 is left at Salyut 7, then Soyuz T-9 would simply be docked with the front of the space station."

Flight International's speculation is illustrative of the paucity of knowledge and lack of reliable information from the Soviets at this time. In fact, as Phillip Clark wrote in 1988, predictions that Cosmos 1443 had either a Soyuz-T-compatible port or some sort of axial docking hub were both incorrect. "The Heavy Cosmos," he explained, "only carries the single docking unit ... The tunnel connecting the main module [the FGB] and the descent craft [the VA] does not incorporate such a system." The only means for Lyakhov and Alexandrov to accept future visitors, including much-needed Progress deliveries and the arrival of a new crew, was to jettison the *entire* Cosmos 1443. Consequently, on 16 August, the two cosmonauts boarded Soyuz T-9, undocked from the aft port and the station was rotated 180 degrees. They then redocked their craft onto Salyut's now-free forward port. Twenty-four hours later, on the 17th, a Progress freighter was launched from Tyuratam and docked on the 19th. Unknown in great depth at the time, word quickly trickled into the Western press that a propellant line had sprung a leak during the transfer of oxidiser from Progress' tanks into the station's tanks on 9 September, raising the very real possibility that Salyut 7 might need to be abandoned. Some accounts have since reported that Lyakhov noticed the pressure of one fuel tank drop close to zero, whilst the cosmonauts were reorienting the station for a radiowave transmission experiment, and Alexandrov, looking through the aft porthole, observed that fuel was escaping into space. An EVA repair by the crew was impossible, due to a lack of appropriate tools, and the leak would eventually be rectified by Leonid Kizim and Vladimir Solovyov in 1984. In the meantime, despite only having half of its 32 thrusters operational, mission controllers elected to keep Lyakhov and Alexandrov aboard to work through the problems. Three weeks later, those 'problems' would be eclipsed by something far worse.

A MOTHER'S WARNING

Gennadi Strekalov's mother did not want him to fly into space. It seemed a natural emotion to experience in the hours before her son chose to sit atop thousands of kilograms of volatile propellants and be blasted into the heavens, but there was something inexplicable about Praskovya Mikhailovna Strekalova's words of warning: even with two previous space missions under his belt, she had an uncanny premonition that *this* one – his third – would end unhappily. Perhaps the traumatic two days of Soyuz T-8 weighed heavily on her mind. Perhaps it was something else. Whatever the reason, Mrs Strekalova's hunch that 26 September 1983 would be a bad day turned out to be entirely accurate. It was the day that her son nearly died. "As the prospective 94th manned space launch to orbit," Rex Hall and Dave Shayler wrote, "the ascent was not expected to be an outstanding event in space history. However, what transpired was a 'first' which both cosmonauts would no doubt have gladly foregone." It was late in the evening when Strekalov and his commander, Vladimir Titov, boarded the Soyuz T-10 spacecraft and took their seats in the descent module. Launch was scheduled for around 10:38 pm Moscow Time and, after docking with Salyut 7, the two newcomers would spend a couple of weeks with Lyakhov and Alexandrov, performing their EVAs, before the outgoing crew departed. Titov and Strekalov would then settle down for their own three-month expedition, which Phillip Clark expected to end during a Salyut landing window in late December or early January 1984.

Preparations for launch seemed to be proceeding normally, until around two minutes before the giant rocket's engines were expected to ignite. Temperatures at the pad had fallen from around 27 degrees Celsius, during the day, to just 10 degrees Celsius that night, whilst winds were gusting at upwards of 40 km/h. Inside Soyuz T-10, Titov and Strekalov listened to music, piped in through the communications system, as they carried out their final systems checks and verified the pressure integrity of their Sokol suits. Ninety seconds to go. All at once, a fuel valve, supplying the first-stage strap-on boosters, failed to close properly, spewing propellant across the pad surface. Within a minute, a fire had erupted and began to spread rapidly, soon licking the sides of the launch vehicle ... which, of course, was fully loaded, its pyrotechnics were armed and its engines were seconds away from ignition. At the blockhouse, a few kilometres away, Launch Director Alexei Shumilin watched with mounting horror as the fire increased in ferocity. Titov and Strekalov had no windows to witness the progress of the fire, although they were very quickly made aware of the situation. Strekalov, who had ridden *two* previous Soyuz-T missions, could tell that something was amiss, judging from the strange sounds emanating from the rocket. He tightened his harness and told Titov to do the same. They could clearly hear urgent chatter from launch controllers over their communications loop. This urgency was intensified when it became clear that the raging inferno had burned through the wiring which should have triggered the rocket's escape system to fire, yanking the Soyuz T-10 orbital and descent modules away from the burning booster and parachuting them a couple of kilometres to safety. Titov and Strekalov had no controls within the descent module to permit them to manually activate the escape system. The only chance to override it was to use a backup radio command ... but *that* required a pair of controllers, located in *different* rooms in a *different* building, more than 30 km from the pad, to press their buttons within *five seconds* of one another. Not only that, but in order for them to be in a position to do this, they each had to receive a code word ('Dnestr'), independently, from *both* Shumilin *and* the rocket's technical leader, a man named Soldatenkov. By the time all of these steps had been concluded, *ten seconds* had elapsed and Titov and Strekalov's booster was literally engulfed in flames and dangerously close to an explosion.

Within seconds of the command from Shumilin and Soldatenkov, the pyrotechnics fired and the escape tower's powerful solid rocket engines – totalling around 80,000 kg of thrust – rapidly pulled the Soyuz T-10 descent and orbital modules away from the burning booster. The immense velocity and the angle of departure imparted high-G accelerations on the two cosmonauts. After the event, Titov recalled feeling the entire rocket physically *swaying*, followed by two waves of vibration as the escape tower fired and a sharp jerk as he and Strekalov were boosted to safety. Observers at the blockhouse witnessed an enormous red, yellow and black cloud, surrounding the top of the booster, after which an object could be seen shooting upwards at a high rate of speed. Within seconds, Titov and Strekalov reached a velocity of Mach 1, travelling near-vertically, rose to an altitude of 950 m and endured peak acceleration loads of between 14-17 G. A mere *six seconds* after the capsule had been tugged to safety, the booster exploded in a holocaust of flame ...

In the blockhouse, several kilometres from the pad, images released many years later show a gaggle of uniformed military officers, their backs to the camera, watching the unfolding drama through large picture windows. As the escape tower pulls Soyuz T-10 away from the inferno, one of them nonchalantly rubs his nose.

After this near-disaster, which one official later described as having brought the programme to within "six seconds from a Soviet Challenger", it became clear that the rocket was *already* beginning to lean precariously, *before* the escape tower even fired, showing that Titov and Strekalov probably were even nearer to death ... and *luckier* than perhaps any spacegoing crew in history. Certainly, some accounts note that the rocket collapsed onto the pad within three or four *seconds* of the activation of the escape tower. Others relate that it was as few as *two seconds*. Whether two or three or four or six, though, is immaterial; for the cosmonauts had escaped oblivion, literally, by the skin of their teeth. Some 650 m above the ground, still rising, the orbital and descent modules were automatically separated, dropping free of the escape tower and shroud, and within seconds Titov and Strekalov reached their maximum altitude and began a slow fall towards the Kazakh steppe. The descent module's heat shield was jettisoned, exposing six solid-fuelled landing rockets in the base, and the shaken and badly bruised cosmonauts came to a hard, parachute-assisted touchdown at 10:43 pm Moscow Time, 4 km from the burning launch pad. Their first action after the firing of the escape tower had actually been to switch off their cockpit voice recorder, because, said Titov, "we were *swearing*!"

Half an hour after touching down, the recovery team reached the capsule. Titov and Strekalov's first request was for cigarettes and they were also offered a helpful tumbler of vodka to settle their frayed nerves. The launch pad, meanwhile, continued to burn for no less than *twenty hours*, before being brought under control and extinguished, and a month later the Soviets announced for the first time that there had been "another accident" in the Soyuz programme. Little more was revealed, apart from the successful performance of the escape system and the survival of the two cosmonauts. The mission subsequently became known in the West as 'Soyuz T-10A' ... and Titov and Strekalov's misadventure had lasted barely five minutes and 13 seconds. For years afterwards, the two cosmonauts would remind each other of the events of that fateful night on which they came within seconds of meeting their maker ... by referring to 26 September as their *second* birthday: for it offered them both a new life, another chance, cheated the Reaper and gave them a second opportunity to breathe the fresh air of Earth.

RETURN TO NORMALITY?

The survival of Vladimir Titov and Gennadi Strekalov was a tremendous relief, but an enormous disappointment, since for the *second* time in less than six months, their mission had failed. Moreover, the need to physically evaluate them after their high-G escape from the booster meant that both cosmonauts were grounded and the *next* manned launch would not occur until the early spring of 1984. In some areas of the Western media, this did not bode well for the Soyuz T-9 crew of Vladimir Lyakhov and Alexander Alexandrov, who had already spent several months in orbit. A handful of journalists wondered if Lyakhov and Alexandrov were now effectively 'stranded' in space, their aging Soyuz T-9 vehicle rapidly approaching its maximum operational lifetime. Repairs to the damaged launch pad were anticipated to cost in the region of between a quarter and half a billion dollars and, although a second pad *was* available, it was recognised that significant pressure would inevitably be imposed on the future Soyuz-T and Progress manifest. Yet the effect on the Soyuz T-9 crew was much less stark. Many Western observers supposed that – based on experience from the earlier Salyut 6 era – the spacecraft could remain aloft for around a hundred days, which was not, in fact, the case. Even *Flight International* noted on 29 October that "the longest that previous Soyuz-T capsules have spent in space is 114 days ... a duration which has just been exceeded by Soyuz T-9." Seasoned analysts brought a measure of sanity to the situation. "More experienced Soviet space watchers," wrote Hall and Shayler, "attributed this to scaremongering, observing that the 100-day expiry date of Soyuz could rather be likened to a 'best before' date on a can of food. There was nothing to prevent its use beyond that date, within reason." One particularly outlandish British story even supposed that the

Americans had been contacted, with a request to use their Shuttle to rescue the cosmonauts ... though *how* this could have been done, with no compatible docking mechanism, was left conveniently unanswered. "All these stories did," wrote Phillip Clark, "was to show the world how little British journalists knew about the Soviet space programme." In fact, the upgraded Soyuz-T had the capability to remain aloft for around six months, and Lyakhov and Alexandrov would return home well before the end of that limit.

Having said this, the remaining weeks of their mission, which had by now been extended far into November, were hardly routine. In a post-flight news conference, held in December 1983, Lyakhov and Alexandrov admitted that it *was* originally intended for them to hand control of Salyut 7 over to Titov and Strekalov and that the latter would have performed one or possibly two EVAs to install and deploy additional solar arrays. Soyuz T-9 would then have returned to Earth, wrote Phillip Clark, "during the first ten days of October". Images published in the *Soviet Union* magazine in 1984 also showed a Soyuz-Salyut-Cosmos complex, with three men aboard, and a representation of the additional solar arrays. It was from this arrangement and this crew size that Clark concluded that Soyuz T-8 – with Titov, Strekalov and Serebrov – would have installed and deployed the arrays during their original mission, launched in April 1983. When this failed, Titov and Strekalov, having been trained specifically for the EVAs, were quickly recycled to Soyuz T-10 to complete the work. The 26 September launch abort, and the effective removal of Titov and Strekalov from immediate mission activities for medical supervision, meant that the array work would now have to be assigned to Lyakhov and Alexandrov ... *neither* of whom had trained for this task. Fortunately, Lyakhov *had* performed an EVA, near the end of his first mission, in August 1979, and Alexandrov later revealed his intimate knowledge of spacewalking and the extravehicular suit in a NASA oral history. He considered it simple to use, exceptionally flexible and reliable and capable of supporting its cosmonaut wearer for seven hours or more in the vacuum of space. The need to install the additional solar arrays, which had been transported into orbit aboard Cosmos 1443, prompted the Soviets to extend Soyuz T-9 by about five weeks, ending in late November.

As newspapers continued their scaremongering, Lyakhov and Alexandrov continued working throughout October and were joined by a new Progress freighter on the 20th, bringing much-needed supplies and 500 kg of propellant for Salyut 7. Yet even *Flight International* joined in the fray, with its 29 October headline reading 'Progress ... buys Salyut time', in which it referred to – without explicitly endorsing – the 'stranded' stories. It was noted, however, that the arrival of the new Progress "could have turned the unmanned cargo craft into a lifesaver" and remarked on "two main dangers" facing the cosmonauts: firstly, the Salyut 7 propellant leak, and secondly, the remaining lifetime of Soyuz T-9. *Flight International* commented on television images of Progress' docking, which were shown by Western networks, "possibly with the idea of playing down any danger to the two cosmonauts". This absence of danger was underlined still further on 1 November, when Lyakhov and Alexandrov donned their suits and ventured outside the station on the first of two EVAs to install and activate the new solar arrays. Supported by the Soyuz T-9 backup crew of Leonid Kizim and Vladimir Solovyov, who had rehearsed many of the procedures underwater in the 'hydrotank' at Star City, Lyakhov and Alexandrov set to work at 7:47 am Moscow Time. (Incidentally, Lyakhov became the first cosmonaut in history to perform a second EVA.) Alexandrov set up a movable television camera for the benefit of ground controllers and the two men carefully attached an auxiliary solar panel to one side of Salyut 7's dorsal array. Forty minutes into the EVA, they passed out of range of Soviet ground stations and tracking ships, for almost an hour, and were forced to wait in darkness until orbital sunrise and the resumption of work. They then used a special "compact and convenient" winch to unfurl the new panel to its maximum size of 5 m long and 1.5 m wide. The new addition was much needed, providing Salyut 7 with 25 percent more electrical power. Lyakhov and Alexandrov's first EVA ended after two hours and 50 minutes.

Two days later, at 6:47 am, the duo were outside again, this time completing all of their principal tasks whilst in voice communication with the ground. In two hours and 55 minutes, a second auxiliary panel was attached to the other side of the dorsal array. In total, the two new panels – which could not be installed on the same day because the dorsal array had to be rotated 180 degrees, requiring the cosmonauts to return inside Salyut – increased the station's electrical output by 800 watts and gave Deputy Flight Director Viktor Blagov reason to beam with pride. "Today's installation operations," he explained, "are the very first steps in solving the orbital energy problems facing us ... in the second place, although all kinds of maintenance work has been carried out in open space before, up to now such major installation operations have *not* been conducted." Of course, had Titov and Strekalov been aboard, both auxiliary panels *could* have been fitted in a single EVA, because Lyakhov and Alexandrov would have been able to rotate the array from inside the station. Still, the cosmonauts' success ticked off yet another first on the list of space achievements, *ahead* of the Americans, it would seem. In the months which followed, as Kizim and Solovyov performed their own EVAs in 1984 to repair Salyut 7's oxidiser leak, Lyakhov and Alexandrov lauded their comrades as "the test pilots who trained us" and assigned acute value on their "assistance" and their "recommendations". In the days which followed, the situation aboard the station stabilised. On the 13th, the Progress was undocked automatically and on the 23rd, Lyakhov and Alexandrov followed suit, bringing to a close an eventful mission of 149 days. Touching down at 10:58 pm Moscow Time, their return was well outside the margins of a 'nominal' Salyut landing window – a month earlier, in fact – but the prophets of doom in the journalistic establishment, once so vocal about an unfolding disaster in space, were now notably quiet in their coverage. As a difficult year drew to a close, *Flight International* reported on Soviet 'admissions' of propellant leaks and the Soyuz T-10A launch failure, but concluded the Salyut 7 was "claimed to be in good working order", with future occupants planned for 1984, including an Indian cosmonaut. In fact, 1984 would be the most successful 12 months in Salyut 7's career, featuring a record-breaking *eight-month* residency by a three-man crew and no fewer than *two* visiting expeditions, including the Indian cosmonaut, a second mission for Svetlana Savitskaya (who would also become the world's first female spacewalker) and a long-awaited flight for the would-be Buran pilot, Igor Volk.

The flight of the Indian cosmonaut, Rakesh Sharma, in April 1984, was one of only a few positive events to occur in what was one of his country's darkest and most violent years. Two months after his mission, Prime Minister Indira Gandhi ordered the execution of 'Operation Blue Star', in which Indian troops brutally removed Sikh separatists from the Golden Temple in Amritsar. Led by Sant Jarnail Singh Bhindranwale, the separatists were suspected of amassing large quantities of weapons, including light machine guns and semi-automatic rifles, within the precinct of the sacred building. The army stormed the temple on the evening of 5 June and within 48 hours had killed Bhindranwale and assumed control, claiming the lives of between 500 and 1,500 civilians in the process. It is particularly notable that this coincided with a Sikh annual festival and many pilgrims, including children and the elderly, were trapped inside the temple as the attack commenced. Three decades later, Operation Blue Star continues to be regarded by many Indians as one of the country's gravest disgraces: in its aftermath, Sikh soldiers mutinied, many Sikhs resigned political and administrative office in disgust and Gandhi herself was gunned down by her Sikh bodyguards on 31 October 1984. This latter act provoked further violence, as anti-Sikh sentiment swept the country, causing hundreds more deaths. The violence was centred on northern India and particularly Delhi and, in addition to the killings, led directly to the displacement of maybe 50,000 people. Many Sikhs consider it an act of genocide and, in 2011, WikiLeaks revealed the United States' conviction that the Indian government was complicit in what was actually a well-planned act of mass murder. Sikhs were dragged out of buses and burned alive and a government MP, Sajjan Kumar, and trade union leader Lalit Maken handed out money, iron rods and alcohol to assailants. At one stage, Kumar fanned the flames of hatred yet more by putting bounties on the heads of Gandhi's murderers: "Whoever kills the sons of the snakes," he declared, "I will reward them." Numerous investigations followed, with human rights organisations condemning the Indian government's role in an organised massacre, but the situation remained tense and the matter remained in the popular consciousness. In June 1990, the area surrounding the temple was forcibly emptied of inhabitants, to prevent the outbreak of any future militant activity. It remains a sad stain on the reputation of one of Asia's most stable democracies.

As the first Indian cosmonaut, and a Hindu by religious conviction, Squadron Leader Rakesh Sharma undoubtedly had his own opinions about Operation Blue Star and the violent events which engulfed his nation in its aftermath. Nicknamed 'Rikki' by his Soviet crewmates, Sharma came from the city of Patiala, in the south-eastern Punjab, which boasts a proud history: situated around the ancient *Qila Mubarak* – the 'Fortunate Castle' – it was the seat of the Patiala and East Pubjab States Union in the mid-19th century and it is perhaps fitting that, since Indian independence, the birthplace of the nation's first cosmonaut should have also established itself as a foremost centre of learning and sport. In fact, Patiala houses numerous universities, public schools, medical and law schools and commerce colleges. Sharma himself, born on 13 January 1949 into a Punjabi family, was educated at St George's Grammar School in Hyderabad and entered the National Defence Academy (NDA) in Khadakwasla, near Pune in Maharashtra, in July 1966 as an Indian Air Force cadet. Today, the NDA, which trains army, navy and air force students, prior to commissioning them into the respective service academies, is one of the most respected military schools in the world and its personnel have fought in all major conflicts in which India has been involved. In fact, NDA alumni included no fewer than *nine* recipients of the *Ashok Chakra*, the supreme Indian medal for "conspicuous bravery", valour and self-sacrifice, away from the field of battle. Only a few dozen of these decorations have been awarded since 1952. One of their recipients was Rakesh Sharma himself, in honour of becoming the first of his countrymen to venture into space.

After completing the NDA, Sharma entered the Indian Air Force as a pilot officer in 1970 and, during the bloody conflict with Pakistan, he flew MiG fighters with distinction. A decade later, in September 1982, he and fellow air force officer Ravish Malhotra were selected by the Indian Space Research Organisation from hundreds of hopefuls to become the nation's first cosmonaut ... and the fact that India was only the second non-Communist, Western-aligned nation to accept a space flight offer from the Soviets did not go unnoticed. According to *Flight International*, the idea of a joint mission was first discussed in the late 1970s, whilst the Janata Party was in power, when Soviet Premier Alexei Kosygin visited India. This was tumultuous time for India: Prime Minister Indira Gandhi had been deposed from power, on charges of corrupt political practice, and constant infighting and ideological differences within the Janata regime made it impossible to address many of the nation's problems. The result was a resurgence of support for Mrs Gandhi and in 1980 she won the general election and was again installed as prime minister. Later that same year, Leonid Brezhnev visited India and talks of co-operation in space resumed. "The idea is said to have been readily accepted by the present government," *Flight International* told its readers. "India and the Soviet Union are already signatories to an agreement covering the mid-1980s launch of an Indian remote sensing spacecraft." Many saw the 'co-operation' for what it was. Coming hard on the heels of the flight of Frenchman Jean-Loup Chrétien, few in the West were under any illusion that the Indian mission was little more than a propaganda coup, with NASA having already announced plans to fly Western European scientists on Spacelab flights.

In a 2010 interview with *The Hindu*, Rakesh Sharma was characteristically modest about his accomplishment, summing it up with just three words: *meri lag gayi* – "I got lucky" – and considering himself as 'just' the 128th human to have left the planet and ventured into the fathomless expanse of space beyond, rather than the *first* Indian. "So much had already been done in space before that," Sharma said, "it had all been documented and there were no real surprises in store for us." Even the cataclysmic power of the rocket ride into space held little fear for him; as a military pilot, with first-hand experience in air-to-air combat, he had lived through equally dangerous situations and had lived to tell the tale. Flying was in Sharma's blood. As a six-year-old boy, he had been fortunate to have a cousin who was a pilot in the Indian Air Force and had been shown around aircraft and the layout of their cockpits from this tender age. "If you end up doing what you are passionate about," he said later, "the journey is so easy."

The next stage of Sharma's journey began at 4:09 pm Moscow Time on 3 April 1984, when he blasted into orbit aboard Soyuz T-11, alongside Soviet crewmates Yuri Malyshev – the commander, dropped two years earlier from the French mission – and flight engineer Gennadi Strekalov, now fully recovered from the trauma of his launch abort, six months before. Yet Strekalov was not originally assigned to the Indian mission at all. The plan drawn up in September 1983 had Malyshev paired with veteran cosmonaut Nikolai Rukavishnikov, a 50-year-old civilian engineer who had already flown three times ... and, through two brushes with extreme misfortune, had *never* boarded a space station. On his first mission, in April 1971, Rukavishnikov and his crew had been unable to dock with the Salyut 1 station and, eight years later, an engine failure had prevented a docking with Salyut 6. The backup crew for the Indian flight consisted of veteran cosmonauts Anatoli Berezovoi and Georgi Grechko, teamed with Ravish Malhotra. Unfortunately, early in 1984, Rukavishnikov was removed from the prime crew, due to "medical problems", but rather than transplanting the entire backup team – or, at least, swapping him with his own backup, Grechko – the Soviets opted to bring Strekalov onto the mission instead. "This showed the Soviet reluctance to use an official backup," wrote Phillip Clark, "to replace an incapacitated cosmonaut: a substitute would be brought into the flight crew, leaving the backup crew intact." Little information has been forthcoming about the cause of Rukavishnikov's problem: some sources suggest that he suffered "a bad cold", others that he failed his final medical exams. (Had Rukavishnikov flown the Indian mission, he would have become the oldest Soviet cosmonaut to date, though not the oldest spacefarer, 54-year-old Bill Thornton having flown STS-8 in August 1983.)

Colonel Yuri Vasilyevich Malyshev carried a certain level of expertise with the Soyuz-T spacecraft, having worked on its systems and development for more than a decade. In January 1974, he was one of four commanders, together with a handful of civilian engineers – whose number included Gennadi Strekalov – to be chosen for the Soyuz-T training group. Over the next few years, several of the team moved onto other assignments, but Malyshev remained, and, in June 1980, he commanded the first Soyuz-T mission to the Salyut 6 station. Malyshev came from the village of Nikolayevsk, near the city of Volgograd, on the eastern shore of the Volga River, where he was born on 27 August 1941. He completed high school in Taganrog, a port on the Sea of Azov, at the age of 18 and promptly joined the military, graduating from the Higher Air Force Academy at Chuguyev in 1963. As a Soviet Air Force pilot, Malyshev flew the MiG-17 and MiG-21 fighters. Four years later, in April 1967, he was selected as a cosmonaut candidate and was initially assigned to follow the development of the Soviet Union's Spiral winged spaceplane. Selection to the Soyuz-T training group was followed in January 1976 by assignment as backup commander of Soyuz 22. In 1979, he was given command of the first manned Soyuz-T, teamed with veteran cosmonaut Vladimir Aksyonov. Removed from Soyuz T-6, apparently, due to a personality clash with the French cosmonaut, Malyshev would ultimately make his second space flight on Soyuz T-11.

When Malyshev, Strekalov and Sharma docked onto Salyut 7 on the early evening of 4 April, the station was not unoccupied. It had served as home for eight weeks to the Soyuz T-10B crew, a unique three-man *expeditsya osnovnoi*, comprising commander Leonid Kizim, flight engineer Vladimir Solovyov and a physician, cosmonaut researcher Oleg Atkov. When they were launched into orbit at 3:07 pm Moscow Time on 8 February 1984, the expectation in the West – partly due to Atkov's presence – was that an extended mission of eight or even nine months was the intention. The bad luck of 1983, it was hoped, would be laid to rest in a spectacular 1984. Many observers had also correctly predicted that the next crew would be three strong, particularly when Salyut 7's orbit was *lowered* on 11-13 January. A day after launch, Kizim guided his spacecraft smoothly to dock with the station's forward port and entered its darkened interior with torches, switched on the lights and set to work configuring its systems. By 17 February, Salyut 7 was fully reactivated and shortly thereafter the Progress 19 freighter arrived, carrying much-needed food, water, propellant, experimental hardware – including a portable cardiograph – and parcels and mail for the cosmonauts. By the end of the month, Kizim, Solovyov and Atkov had set to work on an expansive scientific programme, observing Comet Crommelin, whose 27-year orbital period had carried it to a perihelion of 0.7 astronomical units from the Sun on 20 February 1984. Unlike previous Progress deliveries, in which the cosmonauts had set aside three days to unload the freighter, it was decided that they would remove items as they were needed. By the end of March, they had fully unpacked Progress and it was undocked on the 31st, burning up in the atmosphere the following day. This freed Salyut 7's rear port for the arrival of the Soviet-Indian crew. It was a promising start for the cosmonauts of Soyuz T-10B, only one of whom had any previous space flight experience.

Colonel Leonid Denisovich Kizim's chance of being in the exalted position of commander of the longest human space mission in history might have seemed impossible a couple of decades earlier. He was born in the Cossack-founded town of Krasnyi Lyman, in eastern Ukraine, on 5 August 1941, and like many of his fellows in the Soviet spaceflying corps his dream of aviation started at a young age. However, his shortness of stature caused him to be turned down for flying school; only Kizim's tenacity finally enabled him to achieve his goal and in 1963 he graduated from the Chernigov Lenin Komsomol Higher Air Force School and joined the Soviet Air Force, rising to become an accomplished test pilot and parachutist. In October 1965, he was one of 22 cosmonaut candidates, both pilots and engineers, to be chosen for an anticipated flurry of Soyuz and space station missions. Of those 22 selectees, only *six* ended up actually flying into the heavens ... and of those six, arguably Kizim would achieve the most: three space voyages in total, one of which spent a world-record-breaking *eight months* in orbit, and command of the first (and so far only) mission to visit *two* space stations in a single flight. Significantly, Kizim would become the first human being to chalk up a full years of his life in orbit, accruing no fewer than 374 days by the end of his third mission in July 1986. The wait for his first flight, though, was long and arduous. Completion of initial cosmonaut training at the end of 1967 was followed by assignment to the Spiral spaceplane effort and in January 1974 he was picked as one of the inaugural members of the Soyuz-T team. A year later, he graduated from

Higher Air Force School and in November 1980 he commanded Soyuz T-3, which carried out much-needed repairs and maintenance on the aging Salyut 6 station. The work of Kizim and his two crewmates prepared the old station for another expedition and *two* international visits by cosmonauts from Mongolia and Romania. Had the events of 1983 worked out differently, Kizim and his flight engineer, Vladimir Alexeyevich Solovyov, might have flown a visiting mission to Salyut 7, along with Buran pilot Igor Volk. All three, however, got their respective chances to fly in 1984.

Solovyov came from Moscow, born on 11 November 1946, and grew up with engineering and the sciences in his blood. In 1970, he graduated from the prestigious Bauman Higher Technical School and worked at TsKBEM (later Energia), the former design bureau of Sergei Korolev, as a space propulsion expert. Selected as a civilian cosmonaut engineer in December 1978, he worked on the development of the Soyuz-T spacecraft and served as the backup flight engineer for the French mission and Soyuz T-9 and T-10A. If Solovyov was an expert in engineering and the sciences, then medicine ran like a strong vein through the life and career of his crewmate Oleg Yuryevich Atkov. Born in Khvorostyanka in today's Samara Oblast, in the western Soviet Union, on 5 May 1949, he entered the Ivan M. Sechenov First Medical Institute in Moscow in 1973 and received his doctorate, with a specialism in cardiology, from the Academy of Medical Science in 1978. Atkov enjoyed one of the shortest periods from selection to flight of any spacefarer: chosen as a cosmonaut from the Academy of Medical Science in March 1983, his basic training – "presumably only a check", noted one website – was complete by September of the *same* year and he began intensive preparation for the long-duration Soyuz T-10B. The importance of a medical specialist on this long-duration mission was also illustrated by the presence of a physician, Valeri Polyakov, on the backup crew. Atkov's role, not surprisingly, was to monitor the physiological and psychological adaptation to the microgravity environment of both himself and his two comrades and during the flight he employed a large complement of medical equipment for the purpose. In their history of Soyuz, Rex Hall and Dave Shayler noted that Atkov's inclusion was part of a deliberate effort to include a medical specialist aboard *every* future record-breaking duration flight. This did not wholly come to pass; Yuri Romanenko, who spent 326 days in space in 1987, had no physician amongst his crew and nor did Viktor Savinykh, whose attempt to undertake a nine-month mission in 1985-86 was curtailed by the illness of one of his colleagues. Having said this, when Vladimir Titov and Musa Manarov undertook a year-long expedition in 1987-88, they were joined for several months by Valeri Polyakov, whilst the latter set an empirical, single-mission record of 430 days in 1994-95. Polyakov's record stands to this day as the longest continuous period of time ever spent in space.

The flight of Oleg Atkov on Soyuz T-10B raises another interesting tale. It has already been shown that there was an obsession amongst Soviet senior leaders to score propaganda triumphs over the Americans – from the flights of Gagarin and Tereshkova and Leonov in the early 1960s to those of Chrétien and Savitskaya and Sharma in the early 1980s – but a handful of sources have also remarked that plans were laid to seize the crown of the oldest man in space. Years earlier, in August 1961, the Soviets had launched the youngest cosmonaut, 25-year-old Gherman Titov, whose achievement still stands unchallenged to this day, but since July 1975 the *Americans* had held the record for the *oldest* man in space, having flown Deke Slayton at the age of 51. In the previous volume of this series, *At Home in Space*, this author made reference to an effort to fly veteran cosmonaut Konstantin Feoktistov aboard Soyuz T-3 in November 1980. Feoktistov was an engineer who had been instrumental in the design and development of the Vostok and Soyuz spacecraft and had flown, aged 38, aboard the first Voskhod mission in October 1964. By the time of Soyuz T-3, Feoktistov was 54 years old and might have been in pole position to seize Slayton's record ... but was medically disqualified and grounded shortly before the flight. *Another* attempt to fly him was apparently made in 1982, although opinion differs as to precisely how it happened. Several years ago, in 2000, the journal *Novosti Kosmonavtiki* revealed that Valentin Glushko – the legendary rocket engine designer who took charge of the Kaliningrad-based Energia design bureau in 1974 – was obsessed with the notion of beating Slayton's record. He apparently 'chose' Feoktistov to fly a mission to Salyut 7, alongside Atkov and Kizim, in order to assess the impact of weightlessness on an older cosmonaut. Other observers counter that it was Feoktistov *himself* who insisted on the flight, suggesting that he should fly into orbit in February 1984, alongside Kizim and Atkov, and return home two months later with the Soviet-Indian crew. More recently, Rex Hall and Dave Shayler wrote that Feoktistov had been "asked" to train for the long-duration mission, presumably by Glushko, but note that his removal from consideration occurred when he "developed a chronic illness, which became acute". Certainly, Feoktistov had vanished from even unofficial Soyuz T-10B plans by September 1983, when the formal announcement was made that Atkov and Kizim would fly with Solovyov in the third seat.

By 4 April 1984, Salyut 7's crew had expanded from three to six, with the arrival of the Indian mission, and Rakesh Sharma's work encompassed not only life sciences, but also materials processing research – including silicium fusing tests – and even traditional yoga exercises as a possible countermeasure to the effects of the weightless environment. He also carried a few delicacies from home to share with his Soviet crewmates: mangoes, pineapples and 'crispy' bananas, and the sounds of Indian music echoed through Salyut 7's cave-like interior. Sharma also undertook multi-spectral photography of 40 percent of his homeland, particularly the north, where planning was already underway to build a chain of hydroelectric power stations in the foothills of the southern Himalayas. Additionally, the remote sensing work allowed him to observe a major forest fire in Myanmar. During his time aboard the space station, Sharma spoke directly to Indira Gandhi, thanks to a 'telebridge' set up through Intelsat and Soviet communications satellites. Mrs Gandhi asked him pointedly what India looked like from orbit; after careful thought, he paraphrased the words of the poet Muhammad Iqbal, explaining that their mutual "land of Hindustan ... is the best in the world". Ironic words, it would seem, when one considers the highly tense situation with militant Sikh separatists in Amritsar, who had already fortified the area around the Golden Temple and whose actions would within weeks bring them into conflict with the Indian military. Years later, Sharma would not be alone in his conviction that war and hostility should never be permitted to extend beyond the thin veil of the atmosphere. "*None* of the paradigms that define us here on Earth," he told *The Hindu*, "the borders, the parochialism, the divide, should mar our presence in space." From his privileged vantage point, as he would later tell children, it was not just violence which posed a problem: it was *neglect* of humanity's duty as steward to the Home Planet. To him, the world looked less blue and much greyer than it should. It did not come as a surprise. With disappearing forests, the drying of water sources and appalling levels of atmospheric pollution, much work *had* to be done. Sharma's words and, by extension, the efforts of India herself, are significant, as the nation pursues its conviction to become the fourth country to place its own astronaut into orbit in 2016. Sharma is aware that another chance may never come his way, although he remains involved in India's plan for a future human space programme, "in an advisory capacity".

Since the next Soyuz-T was not scheduled for launch until the late summer of 1984, the final major objective for Malyshev and his crew was to exchange their craft with the two-month-old vehicle of Kizim, Solovyov and Atkov. The visitors duly returned home aboard Soyuz T-10B, touching down on Soviet soil at 1:50 pm Moscow Time on 11 April, completing an eight-day mission, and starting a new and unexpected 'career' for Sharma ... as a national celebrity, making tours, giving lectures and responding to interview requests. It did not come particularly comfortably for a modest man, who had thus far worked under the anonymous umbrella of the fighter pilot, but in his own words Sharma put his best foot forward and later returned to operational flying. Awarded the *Ashok Chakra* – alongside his Soviet colleagues, Malyshev and Strekalov – he remains, to this day, one of only *two* Indians (the other being the late Kalpana Chawla) to have ventured into space. Every year, Sharma continues to receive three cards from a *paan wallah* in Ahmedabad: one at New Year, another for his birthday on 13 January and a third on the anniversary of his launch. Today, as the Soyuz T-10B capsule in which he returned to Earth gathers dust in the Nehru Planetarium in New Delhi, it is Sharma's fervent hope, and doubtless would have been Chawla's, too, that not many years will elapse before another of their countrymen or women follow in their footsteps.

LONG MISSION

With Malyshev, Strekalov and Sharma safely back on Earth, the Soyuz T-10B resident crew continued their long mission. One of their first actions was to board the Soyuz T-11 spacecraft, undock from Salyut 7's aft port and redock at the front end of the station, thereby opening the way for a series of rapid-fire Progress visitors: one every month, in fact, for *three* months, bringing not only basic necessities, together with hygiene equipment and life-support apparatus, but also tools for an extensive series of EVAs. In a four-week period, from 23 April to 19 May, Leonid Kizim and Vladimir Solovyov conducted no fewer than *five* spacewalks – equalled only by US astronaut Dave Scott at the time – to attend to the station's worrisome propellant leak and augment Salyut 7's solar arrays. They subsequently performed a record-breaking *sixth* outing in early August to finally resolve the leak. In doing so, they jointly became the first cosmonauts to perform more than three EVAs (Vladimir Lyakhov having been the first Soviet to make a second *and third* spacewalk in November 1983) and their work benefitted from many hours spent training underwater, in the Star City 'hydrolab', on the wooded outskirts of Moscow. As backups to Vladimir Solovyov and Gennadi Strekalov on Soyuz T-8 *and* Soyuz T-10A, Kizim and Solovyov had virtually duplicated their training to install the solar arrays on Salyut 7 *and* perform the oxidiser line repairs. Their efforts were put to the test on 23 April, when Salyut 7's hatch opened and the two men floated outside for what would turn out to be four hours and 20 minutes, preparing the work site for the repair operation. Two dozen different tools had already been carried aloft aboard Progress 20 for the task, but there was a problem: the main propulsion system was located in the station's unpressurised aft compartment, with no cosmonaut-friendly hand holds in the vicinity, so Kizim and Solovyov required work platforms and foot restraints, before they could even contemplate a repair. In response to this requirement, engineers had equipped Progress 20 with just such a mechanism and this was remotely extended before the cosmonauts ventured outside. With Oleg Atkov monitoring the proceedings from inside Salyut, the two men made their way along the 15 m exterior of the station, from the airlock to the work site, impeded all the way by the bulk of several tens of kilograms of tool caddies, cutters, wrenches, bypass pipes, a waste container and a ladder. Once there, they quickly set to work: driving anchor pins into the equipment compartment's skin to secure their ladder and tool caddies and unfurling the ladder to a full length of 5 m. They returned to the airlock in readiness for a second excursion, on 26 April, when the repair would commence in earnest.

Under the watchful eye of Valeri Ryumin, the cosmonauts spent almost five hours in open space on their second EVA – nearly an hour longer than scheduled – and required no less than 20 minutes simply getting to the work site from the airlock. Furthermore, since the spacewalk was being carried out in the early morning hours, Moscow Time, the orbital geometry of the station meant that tracking ships in the Atlantic and Pacific permitted communications for up to 50 minutes of each 90-minute circuit of the globe. When the men reached the work site, they assembled a television camera, to allow ground controllers to monitor their progress, and began removing thermal blankets and cutting through layers of plastic to reach the oxidiser plumbing. Whilst attempting to replace a valve on a shut-off part of the reserve line, a nut locked in place by epoxy resin delayed their progress by almost two hours, putting them seriously behind schedule. Eventually, they freed the nut and the oxidiser system was pressurised with nitrogen, confirming that the leak was still present. Three days later, on the 29th, Kizim and Solovyov spent two hours and 45 minutes installing a bypass line between the two fill tubes; this served to create a new conduit to the main oxidiser supply. Alas, when nitrogen was again pumped through the system to check its integrity, *again* it leaked. The cosmonauts could do little but replace the thermal blankets and return, acutely disappointed, to Salyut 7's airlock. A measure of success was achieved on the next EVA, on 4 May, when they

With their record-breaking occupancy of Salyut 7, from February to October 1984, Leonid Kizim and Vladimir Solovyov – together with physician Oleg Atkov – performed and supported more EVAs in a single mission than had ever been attempted in two decades of human space exploration. In their two flights together, Kizim and Solovyov performed a total of eight EVAs, spending more than 31 hours in the vacuum of space. Here, Kizim salutes through one of Salyut 7's windows.

demonstrated their adeptness at translating the length of the station, again removed the thermal blankets and fitted a second conduit to the oxidiser system. This enabled troubleshooters on the ground to identify the precise location of the ruptured line . . . but the euphoria was short-lived when it became clear that Kizim and Solovyov lacked the tools with which to perform a repair. Progress 20, with its extension and foot restraints, undocked on the 6th and burned up in the atmosphere shortly thereafter. There was little more to be done until the required tools were delivered by a subsequent Progress, although the cosmonauts *did* venture outside, a fifth time, on 19 May, spending three hours installing new a pair of gallium arsenide solar arrays, to augment one of those already aboard Salyut 7, and winching them open, like an accordion. Gallium arsenide was considered to be more efficient than the silicon cells of Salyut 7's main arrays, increasing the electrical generation capability by more than a quarter. From the interior of the station, Atkov was able to assist by rotating the main array by 180 degrees, enabling them to fit the two extensions. It was intricate, fiddly work, exacerbated by the fact that the spacewalkers were encumbered by their bulky suits. Vladimir Solovyov compared the operation to threading a needle in a pair of boxing gloves. With the new additions, and those panels installed by Vladimir Lyakhov and Alexander Alexandrov, six months before, Salyut 7 now benefitted from an extra 4.6 m^3 of solar cells and an extra 1.2 kilowatts of power. The rate at which EVAs were being performed partially accounts for the remarkable rate at which Progress missions were being despatched toward the station during this period – Progress 20 in mid-April, Progress 21 in early May and Progress 22 in late May – since a quantity of oxygen was being expelled overboard at the start of each spacewalk and required replenishment. Additionally, this expedition was the first to house three resident crew members and their demand on the life-support system was correspondingly higher.

With the completion of the fifth EVA, Kizim, Solovyov and Atkov returned to the normal routine of space station operations. The arrival of Progress 22 at the end of the month brought with it supplies – including mail, parcels ... and the news that Kizim's wife, Galina, had given birth on 24 May to a daughter, Tatiana. It must have been difficult to come to terms with the realisation that father and daughter would not meet for more than *four months*, until October 1984. Still, there was much to keep the cosmonauts busy. Progress 22 brought with it a large complement of medical equipment for Atkov and an extensive series of instrumentation, cameras and film to support six weeks of detailed geophysical research. Not until the middle of July did the freighter undock, by which time a Salyut landing window approached and the launch of a Soyuz-T visiting crew could be anticipated. The size and nature of this crew, however, was *not* anticipated, for the much-used 'rule' of predicting crew size, based on the station's operating altitude, turned out to be wrong: a two-person team was expected, but Soyuz T-12 actually carried *three* cosmonauts. Phillip Clark, in his analysis of the early Soviet programme, did not offer an explanation for this anomaly, merely noting that the launch "disproved the 'rule'". Years later, after the era of *glasnost* and the fall of the Soviet bloc, it would become apparent that Vladimir Dzhanibekov, Svetlana Savitskaya and Igor Volk were assigned to Soyuz T-12 in December 1983, backed up by Vladimir Vasyutin, Viktor Savinykh and Yekaterina Ivanova. With American plans for Kathy Sullivan to make a spacewalk on Shuttle flight 41G in October 1984, it came as little surprise when the Soviets revealed that Savitskaya would perform an EVA with Dzhanibekov during her time aboard Salyut. With the exception of flying steadily increasing duration missions over the next few years, the Savitskaya EVA would be one of the last Soviet propaganda triumphs.

Rex Hall and Dave Shayler explained that Savitskaya had been chosen for the EVA because, of all the women cosmonauts, she obviously had the experience and also the "physical strength" to perform in a bulky suit for several hours in the vacuum of space. Yet the propaganda value of the mission was inescapable. The inclusion of Volk, an experienced test pilot who was then training to command the top-secret Soviet Shuttle, Buran, on its maiden voyage, was apparently done to provide him with space time ... but there was no other Buran pilot on the backup crew! This reinforced in many minds that Savitskaya's EVA took priority over the flight of a future Buran pilot. In his diary, Viktor Savinykh, who shadowed

Savitskaya's EVA training, confirmed what many in the West had already guessed: that the choice of a woman to perform a spacewalk, only weeks ahead of the Americans, was no coincidence. Apparently, the decision was made by Valentin Glushko, head of Energia since 1974, who ordered it explicitly to upstage Kathy Sullivan. Having said this, Yekaterina Ivanova was apparently untrained in EVA and, if the backup crew had been substituted at the last moment, it must be assumed that Vasyutin and Savinykh would have performed the excursion.

Little was known about the Soviet Shuttle in the West in 1984 and the presence of Volk raised more than a few eyebrows. Why, asked the British Interplanetary Society in its journal *Spaceflight*, was a test pilot occupying a Soyuz seat, normally reserved for a researcher or a foreign cosmonaut? Years later, the intimate involvement of Igor Petrovich Volk in the development and testing of Buran would become clear. Born in Zmiiv, on the banks of the Seversky Donets River, in Kharkiv Oblast in eastern Ukraine, on 12 April 1937, he attended primary school in his home city, then moved to the Soviet Far East – to Ussuriysk, midway between Lake Kankar and Vladivostok – and later Kursk, in his native Ukraine, to receive his high schooling. Whilst in Kursk, he spent time at a local aero club and made his first flight in 1954. Aviation hooked him like a fish on a line, yet he had no awareness of where it might lead. "I did not really try to become a test pilot," Volk told an interviewer, years later. "It was made by chance. In the Fifties, I didn't know that there *was* a school of test pilots ... In this time, I was in the armed forces in the division of Baku's air defence." One day, a group of test pilots appeared on the base, participating in high-temperature trials of the MiG-21 jet, and Volk had the opportunity to meet the instructor of the school, Mikhail Agafonov. It was the latter who guided him to enter the test piloting school. In 1965, Volk graduated as a 'Test Pilot, Fourth Class', then progressed rapidly to third class in 1966, second class in 1969 and first class in November 1971.

As a pilot, Volk's credentials were unquestioned and he flew fighters, bombers and transport aircraft during his career. In later life, he would hold the presidency of the National Aero Club of Russia and the vice presidency of the FAI. (Interestingly, when Svetlana Savitskaya took one of her examinations at test pilot school, Volk had been one of her instructors.) On one occasion, with fellow Buran candidate Anatoli Levchenko, he glided a Tu-144 supersonic transport – the Soviet Concorde – to a safe landing ... from an altitude of *twenty-two kilometres ... without* engine power! His involvement in Soviet Shuttle concepts originated with the Spiral spaceplane and, in May 1976, Volk flew a MiG-105 test model and became a cosmonaut in July 1977 to work on Buran. He was one of five test pilots selected by the Gromov Flight Research Institute, based in Zhukovsky, near Moscow, for the Soviet Shuttle and would be the first of their number to fly into space. The other candidates in his group, sadly, would not share much of Volk's fortune: although Anatoli Levchenko *did* fly a Soyuz mission in December 1987, he succumbed to complications from a brain tumour only months later. The others, Oleg Kononenko, Rimantas Stankyavichus and Alexander Shchukin, would all die in aircraft accidents, before ever getting a chance to fly in space. By September 1982, Volk had been assigned to a Salyut 7 visiting mission, with Kizim and Solovyov, but the team was disbanded in May of the following year, after the failure of Soyuz T-8. Eventually, in December 1983, Volk was assigned the third seat aboard Soyuz T-12, with the intention that after his return to Earth he would fly the Tu-154 aircraft, whose instruments had been modified to mirror the displays aboard Buran, from Tyuratam to Akhtubinsk, *and* then return to Tyuratam at the controls of a MiG-25 fighter. The idea was to evaluate a cosmonaut's ability to pilot a high-performance aircraft to a smooth touchdown after experiencing several days in the disorientating weightless environment. After Soyuz T-12, Volk did just that . . . and completed *both* flights perfectly.

Before those flights, he had to complete his first mission into space. Soyuz T-12, carrying tools for the oxidiser leak repair, was launched from Tyuratam at 8:41 pm Moscow Time on 17 July 1984 and arrived at Salyut 7 late the following evening. The highlight came at 5:55 pm on the 25th, when the station's hatch opened and Dzhanibekov and Savitskaya emerged into the brilliance of orbital sunlight for a historic EVA which would last three hours and 55 minutes. Dzhanibekov set up foot restraints, a work lamp and an external power outlet, in order to test a piece of hardware known as the *Universalny Rabochy Instrument* ('Universal Hand Tool', or 'URI'). This consisted of a portable electron beam device, weighing some 30 kg, for cutting, welding, soldering and brazing. Before the flight, the URI had caused great concern on the ground: some engineers felt that its high operating temperatures and heat production might damage the cosmonauts' space suits. However, the instrument worked perfectly. Savitskaya began work with the URI just as Salyut 7 drifted out of radio contact with the ground; after the reacquisition of signal, she used the device to cut a 0.5 mm titanium sample and, in total, she performed six cuttings of titanium and stainless steel, two silver spray coatings of anodised aluminium and six soldering tests of tin and lead solder. During the EVA, her heart rate peaked at close to 140 beats per minute. At length, Savitskaya handed the experiment over to Dzhanibekov, who found it "handy" and expressed his conviction that "I'm sure we'll be using it a lot". Shortly before re-entering the station, they retrieved some exposure sample cassettes for return to Earth and installed a 'Medusa' biopolymer cassette in their place.

After the mission, it would be revealed that Dzhanibekov and Savitskaya had specifically trained to finish the oxidiser line repair, and that Kizim and Solovyov, having already made four arduous spacewalks on this task, had asked flight controllers to allow them to take the responsibility. Their request was accepted. Whilst aboard Salyut 7, Dzhanibekov, who had a wealth of experience in the Star City hydrolab, actually provided instruction to Kizim and Solovyov on techniques for their sixth EVA. He also brought along a model engine, which was fastened onto one of the internal walls of the station, allowing him to demonstrate the repair method, together with video tapes and methodical instructions. The Soyuz T-12 crew returned to Earth on 29 July, their descent module touching down in Kazakhstan at 3:55 pm Moscow Time. Ten days later, Kizim and Solovyov swung open Salyut 7's airlock hatch. Dzhanibekov's crew had delivered a portable pneumo press and, after reading manuals, watching videotapes and studying photographs, they were able to move quickly to their work site and begin the task. By now, translating along the length of Salyut 7 had become old hat; Solovyov compared it to "almost *running*". They pulled back the thermal blankets and used the press to squeeze a stainless steel pipe and successfully seal the leak. Despite Solovyov experiencing a failure of the cooling water pump in his suit, and both men suffering heavily bruised hands, "as if they had been in a fist fight", according to Oleg Atkov, the EVA ended triumphantly after precisely five hours. Upon their return to Earth, the cosmonauts were showered with praise for their achievement.

That achievement extended far beyond the repair itself and a demonstration of the men's abilities in the harsh environment of space; in their six outings, Kizim and Solovyov had made more EVAs than any astronaut or cosmonaut in history and this was made all the more remarkable when one understands that, before Soyuz T-10B, only six spacewalks had *ever* been performed by the Soviets ... and only as recently as November 1983 had as many as *two* EVAs been performed in a *single* mission. Kizim, Solovyov and Atkov had *tripled* the achievement of their Soyuz T-9 predecessors, Vladimir Lyakhov and Alexander Alexandrov, and had performed as many spacewalks as had previously been performed in the *entire* Soviet manned space programme. Across those six excursions, Kizim and Solovyov spent 22 hours and 50 minutes outside the space station, marking a considerable expansion in EVA activity. The mission continued, with the three men quietly exceeding the 211-day record of Anatoli Berezovoi and Valentin Lebedev on 5 September and over the next three weeks, as the next Salyut landing window approached, they steadily ramped up their exercise regime to prepare for a return to Earth. On the first of October, it was announced that the cosmonauts would be back home within 24 hours and the Soyuz T-11 capsule touched down perfectly at 1:57 pm Moscow Time on the 2nd. Leonid Kizim, Vladimir Solovyov and Oleg Atkov were the new world record-holders for the longest single space mission in history, at almost 237 days. The cosmonauts had worked well together, in spite of Kizim's penchant for sticking, inflexibly, to the schedule and refusing to let himself, his crew or flight controllers rest until all daily tasks had been completed. In 1985, the Soviets expected to take still greater strides, with plans to occupy Salyut 7 for up to *nine* months, with the possibility of an *all-female* visiting crew. None of those plans came to pass, but what 1985 *did* demonstrate was the remarkable abilities of cosmonauts in space – their capability to snatch triumph from the fangs of defeat – and would surely test the physical and psychological preparedness of the men sent to explore the final frontier.

Naturally, in the days after the landing of Kizim, Solovyov and Atkov, the Soviets were quick to point out that eight months in space was roughly equivalent to a one-way trip to Mars ... although, three decades later, such voyages with humans remain a distant vision for the future. *Remaining* in orbit for eight months established a baseline of valuable physiological and psychological data, it was true, but the hardware, the life-support mechanisms, the propulsion technology, the infrastructure and the ability to 'live off the land', to a degree, whilst on the Martian surface – all needed for a human expedition to the Red Planet – were, and still are, at the very edge, or beyond, our present capabilities. As Leonid Kizim looked into the eyes of his baby daughter, Tatiana, and cradled her in his arms, he might have wistfully hoped that a member of *her* generation might be the first to set foot on Mars. Sadly, progress beyond Earth orbit with people has been notoriously slow and only now, almost five decades since Apollo, is the talk of missions to the Moon and the planets turning steadily from mere talk and presidential speeches into hardware and launch targets. Kizim, who died in 2010, would never get to see it happen. In reality, it will most likely be one of his grandchildren's generation who will someday take those next bold steps into the Solar System.

3

An age of innocence

TRISKAIDEKAPHOBIA

Thirteen has for hundreds of years been regarded by many cultures as an unlucky number. One person who apparently *did* believe strongly in 'triskaidekaphobia' – an irrational fear of the number thirteen – was NASA Administrator Jim Beggs and it was from the agency's Washington headquarters that an edict came, sometime in the early summer of 1983, announcing a change to the Shuttle numbering system to avoid STS-13. Instead of straightforward numbers, missions would be assigned a cryptic and somewhat clumsy alphanumeric combination, which astronaut Vance Brand described as "a neat new way of designating missions … that *confused* everyone!" Firstly, there was the system itself. Brand's mission came to be known as 'STS-41B'. The first number denoted the last digit of the fiscal year under which the flight was funded (1984 in this case), whilst the second number identified the launch site (with '1' for the Kennedy Space Center and '2' for Vandenberg Air Force Base in California). Finally, the letter described its position in the sequence of flights for a given year; thus, 41B would be the *second* ('B') mission of fiscal year 1984, manifested for launch from the Kennedy Space Center. It is possible to date Beggs' edict to sometime in the early summer of 1983, based on the course of Shuttle flight assignments; by February, full crews had been named through to STS-13 and partial crews for a pair of Spacelab missions, STS-18 and STS-24. By the time the *next* assignments were made in September, Karol 'Bo' Bobko and Rick Hauck were named to command flights known under the new system as '41E' and '41G'. Astronaut Terry Hart, one of the STS-13 crew members, narrowed the timeframe still further in his NASA oral history. He recalled that the decision came "three or four months" after his assignment, thus May or June 1983. In his memoir, *Riding Rockets*, Mike Mullane attributed the final decision over the new system to JSC Director Gerry Griffin. "It was hoped," he wrote, "that this code would sufficiently blind the god of bad luck." The change prompted a retroactive redesignation of earlier flights still waiting to be launched, but caused the plot to thicken still further.

It should be understood that Shuttle payloads for many missions were in constant flux during this period, with some being delayed, others moving forward or backward in the 'pecking order' and frustrating problems with the Air Force's Inertial Upper Stage (IUS) booster having caused a few to be cancelled outright. For example, when astronauts Ken Mattingly, Loren Shriver, Ellison Onizuka and Jim Buchli were assigned to STS-10 in October 1982, they confidently expected to fly Challenger in the autumn of the following year on the Shuttle's first classified mission for the Department of Defense. Their military payload would have utilised an IUS, but when the booster failed to properly deliver NASA's first Tracking and Data Relay Satellite (TDRS) into geostationary orbit in April 1983 it prompted a lengthy inquiry and the cancellation of several missions which relied upon its services. By the end of May, a second TDRS had been deleted from STS-8 and the STS-10 crew soon found their flight removed entirely from the manifest. (They remained together as a crew, however, and in September 1983 were reassigned to 41F.) It was not unexpected, but unsettling, nonetheless. "As we started to do some of the background work and early training," Loren Shriver recalled years later, "it became apparent that STS-10 wasn't going to go in sequence … *or* on time. *That* was when we started to learn that the numerical sequence of the missions didn't mean a lot in those early days!" Since NASA's fiscal year begins on 1 October, the first mission to launch under the agency's 1984 budget allocation was actually STS-9, which took place in November 1983; internally, therefore, this flight was retroactively redesignated as '41A'. Since STS-10 had been cancelled, the *next* mission on the manifest, STS-11, commanded by Vance Brand, became the 'new', tenth Shuttle flight and was redesignated '41B' … and *this* brings us back to Jim Beggs' fear of thirteen. STS-12, another IUS mission, led by Hank Hartsfield, had *also* been deleted, which meant that the 'original' STS-13 – the mission whose unlucky number started the ball rolling in the first place – actually wound up as the *eleventh* Shuttle flight, *not* the thirteenth, and was correspondingly named '41C'. Hartsfield's crew moved a couple of months downstream, picked up a new payload and became 41D, Mattingly's team became 41F and Bobko and Hauck were assigned to command 41E and 41G respectively.

Officially, the decision to redesignate Shuttle flights in such a peculiar fashion was explained by NASA as serving as a sort of 'anchor' for payloads, to prevent them from jockeying around quite so much on the manifest. "Since some reordering of flights may occur, crews will be assigned by payload assignment, rather than an STS number," the agency announced in September 1983. "If a launch moves in the sequence, the mission designator will not change." If this was a partial reason for the change, it actually ended up failing entirely. In the words of astronaut Steve Hawley, who flew on 41D and 61C, it was not unusual for astronauts to change flights on several occasions and it would appear that the idea of the new system acting as a 'payload anchor' actually made little difference. "We were assigned to two or three different flights before it stabilised out," he told the NASA oral historian. "Today, you get assigned to a flight and *that's* the flight you'll fly. If the payload's delayed, then *you* delay with the payload. Back in those days, a lot of the flights were similar; they were launching satellites or running some experiments that could be quickly learned and it wasn't as important to stick with your payload."

Unofficially, in the minds of Hart, STS-13 commander Bob Crippen and lead flight director Jay Greene, the decision to redesignate Shuttle missions was chiefly a response to an ingrained triskaidekaphobia on the part of the NASA Administrator. At first glance, it may seem surprising that an agency whose focus lies in science and technology should devote such emphasis to ancient superstition ... but for one thing: the unlucky voyage of Apollo 13. More than a decade earlier, in the spring of 1970, the Moon-bound mission had ticked each and every one of the 'unlucky' boxes; in addition to its number, it had blasted off at 13:13 Houston time, had been scheduled to arrive in lunar orbit on 13 April and one of its crew members had been replaced due to illness only days before launch. During Apollo 13's journey to the Moon, an explosion destroyed one oxygen tank and crippled another, placing the astronauts in dire peril. Although the crew returned safely to Earth, the legacy of their ill-fated voyage would endure as one of the most harrowing near-disasters ever experienced in the history of human space exploration and would haunt NASA's senior leadership. "The Apollo 13 experience," Hart told the oral historian, "gave NASA a *bad* case of triskaidekaphobia, because there were a whole bunch of thirteens in that." Nor was STS-13 an ordinary mission. It was, in fact, an extremely high-profile flight, involving a complex orbital ballet of rendezvous and spacewalking to retrieve and repair NASA's damaged Solar Max satellite ... and early plans, dating from November 1982, called for it to launch on Friday 13 April 1984! At length, this date was moved forward and by the beginning of 1984 NASA had rescheduled it for 4 April. It would eventually launch on the 6th, fly for seven days ... and land on Friday the 13th! For his part, Jay Greene was also convinced that triskaidekaphobia was the most likely catalyst for the Byzantine numbering system imposed on the Shuttle. Indeed, when Vance Brand's 41B crew released their mission patch towards the end of 1983, it perfectly illustrated the suddenness with which the change took effect. Around the edge of their patch were 11 stars ... for the flight which should have been STS-11.

HUMAN SATELLITES

Also in pride of place on Brand's patch was a snazzy, jet-propelled space suit backpack, known as the Manned Manoeuvring Unit (MMU), together with the surname of an astronaut who had waited longer than most for his first orbital voyage. Bruce McCandless had joined NASA in April 1966, alongside Brand, but his patient wait for a journey into space had exceeded by more than a decade that of many of the newer astronauts. In fact, even old-timers from the Excess Eleven had not waited quite as long as poor McCandless. One of the reasons for his lengthy status as an astronaut-in-waiting was that he had been instrumental in the development of the MMU and had long been tapped to test it on its first outing in space. "In retrospect," he told this author, in an email correspondence from March 2006, "I probably lavished too much attention on scientific and engineering interests, as opposed to the flying, flying and more flying." His long wait – close to two full decades by the time Challenger's main engines ignited at precisely 8:00 am

EST on 3 February 1984 – would be worth it and his famous photograph, snapped by Hoot Gibson, has since graced many a space book, magazine cover, wall poster and screensaver ... and even sparked a bizarre copyright case, involving the singer Dido.

It is often said that great men come from great families with great and illustrious backgrounds and McCandless, born in Boston, Massachusetts, on 8 June 1937, certainly fulfilled much of this criteria. His great-great-grandfather, David Colbert McCanles, was a Nebraska rancher, a former sheriff and, as a member of the 'McCanles Gang', was infamously killed by 'Wild Bill' Hickok in December 1861. The family subsequently changed their name to 'McCandless' and moved to Colorado. McCanles' grandson, Byron McCandless, rose to become a naval commodore who helped to create two separate designs for the Flag of the President of the United States. His son, the 'first' Bruce McCandless, also followed a military career and graduated from the Naval Academy, but made his name and reputation during a bloody naval battle off Guadalcanal in the Solomon Islands. In the early hours of 13 November 1942, Lieutenant-Commander Bruce McCandless was serving as communications officer aboard the USS *San Francisco* when the ship's navigation bridge sustained a direct hit from the Japanese cruiser *Nagara*. The incident killed the admiral, his captain and most of the other senior officers. McCandless himself was rendered unconscious and, upon reviving, he and another lieutenant-commander, the damage control officer Herbert Schonland, took command. Steering and control of the *San Francisco* were lost and regained on several occasions and the ship endured 45 direct hits and sustained major structural damage, yet survived to fight another day. McCandless and Schonland both received the Congressional Medal of Honor, with citations which praised their extreme heroism, outstanding leadership and personal bravery. Both continued their naval careers and both retired as rear-admirals; McCandless would die in 1968, two years after NASA selected his son as an astronaut. Both Byron and the first Bruce McCandless would have streets at naval bases named in their honour, as well as a frigate, the USS *McCandless*, launched in 1971.

Bruce McCandless II was barely five years old when his father fought through the night to save himself and his shipmates in November 1942, but a naval career beckoned and a parent with a Medal of Honor virtually guaranteed him an appointment to a prestigious military academy. McCandless completed high school in Long Beach, California, and entered the Naval Academy, receiving his degree in 1958 and graduating second in his class of 899 students. (One of his contemporaries was John McCain, later a senator and Barack Obama's Republican opponent in the 2008 presidential election.) McCandless underwent flight instruction at the Naval Air Training Command in Pensacola, Florida, and Kingsville, Texas, and was designated a naval aviator in 1960. Following weapons systems and carrier landing training, he flew the F-4D Skyray and F-4B Phantom II and served aboard the USS *Enterprise* during the Cuban Missile Crisis of October 1962. McCandless worked as an instrument flight instructor and in 1964 reported to the Naval Reserve Officer Training Corps Unit at Stanford University to begin a master's degree in electrical engineering. Two years later, he was selected by NASA and in July 1969 entered the

In one of the most iconic images of the 1980s, Bruce McCandless puts almost two decades' worth of development work of manoeuvring units into practice as the world's first untethered spacewalker. During two EVAs with the Manned Manoeuvring Unit, he ventured more than 90 m from Challenger.

headlines as the capcom in Mission Control when Neil Armstrong set foot on the Moon. McCandless was later assigned as backup pilot for the first Skylab mission and participated in the development of an astronaut manoeuvring unit known as Experiment M-509. This nitrogen-fed backpack was tested inside the space station in 1973 and served as a forerunner of the MMU backpack. In fact, the spectacular success of the MMU would win NASA and prime contractor Martin Marietta the coveted Collier Trophy for 1984. McCandless, together with NASA's Charles 'Ed' Whitsett and Martin Marietta's Walter 'Bill' Bollendonk, would be granted special recognition for their contributions to its development and orbital checkout. Whitsett in particular would pay tribute to McCandless' work. "Nobody has left his stamp on any instrument in space," he told the *Washington Post*, "like Bruce has left his mark on the backpack."

In the wake of the Columbia disaster, it seems ironic that the original purpose of the MMU was to enable spacewalking astronauts to inspect and possibly repair damaged thermal protection materials on the Shuttle's wings and lower surfaces. Moreover, in the words of a NASA news release from October 1979, it would also permit rescue operations and even the servicing of disabled satellites. Despite the hype that accompanied its first flight, however, it was actually the latest in a long line of jet-propelled 'guns' and manoeuvring packs whose heritage dated back to the early 1960s and originated from both NASA and the US Air Force. A decade before McCandless undertook his MMU sortie outside Challenger, astronauts performed Experiment M-509 aboard Skylab and, still earlier, in June 1965, the first American spacewalker, Ed White, employed a hand-held pressurised gas 'gun' to manoeuvre himself around the exterior of his Gemini capsule. A year later, in June 1966, Gene Cernan encountered problems with his space suit and was unable to fly the Air Force's Astronaut Manoeuvring Unit jet-propelled backpack. Cernan's difficulties "led to a retrenching to develop EVA technology" and McCandless, together with Whitsett and NASA's David Schultz, began to study the concept which evolved into Experiment M-509. "I hoped to be the first to fly it," McCandless continued, "but that was not to be." Unfortunately, during Skylab's ascent to orbit on 14 May 1973, a solar panel and the micrometeoroid shield were torn away. Temperatures inside the station soared and were only stabilised by the stoic efforts of the Skylab 2 crew in a complex spacewalk and a gruelling programme of emergency repair work. "They, however, were prohibited from trying the manoeuvring unit out due to fears that its nickel cadmium batteries had been damaged by the high temperatures inside the workshop," said McCandless. "The two subsequent Skylab crews did use the M-509 and gave it glowing reports, thus enabling us to sell NASA management on building an MMU in connection with the Shuttle, initially planned for the conduct of tile inspection and repair. Ultimately, those tasks were scrapped and it was built and tested to support the Solar Max repair mission." Before it could be committed firmly to this repair mission, however, a thorough test of its systems and performance in orbit was required...and this was the task of McCandless and Bob Stewart.

If Buzz Aldrin was overshadowed by Neil Armstrong as the second man to set foot on the Moon, then it is a regrettable quirk of history that the achievements of the second man to fly the MMU, Lieutenant-Colonel Robert Lee Stewart, should have been similarly overlooked by the media. It was McCandless' image which became world-famous and appeared on the covers of *Aviation Week* and *National Geographic*, yet Stewart would make history in his own way, becoming the first active-duty aviator from the US Army to be selected as a NASA astronaut. He was born in Washington, DC, on 13 August 1942, but was raised in the Deep South, receiving schooling in Alabama and Mississippi. After high school, he attended the University of Southern Mississippi and earned a degree in mathematics in 1964. Stewart joined the Army in May of that same year and was assigned as an air defence artillery director at Gunter Air Force Base in Alabama. Two years later, after rotary-wing training, he became an Army aviator and flew more than a thousand hours of combat time in assault helicopters in Vietnam. "In Vietnam, it got to where I feared the words, 'Sir, you've got to see this', because one time I had a crack in the pitch change horn, the linkage which controls the main rotor," he told interviewer Elaine Ervin in October 2000. "That pitch horn should never have held together, but it *did*. If I had ever lost that pitch change horn, I would have lost control of the vehicle and we would have all died. Another time, the 42-degree gearbox was shot out. The gears should have seized, but didn't." Years later, Stewart turned to Christianity and credited God as having protected him during these troubled times. Certainly, his awards and decorations speak volumes for his achievements during that war; they include two Purple Hearts – awarded only to military personnel wounded or killed fighting for the United States – and a Bronze Star for bravery and meritorious service.

"During the war," he explained to Ervin, "I was determined that I would not be a 'foxhole Christian'. I was at such a point in my life that I said that this is *not* going to drive me to believe something just to take out a fire insurance policy on my own soul." In addition to his combat time in Vietnam, Stewart served as an instructor pilot at the Army's Primary Helicopter School and completed the Air Defense School's advanced courses for air defence officers and guided missile systems officers. He also received a master's degree in aerospace engineering from the University of Texas at Arlington in 1972. Shortly thereafter, he was sent to Seoul, in South Korea, as a battalion operations officer and executive officer and, two years later, attended Naval Test Pilot School at Patuxent River, Maryland, to complete the Rotary Wing Test Pilot Course. Stewart later moved to Edwards Air Force Base in California as an experimental test pilot for the UH-1 Iroquois and AH-1 Cobra helicopters and the U-21 Ute and OV-1 Mohawk fixed-wing aircraft. He was also the project officer and senior test pilot for the Hughes-built YAH-64 Apache advanced attack helicopter and worked on the electronic flight controls for the UH-60 Black Hawk. Stewart was serving in this capacity when NASA announced its intention to select a cadre of new astronauts in the summer of 1977; he arrived in Houston for a week of physiological and physical tests in late October and was the only Army applicant to be selected in January 1978. After completing his initial training, he worked on the Shuttle's re-entry flight control systems, was on the support crew for STS-4 and served as a capcom during STS-5. During 41B, Stewart assumed the role of the flight engineer, seated behind and between Brand and Gibson on Challenger's flight deck and assisting them with the monitoring of the systems. Yet the real thrill of the mission would undoubtedly be his two EVAs with McCandless and the pair jokingly nicknamed each other 'Buck' (Rogers) and 'Flash' (Gordon) during their time together in orbit.

Although the need to repair portions of the Shuttle's thermal protection system was one of the main reasons for the MMU, its development – which began in earnest in 1975 – was still hampered for some years by management apathy and lack of firm funding. Then, in the spring of 1979, as Columbia was being moved from California to Florida, several heat-resistant tiles were lost from her airframe and renewed vigour was injected into developing the backpack. By the time STS-1 took to the skies in April 1981, most of the tile problems, seemingly, had been solved and no MMU was aboard. It would instead be used, said NASA, for satellite repairs and maintenance and its usefulness was enhanced by the provision of electrical sockets for tools, portable lights and cameras. The device measured 1.2 m high, 81 cm wide and 66 cm deep and on a typical mission, two MMUs were stored on Flight Support Structures (FSS) on opposing walls towards the front of the payload bay. An astronaut would back into it and secure two spring-loaded latches and a lap belt into place, then release the unit from its support structure and float free. After more than four years in the design definition stage, in February 1980 NASA awarded the $26.7 million MMU fabrication contract to Martin Marietta of Denver, Colorado. The first two operational flight units, valued at around $10 million apiece, arrived at JSC in Houston in September 1983 to support astronaut training. Two months later, they were installed aboard Challenger. Each weighing 140 kg, they were painted white to achieve adequate thermal control in the harsh environment of low Earth orbit and were fitted with electrical heaters to keep their components above minimum temperature levels. Affixed to the back of each MMU were two propellant tanks, which supplied 24 tiny thrusters with a total of 18 kg of high-pressure gaseous nitrogen. To operate the thrusters, the astronaut used hand controllers at the end of two armrests: one provided roll, pitch and yaw control, while the other allowed him to move forward, backward, up, down and from left to right. Furthermore, by using both in unison, he could achieve very intricate movements. Particularly useful for repair missions, when a desired orientation had been reached, he could activate an automatic, 'attitude hold' function to free his hands for work. Electrical power came from a pair of silver zinc batteries, capable of supporting the unit for up to six hours of autonomous flight as far as 140 m from the Shuttle. In fact, one of the MMU's widely publicised features was that its wearer did not need to remain attached to the spacecraft by a safety tether. Of course, in the event of problems, most of its systems were redundant and neither McCandless or Stewart ventured so far from Challenger that the pilots would not be able to rescue them if necessary. "We didn't want to come back and face their wives if we lost either one of them up there," joked Vance Brand.

The MMU's controllability was crisp and precise. "The minimal training and precision flying features," said one magazine editor, who flew a model of the MMU at Martin Marietta's Space Operations Simulator (SOS) in Denver, "were demonstrated by my ability, with only a few minutes' practice, to manoeuvre safely in close proximity to fixed objects." Joe Allen, whose own MMU sortie salvaged an errant communications satellite, also remarked that in space it "glided" and displayed none of the idiosyncratic jerks, jolts, bumps and grinding sounds that were characteristic of Martin Marietta's simulator. For Bruce McCandless, who backed himself into the device early on 7 February 1984, it represented "a heck of a big leap", in terms of spacewalking technology and the culmination of his own personal odyssey. Just like the EVA of Musgrave and Peterson a few months earlier, preparations for this excursion had begun soon after Challenger reached orbit. The Shuttle's cabin pressure was lowered from 101.3 kPa ('normal' atmospheric pressure at sea level) to 70.3 kPa in order to reduce McCandless and Stewart's 'pre-breathing' routine from three hours to less than one hour. It was familiar ground for McCandless, who admitted that he was "probably not a representative EVA trainee" and had been "grossly over-trained". Throughout the 1970s and 1980s, he took every opportunity to get into a space suit, an altitude chamber or a water tank and in addition to his work developing Experiment M-509 and the MMU, he had also participated extensively in EVA simulations on Skylab and the Hubble Space Telescope. At Martin Marietta's Denver facility, he and Stewart had 'flown' mockups of the MMU within the SOS, a large room which measured 15 m long by 4.1 m wide and 4.5 m high. "It was quite effective," McCandless concluded, "and could accommodate a fully-suited astronaut and reasonable sized mockups of 'target' objects, such as the underside of the orbiter for tile repair. It also had the capability for introducing malfunctions for training purposes."

In spite of their complexity, McCandless and Stewart's excursions proved successful and the space suits and MMUs performed admirably. The only 'nuisances' were static on the communication channels and difficulties attaching checklists to the suits' arms. "In spite of the 'sound-does-not-travel-through-a-vacuum' tenet of physics," McCandless explained, "it *was* noisy up there, thanks to two independent radio channels and plenty of people wanting to talk to me!" Then, just before leaving Challenger's airlock, Stewart reported a caution and warning alarm, which indicated the pressure of his suit's sublimator had risen to 27.6 kPa. However, after being switched off and back on again, it performed normally. These subtle problems did not distract from the triumph of McCandless' Buck Rogers-style flight that day. Despite the sci-fi analogy, said Vance Brand, the MMU "didn't have the person zooming real fast. It was a huge device that was very well-designed and redundant, so that it was very safe, but it moved along at about one to two miles per hour." At his furthest distance from the Shuttle, McCandless was 91 m away, politely offering to clean Challenger's cockpit windows as he floated over the flight deck. Watching intently from inside, an admiring Brand declined the offer.

Also watching intently, camera in hand, was Robert 'Hoot' Gibson, Challenger's pilot, and the person who snapped the photograph which would make history as one of the top five most-requested images from NASA. In an interview for the Smithsonian in 2001, he recalled the astonishing sight of McCandless flying the MMU. "Bruce first did a couple of brief test flights in the cargo bay, staying very close in case anything should go wrong," Gibson explained. "As we were approaching sunrise on one of our daylight passes, he was cleared to make the translation out to 300 feet from the Shuttle." Grabbing his Hasselblad camera,

Gibson began shooting frame after frame. Since Challenger's orientation was some 30 degrees from the vertical, McCandless appeared at a similar angle with respect to Earth's horizon. Gibson knew that any one of his images could easily make the cover of *Aviation Week* – they actually made *two* – and remembered taking multiple light settings and tweaking the focus five or six times for each photograph, before squeezing the button. The famous shot would come to be known as 'Backpacking' and, even today, McCandless possesses a goofy version in his home, in which his grown daughter pokes her head through the cut-out visor in a life-size reproduction at a Seattle museum. In 2005, McCandless explained that what he liked most about the image was its lack of identity; with his sun visor closed, it is impossible to see his face, "and *that* means it could be *anybody* out there ... sort of a representation, not of Bruce McCandless, but mankind". This assertion gained a measure of irony a few years later, when, in September 2010, he sued the singer Dido for her unauthorised use of the image on the cover of her album, 'Safe Trip Home'. Although NASA images are not bound by copyright, McCandless argued that the image infringed his 'persona', but the case was settled amicably early the following year.

The man who took the photo, Robert Lee Gibson had grown up with the moniker 'Hoot', as had his father. "I always tell people that it comes from 'not worth a hoot'," he once told an interviewer, but in reality it originated from Edmund Richard Gibson, a famous rodeo champion who turned into a cowboy film star in the 1920s and 1930s; *his* nickname of 'Hoot-Owl' came from co-workers and, later, evolved simply into 'Hoot'. "So after that," his astronaut namesake continued, "*everybody* whose name is Gibson usually picked up the name 'Hoot'." In fact, when he progressed into the Navy Fighter Weapons School – the famous 'Top Gun' – Gibson chose the nickname for his radio callsign. Born in Cooperstown, New York, on 30 October 1946, he completed high school in Huntington and entered Suffolk County Community College on Long Island to study for an associate degree in engineering science. He later earned a bachelor's degree in aeronautical engineering from California Polytechnic State University in 1969. Today, he is renowned in space circles as a five-flight veteran, commander of the first Shuttle-Mir docking mission and former chief of the astronaut corps, but underpinning each of those accomplishments has been Gibson's lifelong passion for aviation. His father was a test pilot and inspector for the Civil Aeronautics Administration and built his own private aircraft in the garage, whilst his mother was one of the few women to fly general aviation aircraft in her day; in her youth, she and two friends had bought a J-2 Taylor Cub. With such an impressive pedigree, it is hardly surprising that their son should have charted his own course for the skies and beyond. As a boy, Gibson travelled frequently with his father on CAA business and, on one occasion, the pair were at an airport in Phoenix, sitting in a Beechcraft Bonanza with just a single yoke, and Paul Gibson handed the controls to his son to perform the takeoff. The boy was just ten years old. "I was so proud that he trusted me," Gibson recalled years later. "He was my inspiration." That was just the start. Gibson soloed in a Piper Colt on the "windy, rainy, solid overcast" day of his 16th birthday and gained his private pilot's licence at 17.

After completing his bachelor's degree, Gibson entered the Navy and received basic and primary flight instruction in Florida and Mississippi, then advanced training in Kingsville, Texas, and eventually moved to Naval Air Station Miramar in California for assignment to the F-4 Phantom II fighter. "I was in awe of the F-4," he told Robin White in an interview for *Air & Space* magazine. "It looked so big and heavy and the wings seemed so small. I was reluctant to slow it down. I was sure it would fall out of the sky, but it was just totally rock-solid on approach to the carrier." From April 1972 to September 1975 he served aboard the USS *Coral Sea* and the USS *Enterprise*, flying the F-4 over Vietnam during two tours of duty. When his commanding officer asked him if he wanted a *third* tour, Gibson was not enthusiastic – "I was extremely ready to hit the beach," he told White – until he learned that the tour would involve operational deployment of the new F-14 Tomcat. If Gibson was in awe of the old F-4, then this new fighter functioned on a totally different level. On one occasion, with just 30 hours' experience in the Tomcat, he faced a thousand-hour F-4 veteran for a training dogfight. "We called *Fight's On*," Gibson recalled, "and 30 seconds later I was sitting in his six [behind him]. We ran the engagement three times. The results were *always* the same. An F-14 with a *nugget* at the stick could out-manoeuvre, out-turn and out-fight a Phantom flown by an old hand!"

Completion of the Top Gun course was followed by assignment as an F-14 instructor pilot and graduation as a test pilot in June 1977. Barely two months later, he was summoned to Houston, part of the second group of astronaut applicants to undergo a week of intensive physical and psychological testing, and was selected in January of the following year. In addition to his military aviation, Gibson also participated in air races – one of which would lose him the command of a Shuttle crew in the summer of 1990 – and secured a number of world speed and altitude records. In May 1981, Gibson married fellow TFNG Rhea Seddon and in July of the following year, their first child, Paul, was born. *Time* magazine promptly labelled the first child born to two American astronauts as 'Astrotot'.

One of his reasons for joining NASA, obviously, was to fly the Shuttle, which he considered to be the highest and fastest form of aviation possible. He also hoped that after several tours of Vietnam, landing on aircraft carriers and making it through Top Gun and test pilot school would give him plenty of stories for his grandchildren. "But, man, when I went into space, *that* wiped out everything," he told an interviewer in July 2005. "*That* was the biggest thrill that you could ever have. I wasn't scared, but about a week before I was set to launch, I asked myself, 'Are you *sure* you wanna do this?'" Naturally, after training for more than five years, the answer was a resounding *Yes*, regardless of the risk.

For Bob Stewart, seated behind Gibson and Brand for Challenger's thunderous climb into orbit, the sensation of launch was unlike anything he had ever encountered in his terrestrial existence. "You're enveloped in a noise and vibration far beyond your training," he told Elaine Ervin, "as over seven *million* pounds of thrust hurl you off the pad. Astronauts are trained to know the character of the event, but cannot be trained to its true magnitude. It's a noise that is *felt* more than *heard* – a sharp, staccato noise that *hammers* directly at the core of your being." In an aircraft, flying at several hundred kilometres per hour, it hardly seemed like he was moving . . . but aboard the Shuttle, travelling at orbital velocities of 28,000 km/h,

Backdropped by the inky blackness of space, Bob Stewart flies the MMU.

"you *know* you're smokin' along!" Vance Brand, of course, had seen the views of Earth before, but for his four rookie crewmates, the vista of the Home Planet in all her grandeur drew them to the windows like moths to a flame. *Breathtaking* was one adjective used by Stewart, "how God intended it to be viewed". The deep azure blues of the Atlantic Ocean, overlaid by the pure whites of the clouds, then the greens and browns of Africa, sharply contrasting with the velvety blackness of space. "The colours come *alive*," said Stewart, "and the visible detail is far beyond that which can be brought back on film or tape. It is truly an experience that borders on indescribable."

Four days into the mission, flying the MMU carried that experience to a still greater level; in fact, many spacewalkers have agreed that the sensation of floating above Earth with nothing between themselves and the fathomless expanse of the Universe is a decidedly ethereal one. They savoured the experience as best they could, and in their own ways, with Story Musgrave specifically setting aside time in each of his EVAs to contemplate the enormity of *where* he was. On their first EVA, less than six hours expired between the depressurisation and repressurisation of the airlock. Yet, there was still work to do and for Bruce McCandless it encompassed testing the MMU, on whose design and development he had invested almost two decades of his professional life. During their tethered work in the payload bay, McCandless and Stewart removed a failed television camera for replacement with an in-cabin unit and later installed it during their second EVA on 9 February. The MMU performed admirably, but ironically, it was Brand who undermined its *raison d'etre*. The backpack had long been touted as being capable of more precise and intricate movements than the Shuttle, but on 41B and 41C the real value of Challenger's manoeuvrability and her Canadian-built mechanical arm were demonstrated – by retrieving a lost foot restraint.

"I don't recall now whether it was before or after he went out with the backpack," Brand added, "but he was trying to reposition his foot restraint, so that he could get into it to do work. Our EVA equipment was generally tethered, but it somehow got away from him. I looked back and saw it floating away. I thought about it for a second or two and decided that the ground wouldn't have time to come up with a decision whether we ought to chase it and go after it. It was going to get away from us very quickly, so I couldn't see anything wrong with going after it. We chased it, Bruce caught it and we didn't have to worry about encountering that as 'space junk' the next time we came around the world. By so doing, Brand showed that the Shuttle was capable of the same intricate motions as the MMU and, on the next flight, 41C in April, when a task involving the MMU was frustrated, the RMS mechanical arm would prove itself equally capable. Despite the MMU's success during two satellite recoveries in November 1984, the superb manoeuvrability of the Shuttle contributed to its ultimate demise. In fact, the year immortalised by George Orwell would be the only time the MMU was ever used in space. By the end of 1984, it had seen service on three Shuttle missions, flown by six astronauts for a total of just ten and a half hours, spread across six spacewalks. Other assignments were expected but, in the wake of the Challenger disaster, safety upgrades imposed by the Rogers Commission proved costly and the two flight units were mothballed. A device known as the Simplified Aid for EVA Rescue (SAFER) was developed for the International Space Station and the only other remotely viable role for the MMU – repairing the Hubble Space Telescope – was dismissed, due to fears that plume impingement from its nitrogen-gas thrusters could cause serious damage. Today, the MMU flown by McCandless hangs in the Smithsonian.

LOST IN SPACE

Despite the success of the MMU, two embarrassing failures characterised 41B, together with another problem which directly impacted McCandless and Stewart's second spacewalk on 9 February. Nestled inside the payload bay were the Indonesian government's Palapa-B2 and Western Union's Westar-6 communications satellites; the former was part of a $79 million deal between Perumtel and NASA to launch a pair of satellites, whilst the latter had originally been scheduled for launch aboard the European Ariane booster. Also aboard was the West German-built Shuttle Pallet Satellite (SPAS), which previously flew aboard STS-7. During Brand's mission, this 1,448 kg free-flying platform would be equipped with the same experiments that it carried the previous summer, together with a dummy main electronics box, akin to that of Solar Max. The experiments themselves performed satisfactorily, with the only problem being a failed microswitch on SPAS' mass spectrometer. However, McCandless and Stewart adjusted this switch during their first spacewalk, achieving partial operating capability in the instrument. Next, it was expected that on 9 February, Challenger's fifth crew member – Ron McNair, the second African-American to ride the Shuttle – would grapple SPAS with the RMS arm, raise it to a position a couple of metres from the forward bulkhead, and then rotate it at about one degree per second in order to simulate the attitude and dynamics of the slowly spinning Solar Max. The arm, whose wrist was capable of rolling to 'plus' or 'minus' 447 degrees, was expected to require about 15 minutes to reach the roll 'stop' points.

It was a complex task and entirely fitting that it was given to an astronaut who had risen from the most difficult of beginnings to become a highly regarded physicist, a fifth-degree karate black belt instructor and an accomplished saxophonist. During his early years as an astronaut, he played in an 18-piece swing band at JSC. In fact, Ronald Ervin McNair took his saxophone with him on 41B and, two years later, planned to play it again on his second mission for a very special purpose. In the autumn of 1985, he began working with musician Jean-Michel Jarre on a piece of music for the album 'Rendez-Vous'; the plan was for McNair to record a saxophone solo and, at one point, even for him to participate in one of Jarre's concerts, via live feed. Sadly, McNair's second mission, 51L in January 1986, ended in tragedy, but Jarre would later honour his fallen friend: the final piece on the 'Rendez-Vous' album came to be known informally as 'Ron's Piece'.

McNair came from Lake City in South Carolina, where he was born on 21 October 1950, and developed a passion for learning – particularly science – from his parents. "My parents were not pushers," he once said. "They never *told* us to do anything, but somehow they created an atmosphere, an environment, where it was the thing to do." Growing up in an America where blacks were still persecuted, McNair found himself in a tricky situation as a young child. One day in 1959, he visited a library, a mile from his home, to borrow some science books. His elder brother, Carl, later recalled the story, as the librarian refused to serve him – "*This* library is *not* for coloureds" – and the indignant young boy elected to wait whilst she called the police. When two white officers arrived, along with McNair's mother, Pearl, they could see no apparent disturbance, only a nine-year-old boy sat on the counter.

"Ma'am, what's the problem?" one of the cops asked.

"He wanted to check out the books," the librarian replied, and turning to Pearl, continued: "You *know* your son shouldn't be down here."

After a moment's silence, common sense prevailed and the officer asked: "Why don't you just give the kid the books?" Pearl assured the librarian that young Ron would take care of them ... and insisted that her son thanked the woman on his way out. To Carl McNair, it was a clear indicator of his brother's indomitable spirit and an uncommon ability to see opportunities where others saw only closed doors.

McNair's desire for learning continued and he left Carter High School in 1967 as that year's valedictorian – the highest-ranking student and the one chosen to deliver the farewell speech for his class – before moving on to North Carolina A&T State University to pursue a physics degree. Four years later, he graduated *magna cum laude* (with high honours) and immediately plunged into doctoral research at

Massachusetts Institute of Technology, under the supervision of Professor Michael Feld, famed in the fields of quantum optics and the medical applications of lasers. During his five years with MIT, McNair performed some of the world's earliest development work on hydrogen and deuterium fluoride chemical lasers and high-pressure carbon dioxide lasers. This research contributed to a new understanding and potential applications for highly-excited polyatomic molecules. Completion of his PhD in 1976 was followed by appointment as a staff physicist with Hughes Research Laboratories in Malibu, California, during which time McNair worked on the development of lasers for isotope separation and photochemistry using non-linear interactions in low-temperature liquids and optical pumping techniques. A year later, in November 1977, McNair was one of the last candidates to be invited to Houston to be interviewed by NASA. His selection made him one of only three African-Americans to make the final cut. To many, his achievements may have seemed inconceivable a decade earlier, but to Carl McNair, his brother's own efforts and endless determination had paid off. "Ron ... didn't accept societal norms as being *his* norm," he told a National Public Radio interviewer in January 2011. "*That* was for other people. And *he* got to be aboard his *own* Starship Enterprise!"

In the early morning hours of 9 February 1984, the view through the aft flight deck windows was perhaps not quite as exotic as the Starship Enterprise, but impressive, nonetheless. The plan was for McCandless and Stewart to use a piece of hardware known as a Trunnion Pin Attachment Device (TPAD) to duplicate 'docking' onto SPAS. This would provide NASA with 'real-world' data before the actual Solar Max repair in April. Preparations for SPAS operations with the RMS duly began at 3:40 am EST when the satellite was transferred to internal battery power. However, as McNair worked through his procedures to check out the mechanical arm, the wrist yaw joint experienced a failure – refusing to move when commanded, even though it had worked perfectly during the first spacewalk – and the test was cancelled. Despite efforts to recover full capability in the RMS, including cycling power to the arm to clear the failure indication, it became obvious that SPAS would have to remain secured in the payload bay for 41B. These troubles raised a more serious problem for 41C, because the arm was essential to the planned repair of Solar Max. Worryingly, the difficulty encountered by McNair could not be duplicated using engineering mockups at KSC and, after Challenger's landing, the wrist joint and motor were shipped back to their vendor for further investigation. Some minor corrosive effects were found, but neither thermal or vibration testing was able to identify the problem. The cause remained unknown, so NASA decided to install a 'new' RMS for the Solar Max mission. Fortunately, these difficulties did not significantly hamper the other tasks on 9 February, with MMU tests also being undertaken by Stewart, who scored double firsts on this mission, not only as the first Army astronaut, but also as the first Army spacewalker. During the six hour and 17 minute excursion, he and McCandless performed several successful TPAD 'docking' exercises with the berthed SPAS platform (minus the rotational aspect), replaced the failed television camera and repaired a loose payload bay slide-wire link. The failure of one of their helmet-mounted cameras also required them to make verbal comments to their colleagues inside Challenger for thruster firings in lieu of visual cues.

Both MMU evaluations, read Martin Marietta's post-mission report, "performed as expected and no anomalies were reported". Overall, McCandless flew the MMU for three and a half hours and Stewart for just under two hours. Yet, as has been noted, it was Challenger's own manoeuvrability, demonstrated by Brand, which rendered its future much less certain. "We used the autopilot a lot," Brand said later. "We had the capability to manoeuvre the ship in rotation – roll, pitch, yaw – with a hand controller, but more often than not, we just punched something into the computer and set up the digital autopilot such that we got an automatic manoeuvre. That saved fuel, as we could move at very slow rates. We tested the RCS jets on orbit for translation up or down, sideways or forward and back. On the night side of the Earth, when we translated the ship down, the upward-firing RCS jets were used to do that. At night, it looked like a Fourth of July display because you could look out over the nose and you could see these tubes of fire going up. They were fantastic visual effects."

During neither excursion were McCandless or Stewart 'lost in space', although both of their communications satellites were not so lucky. The first, the 580 kg Westar-6, was nearly identical to the two Hughes-built satellites placed into orbit by the STS-7 crew the previous June. Measuring 6.8 m tall and 2.1 m wide when fully deployed in geostationary orbit, and equipped with 24 C-band transponders, each facilitating either 2,400 telephone circuits or one colour television channel, it was twice as powerful as previous satellites in the Westar series. Since the construction of the first American transcontinental telegraph system in the second half of the 19th century, the Western Union Telegraph Company had closely followed the development of new communications technologies, through the Morse key and sounder to the teletypewriter, microwave transmissions and message-switching computers. By the time that its sixth Westar satellite – destined for use exclusively by business customers – rode into orbit aboard 41B, the company was one of America's primary carriers of voice, data, video and fax telecommunications traffic. During the early 1980s, Western Union contracted with both NASA and the European Arianespace concern for commercial launch services. However, when an Ariane rocket was lost in 1982, one Westar launch had to be rescheduled and the company began to reconsider its future dealings with the Europeans. Further, Western Union felt more confidence in the Shuttle, believing it to be less expensive and more reliable. By April 1983, they had opted in favour of the reusable spacecraft, rather than Ariane, to launch their satellites. Already, efforts had been made to open negotiations with McDonnell Douglas, which had agreed to provide the PAM-D booster for Westar-6. Under the terms of an agreement signed in March 1983, Western Union would hold McDonnell Douglas "harmless" for any damage to their satellite and instead obtained insurance from Lloyds of London to cover potential losses.

After deployment from Challenger's payload bay, it was expected that the PAM-D would insert Westar into geostationary transfer orbit. The satellite's own, Thiokol-built Star-30 apogee motor would then circularise its orbital path at an altitude of some 35,600 km. Supervised closely by the 41B astronauts, Westar's sunshade was opened, it was spun-up to 50 revolutions per minute and ejected from the payload bay at 3:59 pm on 3 February, almost exactly eight hours after launch. Fifteen minutes later, as planned, Brand and Gibson pulsed Challenger's RCS thrusters to manoeuvre to a safe distance before the ignition of the PAM-D. "The impression," said Joe Allen, who watched the proceedings intently from Mission Control that day, "was that the rocket did indeed ignite. Then, somehow, they lost sight of the engine fire, but they weren't sure it was anything out of the normal. The ground controllers, however, detected that the rocket had ignited, the satellite had moved, but then the rocket had extinguished itself. Thus, it was in only a slightly higher orbit. It was a long way from geostationary orbit; a terrible disappointment." In fact, Westar was left in a lop-sided orbital path, with an apogee of barely 1,000 km and a perigee matching the Shuttle of around 250 km. The question now posed was whether Palapa-B2 – identical to Westar and mounted atop a similar PAM-D booster – might be subject to a similar failure. It was scheduled to be sprung free 27 hours into the mission, but that was postponed by a day or so as troubleshooting of the Westar incident got underway.

The decision fell in favour of taking the risk. The Indonesian government concurred with NASA and, at 10:13 am on 5 February, Palapa popped into space. *Surely*, the Westar fault could not be common to both PAM-Ds, it was thought. Unfortunately, the lightning of bad luck, in this case, struck twice and Palapa's booster fired briefly, but also fell silent and its motor nozzle went dark after just a few seconds.

Consequently, Palapa, too, was virtually useless, with an apogee of just over 1,100 km and a perigee of 240 km. Its customers, not just from Indonesia, but also the member states of the Association of South-East Asian Nations (ASEAN), which included the Philippines, Thailand, Malaysia, Singapore and Papua New Guinea, would have to rely upon the services of Palapa-B1 alone. The owners of Westar and Palapa filed insurance claims of $180 million in total, although that of the former was later dismissed as Western Union had already signed a disclaimer with McDonnell Douglas to cover PAM-D failures. The Indonesian government's case, on the other hand, went forward before a jury and the court ultimately agreed that an action in negligence would be allowed to go ahead. During the proceedings, the court heard extensive evidence of possible negligent design in the construction of the PAM-D booster and concluded that the only liable defendant was Thiokol for supplying a 'bad' rocket motor. Efforts were underway, meanwhile, to retrieve both Westar and Palapa and return them to Earth. Within three weeks of Challenger's landing, the satellites' manufacturer – Hughes Aircraft Corporation – had presented NASA with an option to attempt a salvage operation. By the first week of September 1984, the increasingly confident space agency agreed with Hughes and the satellites' majority insurers to commit the Shuttle and MMU to a risky recovery mission.

With two embarrassing satellite failures and an RMS problem that prevented operations with a third, it seemed that little else could go wrong on 41B. Sadly, however, it did! NASA's preparations for the vital resuscitation of Solar Max had shifted into high gear since the spring of 1983, with virtually every Shuttle flight conducting one test or another in direct support of the landmark 41C mission. In addition to the MMU and SPAS demonstrations, Brand's crew was assigned to deploy a large, inflatable balloon and conduct a simulated, 'closed loop' rendezvous with it. Known as the Integrated Rendezvous Target (IRT), the 2 m aluminised Mylar balloon was ejected, along with its Getaway Special canister, from a longeron attached to Challenger's port side payload bay wall. Despite the Palapa deliberations, which had impacted several other mission objectives, the deployment of the 91 kg combination got underway as planned at 6:50 am on 5 February, drifting serenely away at 45 cm per second. Very soon, however, it encountered difficulties. The intention was that, a minute after leaving Challenger, the GAS canister would split in half and the balloon would inflate with nitrogen. Rotating at three revolutions per minute, the balloon would then have been used by Brand and Gibson to firstly practice rendezvous manoeuvres from 9.2 to 14.8 km and, later, from a distance of more than 220 km, using the Shuttle's Ku-band antenna and other ranging systems to provide navigational data to the five General Purpose Computers.

Post-flight analysis would confirm that the IRT did start to inflate, but due to a fabrication defect a series of 'staves' enclosing the balloon failed to release correctly and it burst when its progress was halted by the faulty staves. Still, some success was achieved as the crew tracked several fragments of debris using the Ku-band antenna, their own eyes and Challenger's star trackers. The Ku-band dish, mounted on the starboard wall of the payload bay, had already caused minor headaches earlier in the mission when it failed to conduct a 'self test' and did not properly radar track McCandless as he flew the MMU. It also seemed to be susceptible to interference by external electromagnetic radiation. Nevertheless, it later tracked Stewart and post-flight analysis determined that a single self-test failure out of 19 attempts was "acceptable". With the exception of the spectacular MMU excursions, Gibson remembered, years later, that the entire crew felt "positively snakebit" at this point. Added Joe Allen: "They were now zero for three in satellite deployments! I've never asked him, but I wondered what Bruce was thinking at that point, because he was going to be the fourth satellite!"

Luckily for the spacewalkers, McCandless and Stewart's MMU sorties turned out to be the only fully successful 'satellite' deployments of 41B. Overall, 91 percent of the mission's Detailed Test Objectives were satisfactorily accomplished, with the RMS fault preventing them from conducting MMU docking demonstrations and the burst balloon providing only limited opportunities to evaluate Challenger's rendezvous and laser ranging gear. In spite of this, on the morning of 11 February, Brand completed one of the mission's most important test objectives by landing, for the first time, in Florida. A KSC homecoming had originally been planned for the end of the STS-7 flight, but bad weather forced a 'wave-off' to California. Since STS-1, a landing in Florida had been regarded as a key milestone in achieving truly 'routine' Shuttle missions, as well as saving an estimated $1 million and five days' worth of processing time. Unlike previous flights, Challenger would not be subject to this cost of being ferried from the West to the East Coasts atop the heavily modified Boeing 747 airliner. For Vance Brand, the 41B touchdown provided him with a unique opportunity to have landed in three very different locations during his three-flight astronaut career. He had splashed down in the Pacific Ocean at the end of the

Apollo-Soyuz mission in July 1975, made landfall at Edwards Air Force Base at the close of STS-5 in November 1982 and now returned to Florida.

Following the completion of the 168 second de-orbit 'burn', executed at 6:16 am, Challenger's re-entry flight path took her across the Pacific Ocean to the Baja peninsula, over Mexico and southern Texas and towards Florida. After passing squarely over the Titusville area she flew out over the Atlantic Ocean, preparatory to lining up to approach Runway 15 from the north-west. Later missions would also approach Runway 33 from the south-east. Touchdown at 7:15:55 am, completing a mission just shy of eight full days – Challenger's longest to date – and covering 5 million km, was perfect. For a flight with two failed satellite deployments, an RMS problem that prevented operations with a third and the burst rendezvous balloon, Challenger's fourth voyage had concluded triumphantly. In fact, referring specifically to Brand's landing, the official report remarked that "the precision with which this objective was accomplished shows that all areas of the National Space Transportation System were at their peak of readiness". Indeed they were, for on Challenger's next mission, flown barely eight weeks' later, she would be tested on her most ambitious assignment so far: the long awaited repair of Solar Max.

'LUCKY' THIRTEEN?

Despite NASA's seemingly ingrained case of triskaidekaphobia, which obliged managers to impose the bizarre, '13-free' numbering system on its flights, the crew of perhaps the most important Shuttle mission to date clearly were unsure if 41C was supposed to be unlucky or not. Still internally dubbed 'STS-13', it would actually be the reusable spacecraft's 11th orbital journey; the decision having been taken several months earlier to cancel Ken Mattingly's STS-10 flight *and* Hank Hartsfield's STS-12, both of which had been destined to deploy payloads using the Air Force's troubled Inertial Upper Stage. Perhaps by design or pure circumstance, the scheduled launch date for 41C had moved forward from Friday 13 April (according to the November 1982 Shuttle manifest) to Wednesday 4 April by the end of 1983. Eventually, Challenger launched on Friday the 6th. In fact, the absurdity of NASA's efforts to avoid misfortune befalling the mission had inspired a number of practical jokes from the crew. Dick Scobee, the 41C pilot, designed the 'official' crew patch for the mission ... although a distinctly *unofficial* one also lurked outside of the approval of NASA Headquarters: an insignia of a menacing black cat, emblazoned with the number '13', surrounded by lightning bolts and a Shuttle hurtling from beneath its belly. Scobee's crewmate, Terry Hart, would later admit that they even had coffee mugs made, bearing the 'official' patch on one side and the 'unofficial' patch on the other.

When the names of the crew were announced by NASA in mid-February 1983, one of them was unavailable to begin direct training until later that year. Commander Bob Crippen was preparing to lead STS-7 and his stint on the Solar Max repair would make him the first person to fly the reusable spacecraft three times. He would be joined by pilot Scobee and mission specialists Hart, James 'Ox' van Hoften and George 'Pinky' Nelson. There are many among us who doubt the veracity of 'luck' – good or bad – and although Mission 41C, still carrying its moniker of Unlucky Thirteen, would actually prove hugely successful, in a tragic sense, bad luck *would* eventually catch up with Francis Richard Scobee. The story of his life is an impressive account of a man who worked his way doggedly from the ground upwards, from an aircraft mechanic to an Air Force test pilot to a civilian who flew both the Shuttle Carrier Aircraft *and* the Shuttle itself. Born in Cle Elum, Washington, on 19 May 1939, Scobee attended high school in Auburn and enrolled in the Air Force immediately upon graduation. Aviation had been a fascination from a young age. "Ever since I was a little kid," he told an interviewer, "I had been enamoured of airplanes and that's why I ended up working on them. They generally fascinated the heck out of me and I'd never done any flying. I used to go out to the airport and watch them fly." His earliest work was as a reciprocating engine mechanic at Kelly Air Force Base in Texas and, whilst there, he attended night school to gain college credits for a scholarship into the Airman's Education and Commissioning Program. This enabled Scobee to enter the University of Arizona to study aerospace engineering. He received his degree in 1965, gained his pilot's wings the following year and undertook a combat tour in Vietnam. Returning to the United States, he was chosen for test pilot school at Edwards Air Force Base, from which he graduated in 1972. Prior to his selection by NASA, Scobee participated in test work on the Boeing 747, the experimental X-24B lifting body, the F-111 Aardvark ground-attack aircraft and the C-5 Galaxy military transport. Summoned to Houston by NASA in September 1977, he successfully completed the application process and was selected as an astronaut candidate in January of the following year. He resigned from the Air Force as a major and remained with NASA as a civilian. It underlined another facet of his character: to commit himself totally to worthwhile goals. "When you find something you really like to do," he explained, "and you're willing to risk the consequences of that, you *really* probably ought to go do it."

In contrast to several other crews, whose payloads were being endlessly juggled, the two main objectives of the 41C mission – repairing Solar Max and deploying a huge, bus-sized satellite called the Long Duration Exposure Facility (LDEF) – had remained static, ever since the flight was approved by NASA Headquarters in August 1982. "Our mission was so specialised," explained Bob Crippen in an oral history, "that when we were going up to get Solar Max ... it was not reasonable that we could change it." LDEF, too, was so large that it would have been difficult to remove it from the manifest and put it onto another flight. In fact, the satisfactory performance of the Canadian-built RMS arm was critical, not just for the Solar Max repairs, but also for the deployment of LDEF. The satellite took the form of a 12-sided structure and, as its name implied, it was designed to house experiments which required long-term exposure to the hostile environment of low-Earth orbit. No one could possibly have foreseen, at the time of LDEF's launch, exactly how long it would remain in space before being retrieved by another Shuttle mission and returned to Earth.

NASA intended to collect the satellite during the 51D mission in February 1985, but that was repeatedly delayed, and by the time Challenger exploded, the retrieval had been rescheduled for the autumn of 1986. In fact, it would not be recovered until January 1990, by which point it was only weeks away from an uncontrolled and fiery re-entry. It was a peculiar object, measuring 9.1 m long by 4.2 m wide and weighed 9,520 kg. At its most basic, it consisted of an intricate frame of aluminium rings and longerons, loaded with trays for 57 scientific experiments (some of which occupied more than one tray). Shortly after the formation of NASA in October 1958, researchers began to seriously consider building a satellite that could carry material samples in order to assess how the harsh environment caused them to degrade over time. By the early 1970s, these ideas had acquired a name, the Meteoroid and Exposure Module, to be carried aloft by the Shuttle and retrieved a few months later. As the name implied, its focus was upon the impact of micrometeoroid damage on satellites and how best to protect them. Subsequently renamed LDEF, the contracts for its design and development were granted to NASA's Langley Research Center in Hampton, Virginia.

The structure was complete by 1978 and, after tests, was kept at Langley until a Shuttle flight became available. By this point, its objectives had expanded from micrometeoroid research to studies of changes in material properties over time, performance tests of new spacecraft systems, evaluations of power sources and conducting crystal growth and space physics investigations. The satellite was designed to be reusable and adaptable for differing lengths, though ultimately it would fly only once. Its length was divided equally between six bays for the experiment trays, with a central 'ring' at the midpoint connected by longerons to the end frames. Aluminium 'intercostals' linked each longeron to adjacent rows of longerons on each side and removable bolts joined the longerons to the end frames and intercostals. This meant LDEF could be made 'shorter' or 'longer' if a mission required it. Experiment trays were then clipped into the rectangular openings between the longerons and intercostals. Two RMS grapple fixtures were provided: one to allow it to be picked up by the robot arm for deployment and subsequent retrieval and a second to send signals to initiate the experiments. It had no attitude control system and, said one engineer, "what you saw was what you got": a passive container with no manoeuvring capabilities. It was designed to remain in orbit by being placed into a 'gravity gradient' attitude, with one end facing Earth, making an on-board propulsion system unnecessary. This also freed it from acceleration forces or contamination caused by thruster firings. The orientation of LDEF also meant that the two 'ends' would be subjected to a unique thermal environment, although all parts of the satellite were subjected to daily temperature changes as the Sun 'rose' and 'set' every 90 minutes and solar angles changed annually. Heat management was accomplished by coating the interior surfaces with high emissivity black paint, which kept thermal gradients across the structure to a minimum and maximised heat transfer across LDEF's body. The experiments were also spread evenly to equalize thermal properties across the satellite. Eighty-six trays – 72 around the circumference, six on the Earth-facing end and eight on the space-facing end – accommodated 57 investigations. The 1.3 m × 86 cm trays came in several different depths and housed experiments weighing up to 90 kg. These covered four disciplines: materials and structures, power and propulsion, science and electronics and optics.

They captured interstellar gas atoms to better understand the Milky Way galaxy's formation, observed cosmic rays and micrometeoroids, studied shrimp eggs and tomato seeds and investigated the impact of atomic oxygen on different materials, including solar cells. Originally scheduled for launch in December 1983, delays to several Shuttle flights that year pushed the satellite into the following spring. In June 1983, encased in a specially constructed LDEF Assembly and Transportation System crate, it was transferred aboard a Second World War-era landing craft from Langley down the coast to KSC in Florida. Upon arrival, it was ensconced in the Spacecraft Assembly and Encapsulation Facility and its experiments were prepared and integrated. Eventually, it moved to the Operations and Checkout Building for final processing, before being loaded aboard Challenger on 20 March 1984.

Deployment occurred 24 hours into the mission, on 7 April, and, although Terry Hart admitted "*that* was exciting", it hardly compared with his first and only Shuttle launch a day earlier. "It was a clear, cool morning," he said of Friday the sixth, "and we went through the traditions of having breakfast together and there was always a cake there for the crew before they went out. Next, we went out to the launch pad and up the elevator. As usual, people don't say much in elevators – whether you're in a hotel or on the launch pad – and you watch the numbers tick by and, instead of floors, they do everything in feet in the launch pad elevators. When you walk across the gantry to board the Shuttle, you can look down into the flame trench. The obvious thing that's striking you is that this is for real: we're going to go! Everything was pretty smooth on our launch countdown. We got strapped in and, again, the guys strapping us in were a lot of the same guys that strapped in Al Shepard on his Mercury flight [in May 1961]."

By the time of launch, the 41C crew had spent 14 months together as a team and all five men would agree that *this* mission was the mission that everyone wanted. With a background as a pilot in the Air Force Reserve and master's degrees in mechanical and electrical engineering, Hart had spent his first few years as an astronaut working on the Shuttle's main engines and on developing new rendezvous procedures; although rendezvous had been accomplished during the Gemini and Apollo programmes, bringing the orbiter alongside a giant satellite for repairs was quite different. "The Shuttle was this big truck," he told the oral historian, "and it had a very limited amount of fuel on board, whereas the Apollo command module and Gemini were like sports cars. They could just zip around and change orbits much more readily, especially in close, around an object, they could just kind of move right around with great ease. If *we* started to do much of *that*, we'd very quickly run out of fuel and have to de-orbit, so we had to come up with new design trajectories and procedures to accommodate that difference to ensure that we were flying the most fuel-optimal approach during a rendezvous." When he was assigned to 41C, Hart's initial suspicion was that he would be involved in the rendezvous, and not the EVAs, but the word on the astronaut office grapevine was that James van Hoften – nicknamed 'Ox', due to his physical height and size – was too big to perform a spacewalk. Nevertheless, one day Bob Crippen approached Hart and told him that his expertise was needed to perform the RMS work and the rendezvous. The EVAs would be done by van Hoften and Nelson.

Terry Jonathan Hart, of Croatian-American ancestry, was born in Pittsburgh, Pennsylvania, on 27 October 1946 and after high school entered Lehigh University to study mechanical engineering. He received his degree in 1968 and a master's credential from MIT the following year, then entered the Air Force Reserve on active duty. Hart completed his pilot training at Moody Air Force Base in Georgia in December 1970 and spent three years flying the F-106 Delta Dart interceptor, then joined the New Jersey Air National Guard. In the years since receiving his first degree, he had served on the technical staff of Bell Telephone Laboratories, working on the design of electronic power equipment, for which he received two patents. As an Air National Guardsman, he flew most weekends ("so I had two parallel tracks going") and in the summer of 1977 he noticed a NASA advert in one of the National Guard magazines, calling for astronaut candidates. Hart was invited for interview, but doubted his chances of selection, since many of the other applicants held doctorates or were test pilots. When he received the call from George Abbey in January 1978 that he had been successful, he was, in his own words, *floored.*

"I think one of the reasons I didn't think I was going to be selected," he told the oral historian, "was because even though I had a *breadth* of experience in engineering and flying ... I didn't have the *depth* that a lot of people had, and particularly the mission specialists, who were often very deep in their area of science." As circumstances transpired, this actually worked in Hart's favour. His military training meant that he was able to adapt quite readily to the T-38 flight instruction and many of the civilians found it easier to bond with Hart in the cockpit than with some few of the military test pilots. He picked up another master's degree, in electrical engineering, from Rutgers University in 1978 and continued flying weekends with the Air National Guard – transferring to Ellington Air Force Base in Houston – until 1985.

Hart initially anticipated that the Shuttle would commence operations in 1979 and members of his class would begin to fly a year or two later; he expected to fly maybe twice, then return to his previous career with Bell. Unfortunately, he could not have foreseen that the spacecraft's maiden voyage would be severely delayed and it would be *five years* before he received his *first* flight assignment. Although he was later approached by George Abbey with an offer of a second mission – the German Spacelab-D1 – Hart turned it down in favour of returning to Bell. Throughout 1983 and into the spring of 1984, his attention was entirely devoted to preparing for the intricate RMS operations in support of the Solar Max repair and the LDEF deployment. "The arm engineers wanted to make sure we properly tested the arm moving such a large object," he explained, "so they could understand that it was going to be able to do what it was designed to do, so I spent a lot of time working with the engineers to make sure that I was doing everything that they wanted done during testing." Much of the simulation work was done at JSC and at Spar Aerospace's facility in Toronto and Hart found himself routinely testing the flex of the arm with the huge LDEF attached and capturing Solar Max in a rotating mode.

One of Hart's responsibilities during ascent was to act as a 'second flight engineer'; seated behind Dick Scobee, he assisted van Hoften with checking off the milestones and monitoring the procedures needed in the event of problems. There were none. "Off we went," he said of the 8:58 am liftoff, "right on time on a perfectly clear day. I had a couple of surprises: the shake, rattle and roll of the Solid Rocket Boosters for the first two minutes – is a very low frequency rumble; just a tremendous sense of power. You can look back over your shoulder or look out the top window when you're in the flight deck and watch the world disappearing behind you. Very quickly, the SRBs taper off and separate and that was the surprise I had, because your g-loading builds up close to two and a half G's as the boosters reach their peak thrust. As the solid rockets burn off and separate, the sensation that you have at that point I wasn't quite prepared for, because you go from two and a half G's back to about one and a half. The sensation you have is that you're losing out, that you're falling back into the water! You don't think you're accelerating as much as you should be to get going and, of course, I'd worked on the main engine programme anyway, so I was very familiar with what the engines could do or not do. I think in the next minute I must have checked the main engines to make sure they were running, because I'd swear we only had two working: it just didn't feel like we had enough thrust to make it to orbit! Then, gradually, the External Tank gets lighter and as it does, of course, with the same thrust on the engines, you begin to accelerate faster and faster. After a couple of minutes, I felt like – yes – I guess they're all working." Indeed, Challenger's fifth launch had proceeded without incident. The External Tank behaved superbly and the performance of the main engines, read NASA's post-mission report, "appeared to be normal". The only deviation was when the engines throttled down to 67 percent, rather than the predicted 71 percent, as Challenger passed through maximum aerodynamic pressure a minute into the flight; this lower level was later attributed to a higher than anticipated SRB impulse during the first 20 seconds. Chase aircraft also revealed that one of the main parachutes on the right hand booster failed to inflate, although both boosters were recovered successfully.

Experiencing the launch from a somewhat different perspective, seated 'down-stairs' on Challenger's darkened middeck, next to the side hatch, George 'Pinky' Nelson did not have the luxury of viewing the ascent from Walker's point of view, nor through the wrap-around windows of the cockpit as Crippen, Scobee, Hart and van Hoften could. Still, he recalled the rapidity – and loneliness – of his first ride into orbit. He was also able to peer through a tiny circular window in the side hatch and capture a fleeting glimpse of the enormous, controlled explosion that was underway outside. "I could see the tower go by and the sky and horizon as we ascended," he told this author in an email correspondence from March 2006. "It was a bit lonely down there, but Crip kept a running commentary on how the launch was going, since we were all rookies, but him. That helped keeping up with the events. My first experience with weightlessness was problematic. I'd had many flights on the KC-135 aircraft and hundreds of hours in the water tank, so was familiar with the sensations of weightlessness. I remember how pleasant a sensation it was and how surprised I was that I didn't get sick!"

George Driver Nelson acquired the nickname 'Pinky' as a child, apparently on account of being a pink baby with a mop of reddish-blond hair. He was born in Charles City, Iowa, on 13 July 1950, but grew up and attended high school in

Willmar, Minnesota; a distinctly rural life which he described as "basically a Lake Wobegon kind of experience", after the fictional Garrison Keillor town in which all children were above average intelligence. From at least the age of four, Nelson discovered a fascination for astronomy, the sky and the planets, and at school was interested in mathematics and physics, "but soon discovered that I was a better *physicist* than I was a *mathematician* and my real interest was in nature, not in mathematics". He studied undergraduate physics at Harvey Mudd College in Claremont, California. After receiving his degree in 1972, he moved to the University of Washington in Seattle for a master's qualification (1974) and a doctorate (1978). It was whilst finishing up his PhD research, working on a fellowship in West Germany at the University of Göttingen, that he saw a NASA flier on the bulletin board, calling for astronaut applicants. Nelson had already learned to fly at Harvey Mudd "and the job looked like it combined the three things I was really interested in: space, astronomy – and the intellectual challenge of it all – and then the flying aspect". The physical challenge was important, too, since "I'd always been a jock", but privately Nelson doubted that he had a chance of being accepted. He typed up his application on an old German typewriter, sent it to NASA and, for a few months, forgot about it.

In the summer of 1977, he returned to the United States and worked for several months at Sacramento Peak Observatory in Sunspot, New Mexico, with plans to return to Seattle in November. "Just as I was about to leave Sacramento Peak," Nelson explained, "I got a call from George Abbey, saying they wanted to do an interview." He asked Abbey to postpone the interview, to give him time to drive his family back to Seattle. He arrived in Houston in early October and Abbey quickly found out about his love of baseball ... and, half an hour after his interview with the NASA psychiatrist, Nelson found himself on the softball diamond, playing alongside a group of astronauts and space centre employees. "So I went out and played," he continued, "and drank beer with them afterwards. I figured *that* was my *real* interview!" (Interestingly, Nelson still has a photograph of himself and fellow candidates Jeff Hoffman and Sally Ride, sitting together at a table, eating breakfast during the interview week. "It turns out," he said, "that those are the *only three* of our group that were selected!") Returning to Seattle, Nelson was offered a post-doctoral appointment in the Joint Institute for Laboratory Astrophysics at the University of Colorado at Boulder. He had not yet finished writing up his PhD thesis and remembered spending much of October and November doing "nothing but write all day and compute all night". He defended his doctorate in the second week of January 1978 and moved to Boulder the next day ... to be greeted by a message from the office secretary: somebody called George Abbey had called for him. Abbey told Nelson that if he was still interested in becoming an astronaut, he was in. "So the *first* thing I did at my post-doc job," Nelson laughed, "was *quit!*"

It has often been said that there are very few reliable predictors of what kind of mission assignment an astronaut may draw ... but for Nelson, a position on an EVA crew seemed likely, for one of his earliest duties was to work with Story Musgrave on the development of the space suit. He spent much time in the neutral buoyancy water tanks in Houston and at the Marshall Space Flight Center in Huntsville, Alabama, as well as attending design reviews at Hamilton Standard. Nelson also worked with fellow astronaut Kathy Sullivan as a scientific instrument operator on the WB-57 aircraft (a converted Canberra bomber), used by NASA for high-altitude atmospheric research. The pair travelled to Miami and to various destinations throughout South America – Panama, Lima, Montevideo and out over the Atlantic to the Falkland Islands – for atmospheric sampling and microwave observations.

On his first voyage into space, Nelson adapted well to the new environment. Unfortunately, the same could not be said for Terry Hart. Mission 41C marked the Shuttle's first 'direct insertion' ascent. In other words, only one OMS firing – rather than two – was needed to circularise Challenger's orbit at an altitude of around 530 km. Previously, when less performance data was available for the main engines and some targeting precision was lacking, an initial OMS burn was made to raise the apogee, followed by another, half an orbit later, to raise the perigee and circularise the orbit. On 41C, however, the ascent was to achieve an initial orbit with an apogee at the desired altitude without performing the OMS-1 burn, and the OMS-2 burn provided circularisation. This enabled the engines to provide more energy and permit the easier use of on-board software. The high orbit was needed for the rendezvous with Solar Max and also simplified the procedures for the astronauts. "It's a much easier task from a crew standpoint," explained Bob Crippen, "because you're pretty busy there right after main engine cutoff and this took away some work, so it was a neat thing to try." As Crippen, Scobee and van Hoften busied themselves with readying their ship for orbital operations, Hart was granted the opportunity to unstrap and leave his seat to photograph the jettisoned External Tank as it tumbled Earthward. It was perhaps fortuitous that the LDEF deployment was still a day away, because Hart's initial euphoria turned rapidly into a severe dose of space sickness.

"I had *never* had any motion sickness," he recalled years later. "I was a fighter pilot and could do anything in an airplane. I had a light airplane I used to do aerobatics in and nothing ever bothered me in terms of flying or riding a boat or a train or a car or whatever. I wasn't weightless for more than three minutes and I knew I was in trouble! I could just tell my whole gastro-intestinal system was going into high-speed reverse and I didn't understand it because, psychologically, I was elated. Maybe I got up too quick and started moving around or started looking out the window too soon, but for the whole first day I was really out of it. There were some things I had to do that first day, but they were minimal. I had to unstow the RMS and barely made it through that. I really was totally incapacitated for the first day and I tried the usual drugs that they give you to help, but I had it so bad that nothing helped at all. That night, when we got ready to go to sleep, I was exhausted, really depleted. I remember falling asleep and was asleep for maybe a half hour, when I dreamt that I was falling – I had a visceral reaction to a fear of falling – and I remember reaching out to grab something and I did it with such force that I ripped my sleeping bag. I don't think the other guys were asleep yet, but if they were, I woke them up when I yelled out. That was kind of a low spot and, after that, I acclimatised. I had some kind of fundamental neurological brainstem reaction – totally subconscious – to a fear of falling. I think my initial sickness, after three or four minutes of weightlessness, was something that triggered my basic instincts of falling, even though it wasn't conscious. I couldn't detect it consciously and I think it stayed with me for that first night. The next day, I was able to do all my duties, but it was just a terrible experience. I never heard anyone else relate such a bad experience."

Fortunately, by Day Two, Hart had recovered sufficiently to take the lead in the LDEF deployment, successfully releasing it into space at 12:19:27 pm EST, as Challenger travelled 'upside down', her open payload bay facing Earthward. To activate its many investigations, he firstly grappled the satellite by its so-called Experiment Initiation System fixture and then regrappled the second fixture to actually pick it up and deploy it. "The concern," he remembered, "was that I was going to get it stuck, then we couldn't close the payload bay doors and couldn't come home. Crippen and I were trained on the RMS, with him watching and making sure everything was going well. First, I had to lift it out 'straight' and then the arm did everything it was supposed to do. I put it back in the payload bay, just to make sure it would go back in before I lifted it out one more time to deploy it. We left it out on the arm and did some slow manoeuvres to verify all the dynamics that the engineers wanted to understand about lifting heavy objects out of the Shuttle. Then, we very carefully deployed it. It wasn't detectable at all when I released it; totally steady and we very carefully backed away and got some great photographs." As LDEF drifted serenely into the inky blackness, Crippen and Scobee pulsed the RCS thrusters to increase their distance from the satellite, confirming the separation rates using the Ku-band radar. One problem they highlighted was that their view of LDEF's trunnion pins and berthing guides using the television system was not satisfactory and they expressed concern about its effectiveness during the retrieval, which, at the time, was planned for the following spring.

Pinky Nelson's assignment to perform two EVAs, with van Hoften, in support of Solar Max had come at a restless time for himself and other members of the Thirty Five New Guys, as each waited impatiently for a flight. "This was the mission I wanted," he said of 41C, "because it had EVAs. I remember meeting with Crippen shortly after that, in one of the little conference rooms at JSC, where he doled out the assignments and gave me the role of flying the MMU, which made my year! Here was a mission with four military pilots and they decided to let me fly the manoeuvring unit. Training for that mission was really fun. We were involved quite a bit with Vance Brand and Hoot Gibson. The mission before us was going to test out a lot of the equipment, so we worked closely with what they were doing and watched that flight closely. Ox and I were a great team. It was really the most complicated spacewalk that had ever been conceived and a real precursor to the much more complicated work they've done on the Hubble Space Telescope. We worked hard to choreograph this repair and we had it down to a dance. We knew all the steps and who was where when, what tools were needed and how we moved things."

James Douglas Adrianus van Hoften, nicknamed 'Ox' because of his bulk and 1.93 m height – some six feet and four inches – came from Fresno, California, where he was born on 11 June 1944. After high school in Millbrae, he enrolled in the University of California at Berkeley to study civil engineering, and on graduating in

1966, then moved to Colorado State University for a master's in hydraulic engineering. Less than a year later, in 1969, van Hoften entered the Navy and received initial flight instruction in Pensacola, Florida, then jet training at Beeville, Texas, and finally assignment to fly the F-4 Phantom II at Miramar, California. Aboard the aircraft carrier USS *Ranger*, he completed two deployments to Vietnam in 1972, during which time he flew 67 combat missions. "Right about that time," he told NASA's oral historian, "the Vietnam War was kind of winding down, but not significantly. We didn't know what to expect, but it turns out the war kind of picked back up again and I had a whole war cruise over there, with about nine months on the line." Upon his return to the United States, van Hoften went to the Navy Fighter Weapons School – 'Top Gun' – and completed another Vietnam cruise, but was refused the chance to enter test pilot school because he did not have the requisite flying hours and was over-qualified in terms of his academic education. Years later, he found this snub rather humorous. (In fact, even after his selection by NASA, van Hoften would exhibit a measure of unhappiness that he had been selected as a mission specialist, when "I was equally qualified as a pilot and 90 percent of the guys there.")

Rejection from test pilot school prompted him to resume his academic studies and he left the Navy and returned to Colorado State University to complete his PhD, with a research focus on fluid and wave mechanics. As his studies neared an end, van Hoften began looking for career openings and in the summer of 1976 spotted a vacancy for an assistant professorship of civil engineering at the University of Houston. Whilst there, he taught fluid mechanics and conducted research on biomedical fluid flows in artificial internal organs and valves. "He was just a terribly interesting person to me," recalled fellow astronaut Dick Covey. "One, he was *bigger* than the rest of the guys. Two, he had been a Navy fighter pilot and then had gone and gotten his PhD ... doing research on putting artificial hearts in calves and stuff like that." Van Hoften maintained his ties with the military, joining the naval reserves in Dallas and flying the F-4. Interestingly, it was through one of his PhD students, John Cox, then working for George Abbey at JSC and later to become a flight director, that van Hoften learned of NASA's call for astronauts. Cox encouraged him to apply. Van Hoften's two main worries were his relatively advanced age and his height, which he imagined would immediately disqualify him. Nevertheless, he submitted his application – "I think mine was one of the last ones to go in" – and he was interviewed as part of the ninth group of applicants in November 1977. His daughter, Jamie, had been born prematurely a few months earlier and had spent considerable time in hospital, "so it was pretty tough". He remembered interviewing alongside candidates who were convinced that the most important thing was to achieve the best results on the treadmill or in the other medical tests. "Listen," van Hoften told them, "I've had about an *hour's* sleep and I'm *not* going to run very far!" He was not enjoying his time at the University of Houston and was keen to leave the heat and humidity of the place; so too was his wife, Vallarie. When the call came in January 1978 that he had been successful, her response was not quite what he had expected: "Does that mean we're going to live in Texas *forever?*"

When van Hoften was named to STS-13 – later to become 41C – in mid-February 1983, he had already learned from another astronaut, Judy Resnik, that an assignment might be just around the corner. Van Hoften and Resnik were good friends and frequently flew across country in the T-38 together to training or other locations. She was assigned to STS-12, with the official press announcement on 3 February, and around this time the pair were aboard a T-38 when Resnik told him, "I think you're going to get some good news". Sure enough, when they landed back in Houston, van Hoften was called into George Abbey's office. Fifteen days after Resnik's assignment, the crew for the Solar Max repair was officially named by NASA. Like Nelson, van Hoften *knew* that 41C was one mission that *everyone* wanted to be aboard. Years later, he would speculate that Bruce McCandless, who had worked extensively on developing procedures for the Solar Max repair, was "unhappy" that he did not receive assignment to 41C. "He went up and did the first flight," van Hoften told the NASA oral historian, "but the *real* mission was not to just go out and fly the MMU around; it was to go out and fix the satellite." At first, it seemed that van Hoften himself might not be making the EVAs. "About that time," he explained, "they started running into money crunches and they said that they wanted to limit the space suit sizes, because originally the EMU that they made was meant for everybody. It was supposed to be from the 5th percentile female to the 99th percentile male. Well, I was the biggest guy they ever had, so at some point they decided they weren't going to make extra-large suits anymore; they were just going to make small, medium and large, and those of us on the fringes weren't going to get to do it." It was a disappointment, but van Hoften still had a spot on what he described as "one of the premier missions". At length, it was George Abbey who came to the rescue, insisting that van Hoften should do the EVAs with Nelson and ordering that a pair of extra-large suits should be manufactured.

For the next 14 months, the crew trained with an almost obsessive focus, totally immersed in the minutiae of the mission, and van Hoften is not alone in having lamented the effect that this imposed on families, particularly wives and children. "But the nice part," he told the oral historian, "is that everyone is kind of in that boat, so there's a whole community there to support everybody." Every week, Nelson and van Hoften would don training versions of their suits and descend into the water tank at JSC to simulate removing equipment from Solar Max and installing new components . . . and their families were invited into the space centre to watch them at work. By the late spring of 1984, they were ready.

Not only had they nailed down the mission to perfection but, in Bob Crippen's case, even the pre-flight photographs turned into something of an art form. "I remember the day we posed for our crew picture," recalled Terry Hart, "and all put our blue flight suits on and took maybe 20 pictures, trying to get the right expressions on our faces! Then, the tradition is that you bring them down to the astronaut office and ask the secretaries to pick which one is best. In one of them, one of us would be winking or our smile would be crooked or something like that. Every one of us had maybe a 50 percent 'hit' rate on the pictures, having the right expression on our face. Then we looked at Crippen, who'd been in the public eye from STS-1 until this mission. Every photograph had the same expression on Bob

Crippen's face! He had it down pat. He knew exactly how to smile!" Like his STS-7 flight, Crippen almost gained an extra crew member for 41C. Although never 'officially' confirmed, the Air Force briefly considered a 40-year-old naval engineer named David Vidrine for a payload specialist seat aboard Challenger. For some time, efforts had been underway to train a cadre of manned spaceflight engineers to accompany military payloads on the Shuttle. Although 41C was a 'civilian' mission, the rationale behind flying Vidrine seemed to be that observing the Solar Max repair with a 'satellite servicing specialist' could lead to opportunities for refurbishing important Department of Defense spacecraft in orbit. Indeed, by the time of the Challenger disaster, plans were afoot to repair and refuel the military-funded Landsat-4 Earth resources spacecraft, which had a similar 'bus' to Solar Max. Since Landsat was in polar orbit, that mission would have launched from Vandenberg Air Force Base in California. In the autumn of 1983, Vidrine sat in with the 41C crew for several simulations, but by the following March his assignment had been terminated by the head of the MSE group, Major-General Ralph Jacobson, as having "no value" to the Air Force.

As Nelson and van Hoften worked in the WET-F tank to perfect their orbital repair work, Crippen, Scobee and Hart busied themselves with rendezvous procedures in the Shuttle simulator. In Hart's case, the RMS was another of his responsibilities. This had given trouble on its previous mission, when the wrist yaw joint failed. Although the cause of that failure was still unknown when 41C lifted off, the faulty arm (serial number 201) had been replaced by another (serial number 302) and verified on the ground. Other work performed on Challenger between her missions included replacing her left-hand OMS pod with one from sister ship Discovery, after significant damage had been identified during post-flight inspections after 41B. This was caused, apparently, by ice from the potable and waste water dump nozzles. Although these nozzles were situated close to the side hatch in the Shuttle's forward fuselage, the ice apparently detached 22 minutes after re-entry interface and hit the OMS pod as Challenger flew at Mach 4.5 – some 5,500 km/h! Tile damage had also been caused by debris falling from the External Tank during ascent. In addition, further, albeit less serious, problems were encountered with the brakes during the 41B touchdown and rollout, chipping carbon liner edges and causing retaining washers to fail. Following the meeting of an industry-wide committee at JSC in January 1984, it was concluded that, in view of stresses imposed on the orbiter's brakes, such problems were "normal" and not safety-of-flight issues. 'Hard' braking on the runway had been demonstrated safely by Paul Weitz at the end of STS-6. Nonetheless, NASA opted to install extra instrumentation aboard Challenger for her 41G flight to better understand the dynamic interaction between the brakes and hydraulic systems.

Despite the importance of deploying LDEF, it was overshadowed by the repair of NASA's malfunctioning Solar Max satellite. In fact, virtually every Shuttle flight since November 1982 had helped to lay the groundwork for the reusable spacecraft's most ambitious mission so far. Extensive tests had been undertaken to validate the Canadian-built mechanical arm, requiring it to manipulate larger and more bulky payloads, and three spacewalks had verified the performance of the suits, tools and

MMUs, together with the ability of astronauts to work effectively with them. In particular, Nelson and van Hoften paid a great deal of attention to the MMU assisted spacewalks undertaken on Challenger's 41B mission. Crippen and Scobee, meanwhile, perfected the challenging rendezvous technique.

Their quarry – Solar Max – had launched atop a Thor-Delta rocket from Cape Canaveral Air Force Station in Florida in February 1980, and was to spend a decade (essentially a full 'solar cycle') utilising a battery of gamma ray, X-ray, ultraviolet and other instruments to provide broad spectral coverage of the mechanisms responsible for causing solar flares. Ironically, only months after the MMU fabrication contract had been awarded to Martin Marietta, an unfortunate series of circumstances conspired to lead to the backpack's first operational use. One of Solar Max's instruments, a white light coronagraph and polarimeter, provided by the High Altitude Observatory of Boulder, Colorado, worked satisfactorily from March to September 1980, then suffered an electronics failure which left it inoperative. Then, in December, a fuse blew in Solar Max's attitude control system, causing it to 'wobble' and rendering it incapable of pointing precisely towards the Sun. All was not lost, however, because it had been designed as one of several Multi-Mission Modular Spacecraft (MMS), part of NASA's vision to permit certain satellites to be serviced by the Shuttle. Measuring 4 m long and fitting into a circular envelope some 2.3 m in diameter, the 2,315 kg Solar Max had two sections: a payload module, laden with eight powerful solar instruments, and the MMS for attitude control, power, communications and data handling functions. Connecting the sections was a transition adaptor, which supported two fixed solar array panels to provide between 1,500 and 3,000 watts of electrical power. This modular 'bus' was also being used for a number of other spacecraft. In view of its problems, Solar Max was placed into a slowly spinning 'safe' mode, which it maintained for three years, and although three of its instruments returned valuable data, the primary mission was effectively suspended. Its Hard X-ray Imaging Spectrometer – a 'flare alarm' to alert the other instruments to major solar events – malfunctioned in June 1981 and was left useless. However, by keeping the spacecraft rotating at one degree per second and aiming its solar panels constantly in the direction of the Sun, NASA engineers kept it alive in a dormant state. In the meantime, as efforts got underway to build a replacement electronics box for the white light coronagraph and polarimeter, it was decided to incorporate changes into the 'new' device to improve the instrument's imaging resolution (which had begun to degrade as early as July 1980) and permit space-to-ground communications through the TDRS network. Construction of the new box got underway in November 1982 and, after extensive tests, was complete by the following October. Shortly before Christmas 1983, the box was declared 'flight ready' and transferred to KSC for final checkout.

By the morning of 8 April 1984, after executing a series of thruster firings to set up an approach to their target, the 41C crew glimpsed Solar Max as a steadily brightening star. Crippen halted Challenger about 70 m from the slowly spinning satellite, as Nelson and van Hoften completed donning their spacesuits. They entered the payload bay at 9:18 am EST. "There's *no* comparison to getting into a suit and being outside," said van Hoften. "It was just a whole new world. You open the hatch

Secured into the MMU, Pinky Nelson approaches the slowly-spinning Solar Max.

and you look out and ... *there goes Africa*! It's just so distracting that it took probably 20 minutes until you start saying 'Hey, we've gotta get back on track here'. But it was just amazing. We had done so much work in the water tank, but in zero gravity it's really a *lot* different." The plan was for Nelson, designated 'EV1', to fly the MMU out to the satellite and dock himself to its mid-section using a Trunnion Pin Attachment Device, which had been tested on 41B. Although Solar Max was not spinning too fast for Hart to grapple it with the RMS, "we felt it was more prudent to have Pinky fly over with a backpack, dock himself to the satellite, stabilise it and then I could grab it with the arm". "Donning the MMU went very smoothly," Nelson recalled two decades later, "just like training. Ox and I had practiced so intensely that it was more like a well-choreographed dance than anything else. Once I left the docking station in the payload bay, the MMU flew just like the simulator at Martin Marietta in Denver, where we trained. I had been very well briefed by Bruce McCandless about the few differences between the simulator and the real unit, such as 'chatter' when accelerating, so I didn't experience anything unexpected." Precisely on time, after a ten-minute solo flight, Nelson arrived in Solar Max's vicinity and used the MMU's thrusters to gently match its rotation. Unfortunately, when he moved in to mate his TPAD with the satellite, it did not clamp properly into place. "We didn't know what was wrong," explained Hart, "but, being mechanical engineers, we said 'If a *small* hammer doesn't work, use a *bigger* hammer!' So Pinky went in twice as fast the next time and he hit again and bounced right off again." A third try, which imparted yet more force, also failed. Had the TPAD been affected by the cold of orbital 'nighttime', Mission Control wondered? Its temperature after removal from the payload bay storage locker had not been maintained, but pre-flight tests – and real-world experience on 41B – determined that it was capable of withstanding at least a few hours in the frigid darkness. In fact, on the ground, it had shown that it could operate satisfactorily for up to six hours at temperatures as low as -40 degrees Celsius. So far, in 'real' space, it had been outside for less than two hours and subjected to a balmy -12 degrees Celsius.

Low temperatures, therefore, did not seem to be a contributory factor. Furthermore, when Nelson pushed the TPAD against Solar Max's mid-section, its 'trigger' activated and released a pair of 'jaws' in an attempt to grab onto its quarry. This ruled out any kind of malfunction in the docking hardware. However, as the first EVA continued, the crew saw another problem brewing: Nelson's efforts had 'jostled' Solar Max out of its previously slow spin and Crippen asked him to grab a solar panel to steady it. The gyroscopic effect of this action worsened matters and, with his MMU's nitrogen supply running low, Nelson returned to Challenger. Instead of revolving gently, like a top, Solar Max was now tumbling unpredictably around all three axes.

Four tries by Hart to grapple it with the RMS proved fruitless and Crippen opted to withdraw to a distance of about 160 km until a new strategy could be thrashed out. "The grappling pin I had to grab was underneath one of the large solar panels, so I could only get [the arm] there under certain conditions," recalled Hart, "and it was very hard to predict how it was doing. I got close to it and I was maybe a foot away from getting it, but I'd reach some limit on the elbow or the wrist. I couldn't go far enough or fast enough to get it. It may be a good thing, because the satellite was tumbling so much that if I had gotten it, it may have actually broken the arm! Crippen, rightfully, said 'King's X. Let's go back'. We got the Shuttle back in position in front of the satellite and then we stabilised everything. We had fuel left, but not enough to do what we were doing anymore."

Privately, the astronauts were convinced that they had blown it and that the mission was a failure. "I could see myself spending the next six months in Washington," Crippen told the NASA oral historian, "explaining why we didn't grab that satellite!"

Overnight, as the astronauts slept, engineers at NASA's Goddard Space Flight Center in Greenbelt, Maryland, which operated the satellite, battled to regain control, but since its solar panels were no longer pointing towards the Sun, battery power was gradually dwindling away without recharging. The engineers switched off as many systems as possible, including heaters, but still had only six to eight hours of battery life left. When it became clear that Solar Max's magnetic torquer bars were now slowing the rotation, Goddard implemented a new technique, using a different method of sensing its position. This made the bars more effective in 'pushing off' against Earth's magnetic field and the satellite quickly stabilised itself. Then, just as battery life was running out, it came around in its orbit in such a way that the electricity generating panels faced sunward once more and began to recharge. When the crew awoke on the morning of the 9th, the satellite's batteries were powered and it was rotating serenely at half a degree per second.

Fuel reserves in Challenger's RCS tanks were low, at 13 percent, prompting

Crippen to quip that it *really* was STS-13 now! "Then we talked about what we had to do and Mission Control worked out the available fuel," said Hart, "but we took an extra day and decided we would do a second rendezvous. This time, Pinky and Ox would stay inside the orbiter and I would try to capture it with the arm." It also became clear during the ensuing spacewalk precisely why Nelson's attempts to capture Solar Max had been thrice frustrated: a small grommet, just 20 mm high and 6.4 mm thick, had obstructed the full penetration of the TPAD onto the satellite's trunnion pin. The grommet, which was installed near the pin, helped to hold part of Solar Max's gold-coloured thermal insulation blanketing in place. "What no one noticed," explained Hart, "is that one of the blankets had been put on with a little fibreglass standoff that the grommets would fit over. The engineering drawings didn't specify where those standoffs could be, so when they assembled the satellite, the technicians just put one wherever the grommet was. They glued it onto the metal frame, then stuck the blanket on. That was the correct thing to do, because no one envisioned using that pin for anything." A use for the pin did emerge, however, a year after Solar Max's launch, when the option of a Shuttle repair was first explored in depth, "but when they were designing the TPAD," Hart continued, "no one noticed that there was a grommet there. When Pinky went to dock, it interfered with the docking adaptor." It turned out that, if Nelson had made his approach to the pin within a very narrow pitch angle 'corridor', he might still have succeeded and captured Solar Max. However, during his second spacewalk, he took measurements of where the grommet was and the obstruction it posed, and found that it stuck out 1.5 cm too far ...

The TPAD, clearly, would not work. Either way, Challenger's on-board fuel was now too low to support a rescue if Nelson's MMU happened to fail. Instead, Crippen would fly close enough to Solar Max for Hart to grapple it with the mechanical arm.

As the pilots manoeuvred to re-rendezvous with the satellite, the off-duty EVA crewmen tended a couple of experiments in the middeck, including a student investigation into how well a colony of 3,300 bees made honeycomb cells in space, which Nelson later called, somewhat half-heartedly, "goofy science". Devised by student Dan Poskevich of Tennessee Technological Institute, the experiment theorised that by comparing bee-built structures on Earth and in space, generalisations may be formed for studies of other populations of the order hymenoptera, including wasps and ants. For Poskevich's investigation, two frames were enclosed in an environmentally controlled box, which provided lighting and temperature to simulate terrestrial conditions. Despite noting some disorientation in the bees, they ultimately proved that they could walk, fly and float without difficulty. Moreover, they built a sizeable, structurally 'normal' honeycomb cell and the queen bee laid around 35 eggs. Only about 120 of the insects died during Challenger's flight, representing a little over three percent of the population and significantly fewer than anticipated by Poskevich.

Early on 10 April – on his first attempt – Hart successfully grappled the Solar Max satellite with the mechanical arm and anchored it onto a Flight Support Structure (FSS) at the rear end of the payload bay. "It was a dramatic moment for

Mission Control," he remembered later. "We were euphoric when we succeeded. We really felt that the mission was at risk, which it was, and we were really on a mission that was demonstrating the flexibility and usefulness of the Shuttle to do things like repair."

The spectacular success, sadly, would prove to be the MMU's death knell.

An umbilical line was connected to the satellite to feed it with power from the orbiter and it was pivoted around so that Nelson and van Hoften, during their second spacewalk, which began at 3:58 am on 11 April, could reach and fix its broken attitude control system and the main electronics box of the disabled coronagraph and polarimeter. These repairs were originally meant to occupy one EVA apiece, but with the condensed and rescheduled flight plan, it was decided to attempt both during the same excursion. Replacement of the attitude control box – responsible for crippling the $240 million project more than three years earlier – took barely 45 minutes. Standing on the end of the RMS, his feet anchored in restraints, van Hoften removed a pair of screws, pulled the box out smoothly and plugged in a new unit. The second procedure of fixing the main electronics box to the coronagraph and polarimeter, which was not designed for replacement in orbit, was expected to be a longer and trickier task. Nonetheless, with surprising dexterity and outstanding skill, van Hoften pulled back a panel covering the box, cut and taped back a layer of insulation, removed two dozen screws and cut several wires. Nelson then took over, installing the new electronics box using large, gold-plated beryllium clips, instead of tiny screws, for the connectors. An hour after their second task had begun, the two men were finished and were able to place a 'baffle cover' over the X-ray polychromator to vent its exhaust gases away from Solar Max's other instruments. The second excursion had lasted six hours and 44 minutes, which, together with their two and a half hour outing on 8 April, brought Nelson and van Hoften's spacewalking time to more than nine hours. "The repair itself was a kick," Pinky Nelson recalled years later. "It was so much easier to work in space than it is on the ground. Ox and I, and TJ Hart running the arm, just kind of 'did' this repair. It was a piece of cake! It was so much fun riding on the end of the arm and much easier than working underwater." The pilots, too, were just as excited, particularly Scobee, who persuaded his crewmates to don T-shirts for the space-to-ground press conferences, emblazoned with the legend, 'Ace Satellite Repair Company'.

By this time, of course, Hart and his crewmates had long since found their 'space legs' and had adapted exceptionally well to the peculiar microgravity environment. "The first day or two," said Hart, "you tend to 'over control' your body a little bit and you tend to use your feet too much, so you flail and bounce into things. By the third day, you really get the hang of it, so you just use your fingertips to pull yourself around. It's almost like swimming underwater ... a graceful motion."

Finally, after a day of checkout in the payload bay, on 12 April Hart regrappled the satellite and deployed it back into space. By this time, in view of the additional day of planning needed to retrieve Solar Max, the mission had been extended by 24 hours and rescheduled to land on 'unlucky' Friday 13th after all. Yet, despite the huge success of the repair, bad luck had one more card to play. Much to Crippen's chagrin, the planned landing at KSC was postponed and finally cancelled due to showers in Florida, obliging the crew to land at Edwards Air Force Base. Shortly after the Challenger accident, the Rogers Commission would hear evidence that eight KSC homecomings had been planned between June 1983 and January 1986, with three having been diverted to California at short notice, entirely due to Florida's unpredictable weather. In fact, Crippen himself would testify to the commission in April 1986 that, despite his eagerness to land at KSC, he felt convinced "that you are much safer landing at Edwards". The predictability of 'favourable' weather conditions was essential for returning Shuttle missions because, after performing the 'de-orbit' OMS burn, the crew are committed to touch down approximately an hour later, with no option of returning to space or choosing an alternate landing site. Consequently, weather officers had to be certain, nearly one and a half hours before the burn, that conditions would be acceptable for the Shuttle to land. Thunderstorms in Florida, which build and dissipate quickly in the summer months, together with early morning fog, have made Cape Canaveral a notoriously difficult region to forecast. Edwards, on the other hand, has more stable weather and, in the case of the 41C return, proved the more reliable and safest option.

For Terry Hart, who had already decided before the mission that this would be his only space flight – despite having been offered the chance to fly on Spacelab-D1 – the re-entry came after a sleepless last night in orbit, soaking up as much of the experience as possible. Pinky Nelson, who had ridden into orbit on the middeck, would change places with Hart on the flight deck for the return to Earth. "I didn't have a lot of time, since we were busy on the flight, but the last night on-orbit, I had no duties at all," recalled Hart. "I just figured that I wasn't going to sleep at all. I'd turned down a second mission and was going back to AT&T Corporation, so I was damned if I was going to sleep. I stayed up all night and looked out the window while the rest of the crew was sleeping and watched the Himalayas go by and other parts of the world that I didn't see during the regular shifts."

Re-entry the following morning, he remembered, "was a wonderful thing. Watching the fireball around the vehicle was breathtaking. The 'engineer' side of me wanted to see the G buildup, but I had a camera and remember just letting it go and it would sit there, of course, when we were weightless. As we started to hit the upper parts of the atmosphere, I watched the camera accelerate forward as I let go, because the vehicle was decelerating. I was downstairs, but I was able to stick my head up every once in a while before I strapped in and looked out and could see the fireball overhead 'flickering' – a very impressive experience coming through that, but very smooth and quiet all the way down." True to the weather forecasters' predictions, an ominous thunderstorm arrived over KSC's Shuttle runway at precisely the time that Bob Crippen might otherwise have been landing there. "My family were in Cape Canaveral," continued Hart, "and we were landing in California, but it was beautiful. When I got out of my seat, I felt like I was using almost all of my strength just to get up! I was used to moving my body around with just my fingertips and now, all of a sudden, I had to exert all this force to get up. We didn't want to fall down the stairs on national television, so we were all doing deep knee bends to make sure we got our blood flowing again." Van Hoften, in the flight engineer's seat, recalled watching the effects of the super-heated plasma, whilst Nelson's enduring memory was how *strong* the sense of gravity felt on his body after a week of weightlessness. Yet Crippen's touchdown was perfect. "I'd done so many approaches and landings in the simulator, the Shuttle Training Aircraft and the T-38 jets," Nelson said, "that the landing felt very normal. It was disappointing that my family could not be there." (Unfortunately, Nelson's next mission, in January 1986, would also suffer from having its landing site diverted; his family would thus 'miss' seeing two of his three Shuttle touchdowns.)

Challenger made landfall at 5:38:06 am PST (8:38:06 am in Florida) on the Edwards dry lakebed, Runway 17, completing a mission of almost exactly seven days. As 41C's epic voyage ended, Solar Max's rejuvenated voyage of exploration had scarcely begun. After a four-week checkout, it set to work on what would turn into more than five years of observations of changes in the Sun's energy output. Its coronagraph and polarimeter, repaired by Nelson and van Hoften, resumed work in June 1984 and, despite a few interruptions, continued to capture images of the solar corona during the 'daytime' portions of the satellite's orbit until the end of the mission. Minor problems arose for most of January 1986, when Solar Max suffered a loss of memory in its on-board computer. Observations were again interrupted in December of that year, when the coronagraph's dedicated tape recorder failed, only coming back online in March 1987, thanks to the use of a backup device. Nevertheless, the satellite continued operating – in spite of atmospheric friction gradually dragging its orbit downwards – almost until the end of the decade. Other scientific results included the surprising discovery that the Sun is much brighter during periods at which 'sunspot' activity on its surface reaches its peak. Solar Max's instruments confirmed that, although the sunspots themselves are dark, they are surrounded by bright 'faculae', which more than offset the dimming effects of these Earth-sized blotches. By the time the satellite's mission ended in mid-November 1989, just two weeks before it re-entered the atmosphere, it had chalked up an impressive tally of a quarter of a million images of the Sun's corona and over 12,000 recorded solar flares. Additionally, its on-board gamma ray spectrograph made important contributions to the international study of Supernova 1987A, which had provided astronomers with their first 'local' opportunity to examine such a major stellar event since 1604. It also detected 15 observed gamma ray bursts from 'deep space'. Solar Max's ultraviolet spectrometer and polarimeter, despite suffering from a jammed grating drive mechanism in April 1985, managed to provide pointing and timing information for the other instruments and even conducted four years' worth of ozone-concentration measurements in Earth's atmosphere. Plans nurtured both prior to and in the wake of the Challenger tragedy to retrieve Solar Max once more, bring it back to Earth and refit it as the Extreme Ultraviolet Explorer came to nothing. Pre-Challenger plans from as late as December 1985 called for a retrieval of Solar Max in 1987 or 1988 for a series of EUVE upgrades and re-launch by the Shuttle in 1989 or 1990. Another bus was built for EUVE, which reached space atop an expendable rocket in 1992. Interestingly, part of the original Solar Max remained in orbit for much longer. When the broken attitude control system was returned to Earth by Crippen's crew, it was refurbished and placed aboard NASA's Upper Atmosphere Research Satellite. From September 1991 until its re-entry in September

2011, the latter mission closely monitored the chemistry of Earth's middle and upper atmosphere.

Despite the immense success of the Solar Max repair, plans were already afoot to improve future servicing missions. In November 1985, a satellite services workshop held at JSC, part of which was chaired by 41B's Bruce McCandless, heard proposals to install not one, but *two*, RMS arms aboard future Solar Max-type repair missions. Although the instrumentation on the Shuttle's aft flight deck was not capable of operating both arms simultaneously, it was suggested that one could be employed 'passively' to hold a target satellite steady whilst its 'active' sibling manoeuvred tools, replacement units or spacewalking astronauts into place. The proposal noted that, if two RMS devices had been available to Crippen's crew, the need to fully berth and latch Solar Max into the payload bay might have been unnecessary and not carrying the cradle would have greatly improved the propellant margin. Moreover, the potential risk of damaging the satellite's solar panels or the orbiter during berthing or deployment would be sidestepped and the entire servicing could have been undertaken 'outside' the bay. An orbiter thus equipped would, the paper's author pointed out, have greater ability to reposition the satellite throughout its repair than the more limited 'turntable' option offered by the Flight Support Structure.

If 41C was stricken with bad luck at all, the greatest victim must have been the MMU jet backpack first tested by Bruce McCandless two months earlier. Admittedly, it had performed admirably under the control of McCandless, Stewart, Nelson and van Hoften. It could hardly be blamed directly for the failure of the TPAD. Indeed, it would prove its worth in November 1984 when, flown by Joe Allen and Dale Gardner, it was instrumental in the retrieval of the errant Palapa-B2 and Westar-6 communications satellites. However, what 41C *did* prove was the crisp manoeuvrability of the orbiter itself and the precise handling characteristics of the RMS. It was, observers said later, 'Hart's small grab', rather than 'Nelson's free flight', which had pulled success from the jaws of what might have been an ignominious defeat. Ultimately, the root cause of the TPAD failure was imprecise knowledge of the configuration of the satellite. It is perhaps significant that on the reusable spacecraft's next rescue mission (that of the crippled Leasat-3 communications satellite in August 1985), astronauts relied on their own space suits and an RMS to perform a breathtaking repair. Missions since Challenger have also demonstrated that other equipment such as tethers, safety grips, hand bars and foot restraints can allow astronauts to conduct a multitude of tasks without the need of a bulky, jet-propelled backpack.

Similarly, Vance Brand's ability on 41B to fly the Shuttle with pinpoint accuracy to collect Bruce McCandless' lost foot restraint and the use of the RMS on a subsequent mission to knock a chunk of ice off a waste water port removed the need for additional risk. Despite much criticism of NASA's cavalier attitude towards Shuttle operations before the Challenger disaster, there was also, said McCandless, "a sort of creeping conservatism and EVAs came to be regarded as hazardous, to be scheduled only if absolutely required." More tellingly, Pinky Nelson doubted, even in the heady days before January 1986, that the jet backpack would have flown again. Like McCandless, he stressed that it was "well conceived and engineered but, unfortunately, the planned uses of the MMU were superseded by other capabilities that we developed, but couldn't anticipate".

Finally, when the Rogers presidential inquiry into Challenger's loss presented its findings, renewed emphasis was imposed on increasing the safety of other Shuttle components. "The cost of recertification," recalled McCandless, "eventually killed it off. In the fall of 1989, a proposal was solicited from Martin Marietta for recertification and refurbishment for one Shuttle mission. It came in at $6.1 million, which was deemed too expensive, and despite some small sums for clean room environmental storage, 'just in case', they were eventually retired." Today, the MMU flight unit first used by McCandless hangs in the Smithsonian. The other backpack was loaned to NASA's Marshall Space Flight Center for possible future use as a 'flying testbed' for autonomous rendezvous and docking systems – "subject," said McCandless, "to the constraint that it be maintained in a condition that could be restored to flight configuration." Both units, therefore, were mothballed until such time as their unique capabilities were needed again.

They never were.

Perhaps, through the drawbacks it uncovered with the multi-million-dollar backpack, 41C proved to be an unlucky mission after all.

FIRE ON THE PAD

"T-31 seconds ... We have a Go for autosequence start ... Discovery's computers now taking over primary control of vehicle critical functions until liftoff ..."

The calm, measured tones of NASA commentator Mark Hess provided an assurance that Shuttle launches had become the stuff of routine. It was 26 June 1984 and after a false start the previous day, all seemed to be proceeding normally as the final seconds ticked away to the maiden voyage of the new orbiter, Discovery. Strapped into the flight deck were astronauts Hank Hartsfield, Mike Coats, Mike Mullane and Steve Hawley, whilst downstairs on the middeck were Judy Resnik – America's second female spacefarer – and McDonnell Douglas engineer Charlie Walker, flying as part of a commercial contract with NASA. Walker had been training with the crew since the previous summer, but the others had been assigned in February 1983. It had been a long 16 months. When they were first assigned, they expected to be launched in March 1984 on a mission designated 'STS-12' to deploy the third of NASA's network of Tracking and Data Relay Satellites. However, the real icing on the cake, in Mullane's mind, at least, was that they would be flying the maiden voyage of Discovery, the third orbiter, whose construction had begun under a $1.9 billion contract with Rockwell International in February 1979. Named in honour, primarily, of James Cook's HMS *Discovery*, but also offering a nod to vessels commanded by Henry Hudson to search out the Northwest Passage, by George Nares to reach the North Pole and by Robert Falcon Scott to conquer Antarctica, the spacegoing Discovery's fabrication had begun in August 1979 and the vehicle was structurally complete by February 1983. Several months of testing followed and Discovery was finally rolled out of Rockwell's Palmdale plant in

California in October, commencing an overland trek to Edwards Air Force Base and delivery to Florida atop the Boeing 747 Shuttle Carrier Aircraft on 9 November. Thanks to manufacturing changes to the internal structure of the airframe and the inclusion of newer thermal protection materials, including the quilt-like Advanced Flexible Reusable Surface Insulation (AFRSI) in place of tiles at various points on the upper wings, fuselage, payload bay doors and vertical stabiliser fin, Discovery's dry weight of 67,100 kg was some 300 kg less than her sister ship, Challenger, and more than 2,000 kg less than the queen of the fleet, Columbia. She was undoubtedly the most advanced orbiter yet built. "Aviators live for the day they might be the first to take a new jet into the air," Mullane wrote, "and *we* were being offered the first flight of a Space Shuttle."

If it all seemed too good to be true, it was. It did not take long for the gremlins of misfortune to hit the mission. Within six weeks of the crew announcement, the first TDRS had been left stranded in a useless orbit, thanks to the failure of its IUS booster; and by the end of May 1983, a second TDRS had been deleted from STS-8 and a third from STS-12. According to Boeing, the prime contractor for the IUS, repairs and modifications would require at least a year. As we have already seen, these woes led directly to the cancellation of both STS-10 *and* STS-12 ... but not to the dissolution of their crews. "After many tense weeks of worry," wrote Mullane, "we acquired a new payload of two smaller communications satellites with different booster rockets. Best of all, we still retained the first flight of Discovery." Instead of a TDRS, they were given Anik-C1 – also listed as 'Telesat-I' in NASA's November 1983 manifest – and a military communications satellite for the US Navy, known as Syncom 4-1. The two payloads could not have been more different. Anik-C1 was virtually identical to its siblings placed into orbit on STS-5 and STS-7; a solar-cell-coated drum, mounted atop a PAM-D. Syncom was also drum-shaped, but would be carried aloft horizontally in a 'cradle' and spring-ejected from the payload bay, departing 'sideways', like a frisbee. In addition to Anik and Syncom, according to the November 1983 and January 1984 manifests, Discovery was also to carry the OAST-1 experimental solar array, sponsored by NASA's Office of Aeronautics and Space Technology, and a Large Format Camera in the payload bay for topographical research. In keeping with Jim Beggs' edict, the mission would be redesignated '41D'.

" ... T-15 seconds and counting ..."

Losing their TDRS payload was disappointing to the astronauts, particularly Mullane and Resnik, who would have taken the lead role in its deployment. They had spent a great deal of time at Boeing's plant in Seattle, learning the intricacies of the Inertial Upper Stage. "At the contractors' factories, we also did some 'widows and orphans' appearances," wrote Mullane, referring to NASA's deliberate attempt to present a human face on manned space exploration and hence raise awareness of the deadly consequences of mistakes, "passing out 'Maiden Voyage of Discovery' safety posters to the workers." Judy Resnik had joked that there were certainly *no* maidens on *this* mission! In *Riding Rockets*, Resnik herself is presented as a tragi-comic figure – tragic in terms of her estrangement from her mother and, of course, her untimely death aboard Challenger, but intensely witty in her interactions with others; her romantic crush on actor Tom Selleck became the stuff of banter among fellow astronauts. "Flirtatious, funny ... just a live wire," was classmate Rhea Seddon's summary of her. Yet she was an outstanding over-achiever. Born Judith Arlene Resnik in Akron, Ohio, on 5 April 1949, she was the progeny of a first-generation Jewish-Russian family. Her father, Marvin, was an optometrist and part-time cantor, whilst her mother, Sarah, was a former legal secretary. Soon after entering kindergarten Resnik was able to read and solve simple mathematical problems. She and her younger brother, Charles, received Hebrew schooling and her teachers described as bright, disciplined, personable and a perfectionist. It was whilst at school that she developed her love for mathematics and classical piano. She achieved the highest possible score – 800 – on the mathematics component of her SAT test, graduated from Firestone High School in Akron in 1966 and was accepted into the Carnegie Institute of Technology to study electrical engineering. (By the time she completed her degree in 1970, the institute had been renamed 'Carnegie-Mellon University'.) She was initially hesitant about entering engineering, since it was not traditionally a female career path, but realised that her aptitude for mathematics and the sciences would carry her through. "Maybe I liked it," she once said, "because I was *good* in it." Shortly after graduation, she married a fellow engineering student, Michael Oldak, but the pair divorced in 1974. During her short married life, she was employed by RCA, working on custom integrated circuitry for phased-array radar control systems and the specification, project management and evaluation of control systems. She also undertook work for NASA sounding rocket and telemetry programmes. Resnik later joined the National Institutes of Health as a biomedical engineer and staff fellow, working in the neurophysiology laboratory in Bethesda, Maryland, and at the same time commenced work on her doctorate in electrical engineering. She received her PhD from the University of Maryland in 1977 and joined Xerox as a senior systems engineer.

In *Riding Rockets*, Mike Mullane noted that Resnik's first exposure to the space programme and the idea of becoming an astronaut appeared that same year, when she first saw an announcement on the Xerox bulletin board. To Mullane, it underlined the reality that many women had grown up in society, totally closeted and unaware of the possibility that such careers were available, on the basis of gender or colour. In fact, one of the recruiters who drew Resnik to NASA was the African-American actress Nichelle Nichols, famed for her role alongside William Shatner in 'Star Trek'. Nichols began by affiliating her company, Women in Motion, with the space agency, although its focus extended to ethnic minorities, too, and led to the selection of Resnik and Sally Ride, Guy Bluford and Ron McNair and inspired a number of others, including Mae Jemison, who in 1992 became the first black American woman in space. When Resnik was selected as one of the first six female astronauts in January 1978, it was widely expected that either she or Sally Ride would be the first to fly. In fact, Rhea Seddon remarked in her oral history that Resnik and Ride received "the sorts of technical assignments that really prepared them for flight", such as RMS work and capcom duties. "I think most of us felt it would be Sally or Judy." To Resnik, as one of the 41D crew members, it mattered little. Being a *woman astronaut* or the *second American woman astronaut* or the *first*

Jewish woman astronaut was as insignificant as saying that she was "the 40th or 45th ... *American astronaut* to go on the Space Shuttle in a period of a couple of years". Resnik was simply amazed at how far the space programme had evolved in just a handful of years.

"... *ten* ..."

By the early summer of 1984, as 'Ghostbusters' smashed cinema box offices across the world, the Shuttle seemed to be prospering. The MMU had been tested and Solar Max had been triumphantly repaired. Six more missions were scheduled before the end of the year and, on 4 June, Discovery's three main engines were test-fired in readiness for her first launch. The juggling of payloads remained a serious issue, though, and at some stage between the January and May 1984 manifests Anik-C1 was removed from 41D and rescheduled for another flight early the following year. Anik was a PAM-D payload and the failure of this booster to deliver Westar-6 and Palapa-B2 into orbit in February had led to delays until the definitive cause could be identified and corrected. Hank Hartsfield's crew were left with a relatively spacious seven-day flight to deploy the first Syncom and run the Large Format Camera, OAST-1 and a series of middeck experiments. (In fact, OAST-1 operations would dominate the mission, with the first deployment of its mast on Day Three to support a series of tests of its performance and structural dynamics. The camera, too, would be operated on its payload bay truss throughout the flight.) After 41D, the rest of the 1984 manifest remained largely unchanged, although the numbering system was adjusted slightly. Mission 41E aboard Challenger, previously scheduled for July, would have carried Ken Mattingly's crew on their already-long-delayed assignment for the Department of Defense, but this had been cancelled earlier in the year, due to the IUS difficulties. Next up after 41D, therefore, would be 41F in August, aboard Challenger; Karol 'Bo' Bobko's seven-day flight would deploy a record *three* satellites – a second Syncom (4-2) and a pair of PAM-D payloads known as SBS-4 and Telstar-3C – as well as a retrievable platform called 'Spartan'. Then, in early October, Bob Crippen would fly Columbia on mission 41G to deploy the Earth Radiation Budget Satellite and operate a payload for NASA's Office of Science and Terrestrial Applications. Another mission previously on the 1984 manifest (41H, commanded by Rick Hauck) was cancelled and the first pair of '51-series' flights were scheduled for November (51A) and December (51C).

So it was that the 41D crew arrived in Florida in a fleet of four T-38 jets on the afternoon of 22 June 1984, with liftoff anticipated three days later. They circled over the launch complex and alighted on the runway of the Shuttle Landing Facility. Flying, relaxing, spending time with family members and running through checklists consumed their final hours on the planet. On the morning of the 25th, they suited up – Mullane jokingly offered Resnik an emery board to do her nails during ascent – and headed out of the Operations and Checkout Building for the pad and Discovery. "The pad was eerily deserted," Mullane wrote. "A vapour of oxygen swirled around the [main engine] nozzles. A flag of more vapour whipped from the top of the [External Tank] beanie cap. Shadows played upon that fog ..." Hartsfield and Coats were the first to board Discovery, followed by Mullane on the flight deck and Resnik and Walker on the middeck; the last to be strapped in was Steve Hawley, seated

Discovery's main engines roar to life on the morning of 26 June 1984 ... only to be shut down, seconds later. From the commander's seat, Hank Hartsfield faced a life-or-death gamble: to order a 'Mode One Egress' from the vehicle, and risk exposing his crew to invisible hydrogen flames, or to sit tight. He elected to sit tight and the 41D crew lived to fly another day.

behind and between the commander and pilot in his role as flight engineer. During the wait, Walker recalled being asked to help verify the integrity of a pressure seal. The countdown continued ... until a problem was detected with Discovery's backup General Purpose Computer, which failed to synch with the four primaries. The clock was stopped to permit troubleshooting, but when it became clear that the problem could not be solved, the crew were notified that the launch would be scrubbed and rescheduled for the next day, 26 June. After more than two hours lying on their backs, against the hard aluminium frames of their seats, it was not welcome news that they would be forced to endure the same discomfort tomorrow.

Darkness still covered the Kennedy Space Center when the crew repeated the time-honoured ritual in the small hours of the following morning. "The van starts out slowly," Charlie Walker told the NASA oral historian of his journey to the pad. "We wind our way through the parking lot and out onto the access road to the pad. There's this police escort, of course, in front of you, with lights flashing and everything, so you feel like you're on top of the world already ... Then we arrive out at the launch pad and, at the base of the pad, security waves you on. You don't have to show all your badges and everything this time; they know who you are. At the bottom of the crawlerway, you can get the view ... the first time we'd really seen it clearly without the rotating service structure around it and so it's an awesome experience, knowing that you're going to ride this thing into space and there's your spaceship waiting for you out there. Go up the ramp, the police car leading the way, and you just get to the top of the ramp, and the thing that I noticed first was really different ... was that there was like only one or two other people there. Now, every other time you've been out there, there's dozens and dozens of people all around and conversation going on; but now it's silent, except for the wind, maybe some seabirds now and then. As you get up closer to the vehicle, you begin to hear the vehicle. You begin to *hear* the Shuttle and the External Tank." Months of training had reached their climax; they were *really* going to fly *today*.

" ... *we have a Go for main engine start* ..."

The preparations proceeded remarkably smoothly for the second attempt. The backup computer had been replaced and tested and showed no discrepancies. The countdown moved crisply: built-in holds at T-20 minutes and T-9 minutes were passed and, at 8:39 am EST, with five minutes to go, Mike Coats reached down and switched on Discovery's three Auxiliary Power Units. The trio of hydraulic pumps hummed perfectly to life. "The meters showed good pressure," recalled Mullane, who was seated directly behind Coats. "Discovery now had muscle." The computers automatically commanded a final series of checks of the main engines and the elevons on the wings. With two minutes to go, the astronauts closed their visors. Hartsfield shook Coats' hand and wished them all good luck; reminding them to stick to their training and keep their eyes focused on the instruments. Thirty-one seconds before liftoff, the ground launch sequencer handed over control of the countdown to the Shuttle's computers. In the darkened middeck, Resnik and Walker clasped hands. The Solid Rocket Boosters underwent their final nozzle steering checks and at ten seconds a flurry of sparks from the hydrogen burn igniters gave way to a familiar low-pitched rumble.

" ... *seven, six, five ... we have main engine start* ..."

Inside Discovery's cabin, the astronauts felt the immense vibration as turbopumps awoke, liquid oxygen and liquid hydrogen flooded into the engines' combustion chambers and they roared to life ... and then, suddenly and shockingly, were arrested by the blaring sound of the master alarm. Something had gone badly awry. Two of the main engines – Number Two and Three, closest the aft body flap – had blazed to life, but the Number One engine, directly at the 'top' of the pyramid, had failed to ignite. "The vibrations were gone," wrote Mullane. "The cockpit was as quiet as a crypt. Shadows waved across our seats as Discovery rocked back and forth on her hold-down bolts." From the pilot's seat, all Coats could hear was the screeching of disturbed seagulls outside. Two red lights on the instrument panel indicated that the Number Two and Three engines had indeed shut down, but the indicator for the Number One remained dark. Instantly, Coats, whose responsibility as pilot was to monitor the engines during ascent, jabbed his finger repeatedly onto the button to shut it down. The status indicator did not change; it remained dark. Downstairs, Walker's eyes were focused intently on the procedures for 'Mode One Egress', the instructions for opening the side hatch and evacuating the vehicle.

Several kilometres away, on the roof of the Launch Control Center, the astronauts' families were watching the unfolding drama ... and *they* were perplexed. "A thick summer haze had obscured the launch pad," wrote Mullane. "When the engines had ignited, a bright flash had momentarily penetrated that haze, strongly suggesting an explosion. As that fear had been rising in the minds of the families, the engine-start sound had finally hit ... a brief roar." The sound echoed off the walls of the Vehicle Assembly Building and was gone. Within seconds, it became clear what had happened.

"... *we have a cutoff ... we have an abort by the on-board computers of the orbiter Discovery ...*"

Over the intercom, the astronauts heard the worrisome words 'RSLS Abort', meaning a 'redundant set launch sequence'. This pointed inevitably to a main engine problem, which had forced their automatic shutdown by the General Purpose Computers. Intuitively, the crew knew that safeguards existed to prevent the SRBs from igniting – if *that* had happened, it would have killed them all – but they also knew that only a few seconds existed on the countdown clock. "A couple of seconds in the world of electronics is a *lifetime*," said Mullane, "and I'm sure that all the safety devices had rotated to prevent [the solids] from igniting ... but in the back of your mind, you're thinking *What happens if those ignite?*" The situation was by no means under control. As if the indication that the Number One might still be burning was not enough, Launch Control now told the crew that there was a fire on the pad and the suppression equipment had been activated. The decision over whether to unstrap and make an emergency evacuation of the orbiter was now in the hands of Hartsfield; downstairs, on the middeck, Judy Resnik had unstrapped and was peering through the window in Discovery's side hatch. She could see no fire. The astronauts would have to run across the access arm to a set of seven slide-wire baskets which would whisk them from the pad to safety. Listening to the communication loop, Hartsfield elected to sit tight. It was a decision which probably saved their lives. Hydrogen burns 'cleanly', invisible to the human eye, and it had already begun to ignite combustible materials on the pad surface. Subsequent inspections would reveal scorched paint all the way up the launch pad structure, as far as the crew access arm. Years later, Walker would praise Hartsfield for *not* having ordered a Mode One Egress that day. In his conversations with the launch director after the abort, Hartsfield realised that a lot of doubt also existed over the reliability of the slide-wire baskets and *that* had informed their judgement to keep the crew aboard the orbiter.

At the press site, Mark Hess' commentary continued: *"We have an indication two of our fire detectors on the zero level; no response ... They're side by side, right next to the engine area ... The engineer requested that we turn on the heat shield firewall screen between the engine valve and Discovery's three main engines ..."*

From the flight engineer's seat, Steve Hawley injected a spark of humour into the proceedings. "Gee," he said, in his thick Kansan drawl, "I thought we'd be a lot higher at MECO!"

There *had* been a Main Engine Cutoff, but *not* at the edge of space; Discovery remained firmly shackled to Earth. T-zero had not been reached and so the SRBs had not been commanded to ignite. Hawley's joke "broke the ice and got everybody laughing", said Mike Coats, but did little to dissipate a pervasive sense of gloom that their mission had been aborted just *four seconds* before liftoff. Gloom is often associated with bad weather and rain and when the 41D crew finally saw the light of day and made their way out of the orbiter, they did so in a torrential downpour ... not of rain, but of the waters of the fire suppression system. The entire gantry was soaked, drips from every pipe and platform, the white room ankle-deep in water. As Mike Coats walked out of the elevator at the base of the pad, it was like walking underneath a waterfall. "We got completely soaked to the skin," he explained years later. "Then we got in the astronaut van, which was air-conditioned and *very* cold. As we were driving away, there's a window in the back of the van and all of us were looking back at the Shuttle on the launch pad, shivering and soaking wet, like drowned rats."

Hawley's joke revealed something else. In the inimitable words of Mike Mullane, it showed that some of the civilian astronauts "had steel balls". In fact, Steven Alan Hawley, the youngest of the Thirty-Five New Guys, did not intend it that way; to him, despite the unexpected shock of the engine abort, it felt very much like one of the countless ascent simulations he had performed with Hartsfield and Coats over the course of the past year. In the first tenth of a second after the 26 June abort, as the cabin fell silent and the master alarm blared, Hawley was convinced he was back in the simulator. *Then*, he checked himself. "Yes, except it's not *really* supposed to do that in *real* life!"

Hawley came from Ottawa, Kansas, where he was born on 12 December 1951, but regarded Selina, Kansas, as his hometown. His love of learning traced its origins back at least two generations; *both* of his grandfathers were teachers – on the maternal side, in physics, and on the paternal side, in theology – whilst his father was a minister. (In fact, when Hawley wed fellow astronaut Sally Ride in July 1982, it was Dr Bernard Hawley and Ride's sister, the Reverend Karen Scott, who married them.) As a child, Hawley found himself surrounded by physics books and his fascination with astronomy derived from the 'unconventional' nature of the subject; "all they were able to do," he told the oral historian, "was look at whatever was out there and try to figure out what was going on by looking in clever ways or different ways". The fascination ran through at least one other family member, with his brother, John, becoming a theoretical astrophysicist. Hawley completed high school in Salina in 1969 and entered the University of Kansas to study physics and astronomy. Whilst there, he worked summers as a research assistant at the US Naval Observatory in Washington, DC, and at the National Radio Astronomy Observatory in West Virginia. Graduation in 1973 led him to the University of California at Santa Cruz for his doctorate, working at Lick Observatory on the spectrophotometry of gaseous nebulae and emission-line galaxies, with emphasis on chemical abundance determinations. At around the time that Hawley received his PhD in 1977, he spotted NASA's announcement for astronaut candidates and submitted an application. "Why in the world would they pick me?" he asked later. "I still think, perhaps they didn't mean to, and one day they'll come and tap me on the shoulder and say, 'Excuse me. You've got this guy's desk, coincidentally named *Steve Hawley*, and *he's* the one we meant!"

Hawley was interviewed in Houston and headed off for a two-year post-doctoral assignment at Cerro Tololo Inter-American Observatory in Chile. Several months into the post, in January 1978, he was at the observatory's headquarters, in the foothills of the Andes, when he received a call from George Abbey. He was in.

Despite his youth and baby-faced appearance, Hawley very quickly cemented his credentials as one of the brightest and most outstanding of the civilian astronauts and earned the respect of the military test pilots, who came to nickname him 'The Attack Astronomer'. Mike Mullane explained that this came from the military astronauts, many of whom had their own titles – fighter pilot, attack pilot, gunship pilot – and that Hawley's title was purely honorific. Hawley's own recollection of the nickname has a slightly different slant. During training, the Thirty-Five New Guys were in Florida and Senator Adlai Stevenson of Illinois asked them to introduce themselves: Sally Ride was "a physicist", Dan Brandenstein "a Navy attack pilot", Hoot Gibson "a Navy fighter pilot" ... and when it was Hawley's turn he could not resist identifying himself as "an Attack Astronomer". The moniker stuck. Years later, Mullane paid homage to Hawley's immense brainpower: "Every checklist, including the two-volume malfunction massif," he wrote, "was in a virtual file drawer in his brain, ready for instant retrieval at the sound of a cockpit warning tone." He even did professional baseball umpiring during his summer vacations. It was not idle praise. Some astronauts would consider that having Hawley in the cabin was equivalent to having the benefits of an extra General Purpose Computer.

Another nickname earned by Hawley, quite soon into his training for 41D, was 'Cheetah' ... but *that* nickname actually encompassed the whole crew. During one NASA trip, he and Mullane happened to meet husband-and-wife team John and Bo Derek, who had recently worked together on the 1981 movie, 'Tarzan the Ape Man'. After this, Judy Resnik dubbed Mullane and Hawley as 'Tarzan' and 'Cheetah'. Inevitably, she received the title 'Jane', whilst the entire 41D team became known by office secretaries as 'The Zoo Crew' and Hank Hartsfield, as the flight's non-politically-correct commander, became 'The Zoo Keeper'.

Henry Warren Hartsfield was the only flight-experienced member of the 41D crew. He was born in Birmingham, Alabama, on 21 November 1933. "When I was a kid, all I thought about was *flying*," he told the NASA oral historian. "The guy my dad worked for smoked Wings cigarettes; an old brand that has a little picture of an airplane on every pack. He used to bring them home to me [and] I collected those cards." Hartsfield graduated from local high school and was accepted into Auburn University, majoring in chemical engineering ... until he found out that he was "a natural disaster in the chemistry lab, always blowing things up and catching things afire". He switched to physics, which he thoroughly enjoyed, and received his degree in 1954. Then, as a member of the Army's Reserve Officer Training Corps of Engineers, he noticed a call by the Air Force for pilots. After tests at Maxwell Air Force Base in Montgomery, Alabama, Hartsfield entered the Air Force in 1955 and immediately began theoretical physics work, hoping to achieve a master's qualification. The service agreed to delay his entry into active duty until the following year, but, notwithstanding a strong letter of support from the head of his physics department, refused to grant him additional time to complete his degree. Thankfully, Hartsfield loved flying and decided to remain in the Air Force.

Primary training in Georgia and Texas was quickly followed by gunnery training at Williams Air Force Base in Arizona, at which time he flew F-86 Sabre jets, then the F-100 Super Sabre and eventually the F-105 Thunderchief. During this period, Hartsfield qualified as a fighter pilot and in 1959 decided to pursue a route that would lead into the space programme. He was selected to attend the Air Force Institute of Technology for a master's degree in the new field of 'astronautics', but left partway through his course to continue his flying career and build up enough hours to attend test pilot school. Allowed to resign from his degree without prejudice, Hartsfield found himself, within a month, flying an F-105 from the large air base at Bitburg in West Germany. The assignment lasted for three years and as soon as he passed the magic mark of 1,500 hours in high-performance jets, he applied for test pilot school..."and was selected, *right off the bat*. It was *first try*, so I was fortunate."

During his time at Edwards Air Force Base, Hartsfield met three other students, named Al Worden, Charlie Duke and Stu Roosa, all of whom had their sights firmly set on the space programme. At that time, there were two possible routes: either through NASA or the Air Force. If they applied for the latter, the Air Force would nominate them for either NASA or its own Manned Orbiting Laboratory project. Worden, Duke and Roosa applied only to NASA and were selected by the agency in the summer of 1966. Hartsfield applied to both ... and was picked for MOL. Shortly after the crushing disappointment of MOL's cancellation in June 1969, he opted for a master's degree in engineering science at the University of Tennessee and graduated in 1971. By that time, he had been selected as an astronaut candidate by NASA and throughout the next decade he worked on the development of the Shuttle, focusing specifically on its flight control systems.

By the time the first crew announcements for the Shuttle's Orbital Flight Test (OFT) missions were made in March 1978, Hartsfield had retired from active duty in the Air Force and had been hired by NASA as a civilian astronaut ... yet a flight opportunity still seemed distant. He knew that John Young and Bob Crippen were in line for the first mission (known as 'A') and that Joe Engle and Dick Truly would fly the second ('B'). By the end of 1979, Jack Lousma and Gordon Fullerton were training for the third ('C') mission, when, all at once, Hartsfield and Ken Mattingly were called up to join them for the *same* training. "It was kinda funny," Hartsfield recalled, "because it *scared* them. Lousma made a panic call to Houston; [he] thought *we* were going to *replace* them." In fact, Mattingly and Hartsfield would *back up* Lousma and Fullerton. "It was a little bit confusing as to the way the crews were announced, because *no one* really knew; it was a standard joke ... around the office, trying to figure out this crew structure and *how* it was going to work." Mattingly and Hartsfield need not have worried. On 1 March 1982, NASA formally announced their names to fly STS-4. It was a hugely successful mission and concluded in spectacular style with a touchdown at Edwards Air Force Base on the Fourth of July, witnessed by President Reagan.

Mike Mullane has written a great deal about Hartsfield's decidedly right-of-centre

In this image of Discovery's three main engines in the wake of the 26 June pad abort, the evidence of scorching from the hydrogen fire is clearly visible. Note the discolouration of the smaller OMS engines. Of interest, the base of one of the Solid Rocket Boosters, with its explosive launch pad separation bolts, can be seen in the background.

political stance. At Hartsfield's 50th birthday celebration in November 1983, the astronaut office presented him with outrageously satirical gifts: a copy of *Ms* magazine, signed by feminist journalist Gloria Steinem, a fake congratulatory message from the American Civil Liberties Union and fake letters from Soviet General Secretary Yuri Andropov (thanking him for promoting global communism) and from Senator Ted Kennedy (thanking him for his 'recent donation' to the Democratic Party). "A final gift," concluded Mullane, "a box of Ayds diet candy, was from the gay rights political caucus, acknowledging Hank's support for their cause." The fake gift card read: *Here are some AIDS for you.* "Nothing was out of bounds when it came to astronaut humour," Mullane wrote.

Now, on the morning of 26 June, as Hartsfield's disappointed crew was extracted from Discovery, they were exhausted. "After a launch abort," Mullane told the oral historian, "you could take a gun and point it right at somebody's forehead, and they're not even going to *blink*, because they don't have any adrenaline left in them; it's all been used up." Mike Coats took his wife and children to Disney World, where, later that afternoon, they found themselves queuing for the Space Mountain ride. Replacement of the troublesome main engine would require a return to the VAB and *that* prompted a delay until August at the earliest; Discovery was destacked and returned to the Orbiter Processing Facility by 17 July. The lengthy down time forced NASA to make a number of uncomfortable decisions about the schedule for the remainder of the year. "Payloads were stacking up," Mullane wrote. "Every day a communications satellite wasn't in space meant the loss of millions of dollars of revenue to its operators." The focus was on combining the payloads of two missions into one and deleting the other from the manifest, thereby providing for the minimum distortion of the launch schedule and maintaining NASA's commitment to its commercial customers. Mission 41F had been due to launch on 9 August and was cancelled; its entire payload – with the exception of Spartan – would be shifted onto Hartsfield's flight. Since 41F included the *second* Syncom (4-2) as part of its payload, *that* also ended up on 41D – with the result that the *second* Syncom actually launched ahead of the *first*. (Syncom 4-1 was ultimately launched in November 1984.) Hartsfield's crew retained OAST-1, but lost the Large Format Camera, which moved onto Mission 41G, the dedicated Earth resources flight, in October. In Steve Hawley's mind, training for this change was not a big deal, for the crew had already spent months working on OAST-1, Syncom deployment procedures and PAM-D deployment procedures. By early August, Discovery was back on Pad 39A, with launch anticipated at the end of the month. With SBS-4, Telstar-3C, Syncom 4-2 *and* OAST-1 in her payload bay, she would be carrying the heaviest load – at 18,680 kg – ever taken into orbit by the Shuttle at that time.

The crew returned to the pad on 29 August for their third attempt. It was Walker's 36th birthday and *that*, at least, seemed a lucky omen. It wasn't. The launch was scrubbed when a timing discrepancy was noted between the flight software and Discovery's master events controllers. This was particularly unnerving. Tests had shown that, under worst-case timing conditions, the controller might be unable to process certain critical events commands, such as separation of the SRBs or the External Tank. If the solids failed to separate properly, the crew would be dead. If the External Tank failed to separate properly, the crew would be dead. It was *that* straightforward. Hartsfield's crew assembled for a debriefing later that afternoon. The problem was already known – in *Riding Rockets*, Mullane mentioned that it had been a topic of discussion amongst the astronauts as long ago as April – and it was decided to effect a software 'patch'. This was put in place by engineers overnight and the countdown was recycled for another launch attempt at 8:35 am EST on the 30th. Despite a problem with the ground launch sequencer, which forced an extended hold at T-9 minutes, events ran smoothly ... until a pair of private pilots accidentally strayed into the closed airspace of KSC. A hold was called whilst the pilots were shooed away, but for the crew of 41D, who had already sat through two uncomfortable countdowns, plus a harrowing pad abort, there was a black mood. (*Shoot the fucker down* was the general opinion; Charlie Walker remembered some other decidedly 'colourful' language, aimed specifically at the pilot's *parentage*!) "After nearly a seven-minute delay," wrote Mullane, "its pilot pulled his head out of his ass and flew off. We all wished him engine failure."

Five minutes to go. Mike Coats flipped the APU switches and the three Auxiliary Power Units hummed to life.

Two minutes. The astronauts were instructed to close their visors.

One minute. "Eyes on the instruments," Hartsfield told his crew.

Thirty-one seconds: Go for autosequence start. Discovery's on-board General Purpose Computers assumed primary control of the countdown.

Ten seconds: Go for main engine start ...

Six seconds: For the second time in two months, the orbiter's engines roared to life. This time, all three did so crisply and to perfection. The manifold pressures shot up on the data tapes in front of Coats' eyes. They had three good engines, all running at full power.

" ... *three, two, one ... we have SRB ignition ... and we have liftoff! Liftoff of Mission 41D, the first flight of the orbiter Discovery ... and the Shuttle has cleared the tower ...*"

RIDE OF A LIFETIME

Many astronauts have described the immense power of Main Engine Start, but if anyone was uncertain as to precisely what was happening, T-zero changed that uncertainty forever and there was no doubt that Discovery was heading out of town in a hurry. From his perch on the middeck, Charlie Walker glanced to the left and could see the steel structure of the launch pad tower *move* visibly as the engines shook the 41D stack in a phenomenon known as 'the twang', then a cacophony of noise – he estimated 170 decibels in the cabin – from the crackle of the SRBs. Within a fraction of a second, the tower was *gone*, to be replaced by daylight as Discovery climbed away from Earth and began her GPC-controlled 'roll program' to establish herself on the proper flight azimuth for a 28.5-degree-inclination orbit. Walker was instantaneously pushed 'down' into his seat and he had the feeling that he was in some sort of pickup truck, rumbling down a gravel country road at high speed. As Discovery accelerated faster and faster, he could see the VAB, then the marshy KSC landscape, then the countryside of central Florida and finally the whole state and the offshore Keys. After the roll program manoeuvre had been completed, the window seat enjoyed by Walker and Resnik was facing due south, offering them glorious views of the entire south-eastern portion of the United States. Every second, it seemed, the world was falling further and further away from them. Separation of the SRBs, a little more than two minutes into the climb, was accompanied by a loud bang, a bright flash and an acknowledgement from the flight deck that residual 'gunk' from the boosters had deposited itself on the forward windows.

The view was somewhat different for the four men on the flight deck, who had the six wrap-around windows at the front and two overhead windows, just behind Mullane and Hawley, to behold the controlled explosion that was occurring all around them. "After the boosters separated," Mullane told the oral historian, "I craned my neck back, because the rocket's still going into orbit *upside down*." He was rewarded with his first glimpse of Earth from extreme high altitude – not quite the edge of space, yet, but around 50 km – and could only describe it as "breathtaking". With the SRBs gone, the remaining six minutes of the ride to orbit was, in the words of both Walker and Mullane, "glass-smooth", with scarcely any vibration and very little noise, apart from the ventilation fans, the crackling of the intercom and the astronauts' own breathing. Mullane described the transition from the harsh rattling of the boosters to flying solely on the liquid-fuelled main engines as "just dead quiet". Eighty percent of the thrusting was gone and the astronauts suddenly felt lighter in their seats. Outside, the blue sky turned to black and the curvature of the horizon became more obvious. By MECO – Main Engine Cutoff – at eight and a half minutes after launch, Steve Hawley felt like he was sitting through another run in the simulator … with the exception that no anomalies had arisen. In fact, the majority of their training had been devoted to ascent contingencies. Now, after "all the things you trained for in the sim," he told the oral historian, "*ninety percent* of your training is now *irrelevant!*" With the onset of weightlessness, the astronauts saw and felt things never before experienced: a mosquito, which accidentally found its way into the cabin, struggled to acclimatise to its new environment, whilst screws, nuts and washers drifted out from various nooks and crannies.

The *view* from orbit was astounding. "Your eye can pick up a *lot* more than any camera can," said Mullane, "and it was just so glorious to see the horizon of the Earth, the blackness of space, the blue of the oceans, the white of the clouds." For Walker, the euphoria of reaching space was arrested by Judy Resnik, who gave him a high-five and then told him to stay in his seat until the rest of the crew had completed their final checks. It reminded Walker that although Resnik, Coats, Mullane and Hawley were also rookies, *they* were professional astronauts; and although he had important research to do on 41D, he was effectively a passenger. Although he did not remember any outright belligerence from other members of the astronaut office, it *was* made clear to Walker that he was not fully 'one of them'. Years later, he would express astonishment at how a combination of luck and good timing had put him in the right place at the right time. Born Charles David Walker in Bedford, Indiana, on 29 August 1948, he attended high school in his hometown and entered Purdue University to study aeronautical and astronautical engineering. He gained his degree in 1971 and initially worked as a civil engineering technician and land acquisition specialist and firefighter for the US Forest Service, before moving to Bendix Aerospace to work on aerodynamic analysis, including the design and flight testing of weapons subsystems. Walker joined McDonnell Douglas – the aerospace giant which would eventually lead him to space – in December 1977 as a test engineer on the Shuttle's OMS system, part of a subcontract from Rockwell International. Within a year, he became one of the first members of the company's space manufacturing team and eventually rose to the post of chief test engineer for the Electrophoresis Operations in Space (EOS) project. "Electrophoresis … is really applied to a pretty basic process, as it's been used in laboratories around the world for the past hundred years," Walker told the oral historian, "in which a compound, like a gel or a liquid that has an electrical conductive nature to it … is exposed to an electric field. Within an electric field, they will all move as a group toward the attracting electrical pole and they'll move at different rates, so if you expose, *within* a sample, that sample to an electric field for a period of time, when you shut the field off, you'll have groups of compounds all separated from one another." In précis, electrophoresis encompassed the purification of individually obtained groups from an original mixture. Already, several Shuttle flights had carried the Continuous Flow

Electrophoresis System (CFES) apparatus in the middeck and demonstrated that these separation processes could be accomplished with far greater precision in the microgravity environment; moreover, with increased emphasis on the purification of hormones and enzymes, useful for the treatment of various diseases, McDonnell Douglas anticipated a vast new opportunity to open in the pharmaceutical industry. When the electrophoresis project was first proposed to NASA, the intent was to fly on the Spacelab 3 mission, but as that flight was pushed further and further back on the Shuttle manifest, it was decided to fly a middeck version instead. "The investment of private capital," explained Walker, "could not stand that kind of uncertainty and neither could our pharmaceutical investment partners." By 1983, agreement had been reached for six proof-of-concept missions, followed by two flights of a large production plant, to be mounted in the payload bay.

On 29 June 1983, Walker was assigned to serve as the first 'industry' payload specialist, flying on Hartsfield's crew, specifically to operate an upgraded version of the CFES hardware. It was not his first brush with the space programme. He had long harboured an interest in aviation and had worked towards gaining his private pilot's licence in the early 1970s, but admitted to being "a poor student and I couldn't afford to fly". Still, he applied for NASA's 1978 astronaut intake and it was this experience which guided him whilst at McDonnell Douglas. He failed to make the cut as a pilot or a mission specialist, but knew that plans were afoot for payload specialists to be hired from universities, research institutions ... and from within industry. "I obviously don't work for a university," he said. "I'm not a PhD in any one speciality, so maybe the industrial part." Years later, Walker would admit that, when all was said and done, he was basically very lucky to be selected. "I found the company and the project that – at the time – was virtually the *only* thing in this country in the aerospace arena that was being proposed to NASA from the private side that looked like it was really going to produce something of real benefit to the country and to the commercial side of our economy, that could be done in space, and managed to get into an early key position with that project." It was McDonnell Douglas' EOS Program Director, Jim Rose, who had suggested Walker's inclusion on a Shuttle crew in the summer of 1982, and the request was passed up through the chain of command to NASA Headquarters as a 'special case'. Although CFES had flown previously on the Shuttle, the version on 41D had been upgraded to support continuous operations for around 100 hours, adding further weight to the request for a dedicated payload specialist. Interestingly, the first word of Walker's selection came not from George Abbey or Jim Rose or even NASA's head of the Office of Space Flight ... but from an *Aviation Week* journalist, who had talked to a source in Washington. A few days later, it was confirmed. McDonnell Douglas would pay NASA $40,000 for each of Walker's flights; a trivial sum, by any standards. "If you could get *that* today," he told the oral historian, "you'd be *booked up!*"

Training was quite different from the pilots and mission specialists, of course, and it was made quite clear to Walker that he was *not* a NASA employee, *nor* a civil servant, but remained attached to his parent company. He did not undertake any survival training, but was given a few T-38 flights by Hartsfield to prepare him physiologically and psychologically for high-performance flight, and he participated in integrated simulations with the rest of the crew in Houston. "I knew what systems did," he explained, "like the electrical systems [and] environmental systems. I knew the computer interfaces." Walker wanted to integrate himself into the crew more thoroughly, to 'feel' part of the team, but found opposition from NASA leadership; indeed, he was even barred from participating in the more mundane activities, such as executing waste water dumps in orbit. He *did*, however, participate in a number of NASA medical experiments and wore sensors for heart and blood pressure measurements during ascent. In Walker's mind, 41D was probably going to be his only space mission and he had nothing to lose by volunteering to provide a few points of medical data along the way. Little could anyone have foreseen that Walker – an engineer who had not even reached the *interview* stage for the 1978 astronaut selection, whose flying experience was less than a hundred hours and whose academic credentials rose no higher than a bachelor's degree – would actually fly more missions, and *more often*, than any other astronaut during this period of Shuttle operations.

Less than eight hours after launch, at 4:40 pm EST, the first communications satellite, SBS-4, was sent spinning out of Discovery's payload bay; the wrist camera on the RMS captured the successful 87-second firing of its PAM-D booster. The gremlins of Mission 41B had evidently been banished and the PAM-D was back in business. Early on Day Two, however, a quite different satellite – Syncom – was deployed, in a quite different way. Instead of sitting 'vertically' in the payload bay, it lay horizontally, and was spring-ejected in a manner not dissimilar to a frisbee. Since NASA's fee to its customers was proportional to how much length a satellite occupied in the payload bay, the Syncoms were deliberately designed as full-width, stubby drums; this enabled them to precisely fit the *width* of the bay and minimised their *length*. In shape, they took the form of 1,400 kg cylinders and were designed to provide worldwide, high-priority communications between ships, aircraft, submarines and land-based stations for the US military, as well as the Presidential Command Network. This 'synchronous communications satellite' measured 4 m tall and 4.6 m wide when stowed in the Shuttle's payload bay, but after full deployment in geostationary orbit – when its UHF and omni-directional antennas had been swung out – its height would increase to 6 m. It was spin-stabilised, covered with solar cells capable of generating 1,500 watts of electrical power. A pair of helical antennas on top of Syncom provided receive-and-transmit capabilities in the 240-400 MHz UHF band and it also carried two X-band 'horns' and the omni-directional antenna for tracking and control. All of the antennas were separately attached to a collapsible boom to enable them to be folded during ascent and sprung open in space; this helped to keep the satellite's costs to a minimum. When operational, Syncom provided 13 UHF communications channels and was equipped with its own attitude-control subsystems and nickel-cadmium batteries for backup power.

Syncom was also known by another name – 'Leasat' – because its services were 'leased' by the US Navy from prime contractor Hughes. The project had begun in September 1978, when the Navy awarded contracts to Hughes to build five Shuttle-deployable satellites (one of which would serve as a spare) for the Department of Defense. They would represent the fourth generation in the Syncom family, which traced its origins back to the 1960s. It was intended that this new network would effectively augment the military's Fleet Satellite Communications (FLTSATCOM) system and the Hughes contract also provided for the construction of a control centre at El Segundo in California and a series of fixed and movable ground stations. The Navy acted as the project's executive agent, working on behalf of the Department of Defense. The 'lease' stipulated that the US military would pay to use communications channels, at a cost of $84 million per annum, per satellite. Syncom 4-1, the first of the new series, was originally assigned to 41D, whilst Syncom 4-2 was due to launch aboard 41F. By the time of Discovery's pad abort, the 41F payload – which also included SBS-4 and Telstar-3C – was in the final stages of processing for an early August launch. In the wake of the abort, NASA considered it more straightforward to switch the *entire* 41F commercial payload onto 41D, and reschedule Syncom 4-1 for launch later in the year. Mechanical interface with the orbiter was achieved through a reusable 'cradle', which provided five attachment points, and Syncom was ejected from the payload bay by the release of a series of locking pins and an explosive spring. This provided the satellite with both its separation velocity and its gyroscopic stability. During the deployment procedure, a mechanism on the drum's surface was triggered to activate the post-release sequencer.

The Syncom deployment came at 9:16 am EST on 31 August and occurred perfectly; the satellite departed the bay at a velocity of 0.7 m/sec, spinning at a couple of revolutions per minute. Very soon, it was in the process of executing the required manoeuvres to insert itself into geostationary transfer orbit. What did not work quite so well was the IMAX motion-picture camera, which had previously flown on 41C and would fly again on 41G and whose footage subsequently formed the basis of the 1985 movie 'The Dream is Alive'. Operation of this large and unwieldy camera was Mike Coats' responsibility and one of its earliest uses was to film the deployment of Syncom; unfortunately, during this time, in the words of the official NASA mission report, it "jammed". The background to what *caused* this has been explained with some colour by Mullane and Hawley. As Syncom left the bay, Hartsfield was filming the deployment when, all of a sudden, its belt-driven magazine *sucked up* a shank of Resnik's black hair, which weightlessness had liberated into flowing tresses. The IMAX ground to a stop, blowing the circuit breaker in the process. Eventually, with scissors, they freed Resnik's hair and Mike Coats took the camera down to the middeck to repair it. When the time inevitably came for Hartsfield to report the incident to Mission Control, Resnik gave him a threatening stare and he duly respected her embarrassment, telling the ground simply that IMAX had jammed and Coats was working to fix it. Resnik was only the second American woman in space and keen to cement her credentials; she did *not* want to be remembered as a woman whose hair caused a multi-million-dollar camera to fail. IMAX *did* have a belt guard but, as Hawley told the oral historian, "for whatever reason we decided we didn't need to fly it. I don't know if we were trying to save weight, but we decided we didn't need this belt guard". Coats worked for hours, methodically removing shreds of hair from the camera, and for a time it seemed unlikely that IMAX would work. At length, he restored it to operation and it was used for a variety of tasks, including the deployment of OAST-1. It was an amazing machine. "The thing pulled so much film, so fast," recalled Hawley, "and it's [so] *big* that, in zero-G, it will actually *torque* you like a gyroscope! To use it, you really have to be affixed to something, because it will *rotate* you!"

Certainly, Mike Coats was the hero of the hour, not only for helping to save Judy Resnik's tresses, but also for helping to protect her dignity and saving the IMAX camera. Whereas the other astronauts had received their own Zoo Crew nicknames – Zoo Keeper, Tarzan, Cheetah and Jane – there had been only one name which could best sum up Michael Lloyd Coats: *Superman*. Right from the start, he had been noted for his close resemblance to Christopher Reeve's character. "A curl of black hair would periodically fall across his forehead," wrote Mullane, "making the Clark Kent appearance complete." Coats was born in Sacramento, California, on 16 January 1946, and attended high school at Riverside. He entered the Naval Academy, received his degree in 1968 and was designated a naval aviator in September of the following year. Coats initially trained on the A-7E Corsair II attack aircraft and served two years aboard the USS *Kitty Hawk*, flying more than 300 combat missions over Vietnam. Upon his return to the United States, he became an A-7E instructor pilot in California and was selected for test pilot school in 1973. Graduation the following year led to duties as project officer for the A-7 and for the A-4 Skyhawk attack aircraft. Coats' final flying assignment before selection into NASA's astronaut corps was as an instructor at the Naval Test Pilot School in Patuxent River, Maryland, from April 1976 until June 1977. At the same time, he completed a master's degree in the administration of science and technology from George Washington University. He was attending the Naval Postgraduate School in Monterey, California, studying for another master's in aeronautical engineering when he was summoned to Houston in August 1977 as one of the *first* group of candidates. In Mike Mullane's mind, at least, Coats was quite different from the stereotypical military aviator: a quiet family man, devoted husband and father of two. To him, the fondest memory of Discovery's first flight was getting home to hold his wife and kids again. It was Coats' misfortune, though, to be one of the first to suffer space sickness on 41D ...

Like his crewmates, Coats was enthralled by the view of Earth. "The feeling that is still so vivid in my mind," he once told an interviewer, "is looking out the windows when you get into orbit and looking at the Earth for the first time from space and seeing this *living* planet. The *colours* – the blues of the oceans, the greens, the browns – it's just really obvious this is a living planet ... and this living planet is going through this big, black void. *That's* the perfect word, because there is *nothing* out there. The stars don't twinkle out there; they're just harsh points of light. The blackness of space and the bright, living beauty of the Earth just amaze me and I think they have the same effect on anyone who has been up there."

Following the Syncom deployment, the third and final satellite, Telstar-3C, was sent spinning out of the payload bay at 9:25 am EST on 1 September. Like SBS-4 and most previous PAM-D payloads, it was a cylindrical, spin-stabilised satellite, coated with solar cells for electrical power. Based on Hughes' HS-376 bus, it was 2.16 m in diameter and stood 2.7 m high in its stowed configuration in the payload bay, telescoping out to a height of 6.8 m when fully deployed in geostationary orbit. The name 'Telstar' has gained almost universal recognition, having formed the name of the first active-repeater communications satellite, launched in 1962. Almost two decades later, in August 1980, the American Telephone & Telegraph Company (AT&T) signed a $137 million contract with Hughes to build a trio of Telstar-3 satellites. The first of these (Telstar-3A) was launched atop an expendable Delta rocket in July 1983, whilst the second (Telstar-3C) was assigned to 41D and a third (Telstar-3D) was allocated to a Shuttle mission in the summer of 1985. (None of NASA's manifests from November 1982 onwards make any reference to a Telstar-3B.) In fact, the only solid reference to its existence came from *Flight International* on 30 July 1983, which noted that both 3A and 3B would be launched aboard Deltas. Some sources have assumed that confusion still exists over whether the third satellite should be properly designated '3D' or '3B'.) Each Telstar-3 was equipped with 24 active transponders, permitting 21,600 simultaneous telephone calls or 24 simultaneous TV channels – four times higher than was previously available – and half a dozen 'spare' transponders. Most of the transponders were relatively new, solid-state power amplifiers, whilst a few were conventional travelling wave tube amplifiers; the decision to include this mixture was a compromise between solid-state reliability and the better efficiency of the travelling wave tubes. The provision of single-sideband equipment at ground stations also allowed for the increase in the satellite's capacity. Telstar-3's nickel-cadmium batteries provided for an operational life of up to ten years (three years longer than previous satellites) and they supported communications traffic between the continental United States and Alaska, together with Hawaii and Puerto Rico. To Steve Hawley, Telstar-3C was also notable in that its purpose-built ground station lay at a place called Hawley in eastern Pennsylvania. In his oral history, he admitted that he had previously never heard of the place, "but it's spelled the same way. After the flight, I got to go to Hawley, Pennsylvania!"

With three satellites successfully deployed and on their way to geostationary orbit, the final days of 41D were to be spent on their other payloads, most notably OAST-1, which was primarily Judy Resnik's responsibility. Its hardware sat atop a Mission Peculiar Experiment Support Structure (MPESS) and consisted of a number of experiments, sponsored by NASA's Office of Aeronautics and Space Technology. Of these, the most notable was the solar array, which, when unfurled, projected 31.5 m 'above' the payload bay; taller than a ten-story building. Although it was not designed to generate electricity, the intent was to demonstrate the structural dynamics of a large array, placing it under different levels of stress, including RCS thruster firings, to understand its behaviour. The primary structure was Kapton and the array comprised 84 panels, which folded out, accordion-like, on an epoxy-fibreglass mast from a 1.5 m high canister on the MPESS. After full deployment, the array had a triangular cross-section, longitudinally stabilised by short guide wires between interconnecting battens. On Day Three, soon after the Telstar deployment, Resnik began the first part of the experiment, extending the array to 70-percent-open. To do this, a 'nut' with internal threads rotated at the top of the storage canister, causing the rollers to move up through the threads and allowing the longerons to straighten and the guide wires to keep the structure rigid. The initial few centimetres of mast extension unlatched the containment box lid, gradually unfolding the solar array blankets. In the first test, when the array reached the 70-percent-open position, a tension bar was deployed to pull the unfurled section 'flat'. It was at this stage that the first series of structural dynamics tests were conducted, and the following day, 2 September, the array was extended to its full height.

"Surprisingly," recalled Hank Hartsfield, "once the thing is deployed, it's fairly rigid. In fact, one of the big experiments we did was to turn off the control systems on the orbiter and let it get really stable and then fire the thrusters and ... measure how the array oscillated." The array turned out to be much 'stiffer' than engineers had anticipated and by the time Hartsfield and Coats prepared for the second set of thruster firings, which were intended to increase the motions, the array had almost stopped moving completely. On a couple of occasions, Steve Hawley was given the chance to perform the firings. "Some of it required inputting jet firings simultaneously in different directions," he told the oral historian, "so *that* was where I got to play. Mike Coats was primarily responsible for making the input, but some of the times when we had to make a yaw and a pitch or roll simultaneously, it's easier for two guys to do that." The array had only a few transducer cells for test purposes, but, had it been operational, a structure of its size would have been capable of generating up to 12 kilowatts of electrical power. For the 41D crew, the *size* of the array was one of its most impressive features. "At sunrise or sunset," remembered Hawley, "the first thing that either the last bit of sunlight or the first bit of sunlight would hit was the solar array and it would make it almost look like it was lighting up with its own source of internal illumination. Everything else would be *dark* and the solar array would be *glowing gold*."

Also spectacular, although for different reasons, was another 'structure' which extended not from Discovery's payload bay ... but from a nozzle normally use to dump the crew's urine overboard. Early on 1 September, unusual temperature data from one of these dumps suggested that a large icicle had formed around the nozzle on the exterior of the vehicle. "Heaters on the exit nozzle are supposed to ensure the fluid separates cleanly," Mike Mullane wrote, "and does not freeze to it." Mission Control suspected that an icicle might have formed, but since the exit point of the waste nozzle was located on the port side of the middeck, it was impossible to view it with anything other than the camera on the RMS. Hank Hartsfield duly used the arm to take a look and, lo and behold, there was an icicle, measuring three-quarters of a *metre* in length and about 30 cm wide. If left to its own devices, the icicle might break off during re-entry, potentially damaging the critical thermal protection system. Ice from dump nozzles had caused damage to 41B during re-entry. Initial attempts to dislodge it focused on using the RCS thrusters were unsuccessful. On the 2nd, Discovery's cabin pressure was reduced to 70.3 kPa, preparatory to a contingency EVA by Mullane and Hawley the following day. Mullane was thrilled, but Hawley was cautious; the *location* of the icicle – just aft of the middeck entry hatch – was virtually inaccessible to a spacewalker. "There's no translation path down there," Hawley recalled. "I guess they were talking about taking the CFES unit apart, using some of the poles ... to maybe grab one of us by the boots and hang him over the side [of the payload bay] and knock it off. *That* all sounded like a *bad* plan to me!" Still, the EVA remained an option and as the crew prepared for bed

"Glowing gold," as Steve Hawley described it, the OAST-1 experimental array is extended to its full height.

that evening, they reckoned that it was the only solution. Just before bedding down, Jerry Ross, the capcom in Mission Control, told them of a sudden change. There would be *no* EVA, but they would be asked to dislodge the icicle with the tip of the RMS. Although Hartsfield was not the primary operator of the mechanical arm, he insisted on doing the job. If anyone dinged a brand-new orbiter on its maiden voyage, it had to be the commander's responsibility. Fellow astronaut Sally Ride had worked a procedure in the simulator and she read the instructions up to them. Hawley felt that it was a good plan. Mullane, on the other hand, was upset to have lost his chance to do a spacewalk.

Richard Michael Mullane was born in Wichita Falls, Texas, on 10 September 1945 and attended Catholic high school in Albuquerque, New Mexico, before entering the US Military Academy at West Point. Aviation and space exploration had been his calling from a young age – drawing aircraft, building models, fashioning gliders out of balsa wood – and his dream was to become a pilot. "That never happened," he told the NASA oral historian, "at least not as a military pilot. My eyes were too bad." He earned his degree in military engineering in 1967 and entered the Air Force, training as a backseat weapons system operator in the RF-4C reconnaissance version of the Phantom II. Mullane undertook 134 combat missions over Vietnam between January and November 1969 and spent four years stationed with NATO forces in England. He received a master's degree in aeronautical engineering from the Air Force Institute of Technology in 1975. Mullane subsequently completed the Flight Test Engineer course at Edwards Air Force Base in California and served as a flight test weapons systems operator at Eglin Air Force Base in Florida. "It was while I was nearing graduation from ... the flight test engineer course," he explained, "that NASA announced they were selecting mission specialist astronauts." Previously, Mullane had convinced himself that, based on historical precedent, space was the preserve only of the fighter pilot or the test pilot. *Now*, the space agency was casting a far wider net, to include scientists, engineers, physicians ... and backseaters like himself. He was based on temporary duty at Mountain Home Air Force Base in Idaho when he received the call from George Abbey to inform him of his selection into the world's most elite flying fraternity. "I just went out," he told the oral historian, "and *screamed* with joy. I remember that night I bought some beer for the rest of the people that I was working with there at Mountain Home in the hangar and we had a little party."

Six years later, Mullane was in space, but had just lost the chance to perform an EVA. Yet the option of knocking the icicle free with the RMS was itself fraught with difficulty ... and danger. "One of the rules in those days," said Hawley, "was you're not allowed to operate the arm in a place where you can't see what it's doing ... I was supposed to go down to the side hatch window and watch as best I could, while Henry drove the arm to knock the ice off. He would move it through a pre-determined trajectory and, if he did it properly, they knew from the ground simulations that it would hit the ice." Hartsfield's jab at the ice was successful and, within minutes, Mike Mullane, on the flight deck, was able to photograph the spear-like icicle drifting away. With 'Ghostbusters' having arrived in cinemas in June, the 41D crew earned themselves a new moniker: *Icebusters*. Although the icicle was gone, its presence had caused another problem. "We were told we could not use the urinal for the rest of the mission," Mullane wrote, "for fear *another* ice ball could jeopardise us." They *would* still be able to use the Shuttle's toilet for 'solid wastes', but were forced to use 'Apollo-style' plastic bags for liquids; Mission Control told them that there was sufficient remaining volume in the waste water tank for about three man days' worth of urine, meaning that Resnik could use it, if needed. However, she was keen to hold onto her feminist sensitivities, it seems, and elected to use the plastic bags. At length, the pragmatic gentleman in Hank Hartsfield came to the fore: "Judy ... had a hard time with the bag," he told the oral historian. "I said, 'I don't care *what* the ground says. *You* use the bathroom. The *rest* of us will do the bag trick.'"

Another problem quickly arose. In microgravity, fluids do not readily remain at the bottom of bags, but begin floating around, and it became clear that the astronauts would need to put something in place to soak up the urine. They stuffed dirty towels, underwear and socks into the urine bags to act as an impromptu absorbent. As the bags gradually filled up the waste tanks under the middeck floor, the unsanitary environment was distinctly unpleasant. Randy Stone, the flight director in Houston, asked his superiors for permission to tell the crew to reconfigure one of their water tanks into a waste tank. "It was easy to do," Hartsfield recalled, "just a quick plumbing thing ... but there was a big concern about turning the orbiter around. They had the idea that if they converted a water tank to a waste tank, it would add another week to the [processing] flow at the Cape to get [Discovery] ready to go again." Stone's request to senior management was rejected, *twice*. Years later, Hartsfield would wonder whether he ought to have requested a private medical conference and enforced a decision. After the flight, he spoke to several engineers and managers, who offered their apologies; ironically, replumbing a water tank into a waste tank *would not* have affected Discovery's processing schedule at all.

Whilst all this was going on, Charlie Walker had focused his attention intently on the CFES unit, which operated for more than a hundred hours in total, yet processed only 85 percent of its samples. It stopped working during Discovery's first night in orbit and, although successfully restarted the following morning, experienced a number of other niggling glitches: instrumentation failures, two faulty carrier degassers and also shut itself down when the cabin pressure was lowered for Mullane and Hawley's EVA preparation. Like Mike Coats, Walker had not felt well for most of his first day in space and, thankfully, his flight plan was not too full at the start of the mission. At times, he would venture upstairs to the flight deck for a glance at the Earth through the overhead windows – "just the most marvellous geography lesson one could *ever* have" – and by Day Two he had begun CFES operations in earnest. Since there was not continuous TDRS communications coverage, he found that he had to resolve problems with the hardware, alone, without much ground interaction. Overall, Walker was satisfied with the performance of the CFES, although subsequent investigation would reveal traces of biological contamination in the processed samples.

Offering two quite different perspectives of Discovery's return to Earth on 5 September were Walker, seated in the darkened middeck, and Hawley, riding in the flight engineer's seat and facing into the forward flight deck windows. "There's a bright kind of plasma plume that seems to form above the orbiter during re-entry," Hawley recalled, "and it comes in through the overhead windows. You get this soft kind of pink, whitish, orangish glow that develops and encircles the orbiter." Mike Mullane remembered seeing ribbon-like vortices of white-hot plasma streaming past the overhead windows. Downstairs, Walker used a small water container to convince his eyes that gravity was steadily establishing its grasp on the descending orbiter. He found that, at first, he could let go of the container and it would remain weightless. After a while, when he began to sense the onset of gravity, he tried again and saw it *drift* down to the floor. A third attempt later in the descent would see it fall faster. "I played this little game over the next five minutes or so," he recalled. "I'd catch it before it hit the ground and put it back up in the air and it seems like after about three or four minutes of that, it was dropping so fast, I *couldn't* catch it before it hit the floor and went skittering away somewhere!" The noise outside intensified as they slowed below the speed of sound, although the touchdown at Edwards Air Force Base at 6:38:54 am PST (9:38:54 am in Florida) was perfect and smooth. After two months of delays and a hair-raising pad abort, the Shuttle was back in business ... and for Discovery, whose career would eventually turn her into the fleet leader, her *next* mission in November 1984 would be one of the most ambitious ever undertaken.

EXPERIMENTAL CREW

Who says 'no' when offered the chance of a ride into space?

Bob Crippen certainly didn't, when George Abbey approached him, late in 1983, with an offer to command a dedicated science mission in August of the following year. Mission 41G would involve the deployment of the Earth Radiation Budget Satellite (ERBS) and operations with an expansive payload, sponsored by NASA's Office of Space and Terrestrial Applications and called 'OSTA-3', which included a large synthetic aperture radar for imaging of terrestrial features. It was an exciting offer, but caused a problem, because Crippen was midway through training to lead Mission 41C, the ambitious Solar Max repair effort, which was scheduled for launch in April 1984. Abbey rationalised the decision: with an anticipated increase in the Shuttle flight rate to perhaps two dozen missions per annum by 1987, NASA was keen to determine how quickly it could turn astronauts around between flights. Crippen agreed to Abbey's request, but cautioned that he would be unable to spend much time with the 41G crew until his 41C duties were done. As a result, Sally Ride – who had flown with Crippen on STS-7 and knew how he ran a mission – was assigned as 41G's flight engineer and she would oversee the training of the rest of the crew until the commander became available. Crippen and Ride would be joined by pilot Jon McBride – a burly, sandy-haired naval aviator, nicknamed 'Big Jon' – and mission specialists Kathy Sullivan and Dave Leestma. The latter pair would perform an EVA, which would itself make history, since Sullivan would become the world's first female spacewalker. Or so it seemed. As noted in Chapter 2, it was a Soviet female cosmonaut, Svetlana Savitskaya, who seized this distinction, just three months ahead of Sullivan. Crippen is not alone in his conviction that it was another attempt by the Soviets to score political points. "As *soon* as we named her to do it," he said, "the Russians put up a woman and had her do a spacewalk, just so she could beat Kathy."

By the time 41G took place, in October 1984, the crew had also picked up a pair of payload specialists – Paul Scully-Power, an Australian-born oceanographer, employed by the US Navy, and Marc Garneau, who became Canada's first astronaut – to participate in specific experiments. The 41G crew was an unusual one. Although Crippen was responsible for the success of the mission, his Solar Max commitments during the first half of 1984 meant the crew completed a sizeable portion of their training without him. Sally Ride took the mantle of 'surrogate' commander to nurse three rookies and two payload specialists through their preparations. Only at the end of April did Crippen join them on a permanent basis. "I was the only one on the crew who had flown before," Ride remembered years later of what, unofficially at least, and only on the ground, made her the 'first' woman to lead a mission. "I tried to give the rest of the crew some indication of the way that Crip liked to run a flight and run a crew. Then, thankfully, he landed from 41C and joined us. He's very easygoing, so I don't think our group dynamics changed much when he joined the crew. Everyone knew him really well and had worked with him in the astronaut office for years, so he wasn't an unknown presence joining us." Ride recalled that, since the 41G crew had a lengthy training regime ahead of them, her personal burden was not excessive. "The role I was playing," she said, "was really to talk about the basics of being in space and giving them some familiarity with the space environment."

Crippen's assignment to command yet another (almost)-all-rookie team of astronauts was, said oceanographer Bob Stevenson, who worked with him during the months prior to 41G, possibly the deliberate intention of George Abbey. "Bob is an exceptionally able astronaut," Stevenson explained in a January 2000 interview, "but also an exceptionally able leader of people and I think George, at the time, wanted to use Crip as much as he could to help train and pass on what he could to the new crews. Crippen had a lot of duty in terms of flights!" Among his multi-faceted 41G crew was the first American woman to make an EVA (Sullivan) and the first member of NASA's 1980 astronaut intake to fly into orbit (Leestma). Years later, Leestma remembered fondly the arrival of his astronaut group, who wryly dubbed themselves the 'Needless Nineteen'. "We earned that nickname because there were 35 ahead of us, plus all the other astronauts that were there from the Apollo days that either hadn't flown or were still there, and they were all waiting to fly the Shuttle, so we were way down the line!"

David Cornell Leestma came from Muskegon, Michigan, where he was born on 6 May 1949, although he attended high school in California and entered the Naval Academy to study aeronautical engineering. He graduated first in his class in 1971 and was assigned to the frigate USS *Hepburn*, before receiving orders to report to the Naval Postgraduate School for a master's degree, also in aeronautical engineering. Leestma subsequently moved into flight instruction, received his aviator's wings in October 1973 and underwent training on the F-14A Tomcat fighter. During the next few years of his naval career, he participated in three overseas deployments to the Mediterranean and North Atlantic, flying off the USS *John F. Kennedy*, and in 1977 was sent to Naval Air Station Point Mugu in California as part of the air testing and evaluation squadron. Leestma served as an operational test director on the Tomcat, with his particular focus on the development of new tactical software and programmable signal processors. Whilst aboard the *Kennedy*, his operations officer advised him to apply for NASA's 1978 astronaut intake – and even sent off for the application form on his behalf – but Leestma was too busy with his other duties and put such thoughts to one side. By the end of 1978, he was back on shore duty and when the space agency announced its intent to select more astronauts, he dusted off the application form and submitted it. He was summoned to Houston, part of the very first group to be interviewed, in February 1980, and heard no more for several months. "And then, in late May of 1980, I got this call, early in the morning," he told the NASA oral historian. "I was in California, so it was two hours earlier than here [in Houston]. As I found out later, they start calling around eight o'clock in the morning and it was around six-thirty or something ... and it was George Abbey on the phone. Now, I didn't know that if George calls you, then *you're in*, and if somebody else calls you on the board, then you're *not* in!" Leestma was told to report to JSC in early July, but remembered that selling his house in California and getting his military orders from the Navy took longer than expected. He ended up reporting a week late, as did several other members of his class, including pilots Guy Gardner and Ron Grabe.

It was not just Leestma who remembered the 'Needless Nineteen' handle with which his class was bestowed. In *Riding Rockets*, Mike Mullane recounted a conversation with Jerry Ross, also selected in Leestma's class, who told him that JSC Director Chris Kraft had implied that the new group was not needed at all; there were *enough* astronauts already. Still, Leestma was excited by the enormity of it all: on one training trip, he found himself on a commercial airliner, seated next to Al Bean, who regaled him with tales of his experiences on the Moon during Apollo 12. "The stewardess came by and served us our Cokes," Leestma recalled, "and she doesn't have a *clue* who she just talked to! This man next to me walked on the *Moon!*" However, his *own* experience assured him that a flight assignment would not come along for some considerable time; by the summer of 1983, barely half of the Thirty-Five New Guys had been given missions. The Needless Nineteen would wait even longer. Until ... a chance phone call from George Abbey, one Saturday morning, telling Leestma to come into work. When he arrived, he found Crippen, McBride, Ride and Sullivan and was told of his assignment 'STS-17'. Several flights were pushed back on the manifest due to IUS and PAM-D woes over the next few months, but STS-17 – later renamed 41G – kept its cargo *and* kept its place on the launch schedule, because ERBS and OSTA-3 were unique payloads and required neither of the problem-plagued boosters. "So I ended up flying quite a bit earlier than the other people in my class," Leestma said. "I don't know if that's good, bad or indifferent, but it was sure exciting to see our flight kind of hold its position while others slipped, because we figured, as *they* slipped, then *we'd* slip. But because of the requirements of our payload, we *had* to fly at a certain time of the year and a certain inclination." To this day, he has no idea why he was selected as the first of his group to fly, but Henry Cooper, in his book *Before Liftoff*, suggested a possibility. One of Leestma's earliest assignments in the astronaut office was to devise a checklist for operating the orbiter's Auxiliary Power Units and Cooper believed it was this outstanding work which assured Leestma of a seat on Crippen's mission.

Leestma recalled the training well. "We asked Sally to be our training co-ordinator," he said, "and she became the *de facto* commander, at least for organising our training and assignments and making sure that we were progressing. We trained as a crew of four for a long time. Under those circumstances, there was a lot of pressure on us to know what we were doing and not screw up, because Mission Operations were looking at us carefully to see if this was something that could be done or not. Can you train without one of the crew members, who is doing another flight?" Even at this early stage, NASA was looking at flying astronauts more rapidly than ever before – two or three times per year – and 41G's training cycle gathered valuable, 'real world' data to support this. In his 1987 book, Cooper pointed out that using Crippen so often "went against NASA's policy of building up a pool of experienced astronauts – essential if [the agency] is ever to achieve a rate of one flight a month – and using him again seemed even more unusual in light of his late arrival". However, with a hundred astronauts and between five and seven seats available on each mission, people would indeed be flying relatively frequently.

Since the crew included Sally Ride, who had made history the previous year as the first American woman to reach orbit, and Sullivan, who would make an EVA, the two female astronauts became unofficial 'spokespeople' for the mission. "Jon and I could easily just stand in the background," chuckled Leestma. "It took a lot of the spotlight off us, which was fine." It was also something of a miracle for thick-set Jon Andrew McBride to fade into the background. Late in 1978, the Thirty-Five New Guys had been invited by the Houston Hurricanes to be introduced to the crowd at the Astrodome. Each of the new astronauts felt decidedly ill-at-ease ... except, that is, for 'Big Jon'. "Instead of a nervous wave and a quick step backward," Mike Mullane wrote, "Jon seized the startled cheerleader, swept her backward off her feet and planted a kiss on her. Then he pulled on the Hurricanes T-shirt and waved a greeting to the crowd. Jon was a man for the masses." Born in Charleston, West Virginia, on 14 August 1943, McBride attended high school in his home state and initially entered West Virginia University, then joined the Navy and underwent flight instruction in Pensacola, Florida. After winning his wings, he trained on the F-4 Phantom II and spent several years as a fighter pilot and division officer, flying 64 combat missions in Vietnam. Upon his return to the United States, McBride resumed his education, gaining a bachelor's degree in aeronautical engineering from the Naval Postgraduate School in 1971, with additional work in human resource management. Graduation from test pilot school at Edwards Air Force Base came next and he worked as a maintenance officer and project officer for the AIM-9 Sidewinder air-to-air missile. He was one of the second group of applicants invited for interviews in Houston, arriving in August 1977, and during this time McBride also piloted the Navy's bicentennial-painted F-4J Phantom in various air shows.

As training reached its peak, in June 1984, with Crippen finally dedicated to 41G full-time, the astronauts averaged 80-90 hour working weeks in the JSC simulators. During the course of their training, the simulators were of pivotal importance. Not only did they enable the astronauts to hone their skills by responding to literally hundreds of abort scenarios, but they also allowed them to begin working together as a cohesive unit. Crippen would comment later that it was vital he understood exactly how each of his crewmates would respond to certain malfunctions and under specific conditions; that a bond of mutual trust should develop between them. This was made more difficult by his absence until the early summer. In the wake of the 41D engine abort on 26 June and the cancellation of 41F, Crippen's mission was now at the head of the queue, with a launch date rescheduled for early October. "We knew that our training was going to be hot and heavy that summer," grinned Leestma. It was also an *Olympic* summer, with the Games being held in Los Angeles. When Joan Benoit Samuelson won gold in the first-ever women's marathon, it caused great excitement, particularly for Sullivan, Ride and Leestma. "We really admired that kind of accomplishment," Sullivan told the NASA oral historian, "so we'd come in almost every day, just amazed at what someone had done or pleased with the US result or whatever the particulars might have been."

When the five astronauts were named to 41G in November 1983, their mission was scheduled to begin on the penultimate day of the following August. However, for a time, the identity of their orbiter was still in question. "When we first got assigned, it was on Columbia," Leestma recalled. "We ended up flying on Challenger. NASA always told us when we got selected for a space flight, not to

fall in love with our orbiter or our payload, because they were liable to change!''
Columbia, the 'queen' of the fleet, had already flown in space six times, including the
first Spacelab mission at the end of 1983; in fact, the inclusion of additional
cryogenic oxygen and hydrogen tanks under her payload bay floor in support of that
flight enabled her to remain comfortably aloft for ten days, far longer than her
sisters. Indeed, NASA's manifests from November 1983 and January 1984 list her as
flying 41G for a ten-day mission. Unfortunately, problems arose. Following her
Spacelab flight, Columbia was transported to Rockwell's Palmdale plant in
California in January 1984 for modifications. These included the removal of her
pilots' ejection seats and the installation of new equipment to gather aerodynamic
data during future hypersonic re-entries, together with a Heads-Up Display (HUD)
to match those aboard Challenger and Discovery. These upgrades, coupled with the
need to attend to wear and tear from her previous missions, meant she spent longer
in California than timetabled. Dave Leestma had been detailed by Crippen to follow
Columbia's progress, but he quickly became aware that the flagship would not be
ready in time for 41G. NASA managers were already considering the roomier
Challenger as a more attractive alternative to Columbia, particularly in view of
Crippen's large crew. For the astronauts, however, the 'old girl' was preferable in
terms of her ability to stay longer in space; but without a HUD, she could not yet
confidently support a precision landing on the swamp-fringed, relatively narrow
KSC runway. (In fact, *Flight International* noted in May that 41G would *have* to
land at Edwards Air Force Base, due to Columbia's lack of a HUD.) Not only were
touchdowns at the Florida spaceport highly desirable as NASA sought to make
Shuttle missions 'routine', but Crippen himself wanted a shot at landing there,
having been thwarted twice by bad weather on STS-7 and 41C. By the time NASA
published a revised Shuttle manifest in May, 41G was officially reassigned to
Challenger and, due to her lack of sufficient cryogenic tanks, the mission duration
was accordingly reduced from ten to eight days.

Still, even the HUD-equipped Challenger needed lengthy refurbishment before
she could fly again and it soon became clear that she would not be in a position to
conduct 41G until the beginning of October 1984. During her touchdown at
Edwards, following the Solar Max repair, all of her brakes had suffered varying
degrees of damage: cracked rotors, chipped carbon edges, missing washers and
contamination by surface debris. Tile and thermal blanketing discolouration also
demanded attention. The May 1984 manifest made it official: Crippen's crew would
use Challenger, thereby allowing Columbia to spend more time in California. In
addition to the removal of the ejection seats and installation of a HUD, the veteran
orbiter would also be outfitted with a new nose cone, fitted with air data
instrumentation, and a tail-mounted infrared sensor. "There were a *lot* of upgrades,"
Leestma said. "They weren't going as fast. There's *always* money problems and so
my reports coming back were probably a little bit more negative." Liftoff of 41G was
pushed back slightly to 5 October. To demonstrate the readiness of himself and his
crew, one of Crippen's first actions had been to ask the instructors for a fully
integrated simulation on 8 May. Normally, such 'sims', which involve not only the
astronauts, but also the entire flight control team for the 'real' mission, were

undertaken only in the last eight weeks or so before launch. Despite reservations on the part of some 41G trainers, the fully integrated sim went ahead and proceeded perfectly. By this point, Crippen had three Shuttle missions under his belt and was by far the most experienced, 'in-training' astronaut at the time. "With Crippen there, we had a harder time fooling the crew," lead instructor Ted Browder told Henry Cooper. "Crippen has seen about every training scenario there is!"

Training was complicated yet further by the inclusion of Garneau and Scully-Power as payload specialists in May 1984. Although the astronauts welcomed the newcomers, with Crippen proudly telling a press conference that (with the exception of Sally Ride) his entire crew had a naval, or at least 'nautical', background, there was some concern at the sheer number of people and available room aboard Challenger. Window space, explained Dave Leestma, was a precious resource, together with very real concerns about whether the trouble-prone multi-million-dollar toilet could handle the additional stress. Like Leestma, Paul Desmond Scully-Power – nicknamed 'PSP' by the crew and labelled "a little bit of a loose cannon" by Crippen – had also been bitten by the bug of exceptional good fortune ... but in this case it was a good fortune tinged with a tragedy which afflicted the man who should have flown 41G in his place. Born in Sydney, Australia, on 28 May 1944, Scully-Power received a bachelor's degree and postgraduate diploma in education in applied mathematics from St John's College at the University of Sydney. After graduation, in January 1967 he was approached by the Royal Australian Navy to establish an oceanographic group, which he headed for five and a half years. From July 1972 until March 1974, he then served as a naval exchange scientist with the US Navy, working at the Naval Underwater Systems Center in Connecticut and at the Office of Naval Research in Washington, DC. Scully-Power assisted the Earth observation team on the Skylab space station and, upon his return to Australia, he planned and orchestrated a joint effort with the United States and New Zealand, called 'ANZUS EDDY', a combined measurement of the morphology and acoustics of oceanic eddies. He later served as a foreign principal investigator for NASA's Heat Capacity Mapping Mission, launched in April 1978, and by this time had emigrated to the United States to accept a position as a senior scientist at the Naval Underwater Systems Center, going on to gain US citizenship in 1982.

Originally, Scully-Power's seat on 41G was assigned to another scientist, Bob Stevenson, of the Institute of Oceanography at the University of California in San Diego. Stevenson's involvement with the Shuttle had begun several years earlier, when Kathy Sullivan asked him to give oceanography classes to her astronaut group. Later, during the early Shuttle missions, Stevenson and Scully-Power supported crews with their observations of Sun 'glint' on the oceans, sea water temperatures, photography, analysis of ship wakes and elaborate spiral eddies. During STS-8, Dick Truly had expressed astonishment at spotting spiral eddies "as far as he could see," recalled Stevenson, "either side of the flight path, from the western-southern part of the Indian Ocean, all the way past New Zealand", for five continuous days. Truly's interest provided the oceanographic community with an ally and he pushed for the inclusion of either Stevenson or Scully-Power on a Shuttle flight. In fact, George Abbey had considered such a move in 1982, with

A stunning aerial view of Challenger's ascent.

seats available for a payload specialist on STS-7 and STS-8, but the space sickness episode suffered by Bill Lenoir and Bob Overmyer obliged NASA to add physicians Norm Thagard and Bill Thornton to those missions. Then, in March 1984, Stevenson received a call from Scully-Power, informing him that Abbey had proposed him as a payload specialist for 41G. Stevenson's wife was battling breast cancer at the time and he declined the offer, suggesting to Abbey that Scully-Power fly in his stead. Jovially, he offered to take over if his wife recovered in time or if Scully-Power happened to fall down a hole in the final days before launch. As circumstances transpired, Scully-Power did not break his leg but, tragically, Stevenson's wife passed away a few days before 41G lifted off. "The funeral was the day of the launch," he remembered. "The whole crew called me from the crew quarters that morning. They sent a beautiful arrangement; all orchids. Kathy chose those." For her part, Sullivan felt uncomfortable about the late arrival of the payload specialists – Scully-Power's name was announced by NASA on 29 June 1984 – and the need to assimilate two more members into a crew which had by now knitted itself into a team. "I felt like the standing that I was building with ... the Earth sciences community was pretty stable and gelling," she told the oral historian, "and then here comes this other guy, who's being advertised as *the oceanographer*. Excuse me, I have a PhD in this subject. *This* guy has a *bachelor's degree!*" Bob Stevenson also noted that Scully-Power was the first American astronaut to venture into orbit with a beard.

It was the nature of the on-board Earth resources instrumentation, rather than a question of good or cruel luck, that kept Crippen's crew close to their original schedule. Henry Cooper has also hinted that 41G's testing of an orbital refuelling system, which NASA planned to use on several lucrative satellite servicing missions for the Department of Defense – Landsat-4 being the first – also played a part in averting its cancellation or postponement. Not only the high inclination (57 degrees to the equator, enabling the crew to 'see' much of Earth's surface) but also the altitude (350 km) would be unusual, for the astronauts would adjust Challenger's orbit in order to support two very different payloads. The first, the Earth Radiation Budget Satellite, was part of a series of three platforms to explore the impact of solar radiation on our planet, including its absorption and re-emission by the atmosphere. When NASA began developing ERBS in 1978, the agency hoped it would provide a clearer understanding of the radiation 'balance' between the Sun, Earth, atmosphere and space. This, in turn, would expand knowledge of the mechanisms responsible for terrestrial weather and climatic change. Following the announcement, two remote sensing instruments were identified: an active 'scanner' and a passive 'non-scanner', which used a total of seven radiometers to measure energy intensities in the atmosphere and one sensor to determine solar intensity. Both were fabricated and calibrated by TRW at its Redondo Beach facility in California and flew not only aboard ERBS, but also two other Earth resources missions – the National Oceanic and Atmospheric Administration's NOAA-9 and NOAA-10 satellites, launched in December 1984 and September 1986 – for a comprehensive study. Collectively, this was dubbed the 'Earth Radiation Budget Experiment' and its data has improved scientists' comprehension of how clouds, aerosols and 'greenhouse gases' contribute to the planet's daily and long-term weather. In particular, their results have highlighted how clouds that form over water differ from those that form above land, which, in turn, affects their ability to reflect sunlight back into space. New insights have been provided into Earth's upper atmospheric radiation levels and, in fact, have prompted further satellite projects, including the Tropical Rainfall Measuring Mission and subsequent Earth Observation System, launched in 1997. In addition, data from ERBS and its two NOAA partners has provided a better awareness of how the amount of energy emitted by our planet varies between 'daytime' and 'nighttime'. Furthermore, the amount of radiation emitted by the Sun has increased by nearly 0.05 percent per decade since the late 1970s, which, "if sustained over many decades, could cause significant climate change", said Richard Willson, a researcher affiliated with NASA's Goddard Space Flight Center and Columbia University's Earth Institute in New York.

In addition to its scanner and non-scanner, ERBS carried the 29 kg Stratospheric Aerosol Gas Experiment (SAGE)-2, designed to assess the effects of human and 'natural' activities on Earth's radiation balance – ranging from the burning of 'fossil fuels' and use of chloroflurocarbons (CFCs) to volcanic eruptions. During every orbital sunrise and sunset, SAGE-2 measured the solar radiation passing through the atmosphere at our planet's limb to produce a spectrum that would identify the chemical species along the line of sight. These measurements focused specifically on the lower and middle stratosphere, some 15-25 km above the surface, although the aerosol, water vapour and ozone profiles often extended down into the troposphere. Significantly, the role of nitrogen dioxide, whose concentration in the atmosphere has grown substantially due to increased industrialisation in the 1980s, as a major player in the destruction of stratospheric ozone was first traced by SAGE-2. It also measured the decline in stratospheric ozone over the Antarctic since the much-publicised 'hole' was first identified in 1985. Additionally, the instrument highlighted natural contributors, included the Mount Pinatubo eruption in the Philippines in June 1991, where the airborne dust warmed the 'local' stratosphere by 3 degrees Celsius, but the resulting aerosols spread to middle and high latitudes within a matter of months and served to cool the planet overall by reflecting sunlight back to space. As its name implies, SAGE-2 was the second such device to reach orbit; its predecessor had provided near-global observations of aerosol extinction, together with ozone and nitrogen dioxide concentrations, from 1979 until 1981. This long-term, stable, data-gathering capability spanning more than two full decades was continued by SAGE-3, launched aboard the Russian Meteor-3M satellite in December 2001 and established trends in global ozone levels. In fact, today, SAGE provides key evidence for the United Nations' ongoing assessment of environmental change. Its impact has been profound: the instrument's ozone studies ultimately spurred the international community into action with the 1987-signed Montreal Protocol agreement that led to the virtual elimination of CFCs and the development of new, low-emission technologies in air-conditioning, refrigeration and industrial systems. Added the instrument's principal investigator, Patrick McCormick, of the Center for Atmospheric Sciences at Hampton University in Virginia: "The public should appreciate the investment they made in a satellite mission that has exceeded all predictions and hopes of a long life and for its contribution to making Earth a better place now and for subsequent generations." This enormous success was demonstrated by ERBS' other instruments, too. Originally designed to operate for just a couple of years, the scanner did not expire until February 1990 and both the non-scanner and SAGE-2 continue to return valuable data. Today, the satellite continues to monitor total solar irradiance using its surviving instruments, although budgetary constraints have limited the extent of its observations. At least once every fortnight, the Sun is examined for several, 64-second-long measurement intervals. This accumulation of data has proceeded, virtually unbroken, with the exception of brief attitude control, telemetry or battery cell failures over the years. By July 2002, efforts were afoot to lower ERBS' orbit to permit a controlled, destructive re-entry, although the satellite remains operational to this day.

It was a strange-looking machine: built by Ball Aerospace, it measured 4.6 m wide, 3.8 m high and 1.6 m long. Yet it was, said Dave Leestma, "a beautiful satellite, coated with gold foil insulation and dark, purplish-blue solar arrays". It comprised a keel, base and instrument modules, which provided, respectively, structural support, an interface with the Shuttle's payload bay and mounting points for the scanner, non-scanner and SAGE-2. Two large solar arrays generated up to 2,164 watts of electrical power, supplemented by a pair of nickel cadmium batteries and a hydrazine propulsion system to perform necessary station-keeping man-oeuvres. The deployment – intended to occur eight and a half hours after Challenger's liftoff – did not go entirely to plan.

After a perfect, on-time launch at 7:03 am EST on 5 October 1984, which, according to Kathy Sullivan, was a "huge, *huge* vibration spectrum" with "*tons* of noise", the deployment effort was under Sally Ride's supervision, with Dave Leestma backing her up. Like LDEF before it, ERBS would rely on Challenger's Canadian-built RMS arm for removal from the payload bay. First, though, was the experience, not only of weightlessness, but of the view of Earth from more than 350 km. Sullivan was seated on the flight deck, shoulder-to-shoulder with Ride, for ascent, and her eyes were drawn like a magnet. Momentarily, there was confusion as she *felt* her abdominal diaphragm floating upwards, but the notion of space sickness did not manifest itself. "*This*," she breathed, "is *way* too much fun to be sick over!" Preparing for orbital operations on her first flight was far more straightforward than her two missions after Challenger, when she wore a bulky partial-pressure suit for ascent and re-entry. Consequently, after unstrapping, she, Leestma and Ride were able to break away, floating off in their own directions, to begin activating workstations and readying for the opening of the payload bay doors and deployment of the RMS. "We trained a lot together and spent a lot of time in the simulators and going to Canada," Leestma recalled years later. "It became a little bit of a contest of who could do it quicker or better. All those competitive games were played in everything we did. Sally was very good at the arm, so I learned an awful lot by just watching how she went through the training. When it came time to deploy the satellite, she had let me actually pull the arm out, do the checkout and then grapple it." During this time, ERBS' systems were activated, pre-deployment checks executed and, with the satellite held high above the payload bay, the procedure to extend the solar arrays and other appendages – including communications hardware – duly got underway. "The solar arrays," said Leestma, "were folded up to the sides of the satellite, so we were getting ready to put them out and the ground checked to make sure they were getting current and everything was powered up and looking good." All five NASA crew members, by this time, were crowded into Challenger's tiny flight deck: Ride and Leestma at the RMS controls, Sullivan handling the cameras, McBride flying the spacecraft and Crippen, in his own, rather understated, words, "sitting back and managing". The two 'non-career' payload specialists were confined to the middeck during the deployment effort. Through their headsets, the crew heard the voice of fellow astronaut Dave Hilmers, the capcom in Houston, telling them to release the arrays. "We sent the command for the first solar array to deploy and it went up," said Leestma, "but when we hit the command for the second one, nothing happened!" Several more tries, including one initiated from the ground, to unfurl the stubborn array were also fruitless. Next, they attempted to 'jostle' ERBS by rolling the mechanical arm's end effector, without success, and finally McBride oriented Challenger's payload bay towards the Sun to thaw out possibly frozen hinges. We were talking inside the cabin, of course, about what we could do to free this solar wing. This was back before we had all the TDRS coverage, so we went through long periods of time where we didn't have to talk to the ground or they couldn't see data. We were getting ready to come up over Australia and through the Canberra station and talk to the ground and then we would have a 15-20 minute period before we'd come up over the States; a big loss of signal time."

As the crew awaited the acquisition of signal, Ride and Leestma considered trying again to shake the array open with the RMS. "We changed the payload identification, which tells the arm what's on the end of it," said Leestma, "and changed the payload in the software to 'zero', which meant there was nothing on the end of the arm. Now we could go to the max rates on the arm and play with it." After receiving authorisation from Crippen – on condition that they did not *break* ERBS – Ride jerked the arm as sharply as possible from left to right and back again. During her second attempt, Leestma suddenly noticed something move. Ride put the satellite back into its deployment position and the balky array slowly juddered, then stopped, juddered again and finally sprung open. "We came up over the States," Leestma exulted, "and the ground said 'Okay, we're with you'. I don't remember the exact quote, but they asked 'What did you guys do?' We said 'We aren't going to tell you, but just check it out and make sure that it's ready to deploy'." The glitch, which was later attributed to 'thermally induced problems', delayed ERBS' deployment from Challenger's sixth orbit to her ninth circuit of the globe. During this time, Goddard controllers uplinked new telemetry data to the 2,307 kg satellite to activate its attitude control system. For Bill Holmberg, the keeper of 41G's crew activity timeline for the mission, their carefully choreographed, minute-by-minute schedule had been swept into disarray. However, he later told Henry Cooper that it was easier to rewrite an already extant plan than to write one from scratch.

After the satellite left the RMS at 6:18:22 pm EST, Crippen and McBride pulsed the OMS thrusters to separate and, two days later, ERBS made the first of a series of thruster firings to raise its orbit to 560 km. The next problem for Mission 41G cropped up almost immediately, when the crew began operations with their other major payload, the second flight of the Shuttle Imaging Radar (SIR-B), part of the OSTA-3 package, which experienced difficulties transmitting data through Challenger's Ku-band antenna. This radar, resembling an eight-panelled rectangular dining table 11 m long by 2.1 m wide, had already proved to be something of a headache in pre-mission simulations because of its flimsy nature. Some scientists wanted it to commence radar observations of Earth while Challenger was at her 350 km ERBS-deployment altitude and continue doing so while she lowered her orbit to SIR-B's 260 km operating altitude. However, reducing the Shuttle's orbit required two OMS burns and it was feared that the shock could impart structural damage to the radar. On its first flight, STS-2 in November 1981, SIR had amply demonstrated its unique ability to gather data in support of geographical, geological, hydrological, oceanographic, vegetation and ice-monitoring applications, by acquiring imagery of more than 40 million km^2 at resolutions as fine as just 40 m. It consisted of a side looking synthetic aperture radar, which illuminated Earth's surface with horizontally polarised microwaves transmitted at the L-band wavelength of 23 cm.

During typical science-gathering activities, it radiated pulses of energy and measured the characteristics of the reflected 'echoes'. Subsequent analysis would enable scientists to determine surface textures and types. SIR-B was engineered to 'tilt' at angles between 15 and 57 degrees and its resolution was enhanced to 25 m. Consequently, on 41G, the device was no longer restricted to simply recording the ground-track directly underneath Challenger; by varying its 'look' angle, it became possible to assemble 'mosaics' of adjacent surface features observed over several days. Indeed, when the newly refurbished SIR returned from its second mission, it had greatly outperformed STS-2: yielding data to build three-dimensional models of subtle geological features on California's Mount Shasta, permitting contour modelling of parts of eastern and southern Africa and mapping intricate structural features such as faults, folds, fractures, dunes and rock layers. NASA envisaged that SIR-B would acquire 42 hours' worth of digital data and eight hours of optical measurements during the course of the mission; however, although the optical requirements were met, unforeseen problems limited the digital data to only seven and a half hours. Three main obstacles were responsible for this: problems with Challenger's Ku-band antenna, lost communication links to the Tracking and Data Relay Satellite and depleted power in SIR-B's own transmission system.

Shortly before ending their first day in space, the crew deployed the radar and, for about two minutes, it began transmitting scientific data through the Ku-band antenna via TDRS-1 to the White Sands ground terminal in New Mexico. Then, abruptly, at 7:54 pm, it stopped. Engineering analysis quickly determined that the antenna had lost its 'lock' on the geostationary-orbiting TDRS, due to a failed motor in its 'beta' gimbal. One axis by which the Ku-band dish could move was effectively dead and the other – the 'alpha' gimbal – swung backwards and forwards, operating sporadically. Affixed to the starboard payload bay wall and able to lean out over the sill, the antenna provided space-to-ground communications between the crew and Mission Control, but also acted as a 'rendezvous radar', as during the Solar Max retrieval. However, it could not accomplish both functions simultaneously.

A glorious vista of Earth from Mission 41G. The deployed SIR-B antenna can just be seen at the lower left of the image.

During ascent, an S-band link supported voice and data communications, after which the higher-rate Ku-band antenna was deployed to support the remainder of the mission. Unfortunately, it was often difficult for TDRS to acquire the antenna's narrow beam, so the orbiter employed the wider S-band link to 'lock' the Ku-band into position. To correct the gimbal problem, early on 6 October, Mission Control directed Ride and Leestma onto Challenger's middeck to unplug a wire that routed power to the antenna's motors. It was hoped that, if the wire was removed at the correct time, just as it swung out at right angles to the spacecraft, the astronauts could reorient Challenger such that the Ku-band was once more focused on TDRS-1. The wire was situated behind a row of lockers, so after the astronauts had removed them they waited for Crippen to announce the right time to unplug the wire. Peering at the waving antenna through the aft flight deck windows, as soon as it looked to be in the proper position, Crippen told Ride to pull the wire. The Ku-band antenna stopped its erratic motions and could thenceforth only move slightly in response to external forces, such as thruster firings which manoeuvred the entire vehicle.

At this point, Sullivan – who was responsible for SIR-B – retracted the delicate radar to enable Crippen and McBride to lower their ship's orbit to some 260 km. Using controls on the aft flight deck, she attempted to fold the two outermost antenna 'leaves' onto the central section and close the assembly into a storage canister. It should have gone smoothly, but as she watched the clunking radar components through the window it became clear that SIR-B was improperly stowed. Sullivan tried shutting it with backup controls, without success. A third option was to fire pyrotechnics, thus slamming it closed, but rendering it impossible to re-open. This was not ideal on only the second day of an eight-day voyage. Flight rules dictated that the antenna had to be closed before an OMS burn could take place, for fear that it might be damaged during the orbit-lowering manoeuvre. Already, when she first deployed SIR-B, Sullivan had noticed that it wiggled and writhed around in what Henry Cooper later described as "a classic case of dynamic instability". It seemed likely that if they conducted the OMS manoeuvre with the antenna still partially open, this would inflict damage onto it. Bob Crippen wanted to avoid taking this option. Sally Ride's dexterous handling of the RMS proved the saviour of the day, when she employed its end effector to push the antenna leaves firmly into place. Mission Control was unhappy with this technique, because there was no way of accurately gauging the amount of force imposed on the fragile panels, but neither the arm, nor the radar, appeared dented or scraped. Ride also earned brownie points with Leestma and Sullivan, whose three and a half hour spacewalk – scheduled for 9 October, but later postponed until the 11th – might have been cancelled if SIR-B had not been latched back into place.

After the OMS manoeuvre, the remainder of the mission was spent in the lower orbit, where the radar and two other Earth monitoring instruments could acquire their best results. The Measurement of Air Pollution by Satellite (MAPS) measured the abundance of carbon monoxide in the troposphere on a global basis for the first time. Meanwhile, the Feature Identification and Location Experiment (FILE) provided data to automatically classify surface materials into one of four categories: water, vegetation, bare ground or cloud and snow. NASA hoped that this would enable a new generation of Earth-watching satellites to download only data known to be of interest, rather than a continuous flow of data. With the exception of one niggling problem, in which MAPS experienced thermal fluctuations in its coolant loop, these two experiments performed admirably. FILE acquired 240 images across a broad range of different environments, successfully classifying their composition. SIR-B's data, too, proved of high quality and the radar took advantage of an unexpected opportunity to monitor Hurricane Josephine, detecting wave patterns associated with its motion and speed. Soil moisture content was measured as part of efforts to identify new water sources, support agricultural monitoring and crop forecasting and even, during a pass over Bangladesh, highlighted hidden breeding grounds of malaria-carrying mosquitoes. Plant types in Florida and South America were successfully discerned, ocean waves measuring over 20 m in height were recorded and polar ice flows and evidence of oil spills were detected. Despite the importance of these observations for 'real world' applications, SIR-B's claim to fame on STS-41G was its involvement in the 'discovery' of a lost city on the edge of the Empty Quarter in southern Oman. Archaeological exploration of the site later concluded that it was probably the legendary settlement of Ubar, a major ancient hub on the frankincense-trading route, first founded around 3000 BC.

"I was surprised to find that we were able to readily detect ancient caravan tracks in the enhanced Shuttle images," admitted geologist Ronald Blom of NASA's Jet Propulsion Laboratory in Pasadena, California. "One can easily separate many modern and ancient tracks on the computer enhanced images, because older tracks often go directly under very large sand dunes. We could never have surveyed the vast area where Ubar may have been, nor could we be confident of its location without the advantage of computer enhanced images from space." Excavation began in the summer of 1990 and found a remote well, together with towers, rooms and artefacts from at least 2000 BC. A Los Angeles filmmaker named Nicholas Clapp drew the possibility of using SIR-B to find Ubar to NASA's attention in 1983 and the radar and archaeological exploration confirmed the city – which, according to myth, was 'swallowed' by the desert – had indeed collapsed into an underground cavity. Its discovery, 'on the ground', was delayed by the first Gulf War, but in November 1991 archaeological teams returned to the Omani desert and employed subsurface radar to aid exploration. As digging progressed, Ubar's remains closely matched the Koran's description: an octagonal fortified city with 10 m towers and thick walls, containing a variety of buildings, including storage rooms, together with frankincense burners and pottery sherds.

Apart from their efforts to close the SIR-B antenna using the Canadian-built mechanical arm, the 41G crew had little interaction with the radar, MAPS or FILE. "We turned them on and off, changed the parameters and settings and did some fine-tuning," said Sally Ride. "We changed data tapes for SIR-B. Our direct involvement was not really as scientists, but as operators." The three instruments, mounted on a Spacelab pallet, formed the most visible component of NASA's Office of Space and Terrestrial Applications package. Behind it in the payload bay was a truss-like MPESS, housing another important OSTA-3 experiment: the Large Format Camera, which was capable of conducting high quality orbital photography for cartographic mapping and land use studies. In terms of accuracy, the camera's 305 mm focal length was capable of acquiring images at a resolution of 10 m from a 200-250 km orbit. As planned, 2,280 photographs were taken, including high priority coverage of Mount Everest and the Dead Sea, oblique shots of Hurricane Josephine off the United States' eastern seaboard and even contrails left by aircraft travelling between New York and Europe. In fact, its resolution was so good – detecting buildings, houses and streets – that some pictures were classified by the Department of Defense. Moreover, its position in the payload bay eliminated the distorting effects of Shuttle windows. Upon landing, its imagery of the Great Barrier Reef helped update Australian maps and the topography of national forests in Maine were plotted with greater accuracy than previously possible. Fossil fuel deposits were found in the Middle East and water sources identified in southern Egypt and Ethiopia. It even revealed geological evidence that blocks of land in China were being forced into the Pacific Ocean along the Kunlan fault line.

The Ku-band antenna failure would have ruined the SIR-B experiment, if not for the actions of the crew. With the dish now rigidly fixed in one place, Challenger had to firstly point the radar Earthward, record as much data as possible on tape, and then reorient herself to point the dish at the geostationary relay for playback. This repetitive process slowed down how much radar imagery SIR-B could acquire. Seven tapes were aboard Challenger – each capable of storing 20 minutes' worth of data – but playback, unfortunately, took just as long as recording. Although some steps were taken to maximise scientific return, such as 'dumping' data through TDRS-1 whilst over an ocean so that the radar could be faced back to Earth when the Shuttle approached a landmass, it was a laborious process. Then, on 8 October, TDRS-1 lost attitude control for almost 16 hours and, worse, lost its lock on the White Sands ground terminal, preventing it from being commanded. Given these problems, it is quite remarkable that so much valuable data was successfully returned.

HAZARDOUS HYDRAZINE

NASA managers had already postponed Leestma and Sullivan's EVA from 9 October until the 11th to enable the Earth resources instruments to acquire additional data. Moreover, they would attempt a repair job on the Ku-band antenna to enable it to be properly stowed for re-entry. In spite of the problems, the crew's attitude was "good" and Crippen's absence during early training had no detrimental impact. In fact, the agency declared that, as long as Shuttle commanders were experienced, there did not seem to be a problem with them joining crews at a relatively late stage. One of the objectives of Leestma and Sullivan's excursion was to test hardware for the refuelling of satellites in low Earth orbit. Mounted at the rear end of the payload bay was the Orbital Refuelling System (ORS), containing highly-toxic hydrazine, some of which the spacewalkers would transfer between two spherical tanks.

"Satellites have standard refuelling ports that engineers connect up when they're on the ground," explained Leestma. "One at a time, you very carefully have to handle the hypergolic fuels that go into it, because they're pretty dangerous. Hydrazine is very much like water, but it's got different properties, one of which is that it blows up if it's not handled right! Crip and the safety folks were very concerned that we shouldn't do this with hydrazine; we should just do it with water. Crip sent me to White Sands [Test Facility in New Mexico], so I spent about ten days there, watching them do adiabatic detonation tests, watching all kinds of things blow up. I came back with a real appreciation for the capabilities of this deadly stuff. You can't breathe it. If you get it on your skin, you can get poisoned. There were concerns that if we used hydrazine and it sprung a leak or even got on our suits, how are we going to get back in the airlock? We didn't want to bring this stuff back in."

Bob Crippen was not at all happy with using 'real' hydrazine in the ORS tests. He knew that the volatile substance could prove explosive at temperatures above 230 degrees Celsius; temperatures which could easily be reached in the intense sunlight of orbital daytime. On the other hand, in orbital darkness, it could freeze, contract and then flow back, overpressurising and rupturing its fuel lines. If Leestma and Sullivan got hydrazine on their suits, Crippen and McBride would have had to reorient Challenger's payload bay towards the Sun so they could 'bake' it out. In such a dire eventuality, they would have had to scrub their suits with towels and detergent, seal them in airtight bags, purge the airlock's atmosphere and pipe in fresh air, *before* removing their helmets. The effects on the spacewalkers were not Crippen's only concern. In an interview with Henry Cooper, he felt that 80 kg of hydrazine was enough "to take off the back end of the vehicle" if it exploded. He wanted to use water for the demonstration, but his suggestion was rejected because, said NASA, only by using the real thing would it be possible for realistic testing of safety procedures. Crippen was also assured that, although the ORS tanks and fuel lines had not undergone shaking tests equivalent to the stresses of launch and maximum aerodynamic pressure, they *were* designed with stiffness and robustness in mind. His four NASA crewmates were already comfortable with the experiment during their months of training without him and, at length, he was won over. In fact, Leestma and Sullivan had already successfully argued for a manual system to control temperatures and pressures in the tanks, for the simple reason that no one knew the exact parameters of hydrazine under different conditions.

"Crip finally agreed to have us do it with hydrazine," said Leestma, "because he had watched me several times in the WET-F, doing the whole procedure and how careful we were. We had triple containment of all the liquids at all times. It's a very tedious task, using small tools and lots of arm and hand manipulation." To achieve 'triple containment', three independent valves were placed in each of the coupling 'halves' and three seals were provided at the interface between the fluid path and the astronauts during the refuelling operation. Before the flight, they had encountered problems – narrowly averted – with the tools they would use. Leestma insisted on testing them before they were sent to Florida for packing, although the engineers were not anxious for him to do this. When he did conduct his tests, a 'ball valve' had an extra component fitted which made it 2 cm too long! Had it flown unchecked, the tool would have not have worked correctly. Leestma also recalled problems with a new type of grease applied to several tools which was slightly different to that used in previous tests, but he was assured should work in the same way. Leestma took the tools home that night and put them in his freezer; by next morning, the grease had frozen solid; the test was cancelled and the 'original' grease applied instead. "But that was still on my mind," he said, "that if something changes, you'd better make *sure* that those people know that they've looked at all the different things that can go wrong. I don't think it was so smart on my part. It was just the training that they put into you to kind of question everything. A lot of people don't like the astronauts, because they're always asking those silly, dumb questions, but sometimes those silly, dumb questions are appropriate and that one turned out to be okay."

The excursion began at 11:38 am EST on 11 October, when Leestma pushed open the outer hatch and entered Challenger's payload bay. He would later tell Henry Cooper that the difference between being inside the Shuttle and outside on a spacewalk was "like the difference between sitting at a desk in a *big room* and sitting at a desk in the middle of a *prairie*". Both astronauts needed about 30 minutes to fully acclimatise to their surroundings, learning how to move and how different their suits 'felt' in space, compared to the WET-F. "I grabbed the handhold and pulled out and when I first saw Earth, my heart rate went real high," Leestma said later, "and the docs later confirmed that, because my electrocardiogram reading went real high, 'This is when you came out of the hatch'. I said 'Yeah, no kidding!' I had a tumbling sensation – came out of the hatch and felt I was going to fall! I think my handprints are still in those payload bay handholds, because I just stopped for a short period of time and had to get my heart rate back down and then continue."

However, the excursion went perfectly and six hydrazine transfers were completed without incident. Already, on 6 October, a series of automatic transfers of small quantities of the toxic propellant had been conducted between the two tanks, using controls on Challenger's aft flight deck. Then, during their spacewalk, Leestma and Sullivan modified the piping with the ball valve, leak tested it and transferred another 50 kg of hydrazine through the fuel lines. In fact, because there was no weight on the thread, Leestma found it much easier to attach the ball valve in space than during pre-flight training. The overall procedure took somewhat longer than on Earth – a full, 90-minute circuit of the globe, rather than an hour – but that was because Leestma had to stop work periodically so that Sullivan could photo-document the task. Their work closely mirrored what spacewalkers were expected to do when refuelling Landsat-4 in 1987 for the Department of Defense and replenishing NASA's $1.5 billion Gamma Ray Observatory (GRO) in the summer of 1990. The observatory was intended to be launched in May 1988, loaded with 1,800 kg of the highly toxic fluid and fitted with a specially designed standardised refuelling coupling. Refuelling GRO would mark the first time that a fully functional satellite was refilled with propellant in space. Contracts to develop the refuelling coupling were awarded by NASA in December 1984, for delivery in just 15 months. It was optimistically envisaged that subsequent Shuttle missions would replenish high-pressure helium and nitrogen, and even cryogenic fluids to extend the lives of satellites. In spite of the threefold safety mechanisms, many astronauts breathed a sigh of relief when the 51L accident terminated such plans.

With, arguably, the most hazardous portion of the spacewalk over, Leestma and Sullivan's next task was to ensure that the Ku-band antenna could be retracted and stowed for re-entry. To do this, they had to move it by hand, such that a 'pin', activated from the aft flight deck, engaged to lock it in place; if they could accomplish this, they would leave the dish open so it could continue relaying data from the Earth-watching instruments. If, on the other hand, they could not engage all of the locking pins, they were to manually close, deactivate and latch the antenna. Obviously, for the sake of maximum data return from SIR-B, it was hoped that the second option could be averted. Moreover, if the antenna could not be retracted at all, the crew would be forced to jettison it overboard in order to close the payload bay doors for re-entry. That, said Leestma, was equally unthinkable. "The Ku-band assembly and digital avionics was worth a million dollars," he said, "so it would have been a very big loss to the programme if we had to jettison it." The repair involved not only the spacewalkers, but also their colleagues inside the cabin. In fact, because of her role in the effort, Sally Ride 'missed' watching most of the three and a half hour excursion. After she and Jon McBride had unplugged the wire to the antenna's electrical motors on 6 October, they also disabled a mechanism that drove the pins to 'lock' the alpha and beta gimbal axes into place. Early on the day of the spacewalk, they rigged a 'jump wire' that would allow them to reconnect power to the pins, though not the motors. Unfortunately, both plugs in the jump wire were 'female' and they had to quickly rig up a new, 36-pin 'adapter'. As Ride laboured in the middeck, Leestma manually moved the Ku-band in one axis, then the other, while Sullivan radioed her crewmates when the pins were correctly lined up with the holes they were meant to slot into. Crippen, meanwhile, told Ride when to plug in the two ends of the jumper. Working the current in pulses – plugging and unplugging the cable, such that the pins were 'hammered' into position – the attempt succeeded.

Difficulties with SIR-B's data gathering, though, continued. "Now, that caused us problems, orbiter-wise," admitted Leestma, "because to use the Ku-band, which the SIR-B required, we had to reorient Challenger so the antenna was pointed towards TDRS-1 and make the orbiter rotate. We'd take data and then do data 'dumps' and point the orbiter at the TDRS; then we'd go back and do data 'writes', rather than being able to take data the whole time and point the antenna and dump it. The SIR-B scientists didn't get all the data that they wanted, but the mission was not a loss and they got almost everything." Prior to returning inside Challenger's airlock at 3:05 pm EST, Sullivan took a long look at SIR-B, in an attempt to discover why it had been so difficult to automatically latch it into position. It looked, Henry Cooper wrote later, "like an over-stuffed sandwich"; its thermal insulation having billowed in space to make it 'thicker' than it should have been. This pure white blanketing had thus frustrated previous efforts to close it. "The insulation is billowing enough," she told her crewmates, "to interfere with a single motor closing and you don't need to miss by much to keep the latch from shutting."

In the astronauts' minds, the real deal had been the successful completion of the ORS tests, but to the media, back on Earth, Sullivan's achievement as the first American female spacewalker had seized headlines. Speaking to Henry Cooper afterwards, she would admit that she could not care less that the Soviets had cynically beaten her to it by sending their own woman cosmonaut, Svetlana Savitskaya, on an EVA, but admired her counterpart's credentials and abilities. Like Sally Ride, she considered herself an *astronaut* first, and a *female astronaut* a distant second. Kathryn Dwyer Sullivan was born in Paterson, New Jersey, on 3 October 1951, although her family moved to California when she was six years old and she attended Californian schools, finally entering the University of California at Santa Cruz to study Earth sciences. Whilst there, she spent a year as an exchange student at the University of Bergen in Norway. Sullivan received her undergraduate degree in 1973 and a doctorate in geology from Dalhousie University in Halifax, Nova Scotia, in 1978. Her PhD research focused on remote sensing of the seafloor in the Mid-Atlantic Ridge, the Newfoundland Basin and fault zones off the coast of southern California using acoustic and geophysical instruments.

When NASA issued its call for astronaut candidates, Sullivan applied, but she also had an offer of a post-doctoral position in deep-sea marine geology, using the Alvin submersible. "Two fabulous things were in front of me," she recalled, "either of which just seemed tremendous things to get to be involved in. It made my mother a little crazy that I was either going to 10,000 feet *down* in the ocean or 200 miles *up* off of the planet!" She was summoned to Houston for her interview and then accepted into the astronaut corps. By the late summer of 1983, having worked on the development of new space suits, her assignment to 41G as an EVA crew member came as a great pleasure ... but for one thing: Dave Leestma would be EV1, the lead spacewalker, whilst Sullivan would be EV2. "I'm a class *senior* to Dave," she told the NASA oral historian. "I've been in the programme longer than Dave. I've worked in the suits more than Dave. I worked this payload longer than Dave did, and I'm number two to him on the spacewalk. That's *really* bad optics." Intuitively, Sullivan *knew* that Bob Crippen and George Abbey had faith in her abilities, but she was still perplexed that an organisation in which "class rank matters" and "the senior class guy leads" was apparently changing its position. The EV1/EV2 debate was not lost on the media, either, and several awkward questions were directed at both Sullivan and Crippen during this period. "*Don't* be asking *me* to answer this," Sullivan told Crippen, paraphrased from her oral history, "because I don't see any particularly good reason I'm *not* EV1, but it's your call."

Privately, both Sullivan and McBride felt that the primary focus, at least in the eyes of the media and possibly NASA's senior management, was the fact that this would be Sally Ride's second mission. "In the early press stuff, it was very much slanted that way," she recalled. "Sally was still right in the bull's-eye of all the media interest. The flight was announced ... about five or six months after her first landing, so there's still a flood of interest surrounding her." No one seemed interested in Sullivan, only Ride. Sullivan even went so far as to have a new name tag made for her flight suit, reading *Sally*, but with a bar across it. "Evidently, what I am is *not Sally*," she reasoned. "Can't help that. Sally was *less* than thrilled with that line of teasing."

Challenger's sixth mission, in view of the complex tasks already accomplished, nearly detracted from the fact that there were not five people on-board, but *seven*. Paul Scully-Power's addition to the crew had been announced by NASA in late June, but *Flight International* had reported as early as February that a Canadian astronaut would fly on a mission later in 1984 and that the name would be revealed in March. Six astronaut candidates had been chosen by Canada's National Research Council in December of the previous year – Roberta Bondar, Marc Garneau, Steve MacLean, Ken Money, Bob Thirsk and Bjarni Tryggvason. The nature of the Canadian experiments, one of which focused on the development of an advanced space vision system, prompted some sources to predict Garneau or MacLean or Tryggvason, since they had trained extensively on the specific experiments to be flown aboard 41G. The successful finalist, Joseph Jean-Pierre Marc Garneau, came from Quebec City, where he was born on 23 February 1949. His father was an army officer "and we travelled quite a bit when I was growing up", he told a NASA interviewer, "and I thought that I would like to have a military career, although I was drawn more towards the Navy". Garneau would explain this decision in just three words: love of adventure. He was educated in Quebec and received his degree in engineering physics from the Royal Military College of Canada in Kingston in 1970. Garneau travelled to England in pursuit of his doctorate, studying electrical engineering at the Imperial College of Science and Technology in London. After gaining his PhD in 1973, he joined the Canadian Forces Maritime Command as a Navy engineer, serving aboard HMCS *Algonquin* and later as a forces fleet school instructor. During this period,

Garneau, designed a simulator to train weapons officers to use the missile systems aboard Tribal-class destroyers. Whilst at staff college in 1982, he was promoted to the rank of commander in the Canadian Navy. He was transferred to Ottawa the following year to design naval communications and electronic warfare systems and equipment. In December 1983, he was selected as an astronaut candidate and in February was seconded from the Department of National Defence to commence full-time training. "Two months before flight is when I turned up," Garneau said of his arrival in Houston, "and during those two months ... I was familiarised with the Shuttle to the extent that I needed to be familiar with it: knowing how to prepare meals, use the bathroom, use the communication system on-board the orbiter, what to do in an emergency, getting out of the orbiter, that sort of thing."

Now, in space, Garneau and Scully-Power found themselves spending much of their time conducting their own respective experiments. Obviously, they were not 'career' astronauts, although Crippen treated them as team members enough to invite them to help prepare meals or changeout carbon dioxide scrubbing lithium hydroxide canisters. "We got along fine with Marc and Paul," remembered Dave Leestma. "They were great guys, but still, five plus two equals seven, and *that's* a crowd!" Operating mainly on the middeck, the two payload specialists, whom Jon McBride mentored, helped each other out with their work. This proved useful for Garneau, whose ten investigations occupied much of his time. Known as CANEX-1, these Canadian experiments encompassed three major disciplines – technology, space research and life sciences – with two of the major foci being to develop a space vision system and to research how the space environment affected advanced composites. The plan was for Steve MacLean to fly the CANEX-2 mission in early1987. The space vision hardware was amongst the most important, for its descendant is today used routinely to allow astronauts to align and install components onto the International Space Station. During 41G, Garneau used the device to take video recordings of ten targets attached to the Earth Radiation Budget Satellite – four on its solar arrays and six on its sensor base – during the deployment process, which were then transmitted to Mission Control. It was hoped the system would accurately calculate the position, orientation and rate of movement of ERBS, relative to Challenger, 30 times per second. More than 90 percent of the vision system's objectives were accomplished and, later in the flight, Garneau even made further video recordings with the camera on the RMS wrist to trace the outlines of experiments in the payload bay. The Advanced Composite Materials Experiment (ACOMEX) was attached to the mechanical arm itself and comprised a number of samples, which, on 7 October, Sally Ride positioned in the Shuttle's direction of travel to assess their degradation under 'maximum' atomic oxygen exposure. Earlier missions had already shown that composites tended to deteriorate in the space environment, with the originally bright red Canadian flag on Columbia's mechanical arm eventually turning brownish. Furthermore, the shiny film on thermal blankets for payload bay television cameras quickly became 'dull' and 'flat' and the insulation itself lost around 35 percent of its overall mass. For a day and a half, the ACOMEX samples were pointed in the direction of travel and, after every six hours, Garneau examined them with binoculars and reported his observations of their performance.

Elsewhere, the Measurement of Optical Emissions experiment, known as 'OGLOW', provided data in support of a Canadian-designed imaging interferometer, scheduled for launch in 1988 to measure the temperature of high altitude winds. However, it was known that the Shuttle developed a strange, reddish-orange 'glow' around its extremities in orbit and there were concerns about how this would affect sensitive optical detectors. In addition to this Shuttle glow research, OGLOW gathered data on the Southern Lights (the 'aurora australis'), atmospheric airglow at night and the bioluminescence of the oceans. Other experiments included a solar photometer to measure constituents of the atmosphere and, specifically, the extent of sunlight-scattered dust, moisture, pollution and acidic haze with great precision. One of the instrument's main tasks was to examine the density and distribution of the cloud from the El Chichon volcano in Mexico, which had erupted in March 1982, before it fully disappeared. Finally, Garneau tended to a variety of space adaptation investigations. These focused on the effects of head motion, deterioration of sensory functions, awareness of position, space sickness, microgravity-induced optical illusions and changes in the taste of different foods.

The payload specialists worked together on these tasks, with Scully-Power acting as a 'recorder' for many of the adaptation experiments and as a 'subject' to some of the vestibular investigations, while Garneau helped him with his observations of the oceans. Unfortunately, the planet did not entirely co-operate: there was around 70 percent global cloud cover, instead of the 30 percent anticipated before launch, although the Mediterranean Sea was visible in its entirety and Scully-Power was able to view tightly interconnected spiral eddies moving from one end of it to the other. One of the few problems seemed to be Garneau's lack of conversation over the airwaves, which, he being Canada's historic first man in space, frustrated many journalists who were covering the mission from his country. In fact, one member of the Canadian media even referred to him as *The Right Stiff*, a play on words of Tom Wolfe's description of the Original Seven Mercury astronauts as having 'the right stuff'. Yet Garneau was by no means stiff; nor would this be his only chance to fly into space. Eight years later, in 1992, he was selected by NASA to train as a fully-fledged mission specialist for the Canadian Space Agency and went on to make two more Shuttle flights, the second of which, STS-97 in November 2000, brought some of his 41G work full circle, by using the 'operational' space vision system to install the first US-built set of solar arrays onto the International Space Station. Moreover, his backup for 41G – physician Bob Thirsk – is presently Canada's most experienced spacefarer, having chalked up two missions, including a 188-day stay aboard the station in 2009.

Challenger's mission, meanwhile, at just over eight days, was the longest she would achieve in her short life; eclipsing by about six hours her individual record, set on STS-41B. After checking her preparedness for re-entry and successfully stowing the troubled Ku-band antenna, the OMS engines were fired at 11:30 am EST on 13 October, beginning a hypersonic dive to KSC. Touchdown was perfect on the Shuttle Landing Facility, at 12:26:38 pm on Runway 33, with the orbiter rolling 3,000 m to a halt. For Crippen and Ride, it was also a personal achievement. "All the previous entries," Crippen noted in his oral history, "because we were landing at

Edwards ... came in pretty much over the Pacific, so you weren't flying over land that much. This one, we started up in Canada and pretty much came across the centre of the United States, headed for the peninsula of Florida, and it was a nice, clear day across all the states and you could see *everything*." At one stage, taking a quick peek out of his left window, he could see Jacksonville in Florida, whilst still flying over the Kansas-Missouri area! The peninsula of Florida was exceptionally clear. "I often joke that they've got a 15,000-foot runway," at the Cape, "but they built this *moat* around it and filled it full of *alligators* to give you an incentive to stay on the runway!" Crippen's triumphant landfall in Florida was not entirely without blemishes. Astronaut Dave Hilmers, sitting at the capcom's mike in Mission Control, jovially radioed congratulations on finally making it to the East Coast. However, he alerted them that, judging from Crippen's track record for making successful Floridian landings, their 'welcome home' case of beer had been delivered to Edwards Air Force Base by mistake.

VANILLA TURNS TO CHOCOLATE

Early in February 1984, astronaut Joe Allen – veteran of the STS-5 mission – was at the Kennedy Space Center, watching the tenth Shuttle flight rocket into orbit. In command was Vance Brand. "I remember thinking," Allen wrote in his book, *Entering Space*, "as I wistfully watched the spaceship Challenger climb into the clear morning sky, that Vance had left without me. Although I *had* been assigned to another crew that had begun to prepare for a future mission, the *date* of my next opportunity to enter space and the new mission's specific cargo remained very much in question." Allen's uncertainty about his future prospects were shared by many other astronauts who were on active status during this period. Ten months after STS-5, in September 1983, NASA assigned him as a mission specialist on a flight called 41G. According to the news release issued at the time, Allen and his four crewmates – Rick Hauck as commander, Dave Walker as pilot and fellow mission specialists Anna Fisher and Dale Gardner – would deploy the SBS-4 and Telstar-3C communications satellites for Space Business Systems and AT&T and fly a retrievable astronomy platform, known as Spartan. Their launch was scheduled for the following August. Within weeks, however, all that had changed. In its November 1983 manifest, NASA listed Hauck's crew on the 41H mission, now scheduled for late September and assigned *either* a classified Department of Defense payload or (if the problems with the IUS were rectified) deployment of the second Tracking and Data Relay Satellite. The January 1984 manifest remained very much the same, but when the May manifest was released, Mission 41H (and with it Hauck's crew) had vanished!

Much of the uncertainty focused on concerns over the performance of not only the IUS, but also the PAM-D, which had failed to properly insert Indonesia's Palapa-B2 and Western Union's Westar-6 communications satellites into geostationary transfer orbits after their release during STS-41B. "Preliminary investigation suggests that the solid-propellant motor stopped burning prematurely," *Flight*

International told its readers on 18 February, "and that this may have been caused by the separation of the nozzle. The fact that *both* satellites were similarly affected could be explained if both motors were part of a batch involving sub-standard material or components." Within days, the owners of both satellites had verified that they were otherwise healthy and functioning normally, although in the case of Westar a plan to use its own attitude-control thrusters to raise its orbit was described as "unworkable". At NASA, however, a cunning plan was devised. "The seed of an idea had been planted," wrote Joe Allen. "Could the dramatic debut of the MMU lead to the recovery of Palapa and Westar? The technical challenge of devising a means of recovering the satellites was irresistible. Operating on only the *possibility* that the insurance underwriters who now owned the satellites would endorse our plan, we got to work on potential recovery techniques and it began to look as though the crew to which I had been assigned just might be the one that would fly the recovery mission."

It would herald a sharp turnaround in fortunes. "I'll call it a plain vanilla flight," was Hauck's recollection of his mission in its earliest incarnation. "Go up, deploy two satellites, do some on-orbit experiments, come home." By March 1984, a few weeks after the losses of Palapa and Westar, George Abbey was looking specifically at Hauck's crew to perform the recovery. "I think there were a number of things that worked in our favour," Hauck continued. "One was just the timing of our mission. I had flown proximity operations on STS-7. Clearly, prox ops would be necessary to do this mission." Then, the following month, Bob Crippen's 41C crew demonstrated in spectacular fashion that it *was* indeed possible to retrieve a satellite from orbit and bring it into the payload bay for repairs. Unlike 41C, Hauck's crew would not be *repairing* the satellites, but *salvaging* them and returning them to Earth. The MMU with its Trunnion Pin Attachment Device could be used to accomplish this feat, but some sort of mechanism would also be needed to snare the satellites. One morning, Dale Gardner arrived at work in a state of some excitement. He had been awake all night, he told them, thinking about possible salvage methods. Years later, Hauck could not be sure if it was Gardner, alone, or in conjunction with the EVA equipment team, who came up with the idea, but the consensus was to create a probe-like 'stinger' – a 1.8 m long Apogee Kick Motor Capture Device (ACD) – mounted on the arms of the MMU . "The astronaut could then fly the stinger into the satellite's rocket nozzle," wrote Joe Allen. "Once inside, he could release a lever that would allow toggle fingers to expand, much like opening an umbrella inside a chimney. A hand-driven crank would shorten the length of the stinger and pull the satellite against a padded ring at the stinger's base. The satellite would be held securely and the astronaut could then use the MMU's thrusters to stop its tumbling and hold it while the [RMS] grabbed a grapple fixture on the stinger." It seemed fairly straightforward, but for one thing: only the nozzle 'end' of the satellites could be clamped into the payload bay for the return to Earth; therefore, some other technique would have to be employed to temporarily 'hold' the precious objects – each valued at $100 million – whilst the stinger was detached and a cradling adaptor fitted. NASA's solution was for a truss-like aluminium A-frame (properly termed the 'Antenna Bridge Structure') to be placed over the delicate antenna at the 'top' of

The ingenious capture mechanism developed for Mission 51A is clearly illustrated in this image as Dale Gardner manhandles Palapa-B2 into its berth in Discovery's payload bay. Fellow spacewalker Joe Allen is visible at the right.

each satellite. "Next, the arm would take hold of a grappling pin on the A-frame," concluded Allen, "keeping the satellite motionless while two astronauts manually fitted the adaptor at the nozzle end." With the adaptor in place, the RMS would lower the satellite into position in the bay and the spacewalkers would finally remove the A-frame. It was a brilliant plan, which, if it were executed to perfection, would cement the Shuttle's credentials and demonstrate its capabilities in space.

It also turned Rick Hauck's mission, in a matter of weeks, from 'vanilla' to chocolate.

Having already trained for EVAs on their respective first flights, it made sense for Allen and Gardner to be assigned to perform the pair of six-hour spacewalks to retrieve Palapa and Westar. Meanwhile, the other crew members of the mission – which the August manifest had redesignated 51A and scheduled for launch in early November – set to work in the simulators, refining rendezvous procedures. Hauck was joined on the flight deck by David Mathieson Walker, who retained his naval aviator's nickname of 'Red Flash' for his sandy hair. Walker came from Columbus, Georgia, where he was born on 20 May 1944. After attending high school in Florida, he entered the US Naval Academy, received his degree in 1966, and immediately underwent flight instruction at the Naval Aviation Training Command in Florida, Mississippi and Texas. Walker was designated as a naval aviator in December 1967 and served two tours in Vietnam, flying the F-4 Phantom II from the USS *Enterprise* and USS *America*, during which he gained the Distinguished Flying Cross. Following his return to the United States in 1970, he entered test pilot school at Edwards Air Force Base and later served as an experimental and engineering test pilot at the Naval Air Test Center in Maryland, working on the evaluation of the new F-14 Tomcat fighter. Walker also performed tests of a leading-edge slat modification of the Phantom jet and, in 1975, after replacement pilot training on the Tomcat, he served as a fighter pilot on two overseas deployments to the Mediterranean aboard the *America*. As an aviator, Walker was top-notch, but in the words of Mike Mullane, who flew with him as a member of the T-38 chase team in support of STS-1, "he was *too* cocky, the type of pilot who thinks he's bulletproof even when he's sober". Fellow astronaut Bob Cabana, who later flew into space with him, paid tribute to Walker as "kind of a throwback to the 1960s".

Also supporting Hauck with the rendezvous was Anna Fisher, born in New York City on 24 August 1949. As 51A's flight engineer, she would monitor the performance of the orbiter's systems and providing assistance with malfunction procedures. Their work together provided a breeding ground for some banter. During training for 'transatlantic aborts', simulating an engine failure, late in the ascent, that necessitated an emergency landing in North Africa, Dave Walker would jokingly offer to trade her for camels in exchange for the rest of them getting out. "Nowadays," Fisher told the NASA oral historian, "people would think that's probably not very politically correct. Then, Dave gave me this neat collection ... of *camels*; all different kinds. These are guys who are trained in a different area. They were pilots in Vietnam. They saw all kinds of things. I'd gone to medical school. In histology class as they were doing their slide lectures, they would stick in *Playboy* centrefolds. It's just a way of breaking the ice."

Born Anna Lee Sims (the maiden name under which she would register for her astronaut interview in Houston in August 1977), her father was a military man and the family moved frequently, eventually settling in California, where she attended San Pedro High School. A lifelong interest in science and mathematics eventually inspired her to enter the University of California at Los Angeles to study chemistry, although by this time she had done voluntary work in Harbor General Hospital in Torrance and after gaining her degree in 1971 she started graduate work in X-ray crystallography, but opted to change direction and enter medical school instead. She received her doctorate in 1976 and spent a year as an intern at Harbor General Hospital, then specialised in emergency medicine, practicing in Los Angeles. By this time, she was engaged to another young physician, Bill Fisher, and it was a mutual medical friend, Dr Mark Mecikalski, who first noticed NASA's call for astronaut candidates and encouraged them both to apply. Emergency medicine was tough, as she recalled. "Bill was probably working ten 24-hour shifts a month and I was working probably eight 24-hour shifts," she told the oral historian, "which is really gruelling." As August 1977 drew to a close, whilst planning their wedding, she received a call from NASA, inviting her to interview on the 29th. They married quickly at the brick-and-glass Wayfarer's Chapel in San Pedro and the new 'Mrs' Anna Fisher went to Houston for the interview. Bill Fisher was interviewed in November. When the announcement was made, she was selected and he wasn't. "We had both talked about it," she recalled, "and said that if either of us got selected, I probably had the greater opportunity, because I also had the background in chemistry. At the time, they were really looking for people that had a background in two areas. Bill didn't have that." He was selected in the *next* astronaut class, in May 1980, provoking a great deal of media attention; in fact, even the NASA news release mentioned him specifically by name as "the husband" of Anna Fisher. By the early summer 1983, she was pregnant with her first child and was surprised when she and Bill were called into George Abbey's office. "He said he wanted to assign me to a flight," she told the oral historian. "Did we have any reservations? I'm probably the only person who's been assigned to their flight about *two weeks* before they *deliver!*" Within six months of her daughter's arrival, Fisher was working feverishly, around-the-clock, to prepare for her first mission into orbit.

"Rick, Anna and Dave immediately began to log long hours in the Shuttle simulators in Houston," wrote Joe Allen, "practicing rendezvous procedures that would allow our ship, Discovery, to come within a few feet of each satellite without 'pluming' it out of reach with bursts ... from the orbiter's [RCS thrusters] or, even worse, causing it to tumble like a wildly gyrating top." There were other contingencies for which to prepare themselves. If Allen or Gardner experienced an MMU malfunction – a stuck-on nitrogen thruster, perhaps – it might be necessary to manoeuvre the Shuttle in order to perform a rescue. "*Timing* was critical," Hauck told the oral historian, "because orbital mechanics propagates differential velocities *very* quickly and I think we figured that if *that* were to happen and if he were to go off ... if I didn't manoeuvre the Shuttle within 15 seconds and go after him, he was *gone*." Two weeks after Discovery returned from Mission 41D, NASA announced its plans for the salvage flight: Allen and Gardner had been certified as proficient with the MMU, recovery contracts had been signed with Palapa and Westar's insurance underwriters, equipment had passed vacuum-chamber testing and the entire crew and flight control team were comfortable with the intricacies of the rendezvous. Rick Hauck was livid. To him, the announcement that his crew would deploy two satellites and salvage two others did nothing else but *trivialise* the mission; it made it sound *easy*. "If we get *one* of these satellites back," he told the NASA press affairs office, "it'll be amazing, and if we get *both* of them back, it'll be a *miracle!*" In Hauck's mind, the space agency had shot itself in the foot by creating the illusion that Mission 51A would be a piece of cake, a walk in the park. To Allen, who had been with the space agency since the Apollo days, it was another indication that it considered itself to be bulletproof. "NASA was still in its halcyon days," he told the oral historian, "still riding on the coat-tails of the successful Apollo missions, successful Skylab, successful Apollo-Soyuz, successful first tests of the orbiter. NASA continued to be bullish on itself."

Over a period of several weeks, the orbits of Palapa and Westar were lowered from an apogee of around 970 km to some 350 km, thereby enabling the Shuttle to reach them. "During this process," read NASA's press kit, "the spin rates of the satellites will have been reduced to around 1 rpm. The satellites will be in near-identical orbits, with Palapa trailing Westar by about [1,000 km]." Launch requirements were constrained not only by the need to insert Discovery into the same orbital plane as the satellites, but also by the requirements of 51A's other two customers, Anik and Syncom. Liftoff was originally scheduled at the start of a tight, 18-minute 'window' on 7 November, but was scrubbed when it became evident that predicted wind speeds at altitude would impose shear loads in excess of the design limits of the vehicle. Next day, the situation improved and Discovery thundered off the pad at precisely 7:15 am EST. "History now shows we were also possibly very lucky," Joe Allen told the oral historian, "because both of the tragic accidents, that of the Challenger and that of Columbia, involved launching through very high wind shear conditions and there's some thinking now that high wind shears and Space Shuttles do not go safely together." Deployment of their first payload, Anik-D2, also known as 'Telesat-H', followed on the second day of the mission, at 4:04 pm on the 9th, and its attached PAM-D successfully boosted it towards geostationary orbit. The D-series Aniks operated in the C-band (6/4 GHz) and were equipped with 24 transponders, each capable of transmitting 960 one-way voice circuits or a single colour television programme. Like the Anik-C satellites, launched by STS-5 and STS-7, this fourth generation was based on Hughes' HS-376 bus, measuring 2.8 m high and 2.1 m in diameter in its stowed form, but increasing to a height of 6.6 m when fully deployed in geostationary orbit. In fact, so similar, physically, was this satellite to Palapa and Westar, that in the weeks preceding their launch, Allen and Gardner were cheekily told by fellow astronauts *not* to confuse them. "In other words," Allen recalled wryly, "*don't* bring home satellites that we'd just taken there!"

In a similar fashion to its predecessors, it was powered by a combination of solar cells (capable of generating a kilowatt of electricity) and a set of nickel-cadmium batteries. The two Anik-D satellites – the first was launched aboard a Delta rocket in August 1982, the second aboard 51A – were intended to replace the three members of the Anik-A network, each of which had been in orbit for around a decade. The third day of the mission, 10 November, was devoted to Syncom 4-1, originally part of Discovery's payload on 41D at the time of its pad abort, which rolled, frisbee-fashion, out of the payload bay at 7:56 am and was boosted perfectly into geostationary orbit. With two satellites safely deployed, the next task was to rendezvous with Palapa. Hauck and Fisher were first to spot it as a steadily-brightening star, from a distance of more than 150 km, on the morning of the 12th. The rendezvous was officially completed at 8:00 am. By this time, Allen and Gardner were already clad in their water-cooled underwear and Dave Walker was getting their suits ready in the airlock. Years later, Allen would recall Walker's intense focus on ensuring that every aspect of the checklist was followed; the pilot came away with a fearsome headache and was forced to dig into the medical supplies for a pill. At length, with the two spacewalkers in the airlock, Walker was almost ready to pass Allen his helmet, when he stopped.

"Dave, I'm hungry," said Allen. "I *really* need a cookie or something to eat."

"Oh, Joe, how could you? We're slightly . . ."

Allen was indignant. "Dave, I *need* a butter cookie."

It was not idle gossip, nor a minor grumble on Allen's part. According to the timeline, he and Gardner would be outside for at least six hours and the reader will recall that simply operating in their bulky space suits demanded enormous reserves of energy. "So he goes off into the food pantry," Allen recalled, "and comes back with a butter cookie. I open my mouth. Keep in mind, I can't use my hands now. He puts the butter cookie into [my mouth] . . . then he hits my jaw shut!" Walker clicked Allen's helmet into place and then sealed the two men into the cylindrical airlock. "We could *feel* the hatch being sealed," Allen wrote, "and we waited quietly for 25 more minutes, whilst the airlock was depressurised." The plan was for the EVA to begin during a period of orbital daylight to enable them to set up their tools and prepare the MMU for its first flight.

Every person who has performed a spacewalk has come back with their own stories about it, with many regarding it as incomparably surpassing any other experience on a mission. For Allen, when he pushed open the outer hatch and poked his helmeted head into Discovery's payload bay at 8:25 am EST, his first view quite literally took his breath away: for *there* was Palapa, slowly spinning, directly beyond the forward bulkhead. "I fastened myself into the MMU as darkness fell," he wrote, "tested its two propulsion systems, released the lever that held it to the bulkhead and I glided across the bay. Dale helped me to attach the stinger and, once it was secure, I made another short test flight to see how the MMU flew." As the Sun rose on another 45-minute period of orbital daylight, Walker gave the call – "Let's go get it" – and Allen flew crisply over to Palapa. With the stinger mounted on the front of the MMU's arms, he looked not dissimilar to a medieval knight, about to enter a jousting contest. Back on Earth, in the water tank and at Martin Marietta's Denver facility, it had been quite ungainly, but now, in space, it flew magically. At first, as he headed around the 'base' of the satellite, he was struck full in the face by blazing sunlight, but as he drifted closer and closer and finally entered Palapa's shadow, he could instantly see clearly and was able to guide the tip of the stinger directly into the

Both Palapa and Westar were grappled in the same fashion, with one crew member flying out to the satellite in the MMU, docking with a 'stinger' mechanism and returning to the Shuttle to secure it into the payload bay. Here, Dale Gardner extends the capture bar to snare the slowly-spinning Westar.

throat of the nozzle. Allen waited a few seconds as the stinger moved further and further inward, then pulled the lever to open the toggles. It *worked.* "Stop the clock," he yelled, triumphantly. "I've got it tied!"

After stabilising both himself and the satellite with the MMU's thrusters, Allen watched as Anna Fisher guided the RMS grapple fixture over a pin on the stinger. Returning to the payload bay, he doffed the MMU and positioned himself in a portable foot restraint on the end of the mechanical arm, then watched as Dale Gardner proceeded to attach the A-frame over Palapa's fragile antenna. Suddenly, they hit a glitch. A rigid structure, part of the satellite's wave guide equipment, was protruding further outboard than had been expected; the A-frame would not fit. It was Dave Walker who suggested that their only option was for Allen to physically grab the antenna 'end' of Palapa and hold it steady, whilst Gardner single-handedly attached the adaptor. For 90 minutes – a full circuit of the globe – Gardner worked, before finally declaring success by manually tightening nine bolts around the edge of the adaptor. Shortly thereafter, the two men moved Palapa into a vertical position and lowered it into the bay, securing it in place with payload retention latches. Years later, Allen would pay tribute to Gardner's diligence and persistence in getting the job done, working alone. The astronauts returned inside Discovery, repressurising the airlock at 2:25 pm, after an EVA which had lasted precisely six hours. Their next step was to rendezvous with Westar and Mission Control brought the troubling news that it, too, had wave guide equipment in the same place. Consequently, it was decided that Gardner would fly the MMU out to the satellite and Allen would act as a sort of 'human' A-frame, holding the antenna end of Westar, whilst the adaptor was fitted. The second EVA duly got underway at 6:09 am on 14 November and Gardner quickly captured Westar and returned it to the payload bay. The second retrieval would proceed more smoothly than the first and, in total, the astronauts would spend five hours and 42 minutes outside.

On this occasion, Allen spent a considerable amount of time with his feet secured in a foot restraint on the end of the RMS, which gave him a quite different perspective ... and made him feel peculiarly *precarious*. "Flying the MMU," he wrote, "much like piloting an airplane, had not imparted an ominous sense of *height* to me; I was in control and at ease with the responsive machine. But riding the end of the arm, high above the cargo bay, was like standing on the tip of the world's highest diving board – and a *movable* board at that." The limited visibility of his helmet meant that he could not see his feet, nor the rail by his side. "Only my knock-kneed stance kept me in the foot restraint," he continued, "and my ride was as nerve-wracking as anything I had ever done before." Intellectually, he *knew* that he would not *fall*, but the sensation persisted that if he slipped his restraint, he might either plunge into the payload bay or else directly to Earth. "It was a relief to take hold of the satellite when Dale brought it within my reach," Allen concluded. "I felt like a man on a high wire, being handed a balance bar, and the round end of the cylindrical satellite provided some comfort and security." Gardner finished labouring to attach the adaptor and the two men proceeded to secure Westar into the bay. It was a triumphant moment. They were out of radio communication with Houston at the time and, from the flight deck, Rick Hauck told the spacewalkers that he wanted *them* to announce the success at Acquisition of Signal.

Allen and Gardner declined. After all, it was the captain of a salvage vessel who traditionally must assume such responsibilities. "Rick, that's the commander's job," they told him. "When we come AOS [Acquisition of Signal], *you* report that we have two satellites safely aboard and *you* can also use the words *Fucking Miracle*."

Hauck chuckled at this 'inside' joke. It originated in the KSC crew quarters, a few days before launch. As the commander, he was already irritated by the media assumption that 51A would be a piece of cake, with two 'easy' satellite deployments and two 'easy' retrievals. When a high-ranking NASA official from the agency's Office of Public Affairs called a meeting with them, it was with some trepidation that the crew entered the meeting room. Hauck asked what the agenda was. The official responded that there was *no* specific agenda; he had merely come along to wish them good luck. "We were all surprised," recalled Joe Allen, "because this really was occupying a good chunk of our morning and time was very important to us right then." At this point, Hauck remembered the incessant comments from the media that 51A was an 'easy' mission. "There is *something* that you can do," he told the official, and proceeded to cite the news reports which alleged that the crew had labelled the mission 'easy'. "I can *assure* you that none of us said that, nor do we believe it ... and I will personally tell you that my assessment is: if we successfully capture *one* satellite, it will be *remarkable*, and if we get *both* satellites, it will be a *fucking miracle*! You can quote me on that!" Without another word, Hauck ended the meeting. Now, with this achieved, Hauck opted to keep his language somewhat

Many have argued that the 'Golden Age' of pre-Challenger innocence reached its zenith with the triumphant capture of Palapa and Westar in November 1984. In this image, Dale Gardner displays a 'For Sale' sign above the recaptured satellites. For now, NASA and the Shuttle were bulletproof.

more appropriate. "Houston," he radioed at Acquisition of Signal, "we've got two satellites, *locked in the bay*!" All five of them could hear in their headsets the shouts and cheers from Mission Control. They had *done* it.

Hovering above Palapa and Westar, Gardner now untaped and displayed a sign, emblazoned with the statement: *For Sale*. "The satellites would be returned and would then be in the ownership and the possession of insurance companies," wrote Allen, "which had every intention of selling them as brand-new satellites." From inside the flight deck, Fisher manoeuvred them for photographs. The insurers – Lloyds of London and International Technology Underwriters – loved it, although NASA would mildly rebuke the astronauts after the flight. Lloyds actually rang the Lutine Bell in the rostrum of their Lime Street headquarters, marking only the third time since the end of the Second World War that it had been rung to announce good news. (The bell was traditionally struck once in instances of bad news, such as the loss of a ship, or twice to celebrate good news, such as a safe recovery). In recent years, sadly, it has rung for more bad news than good: the deaths of Princess Diana and the Queen Mother, the terrorist attacks on the World Trade Center, the 2004 tsunami and the London bombings. In the months after 51A, Hauck and his crew were flown, first-class, aboard Concorde, to London to address the Lloyds underwriters in the Captain's Room ... and took tea with Prince Charles at Kensington Palace. Hauck was fascinated, when he used the toilet in the palace and found a page from the prince's logbook, detailing his first solo flight in a helicopter. The page was framed on the wall. When he saw the prince, he asked about it.

"Oh," replied Charles with a grin, "you've been to the *loo*, have you?"

The humour and the state dinners and the meetings with presidents and princes and prime ministers did nothing to detract from the truly remarkable accomplishment of recovering two satellites which had *never* been intended for retrieval, let alone retrieval by spacewalkers. The expectation was that Palapa and Westar would be relaunched in a year or so. In February 1985, however, *Flight International* told its readers that Lloyds had been unsuccessful in securing a buyer, although NASA kept Mission 51L – then scheduled for November of that year – as an available 'slot' for Palapa. Meanwhile, Westar was later sold to the Asiasat consortium and successfully placed into orbit by a Chinese Long March 3 rocket in April 1990. Meanwhile, Palapa's operator, Perumtel, ordered a replacement satellite, Palapa-B3, after the 41B failure, which NASA intended to launch aboard 61H. In the wake of the Challenger disaster, the satellite was shifted onto a Delta expendable booster, renamed Palapa-B2P, and successfully launched in March 1987. Meanwhile, the refurbished Palapa-B2 was sold by its insurers to Sattel Technologies and eventually resold back to Perumtel and launched atop another Delta rocket in April 1990, under the new name of Palapa-B2R. Perumtel retained ownership of the satellite until 1993, when it passed to a private Indonesian company. All this was in the future when Hauck and Walker expertly guided Discovery onto the KSC runway at 6:56 am EST on 16 November, just a few minutes short of eight full days since leaving Pad 39A. It was exactly two years to the day since Allen landed from his first flight. "I landed *twice* on the 16th of November," he said, "once on the East Coast and once on the West Coast!"

DEEP SECRECY

If Rick Hauck considered it a miracle that his crew had managed to snatch two errant communications satellites out of orbit, anchor them into the payload bay and return them safely to Earth, then the *next* Shuttle flight in January 1985 represented a miracle of another kind. In a sense, it was quite literally a miracle that Discovery made it into space at all … both metaphorically and literally, as the Challenger accident investigation would later reveal. When Ken Mattingly, Loren Shriver, Ellison Onizuka and Jim Buchli were named as the crew of STS-10 in October 1982, they confidently expected to launch aboard Challenger in September of the following year. For Mattingly, who had been with NASA for almost two decades, flew in lunar orbit on Apollo 16, commanded the final test flight of Columbia in June 1982 *and* served as the astronaut office's DoD representative, it was an exciting opportunity. If STS-10 flew on time, it would only entail a wait of a little more than a year – quite different to the *decade* he had spent waiting for his first Shuttle voyage. "With all the training and all the years we put into the programme," Mattingly told the NASA oral historian, "the idea of turning around and going right away was *very* appealing; get my money back for all that time."

Having said this, by his own admission, Mattingly actually had little interest in the Shuttle when it was first conceived. He had, after all, flown to *the Moon* and felt that a mission to Mars should be the *real* next step. "I believe that we needed to build a space station, first, so we could have hardware which would gather years of lifetime experience," he said. He was drawn back to the Navy by promises from then-Secretary John Warner of prestigious roles – his own squadron, perhaps – but it soon became apparent that Vietnam was producing active-duty officers who had worked their way through the ranks on the *front line*. Mattingly knew that he could not simply walk into a squadron command and opted instead to remain with NASA. His career with the agency had taken several twists and turns of good and ill fortune and one of those episodes would later see him immortalised on the silver screen. In August 1969, just a couple of weeks after the landing of Neil Armstrong and Buzz Aldrin on the Moon, Mattingly was named as the command module pilot for Apollo 13, joining Jim Lovell and Fred Haise for what was to be the third lunar landing mission. With less than a week to go, an exposure to German measles scrubbed Mattingly from the crew and he was replaced by his backup, Jack Swigert. He persevered on the ground, however, supporting a mission which narrowly averted disaster and in April 1972 he finally reached lunar orbit on Apollo 16.

Thomas Kenneth Mattingly II was born in Chicago, Illinois, on 17 March 1936 and received much of his schooling in Florida. Aviation, he would later recount, was in his blood: his father worked for Eastern Airlines and his earliest memories were of toy planes and model aircraft. "I built every model I could find," he explained, "ate every box of cereal that had a cut-out paper airplane on the back, all that sort of stuff." If the young Mattingly stayed out of trouble for long enough, his father would take him aboard an aircraft and fly to the end of a route and back. Weekends would be spent at the airport, watching planes take off and land. He studied aeronautical engineering at Auburn University, graduating in 1958 and securing membership of the Delta Tau Delta fraternity. Whilst there, as part of the Navy's Reserve Officer Training Corps, Mattingly had the opportunity to fly a propeller-driven attack aircraft and his choice of aeronautical engineering did little to hide an ultimate ambition to become a test pilot. Later in the year, he enlisted in the Navy as an ensign and was vocal in his desire for flight training. When his gunnery officer, Lieutenant-Commander Glenwood Clark, asked if he *really* wanted to go through flight training, Mattingly, naturally, replied in the affirmative. "You're the *dumbest* ensign I've ever met," Clark announced. "Out!" More than a quarter of a century later, having flown to the Moon and commanded two Shuttle missions, Mattingly returned to active duty in the Naval Space and Warfare Systems Command . . . and was introduced to his new commanding officer, one Vice-Admiral *Glenwood Clark*! Mattingly could hardly believe it. After briefing his new boss, the two senior officers sat down together.

"Admiral," Mattingly began, somewhat tentatively. "Do you remember me?"

Clark looked him straight in the eye.

"I sure do. You were the *dumbest* ensign I ever knew!" The two men laughed.

So it was that the Navy's dumbest ensign completed initial flight instruction and received his wings in 1960. He went on to fly the A-1H Skyraider for three years and the A-3B Skywarrior for another two years, then his options included postgraduate school for a master's degree in aerospace engineering – which he was not keen on – or test pilot school at Edwards Air Force Base. Mattingly picked the latter. In March 1965, witnessing the first manned Gemini launch planted the germ of an idea to someday become an astronaut. He knew that the Air Force was selecting candidates from the Edwards school for its MOL – so he reckoned he might get a shot at *that*. Another Navy pilot, Ed Mitchell, had the same idea and the pair applied for MOL ... and were rejected. Although they could not ordinarily apply to NASA, a sympathetic senior office at Edwards, Lieutenant-Colonel John Prodan, gave them a chance and submitted their names to the civilian space agency. In April 1966, they were both selected. Three years later, when he was assigned to Apollo 13, Mattingly had gained a reputation as an expert in command module systems. Eleven months would elapse after the unlucky voyage of Apollo 13 before Mattingly was finally named to a new lunar crew ... and a full decade before he would make his second space mission. During that decade, he worked on the development of the Shuttle and supervised many of the Thirty-Five New Guys and the Needless Nineteen in their technical assignments. His obsessive workaholic nature made him a somewhat comical figure.

"He was pretty goofy," recalled James 'Ox' van Hoften in his oral history. "He would keep his little computer and he would sit there and type up lists. Most of them were 'busywork', but we always used to laugh that your list just got *longer*; it *never* got shorter." Dave Leestma added that Mattingly's fanatical attention to detail extended to his so-called 'green zingers' – action items, penned in green ink – which would assure everyone of *who* they were working for. Yet, despite admitting that Mattingly's style "drove me nuts", George 'Pinky' Nelson and others respected his thoroughness. "T.K. Mattingly is probably the most technically capable person who has ever been an astronaut," said Nelson, "just in terms of his capacity to stuff things between his ears. He knew absolutely *everything* and *had* to know everything ... but, boy, he was a hard taskmaster. He just didn't see the *forest*, but he saw every *tree!*"

Unfortunately, the mission quickly ran into problems when the Inertial Upper Stage booster, built by Boeing for the Air Force, failed to properly inject the first Tracking and Data Relay Satellite into geostationary orbit. Mattingly's mission was manifested to use the same type of rocket stage. The flight hung in limbo whilst an investigation board pored over the failure, made its recommendations and Boeing spent a year correcting the problems and certifying the booster. By November 1983, Mattingly's flight had been redesignated as STS-41E and rescheduled for July of the following year, but within a few months it was delayed yet again. When NASA issued an updated manifest in May 1984, the mission had vanished entirely and Mattingly's crew were reassigned to 51C, still with Challenger and scheduled for December of that same year. "That," said Loren Shriver later, "is when we started to learn that the numerical sequence of the numbers of the missions ... didn't mean a lot." For a time, Shriver wondered if he would *ever* fly, but unlike other missions, whose PAM-D payloads were very much interchangeable, *they* were a DoD crew. "You were kind of *linked* to it," he recalled, "as long as there was some thought that it was going to happen, and it never did completely go away. It just went kind of *inactive* for a while, then came back as 51C."

Loren James Shriver was born in Jefferson, Iowa, on 23 September 1944. Unlike many of his peers, who either grew up with military parents or dreamed of aviation and the prospect of travelling into space, Shriver had no such inspiration as a child. "When I was a young boy," he told the oral historian, "we lived on a farm in Iowa ... so I often wonder how I got started in wanting to be a pilot." Not until he was in his mid-teens did the Space Age commence and, even then, it did not drive his ambition. Nevertheless, an interest in engineering and aviation developed out of nowhere at high school. Despite having been accepted to study at Iowa State University, the problem of paying for a college education led him to the Air Force Academy. He initially went through 'prep school' and underwent training at Lackland Air Force Base in San Antonio, Texas, before entering the academy in 1963. "The Air Force Academy can't grant a master's degree," he recalled. "It's only authorised to give bachelor's-level degrees, but I was among a group of several guys that took advance courses and they were credited by Purdue University." Consequently, after receiving an undergraduate credential in aeronautical engineering from the Air Force Academy in 1967, Shriver worked feverishly to complete the remainder of his master's in astronautical engineering at the civilian university. By January of the following year, master's degree in hand, he left Purdue and entered pilot training at Vance Air Force Base in Enid, Oklahoma. After four years there as a T-38 flight instructor, in 1973 he qualified in the F-4 Phantom II and was sent to Thailand. The prospect of becoming an astronaut had still not figured on Shriver's radar; by his own admission, he was simply following areas of aviation which fascinated him, although test pilot school beckoned and he spent much of 1975 at Edwards Air Force Base. He later served on the joint test force for the new F-15 Eagle fighter. When the NASA call for astronauts materialised, around 60 percent of the pilots at Edwards applied. His wife was none too keen, but eventually agreed and he submitted his application. Shriver himself was fascinated by the notion that this winged spacecraft was billed as being capable of launching into orbit on a *weekly* basis. He was called to Houston in late September 1977, part of the fourth interview group, which also included Rick Hauck and Dick Scobee, and remembered being somewhat bemused the following January when George Abbey informed him, matter-of-factly, of his selection. Five years later, when he was assigned to the classified STS-10 mission, Shriver was not surprised that his crewmates were all active-duty military officers. "I think NASA believed that it didn't *have* to do that," he recalled, "but I think it also believed that things would probably go a lot smoother if they did."

Flying a classified mission posed its own problems for Mattingly. Within NASA, and privy to the complex business of space exploration for many years, he had become familiar with the process and practice of sharing information, particularly regarding the Shuttle. With a Department of Defense payload, the crew could not publicly discuss the particulars of their flight and the exact details were made available to only a handful of engineers, technicians and Air Force managers. "I had some apprehension," Mattingly said, "about could we keep the exchange of information timely and clear in this small community when everybody around us is telling anything they want and *we're* keeping these secrets. Security was the challenge of the mission." Cipher locks were placed on training materials, "but then you had to give the code to a *thousand people*, so you could go to work!" They were given a classified meeting room in the astronaut office, a classified safe for their documents ... and a classified phone, with an unlisted number. In the entire span of their training time together, the phone rang *once*. It was a *sales* call, asking Mattingly if he wanted to buy a new long-distance service! The ridiculous levels of secrecy became even more laughable at other times, particularly when the astronauts were obliged to 'disguise' the places where they were doing their training. They would file T-38 flight plans to Denver, then file new ones to the San Francisco Bay area, then rent a car to eventually reach their military destination at Sunnyvale in California. They were asked to do their mission training during the *daytime* and at *night*, to keep the launch time secret from prying eyes, or anyone who could be bothered to put two and two together, but all this furore never convinced Mattingly than anyone really *cared*. On one occasion, their office secretary booked motel rooms for them – 'secretly', of course – but the four astronauts, crammed into a decrepit old rental car, with Ellison Onizuka at the wheel, had a surprise when they arrived. Jim Buchli spotted it first.

"Stop here," he said. "Now, let's go over this one more time. We made *extra* stops to make sure that we wouldn't come here directly ... and they *can't* trace our flight plan. We *didn't* tell our families. We didn't tell *anyone* where we were. And we can't tell *anyone* who we're visiting. Look at that."

Four sets of eyes peered over towards their 'secret' motel ... and beheld an enormous banner, emblazoned with the legend: *WELCOME, STS-51C ASTRO-NAUTS*. "How's *that* for security?" chuckled Mattingly.

When Challenger returned from 41G in October 1984, she was scheduled to be relaunched on 8 December for 51C, but the post-flight inspections had revealed that almost five thousand of the delicate thermal protection tiles had become debonded during re-entry. One tile, located in the vicinity of the left-hand wing chine, had completely separated from the airframe and, although not a catastrophic problem in itself, revealed a far more worrying issue. A vulcaniser material, known as 'screed', used to smooth metal surfaces under tile bonding materials, had softened to such an extent that its 'holding' qualities were impaired. Subsequent investigation revealed that repeated injections of a tile waterproofing agent called 'sylazane' – coupled with the effects of six high-temperature re-entries – had caused degradation in the bonding material. By the time Challenger flew her next mission, the use of sylazane had been scrapped. In the interim she was reassigned to 51E, scheduled for launch in February 1985, and 51C switched to Discovery with a launch date in late January 1985. Years later, Loren Shriver did not remember any significant mission impact, other than the six-week launch delay, from switching orbiters.

Due to the classified nature of the flight, some Air Force officials did not even want the precise launch date, or even the astronauts' *names*, released to the public. Loren Shriver was not alone in his amazement at this excessive insistence on secrecy. "We weren't going to be able to invite *guests* for the launch in the beginning," he told the NASA oral historian. "This is your lifelong dream and ambition. You're finally an astronaut and you're going to go fly the Space Shuttle and you can't invite *anybody* to come watch ... We finally got them talked into letting us invite ... 30 people, and then maybe some car-pass guests, who could drive out on the causeway

On the Shuttle's first fully classified mission, Discovery thunders aloft at 2:50 pm EST on 24 January 1985. Not until nine minutes before launch did the famous countdown clock start ticking, producing a measure of unexpected excitement for the gathered spectators. A little more than a year after this photograph was taken, the launch would come under the scrutiny of the Rogers Commission and O-ring damage to 51C's Solid Rocket Boosters would be recognised as the worst ever recorded.

... but trying to decide *who*, among *all* of your relatives *and* your wife's relatives, are going to be among the 30 who get to come see the launch, well, it's a career-limiting kind of decision if you make the *wrong* decision. You have part of the family *mad* at you for the *rest* of your life!" Fortunately, Shriver's family and most of his wife's relatives were from Iowa, which was sufficiently distant for many to be unable to make the journey to Florida. Privately, Shriver and his crewmates were worried that their inability to discuss the mission openly might compromise their preparedness and the thoroughness of their training. It must have been an unusual sight to behold the 51C stack, sitting on Pad 39A, with only a select number of military and NASA personnel knowing precisely *when* the launch would take place; in fact, the media had been told to expect liftoff within a three-hour 'block' of time, sometime between 1:00 pm and 4:00 pm EST on 23 January 1985. Freezing weather conditions kept Discovery on the ground that afternoon, but the situation seemed to have improved marginally by the following day. For the spectators at KSC, the famous countdown clock – which normally ticks away the final minutes and seconds – showed a blank face and all communications involving the launch controllers and the flight crew were kept quiet. Then, at 2:41 pm EST, the blackout suddenly ended with a statement from the launch commentator:

" *... T-9 minutes and counting. The launch events are now being controlled by the ground launch sequencer ...*"

The remainder of the countdown proceeded normally and Discovery lifted off at 2:50 pm and thundered into the cold blue Florida sky. Ascent was interesting, because communication between the orbiter and Mission Control was kept strictly under wraps, with only the voice of the commentator reading off a string of standard calls pertaining to the performance of the main engines, the fuel cells, the Auxiliary Power Units and the Shuttle's steadily increasing altitude and velocity. No indication was given as to the precise duration of the 51C mission – one report mentioned that NASA would reveal this information a mere 16 hours before the scheduled landing – and, with the exception that the classified payload would be deployed later that day, very few other details were released. Many of the accredited members of the press who were in attendance mocked the effectiveness of the 'secrecy' which enshrouded 51C; one NBC journalist quipped that "a Russian *tourist* on a Florida beach, a hundred miles away, could have called the Kremlin with the *exact* launch time!"

Today, almost three decades later, 51C remains classified, although rumours have emerged over the years that Discovery's crew possibly deployed a spacecraft codenamed 'Magnum' – a signals intelligence satellite, operated by the National Reconnaissance Office for the CIA – which was boosted into near-geostationary orbit by an IUS. Reports have suggested that the TRW-built Magnum weighed somewhere between 2,200-2,700 kg and was notable for its physical size, featuring 100 m wide umbrella-like reflecting dish antennas to collect radio frequency signals from Earth. *Aviation Week* noted that Discovery entered an orbit of 204 × 519 km, inclined 28.45 degrees to the equator, and executed three OMS burns during its first four circuits of the globe. The payload was then deployed during the seventh orbit. Deployment was the responsibility of the entire crew, although *this* crew was unusual in that it included a unique military expert: Major Gary Eugene Payton of the Air

Force, a member of a new cadre of payload specialists, known as manned spaceflight engineers, specifically chosen by the Department of Defense for these classified missions. Born on 20 June 1948 in Rock Island, Illinois, Payton completed high school in his home town and entered Bradley University in Peoria for a year, then transferred to the Air Force Academy in Colorado Springs to study astronautical engineering. He received his degree in 1971 and proceeded to Purdue University for a master's in astronautical and aeronautical engineering, then progressed into flight training at Craig Air Force Base in Alabama in 1973. Payton subsequently worked as an instructor at the base and in 1976 was assigned as a spacecraft test controller at Cape Canaveral Air Force Station in Florida. In February 1980, he was selected by the Air Force as one of its first team of manned spaceflight engineers, a programme with a long and chequered history.

From its earliest conception, the Shuttle was dominated by the ambitions of the Air Force and an assumption had long been made that the Department of Defense would employ the reusable spacecraft to carry many of its classified payloads. A new launch site was being built at Vandenberg Air Force Base in California, for near-polar missions, and efforts also encompassed the design and construction of a dedicated Mission Control, known as the Shuttle Operations and Planning Center (SPOC). However, as the 1970s wore on and DoD budgets withered under Jimmy Carter's Democratic administration, the Air Force opted to delay the SPOC in favour of making modifications to NASA's Johnson Space Center in Houston to support its missions. Parallel plans to permanently assign one orbiter to military objectives and hire a dedicated Air Force astronaut corps to fly the missions were abandoned and it was decided to use personnel already detailed to NASA. "The only opportunity for an Air Force programme," wrote space historian Michael Cassutt, "seemed to be in NASA's new class of payload specialists." It was Air Force Undersecretary Hans Mark – later to become Deputy Administrator of NASA under Jim Beggs – who introduced the new manned spaceflight engineer position in January 1979 and assigned responsibility for its development to Lieutenant-Colonel Robert Christian of Los Angeles Space Division. Early guidelines called for candidates to have between three and ten years' of active military service, to rank between a first lieutenant and a major, to be able to pass NASA's required flight physicals, to hold a degree in engineering or science and to have at least two years' experience in programme acquisition, test and launch support or flight and missile operations. By August, 14 officers had been selected – a dozen from the Air Force and two from the Navy – although two of them declined the invitation and only one was replaced. Consequently, 13 manned spaceflight engineer candidates arrived at Air Force Space Division in El Segundo, California, in February 1980, under Christian's command. Their number included David Vidrine, the naval officer who would later, briefly, be considered for a seat on Mission 41C, as well as Gary Payton and the man who would serve as his 51C backup, Keith Wright.

Their selection was trumpeted by the Air Force as illustrative of the service's bright future in space, although little interest was shown in NASA's offer to invite the 13 candidates to Houston for two years of training and evaluation. "At that time," grumbled one senior officer, "any Air Force guy who went to NASA *never* came back!" The Air Force's rejection led the civilian space agency to close ranks, refusing further assistance for the manned spaceflight engineers and insisting that it had neither chosen them, nor was it able to control them. "I was naïve enough to believe that the payload side would be treated by NASA the same way the Air Force launch people treated us," Gary Payton explained later. "In the world I came from, payload requirements would drive the time of day you launched, the time of year; *everything*. In 1980, NASA was still worried about getting the Shuttle to *fly*, so we were not paid much attention. It was a rude awakening." Some space agency officials felt that the newcomers should be considered as 'engineers', not 'fliers', and should not participate in *any* flight-related training, until they were formally assigned to a Shuttle crew. Frustrations over the excessive secrecy imposed on the Department of Defense missions often boiled over into disputes. Nevertheless, the manned spaceflight engineers proceeded with their duties, working on the development of military payloads, including the Navstar Global Positioning System, the Defense Satellite Communications System and others, and the group completed training in December 1981. By the late summer of the following year, 14 more candidates had been selected, including two women and one black officer, with a broader range of academic credentials, ranging from bioenvironmental research to computer science and weapons engineers to rescue pilots. In June 1982, several classified payloads were carried into orbit aboard STS-4 and several manned spaceflight engineers were involved in the preparation and execution of this mission. Even so, their relationships with NASA astronauts were poor. Ken Mattingly, who commanded STS-4, described them as "sour". At around this time, Gary Payton and Keith Wright were announced as payload specialist candidates for the STS-10 mission and a handful of others – Jeff Detroye, Eric Sundberg, Brett Watterson, Frank Casserino and Daryl Joseph, all from the first MSE group – were assigned to support follow-on flights. In the summer of 1983, Payton was assigned as the prime manned spaceflight engineer on Ken Mattingly's STS-10 crew.

Some sources have speculated over the years that the inclusion of manned spaceflight engineers was a method of preventing the NASA crew – even though it was made up of military officers – from gaining too much knowledge of the classified payload. For his part, Loren Shriver did not see Payton's role in this way; he was very much like any other payload specialist, assigned to the crew to complete his own experiments and tasks. "Gary had a specific purpose," he said, "but I don't think it was to make sure that we didn't learn about what the details of the mission were. As a matter of fact, we all got briefed into the mission and we knew exactly what was going on." Many of their efforts were effectively hamstrung by the failure of the Inertial Upper Stage on STS-6 and, although several payloads were 'dual-configured' and could be launched by either the Shuttle or an expendable Titan booster, it would seem that Magnum was designed specifically for deployment from the orbiter. As a result, it could *not* be cancelled, only moved to the next available Shuttle opportunity. "In any case," Cassutt wrote, "because of Magnum's importance, the DoD exercised its launch-on-demand option, pre-empting the *next* Shuttle-IUS spot on the manifest." According to the January 1984 manifest, Mission 51C was to have been an IUS flight to deploy the second Tracking and Data Relay

Like a menacing bird of prey, Discovery swoops onto the runway at the Kennedy Space Center to conclude one of the quietest Shuttle flights to date.

Satellite (TDRS-B), but by May its slot had been taken by Magnum. The TDRS was moved a couple of months downstream and reassigned to Mission 51E.

If Gary Payton's role was as an 'observer' of Magnum and the ignition of the IUS, then responsibility for the actual deployment of the payload fell to Ellison Shoji Onizuka – the first Asian-American astronaut – and North Dakota-born James Frederick Buchli. Certainly, the deployment itself went perfectly, for Onizuka would soon be assigned to *another* IUS deployment flight, Mission 51L, with a TDRS. Onizuka said in a pre-flight interview for 51L that he was "very familiar" and "very comfortable" with the performance of the IUS, strongly suggesting that the earlier problems with the booster had been overcome by the spring of 1985. (Certainly, changes had been implemented in the nozzle design and four successful altitude-chamber firings were performed.) Onizuka was born on 24 June 1946 in Kealakekua, in the Kona district of western Big Island of Hawaii, the eldest son and second-youngest child of Japanese-American parents. He attended local high schools and studied aerospace engineering at the University of Colorado at Boulder, receiving his degree in June 1969 and his master's only six months later. He entered the Air Force, serving as a flight test engineer and test pilot, and then in August 1974 was accepted into test pilot school. Following graduation in the summer of 1975, he became a squadron test flight engineer at Edwards Air Force Base and was invited to Houston for a week of physiological and psychological evaluation in October 1977. On

Mission 51C, Onizuka became the first Asian-American spacefarer. Seated to his left side aboard Discovery for 51C, and serving as the flight engineer, Jim Buchli was almost exactly a year older, having been born in New Rockford, North Dakota, on 20 June 1945. After attending high school in Fargo, he went to the Naval Academy. On graduating with his degree in aeronautical engineering in 1967, he was commissioned into the Marine Corps. Buchli underwent basic infantry officer training and was despatched to Vietnam as a platoon commander, executive officer and company commander. Returning to the United States in 1969, he completed flight training in Pensacola, Florida, earned his naval aviator's wings and spent two years at Marine Corps air stations in Hawaii and Japan. A tour of duty in Thailand was followed by completion of his master's degree in aeronautical engineering systems from the University of West Florida in 1975 and graduation from the test pilot school in 1977. Buchli arrived in Houston in mid-November of that year for his interview, just two weeks after Onizuka.

To this day, Mission 51C remains the shortest operational flight of the Shuttle; when Discovery touched down at KSC at 4:23 pm EST on 27 January, she chalked up a time of just over three days. It was the final flight for Ken Mattingly, who had already announced his retirement from NASA in July 1984 to return to active duty in the Navy as head of space programmes for the Naval Electronic Systems Command in Virginia. In fact, he took up his new post only two weeks after 51C landed. Years later, Mattingly admitted that only one other mission might have kept him with the civilian space agency. "The only mission that I really thought I could get interested in was the first Vandenberg mission," he told the NASA oral historian, "and [Bob Crippen] was already doing that, so I decided it was probably best to change assignments." It would appear that the Navy originally wanted Mattingly to head up its new Naval Space Command at Dahlgren, Virginia, but the 51C delays meant that he either had to drop the Shuttle flight or lose the assignment. "I wanted to stay and finish the mission," he said, "because we spent so much time on it and it was a particularly good one for me, because those guys [on the crew] were so good."

MANY CHANGES

Two days after touching down in Florida, Ellison Onizuka was named to his second Shuttle flight. He would be a mission specialist aboard Atlantis on 51L, scheduled to deploy the third in NASA's network of Tracking and Data Relay Satellites *and* one of the satellites – Palapa or Westar – retrieved by Rick Hauck's crew. (*Flight International* suggested that Palapa was the most likely candidate.) The mission was scheduled for November 1985. Before the third TDRS could be launched, however, NASA had to place the *second* into orbit. This was the primary objective of 51E, a long-delayed flight with a long-delayed crew. Commander Karol 'Bo' Bobko had seen the woes with the Inertial Upper Stage from a ringside seat, having flown as pilot on STS-6, whose booster failed to properly deliver the first TDRS into geostationary orbit. In September 1983, Bobko was assigned, along with his own pilot, Don Williams, and mission specialists Rhea Seddon, Dave Griggs and Jeff

Hoffman, to deploy a pair of commercial communications satellites (Syncom 4-1 and Anik-C1) and operate the Large Format Camera and test the OAST-2 solar array. Their launch was scheduled for June 1984. Within weeks, all this had changed. In its November 1983 manifest, NASA listed Bobko's crew as 41F, now scheduled for August and carrying the SBS-4, Syncom 4-2 and Telstar-3C satellites, together with the first flight of the Spartan free-flier. Their payload remained reasonably stable for more than six months . . . until the 41D pad abort in June 1984 threw everything into disarray. Mission 41F was cancelled, but on 3 August NASA reassigned Bobko's crew to 51E, scheduled for February 1985. They would retain Anik-C1 and would also deploy TDRS-B. The absence of the Spartan and OAST-2 payloads shortened the duration of their flight from seven days to just four days and it was revealed that they would become a crew of six, picking up a French payload specialist named Patrick Baudry to operate a middeck echocardiograph experiment.

"It was disappointing," Don Williams told the NASA oral historian about their reaction to the delay, "because to go that far and be within three months of flying and then go back to square one was tough." Reassignment to 51E was the saving grace, even though it was scheduled for six months further down the line, but the crew plunged back into training, practicing not only a PAM-D deployment, but also learning the intricacies of the IUS for their NASA payload. Yet the mood remained sombre, and Bobko saw it as his responsibility to keep them together. "I remember Bo had us all over for dinner," Jeff Hoffman told the historian, "because he recognised that he had to do something for morale." In fact, more than one member of the crew has praised Bobko's qualities. "He was our leader and our shepherd and our instructor and a teacher," Williams said, years later. "Bo really knew the systems and the Shuttle backwards and forwards. I had a lot of respect for him . . . as a crew member, but later on as an officer and a person and particularly as a pilot." As 1985 dawned and Discovery landed at the end of her classified mission, 51E was scheduled for launch aboard Challenger on 20 February and the crew were almost ready to head to Florida in the middle of the month. The crew itself had also changed at short notice, with the additional of a *seventh* member, Senator Jake Garn. A Republican from Utah, he was the chair of the appropriations subcommittee responsible for overseeing NASA's budget. The choice of a politician to fly aboard the Shuttle had already caused much ill-feeling, for a number of reasons, not least because NASA already had plans to fly civilians, including journalists and teachers, into space.

The agency argued that the man in control of their budget was an experienced jet pilot in his own right. In fact, Edwin Jacob Garn had been a naval aviator, a long-time Air Force reservist and a member of the Utah Air National Guard and would ultimately accrue more than 10,000 flying hours in military and civilian aircraft. Born on 12 October 1932 in Richfield, Utah, he came from a Mormon heritage and could trace his ancestry back to England, Scotland, Wales, Norway, Denmark *and* Germany, "so I'm more NATO than the NATO Alliance!" His father had been a pilot in the First World War and later worked as a civil engineer and rose to become Utah's first Director of Aeronautics, based in Salt Lake City. Garn learned to fly at the age of 16 and studied business and finance at the University of Utah, graduating in 1955. He joined the Navy and flew transport missions over Vietnam, before becoming Mayor of Salt Lake City in 1971 and being elected to the Senate in 1974. He succeeded Wallace Bennett for Utah and won a second term in November 1980. In the early years of the Shuttle programme, Garn was only half-joking when he threatened to not appropriate "another cent" to NASA if he did not get the chance to fly in space. Not surprisingly, his flight attracted much interest from the voters in his home state – with a 69 percent approval rating – and he admitted that it was important to *fly* something, before agreeing to *vote* for it. (He had already done so during the Air Force's efforts to secure funding for their B-1 bomber.) Others saw Garn's flight quite differently, as a NASA ploy to curry political favour. "If you want to know what NASA is like," wrote Mike Mullane, "you don't ride a Shuttle; you sit in [Mission Control]." In Mullane's mind, gaining a thorough understanding of how the space agency worked, simply by flying the Shuttle, was ludicrous; as nonsensical, in fact, as showing up for a vote in the Senate, without ever having attended any behind-the-scenes meetings or lobbying actions.

However, Garn's inclusion on a Shuttle crew was not entirely unexpected. He had already undergone training in JSC's altitude chamber in January 1985 when his addition to 51E reached the ears of the media in early February, marking "the quickest astronaut selection-to-flight sequence in history", according to *Flight International*. In his oral history, then-Administrator Jim Beggs recalled being "bearded" frequently by Garn into letting him take a payload specialist seat on the Shuttle. Eventually, Beggs agreed. Jeff Hoffman remembered hearing about Garn's assignment as he left the Orbiter Processing Facility with Dave Griggs, following a particularly long day of equipment testing. "We just cracked up," he said. "We thought this is *never* going to happen, this is *crazy*. Then we got home to Houston and, sure enough, not only was it *going* to happen, but he was going to be on *our* flight!"

The reception of Garn by the other astronauts was mixed. Commander Bobko exhibited nothing but praise for the senator's work ethic in those final weeks of training, although he did admit to some uncertainty about how to integrate a new payload specialist into a crew which had already been working together as a close-knit unit for more than a year. George Abbey had approached Bobko initially to ask him what sort of training programme he would need to bring Garn up to speed. "But Jake was a great person," Bobko said. "He had more flying time than *I* did!" Don Williams remembered a somewhat different reaction – "Oh, man, this is *just* what we need, a *Senator* as part of this crew!" – although it was Rhea Seddon who brought them all back to the realisation than Garn would shine a positive spotlight of publicity on their otherwise 'vanilla' mission. "He knew how to be a crew member," Williams concluded. "He knew how to fly airplanes ... He was actually a very down-to-earth individual, even though he was a very powerful individual, one of a hundred of the most powerful people in the country." Perhaps most importantly, for NASA, Garn was a strong supporter of the space programme. In fact, his first words to the crew when Bobko introduced him were: "Call me Jake!"

After a slight delay, on 15 February Challenger was rolled out to Pad 39A. Her twin satellite payloads, which totalled more than 24,200 kg in weight, were installed shortly thereafter and all seemed to be proceeding normally towards a launch which had shifted slightly to 7 March. By this time, the astronauts had already started their pre-flight quarantine. "We spent *one night* in the crew quarters in quarantine," said Williams, "and the *next* morning, we are eating breakfast and we get a call ..." A NASA news release issued on 1 March reported that 51E had been *cancelled*, due to a problem with one of the 24 cells in TDRS-B's battery and the need to attend to "a timing issue" with the satellite. The latter glitch had already been noted with TDRS-1 and initially thought to be little more than a telemetry problem, not an inherent design flaw. By 26 February 1985, however,, the satellite's contractor, TRW Defense and Space Systems Group of Redondo Beach in California, advised NASA that it was sufficiently serious to warrant postponing 51E. With the benefit of hindsight, it is bitterly ironic that such pre-emptive action was taken by the space agency with regard to serious worries about a *payload* and yet no action would be taken on the night of 27 January 1986 with respect to concerns over the safety of a *crew*.

Both satellites and their supporting hardware were promptly removed from Challenger and TDRS-B was transferred to its processing facility for repairs. In the meantime, Challenger was rolled back to the Vehicle Assembly Building on 5 March, destacked from her External Tank and Solid Rocket Boosters and returned to the Orbiter Processing Facility to await a new flight opportunity. "Challenger will be processed as if returning from a successful mission, and will be reconfigured for the third Spacelab flight," *Flight International* told its readers on 9 March. By this time, Discovery was nearing the end of preparations for her own mission, 51D, previously slated for launch on 19 March. Commanded by Dan Brandenstein and featuring McDonnell Douglas engineer Charlie Walker, flying a second time as a payload specialist, together with Hughes engineer Greg Jarvis, this flight was to deploy Syncom 4-3 and retrieve the Long Duration Exposure Facility after nearly a year in orbit. Since the LDEF retrieval required its crew to be rendezvous-trained, it was dropped from the mission and Anik-C1 was shifted over from the cancelled 51E. This prompted some journalists to speculate that Bobko's crew might be kept together for the 'new' 51D mission, now scheduled as a five-day flight in mid-April, since they were trained for Syncom and Anik deployments. "It was a stressful time," Rhea Seddon remembered. "For most of us, it was our first flight and we didn't care *what* they did to us, as long as they *launched* us!" The identities of the flight's two payload specialists were also subject to change. McDonnell Douglas' electrophoresis machine had already been installed into Discovery's middeck on 19 February, so it made sense to keep Charlie Walker on the new mission, leaving one other payload specialist seat. By early April, Patrick Baudry and the NASA contingent of Brandenstein's crew had been officially moved onto 51G in June, which would benefit from a longer duration of seven days for the echocardiograph experiments. Jarvis moved to Mission 51I in August. As for Garn, not surprisingly, he retained the seventh seat on 51D, with NASA citing "his busy Congressional schedule" and the requirement for him to participate in "a priority medical experiment programme". In the words of Mike Mullane, the astronaut office grapevine had it that Garn had not so much *requested* a seat as *told* NASA which flight he would take.

For Charlie Walker, it was his second flight in a little over seven months ... and, had circumstances worked out differently, he might have flown even *sooner*. Within a month of returning from 41D, McDonnell Douglas and NASA agreed to fly the CFES electrophoresis machine *and Walker* on another mission, with all eyes on the 51A flight in November 1984. Walker was concerned about integrating his schedule and procedures with those of the other mission objectives, and attended several meetings with Rick Hauck's crew. Unfortunately, due to continuing problems with CFES, which had returned contaminated samples from 41D, it was decided to postpone its reflight and Walker was assigned to Dan Brandenstein's 51D mission in March 1985 instead. (Whilst Hauck's crew were in quarantine, Walker sent them a telegram, apologising for his absence from their flight and assuring them that he "had something else *important* to do!") Then, only three weeks before he was due to fly, he found himself again reassigned, this time to Bobko's crew.

With the revised mission, the French echocardiograph would move from 51E and eventually fly on 51G to generate two-dimensional images of changes in the astronauts' hearts, lungs and blood vessels during exposure to the microgravity environment. The results would offer insights into the major cardiovascular changes that occur during an astronaut's first 24 hours in space. Most notably, the left side of the heart, which propelled blood through the circulatory system, reached its maximum size, as did the blood volume. The right side of the heart, however, which collects blood returning from the rest of the body, proved to be typically smaller than when imaged on the ground. By the second day, the entire heart had grown smaller, subsequent changes progressed more slowly and the reduction in the left-side volume remained unchanged for at least a week after landing. Based on this data, and that from an American device carried on 51D, investigators concluded that the cardiovascular system adjusted rapidly to fluid shifts and blood volume loss during orbital flight. However, they identified the need for more extensive testing to determine if the decrease in heart volume was associated with any reductions in its performance. Of 51E's other assigned experiments, the most noteworthy in the eyes of the media was a selection of toys, which the crew would have used to demonstrate microgravity on mechanical behaviour. Each member had their own toy: a spinning top and a set of gyroscopes for Bobko, a spring-wound flipping mouse and paddle ball for Williams, a ball and Slinky for Seddon, a yo-yo for Griggs and magnetic marbles for Hoffman. Williams even hoped to do some juggling tricks. The results would be videotaped to become part of a curriculum package for elementary and junior high schools. "Through the proposed filming of simple generic motion toys in the zero-g environment," said Carolyn Sumners of the Houston Museum of Natural Sciences, "students will discover how different mechanical systems work without the constant tug of gravity."

In spite of the disappointment at losing 51E, Bobko found humour in the sheer number of crew patches he and his colleagues had to design, as their orbiter changed from Challenger to Discovery and as their payload specialist complement changed from Baudry alone to Baudry and Garn to Garn and Walker. "Mary Lee used to be the lady that arranged the patches," he told the NASA historian, "and along the top of her office she had different plaques with all the different patches . . . and then you got to a corner, and there were *four* of them, which were *all* for our mission, or its derivatives." The mayhem of flight changes and cancellations during this period is highlighted by these differing patches. Originally, 41F in August 1984 was supposed to have been the 13th Shuttle flight and Bobko's crew incorporated a 13-star Betsy Ross flag as their centrepiece. When 41F was cancelled and renamed 51E (the 16th flight), the flag made little sense, but was retained. Next, when Baudry was added as the mission's sole payload specialist, a small tab had to be attached to the bottom of the patch for his surname. This changed again with the inclusion of Garn in early 1985 and, finally, when 51E was scrapped and redesignated as 51D, yet another patch resulted, substituting the name 'Challenger' for 'Discovery' and *again* modifying the payload specialists' names to their final configuration of Garn and Walker. Additional humour came from the *positioning* of Garn on the crew photographs; whenever he was asked *where* he would like to stand, the politician responded: "I don't care *where* it is ... as long as it's furthest to the *right!*"

Yet as the chaos in the wake of the 51E cancellation was gradually smoothed into a sensible schedule for the rest of the year, NASA's major satellite-launch competitor, the European Arianespace concern, was gleeful. The Shuttle was far from demonstrating itself to be a reliable and routine launch vehicle, and *Flight International* noted that the Europeans were "on the crest of a wave", having successfully launched their new Ariane 3 booster on its maiden voyage in August 1984. Meanwhile, it was revealed in early April that 'spare' Delta expendable rockets would be available to launch Shuttle payloads, if necessary. Another delay to the long-awaited launch of TDRS-B did little to allay the worries of several customers, including West Germany, whose dedicated Spacelab mission was scheduled for October 1985 and was dependent on the presence of *two* fully-functional Tracking and Data Relay Satellites. Publicly, NASA hoped to launch TDRS-B before the middle of the year, but it soon became increasingly clear that the problems would require longer to resolve and there was talk, as early as April, that TDRS-B would instead be reassigned to Mission 51L in November. It was an embarrassing time, particularly as the mission at the centre of the fiasco was carrying perhaps *the* most important senator, with respect to NASA's budget. Moreover, when 51E was cancelled, NASA had succeeded in launching just *one* of the 13 missions it had planned for 1985. NASA Associate Administrator for Space Flight Jesse Moore rationalised that the Shuttle was quite different to Ariane – it was, of course, a *manned* vehicle – and "schedule is a *secondary* priority. Mission safety and success are *top* priority". Whilst Moore's sentiment was undoubtedly sincere and reflected the view of the majority within the agency, this would prove bitterly ironic in 1986 when the Rogers Commission investigated instances of 'cannibalism' of Shuttle parts, a widespread acceptance of critical problems and, in Mike Mullane's words, "the normalisation of deviance". In reality, schedule *was* the driving factor and, as missions continued to fly, seemingly safely, the pressure to ramp up the flight rate would only intensify.

So it was that on 12 April 1985, four years to the day since the first Shuttle launch, the seven astronauts trooped out to Pad 39A, under very murky skies and a light drizzle, to board Discovery. Launch was scheduled for 8:04 am EST, on the opening of a 14-minute 'window', although a second opportunity ran from 8:45 until 9:00 am. "The weather was a little shaky," Don Williams remembered, "because there was an overcast at about 12,000 feet or so. We didn't think we were going to go and we were kind of just chatting around … because we figured we were going to do a scrub turnaround for 24 hours and come out the next day." In fact, Dave Griggs, the flight engineer, had unstrapped and was sitting on the backrest of his seat, talking to Hoffman, Walker and Garn on the middeck. A freighter, the *Ocean Mama*, strayed into the drop zone for the boosters and had to be shooed away by the Coast Guard, causing a delay of almost an hour. As the second window opened with the clock in a planned hold at T-9 minutes, the crew were convinced that the attempt would be scrubbed. Writing in *Flight International* a few days later, Tim Furniss reported that many of the launch controllers were reluctant to go … "until NASA Administrator James Beggs appeared in the firing room". The implication was clear: schedule and managerial pressure was already being acutely felt. All at once, at 8:50 am EST, the call from the NASA test director in the Launch Control Center crackled over the radio: an extension to the window had been granted and the clock was being manually restarted. Quickly, Hoffman strapped Griggs back into his seat and clambered downstairs to his own seat, next to the side hatch. "It was probably T-2 [minutes] before I really knew that I was finally going to go," Hoffman recalled, "and then I was all psychologically set to go. That's a very exciting time. Lots of things are happening. The Shuttle is *really* alive and it's *moving!*"

Jake Garn turned to Charlie Walker to ask what to expect.

"It's gonna be great, Jake," came the reply. "Just stay calm and enjoy it."

Upstairs, from the pilot's seat, Don Williams could see thick grey clouds; intellectually, he knew that they would not normally launch in such conditions, for fear of moisture damage to the orbiter's thermal protection tiles. The launch itself, which came at 8:59 am, was "sort of like a two-minute-long catapult launch". As they burst through the top of the cloud deck, he asked Bobko, half-jokingly, if he should give Mission Control a 'tops report'; the commander responded with a curt: *Shut up and watch your instruments!* From Hoffman's perch, the vibrations were one of the strongest sensations – reaching such intensity at one stage that he was convinced that Discovery's *wings* were about to fall off – but the steady change from blue sky to black and the magical onset of weightlessness were profound. Numerous flights in the KC-135 parabolic aircraft, had attuned his senses to the unusual environment, but *now* he was in a permanent state of free-fall. "That's when it really hit me," he said. "I floated over and I looked out the window … and *there* was the Earth going by. You could see Africa off in the distance. Then I looked in the mirror and there was *me* in space. I just got this big ear-to-ear grin and I just couldn't stop smiling for several hours! It was just such an elation."

Almost anyone who met Jeffrey Alan Hoffman when he was first selected by NASA in January 1978 would have described him as the stereotypical professor: bearded, riding a collapsible bicycle and carrying a lunch pail in hand were three of the attributes noted by Mike Mullane in *Riding Rockets*. "To the very end," he wrote, "Jeff remained an unpolluted scientist." Yet his life had encompassed far more than academia; he was a skydiver, an accomplished mountaineer and a skilled engineer. He came from Brooklyn, New York, where he was born into a Jewish family of physicians on 2 November 1944. "My parents took me all over the place to museums and concerts," he recalled in his NASA oral history, "and among the other places was the Hayden Planetarium." This quickly hooked the young boy on astronomy. He received his schooling in Scarsdale and entered Amherst College in Massachusetts to study astronomy, graduating *summa cum laude* – with highest honours – in 1966. He went to Harvard for his doctorate, which he received in 1971. His research focused on high-energy astrophysics, specifically cosmic gamma rays and X-rays, and he participated in the design, construction, testing and flight operations of a balloon-borne, low-energy gamma ray telescope. Years later, he believed that this probably attracted him to NASA and vice versa. "I did a lot of work with my hands," he said, "building electronics, machining stuff. That probably stood me in good stead with NASA, because when they're selecting astronauts, they want people who know how to work in a lab, who can fix things and build things." He then moved to England for three years to undertake post-doctoral work at the University of Leicester, serving as project scientist for the medium-energy X-ray experiment on the European Exosat mission. Whilst in England, he met his wife, Barbara, and became a father for the first time. The Hoffmans returned to the United States in 1975 so that he could take up a position at MIT as the project scientist in charge of the hard X-ray and gamma ray experiment aboard the first High Energy Astronomy Observatory. "That was probably the most interesting scientific time that I've ever spent, because ... we discovered a new phenomenon called X-ray bursts," he recalled.

By his own admission, Hoffman had been drawn to astronomy and astrophysics through his fascination for space exploration, although the opportunity to do such things himself seemed out of reach. "The early astronauts were all military test pilots," he pointed out. "I was never particularly interested in that career. In fact, I wasn't particularly interested in airplanes, because they didn't go high enough or fast enough. I always liked rocket ships." The chance finally came in October 1977, when he was invited to Houston for a week of interviews and testing; at first, his wife thought he was joking when he told her about the astronaut application. It came as a shock to Hoffman that it would spell the end of his research career in astrophysics. "NASA made it very clear that they were *not* looking for people to come and be research astronomers," he explained. "They were looking for astronauts who had to be generalists, because there were a lot of different things we were going to have to learn how to do." It was disappointing, in a sense, but it marked a change in Hoffman's career.

"The most unusual thing about my application," he continued, was that "I very well could have been the only person who was selected as an astronaut, who admitted in their application to having been convicted of a *crime*." It had happened during his tenure in Leicester, when he and some friends took a converted coastal steamer across the North Sea to explore the Norwegian fjords. Unfortunately, the original captain of the trip cancelled at the last minute and Hoffman – despite lacking the proper certification – stepped in. Upon their return, the coast guard arrested them and charged them a £10 fine, which Hoffman's friend disputed. The case was upheld because not only did Hoffman lack the required certification, to captain a British flagged vessel it was necessary to possess British citizenship. At the time, Hoffman did not have this and the party were convicted. "We actually had to go to Crown Court," he said, "with the wigs and the whole deal." They were fined £250 and when Hoffman came to fill in his NASA application form, he hesitated before deciding to be honest and admit to his offence. Surely, NASA would not delve *that* deeply into his past. They did. Nor could the selection board prevent themselves from making light of the situation. At his interview in Houston, the first greeting from astronaut Joe Kerwin, on the selection panel, as Hoffman walked into the room was: "Here comes the criminal!" This *criminal* look was surely made complete by a beard, which Hoffman quickly needed to remove. "As soon as I got to the altitude chamber in preparation for T-38 flying, it became pretty clear that you can't make a good face seal with a full beard, so off came the beard. My wife *shrieked* when I walked through the door!" (He kept a moustache, however, throughout his astronaut career.)

The Hoffmans grew particularly close to Dave and Karen Griggs, who lived just across the street from them, and the two astronauts often shared T-38 flights. "He was a superb pilot," Hoffman said of Stanley David Griggs. "All the pilots are good, but some are even better than others, and I think he was one of the best." Fellow 51D crewmate Rhea Seddon would label Griggs "a man's man" and, indeed, he came from a background in naval aviation and his stern, moustached countenance would not have been out of place in the portrait of an American Civil War officer. Griggs was born in Portland, Oregon, on 7 September 1939 and after high school entered the Naval Academy. He earned his science degree in 1962 and underwent flight training, receiving his aviator's wings in 1964. He flew the A-4 Skyhawk on three overseas cruises; one to the Mediterranean and two to Vietnam, aboard the aircraft carriers USS *Independence* and USS *Franklin Roosevelt*. He attended Naval Test Pilot School at Patuxent River, Maryland, in 1967 and was later attached to the Flying Qualities and Performance Branch of the Flight Test Division, taking part in the evaluation of various test projects on fighter and attack aircraft. Griggs resigned from active military duty in 1970 and completed a master's degree in administration from George Washington University, but remained a naval reservist, eventually gaining the rank of rear-admiral. With the reserves, he flew the A-4, the A-7 Corsair II and the F-8 Crusader at naval bases in New Orleans in Louisiana and Miramar in California. Having almost 8,000 hours in high-performance jets, an airline pilot's licence, certification as a flight instructor, more than 300 carrier landings and expertise in over 45 different aircraft, Griggs remains one of the most experienced pilots ever chosen for astronaut training.

His relationship with NASA began in July 1970, when he was hired by JSC as a research pilot. Four years later, he assumed new duties as project pilot for the Shuttle Training Aircraft (STA) and participated in the design, development and testing of its systems, which had to be capable of mimicking the orbiter's descent characteristics as closely as practicable. Headship of the STA office came in January 1976 and the following year he applied to become an astronaut. At the time of his selection in January 1978, Griggs was the only civilian pilot candidate amongst the Thirty-Five New Guys. Unusually, Griggs and another pilot from his class, Steve Nagel, were selected to fly their first missions as flight engineers, with the expectation that both would progress to become pilots later in their careers. In Nagel's case, this was indeed what happened, and for a time Griggs also trained to fly as a pilot, but was tragically killed in a vintage aircraft crash in June 1989, only months before he would have flown his second mission.

For now, on 51D, five rookies were able to acclimatise themselves to their surroundings, but it soon became evident that Charlie Walker and Jake Garn had been affected by space sickness. "I could tell that Jake's feeling it about as rough as I am," Walker recalled, "and … this is my second experience and I had anticipated that, yes, I'll probably have the same kind of symptoms and that *was* the case. Jake didn't know what to expect." Both payload specialists kept their movements to a minimum, avoiding so-called 'zinging of the gyros' by moving their heads around too much in the first few minutes of orbital flight, but Garn's reaction to the space environment was one of the most profound ever seen. One flight surgeon later joked that the 'Garn Unit' had been created as a measure of nausea in astronauts. Clearly, Garn experienced severe space sickness and some rumours suggested that he was even incapacitated for a few days. "He has made a mark in the astronaut corps," remembered Bob Stevenson, "because he represents the *maximum* level of space sickness that *anyone* can ever attain … and so the mark of being totally sick and totally incompetent is 'One Garn'. Most guys will get maybe a tenth [of a] Garn."

Meanwhile, the first payload, Canada's Anik-C1, was deployed at 6:38 pm EST, almost ten hours into the mission, and the RMS camera captured the successful firing of its PAM-D motor shortly thereafter. Bobko and Williams performed a separation burn, which also served to slightly raise Discovery's altitude in readiness for the release of the Navy's Syncom 4-3 early the following morning. This did not go well. At 9:58 am on the 13th, the giant satellite rolled out of the payload bay, but as it drifted away its omni-directional antenna failed to unfurl. Its anticipated 'spin-up' manoeuvre to 33 rpm also failed to occur and, 45 minutes after deployment, its solid-fuel perigee kick motor did not fire. "We sort of *looked*," Hoffman recalled. "We moved away, just in case the engines fired, but of course they didn't. Then, already on the ground, they were starting to think about contingencies. They told us to do another burn, because otherwise we would have kept moving further and further away from it."

For now, the crew continued with their other mission objectives. Charlie Walker focused on McDonnell Douglas' electrophoresis machine, which had again been installed in place of the galley on the port-side of the middeck. Endotoxin contamination on its previous voyage, 41D, had rendered the processed proteins unsuitable for testing and stronger sterilising chemicals had been used and cooler operating temperatures imposed for this second flight to retard bacterial growth. Other problems had appeared only days before launch, when the unit started leaking and there had been momentary concern about whether the unit was unusable and whether to scrub Walker from the flight. Fortunately, the leak was fixed on the pad. "In flight, it worked well," Walker recalled. "I was doing a lot of testing of some design changes in the apparatus and in the software that controlled it, as well as conducting the pro-forma separation and purification of some biological materials over … almost a hundred hours of operation." During 51D, he also participated in a pioneering research effort known as Protein Crystal Growth, an experiment provided by two bioscientists, Charlie Bugg and Larry DeLucas, from the University of Alabama at Birmingham. He was assisted on the electrophoresis machine by Rhea Seddon, who also took the lead in the American echocardiograph experiment, with the two payload specialists *and* Hoffman and Griggs as her subjects. Garn's involvement in the medical studies had actually begun before launch, because he rode into orbit wearing a waist belt, fitted with two stethoscope microphones, and a plethysmography stocking to measure his leg volume.

It is not surprising that Margaret Rhea Seddon was assigned responsibility for 51D's medical investigations, for she was a qualified physician and surgeon and had already been named as a crew member on the Spacelab-4 flight, a dedicated life sciences mission. She was born in Murfreesboro in Tennessee on 8 November 1947, the daughter of an attorney, and developed an early love for the sciences, graduating from Central High School and entering the University of California at Berkeley to study physiology. At Berkeley, she first encountered the Free Speech Movement and became aware for the first time that careers previously barred to women – medicine and aviation, for example – were within reach. Seddon received her degree in 1970 and was accepted into the University of Tennessee College of Medicine. "I was pretty sure when I started medical school that I wanted to be a surgeon of some sort," she told the NASA oral historian. In 1973, having gained her medical degree, she completed a surgical internship at the University of Tennessee and became the only woman on their three-year general surgery residency programme. "I think I was probably the *second* woman they had *ever* accepted," she recalled, "so that was interesting!" During this period, her research interests expanded to cover the nutrition of surgical patients – the technologies needed to feed intravenously were still in their infancy – and it was this mix of skills that Seddon believes attracted her to NASA. She did not believe that she stood a chance of making the cut. "They were hiring people for July 1st 1978," she said, "and I was finishing my residency June 30th." If it did not work out, she told herself, she would return to Plan B: a surgical subspecialty or a PhD in nutrition. As events unfolded, Plan A succeeded and she was selected by NASA in January 1978. She continued working part-time in hospitals in the Houston area and married fellow astronaut Hoot Gibson in May 1981. Their first child, Paul, was born in July of the following year, becoming the first child born to parents who were both astronauts.

Aboard her first space mission, Seddon spent a considerable portion of her time on the medical experiments, including the American echocardiograph. It was already known that bodily fluids have a tendency to 'redistribute' themselves in weightlessness and space life scientists were certain that the heart swelled. "There was a question about exactly what was the position of the heart in weightlessness," Seddon explained, "because your diaphragm comes up a little bit [and] your chest expands a little bit. Does the heart move? What does the heart look like over time? Does it swell up and get smaller, as might be predicted? It was pretty basic stuff, but I think I brought home good data." From her own perspective as a physician, it was good to be practicing science in space and offered an excellent opportunity to rehearse some of the work scheduled for her Spacelab mission.

Deftly manipulated by Rhea Seddon, the RMS arm, with its hastily conceived 'flyswatter' at the tip of the end effector, edges towards Syncom's deploy switch. The effort to activate the switch, and set off a timer to ignite the satellite's rocket motor, was ultimately fruitless, but a spectacular repair would be undertaken by the crew of Mission 51I, just four months later.

By the third day of the mission, consensus had been reached that an attempt would be made to revive Syncom. The flight was extended by 48 hours to accommodate this effort. The crew had followed their procedures correctly and the only possible cause of the failure was something called a 'deploy switch', which should have popped open as the satellite rolled out of the payload bay and triggered a timed sequence of events, leading up to the ignition of the perigee kick motor. On the ground, the switch was held in position by a piece of foam rubber and it appeared likely that this had not been removed before flight. Fortunately, it was an *external* switch and Mission Control felt that a contingency EVA by Hoffman and Griggs might be able to trip it. The astronauts barely remembered the switch. They had visited the factory, many months earlier, when their 41F mission still carried a Syncom as part of its payload, but did not recall ever having been shown the deploy switch. *That*, however, was only part of the worry now facing Bobko's crew. They were being asked to re-rendezvous with Syncom in order to enable Hoffman and Griggs to conduct their spacewalk and Seddon to use the RMS to trip the switch. Almost a year had passed since the cancellation of 41F, which originally included a rendezvous with the Spartan free-flier. Their mission had long since changed and their rendezvous skills were now rusty. "We had not done a rendezvous simulation ... in seven months or so," Bobko told the oral historian, "and we didn't have the books to do the rendezvous, so they sent us up this long teleprinted message." The message was then cut up and pasted into their unneeded post-insertion checklist to create a makeshift rendezvous book. Furthermore, Hoffman and Griggs *had* done extensive contingency EVA work in support of Spartan (perhaps 50 hours or more) but none of it was recent and none of it was in direct support of 51D. Since an EVA was not planned, no mobile foot restraints were aboard Discovery and efforts by astronauts in the WET-F to find workarounds were unsatisfactory. It would be a tricky task.

"Rendezvous in space is a fairly complicated process," explained Williams. "It's *not* like formation flight, where you just join up with another airplane, because you have to take orbital mechanics into effect and there are several manoeuvres and burns ... that *have* to be done at very precise times in order to keep from either overshooting it or crashing into or missing the thing entirely." Donald Edward Williams was born a farmer's son in Lafayette, Indiana, on 13 February 1942, and throughout his youth, after school and at weekends, he worked the fields or drove tractors or carried out general repairs or cared for the animals. It was whilst outdoors that he first saw an aircraft's contrail crossing the sky, planting the seed of an ambition in his head. He went to Purdue University to study mechanical engineering – "because I really liked playing with the mechanical things on the farm ... I like to build things and I like to work on things" – as part of the Navy's Reserve Officer Training Corps and graduated in 1964. His initial flight training was completed in Florida, Mississippi and Texas. Williams received his naval aviator's wings in 1966, and then moved over to fly the A-4 Skyhawk and undertook two deployments to Vietnam aboard the USS *Enterprise*. Upon his return to the United States, he served as a flight instructor, then transferred to the A-7 Corsair II. "At that time," he said, "the Navy had a rule that after you made two deployments to

Vietnam, you didn't have to go back. Unfortunately, they ran out of pilots, so that rule was changed in the course of doing business." Williams ultimately completed *four* tours, all aboard the *Enterprise* and, in total, flew 330 combat missions. In 1973, he was assigned first to the Armed Forces Staff College and then to the Naval Test Pilot School, with the latter being 11 months of rigorous mental and physical challenge. "We did half a day of academics and half a day of flying," he told the oral historian, "and they varied from day to day, but the academics were *very* fast-paced." Graduation in 1974 led to work in the carrier suitability branch of the flight test division and headship of the Carrier Systems Branch within the Strike Aircraft Test Directorate. Williams was project pilot for tests of a new nose gear for the Skyhawk and he was required to perform several nose-first landings, "which is *not* recommended," he said, "but *sometimes* happens". With a squadron command as his next possible option, he noticed an advertisement from NASA which would alter his career trajectory markedly.

Seven years later, seated in the cockpit of Discovery, Williams and Bobko were responsible for handling the intricate rendezvous with Syncom. The other members of the crew were hard at work putting together a makeshift 'flyswatter', which Hoffman and Griggs would fix onto the end of the RMS as a way for Seddon to hopefully trip the deploy switch. The ground was unable to uplink images to Discovery's teleprinter – only a typed description – and the astronauts used the plastic covers of procedures books, sections of aluminium swizzle sticks, metal pieces from the interdeck access panel and a quantity of grey duct table to put together the flyswatter. Seddon handled the bone saw to cut up the sticks, whilst Walker vacuumed up the chippings. "It was arts and crafts time," said Seddon. Uncertainty lingered about whether the device would be capable of achieving the proper angles to catch the switch or, indeed, whether it would even be *strong* enough to do so. On the fifth day of the mission, 16 April, the time came to put it to the test. Hoffman and Griggs were suited-up by Bobko and Seddon and packed into the airlock with the flyswatter. "I was the first one out," said Hoffman. "It was just when the Sun was setting, so the whole Shuttle was lit up red and it was just *so* spectacular." The experience felt quite similar to the underwater simulations . . . and *that* brought back *another* memory. Before launch, he and Griggs had told the neutral buoyancy staff that if they *were* called upon to do an EVA, the *astronauts* would pay for the beer. It was a joke, of course, but Hoffman knew that they would have to pay up. After all, it would be worth it.

The two men were outside for a total of three hours and six minutes, fitting the flyswatter onto the end of the RMS, which Seddon then extended towards Syncom's deploy switch. Charlie Walker watched the proceedings with amazement. He had been forced to temporarily shut down his electrophoresis machine when the cabin pressure was reduced in preparation for the spacewalk. "Remembering that *none* of this had been practiced on the ground," he said later, "that this was *all* done just with the skills that the crew had been trained with generically . . . and yet the crew pulled it off expertly." The deploy switch *was* successfully tripped – in fact, Seddon got three good contacts with the flyswatter as the satellite slowly rotated – but, alas, nothing happened. The fix had not worked; obviously, the fault lay not just with the switch, but possibly also with an electronics failure in the satellite itself. Bobko's crew had little option but to leave Syncom in a 'dormant' state in low-Earth orbit, although very soon the possibility of executing a repair on a later mission came to the fore. For Hoffman, who would go on to service the Hubble Space Telescope in 1993, his first EVA was "an extraordinary opportunity". At one point, late in the spacewalk, he had little to do except gaze out at the Universe and the Home Planet and watch as the Sun peeked above the horizon to yield a new orbital morning. It was also a different experience for Dave Griggs. "He was a *pilot*," said Hoffman, "but he had been assigned to fly as a *mission specialist*, so I think he was a little bit bummed-out." Returning to the cabin, Griggs was forced to acquiesce that, yes, life as a mission specialist, on *this* occasion, was not all *that* bad.

There was something else about the mission that Griggs enjoyed: weightlessness and the feeling of freedom, unhindered by gravity. During the welcome-home speeches following landing, he approached the lectern to give his own comments on the flight. "Ladies and gentleman," he began, "I've *learned* something on this space flight." His audience waited expectantly. Then came the punchline: *"Gravity sucks!"* Everyone laughed. From the mouth of a space traveller, he was right.

VIDE, MATER, SINE MANIBUS

If Charlie Walker's first crew came to be known by its office secretaries as 'the Zoo Crew', then Bobko's team had acquired their own nickname by the time they touched down in Florida on 19 April: 'the Swat Team'. Although the flyswatter had not succeeded in reviving Syncom, it had been an outstanding example of the NASA can-do spirit. "The flyswatter became the symbol of our flight," said Rhea Seddon, "and when we got off the flight, we were handed flyswatters!" To this day, for his part, Walker still has in his possession a baseball cap, emblazoned with the legend: *Bo's Swat Team*.

With the EVA behind them, their return home loomed on the horizon. During one of their final nights in orbit, Bobko called all seven of them onto the flight deck after dinner to simply float and watch the Home Planet drift by. Jeff Hoffman read a poem which had been written by his brother, a composer, whose verses reminisced on human tendencies for lofty thoughts and the possibilities of love, life and freedom. His brother had given him the poem during his mountain-climbing days and Hoffman now saw many analogies between flying in space and summiting the world's highest peaks. (The memory would remain alive with the crew. More than two decades later, Rhea Seddon would contact Hoffman to ask for a copy of the poem.) Re-entry on the 19th brought its own surprises for Hoffman, since he had exchanged seats with Seddon and was now sitting on the flight deck. "You're surrounded by this red, then orange, then yellow, then white-hot plasma around the front windows," he recalled, as Discovery cut through the atmosphere, heading for Florida. "Behind you, there's this flickering wave, just like the wake behind a motorboat, but it's *fiery* and it's just awe-inspiring." After ten minutes, the spectacle dimmed and Hoffman *felt* his weight returning. He could let go of a pencil and watch it gradually tumble, with the grace of a snowflake, towards the floor, faster and faster as he repeated the game in the lower atmosphere. As the Shuttle headed across Florida the four men on the flight deck could *see* the sprawling expanse of KSC. Touchdown on Runway 33 came at 8:54 am EST, barely five minutes shy of seven full days since leaving Pad 39A. However, a shock was in store.

In their original 41F incarnation, Bobko's team had been scheduled to perform the first 'automatic' landing of the Shuttle, part of a procedure which might be used in a contingency situation. The astronauts had even created a mocking Latin motto for themselves: *Vide, mater, sine manibus* ('Look, Ma, *no hands!*'). To Bobko, it posed the added difficulty of having to define a 'box' of performance during the final approach to the runway, whereby he could recover from the autoland in the event of problems and execute a manual touchdown. "The problem," he recalled, "was how to recognise when the auto system was diverging and not let it get so far that I couldn't take over and make a safe landing." As a result, 41F had been scheduled to land on the vast dry lakebed at Edwards Air Force Base. A year later, with the preference loaded in favour of landing in Florida in order to avoid cross-country transport expenses, no one in NASA wanted to demonstrate autoland on the swamp-fringed Shuttle Landing Facility. Bobko would land manually. Unfortunately, an eight-knot crosswind, gusting to 12 knots, required him to apply the right-hand brake and rudder more than the left to keep the vehicle on the centreline during the long rollout, and this 'differential' braking caused the inboard right brake to 'lock-up', followed shortly thereafter by the outboard one. The result was a burst tyre.

Don Williams remembered the incident vividly. "We're down to maybe just about walking speed," he said, "and there's this big *bang, thump, thump, thump, thump*. I knew right away what it was. We're almost stopped anyway, so it turned out not to be a big deal. Of course, the only thing to worry about is, since this tyre is blown, there could be some debris problems, which might cause a puncture or might cause some reason to have to evacuate, a fire or something like that." On the middeck, Jeff Hoffman's first thought was that one of the RCS tanks had exploded. Charlie Walker wondered if they had run over an alligator! The capcom in Houston quickly radioed confirmation that someone from the runway landing crew had verified a tyre blowout. "We didn't think anything more about it," Walker said, "until we got off the vehicle." A member of the ground support team told them about the blowout, that a trail of debris stretched some distance along the runway and that the astronauts would *not* be allowed to do their traditional walk-around of the vehicle lest another of the fully-pressurised tyres might blow and cause injury.

Despite the fruitless attempt to revive Syncom, the 51D crew could be justly proud of their accomplishments. Far from being disappointed, Rhea Seddon said years later, "we were really *excited*. We did *everything* we possibly could have done and we had pulled off all that stuff." It did not detract from criticism, levelled at NASA from certain sections of the media, that *another* satellite had been lost. None of the losses so far – Palapa and Westar, the burst IRT balloon, the fiasco with TDRS-1 and now Syncom 4-3 – were directly the fault of the reusable orbiter or her crews, but *Flight International* noted on 4 May that these incidents "focused attention on the Shuttle deployment method". Satellites carried by Ariane, the magazine explained, were injected directly into geostationary transfer orbits, as part of the launch phase, whereas the Shuttle required them to use *two* separate motors for insertion into geostationary orbit, one at perigee and the other at apogee. Certainly, within days of the loss, and still riding the crest of a tremendous wave following the triumphant Palapa and Westar recoveries, NASA had already begun to focus its attention on a possible salvage mission, as early as the late summer of 1985.

"DON'T SCREW THE MONKEYS!"

When Challenger lost her 51E mission, just days before the scheduled launch, her time in flightless purgatory was relatively short. Discovery had almost concluded processing for her own voyage, 51D, and it made sense for her to fly first. Challenger, meanwhile, was reconfigured to support the Spacelab-3 payload on Mission 51B, with an anticipated launch in late April 1985. Unforeseen problems with the development of the Instrument Pointing System (IPS) had led to the Spacelab-2 Verification Test Flight being postponed until *after* Spacelab-3, which utilised the already-proven pressurised module and Multi-Purpose Experiment Support Structure (MPESS). This module and support structure weighed 8,300 kg when they were loaded into her payload bay and on 10 April Challenger returned to the Vehicle Assembly Building to be stacked onto her External Tank and Solid Rocket Boosters. Five days later, she was 'hard-down' on Pad 39A. It was a remarkable month: Discovery had vacated the same pad with the 51D crew only *three days* earlier and, with Challenger's own liftoff on 29 April, a new record of just two weeks between launches would be set. At face value, this achievement was trumpeted by NASA as evidence that the Shuttle could indeed support 'routine', fortnightly launches. Not until the following spring, and the lengthy Rogers Commission, would the grim reality of mismanagement and schedule pressure and poor oversight of quality control be exposed. Nor was 51B immune to that reality. Little did its crew know at the time, but their own ascent to orbit brought them within milliseconds of tragedy.

Mission 51B was to usher in an era of 'routine' Spacelab flights. Planned to run for seven days and support 15 investigations provided by American, European and Indian researchers, Challenger would circle the globe with her tail pointing towards Earth and her right wing in the direction of travel, to ensure a stable microgravity environment and minimal number of thruster firings. It was recognised that particularly vibration-sensitive materials science or fluid physics experiments could be adversely affected by periodic bursts from the RCS jets. This 'gravity gradient' attitude sidestepped this concern. Preparatory work in support of Spacelab-3 commenced in December 1983, when the module returned to KSC following STS-9. Over a period of several weeks, its racks were removed and the few modifications required for its second flight were made. The roof-mounted Scientific Window Adaptor Assembly (SWAA) would not be needed and so was removed and covered with an aluminium panel, whereas the Scientific Airlock (SAL) was retained to house a French very-wide-field camera. In March 1984, a mission sequence test verified the compatibility of the 15 experiments with each other and with simulated Shuttle subsystems. Each of these tasks was performed inside the Operations and Checkout Building and, unlike many of Challenger's previous cargoes, the Spacelab-3 facility was to be installed into her payload bay in a horizontal position, rather than in a vertical configuration out at the launch pad. With each of the experiment racks loaded into the Spacelab module and two additional sensors – the Atmospheric Trace Molecule Spectroscopy (ATMOS) and the Studies of the Ionisation of Solar and Galactic Cosmic Ray Heavy Nuclei ('Ions'), attached to the MPESS platform – the complete payload was moved to the Orbiter Processing Facility on 27 March 1985 and installed aboard Challenger.

Twenty hours before Challenger's scheduled launch, two nameless male squirrel monkeys (*Saimiri sciureus*), described by mission specialist Don Lind as "cute", and 24 "not so cute" male albino rats (*Rattus norvegicus*) were loaded aboard Spacelab-3. Animal welfare concerns, coupled with the requirement to move the primates and rodents during their 'awake' time in order to avoid causing them undue stress, made it important to wait until the final part of the countdown before loading them into their cages. It proved an interesting event, worthy of comment, particularly as the Shuttle was oriented vertically on Pad 39A. Working from Challenger's middeck, two technicians were gently lowered, one at a time, in sling-like seats down the tunnel into the module. One stayed in the joggle section, while the other entered the laboratory to await the cages, which were lowered on separate slings. The delicate, two-hour procedure was problem-free and the cages were installed into dual Research Animal Holding Facilities (RAHFs) on the module's port side wall. The monkeys occupied single Rack Five, while the rats lived in double Rack Seven. Developed by NASA's Ames Research Center of Mountain View, California, the facility was originally intended to be carried in a middeck locker and transferred to the Spacelab module in orbit. However, as its design progressed, it became clear that moving the bulky unit down the connecting tunnel in space would prove difficult; nor could it be mounted easily in the module's centre aisle. Ultimately, it was decided to install individual cages in the rack-mounted RAHF whilst the Shuttle sat vertically on the launch pad, which meant their animal occupants would be resting on the cage 'sides' for launch.

Spacelab-3's primary focus was on microgravity research, specifically fluid physics and crystal growth, but an additional life sciences thrust evaluated how well the RAHF could support animals in an environment comparable to a ground-based vivarium. It had long been recognised that effective studies of primate or rodent behaviour in space was impossible if their health and well-being were improperly maintained. In addition to the provision of water and food – rice-based bars for the rats, banana pellets for the monkeys – the facility supplied lighting, temperature and humidity control functions. During the course of 51B, Challenger's crew were to work in two 12-hour shifts, with mission specialists Norm Thagard and Bill Thornton – both physicians – assigned to separate teams to keep watch on the animals around-the-clock. Depending upon the RAHF's performance on its maiden flight, NASA hoped to use it again to support several rodent-based experiments on the Spacelab-4 life sciences mission, scheduled for early 1987. Also under test was a Dynamic Environment Measuring System (DEMS) to record the acceleration, vibration and noise in the cages during ascent and re-entry, and a Biotelemetry System (BTS) to transmit physiological data to the ground from a series of implanted sensors. "The squirrel monkeys adapted very quickly," said Lind. "They had been on centrifuges and vibration tables, so they knew what the feeling of space was going to be like. Squirrel monkeys have a very long tail and if they get excited, they wrap the tail around themselves and hang onto the tip. If they get really excited, they chew on the end of their own tail! By the time we got into the laboratory, about three hours after liftoff, they were adjusted. They had, during liftoff, apparently chewed off a quarter of an inch of the end of their tails!" Both monkeys were free of various specified pathogens and it was mandated that six months before launch they must also be free of antibodies to the *Herpes saimiri* virus. Although the virus was not known to cause disease in either squirrel monkey or human carriers, problems had been documented in other species and a global search found five *Herpes saimiri*-free primates. Due to time limitations, NASA only had the opportunity to prepare two of them for microgravity exposure and properly train them to reach the food pellets and activate the water taps in their cages.

The possibility, however remote, of all seven men becoming infected by herpes was hungrily pounced upon by their peers in Houston, according to Mike Mullane. Several Navy astronauts suggested that as long as the Marine Corps and Air Force members of the crew – a none-too-subtle jab at the respective military services of commander Bob Overmyer and pilot Fred Gregory – did not "screw the monkeys", they would be fine. Overmyer was making his second flight into space, having served as pilot on STS-5, whilst Frederick Drew Gregory, an Air Force colonel, was one of only three African-American astronauts chosen in January 1978. Gregory came from Washington, DC, where he was born on 7 January 1941. His father was an engineer and took every opportunity to expose his son to new experiences, frequently visiting Andrews Air Force Base in Maryland. "In the late '40s or early '50s, they had sports car racing at Andrews," Gregory told the NASA oral historian. "They would use the taxiway and the runways for these car races. He would always position himself and me across from a hangar and there would always be airplanes ... and though the object was to watch the sports car racing, you couldn't avoid seeing the airplanes in the background." Such sights were undoubtedly influential in the young boy's maturing mind, as were several of his father's friends, who had been members of the Tuskegee Airmen, the first all-black aviation unit in the Second World War. By the time he was in his mid-teens, Gregory was aware of the link between aviation and the military and decided that it would form the basis of his future career. When he met Barbara Archer (later to become his wife of more than four decades), Gregory took her on a first date ... to Andrews Air Force Base to watch an air show. "She was either very patient," he said with a laugh, "or in fact had those same kinds of motivations."

Gregory entered the Air Force Academy to study for his science degree, becoming one of only a handful of African-Americans ever to be admitted to the prestigious institution. (Three others, Charles Bush, Isaac Payne and Roger Sims, were admitted shortly prior to this.) "These were high-quality people," he reflected. "These were *not* tokens. They weren't brought in just to change the colour of the Academy; they were brought in because they were absolutely equal to the other members of the class." Gregory noted that, with segregation of black and white still widespread across the United States, the *military* actually seemed more open to full racial integration than many other sectors of society, having taken serious steps towards this end as early as 1947, the year the Air Force became an independent service. He received his bachelor's degree in 1964 and began training as a helicopter pilot at Stead Air Force Base in Nevada, receiving his wings the following year. In June 1966, he undertook his "very fulfilling" first deployment to North Vietnam, as an H-43 combat rescue pilot, and upon returning to the United States he flew the UH-1F helicopter as a missile support pilot at Whiteman Air Force Base in Missouri. Then, in January 1968, his aviation experience changed direction, when the opportunity presented itself to transfer to fixed-wing aircraft, firstly on the T-38 Talon and later the F-4 Phantom II. In the meantime, he had also been accepted into Naval Test Pilot School and after graduation in June 1971 he was despatched to Wright-Patterson Air Force Base in Ohio, where he tested both fighters and helicopters. His next assignment, in June 1974, was a detail to NASA's Langley Research Center as a research pilot. NASA specifically wanted a test pilot with experience in rotary aircraft *and* fighters and, years later, Gregory would admit that had he *not* chosen helicopter training earlier in his career, "I probably would not be where I am right now, because I would have been just like any other test pilot with a single capability". By 1977, his interest in remaining a research pilot was waning and it was Nichelle Nichols – the African-American actress who had inspired women and blacks to apply for the astronaut programme – who suggested that Gregory apply. His worry was his relative paucity of fighter experience and, briefly, he considered resigning from the Air Force and entering the astronaut corps as a civilian. By this time, he had earned a master's degree in information systems from George Washington University, and in November he was invited to Houston for interview.

As an astronaut, Gregory's experience as a former helicopter pilot made rides with him aboard the T-38 a thrilling experience. "Apparently," wrote Mike Mullane, "helicopter pilots believed they would get a nosebleed if they ever flew above a few feet altitude, or at least I got that impression from flying with Fred ... We would pass over the tops of windmills with just yards of clearance. The only thing that protected us from running into buzzards and hawks was that *they* had sense enough to cruise at higher altitudes." When a power line loomed, Gregory would hop the jet over it. On one occasion, they swooped down into the yawning Rio Grande River Gorge in New Mexico and found themselves looking *up* to see the canyon's rim! "In what is truly a remarkable irony," Mullane concluded, "many years later, Fred was appointed NASA's Associate Administrator for Safety. I guess we *all* eventually grow brains!"

When Gregory was assigned to Spacelab-3 in February 1983, in some quarters it was perceived as something of a disappointment, for science missions were often viewed in a negative light by the military aviators; they involved no glamorous

EVAs, no complex RMS operations and no intricate rendezvous manoeuvres. "Piloting an MMU or operating a robot arm had a lot more sex appeal and generated a lot more personal fulfilment than watching a volt meter on some university professor's experiment," Mike Mullane wrote. "The Untouchables of *our* strange caste system were those mission specialists engaged in Spacelab missions dedicated to life sciences. They collected blood and urine and butchered mice and changed shit filters for primates." Indeed, on 51B poor Bob Overmyer found an unwanted 'gift' floating right underneath his nose as he sat on the flight deck.

Alongside Overmyer and Gregory, the 'Untouchables' included no fewer than *five* scientists: physicians Thagard and Thornton and physicist Don Lind as mission specialists, joined by Dutch-born chemical engineer Lodewijk van den Berg and Shanghai-born physicist Taylor Wang as the payload specialists. Three of these men were intimately involved in several Spacelab-3 experiments as co-investigators: Lind on an auroral imaging study, van den Berg on a vapour crystal growth system and Wang on a drop physics module. In fact, van den Berg was already recognised as an authority on vapour-driven crystal growth methods. He came from Sluiskil in the Netherlands, where he was born on 24 March 1932. He earned a degree in chemical engineering at the Delft University of Technology in 1961, then moved to the United States to continue his studies at the University of Delaware. Van den Berg received his master's degree in 1972 and his PhD in 1975, both in applied sciences, from that institution. He was then hired by the EG&G Corporation, a major defence contractor, in Goleta, California, to work on crystal growth technologies; the sensitivity of this work demanded that van den Berg become a naturalised US citizen and he did so in 1975. Over a period of almost a decade, he worked on the growth of crystals of various chemical compounds and investigated associated defect chemistries and electronic properties, earning international renown as an expert on the growth of mercuric iodide crystals and their application as gamma ray detectors in the nuclear industry. Years later, he would liken the crystal growth process to gardening. Van den Berg worked on the design and development of a vapour crystal growth apparatus to be flown aboard the Shuttle and NASA offered a payload specialist opportunity. As they drew up a list of candidates, van den Berg and his chief, Harold Lamonds, could only come up with seven names, rather than the required eight. Lamonds told van den Berg to add his name, joking that his age, huge spectacles and limited physical strength would probably cause him to be dropped in the first round of the selection process. It didn't. Four candidates were eliminated by the initial screening for scientific competence. He was now down to the final four for a series of physical and mental tests and he and metallurgical engineer Mary Helen Johnston passed with flying colours, whereas two others fell by the wayside due to possible heart issues. In June 1983, the duo began payload specialist training in Houston and in the autumn of the following year, against all the odds, van den Berg was formally announced as the prime candidate.

The finalists for the second payload specialist seat on 51B were both physicists from NASA's Jet Propulsion Laboratory in Pasadena, California: Taylor Gun-Jin Wang and a Vietnamese-born scientist named Eugene Trinh, both of whom had been instrumental in the development of the Drop Dynamics Module (DDM) designed to study the behaviour and mechanics of liquid drops in the microgravity environment. Wang was born on 16 June 1940 in Shanghai, but moved with his family to Taiwan as a boy and received the final part of his schooling in Kaohsiung. He attended The Affiliated Senior High School of National Taiwan Normal University, then travelled to Hong Kong and, later, the United States, where he completed a degree in physics at the University of California at Los Angeles in 1967. His master's credential came the following year and he received his PhD in low-temperature and solid-state physics in 1971. Wang joined the Jet Propulsion Laboratory, part of the California Institute of Technology, in 1972 as a senior scientist, with responsibilities in the inception and development of containerless processing research. He gained US citizenship in 1975 – the same year as Lodewijk van den Berg – and his work on the dynamic behaviour of rotating spheroids quickly drew the attention of NASA. Wang and Trinh were selected as candidates for payload specialist for the Spacelab-3 mission in June 1983. Wang flew 51B and Trinh went on to fly into space aboard the Shuttle in June 1992.

If this crew *was* a crew of Untouchables, then one of their number who had certainly not been touched by the fortune of a flight crew assignment was Don Leslie Lind. In fact, to this day, he retains the unenviable US record for the longest wait from selection to his first mission, having been chosen by NASA in April 1966 and not launching until April 1985: an astonishing 19 years. Lind was born in Midvale, Utah, on 18 May 1930, and has dated the start of his interest in space exploration to the tender age of six years old. "My mother was a school teacher," he told the NASA oral historian, "and she was teaching me to read before I entered first grade. I was practicing on the comics in the Sunday funny paper and I was absolutely fascinated with Buck Rogers and Flash Gordon. I thought *that* was absolutely the most interesting thing I could imagine." After high school, he entered the University of Utah and graduated with a physics degree in 1953, then entered the Navy and was faced with a decision: submarines or flight training. Lind picked the latter and in his opinion, as a practicing Mormon, what seemed to be a fluke went on to change his life. Whilst at Officer Candidate School, his section leader asked him which physical exam he wanted to sign up for and Lind, jokingly, asked to be signed up for flight training. The section leader marked it down in ink. Lind protested that he had only been joking, but the section leader insisted that he had already marked it down and had only one copy. "If he had marked it in *pencil*," Lind recounted in an interview for the Mormon publication, *New Era*, in April 1985, "he would have just erased it. If that guy had had a *pencil* in his hand, instead of a *pen*, I never would have been an astronaut, because flight skills are one of the requirements." As a Mormon, Lind was adamant that God had touched his life at that moment. He spent four years flying high-performance jets, performing carrier landings on the USS *Hancock* ... and volunteering for a unique experiment to take high-altitude photo emulsions of cosmic rays for the University of California at Berkeley. The latter work enabled him to get into Berkeley on a PhD programme in high-energy nuclear physics, which he received in 1964.

"When I got my doctor's degree," Lind remembered, "I wanted to go into space research and Goddard [Space Flight Center in Greenbelt, Maryland] was the most logical place to go. I didn't think I was going to become an astronaut immediately, but getting into NASA was a stepping stone." At Goddard, he participated in experiments to determine the nature and properties of low-energy particles in the magnetosphere and the realm of interplanetary space. He applied unsuccessfully for the third astronaut class in the summer of 1963 for the simple reason that his 850 hours in jets fell short of the minimum requirement of 1,000 hours. "In my warped sense of values," he said, "I thought ... a PhD is certainly worth a hundred and fifty flight hours." It wasn't. Lind appealed and was turned down. He requested a waiver and was also turned down. He tried again for the fourth group – a class of scientists – but his application was rejected again, this time on the basis that he exceeded the maximum age limit ... by 74 *days*. Finally, when the call for the fifth group was issued in the late autumn of 1965, he ticked all of the boxes: he now *had* in excess of 1,000 jet hours, the age limit had been raised and Lind was interviewed, poked and prodded by physicians and had the innermost portions of his mind explored by psychologists and psychiatrists. "It included every medical examination I have *ever* heard of," he told the historian, "and some procedures that I'm sure were first developed for the Spanish Inquisition." Selection in April 1966 was followed by work on what Lind has described as the Golden Age of Space: the Apollo lunar landing programme. "I don't think we've done *anything* since then," he reminisced, wistfully, "that had the sense of destiny, the sense of history, the sense of sheer high adventure, that those first landings did."

With the euphoria of Apollo over, he was assigned to the Skylab branch of the astronaut office, serving as backup pilot for the second and third missions to America's first space station. In the summer of 1973, when the second crew experienced problems with thruster quads on their service module, Lind and Vance Brand came within days of performing the first 'rescue' mission. There was also hope that Brand, Lind and Bill Lenoir might fly a short-duration expedition to Skylab in 1974, but that mission was eventually cancelled. With the first flight of the Shuttle several years away, many of the scientist-astronauts – and, with his PhD, Lind was considered a scientist – were offered the chance to continue their research. Lind went to the Geophysical Institute at the University of Alaska in 1975 and 1976 to continue the work on auroral science that he had begun years earlier at Goddard. Upon his return to Houston, he refocused his attention on preparing the Shuttle for flight. When he was assigned to Spacelab-3 in February 1983, Lind knew that he would be joined by a pair of payload specialists and anticipated launch in September of the following year. At the end of 1983 it had been postponed to November 1984 and by the summer of that year it had not only slipped to January 1985, but it had also been switched to Discovery. Eventually, payload problems and orbiter delays through the autumn of 1984 and into the spring of 1985 returned the crew to their original orbiter, Challenger, with launch targeted for a couple of weeks after the return of Bobko's mission.

As a dual-shift Spacelab flight, the seven-man team were obliged to begin 'sleep-shifting' during their final few days on Earth. On the 'gold' shift were Overmyer, Lind, Thornton and Wang, while their 'silver' counterparts were Gregory, Thagard and van den Berg. "I was responsible for all the support systems that keep the orbiter functioning," said Gregory of his role as shift leader. "Norm and I had respective jobs on board, but we, in essence, were the folks who supported the work of the payload specialists." As the flight engineer, Thagard was technically part of the orbiter crew, but his work tended to cross over with that of the scientists working in the Spacelab module and, as already mentioned, being a medic, one of his main scientific responsibilities was caring for the rodents and primates on his shift. Lind, meanwhile, was in charge of the activation and deactivation of Spacelab-3 and for the bulk of its experiments, one of which had dictated 51B's launch time. Challenger had scarcely an hour available in which to launch on 29 April, with her 'window' to the heavens opening at precisely 12 noon local time. This was calculated to provide the MPESS-mounted ATMOS instrument with the maximum number of viewing opportunities of the atmosphere during 72 orbital sunrises and sunsets. Designed and built at the Jet Propulsion Laboratory, its two main foci were to determine the composition of the upper atmosphere on a global scale and acquire high-resolution, calibrated spectral data in support of future environmental monitoring missions. The programme for the French-built very-wide-field camera, and the ATMOS calibration and observation timeline, had been 'front-loaded' into the first day of the mission. About 18 hours after launch, Overmyer and Gregory would reorient Challenger for almost six full days in a gravity-gradient attitude to provide a suitably quiescent environment for the fluid physics and crystal growth investigations.

With the exception of a hydrogen leak in loading the External Tank with propellants, the countdown proceeded smoothly until 11:56 am, when, four minutes before the launch window opened, a front-end launch processor failed and prevented the liquid oxygen system's replenishment valve and vent hood from closing automatically. The clock was held as the valves were manually repositioned and Challenger's thunderous ascent at 12:02 pm was described by NASA as "nominal". It was not quite entirely nominal, though, because during the Rogers inquiry the following year, Bob Overmyer would discover how close his crew came to death that day. For Fred Gregory, who became one of the last of the Thirty-Five New Guys to fly, the fear – for now – evaporated and gave way to sheer exhilaration. "I was very excited," he recalled. "I think I was probably anxious, but certainly not afraid. It was similar to the simulations, but they left out the five percent, and that was the 'wow'! I remember the feeling inside when the main engines started; how it was almost a non-event. I could hear it and I was aware of it, but I looked out the window and saw the tower move back.

"At least that's what I thought, but then I realised the orbiter was moving forward and then back," Gregory continued, referring to the 'twang' effect of the main engine start sequence, "and when it came back to vertical, that's when those solids ignited and there was no doubt about it: we were going to go someplace pretty fast! I just watched the tower kind of drop down below me and was probably laughing during this timeframe. Since we had trained constantly for failures, I anticipated failures and was somewhat disappointed that there were no failures. That was Challenger and she went uphill, just as sweet as advertised. The sensation of zero-g was like a moment on a roller coaster, when you go over the top and everything just floats. Once we got there, it was business as usual, just as we had practiced and performed on the ground." For

Gregory, Overmyer and Thagard, the first order of business was pulsing their spacecraft's twin OMS engines to position themselves in a 360 km circular path. The orbit was inclined at 57 degrees to the equator to provide greater observation coverage for ATMOS. For Don Lind, who had waited since April 1966 for a space mission, the reality was surprisingly close to the training. "The simulations are spectacularly accurate," said Lind, whose first task was to leave his seat on the flight deck and photograph the just-jettisoned External Tank as it tumbled Earthward. "With the motion-based simulators, you even got some of the visceral sensations, because they can move the machine around and give you the sense of onset of zero-g. You can't hold it indefinitely, but we had flown hundreds of parabolas in the KC-135 aircraft, so we were quite accustomed to those things."

Gregory felt that he was well prepared, "but it took about half a day to adapt to microgravity. The body very quickly adapted to this new environment and it began to change. You could sense it when you were on orbit. You learned that your physical attitude in relation to things that looked familiar to you – like walls and floors – didn't count anymore and you translated floors and ceilings and walls to your head is always 'up' and your feet are always 'down'. That was a subconscious change in your response: it was an adjustment that occurred up there. You also learned that you didn't go fast, that you could get from one place to the other quickly, but you didn't have to do it in a speedy way. The only referencing system that you have are your eyes, so you can look at something and establish it as a reference that you use."

Following launch, the seven astronauts split into their respective 12-hour teams. Very soon, one of the two squirrel monkeys exhibited the same space sickness symptoms – lethargy and loss of appetite, but no observed vomiting – as humans for the first half of the mission, being hand-fed by Thagard and Thornton at one stage, before recovering completely for the final three days. The second monkey displayed no ill effects. The primates proved to be much less active in space than on Earth, although both they and the rodents grew and behaved normally, were free of chronic stress and differed from their 'controls' on Earth only by way of gravity-dependent variables. The monkeys, in particular, were spoiled, too. "I think the environment they had come from was a place where they received a lot of attention," remembered Gregory. "Norm and I would look into the Spacelab and see Bill Thornton attempting to get these monkeys to do things, like touch the little trigger that would release the food pellets. I could tell they expected Bill to do that for them, even though he was outside, looking in. We looked back one time and could see that the roles were kind of reversed and Bill was doing antics on the outside of the cage and the monkeys were watching!" Thornton and Thagard could view the primates through a window in each of their cages, while a perforated opening gave them limited access to the interior. The rodents' enclosures were similar to those of the squirrel monkeys, with the exception that they housed two occupants per cage, separated by a partition. Half of the 24 rats were rapidly-growing, eight-week-old juveniles and the remainder were mature 12-week-old adults. Four of them had been implanted with transmitters three weeks before launch in order to continuously monitor their heart rates, deep body temperatures, muscle activity and other parameters. These readings were transmitted through the

Biotelemetry System. The data actually proved to be of such high quality that it was possible to monitor one of the rats for indications of stress. Neither of the monkeys was outfitted with BTS sensors, although their cages included provision for this to be included on future flights. Typically, implant data was transmitted via a dedicated computer to scientists at the Payload Operations Control Center in Houston. Although the animals were maintained in healthy conditions throughout their seven days in orbit, the rats proved not quite as 'savvy' as the monkeys in terms of their adaptation to microgravity. "They hadn't learned that this was going to last a while," explained Lind, "and, when we got into the laboratory, they were hanging onto the edge of the cage and looking very apprehensive. After the second day, they finally found out if they'd let go of the screen, they wouldn't fall and they probably enjoyed the rest of the mission." In spite of the slowness in adapting to their new environment, the rats showed no obvious signs of space motion sickness, although post-flight dissection and analysis identified a marked loss of muscle mass and an increased fragility of their long bones. Investigators speculated that this was probably caused by the influence of microgravity, rather than the stress of living in the RAHF cages. Nonetheless, all of the animals were recovered in good physical condition, healthy and free of microbiological contaminants. However, the crew returned to Earth with a number of concerns because the animal enclosures leaked crumbs of food, monkey and rodent faeces and unpleasant odours. "The later analysis was that primarily it was food," admitted Gregory, "though there may have been some contaminants in it. Other than interest in watching it being ejected from the holding facility, I think it was just interest. It was a passing issue; not something that would have caused any disruption in the current activities."

On the ground, however, it became a big news story. "One anecdote involved this bit of animal dung that escaped from a cage and made its way from the Spacelab module to the flight deck," Thagard told this author in a March 2006 email correspondence referring to an object that floated past the mission commander. "Bob Overmyer made a comment about it that prompted an editorial page cartoon that appeared in some newspapers. The cartoon depicts a Shuttle astronaut saying to a crewmate words to the effect of: *I'm not upset, I'm just glad we didn't have elephants on board!*" Behind the humour of the incident, such issues needed to be resolved before the RAHF could be declared operational and flown aboard Spacelab-4, a dedicated life sciences mission, scheduled for February 1987. After landing, the rats and monkeys proved to be in good health and good spirits and strikingly calm when handled, although the rats proved to have a lot of dried urine and food powder on their coats. It was believed this had been caused by a variable flow rate in their cages, which prevented some of the urine, faeces and food powder from being properly deposited in their waste collection trays.

Overall, Spacelab-3 demonstrated that the new facility provided a suitable animal habitat. Time was of the essence, however, to resolve the concerns about leaking food, faeces and odours in time for Spacelab-4. NASA hoped to fly at least one RAHF on that mission, housing 24 rodents and transferring them, in space, to a new unit called the General Purpose Workstation (GPWS). This made adequate containment of particulate debris even more crucial. In the wake of the Challenger disaster, the near-three-year downtime enforced on the Shuttle was used by Ames Research Center to modify the animal holding facility and a 12-day 'biocompatibility' test was undertaken in August 1988 to verify a number of adjustments. Its ability to contain debris – in particular, food bar crumbs and faeces – and deal with odours and micro-organisms were identified as key issues. A single-pass auxiliary fan was added to assist the RAHF's environmental control system. Tests in March 1989 confirmed that the main problems experienced on Spacelab-3 had been overcome. When the dedicated research mission eventually flew, under the name of Spacelab Life Sciences (SLS)-1 in June 1991, tests confirmed it could indeed capture crumbs, flecks of rodent hair and faeces (simulated by black-eyed peas) and no noticeable odours or other contaminants were emitted. Moreover, when the SLS-1 crew moved rats from the RAHF to the GPWS, this enabled scientists to observe their behaviour and performance outside their cages.

It should be remembered that Spacelab-3 was a test flight of the RAHF hardware. The main 'operational' focus of the mission was fluid physics and crystal growth and the two payload specialists were chosen as internationally recognised experts in these fields. Taylor Wang operated his own drop dynamics experiment whilst Lodewijk van den Berg focused on the crystal growth investigations. They became the second pair of 'career' scientists to fly as payload specialists aboard the Shuttle when they were chosen by the Spacelab-3 Investigators Working Group in June 1983, along with their respective backups Eugene Trinh and Mary Helen Johnston. All four received two basic types of training, known as 'dependent' and 'independent'. The former was directly associated with the specific Spacelab experiments, supported by their principal investigators, while the latter focused on practical skills required to live and work safely aboard the Shuttle.

"Spacelab was an interesting assignment," Fred Gregory said years later, "because it was a '24/7' assignment. We had two shifts. Bob Overmyer was the commander of a shift and I was the commander of the second shift and, while one shift worked, the other slept. We had enclosed bunks on the middeck of the orbiter and that's where the 'off' shift would sleep, so we never saw them, really. There was a handover period, but once we began working, they were sleeping and we just wouldn't see them. There was a common portion of the training, and that was the ascent and re-entry, so Norm Thagard, myself and Bob Overmyer were always involved in the ascent and landing portion of the training. I'd say 75-80 percent of the training was on ascent and re-entry. The intent was to try to get us three in a kind of mindset like a ballet without music – individual, but co-ordinated activities that resulted in the successful accomplishment of these phases, regardless of the type failures or series of failures that the training team would impose on you. There were two thousand switches and gauges and circuit breakers, any number of which we would involve ourselves with during ascent or re-entry. The intent was for us to learn this so well – understand the system so well – that we could brush through a failure scenario and 'safe' the orbiter in the ascent, such that we could get on orbit and then have time to discuss what the real problem was and correct it." Gregory's words would prove prophetic, for on Challenger's very next mission, 51F in July 1985, the crew would be obliged to do just that.

"Re-entry was a phase that, prior to the Columbia accident, would have been considered the easier part of the training," added Gregory. "In any scenario, you would have a series of failures, but all those failures would allow you to safe it, come home and land." Nonetheless, potential disaster was at the heels of every mission. Only days earlier, as we have seen, Discovery suffered seized brakes and a burst tyre as it completed its rollout at the Kennedy Space Center at the end of the 51D mission. Five days later, NASA announced its intention to change Challenger's prime landing site to Edwards Air Force Base in California. "The decision will provide more safety margin for the Challenger's tyres and brake system," read the news release, "because of the availability of the unrestricted lakebed and the smoother surface. The Spacelab-3 payload will be a heavy return weight for an orbiter. The decision to land at [Edwards] for the next flight will enable engineers to determine what corrective actions are appropriate before returning to KSC for normal end-of-mission landings." Improvements to the brakes during 1985 culminated in a successful nose wheel steering test at Edwards during the 61A landing in November and East Coast touchdowns were expected to resume in January of the following year. However, after the loss of Challenger in January 1986, NASA insisted on using Edwards for safety reasons. As a result, no more orbiters landed at KSC until Atlantis did so at the end of the STS-38 mission in November 1990 and *that* only came about because of unacceptable weather in California.

GRAVITY GRADIENT

There was little time to admire their surroundings, however, for the flight plan took precedence. As Overmyer and Gregory busied themselves with tending Challenger for seven days – and, potentially, up to nine, in the event of weather-related delays to landing – in space, Lind and Thagard set to work opening the hatch to the Spacelab-3 module and activating the first of its research facilities. These included the ATMOS instrument on its MPESS platform in the payload bay. Although only 19 of its planned 72 observations were achieved before its laser was disabled by a power supply failure, each three-minute data-gathering period provided 150 independent spectra of more than 100,000 measurements of atmospheric constituents between the altitudes of 16 km and 280 km. The results proved the presence of five molecules – dinitrogen pentoxide, chlorine nitrate, carbonyl fluoride, methyl chloride and nitric acid – whose existence in the stratosphere had hitherto only been suspected. The instrument's analysis of the lower mesosphere showed it to be considerably more 'active' than previously supposed, with many 'minor' gases typically being split by sunlight to trigger other chemical reactions. The instrument's spectrometer measured changes in the infrared component of sunlight as it passed through the 'limb' of the atmosphere. Since each of the trace gases under scrutiny was known to absorb sunlight at very specific infrared wavelengths, it was possible to determine their presence or absence, concentration and altitude, by identifying which wavelengths had been absorbed from the data. Furthermore, the instrument's sensitivity and ability to detect these trace gases in concentrations of less than one part per billion, meant its data could be exploited reliably to test theoretical models of atmospheric physics and chemistry. Human influence on the atmosphere was one of the primary reasons for the decision to build and employ ATMOS. In the wake of the Challenger disaster and the resumption of missions, the instrument rode aboard three dedicated Earth-watching flights in the early 1990s. On the first, in March 1992, it examined the effects of the previous year's Mount Pinatubo volcanic eruption in the Philippines and detected large amounts of crustal material and sulphur-based aerosols in the stratosphere. Additionally, many of the Spacelab-3 science team's predictions of atmospheric change between the first and second ATMOS missions were vindicated when chloroflurocarbon (CFC) quantities were shown to have increased dramatically and their role in atmospheric photochemistry had become more pronounced. When the two sets of results were compared, they highlighted an increase in inorganic chlorine levels from 2.77 to 3.44 parts per billion, together with a fluorine rise from 0.76 to 1.23 parts per billion; the latter confirmed that the primary source of the increased chlorine level was indeed from industrial CFCs.

Other studies of Earth focused on its aurorae. By examining changes in its form and motion, it was hoped to derive greater insights into the dynamics of our planet's magnetosphere. In the case of Spacelab-3, observations were conducted from much closer range – in low Earth orbit – than had been possible with previous, higher-circling missions. Five hours of video recordings and more than 270 still photographs were acquired in such a fashion that they could be 'overlapped' and viewed stereoscopically. The results included features never seen before, including the first views from beyond the 'sensible' atmosphere of thin, horizontal layers of enhanced aurorae. Previously considered to be rare, these layers were recorded on two of Challenger's three orbital passes over the aurora, thus eliminating earlier suspicions that ground-based observations may have been optical illusions caused by atmospheric refraction. Of the 21 scheduled opportunities for studies, 18 were achieved and 51B marked the first time since Skylab – which orbited at an inclination of only 50 degrees – that such detailed auroral observations had been undertaken. This experiment proved particularly satisfying for Lind, who had proposed it and served as its primary operator. "Before our mission, the aurora had only been photographed by some slow scan photometers," he explained, "which gives you a blurred picture, like trying to take a picture of a waterfall. We found out that there is a different component to the mechanism that creates the aurora, involving microwaves, that was not understood before."

The second time-critical experiment for the first day of 51B was the French very-wide-field camera, which Lind set up in the SAL in the Spacelab module's ceiling for around 12 hours of ultraviolet observations of very young, massive stars at one end of the celestial scale and their ageing counterparts at the opposite end. Such wide-field observations could be more rapidly achieved than by scanning many individual points, and offered the additional advantages of permitting constant comparison with the background sky and 'reference' stars and being easier to interpret. The camera, provided by the Laboratoire d'Astronomie Spatiale in Marseilles, had yielded promising results on Spacelab-1 with 48 exposures of ten astronomical targets, including a superb ultraviolet image of a 'bridge' of hot gas between the

Spectacular view of terrestrial aurorae from Spacelab-3.

Large and Small Magellanic Clouds, which are two satellite galaxies of our own Milky Way. Had it not been for a bent handle on the SAL, it should have duplicated or exceeded this achievement on 51B. Ground controllers examined the airlock and decided that a maintenance procedure by the crew would be inappropriate. This was a pity, because on its initial extension into space the camera acquired its target and took a brief exposure.

Eighteen hours into the mission, as planned, Overmyer and Gregory manoeuvred the Shuttle into her gravity-gradient attitude to support six days of fluid physics and crystal growth research. "I was Laboratory Director," explained Lind, a title that in the post-Challenger era is roughly equivalent to that of 'Payload Commander'. "We had five scientists on the crew: myself, two doctors and two payload specialists, who were visiting scientists." In the Payload Operations Control Center (POCC) in Houston were Mary Helen Johnston and Eugene Trinh and the principal investigators for the individual experiments. These experiments – 12 provided by American scientists, two by European researchers and one by a team of Indian astrophysicists – had been selected for inclusion in the Spacelab-3 payload by a competitive peer review process. After responding to an initial announcement of opportunity and receiving approval, the principal designers for the experiments typically formed an Investigators Working Group, chaired by NASA's mission scientist; in the case of Spacelab-3 this was George Fichtel of the Marshall Space Flight Center. The group then worked with the Shuttle and Spacelab offices to identify the requirements for their experiments, propose candidates for payload specialist positions and help to train the crew members.

Taylor Wang had been a natural candidate, since his studies of the behaviour of rotating and oscillating liquid droplets led to the development of the Drop Dynamics Module. Housed in a double rack on the Spacelab module's starboard side, this offered fluid physicists their first opportunity to levitate and manipulate drops in a microgravity environment. It had already been theoretically demonstrated that space research could lead to advances in materials technology, including glasses, crystals, ceramics and alloys, whose properties exceeded those of their predecessors in terms of overall quality. However, chemical mixtures of some materials were known to be highly reactive to the walls of their processing chambers and contamination levels of a few parts per billion could seriously degrade the final product. The DDM, explained Wang, had potential applications in the development of future 'container-erless' materials processing methods which could significantly reduce such flaws. Certain fluoride glasses – particularly attractive for their infrared transmission properties – could be manufactured in ground-based laboratories, but imperfections introduced by their containers prevented them from attaining their theoretical performance levels. 'Effective' containerless processing, in which acoustic and electromagnetic forces were applied to suspend and manipulate fluid droplets, could only be practically achieved in space: the influence of terrestrial gravity made it impossible to levitate liquids without introducing forces that masked the very phenomenon that physicists were attempting to examine.

For the DDM's first flight, the fluids carried were water and glycerin, but when Wang attempted to activate it during his shift on 30 April, it shorted out and failed. "Not only that, but I was the first person of Chinese descent to fly on the Shuttle," he wrote later, "and the Chinese community had taken a great deal of interest. You don't just represent yourself – you represent your family – and the first thing you learn as a kid is to bring no shame to the family. When I realised my experiment had failed, I could imagine my father telling me, 'What's the matter with you? Can't you even do an experiment right?' I was really in a desperate situation." On the ground, in Houston, Lead Flight Director Gary Coen told the astronauts that it was doubtful that the mission could be extended beyond seven days, since Challenger did not have the additional cryogenic reactant tanks carried by her sister, Columbia. There would be *no* opportunity for time lost on the troublesome experiments. In his memoir, although he did not specifically *name* Wang, Mike Mullane made reference to the incident. "Its failure severely depressed him and he surrendered to episodes of crying," Mullane wrote, "but this was just the beginning of his torture. He turned out to be a cleanliness freak. Living aboard the Shuttle *doesn't* leave its occupants feeling springtime fresh!" In the midst of this discomfort and upset, Wang asked Mission Control for permission to try to repair the DDM and when given the go-ahead he quickly got to work, opening the Spacelab rack, isolating the fault and completely rewiring part of it. Several dramatic photographs, taken by his crewmates, showed Wang's legs sticking out into the module as the DDM rack appeared to completely swallow his upper body. He had already threatened not to return home if NASA refused to allow him to fix the DDM, so it proved fortuitous that his bluff was not called. "I hadn't really figured out how *not* to come back," Wang told a

Smithsonian interviewer years later. "The Asian tradition of honourable suicide – *seppuku* – would have failed, since everything on the Shuttle is designed for safety. The knife on board can't even cut the bread. You could put your head in the oven, but it's really just a food warmer. If you tried to hang yourself with no gravity, you'd just dangle there like an idiot!"

With the facility successfully repaired, there was no time for suicide and Wang worked virtually non-stop to complete almost all of his experiments in the last three days of the flight, assisted by his crewmates. The results confirmed several age-old assumptions about the behaviour of liquids in a microgravity environment, although others proved somewhat unexpected: the 'bifurcation point', for example, when a rotating droplet takes the shape of a dog bone in order to hold itself together, occurred earlier than predicted under certain conditions. Another dog bone returned to a spherical shape and stopped spinning much more rapidly than anticipated, apparently from internal differential rotation. During typical experiment runs, Wang would position freely suspended liquid drops under the influence of its own surface tension and gently manipulate them with acoustic speakers inside the DDM; Challenger's gravity-gradient attitude kept thruster-induced accelerations to a minimum and avoided unnecessary disturbances. After a drop had been observed as 'stable' and spherical, it was set into rotation or oscillation by acoustic torque or modulated radiation pressure force. In spite of its delayed start, the experiment proved highly successful and, seven years later, an improved version was operated by Eugene Trinh on another Spacelab flight in the summer of 1992. Nineteen months after 51B landed, Taylor Wang received NASA's Exceptional Scientific Achievement Medal in recognition of his "contributions to microgravity science and materials processing in space and for his exceptional contributions as Payload specialist on Spacelab-3". Although he would not fly into orbit again, Wang played an important role in the drop dynamics experiments on the two United States Microgravity Laboratory (USML) missions in June 1992 and October 1995. On both flights, a second-generation Drop Physics Module (DPM) employed speakers to assess the response of water, glycerin and silicone oil to external forces and successfully injected droplets of sodium alginate into calcium chloride drops. It was ultimately hoped, Wang said after the USML-2 mission, that such research could lead to improved techniques for employing polymer shells to encapsulate living cells intended for the treatment of hormonal disorders to protect them from immunological attack and to provide timed releases. Instances in which such methods would be useful included the treatment of diabetes, perhaps by injecting a pancreatic cell to secrete insulin into the patient's body. Clearly, the potential applications of Wang's original experiment were far more expansive than materials processing alone.

Elsewhere in Spacelab-3, located its own rack on the port side, close to the module's aft cone, was the Geophysical Fluid Flow Cell (GFFC) experiment, provided by John Hart's team from the Department of Astrophysical, Planetary and Atmospheric Sciences of the University of Colorado at Boulder. This investigation, which flew again on the USML-2 mission, sought to simulate fluid flows and better understand convective processes in terrestrial oceans, and the atmospheres of the Sun and giant gaseous planets, particularly Jupiter. Simulations of atmospheric dynamics were first undertaken in the early 20th century, using oil and water in rotating pan experiments, but since they were cylindrical their effectiveness proved somewhat limited. Super-computers of the 1960s and 1970s offered greater advances by numerical modelling, although even they had severe imperfections. Even in ground-based spherical models, cold fluids flowed 'downhill' and ended the simulation; the only practical method of largely eliminating this effect of terrestrial gravity was to conduct the experiment in space. Before the flight, "there was a question of whether you could get convection patterns and wind distributions that resembled those on a gas giant planet," Hart recalled years later. This question was partially answered through Spacelab-3's research, by creating and observing so-called 'banana cells' – rapidly rotating columns formed as differential heating was increased – which were thought at the time to be a key feature of Jupiter's atmospheric structure. Not all of these phenomena were able to be fully investigated, however, because of time and film limitations, together with an inability to interact on a 'real-time' basis with the experiments. "The first flight of the GFFC was a little like running an experiment in the lab with the lights off," said Fred Leslie, a co-investigator of the device, who operated it as a payload specialist on its second mission, USML-2. "We had no indication how the fluid was responding to the inputs. On the second flight, not only did we have a real-time video camera to observe the flows, but we also had a computer interface through which the crew could interact with the experiment."

Nevertheless, on 51B the facility operated perfectly, completing all of its computer-run scenarios during a period of 84 hours; an additional unscheduled 18 hours' worth of operations were also undertaken, yielding 46,000 images for post-flight analysis. Ten years later, during the 16-day USML-2 mission, the GFFC and Leslie undertook more than 180 hours' worth of experimental runs and revealed that the long-term evolution of convecting flows in slowly rotating spherical shells depends on initial conditions. "Even under the same external conditions, like rotation and heating," said Hart, "small variations in initial conditions can lead to different end states." The 'heart' of Hart's GFFC was a pair of 'hemispheres' – a baseball-sized one, made from nickel-coated stainless steel, mounted inside a larger, transparent one of sapphire – which were both affixed to a turntable. A thin layer of silicone oil filled the gap between the two hemispheres. During typical operations, the temperatures of both hemispheres, together with the rotation speed of the turntable, were minutely adjusted by the experiment's computer, which also introduced thermally driven motions into the oil. This enabled physicists to model fluid flows within the atmospheres of stars and planets.

One of the primary reasons for the success of both the DDM and GFFC was the high quality microgravity environment established by Challenger's gravity-gradient attitude, which was described by NASA as "quite stable and conducive to the performance of delicate experiments in materials science and fluid mechanics". Each of the experiments which required this environment – which also included two crystal growth facilities – were clustered around the Shuttle's centre of mass, roughly from the mid-point to the aft end of the Spacelab-3 module. The first of these crystal growth facilities shared the same Spacelab rack as GFFC and was provided by French researcher Robert Cadoret of the Laboratoire de Cristallographie et de

Physique in Les Cezeaux. His experiment, which also flew aboard Spacelab-1 in November 1983, processed six cartridges of mercury iodide crystal seeds at different pressures for 70 hours at a time, using a two-zone furnace. As with the geophysical flow cell experiment, this facility operated under computer control, with the astronauts monitoring it for problems. Mercury iodide samples were also grown in the Vapour Crystal Growth System (VCGS). This was provided by Wayne Schnepple of the EG&G Corporation and 'grew' crystals at approximately 120 degrees Celsius in a specially designed furnace. In general, the returned mercury iodide crystals – which have considerable practical significance for gamma ray and nuclear radiation detectors – had fewer defects than their terrestrial-grown counterparts. It was also optimistically hoped that space-produced crystals would allow for the construction of such detectors to operate at more ambient temperatures, rather than having to be cooled to near-cryogenic levels. Typically, a crystal the size of a sugar cube was grown from a 'seed' 20 times smaller. The VCGS carefully controlled the growth process at less than 3 mm per day over a period of 104 hours. Its success led to a reflight on the first International Microgravity Laboratory mission (IML-1) in January 1992, confirming the more 'uniform' molecular structure of space-grown crystals over those produced on Earth. Moreover, electronic measurements verified that the IML-1 crystals were more efficient, with better characteristics for use as X-ray or gamma ray detectors. Lodewijk van den Berg, the 'crystallography expert' on the 51B crew, concluded that vapour crystal growth could be effectively employed in space, where higher quality specimens with better electronic properties could be grown.

Two crystals of triglycine sulphate were also produced in the Fluid Experiment System (FES), elsewhere in the Spacelab-3 module, yielding the first 3D laser holograms and video recordings of their growth process in space. Visual observations by the science crew provided invaluable real-time descriptions of the crystals, whose potential applications include detectors for astronomical telescopes, Earth observation cameras, military sensors and infrared monitors. Furthermore, they do not require cryogenic conditions under which to operate and could perform well at ambient room temperatures. On 51B, the crystals were grown by slowly extracting heat at a controlled rate through a seed crystal of triglycine sulphate, suspended on an insulated 'sting' in a saturated solution of the same substance. Variations in liquid density, solution concentration and temperature around the steadily growing crystal were carefully monitored. By extracting heat from the crystal in this manner, it was possible to maintain saturation at its 'growth interface', permitting slow but very uniform processing and a higher degree of perfection than could be achieved on Earth. The astronauts viewed the developing crystals through a microscope and images were relayed directly to the scientists in the POCC. This allowed them to be tracked through each growth stage and scientists were able to make changes to parameters such as temperature in order to adjust the experiment and reduce defects. An improved version of the FES hardware was flown on IML-1, using polystyrene spheres as markers to aid in characterising residual accelerations in the Spacelab module. "Lodewijk van den Berg and I ran the crystal growing experiments, so we would brief each other on what was going on," said Don Lind, who was the Dutch researcher's counterpart on the gold team. "He'd brief me and then he'd go to sleep and when he woke up, I'd brief him on what I'd done during the last shift. That was pretty well worked out ahead of time."

"I don't think there was *competition*," said Fred Gregory of the relationship between the silver and gold teams, "because the two shifts did two different kinds of science. Taylor Wang did a lot of drop dynamics. Lodewijk van den Berg did crystal growing. Each shift had its own area of interest and would pick up any unclosed item from the shift preceding them, but would very quickly transition to the activities on orbit. There were really about four hours a day when there was an interaction between the two. During that time, it would just be a kind of status brief on orbiter problems or issues, any review of notes that had come up from Mission Control or some deviation to the anticipated checklist that we had." For Lind, the first Mormon astronaut, the gravity-gradient attitude provided a unique perspective of his home planet. "For the first two days of the flight, I did not take one single minute away from the timeline to just be a tourist," he recalled, "but, on the third day, I had about ten or 15 minutes with no immediate assignment. I floated down to the flight deck. We were flying in an orientation with the tail always pointed toward the Earth and one wing always pointed forward in the velocity vector. *That* oriented the windows on the flight deck from the zenith to the nadir and from horizon to horizon, so it was like a Cinerama presentation. Both my wife and I are amateur oil painters. The sensation in space is that you are always right side up, no matter how you're positioned – 'up' and 'down' are just meaningless in space! Intellectually, you know you're moving very fast, so that orbital velocity will cancel gravity, but the sensation is that you are stationary and the world is rotating majestically below you." Lind wondered: could he *ever* paint such a scene? The answer came instantly and abruptly into his head: a firm and resounding *No*. There were no paints deep enough to mimic the blues of the great ocean trenches and, looking tangentially through the atmosphere, there were *twenty* or more different layers of varying shades of cobalt and cerulean and ultramarine.

"When you go over the archipelagoes and the atolls in the Pacific and down in the Bermudas, you see the water coming up from the deep trenches and it appears as hundreds of shades of blue and blue-green up to a little white line, which is the surf and another brown line, which is the beach. Nobody will ever paint that. I looked down and was overwhelmed with the sense of beauty. It was so impressive that it brought tears to my eyes. Now, in space, tears don't trickle down your cheeks; that's caused by gravity. In space, they stay in the eye socket and get deeper and deeper and, after a minute or two, I was looking through a half inch of salt water! I had a spiritual feeling, because several scriptures popped into my mind: the nineteenth psalm, 'The heavens declare the glory of God'. One of the Mormon scriptures is, 'If you've seen the corner of heaven, you've seen God moving in his majesty and power'. I thought, 'This must be the way the Lord looks down at the Earth'."

As Lind and the scientists worked inside the bowels of Spacelab, the 'orbiter' crew kept watch on Challenger's systems. In a gravity-gradient attitude, with few thruster firings, this left them with little to do but observe and conduct photography. Fred Gregory found the heavens and Earth fascinating. "You immediately realise you are either a 'dirt person' or a 'space person'," he said. "I ended up being a space person. It was a high inclination orbit, so we went very low in the southern hemisphere and I saw a lot of star formations that I had only heard about and never seen before. "I also saw the *aurora australis*, which is the Southern Lights. If you were a dirt person, you were amazed at how quickly you crossed the ground; how, with great regularity, every 45 minutes, you'd either have daylight or dark – how quickly you crossed the Atlantic Ocean. The sensation that I got initially was that, from space, you can't see discernable borders and you begin to question why people don't like each other, because it looked like just one big neighbourhood down there. The first couple of days, I was a citizen of Washington, DC, but Overmyer was from Cleveland and Don Lind was from Salt Lake City and Norm was from Jacksonville and Lodewijk was the Netherlands and Taylor was Shanghai, so each had their own little location for the first couple of days. After two days, I was from America, and after five days the whole world was our home. You could *see* this sense of ownership and awareness. We had noticed with interest the fires in Brazil and South Africa and the pollution that came from eastern Europe, but it was only with interest. Then, after five or six days, it *was* of concern, because you could see how the particulates from the smokestacks in eastern Europe circled the Earth and how this localised activity had a great effect. When you looked down at South Africa and South America, you became very sensitised to deforestation and how it affected the ecology."

Not only were the astronauts watching countries from space, but India in particular was observing Challenger's seventh mission with interest, for one of the experiments was provided by an astrophysical team from the Tata Institute of Fundamental Research in Bombay (now Mumbai). Led by Sukumar Biswas, it was a study of the ionisation of solar and galactic cosmic ray heavy nuclei and known alternatively as 'Ions' or 'Anuradha'. Like ATMOS, the Indian study was mounted at the rear of the payload bay on the MPESS and examined the composition and intensity of energetic ions from the Sun and galactic sources. It was a refined version of a similar experiment flown aboard Skylab and after Challenger returned to Earth, its data was analysed to identify the cosmic ray ions of carbon, nitrogen, oxygen, neon, calcium and iron and their ionisation states, intensities, energy spectra, arrival times and directions. It initially refused to respond to commands and rotate its detector stack, but a procedure conducted by the crew enabled it to perform normal operations and it completed two-thirds of its planned observations. New data on the ionisation states of solar heavy nuclei was of particular interest in developing a clearer understanding of the acceleration and confinement of energetic nuclei in the Sun. The experiment's detector consisted of stacks of thin sheets of special plastics, such as cellulose nitrate and lexan polycarbonate, which were efficient low-noise receptors for heavy nuclei. It was possible to determine the identity and energy of particles from measurements of the geometry of the tracks and the ranges traversed through the stacks.

Aside from his Earth observations, Gregory had little involvement in the Spacelab research. One of his tasks, however, was to monitor the deployment of two small satellites from a pair of Getaway Special canisters in the payload bay. Unfortunately, only the North Utah Satellite (NUSAT) was actually released into space; the other satellite experienced a battery failure and was rescheduled for another mission in October 1985. Their carriage on 51B, however, marked the first occasion that miniature satellites had ever been deployed from canisters. NUSAT was an air traffic control calibrator, designed to measure antenna patterns for ground-based L-band radars operated in the United States and member countries of the International Civil Aviation Organisation. It was mounted in its canister by means of a V-band clamp and pedestal. At the instant of deployment, at 4:17 pm EST on 29 April, only four hours into the mission, the full-diameter motorised door assembly on top of the canister was sprung open and the satellite was ejected by a compression spring at about 1.1 m per second. The concept for NUSAT originated in 1978, in response to a suggestion by the Federal Aviation Administration in Utah's Salt Lake City to create a means of providing a stimulating educational opportunity for the United States' students and demonstrating a space-based technique for improving the safety of the travelling public. After several years of definition and review, the project finally got underway in 1982. The satellite was built by Morton Thiokol – the Utah-based company also responsible for the Shuttle's Solid Rocket Boosters – and consisted of a 26-sided polyhedron, measuring 48.2 cm in diameter and weighing 520 kg. Its communications payload comprised six antennas, a transmitter receiver and telemetry and tracking command equipment. It also housed photodiodes for attitude control, a probe for potential and electron temperatures and strobe lights. During eight months in orbit, a typical 'day' began with a command sent from Weber State College in Ogden, Utah, to 'code' its on-board processor and enable it to discriminate against all illuminating radars except one selected for calibration. A clock was then started to command NUSAT's six L-band receivers to turn on simultaneously just as it was about to come over the horizon of the selected radar installation. The latter transmitted a unique pulse position code during the calibration interval, which would then permit the satellite to distinguish between its signal and others. After passing below the horizon, its receivers were turned off to await its next tasking. In spite of the successful deployment of NUSAT, the second miniature spacecraft, a Department of Defense payload called the Global Low Orbiting Message Relay (GLOMR), proved a dismal failure due to battery problems. It was retained in its canister, brought home, and rescheduled to fly aboard another Challenger mission later in the year.

It has often been remarked by dual-shift research flights that the only times the entire crew really got together were shortly after launch and just prior to re-entry. "I think on that particular mission, it may have been anticipated that we would prepare a meal and everyone would eat at the same time," said Gregory. "In reality, that's not what actually happened. I called it 'almost grazing'. You would go down and perhaps get a package of beefsteak and heat it and cut it open and eat it. You may stay on the middeck or you may go back up to the flight deck or you would go back into the laboratory and eat as you were doing your other routine duties. The only time I really had a crew in one place eating would have been on some of my later flights, where I spent two Thanksgivings on orbit and all of us had our Thanksgiving meal together with all the food prepared on the trays."

Many of Spacelab-3's results would require months to fully analyse after 51B returned to Earth on 6 May 1985. Their remarkable success would lead to several reflights. However, some scientists have argued that one of the most significant achievements was the mission's contribution to biomedical research, most notably through its studies of the rats' bone and muscle degradation in microgravity. "It is not surprising that it takes astronauts a few days to recover their pre-flight strength and co-ordination after flight," Kenneth Baldwin of the University of California at Irvine said after a life sciences Spacelab mission in the autumn of 1993, "since their muscles are remodelled by microgravity." Moreover, since muscle protein 'turnover' in rats is much more rapid than in humans, a week or two of microgravity exposure in them was roughly equivalent to two months in us. Spacelab-3's biomedical research did not solely focus on the rodents and primates, but also on the astronauts – and upon van den Berg and Wang in particular, who served as 'subjects' for the Autogenic Feedback Training (AFT) experiment, closely monitored by Lind and Thornton. A number of different techniques were used to counteract space motion sickness, including the wearing of electronic monitors to record physiological data such as sweat, pulse, heart and respiration rates. Provided by Patricia Cowings of NASA's Ames Research Center, the experiment provided 'encouraging' results. One of the subjects exhibited a low heart rate and little sweating, which was indicative of a lack of stress, although the other did not fare as well, showed less ability to control physiological responses and experienced one episode of space sickness. Nevertheless, the AFT work did offer clear insights into the effects of crew workload and behavioural responses to environmental stress; 'baseline' information which would prove important when planning future long-term space station missions or shorter, high-productivity Shuttle flights.

With the minor exception of a fluctuating water flow sensor, Spacelab-3 was hailed as a tremendous success and although the pallet-train configuration had yet to undertake its verification flight test, its scientific yield proved more than sufficient to declare the Spacelab 'system' fully operational. In fact, it has been estimated that some 250 million bits of data were obtained in total from the 51B experiments, together with more than three million frames of video footage. For the pilots, ironically, this very success proved almost disappointing. "The only flying would be attitude adjustments," remembered Fred Gregory, "and those are generally keypunched in and then executed. In our training, we would simulate failures where you had to do that manoeuvre by hand, and it was quite possible to do it, but not as efficient as the automatic systems. I don't recall manually flying any of the manoeuvres in orbit and I don't recall Bob Overmyer doing it either. The only time we really put our hand on the stick was in the less-than-the-speed-of-sound descent for landing."

That descent into Edwards Air Force Base in California proved to be among the most dramatic memories of the mission for Gregory. "Though it takes eight and a half minutes to get up to orbit," he said, "it takes more than an hour to re-enter and it feels very similar to an airplane ride. You get an excellent view of the Earth. You're going pretty fast, but you are not aware of it, because you're so high. It's an amazing vehicle, because you always know where you are in altitude and distance from your runway. You know you have a certain amount of energy and so you also know what velocity you're supposed to land, and you watch this amazing vehicle calculate and then compensate and adjust as necessary to put you in a good position to land. We normally allow the automatic system to execute all the manoeuvres for ascent and for re-entry, but as we slow down for landing, it is customary for the commander to actually fly it in, using the typical airplane controls." The de-orbit burn, lasting closely to four and a half minutes, began at 8:04:48 am PST (11:04:48 am in Florida), slowing Challenger sufficiently to drop her out of orbit and set her on course for a touchdown on the west coast of the United States an hour later. After performing a graceful, 193-degree heading alignment circle turn, Overmyer guided the orbiter to a precision landing on Runway 17 at 9:11 am PST (12:11 pm in Florida), slowing to a halt in 59 seconds and a rollout of less than 2,700 m. Post-mission inspections of the Shuttle revealed only superficial damage to her thermal protection tiles. Overmyer and Gregory's apparent ease in setting Challenger down, however, was achieved only following hundreds of practice runs they had undertaken in the Shuttle Training Aircraft (STA) before the mission. "We had participated, in my particular case, in 500 to 700 landing approaches," said Gregory, "and Bob Overmyer, I'm sure, had 400 or 500 more than that! They are flown using the same profile, the same speed, the same sensation of very high sink rates, with a flare about a mile from the end of the runway."

As soon as the residual hypergolic propellants had been drained from the orbiter, the first time-critical items from Spacelab-3, such as data tapes and film, were removed from the module. About three hours after touchdown, the rats and monkeys were removed – the former to be euthanised – and the remaining samples removed by mid-afternoon on 7 May. Once back to Florida, Challenger was ensconced in the Operations and Checkout Building to be readied for her next mission in mid-July. Barely four weeks had elapsed between two Shuttle missions and another was scheduled to be undertaken by Discovery on 12 June. After countless development problems with tiles and main engines, the reusable spacecraft, it seemed, was finally living up to the vision that NASA had promised to Congress in the 1970s: a commercial, reliable, frequently-launched 'space truck'. On her next mission, Challenger would fall victim to the long-feared main engines and then, as euphoric 1985 wore into tragic 1986, Bob Overmyer's crew would finally come face to face with the disaster that 51B very nearly became.

'THE FROG AND THE PRINCE'

The Shuttle hit its stride during the summer of 1985 and experienced a 'Golden Age' of accomplishments, which captivated the interest of the world. Crews rocketed into orbit frequently, deployed satellites, performed experiments and demonstrated American technological prowess. The presence of Senator Jake Garn on the 51D crew, as we have seen, was illustrative of this change, and no fewer than *four* other US politicians had expressed a similar interest in a flight aboard the Shuttle. NASA's fawning response to most of these requests is, in a sense, understandable, for these were the men and women who controlled the purse strings and cast decisive votes on the space budget. A plan for a new space station had been unveiled by President Reagan in the spring of 1984, and to achieve his goal of establishing a permanent human home in orbit, within a decade, NASA would need politicians on its side. The payload specialist programme extended into other areas, too. If a customer's satellite or experiment filled more than a certain amount of the payload bay, NASA offered an accompanying seat into space for one of their representatives. On the next flight, 51G, in June 1985, which carried the French Echocardiograph Experiment (FEE) and an Arabsat communications satellite, it is perhaps with little surprise that Frenchman Patrick Pierre Roger Baudry and Saudi Arabian Sultan bin Salman bin Abdul-Aziz Al-Saud were aboard as payload specialists. Within NASA's astronaut office, opinions were strong, and some cynically labelled 51G as 'The Frog and Prince Flight'. Others had already picked fun at Jake Garn's 'training' by posting a sign-up sheet on the office bulletin board, asking for volunteers to take an *eight-week* course to become a senator! The fury was not misplaced; the professional astronauts had paid their dues through tours of Vietnam, ferocious workloads at test pilot school and years of academic study. With the Republican Garn having flown, a Democrat – Congressman Bill Nelson of Florida – was the next politician due to fly and NASA was already soliciting applications for its Teacher in Space and Journalist in Space programmes. "This was an unhealthy environment," admitted astronaut John Fabian. "We were taking risks that we *shouldn't* have been taking. We were shoving people onto the crews, late in the process, so they were never fully integrated into the operation of the Shuttle, and there was a mentality that we were simply filling another 747 with people and having it take off from Chicago to Los Angeles. This was *not* that kind of vehicle, but that's the way it was being treated at that time."

Of course, that is not to suppose that the two payload specialists on 51G were unqualified. Lieutenant-Colonel Patrick Baudry, certainly, had a long relationship with the space programme, having backed-up fellow Frenchman Jean-Loup Chrétien on a flight with the Soviets a few years earlier. (In fact, for 51G, the two men remained together and exchanged roles, with Chrétien backing-up Baudry.) Baudry was born in Douala, in Cameroon (then a French colony), on 6 March 1946 and attended secondary and military preparatory school at the Prytanée National Militaire in La Flèche. On completing his studies in 1965, he was admitted to the French Air Force Academy – the École de l'Air – from which he graduated two years later with a master's degree in aeronautical engineering. He undertook flight training at Salon-de-Provence and Tours, received his pilot's wings in 1970 and flew the F-100 Super Sabre fighter and Sepecat Jaguar ground-attack aircraft with the French Air Force. Baudry performed several operational missions in Africa and was accepted into the Empire Test Pilots' School in Boscombe Down, Wiltshire, as part of an exchange programme with the Royal Air Force. He then worked at the Flight Test Centre in Brétigny-sur-Orge, participating in flight tests of the Jaguar, the F-8 Crusader and the Dassault Mirage. His first exposure to space came in June 1980, when he and Chrétien were selected by the Centre National d'Études Spatiales (CNES, the French space agency) as candidates for a joint Franco-Soviet mission to the Salyut 7 station. This mission took place in June 1982, and a little more than two years later the Baudry-Chrétien team were despatched to NASA by CNES for a Shuttle flight involving the French echocardiograph. They were attached to Dan Brandenstein's 51D mission, originally scheduled for March 1985, but the delays to the Shuttle fleet in the spring of that year pushed them further into the summer and they were reassigned to 51G.

As for the Saudis, the responsibility of *their* payload specialist was to 'observe' the deployment of Arabsat-1B, a powerful communications satellite to be deployed atop a PAM-D booster. The Arab Satellite Communication Organisation had been established in 1976 by members of the Arab League to serve the telecommunications, information, culture and education sectors and its first satellite, Arabsat-1A, was blasted into geostationary orbit in February 1985. Unfortunately, one of its solar panels failed to extend correctly and, coupled with other problems, it was relegated to a 'backup' status until 1991, when it was finally abandoned. It was therefore with high hopes and much fanfare that the second of three planned Arabsats was installed aboard Discovery for launch. Built by a French-led international team, including Aerospatiale and Messerschmitt-Bölkow-Blohm (MBB), and based on the cube-shaped Spacebus-100 platform, Arabsat measured $1.5 \times 1.6 \times 2.3$ m and weighed around 700 kg at launch. Electrical power for the satellite came from two rectangular solar 'wings', which spanned 21 m and yielded 1.4 kilowatts. It had a pair of S-band transponders and 25 C-band transponders. Attitude control was by a low-thrust motor, fed by hydrazine and nitrogen tetroxide. Yet, according to John Fabian, who flew aboard 51G, Arabsat had failed *all* of its pre-flight safety reviews. "The crew recommended that it not be flown," he told the oral historian, "the flight controllers recommended that it not be flown and the Safety Office recommended that it not be flown, but NASA management decided to fly it." The political embarrassment, not to mention the commercial impact for missions further downstream, was simply too much to bear. (In fact, Fabian would later reveal that NASA's growing laxity in flight safety prompted his departure from the agency.) Shortly before Arabsat was released, telemetry indicated that one of its solar arrays had prematurely unfurled; it was a false reading, thankfully, and turned out to be nothing more than a problem with a microswitch, but required the crew to do an inspection with the RMS camera, nonetheless.

The two payload specialist candidates chosen to oversee the Arabsat-1B deployment were announced in April 1985, just *two months* before launch: Al-Saud was the first member of royalty to be considered for a seat on a space mission, whilst his backup, Abdulmohsen Hamad Al-Bassam, was a fighter pilot in the Royal Saudi Air Force. Al-Bassam must have worked very much in Al-Saud's shadow. It was the first time Al-Bassam had ever met anybody from the royal family, according to 51G pilot John Creighton, and the entire crew was impressed by the entourage surrounding the prince when he arrived in Houston. "It was obvious to all of us," Creighton chuckled, "that Prince Sultan had grown up in different financial circles than the rest of the crew!" Al-Saud was born in Riyadh on 27 June 1956, making him, at 28 years old, the youngest person ever to fly aboard a US spacecraft. He was the son of Salman bin Abdul-Aziz Al-Saud, governor of Riyadh, the nephew of the then-reigning monarch, King Fahd, and the grandson of the first King of Saudi Arabia, Ibn Saud.

At first glance, it would appear that the choice of a member of royalty as the first Saudi spacefarer was little more than an outrageous example of nepotism. Indeed, this very fear had crossed King Fahd's mind when Al-Saud was first proposed as a candidate. The King turned him down, flat. He would *not* allow his nephew to fly in space. "I was really disappointed," Al-Saud told an interviewer from the Smithsonian. "Even if there was only a ten percent chance I'd get to do it, I wanted at least to go through the process. I asked my father to ask again and the King was still hesitant. He thought there might be talk about how the Royal Family had just pushed in their own guy. I appreciated that, but I explained that this was not going to be Saudi Arabia's choice. I wanted to be put to the test by my own merit, so he approved." For the NASA astronauts, having someone of a profoundly different culture – and religion – aboard their mission posed some challenges. Dan Brandenstein was keen for Al-Saud to feel welcome. The Arab-American Oil Company, ARAMCO, had offices in Houston and Brandenstein called a team from their human resources department to give the remainder of his crew an introduction to Saudi customs. He had few concerns about Baudry; he was French and a fighter pilot, and as such they spoke similar languages. But Al-Saud was an unknown quantity.

"They were worried for a while about somebody making a statement for Allah and doing something dumb up there," John Creighton told the NASA oral historian, "so they gave him a psychological profile. They did the same thing to all of us, too. It's a part of the astronaut selection process. The psychiatrist came back and said: 'He's *saner* than most of the rest of the *astronauts*!'" Aside from the religious issue, Brandenstein did not want to make any cultural missteps. "Don't tell any *harem* jokes," Brandenstein told his crew, "or any *camel* jokes." Very soon, he realised that he need not have worried, for Al-Saud had spent much of his adult life in the United States. At his first meeting with the crew, the prince introduced himself with a self-deprecating joke: "I've left my camel *outside*!" The tension was broken. "In fact," Brandenstein told the NASA oral historian, "he was more attuned to the American way than the Frenchman was. A lot of times we'd [tell] ... a subtle-type joke that the Frenchman didn't understand and the Sultan would lean over and explain it to him!"

'Sultan', as he came to be known by the crew, received elementary schooling in Saudi Arabia, then travelled to the United States to study for a degree in mass communications at the University of Denver and a master's in social and political science from the Maxwell School of Syracuse University. Whilst there, he also qualified for his private pilot's licence. Back home in 1982, he became a researcher in the Department of International Communications at Saudi Arabia's Ministry of Information and, two years later, was deputy head of his country's Olympic Information Committee. Immediately prior to commencing 51G training he was acting director of the Ministry of Information's new department of advertising. When the contract was signed with NASA to launch Arabsat-1B aboard the Shuttle, Saudi Arabia won the payload specialist seat from amongst the 22 members of the Arab League and 20 candidates were chosen, all qualified pilots, in exceptionally good health and capable of speaking fluent English. Al-Saud and Al-Bassam arrived in Houston in April 1985 and received 114 hours of 'habitability' training, learning to adapt to the routines of living aboard the Shuttle. Although the national Saudi dress – the flowing *thwb* and *ghutra* – were inappropriate, Al-Saud was permitted to carry traditional foodstuffs with him, including dates from Medina. Both men also trained for a number of Saudi experiments, including one provided by one of Al-Saud's relatives, Prince Turki bin Saud bin Muhammad Al-Saud. The experiment was part of his PhD dissertation at Stanford University and used TV and still cameras to measure the chemical composition of rocket-exhaust gases and their interaction with the ionosphere. Another investigation was devoted to remote sensing of Saudi Arabia itself, as part of efforts to develop new groundwater exploration programmes, further research into sand movement and better define areas of substantial mineral deposits within the kingdom. Al-Saud's final experiment observed the behaviour of water-oil mixtures in microgravity – with samples of Kuwaiti, Algerian *and* Saudi oil diplomatically chosen for the task. All three investigations were conducted under the auspices of the Arabsat Scientific Experiments Team, led by Abdallah Dabbagh, head of the Research Institute of the University of Petroleum and Minerals in Dhahran.

One other 'experiment' which caused some raised eyebrows was labelled simply *Lunar Crescent Observation* or 'LCO'. It was mentioned in the 51G press kit as an attempt "to observe the crescent of the new Moon with the unaided eye from orbiter windows as it becomes visible close to the western horizon" and would occur on either 17 or 18 June 1985. At first glance, this sounded innocuous enough; a request to observe the new Moon. It wasn't. "The LCO was actually religious in nature," wrote Mike Mullane. "The mission was going to occur in the ninth month of the Muslim calendar, the fast of Ramadan. This period of fasting and spiritual contemplation ended at the sighting of the new crescent Moon. Prince Al-Saud just wanted to be a space observer to the end of the fast of Ramadan." It would seem from Mullane's account that NASA approved the LCO task without realising that it was a religious observation, and when Dan Brandenstein found out the truth he was concerned that Al-Saud would be using an American spacecraft as an orbital minaret. "Knowing that he would be at the centre of a shit-storm," continued Mullane, "Brandenstein confronted the prince and made him agree on the exact wording he would use if he discussed the Moon observation on the air-to-ground link, wording devoid of anything religious." As 51G's commander, he had enough to worry about with ensuring that his crew completed all of the mission objectives, without having the added distraction of worrying about what one of the payload specialists might say over the radio. With his flight into space, Al-Saud thus became the first Muslim in space and the first to witness the new Moon from *above* the atmosphere. Two hundred and thirty Saudi VIPs were at the Kennedy Space Center for the launch, including 29 princes. The mission drew much attention in his homeland, some of it demonstrative of the archaic attitudes and principles still prevalent there.

Mike Mullane served as one of the capcoms in Mission Control during 51G and remembered being given a note at one stage of the flight by a Public Affairs official. Amongst the crew was Shannon Lucid, a member of the Thirty-Five New Guys, who had been seen on several occasions in shorts as she moved around the orbiter's cabin.

Sultan Al-Saud is strapped into the back seat of a T-38 Talon by Dan Brandenstein. In the weeks preceding the launch of Mission 51G, Brandenstein was obliged to address Al-Saud's plans for a religious observation and statement from orbit.

The note requested that the entire crew should wear *trousers*, not shorts, during a forthcoming press conference. "When the note came to me, I understood its intent," wrote Mullane. "Public Affairs was concerned that the Arab world might find it offensive for one of their princes to be seen hovering with a woman's naked legs prominently displayed next to him." Mullane threw the note in the bin. In his mind, he was *not* prepared to tell an *American* woman, on an *American* spacecraft, to modify her dress "to accommodate the values of a medieval, repressive society, where women couldn't drive *cars*, let alone fly Space Shuttles."

Despite his concern about the nature of the LCO experiment, Dan Brandenstein had no intention to call for its removal or cancellation; partly because of the possible effect on his career, but also perhaps due to the fact that his mission had changed beyond recognition from the 51A flight to which he, Creighton and mission specialists John Fabian, Steve Nagel and Shannon Lucid had been assigned, way back in November 1983. They were to have been launched in October 1984, deployed Canada's Anik-D2 satellite and operated a payload known as the Materials Science Laboratory installed on a cross-bay MPESS, together with experiments in a 'bridge' of Getaway Special canisters. By August 1984, the flight had changed significantly; redesigned 51D, it was to fly in March 1985 and deploy Syncom 4-3 and retrieve NASA's Long Duration Exposure Facility (LDEF), which added a rendezvous to the training regime. They also acquired Charlie Walker and Hughes engineer Greg Jarvis as payload specialists, becoming a crew of seven. *This* mission was itself scrubbed only three weeks before its scheduled 19 March liftoff, when the cancellation of 51E disrupted the entire Shuttle schedule. "We did our pre-flight press conference about the LDEF," Steve Nagel recalled, "walked back to the astronaut office ... and we didn't have a flight anymore. It was *gone!*" Instead, Brandenstein's crew gained Arabsat-1B, Telstar-3D, Mexico's Morelos-A, received Baudry and inherited the Spartan free-flier from 41F. They were redesignated 51G and were rescheduled for launch on 17 June 1985. Meanwhile, Greg Jarvis, who was flying to observe the deployment of a Hughes-built Syncom, was correspondingly moved to the next Syncom flight slot, 51I in August. Six crew members expanded to seven with the addition of Al-Saud.

Tradition had long since dictated that the evening before launch would be spent with spouses at the deserted 'beach house' on Cape Canaveral's Neptune Beach waterfront, enjoying a barbecue dinner and private time in seclusion, with only the neighbouring launch complexes and the lapping waves of the ocean for company. Even today, the building, officially known as 'The Kennedy Space Center Conference Center', still possesses dusty old wine bottles, emblazoned with crew patches and signed by outgoing astronaut crews. "This cabin is located ... maybe a mile away from the launch pad," explained John Creighton. "Most of the time, when the vehicle sits on the launch pad ... it's covered by the Payload Changeout Room, sort of a cocoon to protect it from the weather. Well, that night before launch, when they rolled that back and they were in the process of fuelling the vehicle, it's illuminated by these bright searchlights ... and you walk up the beach for maybe a half-mile toward the vehicle. You can get a pretty good view of it." Returning to the beach house, the 51G crew and their spouses toasted their flight with a bottle of wine, then parted. The astronauts went back to their quarters, keenly aware that launch was scheduled for early on 17 June and they would be awakened at around two in the morning.

"NASA has everything scripted right down to the *minute*," Creighton recalled, "and they wake you up about four hours and 45 minutes before launch and you ... take a quick shower and get dressed and go in, still half-asleep, into breakfast, and a bunch of photographers run in and take your picture and they bring out a big fancy cake with your patch on it. *Nobody* feels like eating cake at four o'clock in the morning and it's kept and eventually you get a chance to eat it when you celebrate after you get back." Suiting-up came next, followed by the ceremonial walk-out from the Operations and Checkout Building and the bus ride to Pad 39A. Although it was still dark, 800 million candlepower of xenon illuminated Discovery like a torch. Silence reigned across the launch complex, punctuated only by the creaking and groaning of gaseous propellants boiling off from the External Tank. Brandenstein entered the vehicle first, followed in turn by Creighton and Steve Nagel, who took the flight engineer's seat behind and between the two pilots. "Nagel," said John Fabian, as he was being strapped in downstairs on the middeck, "you're in for one *hell* of a ride!"

For Steven Ray Nagel, it *would* be a spectacular ride, but in a quite different *seat* to the one he had expected when he was selected as a pilot candidate by NASA in January 1978. Already, Dave Griggs served as the flight engineer on 51D, despite having also been chosen as a pilot, and much has been written over the years about why certain astronauts were flown in this fashion. Some have seen the practice as favoritism in the selection of Navy pilots over their Air Force counterparts; yet while Nagel *was* Air Force, Griggs was retired Navy. "It was considered a slap in the face," recalled John Fabian. "Here's a guy who's a pilot, just as well-qualified as J.O. Creighton, who is going to fly the [pilot's] seat, just as well-qualified but less experienced than Dan Brandenstein, but he *wasn't* a Navy officer." For own part, Nagel never questioned why he did not receive a pilot slot on his first mission, even though a handful of his contemporaries, including Brandenstein, had not only *already* flown as a pilot, but had progressed to the commander's seat. Privately, Nagel questioned his own abilities: was NASA telling him that he was not good enough to be a pilot? "Nothing against mission specialists," he told the oral historian. "I'd trade my pilot's slot to go be a mission specialist and do an EVA." The most likely rationale seems to be that the 1978 astronaut class was particularly large and there were many more mission specialist places (three per flight) available than pilots and, in order to fly sooner rather than later, NASA management decided to front-load several pilots onto flights as mission specialists. "The *other* way of looking at it," continued Fabian, "was that they were doing Steve a favour. Better to give him a flight, flying the middle seat as a flight engineer, which would mean that he was learning the procedures necessary to fly the ascent and the entry, rather than to keep him sitting on the ground." As the flight engineer, Nagel trained on the same systems as the pilots and, years later, he admitted that it served him well. If nothing else, he gained mission-specific EVA training out of the assignment. His uncertainty was also calmed by words from George Abbey: NASA *would* fly him as a pilot soon ... and he kept his word. In February 1984, Nagel was assigned as pilot of the 51K mission, a joint Spacelab flight with West Germany, then scheduled for September of the following year. At this point, Nagel's *first* mission, then known as 51A, was expected to fly in October 1984 and that would have meant that his two flights would be spaced about a year apart. Problems arose when his first flight slipped into the spring – then the summer – of 1985, whilst his second flight *didn't move*. As a result, Nagel *did* fly as a pilot, in quick succession, but far quicker than he could have anticipated. In fact, he would launch on his second flight only *four months* after landing from his first flight ... a record which would stand unchallenged for more than a decade.

Like so many of his contemporaries, Nagel grew up with a love of aviation. He was born in Canton, Illinois, on 27 October 1946. His father was not a certified instructor, but owned a small Piper Cub and took his son flying from a very early age – as an *infant*, in fact – and after taking lessons the young Nagel soloed on his 16th birthday. By this time, the first two groups of NASA astronauts had been selected and, although he aspired to such a career, Nagel spoke little about it, preferring instead to enter the military and fly jets. He applied to the Air Force Academy, but was placed on an alternate list and eventually entered the University of Illinois at Urbana-Champaign, in his home state, to study aeronautical and astronautical engineering. Whilst there, he enrolled in the Reserve Officer Training Corps and upon graduation in 1969 entered the Air Force. With the war in Vietnam at its height, he said, "the people pipeline was wide open, so the classes were big in pilot training".

Nagel completed his flight instruction at Laredo Air Force Base in Texas, then commenced training in the F-100 Super Sabre at Luke Air Force Base in Arizona and later served in a tactical fighter squadron in Louisiana. By this time, in the late summer of 1971, the war was winding down. Nagel served for a year in Thailand as a T-28 Trojan instructor pilot, then returned to the United States as an instructor and flight examiner for the A-7D Corsair II. The dream of NASA had matured over the years and Nagel wrote to the agency, asking what sorts of aircraft he should fly. "That was a *dumb* question," he admitted to the oral historian. "They couldn't have told me! Nobody answered me anyway." The path which would eventually lead to space took him next to Edwards Air Force Base in California, and test pilot school. On graduating in December 1975 he remained to conduct test work on the A-7D and the F-4 Phantom II. At one stage during this period, the Approach and Landing Tests of Enterprise were being conducted at Edwards and Nagel found himself providing A-7D flight instruction to astronauts Joe Engle, Dick Truly, Fred Haise, Gordon Fullerton and Vance Brand. In a close-knit community of test pilots, Nagel found that when NASA came calling for astronaut applicants, virtually *everyone* wanted to be considered. As each group came back from interview, they had their own stories: some thought the process was a piece of cake, others were mortified that the panel asked them about current affairs. Some of the pilots even went out and bought copies of *Time* magazine to keep themselves updated with world events. Nagel himself was summoned to Houston in September 1977 and at the time of his selection in January of the following year he was in the process of completing a master's degree in mechanical engineering at California State University.

Now, in the small hours of 17 June 1985, Nagel was amongst the last of his class to fly. There were no technical problems, the weather was good and at T-5 minutes Creighton was given the go ahead to start Discovery's Auxiliary Power Units. This was easier said than done. "On a typical airplane," Creighton recalled, "you've got one switch to start the APU. Well, on the Shuttle, you've got *eighteen* switches! You have *three* APUs, instead of one, but you've got six switches to pressurise the fuel tanks and power up the controllers, and *each* of these switches has got an abbreviation of the word 'control' or 'controller' on it ... and it's *hard* to see. On launch morning, you look at *every* switch and you look at your checklist, you look at the switch ... because they're *not* straight in order. You want to make sure you *don't* mess that up." Creighton had barely finished with the APUs when the call came for the crew to close their visors. Autosequence Start commenced at T-31 seconds and after reaching the eight-second point, all voice contact was virtually impossible, as Discovery's three main engines thundered to life at 7:33 am EST. "The *whole* vehicle starts rumbling and shaking," Creighton continued, "and you can't believe that these big bolts are still holding you to the ground. It feels like it's trying to rip itself off the ground." From his seat, Nagel remembered seeing Brandenstein and Creighton *vibrating* in their seats as first the engines, and then the Solid Rocket Boosters, ignited. "And I remember I just was mentally *behind*," he said. "I think I was left on the *launch pad* with my mind, trying to keep up. It just all happened so *fast*. There was such a rush of events and the sights and sounds ... were almost overwhelming." During the ascent, Creighton had chance to glance through his window and was rewarded with a panoramic view along the Florida coastline. It was only fleeting, for his job was to monitor the engines, and he promptly got back to work.

John Oliver Creighton – universally nicknamed 'J.O.' within the astronaut office – came from Orange, Texas, where he was born on 28 April 1943, although his family moved to Seattle, Washington, where he received his schooling. "I've been interested in flying since I can remember," he told the oral historian. "I can remember going and watching the Blue Angels fly … during the hydroplane races and it was just something that I've always wanted to do." After a year at the University of Washington, he entered the Naval Academy to study for a degree in aerospace engineering, graduated in 1966 and began flight training. He received his wings in October 1967 and undertook two combat deployments to Vietnam, flying F-4 Phantom IIs off the USS *Ranger*. Upon returning to the United States, in June 1970 he attended Naval Test Pilot School at Patuxent River, Maryland, and the following year took up a position as a project test pilot at the school. His duties as Propulsion Project Manager focused on the development of engines for the new F-14 Tomcat fighter. "I got in on the ground floor of the F-14 programme," he said, "and got an opportunity to be one of the first Navy pilots to fly the F-14." In July 1973, he began a four-year assignment as a member of the first operational F-14 squadron and completed two deployments in the Western Pacific aboard the USS *Enterprise*. This cruise was meant to be a peacetime exercise, since the United States had virtually ended its offensive operations in Vietnam. At its end, the *Enterprise* dropped anchor in Manila, in the Philippines, but after barely *half an hour* was sent back to the coast of South Vietnam for 30 days and Creighton and his squadron found themselves providing F-14 fighter cover over the evacuation of the US Embassy in Saigon on the night that the city fell to the communist North. "I saw that, first-hand," he said of the events of 30 April 1975, "from about 10,000 feet. We were just there to make sure that no MiGs came down and tried to harass our helicopters that were evacuating the personnel out of there." Creighton returned to the United States in July 1977 and served as operations office and programme manager for the F-14 at the Naval Air Test Center's Strike Directorate, and later that same year was called to Houston for astronaut screening. Most of the testing, he recalled, was physiological: they tested his eyes and heart, looked in every orifice of his body and took X-rays "until you glow in the dark". Years later, Creighton admitted that the selection panel was *not* something that he could realistically have trained himself to handle. "They don't ask you to derive any differential equations," he said. "They just want to talk to you and, I think, get a sense of the kind of person you are. They'll throw you some oddball questions, just to see how you think on your feet."

As Discovery gained velocity and reached the edge of space, Creighton could not believe how quickly the sky darkened from blue to deep indigo to the pitchest black. The separation of the boosters unleashed a bright flash in the cabin, totally engulfing the front windows in flame for about half a second, and the ride continued under the sustained, quiet push of the main engines. At 7:41 am EST, eight minutes since leaving Florida, the magical call of 'MECO' brought a cheer from the crew and a force of three times of terrestrial gravity was gone, to be replaced by … *weightlessness*.

Behind Creighton sat Shannon Matilda Wells Lucid, the last of the six women chosen in 1978 to reach space, and, at the age of 42, by far the oldest female spacefarer. She had been born in war-torn China on 14 January 1943 and her experiences during her formative years make it unsurprising that she grew up with a "zest for life, steely determination and resourcefulness", according to writer Peggy Mihelich. Her parents, Oscar and Myrtle Wells, were Baptist missionaries and they, together with Shannon, her younger brother, Joe, her aunts and an uncle and her grandparents were taken captive by the Japanese army and held in Shanghai's Chapei Civil Assembly Centre prison camp. She learned to walk in early 1944, whilst aboard the Swedish ship *Gripsholm*, which returned her family to the United States as part of an exchange of non-combatant citizens of the warring nations. It was long and arduous voyage and, during a stopover at Johannesburg in South Africa, she received her first pair of shoes. After the war, the family returned to China – living at times in Shanghai, Nanking and Anking – and Shannon found herself the centre of attention, due to her blonde hair and blue eyes. Her fierce desire to learn to read prompted her parents to place her in a Chinese elementary school. Aged five, she took her first flight. As the DC-3 flew over mountainous terrain and touched down on a gravel runway, the young girl was convinced that *flying* was the most remarkable thing for a human being to do ... and steeled herself to do the same when she grew up. Her family was expelled from China in 1949, after the Communist Revolution, and the young girl received her schooling in Bethany, Oklahoma.

She entered the University of Oklahoma to study chemistry and received her degree in 1963. By now, the first teams of astronauts had already been selected and Lucid was astonished that *all* of them were *male*; in fact, she had written to *Time* magazine in 1960, criticising NASA for choosing only men. Space exploration had fascinated her, even since she read about the rocket experiments of Robert Goddard ... but there was another motive. "The Baptists wouldn't let women preach," she once said, "so I *had* to become an astronaut to get closer to God than my father!" During her undergraduate studies, she took flying lessons, gained her licence and encountered another cruel and harsh reality of life. One day, in her final year of study, she sat down with her professor to discuss her options for getting a job. The professor looked at her blankly. "A job?" he asked. "You plan on *working*? But you're a *girl*!" It underlined the reality that women were not taken seriously in many professional careers. Despite having a private licence, her efforts to become a commercial pilot led nowhere, for the same reasons. Fortunately, the Kennedy and Johnson administrations, with their incessant civil rights campaigning, smoothed the road over the next few years and she found work in academia as a teaching assistant and research chemist, firstly at her *alma mater*, then at the Oklahoma Medical Research Foundation and finally at the Kerr-McGee oil and gas corporation. By now married to Michael Lucid, she returned to study at the University of Oklahoma, earning a master's degree in biochemistry in 1970 and a PhD in 1973. With her doctorate, Lucid gained a job as a research associate with the Oklahoma Medical Research Foundation in Oklahoma City and remained in this position until NASA called for astronaut applicants. Lucid "scrambled" to complete and submit her application. In late August 1977, she was invited to Houston as part of the third group of finalists to be interviewed ... a 20-strong group which included a subset of individuals whose presence, a decade earlier, would have been inconceivable: *eight women*. Three those eight women – Lucid, Anna Fisher and Rhea Seddon – would form half of the female component of the astronaut class that was announced in January 1978.

"It's a remarkable story," John Fabian said of Lucid's life. "It's a story of the human spirit and I love to tell it ... because kids don't realise what opportunities really lie ahead of them. Some are very quick to worry about the disadvantages that they have in their own lives, or as they perceive in their own lives, and I think the Shannon Lucid story is just a great story about overcoming obstacles and blasting through ceilings and knocking down doors and never letting anything get in the way of doing the things that you believe are right."

On her first space mission, Lucid's responsibilities were heavy; along with Fabian, she was charged with the deployment of Morelos, Arabsat and Telstar *and* the release and retrieval of the Spartan free-flier, using the RMS mechanical arm. For Fabian, who had flown as part of a smaller crew on STS-7, it posed its own challenges. "We had a crew of *seven*, living inside a volume about the size of a minivan," he told the NASA oral historian, "and so we were good neighbours for a week!" Eight hours after launch, at 3:37 pm EST, the first of Discovery's load of three communications satellites was successfully deployed when Morelos-A was sent spinning into space, affixed to its PAM-D motor. This payload was particularly important, since it represented Mexico's first communications satellite, with a mission to provide educational and commercial television programmes to the most remote regions of the country, together with telephone and fax services and business and data transmissions. Planning started in November 1982, when the Secretariat of Communications and Transport contracted with Hughes to build a pair of satellites, together with ground-based tracking stations. Morelos differed from previous HS-376s in that it was a 'hybrid' satellite, operating simultaneously in the C-band and Ku-band, and possessed a 'planar' antenna array for transmission and reception. Arabsat followed some 26 hours after launch, on 18 June, and Telstar-3C departed the payload bay at 7:20 am on the 19th.

Next came Spartan, a unique retrievable platform with an name that even NASA's best acronym-makers could be justly proud of: the Shuttle Pointed Autonomous Research Tool for Astronomy. It was a 1,000 kg cube-shaped box, equipped with 136 kg of instrumentation to perform medium-resolution mapping of X-ray emissions from extended sources and regions. Built by NASA's Goddard Space Flight Center at the relatively inexpensive cost of $3.5 million, it measured 3.2 × 1.07 × 1.22 m and was intended to be flown repeatedly, at intervals of between six and nine months, with data stored on internal tape recorders and pointing and stabilisation achieved by a three-axis attitude control system. Deployed and retrieved by the RMS, Spartan was reusable and had already been assigned a second mission, in January 1986, to observe Halley's Comet. Specific foci for its maiden voyage on 51G were areas of hot gas in a large cluster of galaxies within the constellation Perseus and within the centre of our own galaxy, the Milky Way.

The rendezvous operations associated with Spartan were straightforward, according to Nagel, for the crew had already spent months rehearsing the retrieval of the Long Duration Exposure Facility and many of the techniques were similar. Shannon Lucid was responsible for the deployment, which occurred two minutes after midday EST on 20 June, whilst Fabian handled the mechanical arm for the retrieval, about 45 hours later, at 9:30 am on the 22nd. The small spacecraft reached a maximum distance of about 160 km from Discovery. As for the pilots, Creighton manoeuvred Discovery to a safe separation distance after deployment, whilst Brandenstein took the lead in the rendezvous and retrieval. "But when we came back in to retrieve it," Fabian told the oral historian, "it was out of attitude. It was supposed to be in an attitude which would be easy for us to just fly up to and grab. It turned out that the grapple fixture – instead of being out-of-plane to the vehicles, so that we could just go in and get it – was *on top*." One solution was for Brandenstein to fly an out-of-plane manoeuvre, which the crew had not practiced on the ground, but it was decided to bring the Shuttle closer to Spartan "and then reach over the top with the arm and grab it from the top". As a consequence, the final grapple of the payload was an 'off-nominal' event, but it left Fabian with a great deal of pride; not only in his training and his capabilities, but in the ability of the crew as a whole to execute an unplanned manoeuvre.

Downstairs, on the middeck, Baudry and Al-Saud tended to their own experiments. The French Echocardiograph Experiment (FEE), which had been bumped from Mission 51E, employed a non-invasive, ultrasonic technique to obtain data on the physiological adaptation to the microgravity environment, including the pooling of blood in the head and upper torso and changes in the size of heart cavities and differing flow rates in major arteries. Located in two middeck lockers – one housing the electronics, the other holding the video recording equipment and control monitor – the FEE was conducted primarily by Baudry, assisted by Lucid. Another pair of lockers were occupied by the French Postural Experiment (FPE), which was designed to better understand changes in muscular tone, posture, orientation and movement in the strange environment. Baudry and Al-Saud worked together on this experiment. The Automated Directional Solidification Furnace (ADSF) – "a pilot's experiment," joked Nagel, since it needed little more than a couple of switch throws to operate – melted composite material samples of bismuth and manganese and resolidified them in a series of four small furnaces. One of the more notable experiments was a study from President Ronald Reagan's Strategic Defense Initiative – popularly known as the 'Star Wars' programme – which sought to evaluate the ability of a ground-based, low-energy laser to track a moving object in orbit. The High Precision Tracking Experiment (HPTE) consisted of a retroreflector, mounted in a cylindrical housing, which was placed on the middeck side hatch window to receive and reflect a four-watt green laser beam projected from a test site in Maui on Hawaii's Big Island. "The first time we tried, it didn't seem to work," remembered Creighton. "We figured it out, actually, about the same time that the ground did, but it turned out that when you put the units into the computer, we used the *wrong* units! As we passed over [Maui], we were actually tracking a point that was *above* us, instead of *below* us. We figured it out, updated the computer co-ordinates and then we weren't passing over Hawaii, so we had to wait about 24 hours and do it the next day."

For the payload specialists, the once-in-a-lifetime opportunity of flying in space was something to be savoured. For Al-Saud, his 'wow' moment was seeing Saudi Arabia from space. "I was woken up by some crew members," he told a television interviewed. "The Earth was 'above' us and I saw the Eastern Province with its lights. It was a *very* moving sight." For the sultan, though, the happiest moment was returning to Earth. Patrick Baudry, who had carried French fayre into orbit – jugged hare, lobster, crab mousse, chocolate pudding and some small bottles of wine – the strangest thing was to have *someone else* (Brandenstein) land the vehicle. Referring to the fact that 51G was Brandenstein's first landing in the commander's seat, Baudry said afterwards, "It's the first time I've *ever* flown on a plane for the first time, with anybody that was landing it for the *first* time, where I *wasn't* doing it!"

"Well," replied Brandenstein, with a grin, "look at that. We were able to walk away from it!"

Touchdown at Edwards Air Force Base on 24 June came at 6:12 am PST (9:12 am in Florida), completing a mission of just over seven full days, and NASA remarked that 51G was its most successful flight to date. For Steve Nagel, his return to Earth very quickly turned into a return to the simulator, as he plunged into full-time training as pilot of the West German Spacelab mission, now redesignated 61A and scheduled for late October. Before that, however, there were the post-flight public relations tours – and with Saudi and French crew members, this meant foreign trips in which they were quite literally treated like kings and queens. "We flew on Saudi Airlines, out of New York into Jeddah, *first class*, on a 747," Nagel recalled. "They *really* took care of us." Spouses were invited along to places normally excluded to women and the astronauts dined on traditional Bedouin fayre and camel's milk, sitting cross-legged on carpets, then visited the Saudi royal family's summer palace in the mountainous Abha region. It was an experience that Nagel could only compare to a Cinderella-type story. John Fabian would remember the trip, not for him being Astronaut John Fabian, but for being the *nephew* of an old ARAMCO oil employee. "I mentioned to the President of ARAMCO," he said, "that my uncle had spent 35 years working in the oilfields of Saudi Arabia." When the top-ranking oil official asked for the name of Fabian's uncle, it turned out to have been his first boss! The Americans returned to the United States with "gifts, gifts, gifts", according to Fabian – leather briefcases, Swiss watches, a Turkish carpet, an engraved watch with King Fahd's signature on its face, to name just a few – although Shannon Lucid almost got herself into diplomatic hot water.

"She had no love for the Saudis," said Fabian. "She had no love for the Muslim religion. She particularly objected to Saudi treatment of young women and disfigurement ... and she didn't want to be a part of any of that. There was a lot of time and effort spent convincing her otherwise, to absolutely no avail." When the other astronauts arrived in Riyadh, they were greeted by Al-Saud, who looked around for Lucid. Her absence did *not* go down well with the sultan or the king ... or with President Reagan. "Well, the King called the President," recalled Fabian. "The President called the NASA Administrator. The NASA Administrator called the Johnson Space Center Director, who called George Abbey ... and Shannon was on the *next* 747!" Lucid may have lost the battle, but she won the war, for she spent barely a day in Saudi Arabia, shook King Fahd's hand, and returned promptly home. Mike Mullane had a slightly different take on the story. "Shannon's husband could not make the trip," he wrote. "Shannon wasn't concerned. She didn't need a man to hold her hand. Wrong. Saudi Arabia did not allow women to enter the country alone. She had to have a male escort. When Shannon heard this, she told [NASA] Headquarters she wasn't going." It was decided to admit Lucid into Saudi Arabia as the *honorary daughter* of Dan Brandenstein or the *honorary sister* of John Fabian or, worst of all, as an *honorary man*! "I immediately went to Shannon's office and congratulated her on having achieved the highest honour a woman could ever hope to achieve: to be designated an honorary man," Mullane continued. "Shannon had a lively sense of humour and laughed at my antics ... but I made certain *not* to walk down the stairs in front of her for the next few weeks!"

The trip to France proved to be upsetting for the wives of Brandenstein, Fabian and Nagel and for Creighton's bride-to-be, Terry, because *all* were barred. "We went without spouses," Fabian told the oral historian, "because the French couldn't make up their mind what they were going to do with the spouses. Patrick Baudry was in the middle of a divorce and he had a Russian girlfriend, who was the daughter of a KGB agent!" In the astronauts' minds, not only did they have to endure the wrath of their wives – "[The news] was *certainly* not popular in *my* house," recalled Fabian, whilst Creighton admitted that it remained "a sore subject" in his home, too – but preventing them from visiting France and touring Paris and Nice and Marseilles was perceived to be an immense slap in the face for spouses who had supported their husbands through more than seven years of astronaut training and preparation. Whilst in France, the crew were inducted into the Bordeaux Wine Society, having presented the organisation with one of the four small bottles which Baudry carried into orbit. Perhaps in the case of Creighton, the disappointment over spouses was sweetened by his wedding ... only a day after landing from the mission. "Originally, I'd planned to get married at the Cape," he recalled, "because we were going to come back and land at the Cape, but then ... about three weeks before the flight [it was] decided [we should] land in California. They flew Terry out for the landing, but the families weren't going to be out there, so we just made a quick snap decision we'd get married in Houston, after I got back, and any of the families that could cycle through Houston on their way back from the Cape. Most of the family made it."

'Adventure', indeed, is not an excessively poetic term for the mission, for Discovery's fifth orbital voyage offered precisely that. Years later, astronaut Joe Engle recalled a conversation with Al-Saud and the sultan offered a considered, "classic", view of what beholding the Home Planet from on high actually meant to him. "The first day or two in space," Al-Saud said, "we were looking for our *countries*. Then, the next day or two, we were looking at our *continents*. By about the fourth or fifth day, we were looking at our *world*!" In Engle's mind, this was exactly right. It was another reminder that national and international boundaries mean little to spacefarers – gazing down, there are no atlas-like lines or visible barriers between the countries. From space, Earth appeared as it should be: as *one world*.

GOING TO SPAIN

The position of flight engineer aboard the Space Shuttle was arguably one of the most important jobs a mission specialist could possibly hold.

Seated behind and between the commander and pilot during the critical periods of ascent and re-entry, he or she was responsible for helping to monitor the orbiter's instruments and offering a vital third set of eyeballs in the event of 'off-nominal' events. For Story Musgrave, the flight engineer on Challenger's eighth voyage, Mission 51F, his ascent to orbit on 29 July 1985 was arguably the most dramatic of his six-mission career, when the Shuttle suffered a hair-raising main engine shutdown 108 km above Earth. "The ground made the call '*Limits to Inhibit*', which is, for us, an extremely serious omen, [because it] means the ground is seeing problems that are going to shut you down. I'm looking through the procedures book and thinking we're going to land at our transoceanic abort site in Spain. I'm rehearsing all the steps and my hands are moving through the book and I'm thinking 'We're going to Spain. Things are bad!'"

The emergency, which occurred five minutes and 45 seconds after launch, arose when temperature readings for the Number One engine's high pressure turbopump indicated 'above' its maximum redline, prompting Challenger's computers to shut down the engine. At this stage of the ascent – with just three minutes remaining to MECO – the Shuttle was too high and too fast to return to an emergency landing back in Florida. This meant one of two things: either a Transoceanic Abort Landing (TAL) in Europe or a tricky manoeuvre known as an Abort To Orbit (ATO), whereby Challenger would burn her twin OMS engines to augment the two remaining main engines and limp into a low, but stable, orbit. Musgrave's initial focus was upon the page of his checklist that dealt with requirements for a touchdown at Zaragoza Air Base – a joint-use military and civilian installation with a NATO-instrumented bombing range – in the autonomous region of Aragon in north-eastern Spain. Zaragoza had been designated as a TAL site in 1983 and was assigned to 51F because the orbital inclination of 49.5 degrees placed it near the nominal ascent ground track, thus allowing the most efficient use of available main engine propellant and cross-range steering capability. As Musgrave flipped through the procedures that he would recite to commander Gordon Fullerton and pilot Roy Bridges for a diversion to Zaragoza, astronaut Karl Henize, seated to his right, was becoming nervous.

Henize had good reason for his nervousness. The TAL mode was the second available abort after the Return to Launch Site (RTLS) contingency option and encompassed the six-minute period from shortly after SRB separation until, theoretically, main engine cutoff. It utilised one of three sites – one in France (Istres Air Base) and two in Spain (Zaragoza and Moron Air Force Base) – which, under international agreement, provided emergency support for just-launched Shuttles following orbital insertion inclinations between 28.5 and 57 degrees. Flight rules dictated that the TAL option would only be selected in the event of a premature main engine shutdown or other major malfunction – for example a significant cabin pressure leak or cooling system failure – which would prevent the vehicle from continuing into space and completing at least one orbit. Had the command been given from Mission Control that day, Gordon Fullerton would have rotated the abort switch on his instrument panel in the TAL/AOA position and depressed the abort push button next to the selector switch; Challenger's computers would then have automatically steered the spacecraft toward the plane of the European landing site.

Eventually, the call came from Mission Control: "Abort ATO; Abort ATO". Challenger had achieved sufficient velocity and altitude to undertake the next available option: the Abort to Orbit. In fact, she had missed the closure of the TAL 'window' by just 33 seconds! At 4:06:06 pm EST, some six minutes and six seconds into the climb and hurtling towards space at 15,000 km/h, Fullerton fired the OMS engines for 106 seconds, consuming 1,875 kg of the orbiter's much-needed propellant, but permitting the Shuttle to continue into a lower than planned orbit. Two minutes later, at 4:08:13 pm, the Number Three main engine data indicated excessively high temperatures. If the 'Limits to Inhibit' had not already been applied, the computer would have it shut down. The 'inhibit' command effectively instructed the computers to ignore the over-temperature signals and prevented them from shutting down the Number Three engine. The two remaining engines, meanwhile, fired for an additional 49 seconds, shutting down nine minutes and 20 seconds after launch. "We never did get the call for the transoceanic emergency landing," said Musgrave, "and we ended up making it to orbit and finishing the mission." According to Flight Director Cleon Lacefield, Challenger could have achieved orbit without the additional OMS burns, but at the expense of having to jettison her External Tank over north-eastern Africa after a greatly extended burn. "We don't do business that way," he said. Their eventual orbital path, with an apogee of 230 km and a perigee of 174 km and requiring three OMS firings, was considerably lower than the 390 km planned and would impair several of the astronomical and solar physics instruments of the Spacelab-2 observatory.

This incident represented the only in-flight shutdown ever experienced by the Shuttle, and came as a surprise because all main engine parameters had been normal during the countdown, ignition sequence and the first few minutes of the flight. At approximately two minutes into Challenger's ascent, at about the same time as the SRBs were jettisoned, data from Channel A – one of two measurements of the Number One engine's high pressure fuel turbopump discharge temperature – displayed characteristics indicative of the beginning of failure. Its measurement began to drift and, at three minutes and 41 seconds after launch, the Channel B sensor failed. However, its sibling continued to drift, approaching and then exceeding its own redline limit some five minutes and 43 seconds into the flight, which triggered the shutdown. The high pressure fuel turbopump discharge temperature data from Channel B of the Number Three engine, meanwhile, began to climb and passed its own redline just over eight minutes after liftoff. Measurements from its Channel A remained within prescribed limits and, said NASA's post-mission report, all other operating parameters relating to the Number Two and Three engines were deemed normal. Post-mission analysis suggested that the problem was not with the Number One engine itself, but faulty sensors that incorrectly indicated an overheating situation. According to Bill Taylor, who was then-head of the main engine project at the Marshall Space Flight Center in Huntsville, Alabama, these sensors were extremely thin wires, whose electrical resistance changed as they heated up. He added that they had already suffered failures on three earlier flights and that upgraded versions were scheduled to fly aboard Discovery on Mission 51I in August 1985. Otherwise, the performance of the SRBs in propelling Challenger into orbit was described as "nominal". However, gearbox nitrogen pressures in one of the Shuttle's Auxiliary Power Units had exceeded their maximum allowable levels and, during a post-launch sweep of the Cape Canaveral beaches, a fragment of spray-on foam insulation, apparently from the External Tank, was discovered. A survey of the orbiter's heat-resistant tiles by the RMS mechanical arm found a large number of debris impacts. Further inspection after Challenger's touchdown identified a total of 553 'hits'.

As they settled into orbit and divided themselves into their two 12-hour teams – a 'red' shift led by Bridges, together with mission specialist Henize and payload specialist Loren Acton, a 'blue' shift led by Musgrave, with mission specialist Tony England and payload specialist John-David Bartoe, and Gordon Fullerton working across both – the seven-man crew barely had chance to reflect on what had been not just been an eventful day, but a crisis-filled month. Originally scheduled to head for space at 3:30 pm EST on 12 July, they had been thwarted by the second on-the-pad main engine shutdown, only seconds before liftoff. "At T-7 seconds," recalled Bridges, "the main engines start with a rumble from far below. As the person in charge of all engines, I watch the chamber pressure indicators come to life and surge towards 100 percent. I think 'Wait, what's this?' The left engine indicator seems to be lagging behind. Before I can say a word, it falls to zero, followed by the other engines. With less than three seconds before our planned liftoff, we have an abort. The groans from the rest of the crew are now audible. I take a quick look around to see if there's anything else to be done and notice Gordon Fullerton turning to look at me. The thought crosses my mind: 'Gordo probably is thinking I've done something to screw it up'. I show him both hands, palms up, and say 'Gordo, I didn't touch a thing. It was an automatic shutdown'." The 12 July shutdown, executed because the Number Two main engine's chamber coolant valve was slow in closing from 100 percent open to the 70 percent required for startup, necessitated a 17-day wait for a second launch attempt. When one of two command channels failed to execute the closure, fortunately, the backup took over without incident. However, flight rules dictated that both channels had to be fully functional for the countdown and liftoff to proceed. These timings, however, posed a problem. The liftoff time on the 12th lasted barely two hours and was calculated to satisfy lighting conditions needed for the plasma physics and astronomical instruments aboard Spacelab-2. For five or six days after the pad shutdown, the launch window opened at roughly the same time – 3:30 pm – before becoming unfavourable due to a requirement for orbital dark skies. Even the 29 July liftoff, originally set for 2:23 pm, was postponed by more than an hour and a half by an erroneous command to Challenger's fifth, backup flight computer.

THE 'COLA' AND 'SOLAR' WARS

The wait would be worth it and, although the Abort To Orbit proved a momentary scare, the sensation of launch was compared by Loren Acton – in terms of its raw, naked power – to the Loma Prieta earthquake of October 1989, which hit the Greater San Francisco Bay area of California and measured 7.1 on the Richter Scale. Loren Wilber Acton, a 49-year-old solar X-ray physicist from the Space Sciences Laboratory at Lockheed's Palo Alto Research Laboratory in California, was flying with 40-year-old John-David Francis Bartoe, a civilian astrophysicist, working for the Naval Research Laboratory in Washington, DC, on a mission whose objectives were primarily devoted to observing the Sun and astronomical targets. Acton came from Lewistown, in central Montana, the site of a famous 19th century gold rush in the neighbouring Judith Mountains. (Years later, in 2006, he would unsuccessfully run as a Democratic candidate for Montana.) Born on 7 March 1936, he attended high school in his home state and graduated from Montana State University in 1959, with a degree in engineering physics. On his election campaign website, www.acton4montana.com, he gave credit to his father and brothers for having instilled a strong work ethic in him. Acton grew up in a ranching family, working the fields as a youth and riding a pony to a one-room schoolhouse. Nevertheless, he devoted himself wholeheartedly to education and in 1965 gained a PhD in astrogeophysics from the University of Colorado at Boulder with a dissertation about solar X-rays. Indeed, Acton noted on his website that he had worked in solar physics since 1961 and today he is a research professor in the solar physics group at Montana State University. Much of his career prior to his space mission was spent at Lockheed's Palo Alto Research Laboratory, where he conducted solar and celestial research and was a co-investigator for one of Spacelab-2's primary experiments, the Solar Optical Universal Polarimeter (SOUP). On 9 August 1978, NASA selected Acton, Bartoe and two other physicists, Dianne Prinz of the Naval Research Laboratory and George Simon of the Air Force Geophysics Laboratory, to train as payload specialists. Five years later, in June 1983, Acton and Bartoe were announced by NASA as the prime candidates. Already, the Spacelab-2 payload was marked out as quite different from most Spacelab missions, since it would feature a pallet-only configuration, a series of powerful astrophysical and solar instruments and the first flight of a complex device known as the Instrument Pointing System (IPS). Gordon Fullerton called Spacelab-2 'The Gangbuster Payload', because there was simply so much work associated with it.

Ten years earlier, in March 1975, the Joint Spacelab Working Group (JSLWG, spoken as 'jizzlewig') reached agreement on the content of the first two scientific flights, with Spacelab-1 consisting of a long module and single pallet and Spacelab-2 comprising a pallet-only configuration to house 7,000-8,000 kg of research hardware, two or possibly even *three* payload specialists, a week in space and up to 100 hours of experimental work. "This second mission would be very ambitious," wrote Douglas Lord, "with experiment resources considerably beyond those provided for the first mission." Unlike Spacelab-1, which involved co-operation from the European Space Agency, the Spacelab-2 flight would be solely a NASA endeavour. The key to its success was the IPS, a device which Lord said was more challenging than any other Spacelab component, in terms of its technical complexity, organisational and schedule difficulties and cost escalation. When ESA named an industrial consortium, headed by ERNO and Fokker, to build the Spacelab system in June 1974, the German Dornier Satellitensysteme GmbH (now EADS Astrium GmbH) was contracted to undertake IPS design studies. The specifications called for a three-axis system, with a pointing accuracy of one arc-second; the instrument should measure up to 2 m in diameter and up to 4 m in length and weigh no more than 2,000 kg. The sheer size of the yoke mounting for such a large device prompted Dornier to propose an end-mounted approach in which the three gimbal systems would be attached to a Spacelab pallet and would support a circular mounting frame onto which the instruments themselves would reside. Authorisation to proceed with the design was granted in December 1974, but within months ESA was discussing options to reduce project costs by holding the contractor to less stringent specifications. NASA management was also concerned that no single IPS design would satisfy all of its users' pointing requirements. By September 1975, Dornier's design had been rejected on the basis of unacceptable cost and schedule risks and in December ESA issued a new Request for Proposals. Early the following year, two new proposals were delivered: a joint bid from Dornier and MBB and a bid from ERNO which covered the integration of the IPS into Spacelab. In mid-1976, the Dornier-MBB proposal was accepted.

However, issues of cost, schedule and technical difficulty continued to escalate and, with them, tensions grew between NASA and its European partners until, in early 1977, some ESA managers suggested removing the IPS from its Spacelab effort entirely, in order to find another means of development. By June, the agency had finally agreed to continue with the project, contracting with Dornier to deliver the first flight unit by the middle of 1980. Yet the problems persisted. Evidence of the susceptibility of certain components to stress corrosion and uncertainty about the IPS' software were flagged in December 1977. When the European members refused additional funding for the pointing system, Dornier was forced to make modifications and delay the first flight. For its part, NASA was keen to demonstrate its confidence that ESA and Dornier would solve the problems and in May 1980 the agency purchased a second IPS for $20 million. Meanwhile, Dornier submitted a proposal for a redesigned IPS in April 1981 and the first flight unit was delivered to the Kennedy Space Center in November 1984. "The last few months of checkout," wrote Lord, "were fraught with debates about the state of readiness of both the hardware and software and the adequacy of documentation and operating instructions." Harbouring reservations, NASA was torn between pushing for the completion of qualification testing in Europe and pushing for the early delivery of the hardware to Florida in order to begin payload integration for Spacelab-2. Dornier offered the Americans "iron-clad assurances" that all open actions, missing data and tests *would* be completed before launch and NASA and ESA accepted the IPS.

Its advertised capabilities included providing precise guidance for a suite of instruments and pointing them to within two arc seconds, holding them on target to just 1.2 arc seconds. One end of the pointing system was mounted on the pallet and the other end to an 'integration ring', to which the three-axis gimbal for the instrument package was affixed. When operational, the 1,180 kg IPS was capable of manoeuvring telescopes and instruments backwards and forwards, from side to side and could even 'roll' them in a 22-degree arc around its 'straight up' position. Its movements were commanded from the Spacelab subsystem computer and a pair of Data Display Units (DDUs) on the Shuttle's aft flight deck. It could be operated in manual or automatic modes and was capable of spending long periods focused on single objects or conducting slow-scan mapping operations. Moreover, its reaction times were much better than those of the Shuttle's attitude control system. When compared to the orbiter's pointing precision of perhaps a tenth of a degree at best, the IPS' ability to achieve accuracies of one-thirty-six-*hundredth* of a degree has been likened to keeping an instrument on the steps of Washington's Capitol Building aimed at a coin on the Lincoln Memorial, some 3.5 km away! Even the effects of crew motions, equipment operations or Shuttle thruster firings could be compensated by accelerometers mounted on the IPS and the Sun-watching instruments kept on target. Solar physics research from the Shuttle was highly desirable for two main reasons. Firstly, being above the turbulence of the 'sensible' atmosphere meant it could acquire much better images. Secondly, since orbital 'nighttime' lasts only 45 minutes, it proved easier to trace the evolution of phenomena on the Sun's surface and in its atmosphere without long interruptions.

The solar instruments aboard 51F were to investigate the chromosphere, the 'transitional' region and the corona in order to better understand the mechanisms which transfer heat from one layer of the Sun to the next. The temperature of the chromosphere, which is an irregular region above the Sun's visible disk (its photosphere), rises from 6,000-20,000 degrees Celsius. This is sufficient to cause hydrogen to emit a reddish light, as evidenced by prominences on the Sun's limb during total eclipses. Sandwiched between the chromosphere and the much hotter corona is the transitional region, in which heat from the latter flows down into the former and produces a dramatic and rapid temperature change from over a million degrees to 'just' 20,000 degrees Celsius. Hydrogen is ionised at such extreme temperatures, making it difficult to see and the light emitted by the transitional region is dominated instead by ions of carbon IV, oxygen IV and silicon IV. Finally, the outermost 'atmosphere', the corona, produces the glow that surrounds the black lunar disk during total solar eclipses. Early coronal observations revealed bright emission lines at wavelengths that failed to correspond with any known materials, prompting some astrophysicists to propose the existence of 'coronium' as the principal gas feeding the outer atmosphere. The mystery endured until it was determined that coronal gases are super-heated. At such extremes, both hydrogen and helium are stripped of electrons and even carbon, nitrogen and oxygen are 'bare' nuclei. In fact, only heavier trace elements like iron and calcium are able to retain some of their electrons in this intense heat. On Spacelab-2, the IPS had four instruments: the Solar Optical Universal Polarimeter (SOUP), the Coronal Helium Abundance Spacelab Experiment (CHASE), the High Resolution Telescope and Spectrograph (HRTS) and the Solar Ultraviolet Spectral Irradiance Monitor

(SUSIM). The assemblage was clamped pointing along the length of Challenger's payload bay during ascent and re-entry and unlatched in orbit. For safety reasons, there was provision to jettison the IPS if it could not be stowed to enable the payload bay doors to be closed at mission's end.

Early guidelines for Spacelab-2 had specified a pallet-only configuration, initially employing a pair of two-pallet 'trains' – the so-called '2 + 2' approach – although this changed in late 1977 to a single train of three pallets, due to the inclusion of the University of Chicago's Cosmic Ray Nuclei Experiment (CRNE), a giant egg-shaped detector on its own support structure at the rear of the payload bay. This caused some concern for ESA's Spacelab director Michel Bignier, who was already aware that NASA was exploring its own in-house payload support structures and feared that the Americans would eliminate the Spacelab pallets entirely. However, the CRNE structure was a unique design, suited to maximising the size of the experiment. The layout was changed again at the recommendation of NASA's Associate Administrator for Space Flight, Jesse Moore, in 1980, when he proposed swapping the three-pallet train with an igloo to a single pallet with an igloo and a two-pallet train. Moore's rationale centred on the fact that the 6,130 kg Spacelab-2 instrument weight exceeded the 5,000 kg limit for a three-pallet train. The complexity of the mission, though, could hardly be underestimated. Indeed, years later, it was remarked by the solar physics and astronomical communities that records set during Mission 51F would probably "stand until the era of the space station, because no payload now under consideration matches the complexity of Spacelab-2, which tested the limits of hardware, software and people everywhere in the system". That success is all the more remarkable in view of the problems experienced with the observatory when finally activated in orbit.

As its name implies, Spacelab-2 should have been the second flight of the European-developed laboratory, but the IPS problems meant it was postponed until after Spacelab-3. It was the second of two so-called Verification Flight Tests of the Spacelab unit and would put the pallet train and igloo through its paces for the first time. 'Pallets' were U-shaped metal frames, measuring 3 m long by 4 m wide and covered with aluminium panels onto which large instruments, telescopes or antennas requiring unobstructed fields of view could be attached. On Columbia's STS-2 mission in November 1981, for example, an engineering version of the pallet had been employed to hold a large synthetic aperture radar and several other scientific sensors. Up to five pallets – three of them bolted together in a rigid 'train' – could fit into the Shuttle's payload bay and the versatile platforms continue in service in today's International Space Station era. The pallet train was held in place by five attachment fittings – four along the walls of the payload bay and one in the floor – and included aluminium ducts and trays on its port and starboard sides to route cables to and from experiments and subsystems. Thermal control was provided by multi-layered insulation and Spacelab's freon-21 coolant loop, which collected excess heat from the pallet-mounted hardware through a series of 'cold plates' and rejected it into space via the Shuttle's heat exchanger. Although the pallets had been tested on both STS-2 and STS-3 prior to being used operationally to support a payload for NASA's Office of Space and Terrestrial Applications on Mission 41G in October

Mounted on the Instrument Pointing System (IPS), the Spacelab-2 telescopes and detectors operated near-continuously during the eight-day 51F mission. The huge gimbal at the base of the IPS is clearly visible, as is the supporting Spacelab pallet.

1984, their carriage aboard Spacelab-2 marked the first time that the 'train' and another device, the 'igloo', had been utilised for a 'full' scientific mission. The igloo was a 2.1 m tall aluminium alloy cylinder and was mounted vertically on a crossbeam at the forward end of the train, providing a temperature controlled container to hold subsystems and equipment for the instruments. Pressurised to 14.7 psi, the 660 kg igloo offered electrical power, cooling and command and data acquisition services for the pallet-mounted experiments; in effect, it supplied many of the services a 'core' Spacelab module would have offered.

On Spacelab-2, the verification hardware comprised a multitude of sensors installed throughout the pallet train and igloo system and Challenger herself, providing data on their combined performance during launch, ascent, orbital flight, re-entry and landing. This equipment verified that the observatory's thermal control system kept temperatures within required limits and prevent condensation or heat leaks. The thermal, acoustic and structural responses of the entire payload during the most critical and dynamic portions of the mission were closely monitored and the astronauts checked their satisfactory operation and ability to communicate and transmit scientific results through the sole Tracking and Data Relay Satellite. Construction of the Spacelab-2 observatory commenced in 1982, when its three pallets arrived at KSC's Operations and Checkout Building to begin pre-flight testing. In February of the following year, two mission specialists – Karl Henize and Tony England – were assigned to the flight, which was then scheduled for September 1984. Additional equipment was attached to the payload to accommodate the CRNE detector and a number of other experiments were also added. By November 1983, when Gordon Fullerton, Dave Griggs and Story Musgrave were named as the commander, pilot and flight engineer, the launch of Spacelab-2 had slipped until March 1985. Griggs was already training for Mission 41F, scheduled for the summer of 1984, and might have anticipated a rapid turnaround of maybe seven months or so between his two flights. This picture changed dramatically in August, when his first flight was cancelled and his crew were reassigned to a new mission, slated for February 1985. It did not take a mastermind to recognise that Griggs would be unable to prepare for, and fly, two missions in a span of just six weeks, so in October 1984 NASA named Roy Dunbard Bridges Jr to take his place aboard Spacelab-2.

Bridges was born on 19 July 1943 in Atlanta, Georgia, and as a youth was an active Boy Scout, reaching its second-highest rank of Life Scout. After completing high school in his home state, he entered the Air Force Academy in Colorado Springs and earned a degree in engineering science in 1965, then pursued a master's credential in astronautics from Purdue University in Indiana, which he gained in January of the following year. Bridges underwent initial flight instruction at Williams Air Force Base in Arizona and, attached to the 416th Tactical Fighter Squadron, flew the F-100 Super Sabre on 226 combat missions over Vietnam between December 1968 and December 1969. Upon his return to the United States, he served as a T-37 instructor and in July 1970 attended test pilot school at Edwards Air Force Base. Graduation in July 1971 was followed by four years of test work at Edwards, completion of the Air Command and Staff College at Maxwell Air Force Base in Alabama and an assignment to the Pentagon in Washington, DC, where he was involved in the development of the new F-15 Eagle fighter. In April 1980, Lieutenant-Colonel Bridges was summoned to Houston by NASA for a week of interviews, part of its selection process for the second group of Shuttle-era astronauts. Also in Bridges' interview group was another Air Force pilot applicant, Lieutenant-Colonel John Blaha, and when the pair were selected by NASA in May, they jointly became the senior-ranking members of the new class. According to Air Force records, Bridges was promoted to colonel in December 1983. In his later career, he would rise to the rank of major-general and be appointed director of the Kennedy Space Center.

For the payload specialists, it was a frustrating time; indeed, by the time Acton and Bartoe finally flew, they had been training for more than seven years. John-David Bartoe came from Abington, Pennsylvania, where he was born on 17 November 1944, and much later in his career would serve as NASA's Research Manager for the International Space Station. He earned his degree in physics from Lehigh University in 1966, and went on to attend Georgetown University for his master's credential (1974) and doctorate (1976). After gaining his undergraduate degree, Bartoe worked as an astrophysicist at the Naval Research Laboratory, and, like Loren Acton, had an intimate involvement with the Spacelab-2 payload by serving as co-investigator for two of its solar physics experiments. Yet the wait endured by Bartoe and Acton surely paled in comparison with the experiences of Karl Henize and Tony England. Mission 51F would be their first voyage into orbit ... almost *eighteen years* after they were selected by NASA in August 1967. In fact, they jointly hold an unenviable second place, behind Don Lind, for the longest wait by an American professional astronaut to reach space.

"A joy to work with" and "total enthusiasm" were phrases employed by Loren Acton to describe his friend and colleague, Karl Gordon Henize. In *NASA's Scientist-Astronauts*, Acton told Dave Shayler and Colin Burgess that it was Henize's infectious excitement for space travel which often prised him away from his work to simply gaze on the grandeur of Earth or find pleasure in the sensations of weightlessness. Aged 58 and securing a new record as the oldest man in space, Henize knew that 51F would probably be his only flight, and saw it as an adventure. Even at the time of his selection, Henize was a well-known academic, with a strong reputation as an accomplished professor of astronomy at Northwestern University. Born in Cincinnati, Ohio, on 17 October 1926, he grew up on a dairy farm with a love of the natural world. In his childhood, he excelled in science and mathematics and his time as a Boy Scout first introduced him to astronomy. (Prophetically, one of his heroes was Sir Edmund Hillary, conqueror of Everest ... and in October 1993 it was during Henize's own attempt to summit the world's highest peak that he lost his life.) When their father died from pneumonia and a kidney infection, the eight-year-old Karl assisted his older brother, Wilson, in running the dairy and icehouse during the difficult years of the Great Depression.

When the United States entered the Second World War, Wilson Henize volunteered for submarine service in the Pacific, whilst Karl entered the Navy's V-12 officer candidate programme, which took him to Dennison University in Cincinnati and the University of Virginia. The war ended before he received his naval commission and he entered the reserves, eventually reaching the rank of a lieutenant-commander. He pressed on with his studies, earning a degree in mathematics in 1947 and a master's in astronomy in 1948, both from the University of Virginia. Henize moved to Bloemfontein in South Africa to carry out a spectroscopic survey of the southern sky for the University of Michigan and, upon his return to the United States, enrolled as a doctoral candidate. He was awarded his PhD in astronomy in 1954 and accepted a post-doctoral fellowship at Mount Wilson Observatory in Pasadena, California, conducting spectroscopic and photometric studies of emission-line stars and nebulae. Later, he served as a senior astronomer at the Smithsonian Astrophysical Observatory, helping to set up a global network of satellite tracking stations, and in 1959 he became an associate professor (and later full professor) at Northwestern University. His first ties with NASA emerged in the mid-1960s, as principal investigator of an ultraviolet stellar spectrometer for use on the Gemini X mission. By this time, Henize had already applied for the first group of scientist-astronauts, but was turned down due to his advanced age. Two years later, the age limit had been relaxed and he was selected in 1967 at the age of 40. During pilot training at Vance Air Force Base in Enid, Oklahoma, Henize passed with flying colours and earned the distinction of being the oldest man to complete the 18-month military course.

In a sense, Henize and Anthony Wayne England were polar opposites on the Spacelab-2 mission. When they were selected together, Henize became the *oldest* astronaut chosen by NASA at that point and England, at just 25 years of age, became the youngest. Today, England retains his record. He was born in Indianapolis, Indiana, on 15 May 1942, the son of a livestock insurer and part-time carpenter, and his formative years were spent on the move from place to place. By the time he entered his teens, the family had settled in North Dakota, a place that England loved and where he received much of his schooling. At around this time, he developed an interest in aviation, although he feared that his less-than-perfect eyesight would prevent him from pursuing this career option. Still, he attended the local Civil Air Patrol and at one encampment at Ellsworth Air Force Base in South Dakota, he met the young girl who would become his future wife. England applied for both Harvard and the Massachusetts Institute of Technology and was surprised to be accepted by *both* colleges; MIT was his preferred choice, because he perceived its scholarships to be better. He received his degree in Earth and planetary sciences in 1965, followed by a master's qualification, and research towards a PhD. In August 1967, he was selected by NASA as one of its second group of scientist-astronauts ... yet was alone amongst in them in not yet having completed the final requirements for his doctorate. "I thought my PhD research was completed when I left for NASA," he told Shayler and Burgess, "but while writing the dissertation during the fall of 1967, I found that some of the experimental work was inconsistent and had to be redone. My first opportunity to repeat the experiments was during the summer of 1969, after flight school." During the summer that Neil Armstrong and Buzz Aldrin walked on the Moon, England worked feverishly to complete his PhD, defended it in the winter and graduated in May 1970. Over the years, he has made no secret of his disappointment at being unable to walk on the lunar surface and, although he supported the Apollo 16 mission, England opted to resign from NASA in August 1972 to join the US Geological Survey as a research geophysicist, later rising to become deputy chief of geochemistry and geophysics. Seven years later, in June 1979, he rejoined the astronaut programme, hoping for a chance to fly the Shuttle. His return to NASA was not quite what he expected. By his own admission, England considered himself to be "kind of an oddball" and never felt that he truly fitted back in amongst either the older astronauts or the newer ones. His return to the space agency came with one self-imposed condition: he would stay around long enough to get a flight, then return to academia. Getting that flight took longer than anticipated.

By the beginning of 1985, the Shuttle schedule was in disarray; Mission 51E had been cancelled a few days before it was due to be launched, and Spacelab-2 had slipped from April to mid-July. At last, however, payload processing for the complex mission could get underway and sequence tests verified that all the components could function as a single, integrated unit. In May, a 'closed-loop' dummy run successfully commanded the entire payload, remotely, from the POCC in Houston, and on 8 June Spacelab-2 was moved into the Orbiter Processing Facility and installed aboard Challenger. Tests of their compatibility, including another POCC run, were successful, even utilising the Tracking and Data Relay Satellite and high-rate data transfer modes. Despite its main focus upon solar physics, the mission also encompassed studies of atmospheric and plasma physics, high energy astrophysics, infrared astronomy, some technological research and life sciences. Akin to the verification flight of the Spacelab module, the mission was a 'free-for-all', covering virtually all possible areas of scientific inquiry for which the system had been designed. Eleven of Spacelab-2's 13 investigations were developed by United States scientists and two came from the United Kingdom, including an X-ray telescope supplied by the University of Birmingham.

"It was about as multi-disciplinary as you could imagine," Loren Acton said later. "One of the things we learned was that we tried to accommodate and carry out a great variety of experiments." One of those experiments, though, caused some consternation. In the eyes of the world, Mission 51F would become 'The Coke and Pepsi Flight'. Even today, Acton admitted, his visits to schools are dominated by questions from children about the intricacies of carbonated drinks in space, rather than the wonders of solar physics. "Coca Cola had gotten permission to do an experiment in space to see if they could dispense carbonated beverages in weightlessness," Acton recalled. "They got approval to build this special can, put significant money into it and were all set to fly it on one of the early Shuttle missions. This was during the 'Cola Wars' when Ronald Reagan was in the White House. Somebody at a high level at Pepsi found out, went to their contacts in the White House and said 'This cannot be allowed to happen – that Coca Cola would be the first cola in space'. So the Coke can was taken off the mission it was supposed to go on and Pepsi was given time to develop their own can so that they could both fly on the same flight. It turned out that our mission ended up getting the privilege of carrying the first soda pop in space." The astronauts were even instructed to photograph each other drinking the carbonated beverages, with the date/time recording features of their cameras activated, to show which was consumed *first*. (It turned out to be Coke, but neither drink met with approval from the crew.) Although subsequent pictures from 51F showed Bridges' red shift drinking Pepsi and Musgrave's blue team sipping Coke from the dispenser bottles, the distraction for Acton from what was actually an important mission for astrophysical and solar research proved an extreme irritation. "The morning before the launch, there is always a briefing, during which all the last-minute things that need to be talked about get talked about," Acton said later. "We were about halfway through a briefing on the latest data concerning the Sun, when who should walk in, but the chief counsel of NASA, who began to brief us once again on the Coke and Pepsi protocols."

Acton hit the roof. "We've been getting ready for this mission for *seven years*," he thundered. "It contains a *great* deal of science. We have a *very* short time to talk about the final operational things that we need to know. We *don't* have time to talk about this stupid carbonated beverage dispenser test. Please leave!"

The chief counsel turned on his heel and walked out.

This came at a pivotal time in the so-called 'Cola Wars', which had reached their zenith. By 1983, Coca Cola's market share had shrunk significantly in the face of competition from the sweeter Pepsi and the company's senior management had undertaken efforts to change the flavour of the brand. Labelled 'New Coke', this was scheduled for release in 1985, to coincide with Coca Cola's centenary, and in late April it was unveiled with much fanfare. Pepsi used the occasion to declare itself the winner of the Cola Wars and, over time, the American public reacted poorly to the new formula. On 11 July, with Coke and Pepsi dispensers already packed into the middeck on the eve of the first 51F launch attempt, Coca Cola performed a strategic retreat and decided to bring back its original, 'classic' version.

SUN WATCHING ...

Aside from the peculiar media interest in the two colas, the verification of the Instrument Pointing System was a critical mission objective for 51F, particularly in light of its most high-profile future use: the ASTRO-1 mission in March 1986 to undertake ultraviolet observations of Halley's Comet. "If the IPS doesn't work," a senior ASTRO-1 manager told *Flight International*, "the whole mission is down the drain." During Challenger's first day in orbit, tests were performed on the IPS, with the crew unstowing it from its horizontal position on the pallet train and aiming it at several solar targets to verify pointing capabilities and overall accuracy. On 51F, sadly, success took some time to achieve. The SOUP experiment, provided by the Lockheed Solar Observatory, turned out to be somewhat irksome. This proved particularly frustrating for Loren Acton, who, as the instrument's co-investigator, had been specifically picked to fly on Spacelab-2. "About eight hours after its activation," he said later, "it shut itself off and would not accept the turn-on command. The crew did everything it could, but ended up having to forget it." SOUP was meant to observe solar magnetic field activity in different wavelengths and polarisations of visible light and, when it finally came fully to life later in the mission, it returned dramatic movies of the active Sun that proved far more consistent in quality from frame to frame than previously obtained by high altitude rocket flights. In particular, it recorded several hours of sunspots and active regions, including 6,400 photographic frames that solar scientists hailed as "unique" in terms of their extreme stability. Not until the seventh day of the flight (4 August) did SOUP awaken. Sadly, Acton had suffered a severe bout of space sickness and the happy news was relayed to him by his blue shift crewmate, John-David Bartoe. "I got sick as a dog," Acton said years later. "Thirty seconds after the main engines shut off, I felt like my stomach and my innards were all moving up against my lungs. I was sick for four days and learned very quickly that you cannot unfold your barf bag as fast as you barf! When Bartoe came to tell me SOUP was alive, I was feeling so bad I didn't even get up to go look." Post-mission analysis and film processing of the results determined that bubble-like convective cells, known as 'granules', on the Sun's surface were in almost continuous motion at speeds of up to 4,000 km/h. These granules, which typically measure about 1,000 km in diameter, cover the entire solar disk, except those portions where sunspots are prevalent. SOUP showed that they 'float' like corks atop much larger convective cells – varying between 10,000 and 40,000 km across – in which hot fluid rises from the interior in the bright areas, spreads across the surface, cools and then sinks inward along the dark 'lanes'.

Adjoining SOUP was the British-built CHASE instrument, whose objective was to improve measurements of the solar helium abundance, which, at the time, was uncertain by a factor of three, relative to that of hydrogen. The ratio of hydrogen to helium was important for helping to verify models of the birth of the Universe, in particular the 'Big Bang'. Developed jointly by the Rutherford Appleton Laboratory in Chilton and University College London, CHASE recorded ultraviolet emissions from hydrogen and ionised helium, both on the solar disk and in the corona above the limb. Accurately accounting for helium in the Universe has proved pivotal to understanding a number of important astrophysical processes. Since all of the helium in the Sun's surface layers is thought to be primordial in origin, data collected by CHASE was of significance to cosmologists, as well as solar scientists. In addition to its primary task, the instrument also examined the properties of the Sun's outer atmosphere, revealing that hot, active-region material typically formed 'bridges' between the corona and the somewhat cooler chromosphere beneath it. Unfortunately, the instrument's capabilities were confounded by Challenger's orbit, since there was sufficient 'free' helium at the lower altitude to interfere with its measurements.

Eight hours of video and more than 500 still photographs of the Sun in hydrogen-alpha ultraviolet light were acquired by the HRTS, which had been developed by Guenter Brueckner of the Naval Research Laboratory in Washington, DC. This instrument, which had ridden several high altitude rocket flights since 1975 prior to being commissioned for Spacelab-2, observed the fine scale structure of the chromosphere, corona and the 'transition' zone between them, recording over 19,000 exposures of sunspots, spicules (high-speed gas jets shooting up into the corona) and explosive events. Its main focus was upon how energy could be transported and dissipated to form our parent star's hotter, outermost regions by 'seeing' spectral lines emitted in the chromosphere, transition zone and corona at the highest resolution attainable at that time. Eighteen instrument-observing sequences were executed – some many times – over 23 orbits devoted to solar studies. Essentially all of the available film was used, exposing roughly 600 full-frame and 18,000 short-frame spectrograph exposures, in addition to the eight hours of video. It "has the ability to zoom in on very small features on the surface of the Sun," said Bartoe in an April 1986 interview. "The primary goal is to try to understand how the Sun makes the solar wind. Some interesting things happen right on the surface of the Sun – for instance, the temperature goes up very dramatically as you go just above the surface. We're trying to look at that region right there, where that sudden transition of temperature takes place. Most of the light emitted there is in the ultraviolet." Unfortunately, reflected sunlight led to higher than expected temperatures in Challenger's payload bay and pointing problems with the IPS caused complications for HRTS and the other three solar physics instruments. Moreover, the instrument had to be powered down on several occasions in order to prevent its computer and film from exceeding maximum temperature limits. By the second orbit of HRTS operations, its spectrograph started to lose sensitivity, impairing its scientific yield to an extent, although excellent results were ultimately achieved. This being a Naval Research Laboratory payload, it was entirely appropriate that solar scientist Bartoe from that institution flew on the eight-day mission. Watching from the ground was fellow NRL physicist Dianne Prinz.

"The position of Alternate Payload Specialist during the mission is at the top of the pyramid over all the experimenters who are trying to get information up to crew," Bartoe said in April 1986. "During this mission, Dianne had to listen to five or six or seven telephone conversations at a time, listen to the crew and then try to sort it all out. In my opinion, this is the toughest job, much more difficult than flying. We only had seven people and two telephone lines up there!" Added Prinz: "I suppose the crew was just as tired as we were on the ground, but it was really fatiguing trying to keep track of everything at once. Each experimenter was very interested in getting dedicated information about his experiment, but I'd try to prioritise what we could ask of the crew without overtaxing them. I was the interface for all the experiments – not just the solar ones – and I'd be called into back rooms to iron out problems that occurred. There was almost no time to think. All of the NASA crew was of a like mind when it came to the purpose of the mission, which was to get the science done. Everybody knew what everybody else was doing and there were certain fatiguing tasks they'd do in order to get better results. For instance, the crew controlled 'free drift' to maintain pointing stability on the Sun. It was that kind of understanding of what the intent of the mission was that made the crew such an outstanding group. We had a fantastic relationship with each other; very, very close." At the time of the Challenger disaster, Prinz expected to serve with George Simon and a British payload specialist on a reflight of the solar telescopes, known as 'Sunlab', in September 1987. Sadly, even after the resumption of flight operations, this mission never transpired.

The fourth instrument affixed to Spacelab-2's pointing system (SUSIM) was designed to monitor long-term variations in solar ultraviolet radiation which, although it is only a small percentage of the Sun's total output, is the main energy source for Earth's upper atmosphere. The advertised capabilities of the instrument, which was supplied by the Naval Research Laboratory, were to measure solar irradiance from 120-400 nm with an accuracy of between 6-10 percent. The difficulty was that solar ultraviolet radiation was also responsible for causing the instruments measuring it to become degraded and lose their accuracy over time. Consequently, long-term solar changes could not be effectively distinguished from instrument changes. This could easily lead to misinterpretations of long-term changes in instrument readings as 'real' trends in the solar-terrestrial relationship. SUSIM, however, could overcome such problems because it was only destined to fly aboard a week-long Spacelab mission, before being returned to Earth for analysis and calibrations to determine its degree of degradation and thereby enable the deterioration of sensors on satellites to be compensated for. Determining its 'absolute sensitivity' was done at the National Bureau of Standards' Synchrotron Ultraviolet Radiation Facility (SURF) in Gaithersburg, Maryland, both prior to delivery of the instrument for integration and testing and again following its return from Spacelab-2. These calibrations, which began in January 1984 and ended in January 1986, were done by measuring the spectrometer's response to an absolute source of radiation produced by SURF. The results from SUSIM indicated an approximately 38 percent loss in sensitivity for the solar spectrometer and a 20 percent degradation in its calibration spectrometer; both also had a declining loss between seven and ten percent at longer wavelengths of 400 nm. Although this magnitude is quite low, considering the 24-month interval between calibrations, it proved significant when compared to the anticipated small-scale changes in solar ultraviolet output. More alarming have been the results of SUSIM and Shuttle Solar Backscatter Ultraviolet (SSBUV) observations showing depletion of stratospheric ozone since the mid 1980s, due partly to solar effects and atmospheric dynamics, but chiefly to the release of man-made carbon fluorides. This could prove catastrophic for the continuation of life on Earth, because the effects observed over a four-year period, if extrapolated over half a century, could lead to a 60 percent ozone depletion.

All four IPS-mounted instruments proved highly successful, but the initial prospects seemed bad on the evening of 29 July 1985, when 51F Flight Director Lee Briscoe told journalists that the pointing system could not track solar targets smoothly with its built-in Optical Sensor Package (OSP). Instead, it tended to 'wander', or move erratically. "It's like a drunk that can hold it between the ditches," remarked one observer, "but *can't* stay between the white lines!" Karl Henize, a member of Challenger's red shift, was able to fine-tune the tracking system – and, for a time, CHASE and HRTS were used as pointing sensors – but he complained that it did not stay 'locked' on target during manoeuvres. Indeed, the *Houston Chronicle* reported on the 30th that, rather than tracking the equivalent of "a dime at a range of two miles", the IPS "would do good to hit the broad side of a barn". Still, Spacelab-2 Mission Manager Roy Lester told journalists of his confidence that "it will work very well before the mission is over". In spite of these problems, and the Shuttle's lower-than-planned orbit, which affected all of the instruments, it is quite astonishing that Spacelab-2 returned to Earth as a tremendous scientific success.

During troubleshooting of the IPS, a number of software changes were uplinked to adjust its optical sensor package and thereafter most of the originally scheduled observations were completed.

Unlike Spacelab-3, the seven men did not have the luxury of a pressurised module and had to work from the relatively small and cramped area at the rear of Challenger's flight deck. "I held a joystick control in my hand, like on a video game," recalled Bartoe, "that permitted us to move the solar pointing telescopes around to point at particular features on the Sun. We would have a conference call – a solar conference – just before sunrise on each orbit. This gave us a chance to talk to the investigators so we'd know what we were trying to do on that orbit. We also received about 20 feet of typed messages over the teleprinter, every orbit. This was the first mission where we had to replace the teleprinter paper, because we had to totally replan the mission. One of the advantages of the fact that this mission took years from [its] formation to [its] launch was that there was a long time to fine tune each instrument's observing plan. That was good, because a lot of things went wrong, but we really understood how the various instruments' observations fit together and it was easy to make quick changes."

Despite their lack of a Spacelab module, the dual-shift system made Challenger's flight deck considerably more roomy, with no more than four people on duty at any one time. However, it was also *noisy*, according to Story Musgrave. "They were banging around all night," he recalled. "It was hard to sleep. I took a pill to help me to sleep and I *forgot* I took the pill, so I went off to sleep, just out, floating around. Your head doesn't fall; there's *no gravity* to make that happen, so I went off to sleep and actually I went floating *upstairs* where my buddies were!" Their astonished reaction as the sleeping Musgrave drifted towards them was to shriek: "Oh! A *monster!*" Using fingertips, they pushed him back downstairs into the middeck. There was time for fun, too, with the crew recording a few video shorts of their antics. In one episode, Musgrave and Fullerton – both bald – comically argued at length over a hair brush ...

The work schedule, though, was punishing. "During your 12 hours *on*, you ran all these instruments," recalled Fullerton. "During the 12 hours *off*, you had dinner, slept, had breakfast and then went to work. Two weeks before launch, we set that up. I anchored my schedule to overlap transitions, so if something came up on one shift, I could learn about it and carry it over to the next shift. I also had to stagger things so I got on the right shift for re-entry, so I was in some kind of reasonable shape at the end of the mission. We had the red team sleeping up till launch time, so that once we got on orbit, they were the first one up and they'd go for it for 12 hours. The last week [before launch], we didn't see the other team or I only saw part of one and part of the other myself." It was a challenging mission for Fullerton and marked just one in a series of peaks in an aviation career which had begun long before – and would continue long after – his days as an astronaut.

Charles Gordon Fullerton was born in Rochester, New York, on 11 October 1936 and an interest in aviation came naturally; his father was in the Army Air Corps, forerunner of the US Air Force. From a young age he could remember receiving an 'educational' instrument panel, cardboard rudder pedals and control stick – "toy is *not* the word," he told the NASA oral historian – one Christmas. Although Mr Fullerton Senior left the military after the Second World War, he *did* take his son for a flight in a small, two-seater Aeronca aircraft. As the young Fullerton matured, his interests were primarily in mathematics and science. He attended California Institute of Technology, and received bachelor's and master's credentials in mechanical engineering in 1957 and 1958. Flying remained a driving ambition, though. As a member of the Reserve Officer Training Corps, he had the opportunity, from time to time, to fly aboard the T-33 Shooting Star at Norton Air Force Base in California, after which he formally entered the Air Force and was admitted into aviation school.

Fighters were the most desirable aircraft, Fullerton recalled, and *that* was what pushed him to choose training on the F-86 Sabre over the F-100 Super Sabre; pilots of the latter, he felt, were very often shunted into flying heavy bombers later in their careers. As circumstances worked out, ironically, quite the *opposite* happened in Fullerton's case and initial instruction in the F-86 led to an assignment to become a B-47 Stratojet bomber pilot at Davis-Monthan Air Force Base in Arizona. "My *whole class*," he said, "out of F-86s were sent to bombers and transporters! In looking back, though, while it seemed like a terrible thing at the time – some of my classmates almost wanted to *slit their wrists* – I went on with it and decided to be the *best* B-47 pilot the Air Force had … and, in the long term, it paid off." Fullerton felt that the assignment gave him both fighter experience and heavy aircraft experience and possibly proved pivotal in his selection to fly the Approach and Landing Tests aboard Enterprise, more than a decade later. Fullerton remained at Davis-Monthan for four years and his air wing was on full alert at virtually all times. "It wasn't like you were on the verge of World War Three every minute," he explained, although the Cuban Missile Crisis in October 1962 *certainly* caught everyone's attention and Fullerton was dispersed to Hill Air Force Base in Utah for a time to ensure that a strike on the home base would not catch all the aircraft.

Selection to attend the Aerospace Research Pilot School at Edwards in 1964 was followed by assignment as a bomber test pilot at Wright-Patterson Air Force Base in Ohio. At this point in his career, the space programme came knocking at his door, when he was selected as one of the pilots for the Manned Orbiting Laboratory … a project which was cancelled in the summer of 1969. At 32 years old, Fullerton was one of only a handful of men to be selected by NASA, in September of that year, for astronaut training. After a spell working in a support capacity for the final four Apollo lunar missions, he began actively pursuing the Shuttle and was heavily involved in the design of its cockpit displays and controls. At length, Fullerton was paired with Fred Haise for the Approach and Landing Tests of Enterprise, then with Vance Brand for the fourth Orbital Flight Test mission and finally, in the summer of 1979, with Jack Lousma for STS-3. That flight took place in March 1982, lasted eight days, and due to unacceptable weather at both of its primary and backup landing sites, became the *only* Shuttle mission to touchdown in a place other than Florida or California. Instead, Lousma and Fullerton made landfall at White Sands in the mountains of New Mexico.

The work aboard Spacelab-2 was intense, but for the science crew, immensely rewarding, since they had either participated in the development of experiments as co-investigators or were flying alongside instruments built at their former workplaces. Henize, for example, spent some time working for the Smithsonian Astrophysical Observatory before joining NASA and one of his tasks on Mission 51F was to operate one of their experiments. Known as the Small Helium-Cooled Infrared Telescope (IRT), it was developed under the direction of principal investigator Giovanni Fazio and examined infrared radiation from a number of celestial sources. In doing so, it worked in conjunction with the Infrared Astronomy Satellite, launched in 1983, which produced an all-sky survey. During Spacelab-2, the telescope produced mixed results, meeting more of its technical objectives than its scientific ones. In particular, the performance of its superfluid helium/porous plug cooling system exceeded expectations and demonstrated convincingly that extremely low operating temperatures – down to 3.1 Kelvin – could be established and maintained. However, saturation of its mid-wavelength detectors by an intense infrared background compromised some astronomical results. Nonetheless, about half of the galactic plane was satisfactorily mapped at shorter wavelengths than were possible using IRAS. Additionally, data from the telescope yielded useful information about the infrared 'background' of Challenger herself and helped to determine the extent to which 51F's experimental plasma physics studies and the 'Shuttle glow' phenomenon affected its sensitivity. "We see the IRT back there doing its hickory dickory dock," Loren Acton told journalists of the telescope's sweeping motion in the payload bay. "It makes you think of one of those birds that you dip in a glass of water and it dips, dips, dips."

... PARTICLE SNIFFING ...

The Smithsonian's IRT was attached to the third, rearmost pallet in the payload bay, together with a 158 kg Plasma Diagnostics Package (PDP), built by the University of Iowa, and the Superfluid Helium Experiment (SFHE). The former previously flew aboard STS-3 and consisted of a small cylindrical canister of electromagnetic and particle sensors to *sniff* the environment surrounding the orbiter. Its data on the atomic cleanliness of the payload bay proved invaluable in allowing NASA to commit sensitive instruments to Spacelab-2. During its first venture into space, it was hoisted aloft by the RMS mechanical arm in order to analyse the electromagnetic and particle conditions within about 14 m of Columbia. Its data provided, for the first time, detailed insights into the strange plasma 'wake' of the Shuttle passing, boat-like, through the ionosphere of low-Earth orbit. This wake might, it was theorised, complicate the measurements of future detectors and the STS-3 results were pivotal in planning and developing the Spacelab-2 payload.

The 'ionosphere' spans the altitude range 60 to 1,000 km and has shown itself to be an excellent location from which to study ionised gases known as 'plasma'. It had been recognised since the early days of human space flight that low orbiting vehicles are immersed in ionospheric plasma. Shuttle-era scientists were able to deploy and later retrieve small satellites, directly expose sensors and disturb it with beams of energetic particles in order to trace, modify or stimulate the environment. For this reason, the PDP flew in conjunction with the Vehicle Charging and Potential (VCAP) experiment, provided by Utah State University to examine the orbiter's electrical characteristics and its effects on surrounding ionospheric plasmas. This included a fast-pulse 'gun' that fired 100-volt bursts of electrons for durations ranging from 500 nanoseconds to several minutes. It was to investigate the extent to which electrical charges accumulated on the Shuttle's insulated surfaces and how 'return currents' could be established using a limited area of surface-conducting materials to neutralise active electron emissions. During STS-3, VCAP data had provided practical experience of using particle accelerators in readiness for the Spacelab-1 mission. Plans were also afoot, in collaboration with the Italian Space Agency, to build a revolutionary 'tethered satellite', which would be trawled through ionospheric plasma on a 20 km conducting cable to generate electricity. Such tethers, it was argued, could offer a steady supply of power for future spacecraft or space stations.

For four days, beginning on 31 July 1985, the PDP was extended by the end of the Shuttle's mechanical arm, before being released into space to acquire wideband spectrograms of plasma waves at frequencies of up to 30 kHz and distances of up to 400 m. Due to the reduced level of RCS propellant, caused by the Abort To Orbit, a plan to conduct a flyaround survey of the package could not be realised in full. Nonetheless, valuable data was gathered. Two types of interference patterns were subsequently identified in its wideband data: one associated with the ejection of electron beams from VCAP and the other with lower hybrid waves generated by interactions between the neutral gas cloud around Challenger and ambient ionospheric plasmas. Whilst Fullerton, Bridges and Musgrave executed manoeuvres close to the PDP over a six-hour period on 1 August, a momentum wheel spun the satellite to stabilise it sufficiently for accurate measurements. Among the notable findings from its plasma wave instrument was a region of intense broadband turbulence around the Shuttle at frequencies from a few hertz to ten kilohertz. The highest intensities occurred in the region 'downstream' of Challenger and along magnetic field lines passing close to the orbiter, tending to increase during periods of high thruster activity.

In general, the joint PDP-VCAP observations showed that thermal ion distributions around the Shuttle were considerably more complex than predicted and, frequently, there was an unexpectedly intense background level of ion current due to incoming hot ions. Surprisingly, these ions often tended to 'change energies'; indicative of high temperatures and turbulent plasma activity and demonstrative of the huge impact of a large, gas-emitting spacecraft on the ionosphere. Indeed, water vapour was detected in the orbiter's immediate vicinity, extending out to a couple of hundred metres, and proved particularly dominant in its wake. Whilst the PDP was held some 10 m away from the vehicle on the end of the RMS, Challenger performed a roll manoeuvre to sweep the sensors through this wake. Measurements showed that ions from the 'ambient' ionosphere were accelerated into the wake from 'above' and 'below' the spacecraft; triggered, perhaps, by a strong electric field created by density differences between the two. Ground-based and PDP observations were made of Shuttle thruster firings, yielding faint red airglow emissions which produced a cloud

300 km in diameter. Further tests indicated that even minute thruster firings affected the ambient plasma in some way. One particularly interesting experiment, conducted on four occasions whilst over the University of Tasmania's low-frequency radio observatories in Hobart, sought to test the feasibility of carrying out astronomical measurements through artificial 'windows' temporarily created in the ionosphere by thruster bursts. Some cosmic signals were received, but radio waves in the band lower than three megahertz were blocked by the ionosphere. In fact, said Eugene Urban after one 30-second firing over Milstone Hill in Massachusetts, the 'hole' produced was "very large and bright" for radio observations. Significantly for future astronomical and plasma physics research, it was determined by the PDP that contaminants released by thruster firings can interfere with measurements of 'natural' plasma made from instruments in the payload bay.

The superfluid helium experiment was to explore the properties of this unusual substance and demonstrate the performance of a reusable cryostat in space. Superfluid helium, in which helium is cooled almost to absolute zero, was tested for its efficiency as a cryogenic coolant in future astronomical or solar physics instruments. Superfluid helium moves freely through pores so small that they block normal liquids and conducts heat a thousand times more efficiently than copper. Prior to Spacelab-2, many subtleties of the substance were unknown because gravitational effects disturbed the superfluid state in terrestrial experiments. Of particular interest was the behaviour of capillary waves and their sloshing motions and temperature variations. Early results suggested that it could indeed be managed efficiently in space using a porous plug cryostat.

On the middle pallet was an X-ray telescope provided by the University of Birmingham in England. It was one of few instruments not to be affected by the low orbit. It comprised two co-aligned telescopes and employed a coded-mask technique to make X-ray images at energies between 2.5 and 25 keV. The two masks contained different sized holes for different angular resolutions in order to enable the higher resolution telescope to make detailed studies of brighter celestial sources and the lower resolution one to examine fainter regions of diffuse emission. By the end of the mission, more than 75 hours of data were obtained, including observations of eight galactic clusters and the Vela supernova remnant.

"It was a great mission," said Gordon Fullerton. "Some of the missions were just going up and punching out a satellite and then they had three days with [very little] to do and came back. We had a payload bay absolutely stuffed with telescopes and instruments." It proved quite remarkable, journalists reported after the flight, that 51F had turned into such a grand scientific success after suffering an on-the-pad main engine shutdown, a hairy abort during ascent and the problems with the IPS. "We even made up for the fuel we'd had to dump on the way up because of the engine failure and eked out an extra day on it," added Fullerton. "We were scheduled for seven and made it *eight!*"

In addition to the solar and astronomical telescopes and plasma physics detectors, possibly the most unusual instrument was the Cosmic Ray Nuclei Experiment, designed to count and analyse cosmic rays as much as a hundred times more energetic than any previously studied. Nicknamed 'the cosmic egg', it was developed by the University of Chicago. Early results showed that the experiment recorded about 24 million particle events, of which perhaps as many as 30,000 had energies in the formerly unexplored range from hundreds of billions to trillions of electron volts.

... AND LIFE SCIENCES

Although Spacelab-2 was a multi-disciplinary flight to continue verifications of the new laboratory configuration, the presence of two life science experiments in the middeck seemed at odds with the astronomical, cosmic ray, plasma physics and solar research conducted elsewhere on this mission. One investigation, provided by the University of Wisconsin at Madison, measured vitamin D metabolite levels of the crew to gather additional data on the causes of bone demineralisation – loss of density – and mineral imbalances during prolonged exposure to microgravity. Astronauts had typically returned from previous missions with evidence of loss of lower body mass, especially in the calves, together with decreases in muscle strength and negative calcium balances. This process has been compared to the initial phases of some bone diseases or the wasting away of muscle observed in bedrest patients. To undertake the research, three vitamin D metabolites were measured in blood samples taken from four crewmembers before, during and after the mission. Although the levels of two metabolites remained essentially unchanged, a third underwent an interesting pattern: showing a rise in the level of blood samples collected early in the flight, dropping around 3 August and returning to normal after landing.

The second life sciences experiment, labelled 'Gravity-Induced Lignification in Higher Plants', studied the effects of microgravity on the growth and lignification in oats, pine seedlings and Chinese mung beans in a pair of Plant Growth Units (PGUs) on the middeck. This research had begun on STS-3 and sought to determine whether 'lignification' was a response to gravity or a genetically determined process with little environmental influence. Lignin is a structured polymer, which gives plants the structural strength to maintain a vertical posture despite the effects of gravity, and thus is highly important for the plant's ability to grow properly. Earlier experiments aboard Skylab and the Soviet Salyut space stations throughout the 1970s had revealed that the strange conditions in low Earth orbit did indeed cause root and shoot growth to become disorientated, as well as increasing their mortality rates. However, little was known about the physical changes within them. Understanding how plants behave and grow in the absence of gravity was – and, with visions of crews flying long-duration missions to Mars or a deep-space asteroid, still is – essential for growing foodstuffs in space. Chinese mung bean, oat and slash pine seedlings were chosen for both STS-3 and Spacelab-2 because all three could grow in closed chambers and under relatively low lighting conditions. Pine is a 'gymnosperm', meaning that it is capable of synthesising large amounts of lignin, and it was believed that its growth was directly affected by gravity. Unlike the mung bean and oat seedlings, which were germinated only hours before Challenger's launch, the pine samples were germinated four to ten days earlier. Preliminary results indicated that the mung beans and oat seeds behaved normally and the pine seedlings grew well in space. Some reduced growth, in the order of 15-20 percent, in the mung beans was observed and both they and the oat roots grew 'above' the supporting medium of the PGUs, indicative of disorientation. The lignin content was significantly reduced in all three species, compared to the ground-based controls, providing direct evidence of the important role of gravity in lignification.

In terms of its solar and plasma physics research, Spacelab-2 proved to be an extraordinary success, snatching triumph from the jaws of what could have been an aborted mission. Furthermore, the decision to extend the flight from seven to eight days increased its scientific yield substantially. Finally, at 10:43 am PST (1:43 pm in Florida) on 6 August, Fullerton and Bridges fired the OMS engines for 172 seconds to begin the hour-long glide through the atmosphere, descending across the expanse of the Pacific Ocean, to Edwards Air Force Base in California. The re-entry profile featured a number of test inputs on the control stick, including a manual manoeuvre of the aft body flap at Mach 18. Touchdown on Runway 23 at 11:45:26 am PST was picture perfect, with the pilots guiding Challenger to a halt in less than a minute on a rollout of just 2,500 m. Although he had flown with many of the payloads – the PDP, the VCAP, the plant lignification experiments – on his previous mission, Fullerton had a very different role on this flight. "The pressure is higher when you're the commander: the pressure of making sure that not only you, but somebody else, doesn't throw the wrong switch! During the re-entry, it's your fault if this doesn't come out right. When you're in the right [pilot's] seat, it's not all your fault; the commander bears culpability even if you make a mistake."

Many astronauts have remarked over the years that the sheer mental and physical stress of preparing for space missions can reach almost overwhelming proportions. "You're *really* tired after space flight," said Fullerton, "mostly because you elevate yourself to this high level of mental awareness that you're maintaining. Even when you're trying to sleep, you're worried about this and that. Flying in orbit is watching a clock. Everything's keyed to time and so you're worried about missing something [and] being late. We had 270 manoeuvres or something like that. Every sunrise and sunset, we had to go to a different attitude to put the right telescopes at the right stars or Sun. Those are all 'typing' exercises – typing long strings of numbers into the computer and the time to start to manoeuvre so it goes to the right attitude. You mess one number and you're going to go to wrong attitude, then you're going to miss that data."

BACK OF AN ENVELOPE

When Karol 'Bo' Bobko's crew returned to Earth in April 1985, they left some unfinished business behind. Some 300 km above Earth orbited the Navy's Syncom 4-3 communications satellite, whose omni-directional antenna had failed to unfurl and whose solid-propellant perigee kick motor had failed to fire. Instead of the intended geostationary location, 35,000 km above the Home Planet, it lingered in low-Earth orbit. An attempt by Bobko's crew to snare a deploy switch on the side of the satellite using a hastily-improved 'flyswatter' installed on the RMS arm during a contingency EVA by Jeff Hoffman and Dave Griggs had proven fruitless, and within weeks NASA was considering the option of a Shuttle repair. The triumph of the previous year's Palapa and Westar retrievals had imbued the space agency with the belief that its astronauts were bulletproof. For James 'Ox' van Hoften, who had performed two spacewalks in April 1984 to repair Solar Max, it would be an exciting mission.

It had not started out that way. When the crew was first announced in November 1983, commander Joe Engle, pilot Dick Covey and mission specialists Jim Buchli, Mike Lounge and Bill Fisher were scheduled to fly 51C in December of the following year and deploy a Tracking and Data Relay Satellite (TDRS). "I remember going to Seattle for a technical meeting," Lounge told the NASA oral historian, "and there was another crew that also had a very similar payload. Rick Hauck was the commander. We had a *race* to get to Seattle from Houston and we *beat* them. We thought about taking off the travel pod on the T-38 to see if we could go faster without our *clothes!*" Winning was still important to every astronaut and even two decades after the Original Seven, the *right stuff* was pervasive. However, problems with the Inertial Upper Stage booster very quickly changed these plans and both crews ultimately lost their TDRS payloads. By the May 1984 manifest, Engle's crew had moved to Mission 51E and picked up French payload specialist Patrick Baudry. The entire schedule was turned on its head when Discovery experienced her pad abort in June 1984 and in another reshuffle in August, Jim Buchli was reassigned to the West German Spacelab mission and van Hoften took his place. "Stale, pale and *male*," was Lounge's summary of the flight, "but we had a great time!" The 'staleness' and 'paleness' of the flight also changed when Engle's crew lost their TDRS and gained a trio of PAM-D payloads – Morelos-A, Arabsat-1B and Telstar-3D, plus Spartan and a Saudi payload specialist – with an expected launch in May 1985. Circumstances changed *again* in March 1985, when Mission 51E was cancelled and the remaining flights adjusted. Engle's crew was moved to August and became Mission 51I with two PAM-D payloads (the American Satellite Corporation's ASC-1 and Australia's Aussat-1) and the Navy's Syncom 4-4, and gained a pair of payload specialists: Charlie Walker, making his *third* flight with the McDonnell Douglas electrophoresis machine, and Hughes engineer Greg Jarvis to operate a fluid mechanics experiment. As an employee of the company which built Syncom, Jarvis would also serve as an 'observer' for its deployment.

Seated behind and between Engle and Covey, serving as the flight engineer, would be civilian astrogeophysicist John Michael Lounge, who had come to NASA's Payload Operations Division in 1978 after a lengthy spell in the Navy. He was born in Denver, Colorado, on 28 June 1946, and graduated from high school in Burlington, then entered the Naval Academy. He received his degree in 1969 and proceeded onto an immediate master's programme in astrogeophysics at the University of Colorado at Boulder. "If you got selected for, and got a scholarship, somewhere," Lounge told the oral historian, "the Navy let you go and spend a year or 15 months getting a master's degree before you reported to your first duty station." As a naval ensign, he wore his uniform whilst at the civilian university and the nature of his master's credential was an overt statement that he intended to someday apply for the astronaut corps. "To put it in perspective," he said, "two days after I reported to the University of Colorado ... is when Neil Armstrong stepped onto the Moon. Every young ensign in the Navy wanted to follow in his footprints, I'm sure, so I did that."

Lounge completed flight instruction in Pensacola, Florida, moved on to advanced training as a radar intercept officer – "the systems guy" – for the F-4J Phantom II fighter and was deployed to Vietnam in 1972 aboard the USS *Enterprise*. He participated in 99 combat missions, then undertook a Mediterranean cruise aboard the USS *America*. Returning to the United States in 1974, Lounge became an instructor in the Naval Academy's physics department and entered the Space Project Office in Washington, DC, in 1976, for a two-year tour as a staff project officer for reconnaissance satellites. On hearing of NASA's plans to hire astronauts he submitted his application, but was not selected in 1978. However, Lounge did garner sufficient interest to be offered a role at the Johnson Space Center, working in mission operations. When he asked his superiors in the Navy if they would consider reassigning him to Houston he was refused because the Navy wanted to offer him a position aboard an aircraft carrier. Consequently, in 1978 he resigned from the service, with the rank of commander, and spent two years in Houston, working Shuttle payloads and serving as a member of the Skylab Re-entry Flight Control Team. Lounge knew that committing to NASA in this way surely helped his chances of selection in the *next* astronaut class. Whilst there, he worked with two other engineers, a civilian named Bonnie Dunbar and Air Force officer Jerry Ross, and they all applied when the agency issued its next request for astronauts. In May 1980, all three were selected. "Our offices," Lounge recalled, "were within shouting distance, so you could *hear* the shouts!"

Lounge and van Hoften had some similarities in their backgrounds, since both served as military reservists after their time in the armed forces. Lounge was already in the Air National Guard and van Hoften was persuaded to join by George Abbey. One night in April 1985, whilst training for Mission 51I, the two men were on 'alert duty' at Ellington Field near Houston, when a spark of change was injected into their flight. "To make money in the reserves," said van Hoften, "you could go out and spend 48 hours sitting in an airplane ... but they'd give you 15 minutes to get airborne to guard the Texas coast. They paid us a couple of hundred dollars to do that." Midway through their alert duty, they heard about 51D's Syncom problem, started swapping ideas and contacted Joe Engle for his opinion. "Back then," continued van Hoften, "there was a much more can-do spirit at NASA and everyone felt like, hey, you can do *anything*." Back at JSC, they gathered the required data about Syncom – its size, mass properties, angular momentum – and with the benefit of paper and push-button calculator watches sat down with Engle and Covey to lay out a plan. At first, they considered retrieving Syncom and bringing it back to Earth, but the satellite was filled with volatile hydrazine fuel and this idea went nowhere. Finally the plan crystallised into a recovery *and* redeployment, involving the RMS arm. The engineers had determined a technique for bypassing the deploy switch and providing power to the electrical buses from the batteries by using test ports that were external to the satellite and NASA managers were sufficiently enthused to send

Spectacular launch, on the cusp of daybreak, for Mission 51I.

the crew to Hughes Aerospace in Los Angeles. In the auditorium, van Hoften – hardly able to speak, due to a bout of laryngitis – sketched out the plan to around a thousand engineers and technicians.

He reminded his audience that if 'Little Joe' Allen had managed to grab Palapa, then 'Ox' van Hoften could grab the much larger Hughes satellite! In all seriousness, he had handled large masses during his 41C mission and knew precisely what his capabilities and those of his space suit and those of the RMS were.

After his address, van Hoften was taken to one side by Hughes' president, who called NASA Administrator Jim Beggs and immediately offered to invest $5 million in the recovery effort. Engle and Covey and Shuttle Program Manager Glynn Lunney were at the home of Jay Greene, 51I's Lead Flight Director, when they got the go-ahead. They promptly drained a bottle of Old Overholt, which was Greene's favourite whisky. "We drank that bottle," recalled Covey, with a measure of glee, "celebrating just the fact that we had *hoodwinked* the whole system into letting us think that we could go do this. It would *not* happen today." Mike Lounge agreed. "We had a pretty narrow view of how NASA worked, as crewmen, back then. I didn't appreciate the big contractor team that was necessary to make all this work. I knew *who* built *what*, of course, but you didn't know what the problems were until they bubbled up to kind of the flight-readiness level." As the crew plunged into training, there were many aspects for which they *could not* train, since they had to prepare and hone their *skills*, rather than rehearse specific *tasks*, because there were no assurances for how the recovery would go. Van Hoften agreed. His first flight, the Solar Max repair, was choreographed down to the final detail. But *this* mission was different: they were winging it.

Winging it was an apt turn of phrase in more ways than one, because payload and crew assignments were also being moved around like pieces on a chess board. "NASA was manifesting in a pretty scat-of-the-pants fashion," remembered Charlie Walker, who had been moved off 51I and onto a later flight, "compared to today ... where everybody knows darn well what *all* the payloads for the next 14 missions are going to be. It was different, then. Things were coming and going off the manifest and *that* caused a lot of dynamic and some fairly short fuses!" Greg Jarvis was similarly reassigned and Engle's crew wound up as a five-man team; in fact, it was the *only* Shuttle mission of 1985 not to include payload specialists.

It is no accident that Colonel Joe Henry Engle of the US Air Force became only the fourth NASA astronaut in history to command his very first space mission ... and also the *last*, for no American spacefarer has since taken the helm of a crew on his or her rookie flight. Whether or not Engle *was* a 'true' first-timer when he rode Columbia into the heavens for the STS-2 mission is open to debate: for the Air Force had already awarded him astronaut's wings back in 1965 for passing an altitude of 80 km on no fewer than *three* occasions in the X-15 rocket-propelled aircraft. To follow this line of thought has drawn some observers to acknowledge Engle as the first man to fly into space three times. Others adhere to the Fédération Aéronautique Internationale (FAI) ruling that only missions which fly above 100 km – the so-called 'Kármán Line' – may be considered 'space flights'. By the end of his astronaut career, Engle had completed three X-15 missions and two Space Shuttle voyages.

Engle came from the small city of Chapman, in Dickinson County, Kansas, where he was born on 26 August 1932, with aviation in his blood. "I don't know, honestly, when I did *not* want to fly airplanes," he told the NASA oral historian. "My mom used to say the same thing to me, that she couldn't remember me seriously wanting to do *anything* but fly airplanes. Of course, I went through the fireman and the cowboy games, but my core desires and my core toys were always airplanes and flying." As a young boy, he recalled his older sister cutting a small aircraft from an old tin can; since it was sharp-edged, Mrs Engle forbade her son to play with it. Years later, the young boy who would go on to become a fighter pilot, a test pilot, astronaut, 'almost' a Moonwalker and ultimately a major-general in the Air Force, would wonder if this 'forbidden fruit' had actually spurred him on to choose aviation as his life's work. For Engle's hometown, Kansas State University was the place to go, since it was primarily an agricultural college and Chapman was a farming community. However, Kansas State did not offer aeronautical engineering and so he went to the University of Kansas instead, graduating in 1955. Shortly thereafter, Engle received his Air Force commission through the Reserve Officer Training Corps and entered flying school in 1957. "Chapman didn't have a runway," he told the historian, "so my only exposure to flying was at the annual Labour Day celebration, [when] all the farmers would bring in their goods and display them. There was a guy who landed on an alfalfa field and was giving rides." Engle and a friend bought a ride. During the summers at the University of Kansas, Engle worked as a draughtsman for Cessna Aircraft and received flying lessons from his supervisor, Henry Dittmer, a man whom he greatly respected, not only for showing him the rudiments of aviation, but also the *responsibilities* demanded of a pilot. Dittmer even got him a job sweeping the hangars at Cessna, in exchange for flying instruction.

The Air Force soon came calling, however, and when Engle completed fighter gunnery school he transferred to the F-100 Super Sabre to begin training as a fighter pilot at George Air Force Base in Victorville, California. This supersonic jet was one of the hottest in the sky at the time, and Engle gleefully recalled his experiences, practising air-to-air combat – dogfighting – high above the inhospitable terrain of Death Valley and Stovepipe Wells. On one particular occasion, he would literally find himself in a head-on confrontation with Chuck Yeager, perhaps the most famous test pilot on the planet and commandant of the Air Force's Test Pilot School. "I had a flight of four up and *that* was where you always found somebody to engage in a mock dogfight," Engle recalled. "I noticed two airplanes coming back from the north-east, heading back toward George, so I called the flight and we set up for an attack. I was salivating, because everything was ideal and rolling in. It turned out that those two airplanes were *Chuck Yeager* and our operations officer, who had been up to Nellis [Air Force Base in Nevada] to check out some advanced gunnery school classes that we were going to do. They were undoubtedly the two best fighter pilots at George and I found that out *real* quick! They *completely* tore up my flight and just scattered us to the winds and I learned then to be a little more cautious when making attacks and not get too over-confident!" This did not prevent Yeager from *recommending* Engle in 1960 for admission into Test Pilot School and only the second class of the new Aerospace Research Test Pilot School – in fact, it may well have *aided* it, for the commandant's philosophy was simple: the *best* fliers are the ones who have the *most* experience (Engle had more than 1,500 hours), who worked the *hardest* and who took it the most *seriously*.

After graduation, Engle moved over to fighter test operations at Edwards, which he described as "a pilot's heaven", and in 1963 he and fellow pilot Mike Collins applied for admission into NASA's third group of astronauts. Collins was accepted, but Engle was asked to withdraw his application by his commanding officer, Major-General Irving 'Twig' Branch, head of the Air Force's Flight Test Center. Branch had other ideas for Engle: in June 1963, he recommended him to replace Bob White on the X-15 project. "*That* just thrilled me to death," Engle remembered, "because it was a chance to get into *space* ... and to do it with a *winged* airplane, with a stick and rudder." Intuitively, he *knew* that he was still young enough to make another application to NASA at a later date, but flying the X-15 was *not* something a pilot *applied* for: it was a gift from the gods, the kind of assignment that one could only wait and hope would someday come their way. Engle made his first flight on 7 October of that same year and would go on to fly the X-15 no fewer than 16 times. Perhaps his greatest accomplishments were exceeding the Air Force's – though *not* the FAI's – requirement for a space flight, surpassing an altitude of 80 km, on *three* occasions in June, August and October 1965. Engle's parents came to Edwards to see him perform his first mission above 80 km and a few weeks later he joined NASA's Jim McDivitt and Ed White at the Pentagon to receive his astronaut's wings. "I was thrilled," said Engle, "and a little hesitant to rain on their parade of getting their astronaut wings. Initially, there was a lot of good-natured and some maybe serious rivalry going back and forth between the Edwards pilots and the NASA astronauts, but I didn't sense any of that at all. I think it was just three Air Force officers had qualified for their astronaut wings and, at that time, you got to go to the Pentagon to get your astronaut wings."

Engle knew that his time at Edwards was not open-ended and that he would shortly be reassigned, so he decided to reapply to NASA in the spring of 1966 and was selected in April. It brought mixed emotions, but he was leaving the best flying job in the world and was hopeful that he might someday be considered for a seat on a mission to the Moon. As circumstances transpired, *he was*. In the summer of 1969, Engle was named as a member of the Apollo 14 backup crew and might have rotated into the prime slot as lunar module on Apollo 17 ... but was ultimately bumped from *that* mission when NASA decided to fly a geologist, Jack Schmitt, instead. Engle was bitterly disappointed at the outcome and once remarked that the hardest part was explaining to his children that he *wouldn't* be going to the Moon ... but persevered and supported Schmitt in his preparations for what turned out to be one of the most remarkable missions of lunar exploration ever undertaken. In the light of this admirable attitude, Deke Slayton offered to support Engle in whatever he wanted to do next. Forever the pilot, Engle opted for the Shuttle. Slayton obliged him. By the summer of 1985, Engle had long since earned himself a reputation as one of the most gifted fliers in the astronaut office and one of the most experienced Shuttle pilots. Not only had he commanded STS-2, but he had also flown Enterprise during two of her Approach and Landing Tests over the California desert.

At one stage, it seemed possible that Engle might launch on his 53rd birthday, as the launch slipped into the final week of August, a time of year often problematic due to thunderstorms. The first attempt, on the 24th, was scrubbed due to local rain showers. "There weren't any big thunderstorms," recalled Dick Covey, "but that violated a criteria and it was interesting to watch how the weather criteria changed, and not necessarily for good reasons." Engle was particularly irritated, because the showers in question were far from the launch complex – some *fifteen kilometres*, in fact – and never came near to Pad 39A. Discovery had only a couple of days in which to achieve her 'rendezvous window' with Syncom 4-3, and the launch was quickly recycled for the following morning. Yet the second attempt also came to nothing when a failure was experienced in the Shuttle's backup flight software and it was decided that, due to concerns about the status of the propellant lines and insulation and two launch cycles, a couple of days break would be needed before another try. Next day, the 26th, was Engle's birthday and the crew presented him with his cake: unstowed from the middeck, where they had previously placed it, hoping to celebrate with him, *in orbit*. It was not to be and Covey was not alone in his frustration. "I can't *believe* we scrubbed for those two little showers out there," he remarked. "Anybody with half a lick of sense would have said: *Let's go. This could be a lot worse.*" John Young, the chief astronaut, came over and told him straight: the crew could *not* make calls on the weather, as their knowledge was limited solely to the view from inside the orbiter. *They* should focus on the flying, he said, and leave others to worry about the weather.

The morning of the third launch attempt, on 27 August, was even more dismal and the astronauts were clad in yellow raincoats over their flight suits as the Kennedy Space Center suffered a torrential downpour. After being strapped into their seats, Lounge and van Hoften – certain that the weather was too appalling for NASA to give the green light – released their harnesses and took a nap. From his perch on the right-hand side of the forward cockpit, Covey was amazed as they moved smoothly and crisply through the built-in holds and was even more amazed when he was given the go-ahead to start Discovery's APUs, with five minutes to go. As he flipped the switches, he could *see* sprinkles of rain on the forward windows. "The reason," Joe Engle explained, "was that they had one more day delay before they had to de-tank and *that* would have been two more days and the weather forecast was not good for the next day, anyway."

Downstairs, on the middeck, the hum of the APUs awakened Bill Fisher. "What's that *noise*?"

"We're cranking APUs," came the response from the pilots. "Let's go."

"Yeah, sure, we're *not* going anywhere today. Why you starting APUs?"

"Damn it, we're *going*! We're going to launch. Get back in your seats and get strapped in."

On the cusp of daybreak, at 6:58 am EST, Discovery's three main engines roared to life, followed by the blazing brilliance of the twin Solid Rocket Boosters as the 20th Shuttle rose on a pillar of glorious golden flame. "It turned out we launched right through the eye of a *hurricane*," van Hoften told the historian. "It coalesced into a *hurricane* and then we spent the whole time looking down at this major hurricane, going around wiping out Florida." He was making reference to what became Hurricane Elena, which originated as a gigantic tropical wave off the coast of East Africa on the 23rd and progressed weakly westwards, paralleling northern Cuba, becoming a tropical storm on the 28th and a hurricane on the 29th. From their vantage point in orbit, the 51I crew would take several stunning images of the hurricane, which expanded across the entire Gulf of Mexico and whose winds reached a peak of 205 km/h on 1 September, yet steadily weakened as it headed north and finally made landfall in Biloxi, Mississippi. All told, in its few days of havoc, Elena would wreak $1.2 billion of damage, but, miraculously, would cause no direct fatalities.

From the flight engineer's seat, Mike Lounge had crafted himself a small 'mirror' of Mylar and affixed it to the back of his checklist, hoping to watch the reflected launch through the overhead window, just over his left shoulder. "If you hold this mirror, right in your lap," he said, "you get this great view as the orbiter lifts up and rolls. You're looking through that window *right down* at the pad and this huge billow of smoke and flame ... and the pad gets smaller and smaller." There was little time to be a spectator, of course, as Lounge paged through his checklist, mentally ticking off each of the major milestones: *Negative return*, then *Single-engine TAL* and the call which had caused his predecessors on 51F such anxiety: *Abort To Orbit*. After what seemed like an age, the moment of main engine shutdown occurred and the sensation of weightlessness was felt through a gentle rising against the straps and the quirkiness of the checklist, *floating*, right in front of his very eyes. That comical sight was arrested very soon by a sight of unimaginable grandeur: *Africa*, looming large and spectacular in Discovery's windows. They were in orbit.

For Richard Oswalt Covey, this instant had been a *long* time coming; in fact, he was the *last* of the Thirty-Five New Guys, chosen in January 1978, to reach space. "I got that distinction," he said, "and that was hard to take." At the time of his assignment, in November 1983, he knew that he would not fly until at least the end of the following year. "At the time, I didn't realise I was going to be the very *last* one, but I knew I was going to be somewhere down there." Yet Covey's first mission would offer drama, rendezvous and spacewalks and so, too, would his *last* mission, which serviced the Hubble Space Telescope. Born in Fayetteville, Arkansas, on 1 August 1946, Covey was the son of a fighter pilot who had flown in the Second World War and Korea. It was clear that the young boy would follow suit. After high school in Florida, he entered the Air Force Academy and graduated in June 1968 with a degree in engineering sciences, majoring in astronautical engineering. In January of the following year he got a master's in aeronautics and astronautics from Purdue. Covey would pay tribute to Professor Jack Arnold, a family friend and lecturer at the Air Force Academy, for having steered him towards astronautical engineering and, years later, would wonder if it was *that* decision which ultimately guided him to NASA.

Flight training at Williams Air Force Base in Arizona came next, followed by operational work in the F-100 Super Sabre, the A-37 Dragonfly and the A-7 Corsair II, then two deployments to Vietnam, during which time Covey flew 339 combat missions in the final years of the bitterly divisive conflict, right up to the end of planned US aerial activity in August 1973. Flying the Super Sabre was a real thrill. "It was an older fighter aircraft at that time," he recalled, "single-engine, single-seat … and single-engine just seemed to imply more danger." Returning home, he was discouraged from entering test pilot school by his superiors, who felt that he would be ruining his career, but Covey submitted his application and was accepted in 1974, at the relatively young age of 28. "I was just six years out of the Air Force Academy," he said, "and again it was one of those things where somebody was in the right place at the right time." Graduation from Edwards' famed school was followed by test work in the F-4 Phantom II and the A-7D at Eglin Air Force Base in Florida and, later, as joint test force director the electronic warfare testing of the new F-15 Eagle fighter. It was whilst at Eglin that Covey applied for the astronaut programme in 1977 and received an invitation to an interview in Houston. This posed a problem, for he had already planned a week's vacation with his wife, Meredith. Whereas most applicants were *jumping* at the interview, Covey was concerned that asking for a delay would jeopardise his chances. Still, he asked, and was told to come to Houston at a later date … but spent his vacation worrying about whether NASA would call him back. "It was a terrible vacation," he recalled, "a *terrible* vacation!" Towards the end of it, he received the call to Houston and in January 1978 he and a flight test engineer from Eglin, named Mike Mullane, were chosen as part of the newest group of astronauts.

More than seven years later, in orbit, Covey began to realise that all of his experience in fighters had done little to prepare him for the arrival of weightlessness. On each of his missions, he would experience some form of space adaptation syndrome. "Most of them were pretty repeatable," Covey told the historian, "and I could predict what I was going to have. There's an awful lot of talk about it, beforehand, and you talk to the guys that have been there and so you know, *sort of*, what to anticipate, but it's really kind of different when you actually get there. On all of my first three space flights, the first day was just extremely busy. That in itself keeps you from feeling a lot of the effects, because you're overpowering anything that your body is telling you by focusing on the jobs you've got to get done that first day." Mike Lounge also described the malaise in some detail. "The first day, you don't feel much like eating," he said, "and you're stuffy and fluids shift to your head, so you get the headache and the kind of stuffy nose and you want to move slowly and you don't like to see things *upside down!*" Watching one of his crewmates 'sitting' on the *ceiling* certainly bothered Lounge, although he did not get sick. "The *next* day, you feel great," he continued. "It's like a switch. It's really an *adaptation*, I think … that the brain has to just reconcile what it sees with what it *thinks* it should see. As soon as you flip that switch, somehow, it's okay and it's just *fun*." For Discovery's crew, their first day in orbit turned out to be busier than most, because they were the first Shuttle mission to perform *two* satellite deployments in those first few hours in orbit.

Originally, the first satellite, ASC-1, was to be deployed some nine and a half hours after launch, followed by Aussat-1 exactly a day into the mission and Syncom 4-4 early on 29 August. However, a little more than two hours after launch, Aussat's sunshield was commanded open in order to perform routine health checks of the

The enormous size of Syncom 4-3 is amply demonstrated in this image as Bill Fisher's helmeted head almost touches the giant satellite. His EVA partner, James 'Ox' van Hoften, is visible at the bottom.

satellite and the port-side clamshell door only partially unfurled. "It was believed," read NASA's post-flight mission report for 51I, published in October, "that the clamshell structure had been deformed and was believed to be binding on the omni antenna bracket, located on the top of the Aussat." Mike Lounge would later blame himself, and a last-minute change to the flight plan, for how this deformation to the sunshield happened. A couple of weeks before launch, he had been assigned the task to activate a payload bay camera to inspect the sunshield. "I did that," he explained, "then I commanded the sunshield open and I had failed to *stow* the camera. If it had been Day Two, instead of Day One, I would have been more aware of it. On Day One, you're just kind of overwhelmed and you're just down doing the steps ... but that was an example of why you *don't* change things at the last minute and why you *don't* do things you haven't simulated – because we'd never simulated that." The crew was advised to uncradle the RMS and push the balky sunshield door open, but unfortunately the Canadian-built arm was suffering its own problems: a failure of its elbow joint meant that Lounge had to command it in a 'single-joint' mode. "Instead of some co-ordinated motion," he said, it was "a little awkward and took a while." Joe Engle agreed and remembered Lounge having the incredibly difficult and intricate job of manually operating electrical switches, selecting each RMS joint in turn and moving them one at a time to position the arm correctly. "That became a concern," Engle reflected, "as to how much that was going to slow us down in the grapple and the capture and then the redeployment of the failed Syncom that we were going to go repair."

The RMS was uncradled from the port-side sill and, at 11:15 am EST, Lounge successfully pushed Aussat's clamshell door fully open. Due to fears of imposing excessive thermal stress on both Aussat and its attached PAM-D booster, NASA decided not to wait until Day Two to deploy it. Under the supervision of Bill Fisher, Aussat-1 was duly released at 1:33 pm EST, on the descending node of Discovery's fifth orbit, whilst ASC-1's deployment was delayed by a couple of hours from the seventh to the eighth orbit. It was sent spinning out of the payload bay by Lounge at 6:07 pm. This marked the first time that *two* major payloads had been deployed on the same day. "We scrambled to get that done," remembered Dick Covey, "so we could get that one that didn't have the protection done, plus the one that we had already planned on. It made for a very, very busy first day for a bunch of new guys up there!" The satellites were quite different from one another in physical appearance: Aussat-1 was based on the Hughes' HS-376 bus, whilst ASC-1 was a boxy satellite built by GE AstroSpace. Aussat represented Australia's first satellite programme and in May 1982 Aussat Proprietary Limited picked Hughes to build three spacecraft and a pair of ground support stations in Sydney and Perth. Like the earlier HS-376s deployed from the Shuttle, Aussat featured a pair of cylindrical, telescoping solar panels and a foldable antenna, which were unfurled as it neared its final orbital position. The three satellites – two flown on the Shuttle and another aboard the European Ariane-3 rocket – provided 15 channels of telecommunications coverage of the entire continent of Australia, together with Papua New Guinea and the south-western Pacific region, including Venuatu and Fiji. Meanwhile, ASC-1, which measured $1.3 \times 1.6 \times 3.15$ m, with solar array 'wings' spanning 14 m, was equipped with 18 C-band and six Ku-band transponders for television news and data transmissions. The *third* payload aboard Mission 51I was already familiar, since three of its siblings had already flown Shuttles into orbit: Syncom 4-4, or 'Leasat 4', was deployed under the supervision of van Hoften at 6:48 am EST on 29 August. Unlike its immediate predecessor, launched by the 51D crew in April, the omni-directional antenna unfolded without incident, the spacecraft spun-up without incident and its integral perigee kick motor ignited without incident, transferring it perfectly into geostationary orbit. Testing got underway on 4 September, but trouble was in store, for Syncom 4-4 suffered a failure of a transmission cable between its UHF multiplexer and transmitter and ground controllers lost contact with the satellite. Nothing could be done to save it, since it was far beyond the reach of even the Shuttle.

When Dick Covey was named to fly as Joe Engle's pilot, it was both an excitement and an honour; for Engle's experience spanned several decades and encompassed not only the Shuttle, but the Apollo lunar module and the X-15. More than that, however, he was delighted when Engle assigned the rookie the task of performing the rendezvous with the failed Syncom 4-3. This was completed in spectacular fashion on the fifth day of the mission, 31 August. Early in training, they had agreed that Covey would perform the 'phasing' manoeuvres and precise RCS thruster firings to approach to about 300 m, and then Engle, stationed on the aft flight deck, would take over and, peering out through the overhead windows would manoeuvre to position the suited van Hoften, his feet secured in a restraint on the end of the RMS, close enough to the satellite to manually grapple it. "When he had completed the rendezvous manoeuvre and had stabilised," said Engle, "I looked up and kind of expected to see [Syncom] *somewhere* in the field of view in the window, but he had flown that rendezvous and perfectly *nailed it*, so the satellite was right ... in the centre of the [crosshairs]."

In addition to serving as the primary RMS operator, Mike Lounge was responsible for helping van Hoften and Fisher into their space suits and into the airlock. It was then time for him to head up to the flight deck, whilst the two spacewalkers ventured outside for an excursion that would last seven hours and 20 minutes. Van Hoften installed a foot restraint on the end of the arm, and stood on it. By his own admission, the biggest challenge was that "I had to go literally up there and *get hold* of this thing, somehow, until we could get hold of it and get the arm on it, because I had to attach a [capture] bar to it ... It was *every* bit as tricky as I thought it would be." The massive Syncom, which weighed 6,818 kg, was rotating very slowly. He fell behind schedule when the capture bar initially refused to fit, but he persevered. The situation was hampered by a lack of continuous communication with Mission Control, since the full TDRS network had yet to be established, and Lounge found that the RMS difficulties meant he had to be very deliberate with each motion. Eventually, after the installation of the capture bar, Lounge was able to manoeuvre van Hoften and the Syncom towards Fisher in the payload bay, who fitted a second, 'handling' mechanism. Van Hoften then attached an RMS grapple fixture. The two men safed the satellite with grounding plugs, then fitted a bypass cable harness to work around the faulty deploy switch. Syncom's batteries had not frozen and the repair showed its first sign of success when the omni-directional antenna popped open. For Bill Fisher, a physician with a background in emergency medicine, it represented something totally new: microsurgery on perhaps the biggest patient he had even worked upon.

Van Hoften has described his relationship with William Frederick Fisher – an astronaut since May 1980 and the husband of Anna Fisher – as not nearly as good as George 'Pinky' Nelson, with whom he had performed two EVAs to repair Solar Max. "Bill's very competent," he admitted, "but he was obsessed with strength. He had gone out and done lots and lots of bodybuilding before we went on this. For some reason, he thought this was important in a spacewalk and I kept telling him that it wasn't." As Fisher built mass and muscle, van Hoften even worried that he might outgrow his finely-sized space suit. Fisher came from Dallas, Texas, where he was born on 1 April 1946. After high school in New York, he attended Stanford

With the rejuvenation of Syncom 4-3 complete, van Hoften strikes an 'Atlas' pose, apparently supporting the world on his shoulders. Ironically, the Syncom 4-4 satellite, deployed earlier in the flight, would suffer a complete failure after insertion into geostationary orbit and would be rendered inoperable.

University to read for a bachelor of arts degree, graduated in 1968 and moved on to perform work in biology at the University of Florida. In 1971, he entered medical school and completed his doctorate four years later. Fisher completed a surgical residency at the University of California at Los Angeles' Harbor General Hospital – working with his future wife – and transferred to emergency medicine from 1977 until his selection into NASA's astronaut corps in 1980.

The second EVA, which occurred on 1 September and lasted four hours and 26 minutes, involved the two men installing an instrumented cover over Syncom's motor nozzle and arming it. The enormous size of the satellite almost caused it to collide with the orbiter – the spacewalkers could not see each other from their positions on opposing sides of the payload bay – and van Hoften resolved that, "if something happens and I'm about to lose it", he would give it "a heck of a push and *bail out!*" At length, they managed to control its motions and van Hoften manually spun it up to three revolutions per minute and released it. The satellite went on to perform its manoeuvres just as planned.

With the successful deployment of Syncom 4-3, the mission had effectively completed its objectives. Only one middeck experiment, the Physical Vapour Transport of Organic Solids, was carried, and it was tended by van Hoften. From time to time, the sounds of Willie Nelson echoed through Discovery, thanks to Mike Lounge's penchant for country music. On 3 September the astronauts prepared for their return to Earth. Re-entry was mostly in darkness, as Discovery headed across the slumbering Pacific towards the California coastline and, further inland, deep in the Mojave Desert, their landing strip at Edwards Air Force Base. From the flight engineer's seat, Lounge took a handful of photographs of the dazzling plasma wake, trailing behind the Shuttle, and Dick Covey remembered that it "crawled", like fiery fingers, along the bottom of his window. "Then, as you get into the thickest, hottest regions," he continued, it turned "into this complete *sheath* of white over the windows." Touchdown on Runway 23 at 6:16 am PST (9:16 am in Florida) concluded a mission of just over seven days. Rising from his seat, Lounge *felt* heavy – he could feel his weight and it required conscious effort to keep his head up – and he remembered that walking and taking corners was awkward, tentative, even, at first. It did not last. Just as he had adapted to weightlessness, within a day, Lounge was once more adapted to life on Earth. When he and Covey next ventured into the heavens, three years hence, the 'picnic' atmosphere surrounding 51I would be gone, friends would have been lost, a sense of innocence destroyed and the Shuttle would never be the same again.

'OPEN' SECRET MISSION

Management consultant Diane Bobko was particularly irritated as her husband prepared for his third trip into space. Unlike his previous missions, this one – 51J, the maiden voyage of NASA's fourth orbiter, Atlantis – was a classified Department of Defense assignment, totally shrouded in a cloak of military secrecy.

"Bo," she said, one morning in September 1985, "you're *not* telling me exactly what day you're going to land, but I think it's going to be pretty close to a day I have a programme in Baltimore."

"Diane, it's the first flight of a new vehicle," her husband replied. "Probably the safest thing you can do is go ahead and schedule that right now."

To this day, Karol 'Bo' Bobko retains a noteworthy, yet often overlooked record as the only astronaut to have flown the maiden missions of *two* Space Shuttles and he was clearly reflecting on the problems experienced getting Challenger ready for her first flight in April 1983 when he assured his wife that the first flight of Atlantis would probably also meet with delay. Ironically, it did not. Atlantis was named in honour of a two-masted ketch, operated by the Woods Hole Oceanographic Institute from 1930 until 1966, which became the first vessel built specifically for interdisciplinary research in marine biology and geology and physical oceanography. During her time at sea, Atlantis and her scientists scored a number of impressive discoveries, not least of which was the identification and description of the first abyssal plain – the 'Sohm Abyssal Plain', located to the south of Newfoundland – in 1947. Today, she is owned by Argentina as a naval research vessel. The construction of the spacegoing Atlantis got underway in January 1979, following a contract award to Rockwell International to configure a structural test article into the future Challenger and build two additional orbiter vehicles, OV-103 and OV-104. The names 'Discovery' and 'Atlantis' were assigned to the new orbiters a few days after the award. In March 1980, engineers started the structural assembly of Atlantis' crew compartment and over the next few years the vehicle grew – construction of her aft fuselage began in November 1981, her wings arrived from contractor Grumman in June 1983 and she was complete by April 1984. Rollout from Rockwell's Palmdale plant in March 1985 was followed by an overland transfer to Edwards Air Force Base and arrival in Florida, atop the Boeing 747 Shuttle Carrier Aircraft, on 9 April. Two days after Discovery touched down from her 51I mission, on 5 September, Atlantis' three main engines burned at full power in a Flight Readiness Firing, one of the last milestones to prepare her for her maiden voyage. All told, Atlantis required less than half as much time to assemble as did the queen of the fleet, Columbia, and historian Dennis Jenkins has pointed to the greater use of thermal protection blankets, rather than tiles, on the youngest vehicle's airframe as one of the main reasons.

Atlantis' crew for Mission 51J, flying the second classified Department of Defense mission, had been in place for some considerable period of time. As early as November 1983, NASA had identified a DoD 'standby' crew, consisting of Bobko in command, pilot Ron Grabe and mission specialists Bob Stewart, Mike Mullane and Dave Hilmers, and in February 1985 this group – save Mullane, who had been named to the first Vandenberg flight – were assigned to 51J. Also assigned was Air Force Major Bill Pailes, a manned spaceflight engineer, as the sole payload specialist. Command of this new flight posed something of a problem in the spring of 1985, particularly when Bobko's 51E mission was cancelled and he ended up leading the 51D crew into orbit in April. The inevitable consequence was that, like Bob Crippen the previous year, his *other* crew was forced to train without him for a time. What *really* made 51J a pain was its classified nature. "But you lived with that," said

Bobko. He was accompanied by a superb team. Ronald John Grabe, born in New York on 13 June 1945, was an accomplished Air Force fighter and test pilot. After high school, he received his degree in engineering science at the Air Force Academy and spent a year as a Fulbright Scholar at the Technische Hochschule in Darmstadt, West Germany, then returned to the United States to complete pilot training at Randolph Air Force Base in Texas. Grabe flew the F-100 Super Sabre and in 1969 was deployed to Vietnam, where he completed 200 combat missions. Later work included operational testing of the weapons systems of the F-111D Aardvark tactical strike aircraft and in 1974 he was selected for test pilot school at Edwards Air Force Base. Grabe was awarded the prestigious Liethen-Tittle Award for Outstanding Student at the school. Graduation brought an assignment as a test pilot for the A-7 Corsair II and the F-111 and he worked as an exchange pilot with the Royal Air Force between 1976 and 1979, flying the Harrier and Sea Harrier at Boscombe Down in Wiltshire. His final assignment before NASA selected him in May 1980 was as an instructor at Edwards.

David Carl Hilmers came from Clinton, Iowa, where he was born on 28 January 1950. He completed high school in his home state and earned a degree in mathematics from Cornell College in 1972, then entered the Marine Corps in the summer of that year. Completion of Basic School and Naval Flight Officer School brought assignment to Marine Corps Air Station Cherry Point in North Carolina, during which time Hilmers flew the A-6 Intruder as a bombardier-navigator. Subsequent positions included air liaison officer with the United States Sixth Fleet in the Mediterranean Sea and assignments to Marine Corps air wings in Japan and California. He completed a master's credential in electrical engineering from the Naval Postgraduate School in 1977, followed by a degree in electrical engineering a year later. Like Grabe, he was selected by NASA in May 1980. Aside from his military background, Hilmers was very religiously conservative – in fact, his devout faith was remarked upon by Mike Mullane, who flew with him – and cultivated a keen interest in medicine whilst working for the space agency. At one stage he intended to take an Advanced Cardiac Life Support Course. "He wanted to be a doctor," remembered physician-astronaut Rhea Seddon in her NASA oral history. "It was like he never got a chance to be a doctor. He would come and ask me medical questions. I know he had a lot of fun with the medical training, because it was something he wanted to know more about." After his departure from NASA in 1992, Hilmers did indeed enter medical school and today works as a paediatrician and nutritionist.

The fifth crew member aboard Atlantis, William Arthur Pailes, was only the second member of the Air Force's corps of manned spaceflight engineers to fly the Shuttle ... and, as circumstances transpired, he would also be the *last*. In the wake of the Challenger tragedy in January 1986, the military would steadily distance itself from the reusable spacecraft, moving many of its payloads onto expendable launch vehicles, and the cadre was gradually disbanded. This could hardly have been foreseen in October 1985, when 51J speared for the heavens: NASA and the Air Force confidently expected to launch several manned spaceflight engineers on classified missions, with Brett Watterson already assigned as a payload specialist for the first Vandenberg flight and others expected to follow him. Pailes had been born in Hackensack, New Jersey – interestingly, also the birthplace of Wally Schirra, a member of the Original Seven Mercury astronauts – on 26 June 1952. After completing his high schooling in New Jersey, he studied computer science at the Air Force Academy, graduating in 1974, and entered pilot training at Williams Air Force Base in Arizona. Pailes served as a pilot in the Lockheed HC-130 search and rescue aircraft, entered Squadron Officer School in 1978 and received a master's degree in computer science from Texas A&M University in 1981. Selection as a member of the second group of manned spaceflight engineers followed in January 1983. After 51J, he would unsuccessfully apply for a place in NASA's 12th astronaut class.

If Mission 51C had generated a measure of humour in that its payload was known, or at least suspected, before it even reached orbit, then the twin-satellite cargo of 51J became an 'open secret' and details were published by *Aviation Week* as early as 7 October 1985, the very day that Atlantis touched down. A single Inertial Upper Stage booster carried a pair of $160 million Defense Satellite Communications System (DSCS)-III spacecraft, stacked one atop the other, images of which were finally declassified in the summer of 1998. They lend credence to Bobko's claim in his oral history that, for all its 'secrecy', 51J was little more than a 'vanilla' deployment flight. The DSCS – nicknamed 'the discus' – has long been an anchor for the US military's global communications network, operating in geostationary orbit with half a dozen super-high-frequency transponder channels for secure voice and data transmission capability and high-priority command and control links between defence officials and battlefield commanders. The Air Force later admitted that it *had* launched a pair of DSCS-IIIs in 1985 and, according to space analyst Dwayne Day, "the *only* launch that year that fit was the Atlantis mission". Subsequent military documentation highlighted that the DSCS-III satellites *had* been deployed during a Shuttle flight, but refused to reveal the name of that flight. "Military secrecy can be bizarre at times," wrote Day, "like acknowledging that there is a sky, and that the sky can be blue, but *never* saying that the sky *is* blue!"

Physically, the satellites were roughly cube-shaped, three-axis-stabilised, with a pair of articulated solar panels which produced 1,240 watts of electrical power. They measured 2 m in height, spanned 11.5 m across their expansive solar 'wings' and weighed 2,600 kg. Day considered it significant that 51J's payload was so readily revealed, but the natures of the other classified satellites launched between 1988 and 1992 have been kept under wraps to this very day. "If the suspected identities of the other classified Shuttle flights are correct," he speculated in an article for the *Space Review* in January 2010, "then they are *intelligence* satellites. Considering the secrecy that remains about American intelligence satellites, it seems likely that these other flights will continue to remain secret for a long time to come."

Launch of Atlantis came at 11:15 am EST on 3 October 1985, following a 22-minute delay to deal with a power controller in one of the main engines' liquid hydrogen prevalves that showed a faulty indication. As with Mission 51C, the assembled spectators at the Kennedy Space Center were only aware that launch was imminent when the blank face of the countdown clock suddenly came to life and started counting from T-9 minutes. Yet there was still a degree of uncertainty about

51J. *Flight International* suggested (correctly) that if the rumours about the presence of DSCS-III satellites were accurate, then an Inertial Upper Stage was the most likely booster. The Air Force cleverly refused to confirm or deny any of the rumours. After four days – one of the shortest Shuttle missions to date – Atlantis touched down at Edwards Air Force in California at 10:00 am PST (1:00 pm in Florida) on 7 October.

As it turned out, Diane Bobko *was* in California to meet her husband on the runway. Astonishingly, Atlantis had met with no significant delays, launched on time and landed on time. "So she was there to meet me in California," Bobko remembered, "gave me a hug and then she had to leave right away to ... drive down to Los Angeles to catch the airplane to go to Baltimore." Later that evening, Bobko was startled out of his sleep by a telephone call. It was his wife. *Surely*, he thought, if the Shuttle could launch and land on time, on its maiden voyage, then her internal flight would have been trouble-free.

"You're in Baltimore?" he asked.

"No," she replied, glumly. "I'm *still* in Dallas, trying to *get* to Baltimore!"

COMING OF AGE

Spectators at the Operations and Checkout Building beheld an unusual sight on 30 October 1985, as a crowd of blue-clad pilots, engineers and physicists from three nations headed for Pad 39A. Led by snowy-haired skipper Hank Hartsfield, astronauts Steve Nagel, Bonnie Dunbar, Jim Buchli and Guy Bluford, together with West German physicists Ernst Messerschmid and Reinhard Furrer and Dutch physicist Wubbo Ockels, were set to make history as the world's first eight-member Shuttle crew. Although the mission was primarily financed by West Germany, the European Space Agency had contributed around 40 percent in return for flying one of its astronauts as a payload specialist. In the wake of the Challenger disaster, crews were restricted to a maximum of seven members; hence, Mission 61A remains the only flight to have launched with eight people. In view of the restricted volume aboard Challenger, it was fortuitous that her ninth and last successful mission carried a Spacelab module and required a dual-shift system of around-the-clock operations. "The red team of Jim, Ernst and I had to do a circadian rhythm shift," said Bluford, whose four-flight astronaut career involved three missions before which he had to 'sleep-shift', "so, for us, the launch was coming near the end of our work day. While in [pre-launch] quarantine, one team was up while the other was in bed. A new lighting system had been installed in the crew quarters to facilitate the shift in circadian rhythm. Once we got on orbit, the blue team activated Spacelab, while the red team went to bed. We had four soundproof bunks to sleep in while the blue team was at work. The two shift operations worked very well on orbit, with both teams up at the same time during breakfast and dinner, when we transferred Spacelab operations. The simultaneous transfer of responsibility – both on orbit as well as on the ground – went smoothly as we exchanged information and updated our flight data files. Each of the crew shared a sleep bunk with a member from the opposite team. Only Hank had a bunk to himself, which gave him flexibility to work on either shift."

By this time, the European-built pressurised research facility – which, housed in the orbiter's payload bay, provided a miniature space station for a week or more – had amply demonstrated its capabilities on two occasions. For this flight, known as 'Spacelab-D1' for 'Deutschland', was almost entirely focused on life and microgravity investigations funded by West Germany. This was, in a way, unsurprising, since the latter had a 54.1 percent financial stake in the reusable laboratory. Even 61A's launch time of midday EST ("banker's hours", Dunbar joked) was carefully timed, said Bluford, "so as to give maximum TV coverage to Germany". By the time Challenger lifted off, he and Bonnie Jeanne Dunbar had been training for the ambitious mission since February 1984. As Spacelab completed its adolescence and began 'operational' flights, it was becoming common practice to assign 'science crews' in advance of the three-member 'orbiter teams', in order to give them additional time to iron out payload-oriented operational and training issues with experiment sponsors and principal investigators. Preparations for Spacelab-D1 were considerably more complex than most previous missions because it involved a great deal of travel between the United States, Holland and West Germany.

Dunbar came from Sunnyside, Washington, where she was born on 3 March 1949. Her grandparents had arrived from Scotland and homesteaded in Oregon and the young girl grew up on a ranch, raising Hereford cattle in eastern Washington State. She often listened as her elderly grandfather told her why he came to the United States: to own his own soil. His perspective was that life was an adventure and the key to unlocking that adventure was education. Her father had similar, yet also different, ideals. "He was a Marine in World War II in the South Pacific," Dunbar recalled, had "fought for my sons *and* my daughters to be able to become what they want to become, if they wanted to work hard enough to do it". Her parents instilled this ethic into her and she remembered that the first books she received as a child were a set of encyclopedias. If anything appeared on the television which the young girl did not understand, her parents would direct her to the shelf and ask her to read the encyclopedia entry to the rest of the family. "It was fun," Dunbar recalled. "Reading was mandatory to be able to participate in this game." The Soviet launch of Sputnik in October 1957 switched Dunbar on to the idea of space travel for the first time and she became engrossed in the work of H.G. Wells and Jules Verne. In eighth grade at school, her principal asked her what she wanted to do in life and the girl, embarrassed to admit that she wanted to be an astronaut, told him instead that she wanted to design and build spaceships. He did not laugh. Nor did he belittle her dream. "You'll have to know algebra," he told her. She had no idea what *algebra* actually was, but in later years would come to appreciate that algebra, geometry, trigonometry and calculus were effectively the 'keys' for an engineering career. Dunbar had little idea how challenging it would be to *enter* such a career.

After completing high school in Sunnyside, she entered the University of Washington to study engineering and was advised by Dr James Mueller, the dean of ceramic engineering, to change her major from aeronautics to either materials or ceramic engineering. It would prove a pivotal moment in her future career and it was Mueller who introduced her to many key engineering minds who were working on research for the Shuttle's thermal protection system. "That then put me on a course of becoming more involved in the real parts of space," Dunbar recalled. "They had summer positions open and so one of the summer positions at the University of Washington was doing X-ray diffraction research on some of the fibres that NASA was considering." Very few women studied engineering during that period and Dunbar was received with difficulty by some students and professors. "My response to any negative attitudes," she concluded, "was to *ace* the class!" Much of her work was not subjective; it involved work whose results were either *right* or *wrong* and, as one supportive professor told her: If you've got it *right*, you've got it *right!* Gender made no difference.

Graduation in 1971 was followed by two years as a systems analyst with Boeing and a call from Mueller to invite her back for a master's degree in ceramic engineering. From 1973 until 1975, Dunbar worked on the mechanisms and kinetics of ionic diffusion in sodium beta-alumina, part of a project to develop a high-energy-density battery, primarily for the automotive industry. In fact, Dunbar would later give a lecture to the Ford Motor Company, who were performing similar research. By the summer of 1975, she had completed the requirements of her master's degree and was made a quite astonishing offer: the rigorous quality and extent of her research meant that she had effectively completed most of the requirements for a PhD, awardable in just *six months*. But Dunbar was broke, and opted instead for a six-month visiting scientist position in England at the Atomic Energy Research Establishment in Oxford, studying the wetting behaviour of liquids on solid substrates. Years later, she would profoundly regret not going ahead and completing her doctorate. Back in the United States, she joined Rockwell International to work on the Shuttle's thermal protection system and later received the Engineer of the Year Award for it. She applied, unsuccessfully, for admission into the astronaut corps, but, in 1978, entered NASA's Johnson Space Center as a payload officer and flight controller, providing guidance and navigation for the Skylab re-entry. She was finally selected as an astronaut in May 1980 and three years later completed her doctorate in mechanical and biomedical engineering at the University of Houston. Her PhD focus was upon evaluations of the effects of simulated orbital flight on bone strength and fracture toughness. Her assignment as a mission specialist on the Spacelab-D1 flight brought back unpleasant memories of the difficulties she had endured as a female engineer. Many of the West German medical experiments were *not* intended to include female blood and there were concerns that it might ruin their data. "I was actually *told*, in front of my *face*," she recalled, "but any time you're at the point of the pathfinder, there's going to be things happening. I was told that maybe NASA had done this intentionally to offend the Germans, by assigning a woman." Other Spacelab-D1 experiments, including the vestibular sled, also did not fit her size. Eventually, George Abbey came to her aid, insisting that the Germans redesign their equipment to the percentile spread which included Dunbar. NASA "was committed to women," she concluded, "and it was a very equal environment." Dunbar plunged enthusiastically into her preparations, learning German through an instructor and developing an exceptionally smooth working relationship with Reinhard Furrer and Ernst Messerschmid. "Reinhard was a well-known physicist," she told the oral historian. "He was a professor. That's a *very* well-esteemed position in Germany. It's better than being president of the company if you're *Herr Professor.*" There was time for banter, too. 'Dallas' was particularly well-received in West Germany at the time and Furrer was astonished that Dunbar, an *American*, had *never* seen it! When he first asked her about it, she thought he meant the Texan *city*. To her credit though, Dunbar's engineering work had consumed much of her time for several years and television was the *last* thing on her mind.

"Our primary training was conducted at Porz Wahnheide in Germany, a small, very picturesque town, south of Cologne," Bluford recalled. "This European astronaut office housed the ground training units for several Spacelab experiments. Bonnie and I trained on Spacelab systems at Marshall Space Flight Center and Shuttle procedures at JSC." Costing $180 million, and including a $62 million Shuttle launch fee, it took five years to prepare Spacelab-D1 from conception to launch and the mission was managed by the Deutsche Forschungsanstalt für Luft- und Raumfahrt (DFVLR, the Federal German Aerospace Research Establishment) on behalf of the Federal Ministry of Research and Technology. Joining Bluford and Dunbar on the science crew were a record-breaking three payload specialists, all of them physicists – West Germans Messerschmid and Furrer, together with Dutchman Ockels – and the quintet were responsible for 76 experiments in several major facilities aboard the Spacelab module. With names such as Werkstofflabor, Prozesskamer, Biowissenschaften and Biorack, the West German-supplied research complement sounded like a fearsome medieval torture chamber, yet encompassed a range of studies of the behaviour and processing of materials and fluids and the functioning of biological organisms in the strange microgravity environment. Werkstofflabor, firstly, was a multi-purpose unit which housed three furnaces, a fluid physics module and a crystal growth device to investigate areas of materials processing, semiconductor growth for electronics applications, fluid boundary surfaces and heat-transfer phenomena. The Prozesskamer (or 'process chamber') was designed to measure flows, mass transport, heat and temperature distribution during the melting and solidification processes of various materials. Another important facility was the Materials Science Experiment Double Rack for Experiment Modules and Apparatus (MEDEA), which comprised three separate furnaces: one that conducted long-duration crystallisation studies, one that used a 'directional solidification' technique to process metallic crystals at extremely high temperatures, and one with a high-precision thermostat that examined the behaviour of metals under carefully controlled thermal conditions.

By the second day of the mission, although in general experiment operations were running smoothly, the medical and biological investigations were progressing better than the materials science ones. In particular, a problem with MEDEA's pressure sensor had to be corrected by an in-flight maintenance procedure and a lamp on the furnace was also replaced. Unfortunately, this meant that many hours' worth of 'run-time' were lost. Discussions to extend 61A from seven to eight days were ultimately turned down because Spacelab's power usage levels could not be reduced

Guy Bluford (right) and Ernst Messerschmid are pictured during Spacelab-D1 training.

enough to provide the required extra day aloft. Elsewhere, the Biowissenschaften and Biorack facilities focused upon life and biological science applications. Results from the latter, built by the European Space Agency, in particular, offered striking evidence of the influence of gravity on bacteria, unicellular organisms and white blood cells. A total of 14 cellular and developmental biology investigations were assigned to Spacelab-D1 and it was the first occasion on which specimens were 'fixed' and thus preserved in orbit for post-mission analysis. Two of these experiments confirmed observations made on several previous Shuttle flights: bacteria tend to reproduce more rapidly in space than on Earth, suggesting that astronauts could be exposed to a higher risk of infection. Of particular note was an investigation featuring the common pathogenic organism *E. coli*, which has demonstrated an increased resistance to antibiotics in orbit. On the other hand, some of Spacelab-D1's bacteriological research indicated that some cells actually exchanged genetic material through physical 'bridges', perhaps leading to novel techniques for introducing human genes into bacteria to synthesise useful products.

Vestibular experiments involving both humans and tadpoles were also conducted and, in the latter case, revealed pronounced alteration in swimming behaviour upon return to Earth. The tadpoles swam in small circles around fixed centres until their behaviour returned to normal a few days after being returned to Earth. Later examinations of the morphology of their vestibular gravity receptors revealed no structural deformities, indicating that they developed normally in space. This was consistent with earlier studies of amphibians and rodents. Running on rails along the centre aisle of the Spacelab-D1 module was ESA's Vestibular Sled, designed to explore the functional organisation of the astronauts' vestibular and orientation systems and adaptation processes. The accelerations provided by the sled, which could accelerate its subjects at up to 0.2 g along the length of the module, was combined with thermal stimulation of their inner ears and optokinetic stimulation of their eyes.

Not only was Mission 61A's large crew unusual, but so too was its distinction of being the first Shuttle flight to be run from outside the United States. Although Mission Control at Houston was in overall command, the German Space Operations Centre (GSOC) at Oberpfaffenhofen, on the outskirts of Munich, managed daily research activities. This proved to work exceptionally well, although Oberpfaffenhofen's limited data-transmission capabilities meant that several functions had to be monitored from JSC. Moreover, due to the presence of only one Tracking and Data Relay Satellite, Spacelab-D1 received only limited communications coverage for approximately 30 percent of each orbit and it was left to the Intelsat V satellite to relay data to an Earth station at Raisting in Bavaria and from there to Oberpfaffenhofen by microwave link. According to Steve Nagel, training in Munich did not differ significantly from the United States. "You could say it's more complex and there are more issues to be resolved when you're working an international programme," he said. "Not having a US mission manager made it more complex, but I see that mission was an early lead-in to the space station. It was hard for Hank to pull together and complicated when you're dealing overseas. We got along fine with the Germans, but we butted heads about things and the long distance part made it more complex."

Despite the focus of the mission and the common language currency always being English, on a few occasions German was spoken over the space-to-ground communications link, including one opportunity for Messerschmid and Bluford to speak to the head of Bavaria. "The conversation was conducted in German with Ernst doing all the talking," Bluford remembered years later. "Although the mission's dialogue was conducted primarily in English, infrequently, the payload specialists would revert to German during on orbit discussions." Hartsfield remembered that the decision was a controversial one, with the Germans insisting that *their* language be spoken to controllers at a German site in Bavaria. "I opposed that for safety reasons," he explained. "We can't have things going on in which my part of the payload crew can't understand what they're getting ready to do. It was clearly up front: the operational language *will* be English. We finally cut a deal ... that in special cases, where there was *real* urgency, that we could have another language used, but before any action is taken, it has to be translated into English so that the commander or my other shift operator lead and the payload crew can understand it."

For Steve Nagel, the year before the Challenger disaster was a pivotal one, for he flew the Shuttle not once, but twice! Matters had been complicated by the fact that Nagel had also been assigned, along with Bluford and Dunbar, to the Spacelab-D1 flight in February 1984. Peculiarly, as Challenger's pilot, he would form part of the 'orbiter' crew, but was actually named in advance of Hartsfield and Buchli, an anomaly that has never been satisfactorily explained, other than as a visible example of George Abbey making good on his promise to recycle Nagel to the pilot's seat. When the rest of the Spacelab-D1 team was named in August 1984, with launch scheduled for October of the following year, it should have given Nagel a comfortable period of just under a year between his first and second missions. It did not quite happen that way. "One kept *slipping*," Nagel said, "and the other one *didn't*. When we lost [51D], my two flights were four months apart! Before I thought about that, Dan said 'You're in trouble here'. He went over and talked to George Abbey and negotiated for me to stay on both flights; that I could train for both for a while, then stop training for the second one and finish the first one. I don't think they'd ever do that today, so I owe Dan for the fact that I was able to hang onto both of those."

SUSPENDED IN A GONDOLA

As a result, by the time Challenger ascended from Pad 39A on 30 October 1985, a mere 136 days separated Nagel's two launches. This record would not be broken until April 1997, when the STS-83 astronauts were forced to cut short their Spacelab mission after only a few days and were reflown in July. Upon achieving a 320 km, 57 degree inclination orbit, however, Nagel had little time to reflect upon his good fortune: as leader of 61A's blue team, he was in charge of configuring the spacecraft for seven days of operations, while Dunbar and Furrer busied themselves with activating the Spacelab module.

Although not strictly attached to either shift, Hank Hartsfield and Wubbo Ockels tended to align their work schedules with that of Nagel's blue team. "Wubbo decided to freelance," remembered Hartsfield. "He didn't have a fixed shift. His shift would overlap the other two shifts. It was kind of a weird arrangement. He chose to sleep in the airlock. He had a sleeping bag – a design of his own – and the only trouble was that people going back and forth would bump him as they went through there." Wubbo Johannes Ockels was born on 28 March 1946 in Almelo, a municipality in the eastern Netherlands, although the university city of Groningen became his home town and it was there that he received much of his higher education. Although Lodewijk van den Berg was Dutch-*born*, he was a naturalised US citizen at the time of his 51B flight, which meant that Ockels became the first Dutch *citizen* in space. He obtained a master's degree in physics from the University of Groningen in 1973 and a doctorate in physics and mathematics in 1978 for research focusing on the decay process of gamma rays in nuclear systems at Groningen's Nuclear Physics Accelerator Institute. In August 1978, he was selected as one of three payload specialist candidates for Spacelab-1 and in May 1980, under an agreement between ESA and NASA, was accepted for mission specialist training. After completing this training, Ockels resumed his duties as Ulf Merbold's backup for Spacelab-1. The pair changed roles for Spacelab-D1, with Merbold serving as the mission's only backup payload specialist. In honour of the traditions of his Dutch homeland, Ockels took a large bag of gouda cheese as part of his personal allowance. "The coolest part of the vehicle," said Hartsfield, "was the tunnel that went from the middeck to the lab. He taped that bag of gouda up in the tunnel. It was so convenient that anybody that went there – on the way back and forth – reached in. About the second or third day, he was upset because two-thirds of his cheese was gone!"

In a manner not dissimilar to Spacelab-3, Challenger was oriented in a gravity gradient attitude to provide a quiescent microgravity environment for the on-board materials processing and fluid physics investigations. "There's a little bit of atmospheric drag, even at those altitudes, and there's a gravity effect from one end of the Shuttle to the other," explained Nagel, "which will cause it to change attitudes, so you get it in a stable attitude before you turn the jets off. This is interesting, because usually you want the long axis pointed at the Earth, either tail to the Earth or nose to the Earth, and the wing oriented in some way that it'll be fairly stable. And we would get it in this attitude, which was nose at the Earth, and the right wing pretty well forward. You 'slide' along like that and get it all stable and turn off the jets, and it would just stay there. It would slowly wander around a little bit and roll over a long period, like half-hour or so, kind of oscillate. It made for very interesting Earth viewing, because you'd go up in the cockpit and look out the front windows – and the Earth is coming by! It's almost like you're suspended in a gondola."

Activation of Spacelab-D1 was complete by a little over five hours into the mission and, towards the end of the first work day of Jim Buchli's red team, 73 of the 76 experiments were up and running. Another key milestone was deployment, at 12:36 am EST on 31 October, of the Global Low Orbiting Message Relay (GLOMR) satellite from its Getaway Special canister on the payload bay wall. This small, 62-sided polyhedron weighed just 68 kg and was ejected by means of a standard autonomous controller on the aft flight deck. Upon receiving the proper command, a full-diameter motorised door assembly on the canister opened and a spring-loaded device pushed the tiny satellite into space at a steady rate of 1.2 m per second. For GLOMR's manufacturer, Defense Systems Incorporated of McLean in Virginia, the deployment was a moment of triumph, because a previous attempt on 51B in April 1985 had been foiled by a battery problem. It was also Challenger's last successful satellite release.

Perhaps demonstrative of the monotony of Spacelab flights for pilots, Nagel's job consisted of periodically purging the fuel cells, dumping waste water, taking photographs and preparing meals for the rest of the crew on his shift. "But the good thing about the mission," he said, "was the high inclination. We flew 57 degrees, which means you cover most of the inhabited part of the world. It was just a bonanza of Earth observations. We shot all of our film." For Hartsfield, the comparatively relaxed pace for the 'orbiter crew' allowed him to indulge in some light-hearted banter, particularly as Halloween coincided with Challenger's second day in space. "I took the back off one of the ascent checklists," he said, "drew a face on it, cut out eye holes, got some string and made a mask! I took one of the stowage bags and went trick-or-treating in the lab. They don't do Halloween in Germany, so they didn't know what I was up to! I decided not to pull any tricks on them, but I didn't get much in my bag. One of the guys took a picture of me with that mask on, and somehow it got released back in the US. About a month after the flight, I got a letter from a congressman who had a complaint from one of his constituents about her tax money being spent to buy toys for astronauts! I had to explain that nothing was done and it was made in flight from material we didn't need anymore. It was just fun. I never heard any more, so I think maybe that satisfied her."

In addition to the research facilities in the Spacelab module, two devices were attached to an MPESS carrier at the rear of Challenger's payload bay. The Navigation Experiments (NAVEX), comprised a pair of canisters and an antenna to develop and test a precise clock synchronisation to evaluate a new method for precise, one-way distance measurements and position determination. The other payload was the boxy Materials Experiment Assembly (MEA), which had previously flown on STS-7, and investigated atomic diffusion and transport processes in various liquid metals.

The final full flight of Challenger passed remarkably quietly and smoothly. One of the few problems experienced was a cabin leak, which triggered alarms on several occasions. "We discovered, later on, the leak was due to one of the experiments inadvertently venting into space," said Bluford. "We also had a false fire alarm go off on us during flight." Nonetheless, despite the hectic, around-the-clock pace, some time was granted to each spacefarer simply to gaze down on the Home Planet; particularly the trio of payload specialists, for whom the opportunity to fly in space would come only once.

"We were flying into darkness, passing over Tasmania," Jim Buchli told a Smithsonian interviewer years later, "and heading down toward Antarctica. The southern aurora was just unbelievable! It looked like an octopus sitting over the South Pole, with tentacles of light coming out. The orbiter was flying upside down, with the nose pointing toward the pole, and the tentacles shimmered a fluorescent blue-pink. It was like the whole nose was bathed in aurora. Even though we were much higher, you could still see the glow off the front of the nose. I knew what was coming, because I had seen the same geometry when we passed over the pole the day before. I went down to the middeck and literally grabbed Reinhard Furrer, who was on the other shift ... and stuffed him up there in the nose of the vehicle. We're lying upside down, with all the switches and circuit breakers next to our chests, and we're peeking out the front windows, straining to look to the side of the orbiter. For probably ten minutes, we watched these shimmering bands coming off the South Pole. Finally, Reinhard said 'Jim, that was fantastic! That was the most beautiful thing I've ever seen'. Then he went back downstairs to work."

Less than ten years later, in September 1995, Furrer was killed whilst flying back-seat in a Second World War-era Messerschmitt Bf 108 Taifun at Berlin's Johannisthal airfield. He was 54. Reinhard Alfred Furrer came from Wörgl in Austria – then part of Germany – where he was born on 25 November 1940. After the Second World War, his father was expelled from Austria and the family made a new home in Kempten, deep in south-eastern Bavaria. Furrer entered the University of Kiel to study physics, then transferred to the Free University of Berlin, where he received a diploma in 1969 and a doctorate in 1972. Whilst in Berlin, he participated in the construction of 'Tunnel 57' under the Wall, enabling 57 East Germans to escape to the West. Furrer became an assistant professor in 1974 and a full professor in 1979, then spent time at the University of Chicago and the Argonne National Laboratory in Chicago. He was an unsuccessful applicant for the ESA Spacelab-1 payload specialist selection, but he and fellow physicist Ernst Messerschmid were chosen in December 1982 for the D1 mission. Ernst Willi Messerschmid was born in Reutlingen, Germany, on 21 May 1945, and after the *Technisches Gymnasium* (Technical College) in Stuttgart he completed two years of military service and entered the University of Tübingen, where he received his diploma in physics in 1972. He moved on to the University of Bonn and earned a doctorate in 1976, working on proton beams in accelerators and plasmas. In 1977, he joined the Deutsches Electronen Synchrotron (DESY), a major German particle research centre, in Hamburg, to work on the beam optics of PETRA, the positron-electron tandem-ring facility, which would later achieve success in 1979 with the discovery of the gluon, carrier particle of the strong nuclear force. Messerschmid later worked on space-borne communications systems and in December 1982 was selected with Furrer for Spacelab-D1.

QUIET TIME

Watching the aurora and the periods of reflection provided some final quiet time for not only the crew, but also, in advance of the calamity that would befall Challenger on her next mission, for the venerable ship herself. The confidence in which the Shuttle was now held is exemplified by Guy Bluford's businesslike description of his return to Earth on 6 November 1985. "We closed up Spacelab and readied the vehicle for re-entry as the blue team was getting up. I rode upstairs in the cockpit, next to MS2 [Buchli], as we came home. Hank Hartsfield and Steve Nagel flew us home to a safe landing at Edwards Air Force Base in California." After firing the OMS engines for 171 seconds whilst on the opposite side of the planet, Challenger began her hour-long glide to Earth, touching down at 9:44:51 am PST (12:44:51 pm in Florida) on Runway 17 and slowing to a stop in 49 seconds. As part of preparatory work to resume landings in Florida, Hartsfield conducted a computerised steering test of the nose wheel during the 3,000 m rollout. Until 61A, the left and right wheel brakes were applied to steer the orbiter on the runway, although this had regularly caused excessive brake and tyre wear. On this mission, Hartsfield had the ability to depress either the left or right rudder pedal, signalling Challenger's computers to direct a hydraulic actuator to turn the nose wheel and steer the spacecraft onto the runway's centreline. As she slowed to around 170 km/h, he deliberately steered the Shuttle off the centreline by just under 7 m, before returning to normal as he braked to a halt. "It went very well," he said later. "I didn't get very far off the centreline." The test was hailed a success and Mission 61C in mid-December was provisionally approved to land at KSC. In spite of the steering test, however, Hartsfield was not so keen to resume Floridian landings so soon. "As a test pilot, I would like to see a concrete landing at Edwards [for 61C]," he added. "One landing does not prove a system." On balance, the veteran astronaut remarked that he had the most fun of his entire spacefaring career, spending most of his time taking photographs through the flight deck windows and nursing Challenger through a virtually trouble-free week-long trip.

As Challenger settled onto the runway from her ninth mission, she and her crew had good reason for pride in their achievement. The West German sponsors of many of her experiments would later describe the mission as "extremely challenging", but would label its results "outstanding". Spacelab, it seemed, had been cleared for ever more ambitious journeys of scientific discovery. In March 1986, Columbia was scheduled to take three ultraviolet telescopes into orbit to study Halley's Comet. Challenger, too, was to play an important role in the sixth year of Shuttle operations, with her heaviest number of mission bookings to date: five. She would finally deploy the second Tracking and Data Relay Satellite in January, followed by a third in July, and also release the joint US/European Ulysses probe on a journey to observe the Sun's polar regions in May. Later that year, in September, she would deploy an Indian communications satellite and then complete unfinished business by retrieving the Long Duration Exposure Facility from orbit and, in December, she would stage her first top-secret Department of Defense mission.

From a public relations standpoint, too, 1986 would be a banner year for Challenger. High school teacher Christa McAuliffe – the first private citizen passenger to fly on the Shuttle – was already listed among the seven-member crew for 51L and a journalist was tipped to ride 61I in September. McDonnell Douglas engineer Bob Wood, Indian astronaut Nagapathi Bhat and Air Force manned spaceflight engineer Chuck Jones would fill payload specialist slots on three other flights. With the completion of Mission 61B by Atlantis in December 1985, the four-strong Shuttle fleet had flown 23 times and was expected to fly 15 missions in 1986.

However, beneath its 'routine' veneer, it was faltering: voraciously consuming man-hours in preparing orbiters for launch and the shortage of spare parts led to an increasingly dangerous practice of 'cannibalism' from one vehicle to equip another.

After several delays and two brushes with disaster, Columbia lifted off on 12 January 1986 to begin 61C. Her landing in Florida was cancelled due to bad weather and she came home two days late into Edwards Air Force Base. This had already pushed Challenger's tenth flight, 51L, to the end of the month. As the launch schedule grew ever more fierce, something, it seemed, was bound to break. On 27 January, fortunately, the only thing that snapped was a drill bit; but on the following day, for Challenger, the Golden Age would be over.

ACCESS ALL AREAS WITH EASE

When President Ronald Reagan announced plans to build a permanent space station – later to be named 'Freedom' – in the spring of 1984, NASA quickly recognised that this on-orbit construction would involve astronauts performing many hours of EVA. Moreover, the complexity of the assembly work promised to be far more arduous and intricate than anything previously attempted. The space agency already had plans for astronauts to perform a pair of tests – the Experimental Assembly of Structures in EVA (EASE) and the Assembly Concept for Construction of Erectable Space Structures (ACCESS) – which would feature the construction of a geometric object, similar to an inverted tetrahedron, and a 13 m 'tower' in the Shuttle's payload bay. The experiments had been jointly devised by NASA's Marshall Space Flight Center and Langley Research Center and the Massachusetts Institute of Technology. All the hardware was mounted atop a cross-bay MPESS carrier, which also served as a work platform. At first glance, the EASE-ACCESS task looked relatively straightforward. No tools were involved and the spacewalkers were required to snap together the prefabricated segments, linking them with a series of nodes, clusters of sockets and lockable 'sleeves' to erect and secure each structure. There were, however, a large number of components. The ACCESS tower had no fewer than 93 tubular aluminium struts, including diagonals, measuring between 1.35 and 1.8 m in length, and the EASE tetrahedron had half a dozen beams, each 3.6 m long. For some time it was open to question *which* crew would be responsible for this exciting and dramatic EVA task. In its May 1984 manifest, NASA listed EASE-ACCESS on Mission 51G, then scheduled for May of the following year. By August, after the 41D pad abort, it had slipped to Mission 61B in November 1985. Then, in the March 1985 manifest, it moved to 61C, scheduled for December, before ultimately reverting to 61B. As a result of this juggling, many astronauts, including James 'Ox' van Hoften and Steve Hawley and, for a time, even Dick Scobee's crew, undertook EASE-ACCESS training and privately hoped for the assignment.

Yet the team who *would* perform the experiment could hardly have guessed this to be the case when they were announced in February 1984. Commander Brewster Shaw, pilot Bryan O'Connor and mission specialists Jerry Ross, Mary Cleave and Woody Spring were assigned to Mission 51D, which was then scheduled for

February 1985 to deploy the Syncom 4-3 communications satellite and retrieve NASA's Long Duration Exposure Facility after almost a year in orbit. For several months, their flight did not change, but by August 1984 the manifest had shifted and Shaw's team were realigned to deploy a Tracking and Data Relay Satellite on Mission 51L in July 1985. In the March 1985 manifest they were reassigned to 61B with a cargo of three communications satellites – one for Australia, another for Mexico and a third for the Radio Corporation of America – and a large payload bay-mounted version of McDonnell Douglas' electrophoresis machine. At this point, Shaw's crew picked up a Mexican payload specialist, flying to observe the deployment of his country's satellite. By their own admission, it was Ross and Spring, both of whom had worked extensively on EVA issues since joining the astronaut office in May 1980, who pushed for EASE-ACCESS to be added to their mission.

For Ross, who would go on to help build a *real* space station later in his career, EVA was an obsession. It was a remarkable achievement for a highly skilled mechanical engineer ... who could not *swim*. "I got assigned to EVA because I was the class *rock*," he told the oral historian. "When we did our swimming training and our scuba training, I was the guy that sank to the bottom. We had a couple of different people in the astronaut office who had Red Cross swimming training ... try to teach me how to swim." One day, Mary Cleave took Ross over to Anna and Bill Fisher's house and showed him a few strokes in their swimming pool. "Either I sank to the *bottom*," Ross remembered, "or I went *backwards*!" In the end, Cleave gave up. Training underwater had long been standard practice for EVAs and Ross eventually learned to scuba dive and remained qualified throughout his time as an astronaut. After qualifying in the space suit he began to focus on satellite servicing procedures, and worked extensively with Bruce McCandless on the Manned Manocuvring Unit. He was a capcom during Joe Allen and Dale Gardner's spacewalks on 51A to retrieve Palapa and Westar. "Once I got into the EVA area and worked EVA," Ross said, "I *never* let it go. Even if I wasn't officially assigned to it, I continued to work in it and to do whatever I could to have opportunities to get in the [water] tank [to] do development or testing work." By the end of his astronaut career, he would have performed a grand total of *nine* EVAs – or possibly *ten*, if the reader believes rumours of a 'secret' excursion on one of his missions, to be discussed in Chapter 5 – and spent more than 58 hours working outside the Shuttle in space. In April 1985, during the effort to develop a 'flyswatter' repair to activate the deployment switch on Syncom 4-3, Ross and fellow astronaut Mark Lee practiced the EVA procedure underwater in Houston. In the autumn of 2011, his experience had secured him third place in the list of the world's most experienced spacewalkers.

Ross and Spring kept track of EASE-ACCESS and eventually succeeded in persuading programme managers to give the assembly task to their flight. Over a period of several months, the two men worked with the experimenters to choreograph a pair of six-hour EVAs, whose primary objective would be to assess the efficacy of assembling the structures in a microgravity environment, encased in a pressurised space suit. ACCESS did not pose insurmountable obstacles. "Both crew members were in fixed foot restraints," said Ross. "It was basically just a matter of bringing a part out, putting it onto this assembly fixture, hooking the components together, rotating to the three faces, then sliding the completed segment of truss up, and repeating the process for a total of ten 'bays'. We knew that that technique would be a very satisfactory way of doing business, because when a crew member's feet are anchored properly, that gives you both hands free to do work." EASE, on the other hand, was more problematic, since one of the men would be positioned, free-floating, without foot restraints, at the 'top' of the structure, holding on with one hand and torquing the beams into position with the other. Lessons from earlier EVAs had already proven that the absence of foot restraints and adequate hand holds made it extremely difficult for spacewalkers to steady themselves and perform tasks. Six months before launch, Ross and Spring were performing a least one long-duration simulation, per week, in the WET-F tank in Houston, deliberately spending up to six hours underwater at a time to mirror their actual timeline as closely as possible. It was tough, physically demanding work. "It's a *lot* of work, doing an EVA," admitted Spring. "It's one of the few times you actually get to *work* on-orbit. For some of the 'tall' work, we had to go up to Marshall, where they had a bigger tank. We knew, for [the] space station, it's absolutely essential to work with the masses and the volumes and do the choreography *exactly* as it will happen on orbit, so you know what to expect. If you don't, you'll regret it!" Both men instinctively knew from their training that EASE threatened to be an Achilles' heel. "We learned to do it," concluded Spring, "but also learned that just free-floating is *not* the way to put things together."

By November 1985, the plans for two EVAs had been finalised. The first excursion would begin with the construction of ACCESS, with one astronaut in foot restraints on the MPESS carrier and the other at the end of Atlantis' RMS arm. They would disassemble the tower and spend the second half of the EVA repeatedly building and dismantling the EASE tetrahedron, perhaps as many as six times. Two days later, the second excursion would involve further construction tests, attaching flexible cables to simulate electrical wiring and practicing the physical movement, by hand, of the structures around the payload bay. In practice, the EVAs went well. On 29 November, three days after launch, Ross and Spring departed Atlantis' airlock and spent five hours and 32 minutes outside. Donning the suits in the airlock presented something of a surprise, for the spine *grows* slightly in weightlessness, and Spring remembered having more butterflies in his stomach *now* than he had in the seconds before launch. "What's going through your mind is: *Oh, I hope I don't screw up*! It's your big chance ... and they've got all the video cameras in the *world* on you! If you screw up, your friends will have photos and video ready for you at the pin party, too." For his part, Ross was so excited that he had to muster all of his strength *not* to let out a "war whoop of glee" when he ventured outside. The ACCESS tower was built in less than an hour – half the time allocated – and so the two men disassembled it and reassembled it a *second* time. EASE also proved somewhat 'easier' than anticipated, with *eight* assemblies completed, rather than the planned six. Aware of the problems encountered in the WET-F tank, the first four of these were done in foot restraints, although Spring remembered a moment of anxiety as Atlantis drifted from orbital daytime into darkness. "I was up on the EASE structure and, all of a sudden, night fell. I just wasn't used to all of sudden going *dark*, so you've got to get your visor up and get your headlights on and then everything was cool. But I remember that little bit of anxiety ... because you're up on this kind of tippy structure and you're thrashing around just a little bit." Both men found that their fingers grew numb over time and tiredness quickly set in. Fatigue was a problem. "Literally, your mind is going a million miles an hour," said Ross, "thinking about what you're supposed to do, thinking about every step of the procedures, what your buddy's doing, how's the suit doing ... and *looking up*, every once in a while, and trying to capture a snapshot of where you're flying over. I was literally *mentally* the most fatigued, even more than *physically*." On 1 December, the duo were outside again, this time for six hours and 42 minutes, and Ross perched himself in a mobile foot restraint on the RMS, whilst Mary Cleave manoeuvred him to the end of ACCESS to assemble its tenth and final bay. Ross installed mock cables and Spring, located on the MPESS carrier, released the tower and allowed his crewmate to handle it manually from the end of the mechanical arm. The EASE tetrahedron was also moved by Spring and 'repair' methods, involving the removal and replacement of components, were satisfactorily completed. So jubilant were the two men when they returned inside Atlantis that they volunteered to prepare the evening meal for the entire crew!

"EVA is *not* a two-person job," said Spring. "The two guys that get to go outside get the glory from it ... The *rest* of the crew was dead tired. Jerry and I were the ones that prepared supper, because everybody else looked so tired from moving the [RMS], so they could get the pictures, holding the orbiter, turning on lights." Although Spring was correct that the spacewalkers got the glory of the EVA, the *real* glory was the sheer grandeur of the Home Planet itself. "I remember one whole pass over Africa," he recalled. "We launched in November, which is about when the monsoons break in Africa, so almost three-quarters of that continent, at least coastal, is just *consumed* with thunderstorms!" The flashes of lightning were almost continuous. Both Spring and Ross took the time to watch in awe. Aside from the anticipated difficulties with EASE-ACCESS, the spacewalkers found that they typically met or exceeded the speeds at which they had been able to assemble and disassemble the structures in the WET-F. (The crew tried to manifest the Manned Manoeuvring Unit onto 61B, thinking that it would prove useful for moving components and attaching cables.) However, they were dissatisfied that the WET-F did not even come close to mirroring the environment for building a space station. Neil Hutchinson, a former flight director who had been named as NASA's head of the station project, attended the 61B post-flight debriefing and the astronauts explained the need for a much larger underwater training facility. "For the next ten years," remembered Ross, "I ended up being one of the delegates from JSC that went to Washington to make presentations on the need of a new facility." He also led tiger teams and served on inspection teams for an effort which eventually produced the Sonny Carter Training Facility at JSC, which is today used for the EVA preparations of all astronauts.

For an Air Force mechanical engineer who could not swim and an Army test pilot who grew up taking apart and repairing bicycles, toys and trinkets, Ross and Spring had succeeded triumphantly.

Jerry Lynn Ross was born on 20 January 1948 in Crown Point, Indiana, and represents one of only a handful of individuals to have dedicated virtually his adult working life to the astronaut business. His EVA accomplishments have already been noted, but in April 2002 he became the first human to chalk up as many as *seven* missions into space. It is a record which was tied a few months later by fellow astronaut Franklin Chang-Díaz and which both men jointly hold to this day. Ross grew up at a time when the Cold War was at its peak and the idea of rockets, whether for carrying explosives or men, was steadily entering the popular consciousness as something more than a facet of science fiction. As a child, he watched television shows about space stations, read articles in *Life* magazine, created scrapbooks of space-related events and watched in awestruck astonishment when the Soviets launched Sputnik and America responded with Explorer-1. Even at this young age, he was introduced to the word 'engineer'. "I truly didn't fully understand what an engineer *was*," Ross told the NASA oral historian, "but I knew that they had to use a lot of math and science. I *liked* math and science, so I thought that's what I wanted to do. I wanted to become an engineer." By his own confession, this gave him a one-track mind, working on farms to earn enough money for a bank account which would someday pay his way through the prestigious Purdue University.

His three-step plan was relatively straightforward: (1) Be an engineer, (2) Go to Purdue and (3) Get into the space programme. In Ross' mind, it was as cut-and-dried as that. Unlike so many others, he stuck doggedly to his plan and not only achieved it, but surpassed it. After completing high school in 1966, he entered Purdue to study mechanical engineering and received his bachelor's degree in 1970 and a master's in 1972. He entered active duty with the Air Force and worked on computer-aided design of ramjet engines and captive tests of supersonic ramjet missiles at Wright-Patterson Air Force Base in Ohio. Ross graduated at the top of his class from the Air Force Test Pilot School's flight test engineer course in 1976 and later served as project engineer for the flying qualities of the B-1 Lancer bomber. Part of this role was the training and supervision of all B-1 flight test engineers but also mission planning for the bomber's offensive avionics. "The B-1 at the time was the Air Force's highest-priority programme," he remembered, "and I was given the opportunity to come on-board as a B-1 flight test engineer and to work in the stability and flight controls areas of the B-1." It was shortly after the B-1 effort that Ross learned about NASA's plans to hire astronauts; he was one of thousands of hopefuls who submitted applications in 1977 and, though summoned to Houston for interview, was not successful. Still, he persevered and George Abbey offered him a position as a payload officer at JSC, working on the integration of military payloads into the Shuttle, with the hint that it might stand him in good stead for possible future selection. Two years later, in May 1980, after reviewing six thousand applications and interviewing 120 of those, NASA selected 19 new astronauts ... including Ross! In fact, two of Ross' office mates – Mike Lounge and Bonnie Dunbar – had also been chosen. "We had a party at Mike Lounge's house that evening to celebrate," Ross said. "My kids were up in Indiana. School had just got out and so we had to call them and tell them that Daddy had been selected into the astronaut office."

For a man who could not swim, Jerry Ross forged an impressive career for himself as not only a record-setting astronaut and exceptional engineer, but also as one of the world's most experienced spacewalkers. In this image, he is perched on the end of Atlantis' RMS mechanical arm, assembling the ACCESS tower.

If Ross went on to fly seven times in space, then Sherwood Clark Spring was quite the opposite, for 61B would be his only mission. He came from Hartford, Connecticut, where he was born on 3 September 1944. "As a child, engineering was just something I always tended to gravitate to," he told the NASA oral historian. "It's where my interest lay." Growing up on a farm in Rhode Island, it was a useful talent and Spring pursued it at the Military Academy, from which he graduated in 1967 with a degree in general engineering. After West Point, he entered the Army and served two tours in Vietnam, firstly in the 101st Airborne Division and later, after flight training, as a helicopter pilot with the 1st Cavalry Division. Returning to the United States, he enrolled on a master's degree programme at the University of Arizona – where he met his future wife, Debbie – and graduated in 1974. Two years later, Spring served as a flight test engineer at Edwards Air Force Base and completed test pilot school at Patuxent River, then served at the Army's flight test facility as an experimental test pilot. Shortly thereafter, he began to show more than a passing interest in NASA. A close Army friend, Bob Stewart, had been selected as an astronaut and in January 1980 Spring was himself approached by the space agency to be interviewed. Four months later, he was in Tucson, Arizona, preparing for his next Army assignment at Fort Monmouth in New Jersey, when he received a call from George Abbey, inviting him to Houston to become an astronaut. During the interview process, Spring had described himself as "a risk-taker" and, indeed, both of his children have risen to great heights. His daughter, Sarah, became an Army paediatrician, whilst his son, Justin, was a member of the US Men's National Gymnastics Team, which won bronze in the 2008 Beijing Olympics.

As a crew, the 61B astronauts grew into an incredibly close-knit team, with Brewster Shaw described as both a "mentor" and a "mother hen" by more than one of the rookies. With a crew of experienced test pilots and test engineers, it was his decision that the only civilian member of the crew – Mary Louise Cleave – would be trained to serve as the mission's flight engineer, seated behind and between Shaw and O'Connor during ascent and re-entry to offer a third set of eyes on Atlantis' instruments. Since she was not an aviator, the others backed her up at first, but Spring admitted that as launch drew closer, "Mary was a *tiger* and had it all down pat". As 61B's primary RMS operator, Cleave also played a critical role in supporting Ross and Spring … although she would later admit to some disappointment that she was not assigned as an EVA crew member herself. "They couldn't fit me in a space suit," she told the NASA oral historian, "because I was too small and they decided not to buy these small hard upper torsoes to save money. They were going to fit the mid-range of the astronaut corps." (Cleave stood only 157 cm, barely five feet and two inches, making her one of the shortest astronauts ever selected.) Years later, she noted that NASA had a tendency to assign RMS tasks to the female astronauts and there has been some speculation that the women were more adept at operating the mechanical arm than some of their male counterparts. Cleave was born in Southampton, New York, on 5 February 1947, and after high school entered Colorado State University to study biological sciences. Flying had grown from an interest to a fascination and into a passion. "I was interested in airplanes before space," she reflected. "I started flying when I was 14. No one's really sure why I was so crazy about airplanes. Nobody else in my family flew, except for my mother's brother, who was a pilot, killed in World War II, so I didn't even know him." She had built model aircraft as a child and after receiving her first flying experience as a teenager was told by her parents that she would have to make the money to earn her pilot's licence. "I gave baton-twirling lessons," she said, "and babysat a lot and got the money and went ahead and took my first lesson." Cleave soloed at the age of 16 and received her licence a year later.

At Colorado State, Cleave aspired to be an airline stewardess, but was a couple of inches too short and when she graduated in 1969 she was rejected. Nevertheless, she continued with her academic studies and received a master's degree in microbial ecology from Utah State University in 1975 and progressed immediately into doctoral research. Cleave received a PhD in civil and environmental engineering in 1979 and was working as a sanitary engineer at the Utah Water Research Laboratory when one of her colleagues approached her with news of a NASA advert, calling for astronauts. "You're the *only* engineer I know that's *crazy* enough to want to do something like that," he told her. Cleave was invited to Houston in March 1980 – "the weather was perfect," she recalled, "and the mockingbirds were up in the trees" – and to her great surprise was selected in May. On her first flight in a T-38 Talon, she must have been struck by the irony of having been rejected as an airline stewardess a dozen years earlier … and *now* she was at the *controls* of a high-performance jet trainer. Her instructor, Bud Ream, had received strict instructions *not* to make the 'weakling' scientists sick … but Cleave spent the entire flight *pleading* with him to do some aerial manoeuvres and *promising* that she would not be sick!

Mission 61B was only the second Shuttle flight to launch in the hours of darkness, blasting off at 7:29 pm EST on 26 November 1985 in what Brewster Shaw later described as an ascent "immersed in light". Since Atlantis had landed from her maiden voyage barely 50 days earlier, her turnaround between flights remains the shortest ever achieved in the Shuttle's history. She returned to Florida on 12 October, spent less than four weeks in the Orbiter Processing Facility and was back on Pad 39A by 12 November. The mission had already attracted much popular attention, not only because of the EVAs, but also due to the presence of Mexico's first man in space, electrical engineer Rodolfo Neri Vela. Interestingly, Mexico City had been recently hit by a massive earthquake, the after-effects of which continued to be felt into the spring of 1986, and, according to Cleave, these were instrumental in adjusting Atlantis' launch time from the hours of daylight to those of darkness.

A MEXICAN IN SPACE

In the early hours of 19 September 1985, an earthquake measuring 8.1 on the Richter Scale hit Mexico City, directly causing the deaths of more than ten thousand people. It was not totally unexpected, for a slightly smaller quake had taken place in May and the main event was followed by a pair of major aftershocks on 20 September and at the end of April 1986, together with a dozen or more minor shocks. Although the main quake occurred off the Pacific coast, several hundred kilometres away, its sheer strength, combined with Mexico City's location on an old lakebed, meant that it inflicted great suffering on the densest population centre in the world: in addition to the devastating death toll, hundreds of buildings collapsed, thousands more were wrecked beyond repair and more than $3 billion of damage was caused. The effects of the quake were felt as far away as Los Angeles and Houston.

Against the backdrop of this calamity, Rodolfo Neri Vela had been training throughout much of 1985 to become Mexico's first man in space. He would be aboard 61B to observe the deployment of the Morelos-B communications satellite, almost identical to the one placed into orbit by the 51G crew in June. Yet the effects of the earthquake and its devastating impact on the Morelos ground support infrastructure in Mexico almost resulted in the satellite being deleted from Atlantis' payload roster, according to *Flight International* on 9 November. "If Morelos ... does fly," the magazine told its readers, "it will be placed in parking orbit until Mexico's telecommunications are rehabilitated following the Mexico City earthquake." Ultimately, Morelos-B remained aboard, as did Neri Vela.

His family lived in Chilpancingo de los Bravo, the second-largest city and capital of the mountainous state of Guerrero, in the far south of the country. He was born on 19 February 1952, of Spanish and Italian extraction, and received a degree in mechanical and electrical engineering from the National Autonomous University of Mexico in 1975. Neri Vela then travelled to the United Kingdom and earned a master's degree in science, with a specialisation in telecommunications systems, from the University of Essex in 1976 and a doctorate in electronic engineering, focusing on electromagnetic radiation, from the University of Birmingham in 1979. He remained at the university as a post-doctoral researcher in waveguides and was selected as one of two Mexican payload specialist candidates, alongside mechanical engineer Ricardo Peralta y Fabi, in July 1985. Less than six months to train with an unknown crew member worried Brewster Shaw. "I'm probably a paranoid kind of guy, but I didn't know what he was going to do on-orbit," Shaw told the NASA oral historian, "so I remember I got this padlock and ... went down to the hatch on the side of the orbiter and I *padlocked* the hatch control, so that you could not open the hatch." Payload specialist Charlie Walker was a 'known' quantity, with whom the astronauts felt comfortable, but Shaw was particularly concerned that Neri Vela might "flip out" during launch or whilst in orbit. Mike Mullane made reference to the padlock episode in *Riding Rockets* and noted that future missions benefitted from a similar arrangement, with only the commander given the key. "I don't know if I was supposed to do that or not," said Shaw, "but that's a decision I made as being responsible for my crew. I don't think Rodolfo noticed it, but some of the other crew noticed it." Fortunately, Neri Vela proved himself to be a model crew member, but the situation demonstrated how nervous the 'professional' astronauts were in flying alongside part-time and relatively unknown payload specialists.

As if to underline their diversity, the 61B crew posed for an unofficial photograph, featuring engineers Mary Cleave and Charlie Walker in their white lab coats, Brewster Shaw wearing a 'Boss' badge, Bryan O'Connor dressed in leather cap and goggles as the barnstorming fighter ace, Jerry Ross and Woody Spring in space suits and construction helmets ... and Neri Vela in a traditional serape and sombrero. "*That* photo could not be the official one," admitted Spring, "because the Mexican government took a little bit of umbrage at Rodolfo being dressed up in a serape and a sombrero, but then post-flight we went down to Mexico City ... and the *first* thing they did was take us to the folk ballet, where *everybody* is dressed up exactly like that!" In the photograph, Spring also posed with a stuffed kangaroo toy in his lap, honouring one of 61B's other payloads: Australia's Aussat-2, a virtual copy of the communications satellite deployed by 51I. Their third satellite was an American payload, Satcom K-2, which was, at the time, the highest-powered domestic communications satellite in service. Its size and weight also required it to be boosted into geostationary orbit using the uprated PAM-D2 booster.

If Woody Spring considered the moments before his first EVA as being more anxious than the launch itself, that did nothing to detract from the excitement of Atlantis' second climb into orbit. "It was a perfect night," he recalled. "We had a full Moon, the day before Thanksgiving, severe, clear, not a cloud in the sky; it was a *gorgeous* night." The festive, picnic-like atmosphere was evident in the number of guests: Spring, for one, had invited as many of his family and friends as possible to witness the launch. As he ascended the elevator to the level of the white room, it was obvious that the Shuttle was *alive*; loaded with cryogenic propellants, it squealed and creaked and moaned its readiness to them. Aboard the flight deck, Spring was seated on the middeck, together with the payload specialists, whilst Jerry Ross sat upstairs, directly behind the pilot. With five minutes to go, Marine Corps aviator extraordinaire Bryan Daniel O'Connor was instructed to activate the three Auxiliary Power Units, and they all hummed perfectly and rhythmically to life.

It is no accident that O'Connor became the first pilot from the May 1980 class to receive a flight assignment, although as the schedule writhed and contorted, he ended up being the *third* to actually launch into space, after Roy Bridges and Ron Grabe. The reader will recall that five pilots had been dropped from the 1978 astronaut selection, due to the requirement to pick more women, and O'Connor was one of them. Two years later, he was accepted. Born in Orange, California, on 6 September 1946, he grew up in a military family – his father was a career Marine Corps pilot – and spent his weekends looking at aircraft on the flight line. After high school, he entered the Naval Academy and received a degree in engineering in 1968. Two years later he got a master's credential in aeronautical systems from the University of West Florida. At first, he had mixed feelings about where he wanted his own military career to take him. "Maybe I would get into submarines or ships," he said, "but the aviation bug was too deep in there and so when it came time to select a service and an occupation, I went for Marine Corps aviation." After Basic School he received his naval aviator's wings in June 1970 and flew as an attack pilot in the A-4 Skyhawk and the AV-8A Harrier. "*That's* when I got bit by the test pilot bug," he said. In 1975, he was accepted at Naval Test Pilot School and later served within the Strike Test Directorate, participating in the evaluation of various vertical takeoff and landing aircraft. Going on to graduate from the Naval Safety School, he was in charge of Harrier flight testing when he applied for NASA's 1978 astronaut selection. Rejected for what he considered to be his lack of experience, he spent 1979 gathering more flight test time in the Harrier. "I don't think I would have flown the Shuttle any sooner had I been picked up in that first class," he said later. "They were going to hire 40 and they cut five pilots out, because they didn't need them, right at the last minute. I found out much later that I was one of those five. The reason they didn't need them is because in that time frame, Columbia was having problems. All the tiles had fallen off on a flight across country. One of the main engines had blown up on a test stand and the whole Shuttle programme was slipping, so if they had picked me up in 1978 and I had joined the astronaut office, I probably would have eventually flown the same time anyway as I did."

Not until relatively close to the flight did O'Connor and his crewmates realise that they would be launching at night. The only astronauts with actual experience were the STS-8 crew and the 61B crew consulted them at length about how it differed from a daylight launch. As the final seconds ticked away to launch, O'Connor glanced across the cabin towards Shaw and noticed that the commander had momentarily removed his gloves to wipe sweat from his hands. "Oh, my God," O'Connor thought. "My commander, who's been through this before ... his *hands* are *sweating*! Why aren't *mine* sweating? I need to be nervous now, if *he's* nervous." (It brought back memories of John Young's remark that if an astronaut was *not* nervous, sitting atop thousands of kilograms of volatile propellants, he or she did not have a full understanding of the enormity of the situation.) The sound of the main engines igniting, said Woody Spring, was like a roomful of lions, roaring, directly behind him. "And you can *feel* it," he remembered. "This vehicle's *alive*. Then the main engines gimbal, getting ready. From the moment of main engine start, you get one and a half seconds where the vehicle actually *swings* about three degrees of arc and then comes back again. *Then*, the Solid Rocket Boosters ignite." With a force akin to a sledgehammer blast, they were propelled away from the launch pad in what Spring could only describe as "a barn burner". Eight minutes later, in a preliminary orbit, he released a pencil and watched it *float* freely. He let out a whoop of delight as Mary Cleave giggled with excitement over the intercom. They were in space.

A busy mission got underway almost immediately, with Spring supervising the deployments of the Mexican and Australian communications satellites and Ross overseeing the release of Satcom K-2 and its PAM-D2 booster. Morelos was sent spinning out of the payload bay at 2:47 am EST on the 27th and, after the crew had slept, was followed by Aussat at 8:20 pm. Finally, Satcom was released at 4:57 pm on the 28th. To say that the 61B crew had a good time together, on the ground and in orbit, is something of an understatement. Jerry Ross would recount that O'Connor and Spring were the comedians, with the latter producing the idle jokes and the former mastering the dry wit. "He will sucker you in on some really serious discussion," Ross said of O'Connor, "and then *hit* you over the head with a two-by-four with some joke or comment!" Before launch, O'Connor had craftily recorded the Naval Academy's song, *Anchors Aweigh*, somewhere in the middle of Army's aviator Spring's Walkman music cassette. A couple of days into the mission, with the lights turned off and the crew ready for sleep, Spring suddenly screamed, "O'Connor, you *son of a bitch*!" The prankster had completely forgotten about it, but was quickly reminded. "It was his Peter, Paul and Mary album," he recalled with glee, "and it was right in the middle of *I've Got a Hammer* ... and suddenly up comes

Woody Spring works in Atlantis' payload bay.

this *really* loud Navy fight song!'' Thanksgiving on the second day of the mission offered the chance for the crew to eat irradiated turkey, pumpkin bread, mashed potatoes, beans and a somewhat tasteless concoction which was labelled 'gravy'. In fact, many foods which tasted fine on Earth were quite different in orbit: shrimp cocktails were like battery acid in sawdust and Spring found the grapefruit juices appalling. Neri Vela also brought along some Mexican foodstuffs, including flour tortillas.

Watching the Home Planet was a frequent and favourite pastime. In a single eyeful, the astronauts could behold the entire history of the world as they knew it: Egypt and the Red Sea, the Holy Land, the eastern Mediterranean, and beyond. Reflective sand dunes and cloud-free deserts and translucent water proved mesmerising and Spring found it difficult to 'burn' a memory of it all into his brain. For O'Connor, the greatest pride was acquiring high-quality images of the oceans for Bob Stevenson and his team on the ground. "Some of the things you can see when you look into the Sun glint in the oceans are just incredible," O'Connor told NASA's oral historian. "To this day, when I'm flying in an airplane over the ocean ... I try to pick up things like ship wakes and circular eddies." As the five professional astronauts worked, Rodolfo Neri Vela tended to a battery of Mexican experiments on the middeck involving bacterial growth, the germination of various seeds, the transport of nutrients through plants, the validation of theories of 'electropuncture' to stimulate disequilibrium in human organs and, of course, photography of Mexico and its recovery in the aftermath of the earthquake.

Charlie Walker was flying for the third time with McDonnell Douglas' electrophoresis machine and produced pleasing results. However, many pharmaceutical companies – including Ortho Pharmaceutical – were beginning to back away from the space-flown experiment, finding that cheaper alternatives existed on the ground, including genetic engineering technologies. Walker recognised this to be "a bump in the road", although McDonnell Douglas announced at around this time that it was producing 'erythropoietin', a glycoprotein which controls red blood cell production, with potential applications in the treatment of anaemia from chronic kidney disease and myelodysplasia from chemotherapies and radiotherapies. As with his previous flight, 51D, Walker had to shut down his equipment when the cabin pressure was reduced in readiness for the EVAs. He could not have foreseen the catastrophe which would engulf the Shuttle and the nation in the coming weeks. With the EOS-1 payload bay hardware scheduled to make its first flight in July 1986, Walker confidently expected a fourth mission, but in the wake of Challenger it became increasingly clear that commercial payloads – even on the middeck – would be 'secondary', rather than 'priority', experiments. By 1988, McDonnell Douglas was beginning to look at ground-based electrochemical or fluid electric purification processes. The commercialisation of electrophoresis on the Shuttle was over.

If Walker expressed confidence for the future as November wore into December 1985, so too did the NASA members of the crew. Normally after a mission, the astronauts celebrated with a party, but with the packed Shuttle schedule and 61C and 51L due to be launched on 18 December and 22 January, Brewster Shaw's crew decided to wait until the next two flights had returned and hold a 'big' joint party.

Obviously, it never happened. In fact, the next time that Charlie Walker saw his 61B crewmates after Atlantis' landing was more than two months later, in February 1986 ... at the memorials for the Challenger astronauts.

"The landing," said Bryan O'Connor, "isn't *nearly* as exciting as the launch." Re-entry was over the darkened Pacific Ocean and although cloud cover restricted their view of the runway at Edwards until Atlantis was a couple of kilometres above the ground. Shaw brought her in to a perfect landing at 1:33:49 pm PST (4:33:49 pm in Florida) on 3 December, an orbit earlier than planned, after seven days in space. There was a slight tail wind and both Shaw and O'Connor were concerned about the integrity of the brakes and tyres, but the vehicle slowed to a stop, right on the centreline. Woody Spring, who had exchanged seats with Jerry Ross for re-entry, remembered struggling to lift himself out of his seat ... and the problem of returning to terrestrial gravity after a week in free-fall continued when he got home. His four-year-old daughter jumped onto the bed and Spring worried that she would ricochet across the room, just as he had seen happen in space. Whenever the telephone rang, he instinctively made as if to *float* over to answer it. Within days, his 'Earth legs' had returned and he felt back to normal. Eight weeks later, the entire Shuttle programme would have been brought to its knees and nothing would ever be truly 'normal' again.

BRINGER OF MISFORTUNE?

By the beginning of January 1986, Steve Hawley had developed something of a reputation within the astronaut corps. The boyish-looking, 34-year-old astronomer had already made one Shuttle flight in the late summer of 1984 and that had been sprinkled with the first seeds of bad luck. Hawley's first flight should have taken place in June 1984, but was halted on one occasion by a computer glitch and, later, by a dramatic shutdown of Discovery's three main engines, seconds before liftoff. Two months later, after extensive repairs, he and his five crewmates returned to Pad 39A to try again and were thwarted by bad weather. On their *third* attempt, they finally launched on 30 August. By October of that year, Hawley had been assigned to a mission known as 51I, with commander Robert 'Hoot' Gibson, pilot Charlie Bolden and mission specialists George 'Pinky' Nelson and Costa Rica-born Franklin Chang-Díaz. Scheduled for August 1985, they were to deploy the ASC-1 and Syncom 4-4 communications satellites and operate the Materials Science Laboratory (MSL-2) in the payload bay. However, by March 1985 their mission had slipped again, this time to January 1986, and was redesignated 51L. They would have deployed a Tracking and Data Relay Satellite and the Spartan-203 free-flier to observe Halley's Comet. Then, in July, they received the new designation 61C, gained two new communications satellites and picked up a payload specialist from the Radio Corporation of America (RCA).

Robert Joseph Cenker was born on 5 November 1948 and grew up close to Uniontown in Pennsylvania. He received bachelor's and master's degrees in aerospace engineering from Pennsylvania State University, together with a second master's in electrical engineering from Rutgers University in New Jersey. In total, he would spend almost two decades of his career with RCA Astro Electronics (which was bought by General Electric in 1986 to become GE Astro Space), and at the time of his Shuttle mission was a senior staff engineer, working in several communications satellite projects. In July 1983, RCA Astro Electronics received a $120 million contract from its sister company RCA American Communications to build three high-powered Satcoms for television broadcasting. Two of these were to be launched in quick succession, late in 1985, with the third manifested for 1987. Known as 'Satcom-Ks' (or 'Satcom-Kus'), each was capable of broadcasting in the 12-14 GHz Ku-band. Since Ku-band frequencies were not shared with terrestrial microwave systems, dishes served by the Satcoms could be located within major metropolitan areas characterised by heavy terrestrial microwave traffic. "The satellites weigh 2,000 kg," *Flight International* told its readers, "and operate at a higher frequency than their C-band predecessors, making them the most powerful domestic satellites in service." For Cenker, who had served as RCA's manager of systems engineering for the Satcom-K, it was 'his' satellite and his technical responsibility. RCA paid NASA $14.2 million to launch Satcom K-1 on Mission 61C and the satellite's 16 transponders were expected to provide coverage of the entire contiguous United States. Confusingly, the *second* Satcom-K ('K-2') was launched aboard Mission 61B, *before* K-1, and physically the satellites differed from many of their predecessors, in that they were cube-shaped, carried wing-like solar arrays and were also much heavier. In fact, the Satcom-Ks were *three times* the weight of the SBS-3 and Anik-C3 payloads deployed on STS-5. This demanded a more powerful version of the PAM-D booster. Enter the new PAM-D2, which could accommodate payloads measuring up to 3 m in diameter, as opposed to just 2 m for its predecessor. Thanks to an upgraded Star 63D engine, it could send much heavier satellites into geostationary transfer orbit, but it required a slightly larger cradle in the Shuttle's payload bay. Already, in March 1985, McDonnell Douglas received a $169 million contract to supply 28 PAM-D2s to the Air Force to launch a network of Navstar satellites from 1987 onwards.

The second payload specialist on 61C *should* have been Greg Jarvis, a Hughes engineer who had been selected by his company in the summer of 1984 and was first attached to Dan Brandenstein's 51D mission. Unfortunately for Jarvis, as the manifest shape-shifted in the wake of the 51E debacle, he moved further and further downstream: since Hughes had built the Syncom, it seemed likely that he would accompany one of those satellites into orbit and he was earmarked to fly on Joe Engle's 51I mission for a while, before inexplicably disappearing from the crew roster and winding up on 61C. Then, in the autumn of 1985, Jarvis was moved again, this time to 51L. His replacement on 61C would be Congressman Bill Nelson, a Democrat from Florida, who had trained alongside Jake Garn. Jarvis' reassignment to 51L, which did not even have a Syncom aboard, reinforced in many minds the reality that he had been moved simply to make room for another politician to fly.

Clarence William Nelson – "Willie Nelson," as chief astronaut John Young called him, "the one who *can't* sing!" – was born in Miami on 29 September 1942, the only son of Clarence Nelson and Nannie Merle Nelson. He spent his early childhood in

Melbourne and attended the University of Florida, then Yale, where he received a bachelor of arts degree in 1965. Whilst an undergraduate, Nelson was enrolled in the Reserve Officer Training Corps and served as an Army reservist for six years. He completed a law degree at the University of Virginia and was admitted to the Florida bar in 1968. Several years of legal practice in Melbourne followed – he served as legislative assistant to Governor Reubin Askew for a time – and in 1972 he was elected to the Florida House of Representatives. Nelson won re-election in 1974 and 1976 and was then elected to the House of Representatives in Washington in 1978, serving until 1991. At the time of his Shuttle mission, he chaired the House's Space Science and Technology Subcommittee. Since 2000, he has served as the senior Senator for Florida and over the years has proven to be one of NASA's staunchest supporters and most vocal political champions.

Nelson's assignment to 61C brought with it the inevitable question of *what* he should do in orbit. Photography seemed the most obvious option, although Nelson wanted to be assigned something more important. Principal investigators would spend many years, and in some cases many millions of dollars, preparing their experiments for flight and training the astronauts to operate them; and were reluctant to have a non-technical politician step in at the last minute and possibly screw them up. Commander Gibson remained firm: *mission specialists* would operate the experiments. Some astronauts jokingly wondered if Nelson's 'important' experiment was to find a cure for cancer. In all sincerity, Nelson proposed that he take photographs of Africa to aid with the humanitarian effort. Then he suggested a communications link between the Shuttle and the cosmonauts aboard the Salyut 7 space station. Hearing of Nelson's three 'missions' – curing cancer, ending famine and fostering better US-Soviet relations – the astronaut office humourists quickly got in on the act.

"Do you know how to ruin Nelson's entire mission?" came the joke.

The response: "On launch morning, tell him they've found a cure for cancer, it's raining a *flood* in Ethiopia and the Berlin Wall is coming down! He'll be *crushed*!"

Yet, as a crew member, Nelson proved a model payload specialist. "He worked very hard," admitted Pinky Nelson. "Physically, he was in better shape than *we* were! He had no experience ... in aviation or anything technical. He was a lawyer, so he had a huge learning curve, but that didn't stop him from trying and I think he knew where his limitations were."

In fact, Pinky and Bill Nelson, though unrelated, became the first two men sharing the same surname to fly together in space. Also aboard were a pair of astronauts from NASA's 1980 selection: Charlie Bolden was the fourth African-American to fly the Shuttle – and only the *second* to fly in the pilot's seat – whilst Franklin Chang-Díaz was the first Hispanic-American spacefarer. With its increased emphasis on the recruitment of minorities in the late 1970s, NASA recognised that whilst blacks and women were applying for the astronaut programme, Hispanic applicants were fewer in number. In September 1979, the agency decided to act and specifically called for interested volunteers. "Many qualified Hispanics are hesitant to apply," admitted Jose Perez, deputy chief of NASA's Equal Opportunity Programs Office. "I would like to encourage those persons, and others, to call or write NASA for an application." One young man who answered this call, Franklin Ramón Chang-Díaz, originated from Costa Rica in Central America. He was born on 5 April 1950 in San José, the son of Ramón Chang-Morales and María Eugenia Díaz De Chang. Though his parents were both Costa Ricans, his father was of Chinese descent, whilst his mother was wholly Hispanic, and the family had lived in Venezuela for a time. After studying at La Salle School he moved to the United States as a teenager in August 1968, with $50 in his pocket, hoping to become an astronaut. "I was captivated by Sputnik as a child," he told an interviewer. "I felt that, someday, humans would travel to distant planets and I decided that I wanted to be one of those travellers. I would be a space explorer."

It seemed an impossible dream for a youth who spoke no English. "My family never prevented me from doing that," he continued, "but they couldn't really help me. We were not a well-to-do family. Even though my parents put us in the best, most expensive schools and we got a first-rate education, my parents were not rich. Neither one of them finished college, so I was expected to make my own way as soon as I finished high school. I couldn't expect to receive a college education on my father's dime, but I was expected to *have* a college education." Encouraged by his parents, and his grandfather, to go to the United States, Chang-Díaz enrolled at the public high school in Hartford, Connecticut, and learned English through sheer, total immersion in the language. He failed his classes in the first two quarters, but his third and fourth quarters were outstanding. One of his teachers, Alan Winter, took notice of his efforts and began coaching the teenager and preparing him for university. Chang-Díaz succeeded in securing a scholarship for the University of Connecticut as an engineering student, but his lack of US citizenship presented an obstacle. His admissions tutor thought he was from Puerto Rico, *not* Costa Rica. "Well, *that* was a bucket of cold water," he remembered. "I went back to the high school and related this story to the teachers, who apparently wrote a *petition*. The Connecticut legislature met and decided to offer me one year of the scholarship and let me pay the lower, in-state tuition, because they had already offered the scholarship." Chang-Díaz was obliged to take loans and work in the university's physics department to support himself through the remaining three years. To him, the story of those years was all about America – "the ability to get ahead by hard work" – and he got his bachelor's degree in mechanical engineering in 1973.

By this stage in his life, he was gravitating towards energy and nuclear fusion, figuring that *this* power source would be critical for getting future astronauts to Mars. It offered a small insurance policy in the event that he did not succeed in his aspiration to become an astronaut. Chang- Díaz entered graduate school at MIT and in 1977 earned a PhD in applied plasma physics, with a research focus on the problems of controlled thermonuclear fusion. After the completion of his doctorate, he joined the technical staff of the famed Charles Stark Draper Laboratory, working on the design and integration of control systems for fusion reactor concepts. Two years later, NASA announced its intention to select Hispanic candidates for the astronaut programme and he tendered his application. When he received the call, in May 1980, that he had been selected, Chang-Díaz "went running out the door and across the street. I almost got run over by a *cab*! I was in a *totally* different world. My life changed completely from that day on." At first, being a scientist, and not a test pilot, posed another hurdle. It seemed to Chang-Díaz that his lack of flying credentials made him a less attractive candidate to draw a mission assignment. "That didn't seem right to me," he said, "and I kept working to remain both a scientist *and* an astronaut. In the end, I won out. I remained a scientist and I flew more than anybody else!" In fact, by the end of his career with NASA in 2005, he had flown *seven missions*, creating a joint record with fellow astronaut Jerry Ross. Their record stands to this day. His role as an active, practicing scientist also continued throughout his astronaut career and beyond. He served as a visiting scientist with MIT's Plasma Fusion Center from 1983 to 1993 and later headed JSC's Advanced Space Propulsion Laboratory, helping to design future plasma-propelled rockets. He was also instrumental in the creation of the Astronaut Science Colloquium Program and Astronaut Science Support Group in January 1987. Today, he runs the AdAstra company, developing the Variable Specific Impulse Magnetoplasma Rocket (VASIMR) as a propulsion system for deep-space missions. Even Steve Hawley and Pinky Nelson were in awe of Chang-Díaz. "Franklin's smarter than both me *and* Steve, put together," Nelson told the NASA oral historian. "He's the one who's going to take us to Mars, I think, in his rocket."

If Chang-Díaz conquered almost insurmountable difficulties in his life, so too did Columbia's pilot for 61C, Charles Frank Bolden Jr, who is today the first – and so far only – African-American Administrator of NASA. He was born in Columbia, South Carolina, on 19 August 1946. "I never wanted to be an aviator," he told the oral historian. "I saw a programme on television, called *Men of Annapolis*, when I was in eighth grade; fell in love with the uniform, fell in love with the fact that they seemed to get all the good-looking girls!" After high school, where his father served as the head football coach, Bolden entered the Naval Academy and graduated with a degree in electrical engineering in 1968. His original intent was to become a Navy frogman or an underwater demolition expert, but after graduation he entered the Marine Corps, inspired by an infantryman, Major John Riley Love, who had been his first company officer at the academy. "I knew that infantry officers *died* real quick when they went to Vietnam," Bolden said. The devastating Tet Offensive had recently occurred and life expectancies for infantrymen were expressed in *months*, rather than years, but he pressed on through Basic School, then decided to change course and enter Marine aviation instead. He underwent flight instruction, received his wings in May 1970 and later flew 100 sorties – many of them in the hours of darkness – over Vietnam in the A-6 Intruder.

The lure of test pilot school was strong, although Bolden met with many rejections. Upon his return to the United States from Vietnam, he served as a selection and recruitment officer in Los Angeles and took various assignments at Marine Corps Air Station El Toro, California. After half a dozen attempts, he was finally accepted by the test pilot school at Patuxent River in Maryland and graduated in June 1979. Years later, he was convinced that a master's in systems management, earned from the University of Southern California in 1977, was instrumental in this success. At around this time, Bolden picked up a NASA application form, but did not fill it in, preferring not to waste his own time and that of the Marine Corps. However, when he *met* some of the group who *had* been selected, Bolden decided to try his hand and was interviewed in Houston. At the end of May in 1980 – on his wife's *birthday*, of all times – he received the call from George Abbey that would change his life.

Assigned as Hoot Gibson's pilot, Bolden would credit his commander with having taught him a great deal about Shuttle systems and aerodynamics ... and, most significantly, *Hoot's Law*. By his own admission, Bolden struggled through his first few months of mission-specific training. "I really wanted to impress everybody on the crew and the training team," he told the NASA oral historian. One day, in the simulator, the instructors threw an engine failure at the crew and, to distract them, quickly piled on more problems. Bolden followed his procedures, safing the engine, and was quickly presented with a minor glitch in an electrical bus. His attention shifted to the new problem, but he picked the *wrong* bus and inadvertently shut down the bus for an *operating* engine. "When I did that," he said, "the engine *lost power* and it got *real* quiet! We went from having one engine down in the orbiter, which we could've gotten out of, to having *two* engines down, and we were in the water, *dead*." Bolden felt awful. If this ascent simulation had been for real, he would have brought death to them all. Gibson, in his infinite wisdom, reached across the cockpit and patted Bolden on the shoulder.

"Charlie," he said, "let me tell you about Hoot's Law."

"What's Hoot's Law?"

"No matter how *bad* things get, you can *always* make them worse!"

A 'NEW' COLUMBIA

On Mission 61C, Columbia was making her first orbital flight in over two years. On 26 January 1984, a month after returning from STS-9, she was mounted atop the Boeing 747 Shuttle Carrier Aircraft and transported from Florida to California for a major programme of modifications. Following a brief stopover at Kelly Air Force Base in Texas, she arrived at Edwards Air Force Base on 30 January and was rolled overland to prime contractor Rockwell International's facility in Palmdale. Subsequent modification periods would involve flying the orbiters directly into Palmdale, but at this time the plant was not equipped with a Mate-Demate Device to remove the spacecraft from the top of the Boeing; removal thus had to take place at Edwards and Columbia was towed, to the amazement of open-mouthed motorists, down the highway to what would be her new home for 18 months.

Many observers have remarked that the result of this modification work was not so much an *overhauled* spacecraft, but a *brand-new* spacecraft. Already, by the time she flew to Edwards, Columbia had changed a great deal since her maiden voyage in April 1981. Parts of her ejection seats had been dismantled, her development flight instrumentation pallet was more or less gone, sleeping bunks had been added for her larger crew, a new Ku-band antenna had been installed, her brakes and tyres were greatly improved and the floor of her payload bay had been strengthened. Much of the data from Columbia during her four test flights had led engineers to the conclusion that they had *over-designed* the Shuttle. "It was *too* strong, *too* beefy," remembered Arnold Aldrich, "and what we could actually do was take [weight] out of the orbiter by redesign. That was very desirable, because that would be directly related to payload. Both Columbia and Challenger were built to this heavier design, but Discovery, Atlantis and Endeavour weren't yet created. They could take advantage of this knowledge of areas where we could take some of the weight out." In addition to structural inspections, Columbia received a Heads-Up Display for her pilots, lost her ejection seats in favour of lighter versions and was fitted with a myriad of sensors to monitor her performance in ascent, orbital flight, re-entry and landing.

One device in particular was the Shuttle Infrared Leeside Temperature Sensing (SILTS) instrument, which took the form of a 50 cm cylindrical pod, attached to the top of Columbia's vertical stabiliser tail fin. It was intended that SILTS would acquire high-resolution infrared imagery of the upper surfaces of the port-side wing and fuselage during the high-temperature portion of re-entry. This was expected to highlight those parts of the spacecraft which experienced maximum amounts of heating during the fiery plunge back to Earth. A hemispherical dome at the forward-facing end of the cylinder contained two windows – one forward-facing, the other at a slightly oblique angle – for the infrared camera. Through re-entry, a constant stream of room-temperature nitrogen gas would flow across the windows to protect the windows from the onslaught of atmospheric heating. If they had not been, the camera would only have been able to see the window itself, rather than what lay beyond. To accommodate SILTS, the pod and the upper couple of metres of Columbia's tail fin were coated with black High Temperature Reusable Surface Insulations (HRSI) tiles. It was planned for SILTS to be activated by the General Purpose Computers at an altitude of around 120 km, at the cusp of 'entry interface', when two plugs would be jettisoned from the windows. For the next 20 minutes or so, the camera would alternate every 11 seconds between the port-side wing and the fuselage, until the spacecraft had descended below 25 km and was through the worst of aerodynamic heating. It would then be deactivated. With the benefit of hindsight, it is a pity that SILTS did not remain operational for the remainder of Columbia's career, for it might have provided additional insight into her loss in February 2003.

Other new devices included a brand-new nose cap for the Shuttle Entry Air Data System (SEADS), which consisted of pressure sensors to assess the spacecraft's aerodynamic performance at various stages in the high atmosphere. The new nose cap contained 14 tiny holes, through which the pressure of the outside airflow could be determined. This provided data about Columbia's attitude in relation to the airflow, and enabled predictions to be made about atmospheric density at differing altitudes. To ensure that the holes in the nose cap would not generate leaks and risk destroying the vehicle, the interior bulkhead, 'behind' SEADS, was covered with protective HRSI tiles. A third instrument was the Shuttle Upper Atmosphere Mass Spectrometer (SUMS), which took air samples through a small hole on the underside of Columbia's nose, just between the nose cap and the forward wheel well. These samples were then used to measure the quantity and composition of gas at different altitudes and thus the density of the atmosphere through which the vehicle had flown.

With all of these devices fitted, Columbia was returned to Florida on 11 July 1985. After temporary storage in the cavernous Vehicle Assembly Building and the Orbiter Processing Facility, processing began in late September for her next flight, by now redesignated 61C and scheduled for December. Hoot Gibson's crew had retained the MSL-2 payload, but the remainder of their planned five-day flight was relatively roomy, with few middeck experiments to perform and only Satcom K-1 to deploy. MSL-2 was a largely automated materials processing facility, housed on an MPESS structure. Another bridge-like assembly carried a dozen Getaway Special canisters. To say the very least, in the eyes of the astronauts, 61C was a distinctly 'ho-hum' mission.

Steve Hawley would later speculate that Congressman Bill Nelson's presence had probably saved their flight. "Frankly, our payload wasn't very robust," he told the oral historian, "and were it not for [Nelson's] presence on the flight, we might have been cancelled. We had *one* satellite and some other experiments. It was a clearing-house sale!" Such a mission would almost certainly have been cancelled, or at least remanifested onto another flight, but for the presence of the powerful congressman. Mission 61C finally received a firm launch date of 18 December 1985 and should have been NASA's tenth flight of the year, capping a triumphant run of Shuttle achievements and heralding the dawn of an even brighter 1986. However, it did not happen. Launch on the 18th, scheduled for 7:00 am EST, at the start of a 49-minute 'window', was routinely postponed by 24 hours to give technicians more time to finish closing out Columbia's aft compartment. The second effort to get the queen of the fleet airborne, on the 19th, ended dramatically just 14 seconds before launch, when flight controllers received an indication that the hydraulic power unit on the right-hand Solid Rocket Booster had exceeded maximum allowable turbine speed limits. "We were happy as clams," Charlie Bolden recalled of the build-up to that attempt. "All of a sudden, *everything* stopped and the countdown clock went back to T-9 [minutes] and kind of ticked there. We had no idea what had happened. As they started looking at the data, they had an indication that we had a problem with the right-hand booster." The signal turned out to be erroneous, but by then the window had closed and the seven astronauts were obliged to disembark.

Launch was moved to 6 January 1986, and was under considerable pressure to get underway, because the *next* flight – Mission 51L – was due to fly from the newly-refurbished Pad 39B on the 22nd, carrying New Hampshire school teacher Christa McAuliffe as the first private citizen astronaut. NASA also wanted Columbia back from 61C as quickly as possible, because she was booked for the ASTRO-1 mission to observe Halley's Comet in early March. A launch on 18 or 19 December would have allowed just enough time to refurbish Columbia and install a battery of three ultraviolet telescopes in her payload bay. Delaying 61C into the new year was a headache which the space agency could ill-afford, but worse was to come.

For the crew, the down time over the festive period was a chance to relax after more than a year of intensive training in the simulators and uncertainty over when they would ever fly. "We stayed in quarantine a lot of the time," remembered Hawley. "When you're in a launch mode, down in Florida, the pace is not very hectic. You're not in training . . . like you would be if you're in Houston and going to the simulators every day. You're reviewing procedures and checklists and having a

Robert 'Hoot' Gibson regarded Mission 61C as "the end of innocence". Only ten days after this image of Columbia on the runway at Edwards was taken, a catastrophe of unimaginable – though not unexpected – proportions exploded on television screens across the world. After January 1986, the Shuttle would never be the same again. Note the SILTS pod, mounted atop Columbia's tail fin.

nice time, because you have the opportunity to sort of sit back without the pressure of having to be in a sim. I've always enjoyed the time in quarantine, although, because of the launch time, we were getting up at *two* in the morning every day!"

Columbia's launch attempt on 6 January turned out to be one of the most hazardous yet in the Shuttle's five-year operational history. The count was halted at T-31 seconds, right on the cusp of 'Autosequence Start', following the accidental *draining* of more than 1,800 kg of liquid oxygen from the External Tank! The fill and drain valve, it seemed, had not properly closed when commanded to do so. Launch controllers reset the clock to T-20 minutes and efforts were made to reinitiate the liquid oxygen tanking, but it was quickly realised that time was running out and the window would close before the vehicle was ready. Another 24-hour delay was called. The *next* attempt, on the 7th, was scrubbed due to poor weather at two Transoceanic Abort Landing sites in Spain and Senegal, both of which provided lengthy runways to support the Shuttle in the event of an emergency during ascent. Yet another try on the 9th similarly came to nothing when a liquid oxygen sensor on Pad 39A broke off and lodged itself in the prevalve of one of Columbia's three main engines. "*That* would have been a *bad* day," Bolden recalled, grimly, years later. "It would have been *catastrophic*, because the engine would have exploded, had we launched." Heavy rain put paid to the next opportunity on 10 January, but on this occasion the seven-man crew was relieved. "We went down to T-31 seconds," said Bolden, "and they went into a hold for weather and it was the *worst* thunderstorm I'd ever been in. We were really not happy about being there, because you could *hear* the lightning!

You could hear stuff crackling in your headset. You're sitting out there on the top of two million pounds of liquid hydrogen and liquid oxygen and two [SRBs]. *None* of us were enamoured with being out there."

The repeated delays took a financial toll, too. "We spent a long time in crew quarters," remembered Pinky Nelson. Previously, astronauts were responsible for getting their families to Florida, paying their way, finding motel rooms for them and putting them up. Nelson's wife, Susie, spent *three weeks* in a condo at Cape Canaveral, waiting, and their young children ended up missing the launch because they had missed so much school and went back to Houston. "Had the accident occurred on *that* flight, instead of the flight afterwards," said Nelson, "it would have been just a nightmare scene, because the families were scattered all over the place."

As for the *cause* of the delays, there could be only one person to blame: *Steve Hawley*.

When the astronauts left their quarters in the pre-dawn darkness of 12 January, Hawley had ridden the bus to the launch pad on *ten* occasions for only *two* liftoffs. To this day, he is confident that a conversation and agreement he had with Hoot Gibson may have helped to finally get Columbia into orbit. "I decided that if [Columbia] didn't know it was *me*, then maybe we'd launch," he said, "and so I taped my name tag with grey tape and had the glasses-nose-moustache disguise and wore that." It *worked* and 61C roared into space at 6:55 am EST. Yet misfortune was not done with the crew, as recounted dramatically by Bill Nelson in his book about the flight, *Mission*. Shortly after leaving the pad, Bolden noticed an ominous indication of a possible helium leak. An alarm sounded in the cockpit. "I looked down at what I could see," Bolden said, "with everything shaking and vibrating, and we had an indication of a helium leak in the right-hand main engine. Had it been true, it was going to be a bad day." The pilots communicated the information to Mission Control and advised them that they would work their procedures. "The ground didn't see anything" on their data, Bolden continued. "It was a glitch in one of the computers." He and Gibson attempted to isolate the first system, but there was no change. It *still* looked as if there was a leak. Isolating the second system, similarly, had no effect. He told Gibson that it looked like a sensor problem, not a genuine leak. Gibson concurred and they reconfigured the system back to normal. It was an inauspicious start, for *all* this happened within seconds of liftoff.

Other than the malfunctioning helium sensor, Columbia's seventh ascent was normal, although Bolden recalled that his first flight into space "went by *really* fast". He has since downplayed his role in isolating the problem, but Bill Nelson, sitting on the middeck and listening to the communications through a headset, would later praise the pilot for 'saving' the mission. "We *had* a problem," Bolden admitted, "but it was an *instrumentation* problem." As one of the few black astronauts, there was one particular place that Bolden wanted to see after reaching orbit: *Africa*. Years later, he would describe it as "awe-inspiring" and it brought tears to his eyes. However, his first glimpse of his ancestral homeland from space would closely mirror the reflections of many other astronauts: for there were *no lines* or boundaries to demarcate the countries or, indeed, the continents, and Bolden found it difficult to orient himself and realise what he was observing.

Deployment of Satcom K-1 was supervised by Franklin Chang-Díaz and Pinky Nelson and took place near the end of the first day in space. After several hours of checking out the payload, the satellite and its PAM-D2 booster were released at 4:26:29 pm EST, a little under ten hours into the mission. Fifteen minutes later, Gibson and Bolden pulsed Columbia's OMS engines to create a safe separation distance before the perigee kick motor ignited. All went well, and Satcom finally achieved its geostationary 'slot' at 85 degrees West longitude, where it remained operational until April 1997. It marked the last 'major' commercial satellite to be launched by the Shuttle and the remainder of Mission 61C, in Hoot Gibson's words, was "the end of innocence" for the reusable spacecraft. The astronauts settled down to what should have been five days of experiments and observations of Halley's Comet, which was making its closest approach to Earth since 1910. The Comet Halley Active Monitoring Program (CHAMP) required Pinky Nelson to bury himself under a black shroud to eliminate cabin light interference and take high-resolution photographs of the celestial wanderer. Much to Nelson's annoyance, the camera failed due to a battery problem and did not even return *one* decent image of the comet. Although Nelson and his crewmates tried to repair CHAMP, their efforts were in vain and by 14 January they were told to press on with the rest of their mission. Another attempt was scheduled for Challenger later in the month and then, in March, ASTRO-1 would conduct a comprehensive ultraviolet analysis of Halley. However, the loss of Challenger and the stalling of the Shuttle programme meant that NASA lost its chance to capitalise on viewing the comet from space.

Chang-Díaz, on the other hand, was having more luck with the three materials processing experiments under his watch, for he was responsible for the MSL-2 payload. This was similar to the facility which flew aboard STS-7 in June 1983 and consisted of an electromagnetic levitator, an advanced directional solidification furnace and a three-axis acoustic levitator. The electromagnetic levitator examined the solidification processes of six melted materials, suspended within a cusp coil, whilst the three-axis levitator contained a dozen different liquids, suspended by the pressure of sound waves, which were rotated and oscillated to study bubble formation and dynamics. Finally, the furnace – actually *four* furnaces – melted and resolidified several materials samples in microgravity. The three-axis experiment worked well and was complete by 13 January, although the others experienced power problems and were ended earlier than planned.

The second bridge in the payload bay was hidden behind Satcom's sunshield, but this was the first time that as many as 12 dustbin-sized GAS canisters had flown together. A 13th canister was also carried on the payload bay wall. The decision to fly this 'GAS bridge' came about following the cancellation in November 1985 of the Navy's Syncom 4-5 from the mission, which had been removed for inspections following the UHF electronics failure of its predecessor. (By the time of the Challenger accident, Syncom 4-5 was rescheduled for Mission 61L in November 1986.) NASA decided to fly the GAS bridge in Syncom's place, to help to clear a growing backlog of small experiments. The bridge was up and running by 14 January and its payloads included an ultraviolet investigation, an Air Force study, numerous student experiments and an amateur radio package. Like Hawley, Bolden described these payloads as "a year-end clearance sale". There was also a new carrier aboard: the Hitchhiker. Unlike the GAS experiments, Hitchhikers could draw power from the Shuttle and were controlled from NASA's Goddard Space Flight Center, which permitted scientists to actively interact with their payloads. On Mission 61C, the Hitchhiker, attached to the payload bay wall, carried an Air Force particle distribution study, a set of coated mirrors to assess the effect of the Shuttle's environment on sensitive instruments and a capillary pumped loop investigation for heat-transfer experiments. Other activities saw Chang-Díaz provide a live television tour of Columbia, speaking in Spanish for his Latin American audience. Meanwhile, Bill Nelson operated a hand-held protein crystal growth experiment and participated in several medical tests.

If the factors which conspired to keep Columbia on the ground were maddening, those which hampered her return to Earth proved equally trying. Originally scheduled to touch down in Florida on 17 January – marking a return to operational use of the Shuttle Landing Facility after improvements to brakes and tyres – the attempt was brought *forward* by a day to provide more time to prepare the vehicle for ASTRO-1. However, bad weather at the Kennedy Space Center forced a 24-hour delay. *Another* delay was implemented on the 17th, for the same reason. For Hoot Gibson, the time spent in orbital limbo was spent writing a song to the tune of *Who Knows Where or When?* He and Bolden sang it to Mission Control in a two-part harmony:

It seemed that we have talked like this before
The de-orbit burn that we copied then
But we can't remember where or when

The clothes we're wearing are the clothes we've worn
The food that we're eating is getting hard to find
Since we can't remember where or when

Some things that happened for the first time
Seem to be happening again

And so it seems we will de-orbit burn
Return to Earth and land, somewhere,
But who knows where or when?

Mission Control loved it. In fact, they even cobbled together a mock 'Wanted' poster for all seven astronauts, calling for them to be returned to Earth immediately, if found.

The first opportunity to land in Florida on 18 January was postponed by one orbit and when NASA finally ran out of time, Columbia had to be diverted to Edwards Air Force Base in California. "Everything worked, except God!" Charlie Bolden joked. "Finally, on our fifth attempted landing, in the middle of the night, we landed at Edwards, which was interesting, because with a daytime scheduled landing, you'd have thought we wouldn't be ready for that. Hoot, in his infinite wisdom, had decided that half our landing training was going to be nighttime, because you needed to be prepared for anything." Columbia touched down on Runway 22 at 5:58:51 am PST (8:58:51 am in Florida), wrapping up a mission of a little more than six days. Subsequent inspections would reveal severe thermal damage to her right-hand main gear inboard brake and it was decided that major improvements to withstand higher energy wear would be incorporated before her next voyage, Mission 61E. Coupled with the landing in California, this made the scheduled launch of 61E, just *seven weeks* later, increasingly untenable. "The landing," said Bolden, was "uneventful, other than the fact that it *really* upset Congressman Nelson, because he really had these visions of landing in Florida and taking a Florida orange!" The recovery team at Edwards showed no mercy and fellow astronaut Dan Brandenstein greeted the crew with a basket of *California* oranges and grapefruits from the California Growers' Association! It was a joke that Nelson did *not* appreciate. Ten days later, he would appreciate, at least, that he had returned *alive*.

ULTIMATE FIELD TRIP

"Now don't break our airplane," Judy Resnik joked.

Hank Hartsfield promised not to. It was October 1985 when the two astronauts shared their moment of camaraderie. Three months later, he would recall Resnik's light-hearted advice with sorrow; for it was on *her* flight, rather than *his*, that the 'airplane' – Space Shuttle Challenger – finally broke. Hartsfield had just returned from the week-long Spacelab-D1 mission aboard Challenger and had immediately flown to Europe on his crew's public relations tour. He was aboard a commercial airliner, returning to the United States, when Resnik and the rest of Challenger's tenth crew blasted off. Whilst in Europe, he had followed with interest Resnik's seemingly fruitless efforts to return to space in January 1986.

It was a frustration that NASA could not afford. The planned six-day flight, designated 51L, would feature the first private citizen to fly aboard the Shuttle – a social studies high school teacher from Concord in New Hampshire, named Christa McAuliffe. Picked from thousands of applicants for the Teacher in Space initiative in July 1985, she was to teach two lessons from space, providing a much-needed publicity boost for the agency as it sought to demonstrate that its reusable fleet of orbiters were truly the spacegoing equivalents of commercial airliners...and convince senior politicians to support a permanent space station. Years later, her mother would insist that the general atmosphere in the weeks leading up to Challenger's fateful launch was that the Shuttle was far safer than an airliner, simply due to the higher number of precautions taken by NASA. Even McAuliffe herself had expressed confidence that her only 'fear' was a failure of the orbiter's toilet.

Born Sharon Christa Corrigan in Boston, Massachusetts, on 2 September 1948, she was the eldest of five children of accountant Edward Corrigan and teacher Grace Corrigan, with an interesting ancestry which included Irish, Lebanese, German, English and Native American. She completed high school in Framingham and developed an early fascination with the space programme – telling one of her classmates after John Glenn's flight that, someday, people *would* travel to the Moon

Teacher Christa McAuliffe climbs out of a T-38 Talon jet trainer.

and beyond – before entering Framingham State College to study education and history. She completed her degree in 1970 and, within weeks, married her long-time boyfriend Steven McAuliffe. The couple moved closer to Washington, DC, to enable him to attend law school, and Christa McAuliffe found employment as an American history teacher at Benjamin Foulois Junior High School in Morningside, Maryland. A year later, she moved to Thomas Johnson Middle School in Lanham, teaching history and civics, where she remained until 1978. During this period, she completed a master's degree in education supervision and administration at Bowie State University and in 1978 the McAuliffes moved to Concord, where Steve took a position as assistant to the New Hampshire Attorney General. Christa, meanwhile, began teaching American history, law and economics – and a self-designed course, 'The American Woman' – at Concord High School in 1982.

Two years later, in August 1984, President Ronald Reagan announced the Teacher in Space project, requesting NASA to find a gifted educator with the ability to communicate his or her enthusiasm to students from orbit. The non-profit Council of Chief State School Officers was selected by NASA to co-ordinate the selection process and from November 1984 until February 1985 more than 11,000 applications were submitted. These were winnowed down to 114 semi-finalists by state, territorial and agency review panels and McAuliffe was one of only two teachers to be nominated in New Hampshire. In her application, she wrote: "I cannot join the space programme and restart my life as an astronaut, but this opportunity to connect my abilities as an educator with my interests in history and space is a unique opportunity to fulfil my early fantasies. I watched the space programme being born and I would like to participate." A judging panel, including former astronauts, university presidents, actress Pam Dawber, former basketball player Wes Unseld and Robert Jarvik, inventor of the artificial heart, presided over these candidates at interview and eventually narrowed the list to just ten finalists. On 26 June, at the White House, President Reagan remarked to the candidates that "whichever one of you is chosen might also want to take under consideration the opinion of another expert: *The acceleration which must result from the use of rockets inevitably would damage the brain,* so consider yourself forewarned!" Little could he possibly have known that such dire predictions would come awfully true for the Teacher in Space in January of the following year ...

Late in July, it was Vice President George Bush who announced McAuliffe's name as the prime candidate, backed up by Idaho elementary school teacher Barbara Morgan. NASA psychiatrist Terry McGuire told *New Woman* magazine that McAuliffe was the most 'balanced' of the ten finalists, whilst other senior officials found that her endearing manner and infectious enthusiasm set her apart from the others. Training in Houston began in September and by the time the 51L crew arrived at the Kennedy Space Center in January 1986, their launch had already been postponed by delays in bringing Columbia home from her flight and weather concerns at a Transoceanic Abort Landing site in Senegal. More trouble was afoot. Predictions of unacceptable weather in Florida put paid to a second attempt on the 26th and, when Dick Scobee and his six crewmates settled into their seats aboard Challenger on the 27th, they were again thwarted by high winds and a frozen handle on the hatch. In the years that followed, many observers commented that launching on 'Super Bowl Sunday', the 26th, might well have saved 51L. "It's another case of fate, playing tricks on you," remembered Jim Beggs. "They decided not to launch on Super Bowl Sunday ... but Super Bowl Sunday was a *fine* day. If they'd have launched *then*, they wouldn't have had any trouble, so they held it over and caught the cold spell and lost the vehicle and the crew." On the night of Monday the 27th, temperatures at the launch site plummeted, precipitously, to an unseasonal (for Florida) -13 degrees Celsius, forcing technicians to switch on safety showers and fire hoses at Pad 39B to prevent water pipes from freezing. This proved particularly worrisome for the ice inspection team, who began their final 'sweep-down' of the pad area in the early hours of the 28th, and they were obliged to knock a large number of 30 cm icicles away with broom handles as the countdown clock continued ticking towards launch. As the Sun rose, temperatures rose slightly to a few degrees above zero Celsius, producing the coldest conditions under which a Shuttle launch had ever been attempted, a fact that would be investigated in depth during the subsequent presidential inquiry into the cause of the tragic events later that day. The copious amounts of ice on Pad 39B then forced an additional two-hour delay to permit the Sun to thaw it. Nonetheless, many of the astronauts' families, including Scobee's wife, June, doubted that NASA would fly in such conditions. She was partially appeased by her husband's insistence, over the phone that morning, that he felt it was safe to do so. Hank Hartsfield, good-naturedly, had called the crew regularly to jokingly ask what the hell they were up to. Would they *ever* launch, he wondered?

Tragically, as we know, Resnik and her colleagues – Scobee, McAuliffe, pilot Mike Smith, mission specialists Ellison Onizuka and Ron McNair and Hughes payload specialist Greg Jarvis – would indeed launch that frigid Tuesday, with catastrophic consequences. More than two decades later, the world is familiar with the technical and human causes of Challenger's loss, but the disaster also put paid to plans for two important satellite deployments, a range of scientific and engineering experiments and a comprehensive survey of Halley's Comet. Two weeks before Challenger lifted off, problems with the CHAMP experiment had prevented Columbia's 61C crew from making significant observations. Mission 51L would have utilised CHAMP and deployed a free-flying Spartan satellite to study Halley's tail and the gaseous 'coma' immediately surrounding its nucleus. Onizuka was responsible for CHAMP. "I will have about two minutes on four different orbits to photograph Halley's Comet in both the visible and ultraviolet spectrum," he told an interviewer. "The objective is to try to get this data as the comet approaches perihelion, which is just as it goes around behind the Sun and starts to head back out. It's a regime where we do not have any data at the present time, so I've been told we'll probably be the only human beings to see it at that time."

As Onizuka worked, his crewmates would have been involved in preparing Spartan for deployment on 30 January to commence its own series of Halley observations. Built by the agency's Goddard Space Flight Center, the small, boxy satellite had previously flown in June 1985 and was designed to be serviceable and capable of returning to orbit every six to nine months. On 51L, it would have used a pair of ultraviolet spectrometers and two modified Nikon F-3 cameras to study the composition of the comet's dirty-snowball-like nucleus and million-kilometre-long shimmering tail. McNair, with Resnik, was responsible for deploying and later retrieving Spartan with the Canadian-built RMS mechanical arm. Interestingly, the spectrometers, produced jointly by Goddard and the University of Colorado's Laboratory for Atmospheric and Space Physics, were derived from backups for an instrument aboard the Mariner-9 spacecraft which had begun investigating Mars' atmosphere in 1971. "Comets happen to be one of the remnants of the creation of the [Solar System]," McNair said before the mission, "and they're just a big mass of ice – of frozen gases – and the last time [Halley] came around, we weren't sophisticated enough to do the type of things that we're doing now. Scientists will be able to analyse the gases [and] emissions by looking at the Sun's reflection and the absorption of sunlight and give some credibility to some of the theories – or possibly tear them down – about the origin of the Universe."

The agency hoped that from 20 January until 22 February, Halley – then some 225 million km from Earth and only 97 million km from the Sun – would be chemically at its most active and yield the most desirable scientific data. The Spartan mission to explore the comet, codenamed '203', would have gotten underway on the second day of 51L, when Scobee and Smith were scheduled to fire Challenger's OMS engines to adopt a slightly higher altitude, about 245 km above Earth. The assignment of Spartan-203 had already resulted in some changes to the Shuttle's launch window – originally targeted for a morning liftoff, it was moved to the afternoon in order to provide the best lighting conditions for the satellite's observations. However, an afternoon start would delete the option of touching down in Casablanca in the event that a Transoceanic Abort Landing in Morocco became necessary. Ultimately, as Challenger's launch was pushed into the final week of the month, conditions for optimum viewing of Halley, based on an afternoon window, could no longer be met and the liftoff time was shifted back to the morning hours. In a sense, therefore, the delay was actually beneficial.

By 30 January, after Spartan-203's software had been uploaded from Houston and voltage and current checks carried out, it would have been 'hung' over the payload bay wall and released into free flight by the RMS. The satellite would then have executed a slow, minute-and-a-half-long pirouette to prove that it was working properly, after which Scobee would have withdrawn by around 145 km. This would have ensured that sunlight reflected by Challenger's pristine white surfaces did not 'confuse' Spartan-203's sensors. Following two orbits of further tests, the aperture doors covering the satellite's two ultraviolet spectrometers would have automatically retracted to initiate an aggressive, 40-hour-long phase of free flight, of which more than half would have been dedicated to studies of the photodissociation of water in Halley and analysis of its various nitrogen-, carbon- and sulphur-containing molecules. Meanwhile, its cameras would have offered an ongoing record of the 'large-scale' activity of the comet itself, including outbursts from its nucleus and asymmetries in its coma. Retrieval would have followed on 1 February and Spartan-203 would have been repositioned on its MPESS carrier in the forward section of the payload bay.

Had Onizuka and his six crewmates survived their violent climb to orbit on 28

January, however, the delicate and tricky Spartan-203 deployment, two days of stationkeeping and retrieval would have actually been the secondary task of their mission. By far the largest, most expensive and most powerful payload aboard 51L was NASA's second Tracking and Data Relay Satellite, which, together with its predecessor, would enable future Shuttle astronauts to communicate directly with Mission Control for most of each 90-minute circuit of the globe. "That's going to be a big improvement," Smith told an interviewer in the weeks leading up to the launch, "not only for the Shuttle, but also for the space station when it gets up later on." Until the early 1980s, missions had relied on a network of ground stations to relay communications between orbiting crews and Houston-based controllers. The TDRS network of at least two large satellites, positioned in geostationary orbits, 35,000 km high, would bring this era to a close. Onizuka, though, was simply thrilled at having the chance to help deploy "one of the largest communications satellites ever!"

A Boeing-built Inertial Upper Stage with two solid-fuelled sections would have delivered the satellite, over a period of about seven hours, into its operational location. Deployment of the combo would have consumed most of Challenger's first day in space and, although all five 'career' members of the crew would have been involved, the lengthy procedure would have been conducted under the direction of Onizuka and McNair. Shortly after reaching space, the two men would have rotated the 'stack' to a pre-deployment angle of 29 degrees using the ring-doughnut-shaped 'tilt table'. As Scobee and Smith manoeuvred Challenger into the correct attitude, Onizuka and McNair would have switched TDRS-B over from the Shuttle's electricity supply to the IUS' internal batteries. Next, they would have commanded the tilt table to increase the angle to 59 degrees and, precisely ten hours after leaving Earth, spring-ejected it, such that it swept smoothly over Challenger's cabin roof. Nineteen minutes later, Scobee would have fired the OMS engines to create a safe separation distance in anticipation of the IUS' first stage ignition. After computing the stack's correct attitude by taking star sightings, the IUS would have fired its engine an hour after deployment and run for two and a half minutes. An additional burn by the second stage, lasting just under two minutes, would then have achieved near-geostationary orbit.

Whilst still attached to the now-exhausted second stage, the satellite's solar arrays would have opened – "like an insect coming out of a cocoon", astronaut Mike Lounge later remarked – and, eventually, so too would its communications payload. Over a period of several months, during a series of extensive tests, it would have gradually drifted westwards to its final orbital position. It was a complicated task and one for which the IUS itself had made a rather inauspicious start. However, following his first mission in January 1985, which featured the successful deployment of a classified Department of Defense satellite affixed to an IUS, Onizuka expressed confidence in the weeks leading up to 51L in the training and procedures involved with releasing both the TDRS and its booster. "The basic training was the same," he said of his first and second flights. "Once we enter the area of payload and mission operations, there were some differences, [but] I'm very familiar with the IUS; very comfortable with it." By September 1985, reported *Flight International*, TDRS-B had completed its repairs and was ready to go. It would technically bring the system up to fully operational status. Nevertheless, a third satellite was scheduled to be ferried into orbit in July 1986 to replace the degraded TDRS-1. Until the arrival of this third member of the network, TDRS-B would operate from an initial 'spare' orbital slot of 136 degrees West longitude, providing much-needed backup services for its prematurely aging sibling. After the launch of the third satellite, however, TDRS-B was to be moved to its final position directly above the equator south of Hawaii, at 171 degrees West longitude.

Despite its important contributions to astronomy and communications, the 51L mission naturally attracted media attention, as NASA had intended, thanks to the presence of teacher observer McAuliffe. Explorers, journalists and entertainers were considered in the early 1980s as the agency weighed up options for which profession would yield 'the best' private citizen to send aloft on the pioneering mission. Ultimately, it was decided that a teacher would fly first. Dick Scobee agreed that it was the right decision. "Teachers teach the lives of every kid in this country through the school system and if you can enthuse the teachers about doing this, then you enthuse the students and impress on them that's something to expect in their lifetime," he explained. "Man needs to explore and that's part of the thing we have to do to ensure our future. So as far as I'm concerned, it's a good insurance policy for the human race." McAuliffe's tasks included two, 15-minute-long lessons: the first, entitled 'The Ultimate Field Trip', was a guided tour of the Shuttle to familiarise students with on-board living and working conditions, while the second, called 'Where We've Been, Where We're Going', focused on NASA's fledgling plans for a permanent space station. Both were to have been aired by the Public Broadcasting System on 2 February and McAuliffe would have explained the roles of her six crewmates, identified and summarised the experiments aboard Challenger and enthused 'her' students with a vision of the future.

"I think it's going to be very exciting for kids to be able to turn on the TV and see the teacher teaching from space," she said. "I'm hoping that this is going to elevate the teaching profession in the eyes of the public and of those potential teachers out there. Hopefully, one of the secondary objectives of this is students are going to be looking at me and perhaps thinking of going into teaching as professions." McAuliffe and Jarvis both joined 51L relatively late in the training flow. Yet both were quickly accepted and grew to become highly respected members of the team. "It's refreshing to have somebody on board that's really dedicated and enjoys doing what they're doing," Scobee remarked, "but also she goes into the training with a positive attitude and stays out of the way when she needs to stay out of the way, she gets involved when she needs to get involved and does basically all the right things, and so does Greg Jarvis. Both of them, from our standpoint, are good payload specialists. They came aboard with a good, open mind, they're accommodating to our system, we try to be accommodating to theirs and it's a nice tradeoff."

The level of respect was, of course, mutual and Jarvis recalled one particular session as an example of the ability of the astronauts to operate seamlessly together. "When you watch them work through the malfunctions they work through, you get very comfortable that they know what they're doing," he said. "One time when we were in the Motion Base Simulator, the lights went out for the visual for the landing.

Only hours before Challenger's fateful launch, ice hangs in copious quantities from the gantry of Pad 39B.

The commander called down and said 'Aren't the lights out?' [Mission Control] said, 'I think so, we'll get back to you on that'. The conversation went on for about two or three minutes and it turns out they had mistakenly turned the lights out on the visuals. The thing you didn't realise was that he made a perfect landing without any lights!"

Gregory Bruce Jarvis was born in Detroit, Michigan, on 24 August 1944 and completed high school in New York, then entered the University at Buffalo at the State University of New York to study electrical engineering. He graduated in 1967 and received a master's degree from Northeastern University in 1969. Jarvis served for the next four years in the Air Force, reaching the rank of captain, and moved to Hughes Aircraft Company's Space and Communications Group as a communications subsystems engineer for the Marisat programme. Whilst at Hughes, he also completed the coursework for a master's credential in science management at West Coast University in Los Angeles. Later, from 1978, Jarvis worked on the formulation of concepts and formal proposals for the Syncom 4 project and subsequently served as a subsystems engineer for the satellites. He was responsible for managing the testing and integration of the first three Syncom 4s, together with their payload bay cradles, and was selected from over 600 Hughes applicants as a payload specialist in July 1984.

The arrival of Jarvis in October 1985 had come particularly late in the crew's training period. During the mission, he was assigned to conduct a battery of investigations using spinning, fluid-filled plastic models on the middeck to evaluate 'optimum' shapes for future satellite fuel tanks. The reason for his late assignment was primarily linked to the fact that payloads for Shuttle missions were in constant flux; indeed, the cargo for Dick Scobee's flight had changed several times, as had the identities of 'his' payload specialists. Originally assigned to fly aboard Columbia on 61C, Jarvis was transferred to 51L, ostensibly because Syncom 4-5 had experienced technical problems and been delayed. This sounded reasonable, except for one thing: 51L, also, had no Hughes satellite aboard! The more likely reason for reassigning Jarvis was that Congressman Bill Nelson had requested a Shuttle flight and the space agency had hurriedly complied. "NASA bumped the oft-abused Jarvis one mission to the right," Mike Mullane recalled. "The next time he would pose for a crew photo would be for 51L, the mission that would kill him. He would die on a mission that had no Hughes satellite to deploy, the singular event that had been the original justification for his assignment to a Shuttle flight."

Even when the five NASA crew members were assigned to 51L in January 1985, the payload was changing every few months: first they would deploy a Tracking and Data Relay Satellite and one of the satellites retrieved by 51A – probably Palapa – and for a time it appeared that they might launch aboard Atlantis, rather than Challenger. Within weeks, NASA's March 1985 manifest listed them as the crew for 61C, scheduled for December to deploy a pair of PAM-boosted communications satellites, including Satcom K-2, and conduct a pair of EVAs to operate the EASE-ACCESS experiment. By the autumn of 1985, their designation was again 51L with TDRS-B and Spartan-203. One of the issues raised by the Rogers Commission – established by presidential order and chaired by former Secretary of State William

Rogers – in the wake of the Challenger accident was this practice of constantly juggling payloads between missions. In 51L's case, this had led to no fewer than *six* postponements of the Cargo Integration Review, an important meeting at which payload requirements were assessed to enable the development of the final flight products to begin. Although the commission admitted that most payload adjustments were complete by the time of the review in June 1985, it was particularly critical of the late assignment of Jarvis and his experiments. "The Flight Planning and Stowage Review was conducted on 20 August 1985," the Rogers report explained, "to address any unresolved issues and any changes to the plan that had been developed to that point. Ideally, the mission events are firmly determined before the review takes place. For 51L, however, Mr Jarvis was not added until 25 October 1985 and his activities could not be incorporated into mission planning until that time. There were changes to middeck payloads, resulting from the addition of Mr Jarvis, that occurred less than three months before launch. The most negative result of the changes was a delay in publishing the crew activity plan. [This] specifies the in-flight schedule for all crew members, which in turn affects other aspects of flight preparation." Furthermore, Rogers investigators expressed concerns that changes were being made at very short notice – not only payload specialist adjustments, but also satellite swaps and experiments being added, delayed or dropped – which, of course, would directly impact the training time available for crews.

"Had we not had the accident," said Hank Hartsfield in his testimony to the commission, "we were going to be up against a wall; 61H [planned for June 1986] . . . would have had to average 31 hours in the simulator to accomplish their required training and 61K [scheduled for August] would have to average 33 hours. That is *ridiculous*. For the first time, somebody was going to have to stand up and say [that] we have got to slip the launch because we are not going to have the crew trained." Training was also affected by the presence of only two Shuttle simulators at JSC, capable of supporting crews for no more than 12-15 missions per annum. "The flight rate at the time of the accident," read the Rogers report, "was about to saturate the system's capability to provide trained astronauts for those flights."

At length, with everything and everybody in place, the 51L mission-specific training commenced in the late autumn of 1985, with the astronauts averaging 49-hour working weeks to ensure proficiency in RMS operations, Spartan deployment and retrieval activities, IUS systems, ascent and re-entry procedures and each of the experiments on Challenger's middeck. The mission itself was deemed 'moderately complex' in view of the Spartan commitment, although both it and a TDRS deployment had already been 'baselined' on previous flights. Still, despite a hectic six days in space, all seven astronauts intended to spend some moments appreciating the uniqueness of where they were. "We have a fairly busy timeline and it's nice to have time to go look out the windows," Scobee, who had flown Mission 41C, said during one of his last interviews. "I guess one of the things that pleasures me most is to have a quiet time where you can go look out the windows, turn out the lights and look at the stars and Earth and thunderstorms. I enjoy the excitement and thrill of the ascent, because it is really dramatic. Entry is fiery – just an amazing light show – and the fires of hell are burning outside your window and you're sitting there nice and comfortable watching all this go on and it's just a neat feeling." Nonetheless, Scobee had already announced that 51L would be his last space mission. His last comments of encouragement to his crewmates over Challenger's intercom in the final seconds of the countdown were words that conveyed enthusiasm, dedication, professionalism, child-like wonder – and an uncanny, though unknowing, preview of what would happen.

"Everybody strap in tight," he told them cheerily. "We're about to go for the ride of our lives."

THE GOLDEN AGE ENDS

That ride began at precisely 11:38 am EST on January 28th 1986. Six and a half seconds before liftoff, Challenger's three main engines thundered to life and, as the countdown clock touched zero, the assembled spectators at KSC were greeted by the ear-splitting staccato crackle of her twin Solid Rocket Boosters. It proved to be the failure of both primary and secondary O-ring seals at the base of the right hand booster, Rogers investigators would later conclude from photographic, physical and other evidence, that was directly and solely responsible for the destruction of 51L and the loss of her crew.

The response from the astronaut corps was one of astonishment: many knew nothing of the concerns over the integrity of the O-ring seals, a problem which had first drawn engineers' attention during Columbia's STS-2 ascent in November 1981. In fact, the astronauts most feared a failure of the Shuttle's main engines. The boosters, on the other hand, were deemed uncontrollable whilst firing, yet they 'worked' and were regarded as 'big' and 'dumb' ... and totally reliable. Clear evidence of their fallibility, made public for the first time by the Rogers report, occurred serendipitously when, 0.678 seconds after liftoff, a video camera mounted close to Pad 39B captured "a strong puff of grey smoke ... spurting from the vicinity of the aft field joint of the right Solid Rocket Booster". The camera had identified the tell-tale result of both the primary and secondary O-rings failing, disintegrating and streaming away in the moments after ignition. More significantly, the point of failure directly faced the External Tank and its two million litres of highly volatile propellants. Any flame from the compromised booster could now play on the tank like a blowtorch, igniting its contents in a fireball and destroying Challenger, together with the entire launch complex. Years later, Morton Thiokol structural engineer Roger Boisjoly would express astonishment that the vehicle did not explode on the pad; by an incredible sequence of events, a chunk of solid fuel temporarily plugged the O-ring hole and the first minute of Challenger's ascent proceeded normally.

The temporary plug, however, was just that: temporary. It would not hold.

Several more puffs of increasingly denser, darker smoke – further indicative that the products under combustion were indeed the grease, insulation and rubberised O-ring material from the joint seals – were recorded by other ground-level cameras between 0.836 and 2.5 seconds after liftoff, as the boosters' hold-down posts were severed and the Shuttle commenced its climb for the heavens. As each puff was left behind by Challenger's upward trajectory, the next fresh puff could be seen close to the level of the joint. The frequency of these emissions was directly related to flexure within the structure of the SRB as the gap in its joint cycled open and closed. The last incidence of smoke above the joint was timed at T + 2.733 seconds. In the moments that followed, a combination of atmospheric factors and exhaust from the boosters made it difficult to determine if any more smoke was emerging from the failure point.

A little under eight seconds into the mission, as planned, the vehicle cleared the tower and began a programmed roll manoeuvre, moving onto the correct flight azimuth for a 28.45-degree-inclination orbit, then pitching onto her back. Shortly thereafter, at T + 19 seconds, to prepare herself for passage through a period of maximum aerodynamic turbulence, Challenger's main engines were throttled down from 104 to 94 percent, and later 65 percent, of rated thrust. Thirty-seven seconds into the ascent, she encountered the first of several high altitude wind shears, lasting until just past a minute after launch. In its inquiry, the Rogers report noted that the Shuttle's guidance, navigation and control system immediately detected and compensated for these conditions and – although 51L's aerodynamic loads were higher than previous missions in both the yaw and pitch planes – the SRBs, too, responded effectively to all commands.

It is possible that the mission may still have proceeded normally, had the plug of solid fuel remained jammed into the O-ring breach. However, by an incredible stroke of cruel luck, Challenger happened to pass through the most severe wind shear ever encountered by an ascending Shuttle; a shear which dislodged the plug somewhere around a minute into the mission. After passing through maximum aerodynamic turbulence, 51 seconds into the climb, her main engines were throttled back up to full power; shortly afterwards, at 58.788 seconds, a frame of video recorded the first evidence of a flickering flame from the right hand SRB's aft joint. The temporary plug of solid fuel had gone and, although they were oblivious to anything amiss, the crew's fate was now sealed. The flame rapidly established itself, growing into a well-defined plume within barely half a second. Exactly a minute into the mission, downlinked telemetry pointed to an unusual chamber pressure differential between the left and right boosters – the pressure of the latter was some 11.8 psi lower than the other, indicating a leak in its aft joint. As the flame increased in size, Challenger's aerodynamic 'slipstream' deflected it backward and circumferentially by the protruding structure of the upper ring which linked the SRB to the External Tank, focusing the flame directly onto the surface of the tank. Sixty-two seconds into the ascent, the left booster's thrust vector control moved to compensate for the yaw motion caused by the reduced thrust from its right-side counterpart. A couple of seconds later came the first visual manifestation that the flame from the damaged booster had breached the lower segment of the External Tank: an abrupt change in the shape and colour of the flame, indicating that it was now mixing with leaking liquid hydrogen. Moreover, pressurisation data telemetred at around this point reinforced the fact that its liquid hydrogen tank was indeed ruptured.

Fifty-nine seconds into the ascent, a well-formed tongue of flame is readily apparent at the base of the right-hand Solid Rocket Booster. Had it not been for the most severe wind shear ever experienced by the Shuttle, which dislodged a temporary plug of solid fuel in the O-ring breach, Challenger might have survived to orbit ... and the devils in the system would have awaited another unsuspecting crew.

In Mission Control, Capcom Dick Covey relayed a standard call: "Challenger, Go at throttle up."

Commander Dick Scobee came back a second or two later. "Roger," he replied, "Go at throttle up."

In the seconds that followed, an incredibly rapid sequence of events concluded with the destruction of the External Tank, the separation of both boosters and the structural disintegration of Challenger into several large pieces. Seventy-two seconds after liftoff, the flame from the right SRB finally burned through the lower of two struts holding it onto the External Tank; pivoting around its upper strut, the top of the booster impacted the inter-tank and the base of the liquid oxygen tank, breaching them both. Nearly simultaneously, around T + 73.1 seconds, clouds of white vapour were spotted at the top of the tank and around the area of its bottom dome: the former was clearly indicative of the ruptured liquid oxygen tank, the latter conclusive evidence of structural failure. Almost immediately, at T + 73.6 seconds, came a massive – "almost explosive", read the Rogers report – burning of both the hydrogen leaking from the lowermost tank and the oxygen from its uppermost section. At this point, 51L was at an altitude of 14 km over the Atlantic Ocean, travelling at almost twice the speed of sound, and Challenger was lost from view in the resultant explosive burn. Her Reaction Control System ruptured during this period, setting off the hypergolic burning of its propellants, evidenced by a reddish-brown hue around the edge of the fireball. Meanwhile, the two boosters, now released of their loads, rapidly climbed away from the catastrophe, before being remotely destroyed by the Range Safety Officer at 11:39:50 pm, some 110 seconds after launch. "Obviously a major malfunction," was all Steve Nesbitt, the stunned launch commentator at KSC, could remark.

Even as he spoke, most, if not all, of the 51L crew were still conscious and may have remained so as they tumbled to Earth. Now placed under significant structural duress and aerodynamic loads for which she had not been designed, Challenger had disintegrated into several large fragments. Clearly visible, tumbling away from the blossoming cloud of debris that had swallowed the External Tank, were her aft compartment – with main engines, briefly, still firing – together with one wing and a fuzzy, roughly triangular blob: the forward fuselage, containing the crew cabin, trailing a jumble of umbilical lines ripped from beneath the payload bay floor. It continued, for a time, on an upward trajectory on a ballistic arc that reached a maximum altitude of around 18 km, before gravity inexorably pulled it down towards an Atlantic grave. Tumbling at close to 400 km/h at the point of impact, ocean water would have been as unyielding to Challenger as solid ground.

A PREVENTABLE TRAGEDY

Six weeks later, on 7 March, the crew cabin was found by divers in less than 30 m of water, some 27 km north-east of KSC, and recovered by a team from the USS *Preserver*. It "was disintegrated, with the heaviest fragmentation and crash damage on the left side," read the Rogers Commission's final report. "The fractures examined were typical of overload breaks and appeared to be the result of high forces generated by impact with the surface of the water." Tellingly, US Navy spokeswoman Deborah Burnette told a *Washington Post* journalist shortly after the discovery that "we're talking debris, and not a crew compartment, and we're talking remains, not bodies". Mike Coats, one of several astronauts directed by NASA to examine the wreckage, described it as looking "like aluminium foil that had been crushed into a ball". It contained the remains of the crew, but their horrific condition could be guessed from pathologists' difficulty in identifying them: a few strands of Judy Resnik's hair and a necklace were all that was left of Mission Specialist Two. Indeed, in the months after the disaster, all astronauts were required to submit a clip of hair and a footprint to NASA for identification. In the case of the 51L remains, apparently, even *dental* records were insufficient for positive identification ...

In his memoir, Mike Mullane expressed fervent hope that the explosive burn of the External Tank's propellants had been enough to completely destroy Challenger's crew cabin, or at least breach her flight deck windows, thereby causing a rapid depressurisation and a mercifully rapid death. Having said this, when tested to 140 percent of its design strength in Lockheed's Plant 42 rig almost a decade earlier, that same cabin had proved to be extremely hardy, and certainly its wreckage showed little evidence of having experienced an explosive depressurisation. Such an eventuality would have led to an upward 'buckling' of the flight deck floor as air from the middeck rapidly expanded; no such buckling was detectable. Additionally, wrote JSC's head of life sciences, Joe Kerwin, in a 28 July letter to Associate Administrator for Space Flight Dick Truly, the "impact damage to the windows [examined after recovery from the Atlantic] was so extreme that the presence or absence of in-flight breakage could not be determined. The estimated breakup forces would not in themselves have broken the windows. A broken window due to flying debris remains a possibility; there was a piece of debris embedded in the frame between two of the forward windows. We could not positively identify the origin of the debris or establish whether the event occurred in flight or at water impact ... impact damage was so severe that no positive evidence for or against in flight pressure loss could be found". Astronauts Jim Bagian and Manley 'Sonny' Carter, both physicians, speculated that penetrations in the cabin's aft bulkhead – created by the violently severed payload bay umbilical lines – could have led to a slower depressurisation and quick unconsciousness for the seven astronauts, although this was conjectural. More conclusive evidence that at least some of the crew had remained alive and conscious for most of the fall to Earth came in mid-March, when four Personal Egress Air Packs (PEAPs) were recovered. These were to provide each astronaut with a limited amount – some six minutes' worth – of breathing air for use in emergencies. Analysis of the packs led to an announcement on 21 May that at least one had been activated in the seconds after structural breakup and, later, that this activation was not caused accidentally at water impact. Then, on 9 June, investigators revealed that one of the packs belonged to pilot Mike Smith.

This raised an interesting scenario. Smith's PEAP was affixed to the back of his seat, placing it out of his reach, which implied that either Judy Resnik or Ellison Onizuka, seated behind him on the flight deck, had leaned forward and switched it

Minutes after hearing their comrades' final words and still stricken by the horror of seeing seven friends die, veteran flight director and then-head of Mission Operations Gene Kranz (left) talks to capcoms Fred Gregory and Dick Covey. Covey's words were the last ever received by the Challenger crew.

on in a valiant effort to save his life. A second identifiable PEAP belonged to Dick Scobee and had not, apparently, been activated. The owners of the two other packs were never identified. The quantity of air which remained in Smith's PEAP, in particular, led to a suggestion that apparent 'crew inactivity' after breakup could be an indication that they had rapidly lost consciousness. Every scrap of paper from Challenger's wreckage was analysed and it was determined that none of the astronauts had written a note; moreover, Smith's air pack was depleted by barely two and a half minutes – almost precisely the length of time it took for the cabin to fall from the fireball to the Atlantic – which suggested he had kept his helmet visor closed during the descent. If it had remained open, all six minutes of his PEAP air would have leaked out. Immediately after breakup, Challenger's intercom, lights, computers and electronics went dead. Bagian and Carter postulated that, in order to communicate, the crew's only option would have been to raise their visors and speak aloud. Unfortunately, the helmets themselves were obliterated, which rendered it almost impossible to determine how, or *if*, the astronauts communicated during those final frantic minutes. However, Mullane believes from his own experience as a US Air Force navigator, flying in the back seat of F-4 Phantoms in the 1960s and 1970s, that hand signals as a means of communication would have worked perfectly well. Furthermore, Scobee and Smith's years of experience as fighter and test pilots would have taught them to keep their visors down, rather than risk lifting them and suffocating.

The case of Michael John Smith is among the greatest tragedies of Mission 51L; not just that he and his crewmates made such valiant efforts to keep each other alive and functioning to the very end, but also the fact that he came so close to space, yet fell so far short of turning a dream into a reality. Smith was born in Beaufort, North Carolina, on 30 April 1945. He received a degree in naval science from the Naval Academy in 1967, then entered the Naval Postgraduate School, gaining a master's credential in aeronautical engineering the following year. He completed flight instruction in Kingsville, Texas, got his aviator's wings in May 1969 and was assigned to the Advanced Jet Training Command. After two years as an instructor pilot, he was deployed to Vietnam aboard the USS *Kitty Hawk*, flying A-6 Intruders as part of Attack Squadron 52, along the way earning the Distinguished Flying Cross. Smith completed test pilot school in 1974 and was detailed to the Strike Aircraft Test Directorate at Pax River, working on missile guidance systems for the A-6. He later served as an instructor at the school and completed two Mediterranean Sea deployments aboard the USS *Saratoga*.

One factor is almost certain: most, if not all, of the astronauts were aware of their dire predicament. Milliseconds before the External Tank disintegrated, a bright sheet of white vapour flooded across Challenger's nose – probably visible to Smith, sitting in the right-hand seat – and may have prompted him to utter a brief exclamation ("Uh, oh"), which turned out to be the last vocal communication from the orbiter. It is also quite possible that he saw the top of the right SRB pivot into the side of the External Tank. Despite hoaxed intercom 'transcripts' which alleged that the terrified, panic-stricken crew screamed and cursed their way down to the Atlantic, Mike Mullane is confident that Scobee and Smith would have fought to the very end to regain control of their crippled ship.

In the days after the disaster, most of the astronauts in Houston became convinced that a failure or explosion of one or more of the Shuttle's main engines was the most likely cause. Remnants of all three were dredged from the Atlantic on 23 February, each still attached to the thrust structure, and the controllers for the Number Two and Three engines were found, disassembled, flushed with deionised water, dried, vacuum baked and their data extracted. All of the engine debris exhibited burn damage caused, according to the Rogers report, "by internal over-temperature typical of oxygen-rich shutdown". Thus, the loss of hydrogen fuel after the rupturing of the lower part of the External Tank appeared to have caused all three units to begin shutting themselves down within milliseconds of each other at around $T + 73.5$ seconds. Overall, the performance of the main engines was satisfactory and in line with observations from previous missions. They first exhibited 'abnormal' behaviour almost exactly a second before breakup, when their fuel tank pressures dropped and the controllers responded by opening the fuel flow-rate valves. Next, turbine temperatures increased due to the leaner fuel mixture feeding into the combustion chambers from the External Tank. Otherwise, the report continued, "engine operation was normal". They did not contribute to the loss of 51L. Nor did the gigantic tank itself, of which 20 percent was recovered, mostly debris from the inter-tank and the lowermost hydrogen section. Initial speculation that there had been premature detonation of range safety explosives was discounted, partly because the unexploded ordnance was among the debris, as were theories of structural imperfections in the tank's design or damage incurred at liftoff. The possibility of a liquid hydrogen leak at liftoff was also dismissed, since it would immediately have been ignited by the exhaust from the Solid Rocket Boosters or main engines and would have been evident in the downlinked telemetry data.

In total, around 30 percent of Challenger herself was found and inspections revealed that she had disintegrated as a result of massive aerodynamic overloads, with no evidence of internal burn damage or exposure to explosive forces. Chemical analyses indicated that her right side had been sprayed with hot gases from the leaking SRB, but telemetry indicated that all of her systems operated normally until shortly prior to the breakup. No problems were detected with either of her payloads: Spartan-203 was unpowered during ascent and the deployment ordnance for the Inertial Upper Stage and TDRS-B, of which about five percent of debris was recovered, showed no indication of having prematurely activated. The finger of blame pointed squarely at the third component of the Shuttle system – the SRBs – and, in particular, at the leaking right-side booster. Initial suspicion that its range safety explosive charges had been inadvertently fired was dismissed when telemetry data revealed that no such commands were sent to either booster until both were remotely destroyed by the Range Safety Officer at $T + 110$ seconds. For a number of engineers and managers at Morton Thiokol and within NASA, however, the cause of the disaster had been identified more than a year before Challenger's maiden voyage: the primary and secondary O-rings meant to prevent a leakage of hot gases were incapable of properly sealing the gaps between the SRB joints in extremely cold weather. Already, catastrophe had been averted on one previous cold-weather launch in January 1985 and conditions in the hours leading up to 51L's liftoff were colder still. Moreover, an application of zinc chromate putty, intended to act as a 'thermal barrier' and keep the combustion gas path away from the two O-rings, had been shown as early as 1984 to be susceptible to the formation of 'blow holes', which compromised its effectiveness.

"It was intended," read the Rogers report, "that the O-rings be actuated and sealed by combustion gas pressure displacing the putty in the space between the motor segments. The displacement of the putty would act like a piston and compress the air ahead of the primary O-ring and force it into the gap between the [field joint's] tang and clevis. This process is known as 'pressure actuation' of the O-ring seal. This pressure-actuated sealing is required to occur very early during the solid rocket motor ignition transient, because the gap between the tang and clevis increases as pressure loads are applied to the joint during ignition. Should pressure actuation be delayed to the extent that the gap has opened considerably, the possibility exists that the rocket's combustion gases will blow-by the O-rings and damage or destroy the seals. The principal factor influencing the size of the gap opening is motor pressure, but gap opening is also influenced by external loads and other joint dynamics." One of these external factors was the detrimental impact of low launch temperatures, together with the effect of water and ice, on the O-rings. In

Members of the Rogers Commission inspect the O-ring channel in an SRB segment during their investigation.

the case of 51L, on the night of 27 January 1986, ambient temperatures had dipped to the lowest ever recorded for a Shuttle launch: around -13 degrees Celsius. Indeed, at the moment of ignition the following day, the right-hand booster's aft field joint was the coldest part of the stack at -2.2 degrees Celsius. Ground tests had already confirmed that reduced temperatures could cause the O-rings' resiliency to degrade and during the Rogers investigation it was learned that a small quantity of rain water had been discovered in Columbia's SRB joints during preparations for STS-9 in November 1983. It was theorised that 51L, which had been sitting on Pad 39B for a total of 38 days and been exposed to significantly more rainfall than Columbia, could have suffered from the further disruption, and perhaps even 'unseating', of its O-rings by frozen water.

The observed problem with the boosters first arose in November 1981, shortly after STS-2. Routine inspections revealed significant erosion of the right-hand SRB's primary O-ring due to hot combustion gases, yet the secondary seal remained intact and the anomaly went unreported at the Flight Readiness Review for STS-3 in March of the following year. Morton Thiokol believed that the erosion had been caused by blow holes in the zinc chromate putty and began tests to alter the method of its application and the assembly of the booster segments. The manufacturer of the original putty, Fuller-O'Brien, discontinued its use and a new putty from the Randolph Products Company was selected in May 1982; however, after more changes, it was substituted for the original putty the following summer, shortly before the launch of STS-8. Since December 1982, the O-rings had been designated a

'Criticality 1' item by NASA, denoting a component without a backup, whose failure *would* result in the loss of the Shuttle and its crew. Prior to that, they had been labelled by NASA as 'Criticality 1R', meaning that, although "total element failure ... could cause loss of life or vehicle", the presence of primary and secondary O-rings lent 'redundancy' to the design: in effect, the secondary seal would expand to fill the joint if its primary counterpart failed. However, in its Critical Items List of November 1980, NASA acquiesced that "redundancy of the secondary field joint seal cannot be verified after motor case pressure reaches approximately 40 percent of maximum expected operating pressure. It is known that joint rotation occurring at this pressure level ... causes the secondary O-ring to lose compression as a seal". Following a series of high-pressure tests of the O-rings, conducted by Morton Thiokol in May 1982, it became clear that the secondary seal did not provide sufficient redundancy and NASA changed their criticality listing later that year. According to then-Associate Administrator for Space Flight (Technical) Michael Weeks, who signed a waiver to accept the new criticality level in March 1983, "we felt at the time that the Solid Rocket Booster was probably one of the least worrisome things we had in the programme". This view was shared by managers and astronauts, too. But not by Thiokol structural engineer Roger Boisjoly.

By the time Boisjoly inspected severely damaged field joints from 51C's boosters in January 1985, a number of other missions had yielded disturbing O-ring erosion. On 41B almost a year earlier, the left-hand SRB's forward field joint and the nozzle joint belonging to its right-hand counterpart were found to be badly degraded, to such an extent that NASA requested Thiokol to investigate means of preventing further erosion. A week prior to the launch of the next flight, 41C, the company concluded that blow holes in the zinc chromate putty were one "possible cause" and NASA's SRB project office at the Marshall Space Flight Center in Huntsville, Alabama, decided that, as long as the secondary O-ring could survive gas impingement, the mission was safe to fly. It was the beginning of a disturbing chain of thought within NASA and Thiokol, explained the Rogers report, that "there was an early acceptance of the problem" and both organisations "continued to rely on the redundancy of the secondary O-ring long after NASA had officially declared that the seal was a non-redundant, single-point [Criticality 1] failure". One of the members of the Rogers inquiry was the celebrated physicist Richard Feynman, who judged the cavalier attitude of NASA and Thiokol as representing "a kind of Russian roulette ... [the Shuttle] flies [with O-ring erosion] and nothing happens. Then it is suggested, therefore, that the risk is no longer so high for the next flights. We can lower our standards a little bit because we got away with it last time. You got away with it, but it shouldn't be done over and over again like that." Mike Mullane scornfully called it the "normalisation of deviance".

The 51C damage was among the most serious yet seen. Launched in freezing conditions on 24 January 1985, its recovered left and right SRB nozzles showed evidence of 'blow-by' between the primary and secondary O-rings and, moreover, it proved to be the first Shuttle mission in which the secondary seal displayed the effects of heat. "SRM [Solid Rocket Motor]-15," said Boisjoly of one of the 51C boosters, "actually increased concern because that was the first time we had actually penetrated a primary O-ring on a field joint with hot gas, and we had a witness to that event because the grease between the O-rings was blackened, just like coal. That was so much more significant than had ever been seen before on any blow-by on any joint." When the blackened material was analysed, Boisjoly told the Rogers hearing, "we found the products of putty in it [and] the products of O-ring in it". Four days after 51C landed, on 31 January, Lawrence Mulloy, head of the SRB office in Huntsville, expressed concern over the impact O-ring problems may have on the next scheduled mission, 51E, then projected for launch in late February. One of Thiokol's conclusions before the Flight Readiness Review was that, while "low temperature enhanced probability of blow-by ... the condition is not desirable, but is acceptable". It was the first occasion on which a link between cold weather and O-ring damage had been officially acknowledged.

MISSED WARNINGS

Three months after the worrisome 51C boosters had drawn Boisjoly's attention, Bob Overmyer's crew lifted off on the Spacelab-3 mission. Subsequent examination of their SRBs also indicated erosion of the secondary O-ring, clearly pointing to the failure of its primary counterpart. The problem was attributed to leak check procedures. So serious was the episode, however, that "a launch constraint was placed on flight 51F and on subsequent launches," read the Rogers report. "These constraints had been imposed, and regularly waived, by the Solid Rocket Booster Project Manager at Marshall [Space Flight Center], Lawrence B. Mulloy. Neither the launch constraint, the reason for it, or the six consecutive waivers prior to 51L were known to [NASA Associate Administrator for Space Flight Jesse] Moore or [Launch Director Gene] Thomas at the time of the Flight Readiness Review process for 51L ..." In fact, as Overmyer would later discover, his own launch had been milliseconds from disaster. Crewmate Don Lind journeyed to Thiokol in Utah for further explanation. "The first seal on our flight had been totally destroyed," recalled Lind, "and the [other] seal had 24 percent of its diameter burned away. Sixty-one mil[limetres] of that [last seal] had been burned away. All of that destruction happened in 600 milliseconds and what was left of that last O-ring, if it had not sealed the crack and stopped that outflow of gases, if it had not done that in the next 200 to 300 milliseconds, it would have gone [all the way]. You'd never have stopped it and we'd have exploded. That was thought provoking! We thought that was significant in our family. I painted a picture of our liftoff, then [added] two great celestial hands supporting the Shuttle and the title of that picture is *Three-Tenths of a Second*. Each of [my] children have a copy of that painting, because we wanted the grandchildren to know that we think the Lord really protected Grandpa."

Shortly after the analysis of the 51B boosters, on 31 July 1985, Roger Boisjoly expressed his growing concerns over the O-rings in a memorandum to Thiokol's vice president of engineering, Bob Lund. "The mistakenly accepted position on the joint problem," he wrote, "was to fly without fear of failure and to run a series of design evaluations which would ultimately lead to a solution or at least a significant reduction of the erosion problem. This position is now changed as a result of the [51B] nozzle joint erosion, which eroded a secondary O-ring with the primary O-ring never sealing. If the same scenario should occur in a field joint – and it could – then it is a jump ball whether as to the success or failure of the joint, because the secondary O-ring ... may not be capable of pressurisation. The result would be a catastrophe of the highest order: loss of human life." Boisjoly recommended the establishment of a Thiokol team to investigate and resolve the problem and, on 20 August, Lund announced the formation of a task force. However, only a day earlier, in a joint Thiokol-Marshall briefing to NASA Headquarters on the issue, programme managers concluded that the O-rings were a 'critical' issue, but that, so long as all joints were leak checked with a 200 psi stabilisation pressure, were free of contamination in the seals and met O-ring 'squeeze' requirements, it was safe to continue flying. As the year wore on, Thiokol's O-ring team, which had only eight to ten members, found many of their efforts frustrated by senior management. "Even NASA perceives that the team is being blocked in its engineering efforts to accomplish its task," Boisjoly wrote in a 4 October memo. "NASA is sending an engineering representative to stay with us, starting 14 October. We feel that this is the direct result of their feeling that we [Thiokol] are not responding quickly enough on the seal problem."

A little over three weeks later, Challenger flew mission 61A, experiencing nozzle O-ring erosion and blow-by at the field joints; neither of these problems were identified at the Flight Readiness Review for the next mission, 61B, in November. Indeed, that flight *also* suffered nozzle O-ring erosion and blow-by. By early December, in response to these problems, Thiokol recommended that their testing equipment needed to be redesigned. Only days later, on the 10th, the company requested closure of the O-ring critical problem issue, citing satisfactory test results, future plans and work carried out thus far by its task force. This closure request was harshly criticised by the Rogers investigators. One panel member pointed out to the Thiokol senior managers: "You close out items that you've been reviewing flight by flight – that have obviously critical implications – on the basis that, after you close it out, you're going to continue to try to fix it. What you're really saying is [that] you're closing it out because you don't want to be bothered." Part of the problem was NASA's desire, since the mid-1970s, to create a reusable transportation system that would provide regular and routine access to low Earth orbit. Original plans to fly the Shuttle once every fortnight, admittedly, were unrealistic with only four operational orbiters – rather than six or seven – but in its December 1985 launch schedule, the agency envisaged staging up to 24 missions per year from 1987 onwards. In correspondence with the author, one former Shuttle engineer expressed serious doubts that such flight rates werre achievable, even with overtime and three shifts working around-the-clock in the Orbiter Processing Facility. Nine or ten missions in any 12-month period stretched resources to their limits. Overtime and overwork presented their own problems. Numerous contract employees at KSC, the Rogers Commission heard, worked 72-hour weeks and frequently supported 12-hour shifts. "The potential implications of such overtime for safety were made apparent during the attempted launch of Mission 61C on 6 January 1986," read the report, "when

Poignantly, among the fragments of Challenger to be recovered from the Atlantic Ocean
was this piece of debris, bearing part of the fallen craft's name.

fatigue and shift work were cited as major contributing factors to a serious incident
involving a liquid oxygen depletion that occurred less than five minutes before
scheduled liftoff."

Furthermore, the commission discovered disturbing evidence that NASA's
provisions to support the projected 24-flight annual rate were woefully inadequate.
Spares for individual orbiters were in short supply (only 65 percent of the required parts
inventory was in place by January 1986), leading to an increasingly dangerous practice
of 'cannibalism' from one vehicle to equip the next, and resources focused primarily on
'near-term' problems, rather than longer-term issues. An $83.3 million budget cut in
October 1985 necessitated additional major deferrals of spare parts purchases. Prior to
the disaster, the shortage of spare parts had no serious impact on flight schedules, noted
the Rogers report, but further cannibalism was "possible only so long as orbiters from
which to borrow are available. In the spring of 1986, there would have been no orbiters
to use as spare parts 'bins'. Columbia was to fly in March, Discovery was to be sent to
Vandenberg [Air Force Base in California] and Atlantis and Challenger were to fly in
May." Indeed, KSC's Shuttle engineering chief, Horace Lamberth, predicted that, had
51L flown successfully, the schedule would have been brought to its knees that spring
by the spare parts problem alone. "Compounding the problem," the report explained,
"was the fact that NASA had difficulty evolving from its 'single flight' focus to a system
that could efficiently support the projected flight rate. It was slow in developing a
hardware maintenance plan for its reusable fleet and slow in developing the capabilities
that would allow it to handle the higher volume of work and training associated with
the increased flight frequency."

"THE SUN KEPT RISING"

Had 51L flown safely, the frequency with which the Shuttle flew into space in 1986 would have greatly eclipsed all previous years, with 12 more missions planned. One of the most significant was the launch of Ulysses, which, uniquely for the Shuttle, would have been one of two deep space missions launched within only five days. By the beginning of May, it was expected that Challenger would be on Pad 39B and sister ship Atlantis, carrying the Jupiter-bound Galileo explorer, on Pad 39A to support launches on the 15th and 20th of that month. Both missions were fixed within a narrow launch window.

Chief astronaut John Young referred to the missions as the 'Death Star' flights.

Behind the dark humour lay real concern. Even with an increasingly confident outlook on the Shuttle's capabilities, Young instinctively knew that Challenger's 61F mission and Atlantis' 61G voyage would be two of the riskiest ever attempted. Astronauts Rick Hauck and Dave Walker, who would command them, echoed his concern. "As with any flight," said Hauck, "if everything goes well, it's not risky. It's when things start to go wrong that you wonder how close you are to the edge of disaster." Yet Hauck and Walker's flights, scheduled to occur just five days apart, would have carried additional danger. This was partly due to the importance of the Ulysses and Galileo payloads, both of which were equipped with plutonium-powered Radioisotope Thermoelectric Generators (RTGs). This risk was worsened by the fact that, attached to the base of each probe was a thin-skinned, liquid-fed rocket that many astronauts and managers had condemned as unsafe and unacceptable for use in conjunction with a manned spacecraft. Measuring 9 m long and 4 m wide, it was called the 'Centaur-G Prime' and, for Rick Hauck, it was his baby.

Just like a baby, it was both temperamental and unpredictable.

"I was assigned to be the astronaut office's project officer for Centaur," Hauck recalled. "It's pressure stabilised, which means if it's not pressurised, it's going to collapse by its own weight. If it were not pressurised, but suspended, and you pushed on it with your finger, the tank walls would 'give' and you'd see that you're flexing the metal!" Nicknamed a 'balloon tank' because its rigidity thus depended on full pressurisation, the Centaur had long been viewed warily by NASA, whose safety rule of thumb on the Shuttle dictated that no single failure should be capable of endangering the vehicle or crew. The Centaur-G Prime did much more than that. Much of its pressure regulation hardware was not redundant – it lacked a backup facility – and, worse, a failure of its internal bulkhead had the potential to rupture both its volatile liquid oxygen and hydrogen tanks. Additionally, it was recognised that the sheer mass of propellants – which totalled more than 16,500 kg – could cause 'sloshing' and a myriad of other controllability problems that could hinder Hauck or Walker if the need arose to execute an emergency landing shortly after liftoff. In spite of the hazards, the Centaur's key advantage was that its liquid propellants provided considerably more oomph to push large payloads out of Earth orbit and onto trajectories to other planets than solid-fuelled rockets could achieve. It was also known that liquid-fed boosters produced a much 'gentler' thrust than the notoriously harsh impulse of solids. Still, the safety concerns rightly overshadowed and ultimately overwhelmed these benefits.

"The Shuttle was obligated to launch Ulysses and Galileo," explained Hauck. "[NASA] needed the most powerful rockets they could have [and] at some point the decision was made to use Centaur, which was never meant to be involved in human space flight. That's important because rockets that are associated with human space flight have certain levels of redundancy and certain design specifications that are supposed to make them more reliable. Centaur did not come from that heritage, so, Number One, that was going to be an issue in itself, but Number Two is [that] if you've got a Return to Launch Site abort or transatlantic abort and you've got to land – and you've got a rocket filled with liquid oxygen [and] liquid hydrogen in the cargo bay – you've got to get rid of [it], so that means you've got to dump it while you're flying through this contingency abort. To make sure that it can dump safely, you need to have redundant parallel dump valves [and] software that makes sure contingencies can be taken care of. Then, when you land, you're sitting with the Centaur in the bay that you haven't been able to dump all of it, so you're venting gaseous hydrogen out this side [and] gaseous oxygen out that side. This is just not a good idea!"

To support their rocket payloads, both Challenger and Atlantis underwent extensive modifications, costing around $5 million apiece, which included extra plumbing to load and drain the Centaur's propellants and control panels in their aft flight decks to monitor its performance. A Centaur Integrated Support Structure (CISS) at the rear of the payload bay would have held the booster and its payload and oriented them for deployment. As NASA's newest orbiter, Atlantis had been made Centaur-capable during her initial construction and was destined to spend the first part of 1986 out at Pad 39B undergoing validation tests of the new hardware. Challenger, too, had received the Centaur upgrades, which also included an S-band transmitter to handle the booster's telemetred data. During typical, pre-launch loading operations, the Centaur's liquids would have been fed through plumbing 'tapped into' the Shuttle's main propulsion system feedlines. Emergency dumping vents – capable of draining all liquid oxygen and hydrogen from the booster within 250 seconds of an abort being declared – were situated on opposite sides of the aft fuselage, just beneath the OMS pods. As part of her validation tests, Atlantis would have rolled to Pad 39B in February 1986 with a real Centaur-G Prime and a mockup of Galileo in her payload bay. Whilst on the pad, the booster would have been fuelled and a series of 'wet' tests carried out. Atlantis would then have been removed from the pad to enable the real spacecraft to be installed and then she would be rolled out to Pad 39A. By mid-April, she would have been joined on adjacent Pad 39B by Challenger, laden with Ulysses and its own Centaur.

Doubts over the reliability of the Centaur-G Prime riding the Shuttle had already, in the autumn of 1981, obliged NASA to cancel it and opt to install Ulysses and Galileo onto 'safer' – though less powerful – solid-fuelled Inertial Upper Stage boosters. For Galileo, which comprised a Jupiter orbiter and atmospheric probe, the swap from Centaur to IUS meant that its journey time to the giant planet would almost double to four and a half years and most likely require the mission to be split into two 'halves'. When the projected cost soared to $1 billion, Congress pressed NASA in late 1982 to resume work on a Shuttle-borne

Centaur. Not only Galileo, but also Ulysses, required close encounters with Jupiter – the latter in order to alter its trajectory and pass over the Sun's poles – and both missions were allocated the same launch window from 15-21 May 1986. According to the 61F Crew Activity Plan, published on 14 January 1986, Hauck and his men would have lifted off from Pad 39B aboard Challenger at around 4:10 pm EST on the 15th, followed by Walker's team from adjacent Pad 39A aboard Atlantis five days later. The flights had scarcely an hour apiece available in which to launch and, in order to minimise weight, both would carry just four astronauts. Hauck would have been joined by pilot Roy Bridges and mission specialists Mike Lounge and Dave Hilmers, whilst Walker's crewmates were pilot Ron Grabe and mission specialists Norm Thagard and James 'Ox' van Hoften. In van Hoften's mind, the excitement of deploying "this goofy thing to Jupiter" hardly compared to his two previous flights and their dramatic EVAs. "That mission looked honestly *really* boring," he told a NASA oral historian. "I had two of the best flights that NASA ever did and I was just on top of the world. [Mission 61G] was going to be a very short flight and there was going to be nothing to it, other than going up and launching this Centaur." There would be no secondary experiments and the payload bays would be empty, save for the probes and their attached Centaur boosters and support structures. Even some elements of crew equipment in the middeck, including the galley, would have been eliminated to save weight. In January 1986, NASA accepted a recommendation to fly Atlantis with her 'Phase II' main engines running at a hitherto-untried 109 percent thrust: to launch at the standard 104 percent, it was argued, would have meant that the Centaur with the 2,270 kg Galileo could not have accommodated a full load of propellant, which would have reduced its margins for the escape burn. Ulysses, on the other hand, was considerably lighter, at just 370 kg, and Challenger's engines for 61F could run at the 'standard' thrust rating.

Additionally, the two flights scheduled to last four days apiece – were headed for lower than normal, 168 km orbits because, said Hauck, they needed the additional performance simply to get the heavy Centaurs into space. Moreover, assuming an on-time liftoff of 61F, the astronauts would have had no more than about nine hours to get Ulysses out of Challenger's payload bay and on its way to Jupiter, since the Centaur was required to periodically dump its boiled-off gaseous hydrogen to keep tank pressures within mandated limits. After too much time, it would have 'bled' so much hydrogen that the remainder would not be sufficient to perform its trans-Jovian engine burn. Consequently, three deployment opportunities were manifested for both missions: in the case of 61F, the first chance came at 11:10 pm EST, some seven hours after launch. The Centaur-G Prime's twin Pratt and Whitney-built RL-10A-3A engines, each generating a thrust of 7,300 kg, would then have ignited about 45 minutes later and Ulysses would have been on course, first for a Jupiter rendezvous in July 1987 and ultimately for passage over the Sun's polar regions in 1989-1991. One can imagine that Hauck, Bridges, Lounge and Hilmers would have been glad to see the back of both it and the Centaur. Throughout the second half of 1985 and into the spring of 1986, in addition to their rigorous training regimes, both Hauck and Walker found themselves routinely questioning their own judgement over how many potential failure modes and problems they could live with. "In early January 1986," Hauck recalled, "we were working an issue to do with redundancy in the helium actuation system for the liquid oxygen [and] liquid hydrogen dump valves and it was clear that the [senior Shuttle management] was willing to compromise on the margins in the propulsive force being provided by the pressurised helium. We were very concerned about it. We had discussions with the technical people, but we went to a [review] board to argue why this was not a good idea to compromise on this feature. The board turned down the request. I went back to the office and said to my crew, in essence, 'NASA is doing business differently from the way it has in the past. Safety is being compromised and, if any of you want to take yourself off this flight, I will support you'. Two or three weeks later, Challenger blew up. Now, there is no direct correlation between my experience and Challenger, but it seemed to me that there was a willingness to compromise on some of the things that we shouldn't compromise on."

With 20-20 hindsight – or 'hindsight bias', the inclination to see events which have already occurred as being more predictable than before they took place – it has often been argued that the danger signs and the indications of impending catastrophe were visible and apparent, long before 51L. Some Thiokol engineers saw potentially fatal flaws with the SRBs, technicians saw insurmountable hardware problems and astronauts saw flaws in training and overbearing schedule pressure, all of which could lead to disaster. There were other concerns. In the event of a Return to Launch Site abort, during ascent, it was not possible to vent the Centaur's propellant overboard, risking an explosion in the payload bay ... and, according to John Fabian, sloppiness on the part of the processing workforce heaped additional worries. Fabian even remembered seeing one technician clambering onto the Centaur with an untethered wrench in his back pocket, whilst another tried to smooth out a weld and accidentally scarred the skin of the upper stage. Yet Mike Lounge rationalised their thinking. "We assumed we could *solve* all these problems," he said. As if to underline the point, the 61F crew were in a flight procedures meeting on the morning that Challenger exploded, reviewing the techniques to vent the Centaur's propellants in the event of a launch abort. "Until Challenger, we just thought we were bulletproof and the things would *always* work."

Years later, Rick Hauck remained undecided as to whether he would have refused to fly 61F, but admitted that Shuttle managers were taking unacceptable risks. Only days after the tragedy, any lingering doubts were resolved for him. The Kennedy Space Center's safety office refused to approve advanced processing of the first Centaur-G Prime, citing "insufficient verification of hazard controls" from both NASA and the booster's manufacturer, General Dynamics. Additional safety concerns, and cost overruns to the tune of $100 million, ultimately led to the project's cancellation in June 1986. Fortunately, a few years later, the Galileo and Ulysses missions went ahead, reverting to the less powerful IUS to get them successfully to their celestial targets. Both achieved considerably more than expected and revolutionised our understanding of both our parent star and our planetary big brother. Barely 23 *hours* after Challenger's scheduled landing at 4:21 pm EST on 19 May, Atlantis would have lifted off for her own four-day mission to set Galileo on course for Jupiter. Judging from Horace Lamberth's comment to the Rogers panel that a lack of spare parts would have brought the programme to its knees, it will never be known if NASA could have succeeded in launching two missions within the same week. If either had been postponed beyond the May 1986 window, the next Jovian opportunity would not have opened until June 1987. This delay would have also produced an estimated cost penalty in excess of $50 million. The schedule pressure grew so intense that in January 1984, Centaur technicians were even issued with unusual 28-month calendars ... which *ended* in May 1986; the deadline to launch was as immovable and inflexible as a rock. Most observers doubt that these missions could have launched on time, not just because of the short window or the lack of spare parts, but due to ongoing problems with certifying the Centaur-G Prime for advanced processing at the Kennedy Space Center in the spring of 1986. Others have disagreed. "I'm convinced to this day we *would* have made the launch window in May of '86," said General Dynamics' Marty Winkler, "but it was a sprint to the finish. It was like the racehorse that overtakes you at the end." The question ultimately became a moot issue. When the two spacecraft finally launched several years later, *sans* Centaur, they did so in separate windows: Galileo in October 1989, Ulysses in October 1990.

The success stories that Ulysses and Galileo ultimately became could scarcely have been further from NASA's mind on 28 January 1986, as Challenger's wreckage tumbled into the Atlantic. All Shuttle missions were indefinitely suspended until the Rogers Commission, whose staff panel included astronauts Neil Armstrong and Sally Ride, had made its recommendations. Among its conclusions were that NASA and Thiokol's operation of the Shuttle was seriously flawed – concerns from individual engineers were not reaching appropriate managers, 'critical' items were not given the attention they demanded and the need to stick to a 'schedule' was overriding 'safety'. Not only was NASA attempting to accommodate its major customers but, evidenced in a teleconference with managers at the Marshall Space Flight Center and Kennedy Space Center on the evening of 27 January 1986, Thiokol showed that it was prepared to ignore the safety concerns of its engineers in order to accommodate NASA, its own major customer. Worries of potential O-ring failure in the near-freezing weather conditions predicted for the following morning, as expressed by Roger Boisjoly and others, were downplayed and Thiokol collectively voted that Challenger was fit to fly, unwittingly signing the 51L crew's death warrants.

During that fateful teleconference, Bob Lund argued that his team's 'comfort level' was not to fly SRBs at temperatures below 12 degrees Celsius – some 53 degrees Fahrenheit – for fear of catastrophic 'blow-by' of the O-rings and field joints, but he could present no evidence to Marshall that 'proved' it was unsafe to do so. In a lengthy debate, Lawrence Mulloy – based in Florida as Marshall's KSC representative at the time – and other NASA officials challenged Thiokol's data and questioned its logic. At one stage, Marshall's head of science and engineering, George Hardy, remarked that he was "appalled" at the company's decision. So was Mulloy, who scornfully exploded with "For God's sake, Thiokol, when do you expect me to launch? Next April?" Neither man, however, was prepared to ignore the recommendation of their major contractor. Lund stood firm and, had he continued to do so, NASA would have had little choice but to postpone the 51L launch. Shortly thereafter, Thiokol requested a five-minute recess from the teleconference to consider the situation. Five minutes ultimately became half an hour. Throughout this recess, Boisjoly and fellow engineer Arnie Thompson continued to argue that it was unsafe to fly outside of their proven field joint temperature range, but the Thiokol senior executives in attendance felt the O-rings should still seat and function properly, despite the cold weather. "Arnie actually got up from his position and walked up the table and tried to sketch out once again what his concern was with the joint," Boisjoly told the Rogers Commission, "and when he realised he wasn't getting through, he stopped. I grabbed the photos and tried to make the point that it was my opinion from actual observations that temperature was indeed a discriminator and we should not ignore the physical evidence that we had observed. I also stopped when it was apparent that I couldn't get anybody to listen."

Then, executive Jerry Mason explicitly asked Lund to remove his engineering hat and put on his management hat. When the teleconference resumed, Lund changed his vote and Thiokol changed its position on the issue. The company's new recommendation was that, although frigid weather conditions remained a problem, their data was indeed inconclusive and the launch of 51L should go ahead the following morning. None of the engineers wrote out the new recommendation – "I was not even asked to participate in giving any input to the final decision charts", Boisjoly told the Rogers hearing – and only the executive managers signed it. However, when Marshall and KSC managers asked for any additional comments from around the Thiokol table before closing the teleconference, none of them voiced their concerns. Boisjoly, in particular, remained silent; a fact which would later lead some observers to brand him a witness who turned 'state's evidence', rather than a noble 'whistleblower'. When questioned by a Rogers panel member, he emphasised that "I never [would] take [away] any management right to take the input of an engineer and then make a decision based upon that input, and I truly believe that. There was no point in me doing anything any further than I had already attempted to do ... [but] I left the room feeling badly defeated. I personally felt that management was under a lot of pressure to launch and that they made a very tough decision, but I didn't agree with it." Having analysed the results of the teleconference, and interviewed the participants, the Rogers report concluded that "there was a serious flaw in the decision-making process leading up to the launch ... A well-structured and managed system, emphasising safety, would have flagged the rising doubts about the Solid Rocket Booster joint seal." In fact, when brought to testify before the panel, key officials intimately involved with the decision-making process, including Gene Thomas and Jesse Moore, admitted that they had not been privy to the issues raised at the 27 January teleconference.

Over the years, many observers have commented that, had Challenger not been lost, *another* unsuspecting Shuttle crew *would* have fallen victim to catastrophe. Astronaut Bob Parker, who would have flown aboard the next flight, Mission 61E, has expressed his fervent belief that disaster *may* have befallen himself and his crewmates ... for the weather conditions in Florida in the early hours of 6 March 1986 were even *colder* than those on the night before Challenger's fateful flight. Although it seems unlikely that Columbia could have been ready in time, NASA was still aiming to launch 61E at 5:45 am EST on the 6th, kicking off an ambitious flight, during which a 'red' team of pilot Dick Richards, mission specialist Parker and payload specialist Sam Durrance and a 'blue' team of mission specialists Dave Leestma and Jeff Hoffman and payload specialist Ron Parise would have operated the ASTRO-1 payload around the clock. Commander Jon McBride would have anchored his work schedule across both shifts. When the Crew Activity Plan for 61E was published in November 1985, McBride's crew were expected to attempt the second-longest Shuttle mission to date; their landing, scheduled for 3:47 am EST on 15 March, would have concluded a flight lasting two hours shy of nine full days. Jeff Hoffman had been assigned to follow ASTRO-1 several years earlier, in 1982, and he quickly concluded that the mission was sufficiently complex to warrant a pair of payload specialists. "We knew George Abbey didn't like the idea of payload specialists," Hoffman told the oral historian. "Was writing this report going to be *career-limiting?*" It turned out not to be the case and in June 1984 he and Bob Parker were assigned as mission specialists for the flight, which, at that time, was to launch in March 1986 with the large Intelsat 6-1 communications satellite sharing the payload bay. Intelsat was subsequently removed and replaced with Westar-6S, which was itself later removed. By the end of January 1986, ASTRO-1 had completed its pre-launch processing and was ready for installation into Columbia's payload bay. Its instruments would have been mounted on a Spacelab pallet and IPS. On the day that Challenger exploded, the 61E crew was in the simulator, practicing ascents and re-entries, and paused briefly to step outside and watch the 51L launch on television. Years later, Bob Parker would muse as to why unbearable schedule pressure and a cavalier attitude to the Shuttle was repeatedly allowed to override flight safety. "Mission 61E ... must be launched by 10 March," *Flight International* had told its readers in December 1985, "to achieve maximum science return. A slip to March 20 would result in the flight's cancellation." Such immovable targets astounded Parker. "You'd have thought the world was going to end! My favourite expression is: *Guess what?* The Sun kept on rising and setting! The Sun didn't even *notice* if we missed our launch windows."

Schedule pressure and the need to revise managerial communications channels to enable individual engineers to express concerns more openly were only part of the problem. On the technical side, decreed the National Research Council's Shuttle audit committee, the most important requirement was the redesign of the SRB field joints and O-ring seals to prevent future leakages. In its July 1986 response to President Reagan and the Rogers Commission, NASA announced a $680 million plan: to redesign the joint's metal components, insulation and seals, thereby providing "improved structural capability, seal redundancy and thermal protection". New capture latches would reduce joint movements caused by motor pressure or structural loads and the O-rings were redesigned to not leak under structural deflection at twice the expected level. Internal insulation was modified with a deflection relief flap, rather than putty, and new bolts, strengtheners and a third O-ring were added. External heaters with integrated weather seals would ensure that future SRB joint temperatures did not fall below 24 degrees Celsius and prevent water from entering the seals. "The strength of the improved joint design," read NASA's reply to Reagan, "is expected to approach that of the [SRB] case walls."

Concerns over the integrity and strength of booster casings had already raised questions over the planned use of 'filament-wound' SRBs for the planned Shuttle launches from Vandenberg Air Force Base in California. Rather than having solid-steel case segments, remembered astronaut Jerry Ross, who was assigned to the first Vandenberg flight, Mission 62A, the filament-wound segments were composed of a more flexible graphite-epoxy material. "They had the same joint design as the steel cases," he explained, "and since the graphite ones would have been more flimsy, we always were wondering what would have happened to us had we tried to launch with those, considering the Challenger accident." The rationale behind filament-wound SRBs was that they provided a weight saving, enabling the Shuttle to transport several thousand kilograms of additional payload into orbit. The new boosters were built by Hercules Inc., subcontractors to Thiokol, and employed a unique 'capture' feature to eliminate rotation in the joints between each segment. In addition to questions over the reliability of the new boosters, there was also an alarming possibility that hydrogen could become trapped in the exhaust duct, beneath the main engines, perhaps triggering a fire or explosion during liftoff. Only days before the loss of Challenger, these concerns crystallised into additional costs as the Air Force acknowledged its latest technical hurdle in getting Vandenberg operational. By now, the site had swallowed close to $3 billion for its Shuttle operation, more than twice as much as had been predicted in 1977. As the Rogers investigation got underway, scrutiny was cast on the boosters and on more generic safety problems. The Air Force had already put in place plans to fit 54 outward-firing igniters inside each of the Shuttle's main engines to remove concentrations of hydrogen. Eventually, following a $7.3 million study, it was decided to employ steam to solve the problem: storing hot water in pipes, re-circulating it through boilers to keep its temperature stable and spraying it into the main engine duct just before launch. However, flying from Vandenberg posed other difficulties and risks and it is important to consider what was being planned at the California launch site and why.

Since the dawn of the space age, each US manned mission had flown from Cape Canaveral in Florida. That monopoly, however, was set to change in July 1986 with the launch of Discovery from Vandenberg Air Force Base, located on the coast roughly halfway between San Francisco and Los Angeles. It had been chosen to stage a series of military Shuttle missions carrying primarily Department of Defense payloads into near-polar orbits. NASA's agreement to detail one of its orbiters to Vandenberg on a more-or-less permanent basis was the payoff for the Air Force's political support for the Shuttle during its development. One of the most significant satellites to be launched from the Californian base was the KH-12 imaging platform, which many observers felt could not reach its required orbit from Florida. In accordance with nominal safety criteria, Shuttle missions were typically limited to maximum inclinations of around 57 degrees, although higher inclination trajectories were technically feasible by the technique of executing a 'dog-leg' manoeuvre during ascent, travelling far offshore before swinging north or south. However, the additional energy needed for the orbiter and External Tank to 'turn' at such high velocities during the dynamic climb to space was at the cost of a 30 percent reduction in payload capacity. The implications, articulated by a high-ranking Air Force official during testimony to Congress in 1978, was that heavy KH-12 and other polar launches could not be effectively achieved from the East Coast. Additionally, in the run-up to the first Shuttle mission, the Department of Defense was already developing payloads whose size, weight and complexity specifically required the capabilities of the reusable spacecraft to reach orbit.

Safety concerns were yet another issue precluding polar flights from Florida, in that the trajectories of ascending Shuttles would cross heavily-populated South Carolina and the Great Lakes; furthermore, the jettisoned Solid Rocket Boosters would fall within an impact 'footprint' which included Brunswick in Georgia and the External Tank would follow a suborbital path across Canada, the North Pole, Russia, China and perhaps India. Aside from the risk of killing or injuring civilians, the chance of an abortive mission crash-landing in Soviet territory whilst carrying a highly-sensitive national-security payload did not bear thinking about. Although Vandenberg's use for polar launches would have sidestepped many of these diplomatic difficulties, the sheer value of KH-12 in particular obliged NASA and the Air Force to design 'alternate' mission plans in which the satellite could be deployed as early as the first orbit, to avoid overflying Russia. At length, partly due to problems accommodating the KH-12 aboard the Shuttle and chiefly because the first Vandenberg launch was considered a test flight, it was decided that this should carry an experimental spacecraft before committing more sensitive payloads at a later date.

Known as 'Teal Ruby' and originally codenamed 'Air Force Program-888' (AFP-888), it was a bizarre contraption which would have evaluated infrared detectors for future early-warning satellites and featured extremely lightweight instruments and a novel 'cryostat' which utilised cryogenic neon and methane, rather than helium, as a coolant. "It was a staring mosaic infrared sensor," explained Jerry Ross, "that was trying to be able to detect low-flying, air-breathing vehicles – things like cruise missiles – and a way to detect these approaching US territories". Also aboard Discovery would have been an experimental telescope called the Cryogenic Infrared Radiance Instrument for Shuttle (CIRRIS). Teal Ruby required an orbit of 560-720 km, inclined at 72 degrees, which only a Vandenberg launch could safely and efficiently provide. After lifting off, the Shuttle would pitch over to fly a trajectory down the Pacific coast of the Americas. After deployment, the operational lifetime of the satellite would be constrained to just one year by a dwindling on-board supply of cryogenic coolant needed to keep its infrared sensors at sufficiently low temperatures. In shape and cost, it has been described as "a complete mess", with funding originally set at $80 million and three years allocated for its development. In reality, it took more than twice as long to build and cost almost $500 million.

The Vandenberg mission would also have added to the confusion already caused by a cryptic and somewhat clumsy combination of numbers and letters used to identify Shuttle flights since February 1984. Internally, it was variously labelled 62A or 'Sixty-Two-Alpha'. The first number denoted the financial year in which it was due to be flown (1986 in this case), while the second identified the launch site (with the digit '2' reserved for Vandenberg). The letter, lastly, highlighted the position of a particular flight in the launch pecking order to a specific year and from a specific site. By following this peculiar logic, Sixty-Two-Alpha was the first of two classified flights scheduled for launch from Vandenberg in 1986. The second, carrying the long-awaited KH-12, was expected to follow on 62B in late September. The only crew names attached to the mission were a pair of Air Force payload specialists: Captain Katherine Roberts, a manned spaceflight engineer, and General Lawrence Skantze, commander of the Air Force Systems Command at Andrews Air Force Base in Maryland. However, such a short turnaround time for Discovery was considered unlikely because Vandenberg did not possess such sophisticated processing facilities as the Kennedy Space Center. In fact, even before Challenger, NASA anticipated turnaround times between Vandenberg flights to be around eight months, which might easily have pushed 62B into early 1987. The Department of Defense, it seemed, would be lucky to achieve just two missions per year – a far cry from the dozens envisaged annually when the Shuttle began flight operations in April 1981. The reasons for this overestimation are complicated, but can be attributed to problems with both the launch site itself and a number of key design changes needed to haul heavy payloads into high polar orbits.

Vandenberg's Shuttle operation centred on Space Launch Complex (SLC)-6, nicknamed 'Slick Six. The expectation was that upgrades to existing hardware would help to drive down development costs; however, unforeseen problems throughout the late 1970s and into the 1980s pushed the price tag relentlessly upwards. Flame ducts had to be built for the booster exhaust, as did a specialised payload preparation room, an assembly building, a Shuttle runway, a new launch tower and facilities for mating the orbiter to its Boeing 747 carrier aircraft. Unlike the Kennedy Space Center, where the stacking of the orbiter, its boosters and External Tank was undertaken in the Vehicle Assembly Building, preparations for missions at Vandenberg would have been conducted at the launch pad.

The dangers of acoustic, blast and thermal excesses on the base had been questioned as late as the autumn of 1985 and would be revisited during the Challenger inquiry, because at just 300 m from the pad, the launch control centre was extremely close for the comfort of its 175 occupants. Vibrations experienced during launches – exacerbated and reflected by the surrounding mountains – could upset computers and cause major structural damage to an ascending Shuttle, while unpredictable weather conditions, including ice, heavy rain and thick fog, posed additional hazards. In fact, in the summer of the previous year, recommendations had been made to mothball Slick Six until at least 1991, pending the construction of a new orbiter and the final resolution of problem of hydrogen concentrations on the pad. Ultimately, with the reduction in the Shuttle's payload capacity to polar orbit and the suspension of manned missions, it became more prudent to reassign some Department of Defense payloads to expendable rockets. By this point, with a few exceptions, the military was seeking a way out of its dealings with NASA at the earliest opportunity. Ironically, it was one of Jerry Ross' crewmates on the 62A mission, Air Force Undersecretary Edward 'Pete' Aldridge, who pressed for a return to using expendable rockets and ultimately terminated the Shuttle effort at Vandenberg in December 1989.

Aldridge's assignment was an interesting case in itself. Since becoming under-secretary in 1981, he had clearly seen the writing on the wall that the Shuttle would be unable to meet its advertised flight rates. In fact, his actions during those years led to the offer of a seat on the reusable spacecraft. "I believe Jimmy Carter wrote a presidential directive [in 1978] that the Space Shuttle ... would meet *all* the demands of *all* the users," he told the NASA oral historian. When Columbia finally launched on her first mission, it became abundantly clear that promises of weekly flights were a long way off, the turnaround times were far longer than anticipated and *two* of the four orbiters were too heavy for effective use by the Department of Defense. It did not bode well for the future. "We only had two orbiters that could meet the DoD weight and size demands," Aldridge continued. "The cost was *not* one-third of the cost of an expendable [rocket]; it was more likely equal at best, and possibly much higher than that. When we first started to see this, we began to worry that we were not going to meet the demands of the Department of Defense. We had a requirement for 12 flights a year from the Shuttle. Our estimates of what we were seeing as turnaround time said [a total of] 12 to 18 [flights per annum] was more likely the number. If it was going to be at the lower end, or even at 18 per year, we were going to take 12. We had a hard requirement to fly 12 flights. This meant the civil and commercial space business was not going to be as robust as we thought it was going to be. We could *pre-empt* the launch of a commercial satellite in order to get a national security satellite up. It was highly uncertain whether or not any of the commercial or civil programs were going to have much viability if the orbiter flight rate was in the 12 to 18 per year."

In 1983, Aldridge approached Secretary of Defense Caspar Weinberger to advise *against* the termination of expendable rocket production. Weinberger agreed, and so too did President Reagan, and proposals were put in place to continue at least the Titan booster production for another five years. Furthermore, some payloads were reconfigured to fly aboard expendable vehicles. "NASA got *very* upset about it," added Aldridge. "Jim Beggs [who headed the agency from 1981 until 1985] saw that as a ploy of the Air Force to remove itself, ultimately, from the Shuttle." For his part, Beggs noted that he had no concerns about the Air Force opting for expendable launch vehicles, as a "backup" flight capability, but stressed that those rockets were equally as vulnerable as the Shuttle to failure. (In fact, *two* Titans were lost in separate accidents, a few months either side of Challenger.) At length, after discussing the issue with Weinberger, a compromise would be reached between Beggs and Aldridge, in which the Department of Defense would buy up a third of each year's Shuttle missions for its purposes. The seeds of uncertainty had already been sown, however, and Beggs considered it appropriate to re-establish the partnership between NASA and the Air Force. A payload specialist seat for a high-ranking official was one obvious outcome; the 68-year-old Secretary of the Air Force, Vernon Orr, was not interested in flying, so the offer trickled down to his deputy, Aldridge. Consequently, in December 1985, Aldridge arrived in Houston for 62A training. "We were going to fly at a 72-degree inclination, which man has never flown before," he remembered. "For military missions, you want to cover the entire Earth. Well, you don't do that by flying east and west, so you fly north and south. We were actually going to fly [close to] the poles on every orbit, which was unique. It was exciting to have a completely new mission."

The second payload specialist aboard 62A was Brett Watterson, a manned spaceflight engineer. In February 1985, NASA named Bob Crippen as the commander, joined by pilot Guy Gardner and mission specialists Mike Mullane, Jerry Ross and Dale Gardner; interestingly, it was speculated by *Flight International* that an EVA might have been planned, although this remains unconfirmed. "I *really* wanted that polar flight," Crippen told the NASA oral historian. "I lobbied for it and ended up being selected, although not without some consternation. I think, since this was primarily an Air Force mission, there was a big push by the Air Force to have an Air Force commander on the flight." (Some astronauts regarded the assignment of Crippen – a naval aviator – as another example of George Abbey's preferential treatment of Navy pilots.) Mullane was equally excited. "You're basically going to see the *whole world*," he exulted. "In an equatorial orbit, or a low-inclination orbit ... you don't get to see a lot of the world." At the time of the 'core' crew announcement, their launch was set for January of the following year, but by mid-1985 it had already slipped to late March and at the time of the Challenger accident was tentatively scheduled for mid-July. The arrival of Discovery at the West Coast launch site – originally scheduled for September 1985, but postponed by several months, due to technical difficulties – was partly to blame for this delay, although "propellant loading and other problems" at Vandenberg were offered by *Flight International* as contributory factors in December.

The loss of Challenger irrevocably sounded the death knell for polar-orbiting Shuttle flights from Vandenberg. The lightweight filament-wound casings, which had been conceived to carry heavier payloads into higher inclination orbits, would have required additional fixes to meet one of the Rogers Commission's recommendations. These fixes, unfortunately, would have added extra weight and virtually cancelled-out any advantages in terms of their lifting capacity. By February 1987, the Air Force accepted that Rogers-enforced structural changes meant the Shuttle could no longer effectively haul heavy payloads into polar orbits. In terms of its Shuttle operation, Vandenberg and Slick Six were indeed stricken by bad luck, having cost more than $4 billion without achieving a single launch. To this day, some observers feel it was for the best. The catastrophic failure of booster joints during Challenger's final flight has led Jerry Ross to doubt how well the "flimsy" filament-wound cases would have performed during a 'real' ascent. Moreover, Vandenberg's unpredictable weather would have caused excessive ice build-up on the External Tank which could have crippled the orbiter's heat shield and triggered disaster. That, of course, is to leave untouched the other unknowns surrounding the area: the vibration and acoustic effects, the hydrogen-concentration problems and the ever-present, obvious dangers of strapping seven astronauts atop several million pounds of highly-explosive propellants.

For the United Kingdom, 1986 was to be a banner year, with the first British astronaut, Nigel Wood of the Royal Air Force, due to fly on 61H in June as a payload specialist to observe the deployment of the Ministry of Defence's Skynet 4A communications satellite. Planning for the flight had been announced more than two years earlier, in January 1984, when the MoD revealed that it would launch Skynet 4A and 4B on the Shuttle. However, it reserved the right to fly its third satellite – Skynet 4C – aboard Ariane, thereby demonstrating visible support for the European booster. "The £60 million or so that it will cost to fly two Skynet 4s on Shuttle is probably slightly cheaper than flying on Ariane," *Flight International* reported, "and no doubt was a factor in selecting the American launcher." Rumours of shortlisted candidates from the Royal Air Force and Royal Navy proved accurate, when, in February 1984, four finalists were selected. In addition to Wood, they were civilian physicist Christopher Holmes, who was the deputy manager of the Skynet 4 project, together with two military officers, Peter Longhurst of the Navy and Tony Boyle of the Royal Signals. (Boyle was later replaced by fellow Signals officer Richard Farrimond in July.) Wood's selection as the prime payload specialist was revealed in May 1985 and he joined 61H commander Mike Coats, pilot John Blaha and mission specialists Jim Buchli, Anna Fisher and Bob Springer. In addition to monitoring the Skynet deployment, Wood trained to conduct a series of six British experiments, focusing on the effects of cosmic radiation, changes in head-eye co-ordination and adaptation to weightlessness, studies of adhesive bonding, the ability to estimate mass in microgravity, motor skills – including postural control – and an ergonomics experiment. In September 1985, the equipment for Skynet 4A was delivered by Marconi Space Systems to British Aerospace and represented the most advanced communications payload ever built in Europe. Later that same month, a team of candidates were selected by Indonesia, for a *second* payload specialist seat on the flight. The finalists included a journalist, a pilot, a telecommunications engineer and a female microbiologist named Pratiwi Sudarmono, who was chosen to fly. She would monitor the deployment of her nation's Palapa-B3. This crew would also place a *third* commercial satellite, Westar-6S, into orbit.

Another key outcome of the Challenger tragedy was that the Shuttle would henceforth only be used for missions which explicitly required its capabilities and those of its astronauts. Particular focus would be granted to scientific research. More than two dozen commercial and military satellites, that were booked to fly on the Shuttle in 1986 and beyond, were transferred to expendable rockets. As early as April 1986, the Ministry of Defence was in talks with Arianespace to launch Skynet 4A and 4B aboard the European rocket and *Flight International* admitted that the opportunities for 'passenger' payload specialists would probably be low on NASA's list of priorities. With a fleet of only three surviving orbiters, the agency knew that it would struggle to attract commercial interest and when the first post-51L manifest was published in October 1986, it featured primarily NASA or Department of Defense payloads. Apart from a few contracts signed prior to 51L, no further commercial 'primary' cargoes would be carried. A deviation from that policy came in May 1992, when, on her maiden voyage, Challenger's replacement, Endeavour, retrieved the stranded Intelsat 6-3 communications satellite, fitted it with a new perigee kick motor and redeployed it. Although successful and once more demonstrative of the Shuttle's unique capabilities, this mission was an initial, worrying hint that the lessons of 51L were already fading from NASA's mind.

Among the improvements made to increase the survivability of future crews in the wake of 51L were upgraded brakes and tyres, the development of a drag chute to support the Shuttle's high-speed landings and the incorporation of an escape pole which could be used to bail out of the middeck side hatch in the event of serious problems. It was recognised that, without a pole to provide sufficient clearance, astronauts evacuating a crippled orbiter in flight would likely hit the port-side wing. Unfortunately, the pole – which was attached to the middeck ceiling during a mission – could only be used when the Shuttle was in controlled, gliding flight, and not much higher than Challenger when she disintegrated. In the wake of Challenger, each astronaut was provided with a partial-pressure suit – upgraded, in 1994, to a fully-pressurised ensemble – which would provide hyperbaric protection during ascent and cold-water immersion protection in the event of an emergency ditching in the ocean, together with a parachute and life raft.

Other concerns raised by the Rogers Commission were the short periods separating individual missions, which provided insufficient opportunity for flight data from one voyage to be properly analysed before the next one set off. One particular example was a potentially serious problem with Columbia's brakes during 61C, launched only 16 days before 51L. The Flight Readiness Review for the latter was on 15 January 1986, with Columbia still in orbit, and the data from the brake problem – together with further, disturbing O-ring erosion – was not analysed until the 30th. By then, of course, it was too late for Challenger and her crew. When missions resumed in September 1988, NASA mandated that a minimum of three weeks should separate every Shuttle launch, although this was waived in November-December 1990. Certainly, with plans to stage 14 flights after 51L in 1986, close proximity between launches was a fact of life. As we have seen, only a few *days* would have separated the 61F and 61G missions and the remainder of the year was also tightly packed. After Discovery's 62A, Challenger was scheduled to fly Mission 61M on 22 July. During the five-day flight, the crew would have deployed a Tracking and Data Relay Satellite, whilst payload specialist Bob Wood operated the payload bay-mounted version of McDonnell Douglas' continuous flow electrophoresis machine. Wood had designed its software and would have operated it from a middeck control panel. However, according to Charlie Walker, who would have served as backup for 61M, Wood's training "didn't go as smoothly as hoped [and] management was considering putting me up as Number One again. I had my doubts. I was tired of the pace. Of course, it ended up being a moot issue".

In yet another mission which Challenger herself had 'baselined' previously, India's Insat-1C would have been carried into orbit aboard 61I on 27 September, together with Indian mechanical engineer Nagapathi Bhat as a payload specialist. In a similar vein to the Teacher in Space effort, 61I's second payload specialist seat would also have been granted to a journalist – and, at the time of the disaster, the applicants had been winnowed down by NASA to a list of 40 semi-finalists. These included Pulitzer prizewinners John Noble Wilford and Peter Rinearson and veteran CBS anchorman Walter Cronkite. At the time of the disaster, the screening of those

40 semi-finalists was to take place at JSC in April 1986 and the successful candidates would have begun formal training in May. All 40 lost their chance that cold January day when not only the Shuttle was suspended, but so were its commercial ventures. Interestingly, John Noble Wilford later shared a second Pulitzer Prize in 1987 for his reporting of the Challenger explosion and its aftermath.

Other missions included two classified Department of Defense assignments from Florida and a commercial deployment flight, 61L, in November to place the GStar 3 communications satellite and the Navy's long-delayed Syncom 4-5 into orbit. Manned spaceflight engineers Frank Casserino and Chuck Jones were assigned as payload specialists for the DoD missions and Hughes engineer John Konrad as a payload specialist for 61L. The remaining missions, 61K in August and 61J in October, both of which were to be flown by Atlantis, had recently swapped places. The latter was to deploy NASA's scientific showpiece, the Hubble Space Telescope, which required additional time for pre-launch checkout. Underscoring its importance, 61J would have been commanded by chief astronaut John Young. Meanwhile, the 61K flight, led by Vance Brand, would have flown the Earth Observation Mission involving a pallet of experiments and a Spacelab short module and featured amongst its crew the first European mission specialist, Swiss astronaut Claude Nicollier. NASA's November 1985 manifest indicated that the flight rate was expected to increase further, with 19 missions planned for 1987.

So many flights by so few orbiters was hugely ambitious, if not ludicrous, in its scope. Jim Beggs, in discussion with his deputy, Hans Mark, had concluded long before Challenger that the maximum achievable yearly flight rate would be around a dozen. *Nineteen* missions per annum – translatable to a launch every two and a half *weeks* – was simply unrealistic. Moreover, all four vehicles would have been operational, with Columbia, Challenger and Atlantis operating from Florida and Discovery from Vandenberg, which meant that the availability of spare parts would have been limited and the option of cannibalism would have been a difficult temptation to resist. Not until after the resumption of Shuttle flights in 1988 was a co-ordinated effort set in motion to ensure that each orbiter was removed from service after every seven or eight missions for maintenance, inspection and refurbishment. In addition to the logistical and technical nightmare of continuing to prepare and execute missions with just six to eight weeks between each launch of individual orbiters, the production of sufficient SRBs and disposable External Tanks would have driven the programme to its knees. Many observers predicted the Shuttle flight rate would drop precipitously in the spring of 1986. Ongoing problems with O-ring erosion, failures of brakes and tyres, main engine trouble, inadequate time to train crews, a multitude of niggling issues with no time to properly address them and an increasingly aggressive launch schedule would all have taken their toll. When asked if the Shuttle could have even achieved 12 flights per annum, one engineer from those seemingly bulletproof days responded with a resounding "No".

Peculiarly, as late as March 1986, still in advance of the Rogers Commission's final report, NASA was anticipating a return to flight in the spring of the following year. Furthermore, the agency envisaged nine missions in 1987, more than a dozen in the following year and an average of between 16 and 19 per annum by the end of the decade. At around the same time, the plan was to retain commercial payloads – including Westar-6S and Insat-1C – and continue to carry passengers. NASA's dream, even in the wake of Challenger, to continue to fly the Shuttle 'routinely' and 'regularly', was in for a rude awakening. When Columbia disintegrated during re-entry in February 2003, her mission was the first of six flights planned for that year; a pitiful figure when offered in comparison to the numbers quoted above. The reusable fleet reached its peak of operations in 1985, when three orbiters, some of them cannibalising parts from the fourth, undertook *nine* voyages. Never again would the Shuttle duplicate this flight rate, although it did average seven missions per year throughout most of the 1990s. None of these missions were conducted without risk, but thanks to the sacrifice made by Dick Scobee and his heroic crew that tragic January day in 1986, once the Shuttle resumed flying, each launch was immeasurably safer.

4

Road to Peace

LAZARUS FROM THE DEAD

With the return of Leonid Kizim, Vladimir Solovyov and Oleg Atkov from Salyut 7, the Soviet Union concluded 1984 in spectacular style: a new endurance record of almost 237 days, the first female spacewalker, the first woman to make a second mission, the first Indian cosmonaut and the first flight to feature six EVAs. Moreover, in the case of the latter, all six excursions had been complex and challenging and had triumphantly repaired a damaged fuel line and successfully installed extensions to the space station's electricity-generating solar arrays. The United States may have held the technological superiority in terms of the Shuttle's systems, its reusability and its capacity to fly large crews, but Soviet cosmonauts had spent far longer in orbit in 1984 than had the Americans. The balance would change, to an extent, in 1985, when *nine* Shuttle flights took place ... but even those would accrue, in total, barely eight weeks aloft. The Soviets, on the other hand, were planning a mission lasting at least six months, featuring a new 'Heavy Cosmos' module and a visiting expedition, possibly by an all-female crew. How those plans would have played out has led many observers to suggest that the female crew (aboard Soyuz T-14) would probably have comprised Svetlana Savitskaya – the only experienced woman cosmonaut, still on active duty – in command, teamed with engineer Yekaterina Ivanova and physician Yelena Dobrokvashina. According to Rex Hall and Dave Shayler, such a mission would most likely have taken place in November 1985, coinciding neatly with the anniversary of the Bolshevik Revolution. Theirs would be a two-week flight to visit Salyut 7's resident crew, the Soyuz T-13 cosmonauts: commander Vladimir Vasyutin, flight engineer Viktor Savinykh and 'research engineer' Alexander Volkov. The latter were meant to fly for about six months, from May to November, although as circumstances transpired, the 1985 plans changed markedly from February onwards and Savinykh would be prepared for a record-breaking nine-month mission.

Nine months in space – 270 days or thereabouts – was mentioned by Phillip Clark, who suspected that Vasyutin, Savinykh and Volkov were originally scheduled to fly in February 1985, "with perhaps one or two visiting missions", both involving "all-Soviet crews". In addition to the Savitskaya flight, Clark suggested that the 50-year-old cosmonaut Boris Volynov – a veteran of two previous missions – might be in the running to command the other. Also mentioned was Oleg Atkov's backup, the physician Valeri Polyakov, a cosmonaut since 1972, yet still awaiting his first flight. "It was decided," wrote Clark, "that someone who was medically trained to Atkov's degree was too much of a specialist for the routine Salyut operations. A doctor *had* a place in space, but it would probably be in the third seat of an all-Soviet visiting crew." According to Clark, Polyakov was dropped from Soyuz T-13 and replaced by a military engineer, Alexander Volkov. Both Hall and Shayler, and Clark, concur in their assumption that the all-female Soyuz T-14 would have taken place in the latter part of 1985 *and* that the *expeditsya osnovnoi* (EO), the principal expedition, of Vasyutin and his men, would also be visited by a Heavy Cosmos, one of the TKS military spacecraft. The reader will recall from Chapter 2 that plans had been abandoned to fly *manned* TKS missions and the vehicles were being retasked as 'add-on' modules for the front of Salyut 7 … but with a caveat: they only possessed a *single* Soyuz-T-compatible docking mechanism. This meant that, whilst a Heavy Cosmos was in residence, only *one* Soyuz-T spacecraft could be docked at the station and, thus, only one crew could be resident. Moreover, the absence of a second available port meant that the docking of a Progress resupply craft would be impossible whilst a crew was aboard. With these factors in mind, Hall and Shayler remarked that the Heavy Cosmos would have been launched in April 1985, a month before Vasyutin's crew, and would have been undocked and de-orbited in the early autumn, prior to the arrival of Savitskaya's mission.

Dropping Valeri Polyakov in favour of Alexander Volkov is only part of the story. The Soyuz T-13 resident crew was expected to conduct operations with the TKS – later to become known as 'Cosmos 1686' – which was, to an extent, familiar territory for Vladimir Vladimirovich Vasyutin and Alexander Alexandrovich Volkov. Both were Soviet Air Force officers and experienced fighter and test pilots and much of the experimental work aboard the new module was thought to be of a military nature. Vasyutin came from Kharkiv in the Ukraine, where he was born on 8 March 1952, and was a graduate of the Chuguyev Higher Military Pilot School and Test Pilot School, both located in Kharkov. He was selected as a cosmonaut in August 1976 and appears to have undergone training specifically for the original, manned version of the TKS spacecraft. Volkov, meanwhile, would later earn a family place in history, since both himself and his son, Sergei, served as cosmonauts. The Volkovs thus became the first case in which successive generations of the same family flew into space. Yet the elder Volkov's interest in the heavens extended back much earlier. Like Vasyutin, he was of Ukrainian descent, having been born in the city of Gorlovka – also known as 'Horlivka' and today twinned, interestingly, with the Yorkshire town of Barnsley – in the Donetsk Oblast on 27 April 1948. Two weeks before his 13th birthday, Volkov had watched, awestruck, as his fellow Soviet countryman, Yuri Gagarin, became the first man to venture into space and it was *this* which inspired the young boy to go on to train as a cosmonaut. In time, he would not only fly three space voyages, accruing more than a year aloft, but also command the cosmonaut corps from 1991-98 and, of course, produce a spacegoing son, who has himself flown a pair of six-month missions.

In addition to Vasyutin, Savinykh and Volkov, six other cosmonauts had been assembled as alternate crews for the TKS missions. Military pilots Alexander Viktorenko and Yevgeni Salei were teamed with civilian Alexander Alexandrov, whilst military pilots Anatoli Solovyov and Nikolai Moskalenko were joined by civilian Alexander Serebrov. (The reader will recall that Alexandrov had earlier flown on the Soyuz T-9 mission to Salyut 7 with the Cosmos 1443 module present, whilst Serebrov might have done the same, but for the failure of Soyuz T-8 to dock with Salyut 7.) These three teams of cosmonauts were in dedicated training by September 1984. By the early spring of the following year, the second Heavy Cosmos was ready for launch; offering a habitable volume of around 45 m^3, it closely resembled its predecessor, Cosmos 1443, in that it possessed the main FG8 module and recoverable VA capsule, although the latter had been heavily modified to carry scientific equipment, including an infrared telescope and the Ozon ('Ozone') spectrometer. When Cosmos 1686 was launched, atop a Proton rocket, it was expected to deliver around 4,500 kg of supplies to Salyut 7 and almost *double* the amount of living space for the cosmonauts. All of these plans changed abruptly on 11 February 1985, when a complete loss of telemetry downlink from Salyut 7 was experienced and it entered 'free drift'. Its Igla radar transponder was unable to pick up the signal from an incoming spacecraft, so launching an unmanned Progress to dock with the station and stabilise it in preparation for the arrival of a future crew was impossible. (Much later, it would become clear that all eight batteries had run down and a telemetry glitch prevented flight controllers from detecting the problem; in response, Salyut automatically shut down all of its systems and it was *this* which was responsible for the broken downlink.) Early in March, the Soviet news agency, Tass, announced that "in view of the fact that the planned programme of work ... has been fulfilled completely, at present the station is mothballed and continues its flight in an automatic regime". Others hinted that Salyut 7's "primary mission ... was at an end" – perhaps hedging their bets against admitting a catastrophic failure – but after the extensive repairs, conducted by Kizim, Solovyov and Atkov, they did not intend to abandon it to an ignominious demise ... or, from a political perspective, risk an uncontrolled re-entry. Had the fuel line repairs failed, some observers wondered? Certainly, by May 1985, elements of the Western trade press were speculating that Salyut 7 had experienced major electrical problems. A manned mission to bring the 'dead' station back to life was acutely needed. It would mark the first occasion on which Soviet cosmonauts had ever attempted to rendezvous and dock with an inert vehicle. Four experienced commanders – Vladimir Dzhanibekov, Anatoli Berezovoi, Leonid Popov and Vladimir Lyakhov – were chosen to train for the repair mission. Most of Vasyutin's crew, having trained specifically for the Heavy Cosmos, were bumped to a later flight, with the exception of Viktor Savinykh, who had undergone extensive preparation for EVAs in the Star City hydrolab. In March 1985, he was assigned as the flight engineer of the 'new', rescoped Soyuz T-13, paired with Dzhanibekov, who was already one of the Soviet

The dead Salyut 7 space station, viewed from the incoming Soyuz T-13.

Union's most experienced cosmonauts on 'active' status. In fact, he already had four space missions under his belt, including the dramatic July 1984 EVA with Savitskaya *and* he was trained to make a long-duration flight. On Soyuz T-13, he would become the first cosmonaut to chalk up five flights.

Despite only having flown one previous mission, Viktor Petrovich Savinykh was actually *more* experienced, in terms of flight time, than his commander, having accrued 74 days aboard the Salyut 6 station in early 1981. He came from Berezkiny in Russia's Kirov Oblast, where he was born on 7 March 1940. After graduating from the Moscow State University of Geodesy and Cartography in 1969 he joined the TsKBEM design bureau. In December 1978 his goal of becoming a cosmonaut was achieved when he was selected as one of seven civilian candidates. Two years later, he was shuffled several times and ended up as a backup flight engineer for the Soyuz T-3 mission, before being paired with Vladimir Kovalyonok for a two-month flight to Salyut 6 aboard Soyuz T-4. This mission, which launched in mid-March and ended in late May 1981, earned him the accolade of becoming the world's hundredth spacefarer *and* the 50th Soviet cosmonaut. That Savinykh remained aboard Soyuz T-13, alongside Dzhanibekov, owes itself partly to the fact that he was a veteran cosmonaut with a wealth of EVA training (whereas Vasyutin and Volkov were both 'rookies' at the time), but he had also served on the backup crews for no fewer than *four* earlier missions: Soyuz T-7 and T-12, both of which were 'visiting expeditions', together with the long-duration Soyuz T-8 and (with Vladimir Vasyutin and Valeri

Polyakov) Soyuz T-10B principal expeditions. In short, both Savinykh and Dzhanibekov had the all-round knowledge, the training and the experience to handle what the Russians consider to be one of their riskiest manned missions. The backup crew, named in March 1985, was Leonid Popov and Alexander Alexandrov, both of whom had flown long-duration missions. In addition to docking with an unresponsive target – and a *big* one, at that – the repair crew would be required to enter its darkened interior and rectify whatever problems Salyut had suffered. It was not known if the station's atmosphere was still breathable. The cosmonauts would most likely need to stabilise Salyut, realign its solar arrays, recharge its batteries and bring its life-support systems back online, before they could even contemplate an extended stay. Moreover, several months after the departure of Kizim's crew, the temperature aboard the station would certainly be cold, perhaps below freezing. On the eve of their launch aboard Soyuz T-13, on the night of 5 June 1985, Vladimir Dzhanibekov and Viktor Savinykh knew that theirs would be unlike any previous mission. Many journalists compared their task to the rescue of Skylab by its first crew in May 1973. They would literally be bringing the space station equivalent of Lazarus back from the dead.

REOPENED FOR BUSINESS

Soyuz T-13 thundered into the Tyuratam sky at 9:40 am Moscow Time on the 6th, carrying Dzhanibekov and Savinykh towards their quarry. In its summary, Tass made no reference to the dramatic nature of the mission, stating simply that the cosmonauts would be "carrying out joint work with the orbital research station, Salyut 7". In order to conserve propellants aboard their spacecraft, it had been decided that the cosmonauts would adopt an extended, two-day rendezvous profile to reach Salyut 7, rather than the 25 hours normally followed. Although the final approach was done without the aid of the Igla, the cosmonauts were fortunate in that Vladimir Titov had demonstrated on Soyuz T-8 that it *was* possible to draw close to Salyut, without using the radar. In Titov's case, he had failed because he was untrained to perform a fully manual approach and docking *and* had insufficient propellant at his disposal for lengthy stationkeeping and repeated attempts. With this incident in mind, it is easier to understand why mission planners incorporated a far more conservative rendezvous into Soyuz T-13's flight profile. The rendezvous training itself had included *manual* approaches from a distance of 30 km, although the actual docking by Dzhanibekov on 8 June was near-flawless. He sighted Salyut 7 from a distance of 10 km. The station seemed to be 'nutating' – slowly rotating about its main axis, drawing out a circle – at a rate of about a third of a degree per second, and Dzhanibekov found that he could match this motion with Soyuz T-13's attitude control thrusters. The latter had been modified to include specialised control levers for proximity operations and stationkeeping. It soon became clear from the orientation of Salyut's solar panels – which had lost their 'lock' on the Sun and were about 75 degrees apart – that some sort of orientation or power supply failure had been experienced, rather than a communications or commanding failure. Still, the station was technically 'dead', totally unresponsive to the Igla radar, unpowered, unable to orient its forward docking port to face Soyuz T-13 and almost certainly cold and potentially hazardous inside.

Dzhanibekov had luck on his side, though, in two areas. Firstly, he had a comrade who had earned himself a reputation as a highly skilled and competent engineer; in fact, Viktor Savinykh's nickname among his fellow cosmonauts was "the human computer", owing to his astonishing ability to calculate complex problems in his head. During the rendezvous, Savinykh called out a string of closure ranging and rates and tapped data quickly into Soyuz T-13's computer. The second aid for Dzhanibekov had come about following Vladimir Titov's unsuccessful bid to dock with Salyut 7, two years earlier: a laser range-finder – the first time such a device had ever flown aboard a Soviet manned craft – to gather precise data on distances and closure rates, together with a low-light image intensifier, enabling the crew to 'see' the station against Earth's shadow and reveal its orientation and precise position. As the cosmonauts drew closer, reaching just 200 m, they paused to station-keep and allow ground controllers to assess the situation. At length, after describing Salyut's physical appearance, Dzhanibekov received the go-ahead to attempt a manual docking. He masterfully guided his craft into alignment with the forward port, matching its slow roll with a burst from Soyuz T-13's thrusters, and brought the two craft together as they flew into the shadow of Earth. The docking, in darkness, came at 11:50 am Moscow Time on 8 June ... on Dzhanibekov's *first* attempt!

Their gruelling mission, though, had barely begun. The two men had first to ascertain whether the station possessed a breathable atmosphere and, with no power, the only way to do this was by opening a plug valve in the docking mechanism to equalise the pressures between the two vehicles. The result was encouraging: there *was* air inside Salyut 7, but a sample would need to be taken in order to satisfy flight controllers that it could keep them alive. When this sample showed that there were no hazardous contaminants in the atmosphere, Dzhanibekov and Savinykh opened the hatches, clad in fur-lined jumpsuits and gloves, woolly hats and oxygen masks, and floated into its cold, dank, musty-smelling interior. Neither man had any way of knowing if a fire, started, perhaps, by a short circuit, had filled the air with toxins. Everything looked as it should, with nothing awry, until they came to switch on the lights. Nothing happened. The fans were quiet, whilst the voltage indicators on the storage batteries hung limply at zero. The cosmonauts threaded their way through the dead station, the only illumination coming from their flashlights, whose flickering beams danced in the gloom. "Every surface," wrote Bryan Burrough, "was coated with frost and icicles." The inside temperature was estimated at -10 degrees Celsius. Their first task was to connect the solar panels directly to the storage batteries, in order to recharge them; two were beyond repair, but the other six were still functional. As they worked, it became clear that Salyut's lack of proper ventilation to clear their exhaled carbon dioxide was leaving them drowsy and, periodically, they had to retreat back to the Soyuz to rest and regain their strength. The cold clawed its way into their bones, forcing them to take breaks every half an hour or so to warm themselves up. "Finally, after a whole day of arduous work," Dzhanibekov told a journalist in 1986, "we had the thrill of seeing the voltage needle

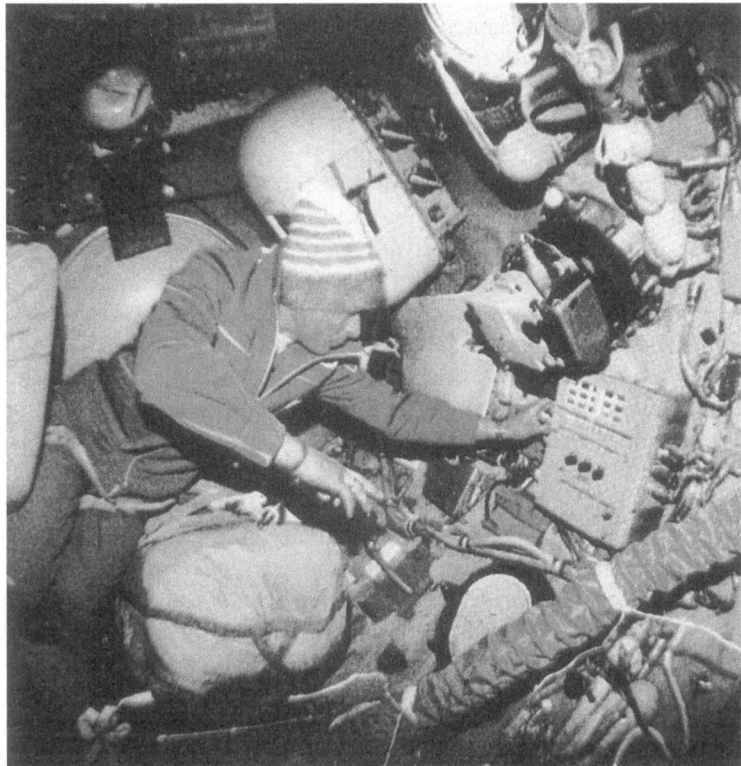

Clad in woolly hat and warm clothes, Viktor Savinykh works to bring Salyut 7 back to life. Not only were the two cosmonauts faced with immense technical challenges, but the frigid conditions and lack of proper ventilation required them to frequently retreat to Soyuz T-13 to rest and warm themselves. Their salvation of Salyut 7 from an untimely demise is today recognised as one of the most audacious and risky missions ever attempted.

tremble, move slowly off zero and begin to climb." Quickly, they were able to switch on Salyut's lights. Gradually, over the next few days, they resuscitated the station. A sensor in the solar arrays' pointing system had failed, thus preventing them from properly recharging, and the cosmonauts were obliged to use Soyuz T-13 itself to 'turn' Salyut and place them back into sunlight. Six of the eight batteries could then be recharged. Over the next few days, they patched up burst pipework, thawed water supplies and, when the frost finally evaporated, switched on the wall heaters. It was a long, difficult slog – "Patience" was one word used by Dzhanibekov to describe the quality which served him best – and the poor quality of the atmosphere meant that headaches and nausea plagued them on a near-constant basis. The only means of heating fluids in the freezing conditions was by using a powerful television lamp!

At length, by the 11th, the station was in a condition from which the two cosmonauts could begin 'demothballing' its systems. They sent a list of items to flight controllers: essentials needed on the next Progress mission – propellant, obviously, but also spare parts and replacement pieces for Salyut's faulty temperature controls – and continued to press on with the reactivation of the life-support apparatus, the air regenerators, the carbon dioxide absorbers, the gas analysers and the heaters. On 12 June, they brought the station's communications system back to full working order; no longer would they be totally reliant upon their radios in Soyuz T-13. The first televised images from Salyut were downlinked the following day and by the 13th it was with much enthusiasm that the first successful efforts were made to command the orientation of the station. On the 16th, running 'water' was finally produced as the ice thawed. The eagerly anticipated Progress – the 24th in the series – arrived on 23 June, docking automatically, and by this time Dzhanibekov and Savinykh had settled down to begin scientific studies. These included the Kursk-85 Earth resources experiment and, in early July, work with the MKF-6M and KATE-140 cameras. By the middle of the month, when Progress 24, loaded with unneeded items and rubbish, was undocked and burned up in the atmosphere, Salyut 7 was back in business and ready to support another long-duration crew.

Today, in an era of greater openness and transparency, the remarkable exploits and accomplishments of Dzhanibekov and Savinykh are better known, but at the time the Soviets kept the true drama of the mission under wraps. At least, that is, until success had been assured. Even *Flight International* told its readers on 29 June only that the cosmonauts had "docked with the mothballed Salyut 7 ... after first checking the station's exterior and solar panels". Reference *was* made to possible damage – but only with respect to erroneous reports that the station had been used as a target for a ground-based high-energy laser test – and it was not until mid-August that the first murmurings of what had actually transpired became clear. Writing in *Pravda*, Konstantin Feoktistov explained that broken batteries had rendered Salyut useless and had frozen the interior. Why the Soviets chose to reveal Salyut's problems may have its roots in the improved attitude of newly-appointed General Secretary Mikhail Gorbachev and his desire to unveil the nation to a wider world ... or it might simply reflect a desire to crow about another space achievement. A year later, in an interview with an Australian publication, Dzhanibekov explained that the rescue mission was intended to prevent the space station from making an uncontrolled re-entry. "There were fears that debris from Salyut 7 could have fallen on inhabited areas," *Flight International* told its readers in June 1986, "causing a serious international incident." (Perhaps in making such admissions, the Soviets were jabbing at America's failure to control the re-entry of its Skylab station, several years earlier. In hindsight, this line of thought is somewhat ironic, considering that Salyut 7 eventually made an uncontrolled re-entry in February 1991, raining debris onto Argentina.)

None of this detracts from the reality that Dzhanibekov and Savinykh's effort was unremarkable. After the flight, Valentin Glushko, head of Energia, recommended that Dzhanibekov be granted a third Gold Star for his expert command of what is recognised – even today – as *the* most challenging Soviet manned space mission in history. Glushko's recommendation was declined, on the basis that even Sergei Korolev had only accrued *two* Gold Stars, but Dzhanibekov subsequently rose to become a general, served as deputy commander of the cosmonaut corps and acted as deputy to the Supreme Soviet of the Uzbek Soviet Socialist Republic. In January 1993, he joined balloonists Larry Newman and Don Moses in a bid to circumnavigate the globe, as part of the 'Earthwinds' project. Unfortunately, the attempt failed when their ballast balloon tore on a mountain peak. Not until March 1999 and the 20-day flight of Bertrand Piccard and Brian Jones aboard *Breitling Orbiter 3*, did a piloted hot-air balloon fly successfully, and non-stop, around the world.

As a cosmonaut, of course, Dzhanibekov had circled the world many times and by the end of Soyuz T-13, the fifth and final mission of his career, he would have accumulated a grand total of 2,303 orbits of Earth. In the late summer of 1985, however, the precise *length* of the mission turned into something of a surprise. It has already been noted that the Soviets were earnestly progressing towards a permanent occupancy of their *next* space station, called Mir, but an attempt to semi-permanently man Salyut 7 was also in the works. (Had Soyuz T-10A not exploded on the pad, this might have been achieved in September 1983.) It came as something of a surprise in the West, when, on 19 July, a spacecraft called 'Cosmos 1669' was despatched towards Salyut. It docked successfully a couple of days later and speculation was rife that the new arrival might represent another 'Heavy Cosmos'. However, it had been launched aboard a Soyuz-type booster – not a Proton – and many observers wondered instead if it was actually a new version of Progress, perhaps a free-flying scientific platform or possibly equipped with additional solar panels or a descent module capable of surviving re-entry. Certainly, the conclusion that Cosmos 1669 was actually a Progress did not take long to reach. One report, quoted by Phillip Clark, even noted that Dzhanibekov and Savinykh were "getting ready for a meeting with the Progress 25 cargo ship", although the Radio Moscow World Service described it, rather ambiguously, as a "support satellite", and admitted that it was "similar to spacecraft of the Progress series". Years later, Rex Hall and Dave Shayler noted that initial telemetry from the new craft indicated that its Igla antenna had not deployed, thus precluding a rendezvous and docking, "and it was therefore decided to assign it a Cosmos cover name, rather than name it Progress 25". Subsequent analysis revealed that the spacecraft's Igla *did* successfully deploy – the telemetry readings reflected nothing more than a faulty sensor – but by this point the Soviets had already named it. Indeed, it had long been Soviet practice to assign a Cosmos number to missions which failed or experienced technical difficulties, in order to screen their precise objectives from prying Western eyes. (More recently, it became clear that Cosmos 1669 represented an upgraded Progress, designed for operations in conjunction with Mir.) Nothing of substance was revealed about the cargo carried to Salyut, although it can be supposed that equipment for a forthcoming EVA might have been amongst it. This 'equipment' included a pair of new 'Orlan-DM' space suits, intended for use aboard Mir, which featured bright lights at the temples of the helmet, to illuminate suit control dials, together with improved systems. The suits were of much sturdier construction than earlier models and included rubberised fabric shoulder belts, rather than the wholly rubber belts of the Orlan-D, and provided better mobility. On 2 August, Dzhanibekov and Savinykh clambered outside the station in their new suits for what would turn out to be five hours of work to augment Salyut's port-side solar arrays with the third and final pair of extension panels carried aloft by Progress 24. One of these arrays was of 'experimental' design and the efforts of the cosmonauts concluded a long series of augmentation work which had begun with Vladimir Lyakhov and Alexander Alexandrov, almost two years earlier. Moscow TV showed parts of the EVA 'live' to its audience and Dzhanibekov and Savinykh installed sample materials and a joint Soviet-French experiment to gather meteoritic dust, together with particles from Halley's Comet.

Late in August, Cosmos 1669 undocked from Salyut 7, then performed a rendezvous and redocking to test the reliability of its systems, before finally departing and burning up in the atmosphere on the 30th. By now, almost three months into their mission, the cosmonauts had settled into a routine of scientific activities, performing observations of high-energy particles in Earth's radiation belts with the Mariya device, whose results were expected to better the Yelena gamma spectrometer, providing data within just a few minutes. Other investigations, including Medusa, sought to understand the synthesis in components of nucleic acids in the microgravity environment, whilst the cosmonauts also participated in technological, biological and medical studies. Other studies included geological and petrological observations of oil-bearing and gas-bearing fields in Tajikistan and the identification of major freshwater deposits in the Pamir Mountains. This work, and doubtless also a difficult few months in orbit, did little to ease their spirits and the longing for home was strong. Televised meetings with family members were enormously important for the two men and they frequently listened to tape recordings of sounds from Earth – rain, rustling leaves, birdsong – which offered another psychological crutch and improved their efficiency. With a Salyut landing window approaching, it came as little surprise in the West when Soyuz T-14 was launched at 3:39 pm Moscow Time on 17 September, carrying a somewhat unusual crew: two members of the *original* T-13 team of Vasyutin and Volkov, together with 54-year-old veteran cosmonaut Georgi Mikhailovich Grechko. The latter had long experience in the Soviet space programme and, it would seem, had been pulled out of semi-retirement to lend his expertise to this crew. It was not the first time that Grechko's skills had been thus tapped. Late in 1977, he had flown with rookie cosmonaut Yuri Romanenko on the first long-duration expedition to the Salyut 6 station, which broke the 84-day Skylab record.

Phillip Clark has noted that Grechko was an experienced spacecraft designer and, with his flight aboard Soyuz T-14, this experience made him the first cosmonaut to fly aboard *three* separate space stations. The bubbly, jocular Grechko came from Leningrad – today's St Petersburg – where he was born on 25 May 1931. As a young engineer, he had worked for Sergei Korolev's OKB-1 design bureau and became one of only a handful of civilians to pass the preliminary screening for a flight aboard the Voskhod spacecraft in 1964. Two years later, after a slight relaxation of the rules on health requirements, he was picked as a cosmonaut trainee. Later in the decade, Grechko seems to have served as something of a pawn between Korolev's successor, Vasili Mishin, and the powerful commander of the cosmonaut team, Nikolai Kamanin. In the days preceding the Soyuz 7 mission in October 1969, for example,

Mishin pushed for Grechko to replace Vladislav Volkov, but Kamanin nixed the idea. Six months after that, it was *Kamanin* who wanted Grechko to fly Soyuz 9 instead of Vitali Sevastyanov and, this time, *Mishin* strenuously objected. From August 1972, Grechko was teamed with Alexei Gubarev for a long-duration Salyut flight and they flew to Salyut 4 for 29 days in the spring of 1975. This was followed, for Grechko, by a 96-day stay aboard Salyut 6 in 1977-78 and, much later, a position as backup flight engineer for the Soviet-Indian Soyuz T-11 mission. In a sense, Grechko is arguably one of the few cosmonauts who *benefitted* from Salyut 7's failure in early 1985, for had it *not* occurred, Vasyutin, Savinykh and Volkov would have flown Soyuz T-13 to the station and Savitskaya's crew would have been aboard Soyuz T-14. Grechko, almost certainly, would never have flown again. In fact, when he and his crewmates rose into orbit on the afternoon of 17 September, he secured a new record as the oldest cosmonaut ever launched into orbit. At 54, he broke Oleg Makarov's previous record by seven years and fell into joint third place, with American astronaut Don Lind, on the world list of oldest spacefarers.

If Grechko was at the older end of the scale, then Vasyutin and Volkov, were at the opposite extreme, as mere thirtysomethings. After docking at Salyut 7's rear port, the population of the station expanded to five men and the Soviets announced that a crew rotation *would* take place, though not quite on a par with earlier practice. Vladimir Dzhanibekov would return to Earth with Grechko, and Savinykh would remain in orbit with Vasyutin and Volkov, effectively reuniting the 'original' Soyuz T-13 crew for a long-duration mission, possibly stretching into the late spring of the following year. In the early hours of 25 September, Dzhanibekov and Grechko undocked from the station and were safely back on Earth at 12:52 pm Moscow Time the following afternoon. After their departure from Salyut, they had spent around 30 hours evaluating "differing methods" of approaching the station, though they did not redock. It was later revealed that these exercises were conducted in support of the forthcoming Mir station. Dzhanibekov completed a relatively 'short' mission of 112 days, whilst Grechko logged just under nine days, but this was expected to be dwarfed by Viktor Savinykh, as he prepared for an extended nine-month residency aboard Salyut. Many observers in the West expected him to return to Earth, alongside Vasyutin and Volkov, in March 1986. Their marathon expedition seemed to get off to a good start when another TKS spacecraft was launched into orbit, atop a Proton, on 27 September 1985. Named, Cosmos 1686, it docked automatically at Salyut's forward port a few days later. Although the Soviets revealed little of substance about its purpose, it apparently carried about as much cargo as a Progress, including the 'Kristallizator' materials processing unit, together with "food, gas regenerators, new scientific apparatus and individual assemblies and parts". One key difference from the earlier Cosmos 1443 was that the VA (normally employed as a recoverable re-entry capsule) had been stripped of its landing systems and environmental and other controls; these were replaced by high-resolution photographic equipment and optical sensors, including an infrared telescope and the Ministry of Defence's Ozon spectrometer. In a sense, Cosmos 1686 formed a prototype for the large research modules in the pipeline for the Mir complex, although it was devoted almost exclusively to military objectives.

Suspicion in the West about the new module's military capabilities was piqued in mid-November, when the Yorkshire-based Kettering radio group picked up encoded voice transmissions from the station and many observers wondered if Alexander Volkov – a military researcher – was engaged in classified operations. Something *was* afoot, but it had less to do with military reconnaissance and more to do with the psychological condition of one of the cosmonauts. "The Kettering Group received normal telemetry and voice from the crew during October and most of November," wrote the Swedish analyst Sven Grahn on his website, www.svengrahn.pp.se. One particularly telling comment came in early November, when Volkov made a strange request. For several weeks, Cosmos 1686 had been used to maintain Salyut's orientation; however, on *this* occasion, such precision was *not* needed, but he asked for it to be maintained anyway, in order to preserve "a normal psychological climate" aboard the station. Then, on the evening of 13 November, Grahn "picked up scrambled voice ... indicating that something confidential was being discussed, probably a medical problem". Similar data was collected by Kettering personnel in the next couple of days ... and something else: something which implied that the 'medical problem' was serious and did not bode well for the continuation of the Soyuz T-14 mission. "The 20.008 MHz telemetry transmission also began to be heard at this time," continued Grahn, "indicating that the status of the *return ship* was of great interest." On the 15th, Tass revealed that medical checks of the cosmonauts were underway. There was nothing unusual about *that*, but it was then revealed that during these tests they were wearing their Chibis ('Lapwing') lower-body negative pressure suits, normally employed to condition themselves for an imminent return to Earth. Within a week, on the 21st, Soyuz T-14 had undocked from the station and touched down safely at 1:31 pm Moscow Time. Savinykh had spent 168 days in orbit, whilst Vasyutin and Volkov had accrued 65.

It soon became evident that Vasyutin had fallen ill, although the precise nature of his condition has become the stuff of legend and subject to much debate over the years. At the landing site, he was described as being "pale" and, when asked how he felt, his response was that "the main thing is to be back on Earth". Clearly, space living had proven incompatible for him. To their credit, the Soviets admitted immediately that Vasyutin *had* been ill and required hospital treatment – he was flown to Moscow, where he remained until 20 December – and the Western rumour mill was rife with speculation: kidney stones, appendicitis and viral pneumonia were just three possibilities aired by the press. Upon his release from hospital, the man himself explained that he had fought "inflammation" for three weeks and had suffered from a high fever. At the end of December, the newspaper *Pravda* included in its pages a few excerpts from Viktor Savinykh's diary, which noted that Vasyutin suffered from "anxiety in his behaviour and loss of sleep and appetite", followed by "pain", in the days following the departure of Dzhanibekov and Grechko. Within weeks, his condition had deteriorated drastically. On 28 October, Savinykh described him as "tense ... a bundle of nerves" and as November dawned the commander was spending much of his time in his sleeping bag, whilst his crewmates carried out his chores.

Vasyutin's illness and inability to work took its toll and he became acutely depressed and withdrawn; this was reinforced by Bryan Burrough, writing in 1998, who recalled conversations with psychologists that there had been "mood and performance issues" among the Soyuz T-14 cosmonauts. The official order to terminate the mission came on 17 November and it was *Savinykh*, the civilian engineer, who took command of the Soyuz during re-entry and landing. Two years later, during a visit to England in October 1987, Oleg Gazenko, then-head of the Soviet Institute of Biomedical Problems, refused to divulge more detail, citing medical privacy, but he did say that Vasyutin made a full recovery. For his part, Alexander Volkov explained that, had Vasyutin not fallen ill, the mission would have ended on 15 March 1986, after 179 days for the two of them ... and a record-breaking 282 for Savinykh! And for Savinykh, the blow was harshest: for not only was there much "regret" over the lost work from the mission – having spent many weeks painstakingly assembling and checking out Cosmos 1686's hardware – but the chance to secure a world record for space endurance had slipped like grains of sand through his fingers. There was only *one* remaining Soyuz-T vehicle, because a new version of the spacecraft was to be introduced with Mir. This almost certainly would have been used for Svetlana Savitskaya's crew. It can be supposed that *this* flight would have taken place a couple of weeks before the return of Vasyutin, Savinykh and Volkov, perhaps deliberately timed to coincide with International Women's Day on 8 March 1986. Had the all-female mission taken place, it would have required the undocking of Cosmos 1686 from Salyut 7. As circumstances transpired, the TKS would remain linked to Salyut for the remainder of its life and the pair would burn together during re-entry in February 1991. As for Vladimir Vasyutin, it is not surprising that he never flew into space again, although, sadly, he did not survive into old age, either. After a lengthy battle with cancer, he succumbed to the horrific disease in 2002, only months after his 50th birthday.

PAST AND FUTURE

If Soviet politics in the mid-1980s offered a Janus-like vista of past and future in a single eyeful, then the final crewed mission to Salyut 7 provided nothing less: for the veteran crew of Soyuz T-15, Leonid Kizim and Vladimir Solovyov, would become the first – and so far only – team of cosmonauts to visit not one, but *two*, space stations in a single mission. According to Phillip Clark, the men and their backups – Alexander Viktorenko and Alexander Alexandrov – had been assigned to the mission in November 1985. The primary destination of Soyuz T-15 was the core module of the new space station, Mir. Its name was translatable as either 'World' or 'Peace' and seems to have formed part of a propagandist Soviet effort to present their space programme as one of 'Star Peace', and thus diametrically opposed to Ronald Reagan's belligerent 'Star Wars'. This theme was reinforced by former cosmonaut Alexei Leonov, speaking shortly after Mir's launch, "by naming the space station in this way, we want to emphasise once again that the Soviet programme for space research and for the use of outer space is intended solely for peaceful purposes". This notion was reinforced by Leonid Kizim, during his stay aboard Mir, when he responded to a question about whether the new station would be used for military purposes. "The programme for our work," he replied, "does *not* contain any experiments for military purposes. As for the statements by US officials, it seems to us that they are being made in order to justify their *own* plans for transferring the arms race to space." *Perestroika* might be steadily moving towards *glasnost*, but even a year into Gorbachev's reformist rule, the mistrust between East and West remained strong.

The roles and capabilities of the new station were at least an order of magnitude higher than anything previously attempted. Unlike the 'monolithic' Salyuts, Mir represented something entirely novel: a first effort to create a 'modular' complex in space. After the insertion into low-Earth orbit of its 'core' module, equipped with a unique, multiple-port docking adaptor, the plan was for it to be expanded over several years into a fully-fledged research facility, with the addition of large laboratory modules, each weighing around 20,000 kg, devoted to astrophysics, remote sensing and the life and microgravity sciences. For a decade and a half, counting from the launch of its core in February 1986 to the end of its life in March 2001, Mir would prove itself to be one of the most successful and influential space stations ever launched. Record after record would be secured by its cosmonauts, many of which still stand to this day, including records for the longest time spent in orbit, and it would be occupied, uninterrupted, for more than ten years.

Mir's story began in February 1976, when a Soviet decree identified it as quite distinct from the other Salyuts: it would possess docking ports at each end of its core module, just like them, but its spherical multiple adaptor would carry two radial ports for additional research laboratories. By the summer of 1978, this plan had changed to a final configuration of a single aft port and no fewer than *five* ports on the multiple adaptor: one at the front and another four radial ports, spaced at 90-degree intervals. These radial ports would accommodate small laboratories, weighing in the order of 7,000 kg apiece, and probably derived from the Soyuz spacecraft but with the orbital and descent modules replaced by a pressurised research compartment. Plans shifted yet further in the spring of 1979, when a government resolution called for the project to be merged with Vladimir Chelomei's cancelled Almaz military project and Mir's ports were reinforced to handle larger modules based on the TKS ('Heavy Cosmos') in size and shape, each weighing around 20,000 kg. Mir would benefit from digital flight computers and gyrodyne flywheels for attitude control, an automated rendezvous system, a dedicated satellite communications network and improved oxygen generators and carbon dioxide scrubbers. Unfortunately for Mir, financial resources during this period were very much directed into the Buran programme and it was not until the spring of 1984 that Valentin Glushko was ordered to orbit the new station by February-March 1986, with the launch intended to coincide with the 27th Communist Party Congress. (It is probably not entirely coincidental that Glushko received his order only weeks after President Ronald Reagan announced plans for the United States to build its own space station.) It would be difficult to get Mir ready for launch in such a short span; in fact, original plans had not envisaged the core module being placed into orbit before late 1986. Static and dynamic testing of the core was completed in December

1984, but the remainder of the processing was far from complete. There were other headaches, too. The weight of electrical cabling pushed the core module a couple of thousand kilograms outside the Proton booster's performance envelope. In response, a sizeable chunk of the experimental hardware had to be removed and remanifested onto Progress supply missions, but the weight overspill was still unacceptably high. At length, in January 1985, the planned 65-degree orbital inclination for Mir was relaxed to 51.6 degrees, like the Salyuts, although this would reduce the photographic coverage of the Soviet Union. Difficulties with the development of the new Strela ('Arrow') flight computers – whose development team included former cosmonaut Nikolai Rukavishnikov – forced engineers to install older-specification units, as a short-term measure, in order to meet the politically mandated launch target. In April 1985 it was decided to ship the core module to Tyuratam to complete systems testing and integration. When it arrived at the desolate launch site in early May, more than *eleven hundred* wires – almost half of the total – required substantial rework, but by October it had completed its final clean room inspections and the focus changed to communications checks (using the Cosmos 1700 satellite in geostationary orbit).

The first launch attempt for the core module came on 16 February 1986, less than three weeks after America had reeled from the loss of Challenger. It was scrubbed when Mir failed a communications test, but the launch occurred successfully at 12:28 am Moscow Time on the 20th, with the Proton booster inserting the 20,400 kg infant station perfectly into orbit, inclined 51.6 degrees to the equator. By 6 March, the very day on which the 27th Communist Party Congress ended, Mir was established in its operational orbit. Physically, the core measured 13.1 m long and 4.2 m in diameter and, with its twin solar arrays fully deployed, had a wingspan of 20.7 m. These arrays each covered 76 m^2 and utilised high-performance gallium arsenide cells, with an initial electrical power yield of around nine kilowatts. (A third, 'dorsal' solar panel, presumably omitted at launch, due to weight considerations, would finally be installed during a spacewalk in June 1987.) Two liquid-fuelled main engines, with an individual thrust of 300 kg, were situated at the rear of the core module, but could not be used after the spring of 1987, due to the arrival and permanent docking of the first scientific module, the astrophysics laboratory Kvant ('Quantum'). From 'front' to 'back', Mir's most visible feature was the spherical transfer compartment – the multiple adaptor – which measured 2.2 m in diameter and 2.8 m long. "It serves as a lobby for spacecraft docking," noted a 1987 US Senate report on Soviet space activities, "and houses the airlock for egress into space." It was intended that arriving modules would firstly dock at the 'forward' port, after which an arm-like manipulator device would robotically separate them and manoeuvre them around to one of the radial ports. A pair of 'sockets' for these devices were situated on the multiple adaptor. Moving out of the multiple adaptor, the cosmonaut entered Mir's cylindrical working and living compartment, 7.7 m in length and comprising two sections, which measured 2.9 m and 4.2 m in diameter, connected by a short conical frustrum. At the rear was a service propulsion area and transfer adaptor, the same diameter as the main compartment and measuring 2.3 m long, which housed a pair of 300 kg main engines, communications and rendezvous

Pictured from the departing Soyuz TM-2 spacecraft, this image of Mir reveals its unique multiple docking adaptor at the left, the two light-coloured cylindrical segments of the base block itself, the large dark mass of the Kvant astrophysics module – added in April 1987 – and, at the far right, the Soyuz TM-3 vehicle.

equipment, and four propellant tanks for the 32 thrusters of the attitude control system.

Inside the core, Mir differed greatly from the earlier Salyuts. With the exception of the station's control consoles, it was largely devoid of experimental apparatus. It had instead been designed as a 'habitat' or living area, which the cosmonauts came to affectionately nickname 'the base block', and could comfortably accommodate a crew of two. Each man was provided with his own private cabin, equipped with a window, a sleeping bag and a desk. It also housed a toilet, an entertainment system for movies and music, exercise equipment, medical supplies, and a galley and dinner table with food warmers and movable 'chairs'. "The crews cook on a hot plate," noted the US Senate report, "and are allowed to select their own food, as long as they consume the required number of calories." *Spatial awareness* was provided by a dark green 'carpet' on the 'floor', light green 'walls' and a white 'ceiling' on which were installed fluorescent lamps, and the installation of equipment was deliberately arranged to mimic this Earth-like orientation. Communications were to be provided via a network of command and control satellites, known as Luch ('Ray'), which could trace their genesis to the very same February 1976 decree which had brought Mir into existence, although at that time there was only one satellite in operation. Similar in function to NASA's Tracking and Data Relay Satellites, the Luch network provided a high-rate space-to-ground link for the station and its visiting Soyuz spacecraft, together with communications support for the Soviet Navy. In total, a trio of three-axis-stabilised Luch satellites were launched between October 1985 and December 1989, under the cover names of Cosmos 1700, Cosmos 1897 and Cosmos 2054. Although much of the hardware destined for Mir would be launched by Progress or aboard subsequent modules, the core did arrive in orbit with some of its research facilities in place: crystal growth apparatus, melting and solidification facilities, electrophoresis equipment, photometers and spectrometers, metallurgy hardware and cameras.

With Mir established in orbit, the Soviets now had *two* operating space stations at their disposal. Yet although Salyut 7 (vacated by Vladimir Vasyutin, Viktor Savinykh and Alexander Volkov the previous November) could still receive visitors, many Western observers doubted that it would be used again. "Almost certainly," wrote Phillip Clark, "the Soviets had not planned to re-man Salyut after the scheduled nine-month mission." Significantly, Mir and the Salyut 7-Cosmos 1686 complex were in almost the same plane. The old complex's orbit was allowed to decay until early February 1986, when a small manoeuvre might have been executed – "on the borderline of the two-line orbital elements error limits", Clark acquiesced, and thus "very uncertain" – and when the two stations passed within a few kilometres of each other on 8 March there was renewed speculation that, perhaps, they might dock. "[Salyut 7-Cosmos 1686] is expected to join up with Mir at some point," *Flight International* emphatically told its readers on 29 March, "possibly after Salyut has been closed down." Although the respective altitudes of the two stations *were* suitable for such a rendezvous to occur, ultimately this did not transpire ... but what *actually* happened on the Soyuz T-15 mission prompted even greater surprise. Judging from *Pravda* reports, there was some thought in the West that, possibly, Soyuz T-15 might fly to both stations, although this remained far from certain.

With the landing of Vasyutin, Savinykh and Volkov, the Soviets possessed only one more Soyuz-T spacecraft (which included the refurbished descent module from Vladimir Titov and Gennadi Strekalov's aborted Soyuz T-10A mission) and planned to commence manned operations with a 'modified' vehicle – the 'Soyuz-TM', whose systems were designed specifically for work with Mir – in the spring of 1987. It would appear that the very lack of available Soyuz-Ts eliminated a whole plethora of plans for follow-on visits to Salyut 7; in fact, Mark Wade, on his website, www.astronautix.com, has remarked on at least three of these. The first, which he called 'Soyuz T-15A', might have been intended to fly in early 1986, crewed by cosmonauts Boris Volynov, Musa Manarov and Buran pilot Anatoli Levchenko. The second, 'Soyuz T-15B', with Alexander Viktorenko, Alexander Alexandrov and Yevgeni Salei, might have continued the Cosmos 1686 work, left undone by Vasyutin's

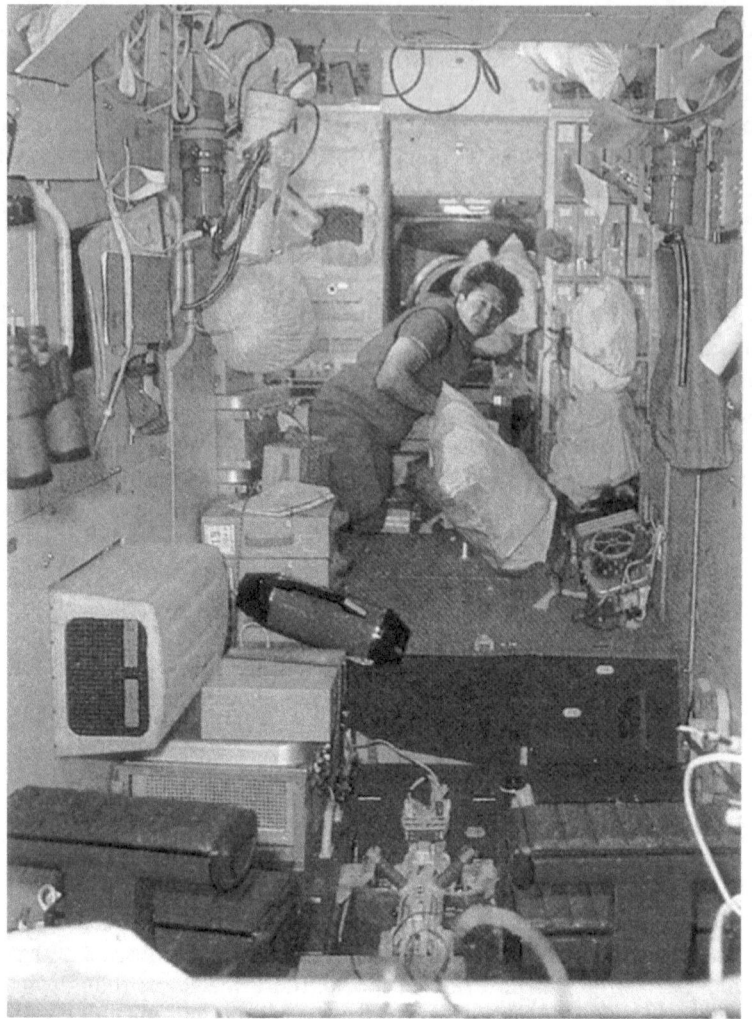

Leonid Kizim, the first commander of Mir, glances up from his work to smile for Vladimir Solovyov's camera. Behind Kizim is the hatchway leading to the station's aft port, whilst in the foreground are the seats for the main control console.

illness. 'Soyuz T-15C' was originally to have coincided with International Women's Day in March 1986, but was postponed to at least September. This might have been the all-female mission, with Svetlana Savitskaya commanding Yekaterina Ivanova and Yelena Dobrokvashina. Officially, it was cancelled due to the birth of Savitskaya's son, Konstantin, in October 1986, although rumour has abounded for years that strong opposition existed in the higher echelons of the Soviet space hierarchy against the idea of an all-female flight. (Ironically, the backup crew for the mission – Alexander Viktorenko, Alexander Alexandrov and Anatoli Solovyov – was *all-male*, further underlining the hypocritical reality that the all-female flight had little practical merit, other than propaganda.) As a consequence, with Mir in orbit, earlier than planned, with barely a fraction of its scientific equipment aboard and only one Soyuz-T available, the decision was taken to fly Leonid Kizim and Vladimir Solovyov on a unique mission to begin work aboard the new station *and* complete the work left undone aboard Salyut 7-Cosmos 1686. All this could be accomplished, wrote the Swedish radio analyst Sven Grahn on his website, by injecting the new station into an orbital plane very close to that of Salyut "and then using the propulsive capacity of Progress freight ships to manoeuvre Mir to drop off and pick up the Soyuz T-15 crew". A pair of Progress vehicles, the 25th and 26th in the series, were launched in March and April 1986 to accomplish these manoeuvres.

The choice of Kizim and Solovyov, both of whom had trained extensively on Salyut systems, may seem odd at first glance. Clark revealed that neither man had specialised with the systems of the new station, but the absence of much of the experimental hardware suggested that the Soviets considered it poor economy to commit a fully-trained Mir crew to a 'barebones' maiden mission. Instead, Kizim and Solovyov, as a highly experienced cosmonaut duo, with long-duration expertise, were ideally suited to activate the station and check its primary systems.

However, there was a problem. Mir's front port could only be used in conjunction with new Kurs ('Course') rendezvous system. The upgraded Soyuz-TM spacecraft had this, but Soyuz-T carried the older Igla instrumentation. Mir's aft port, on the other hand, *was* capable of accepting an Igla-guided approach, but it was preferred to keep this free for Progress freighters. This meant that after Soyuz T-15 had used its Igla to complete the rendezvous, drawing to a halt 200 m from the rear port, Kizim would take control and fly around to the front in order to dock there. The spacecraft was duly launched from Tyuratam at 3:33 pm Moscow Time on 13 March 1986 and, two days later, the Igla completed the rendezvous and was then shut down. Kizim took command, performed the flyaround to the front and docked on the first attempt. He and Solovyov had the experience of their own Soyuz T-10B rendezvous with Salyut 7 as a point of reference, and to undertake the fly around they had available many of the tools used by Dzhanibekov and Savinykh, including the laser range finder and the low-light image intensifier. Yet it was a triumph which enabled the Soviets to start 1986 in spectacular style, only weeks after America had so publicly lost Challenger. (In fact, in an unprecedented move by the Soviets, the Soyuz T-15 launch had been broadcast live, perhaps for the propaganda value it afforded.) The cosmonauts showed television viewers the new station's expansive interior, free of clutter and experimental hardware, and for the next 51 days – until the beginning of May – worked to bring its systems to life and unpack its first pair of Progress visitors.

With many in the West wondering if Mir would dock onto Salyut 7, very little information emerged from the Soviets about what was planned. Then, on 30 April, Deputy Flight Director Viktor Blagov first reported that Mir was "tuned and playing like a grand piano" and then announced that the cosmonauts *were* preparing to transfer to Salyut 7. Phillip Clark agreed that until this time there was absolutely "no hint of any plans for Salyut 7" and related that on 3 May the Soviets announced that the "first stage" of Mir's mission was "nearing completion" and Kizim and Solovyov would shortly "fly over" to the old station. According to Clark, Mir orbited at 309–345 km, whilst Salyut was at 358–360 km, and "in order to counter the distance *around* the orbits ... some manoeuvres by Soyuz-T had to be completed". The simplicity of this description belied a complex ballet of orbital mechanics. In his analysis, Sven Grahn explained that the separation of the two stations on the afternoon of 4 May was fixed at approximately 4,100 km, with their orbital planes a mere 0.07 degrees apart. "Mir lowered its orbit to start approaching Salyut 7 from behind," wrote Grahn, in a manoeuvre which probably occurred before 7:10 pm Moscow Time. This caused the orbits of the two stations to be almost exactly co-planar when the time came for Kizim to dock at Salyut 7. When Soyuz T-15 separated from Mir on the late afternoon of the 5th, the distance between the two stations had been reduced to around 2,000 km. It would appear that Mir passed within about 25 km of Salyut, soon after midnight, Moscow Time, on the 6th, some five hours before Kizim and Solovyov brought their craft to a smooth docking. "During the approach to Salyut 7," continued Grahn, "Soyuz T-15 moved away from Mir by only 125 km. In this way, the Soyuz did not have to make such extensive manoeuvres." In effect, Mir acted as a 'delivery truck' to get them to Salyut 7 and as a 'pickup truck' to get them back. Years after the event, it is quite remarkable, when one considers the constraints imposed by propellant expenditure and orbital mechanics and lighting opportunities, that the transfer was accomplished. However, even taking into account the two-day, propellant-saving approach profile adopted to reach Mir, the craft lacked sufficient reserves to rendezvous with Salyut 7 *and* execute a re-entry burn at the end of its mission; as a result, the engines of Progress 26 would be used to adjust Mir's orbit. Before setting off, Kizim and Solovyov loaded around 500 kg of equipment into their craft's orbital module for the 29-hour journey to Salyut 7 and successfully docked at the aft port on the late evening of the 6th.

Among their tasks for the next seven weeks was the completion of experiments left undone by the Soyuz T-14 crew. One major objective, for which Vasyutin and Volkov had originally trained, was a pair of lengthy EVAs. On 28 May 1986, Kizim and Solovyov opened the station's hatch and ventured outside for a four-hour spacewalk. The objectives were twofold: firstly, to retrieve a series of 'cassettes', which had exposed various samples, including cables, nuts and bolts, metals and biopolymers, together with a joint Soviet-French micrometeoroid collector, and secondly, to install and test a deployable truss structure. They began by setting up a work platform, close to Salyut's hatch, onto which they installed a large 150 kg cylindrical 'drum', from which the truss would be deployed. Next, they deployed the 20 kg tubular metal truss, which measured 12 m long when fully unfurled, connecting it together with hinges and springs. Built by the Paton Institute in Kiev – the very same facility which had created the URI welding tool used by Dzhanibekov and Savitskaya in July 1984 – it would appear to be a Soviet equivalent of the EASE-ACCESS tasks performed by Jerry Ross and Woody Spring a few months earlier. "Kizim operated the three buttons that controlled deployment," explained David Portree and Robert Treviño in *Walking to Olympus*, "then climbed halfway up the truss. He found it sturdy, with oscillations limited to a few centimetres of amplitude." At the top of the truss was the Fon ('Background') device, an instrument built by Leningrad Polytechnical Institute to assess the environment around Salyut 7. After evaluating the performance of the structure, Kizim and Solovyov refolded the truss and returned inside the station. Their work was not over, though, and on 31 May they were back outside on the record-breaking *eighth* EVA of their career. On this occasion, they spent four hours and 40 minutes erecting the truss a second time, then used a visible light communications system, called 'BOSS', to relay stability data from the instruments at its apex to the station and from there to ground controllers. The instruments included a tiny seismograph for tracking low-frequency vibrations imparted by Salyut 7's acceleration and monitored a small orange beacon to measure high-frequency movements. Next, Kizim and Solovyov rigidised the truss by welding portions with the URI tool, then dismantled it for a second time. Before returning inside the station, they installed a device to study the response of aluminium-magnesium alloys to structural loads and collected a sample of solar cell material to enable its degradation after so long in space to be studied.

Less than four weeks later, on the evening of 25 June, having loaded a number of instruments – including the KATE-140 topographical camera, a French atmospheric and astrophysics device, an ultrasound cardiograph, an electrophoresis machine and a heat-transfer experiment, weighing a total of around 400 kg – into Soyuz T-15, the cosmonauts undocked from Salyut 7 for the last time. (According to *Pravda*, the equipment transferred from Salyut 7 to Mir would be utilised "until scientific modules are sent up to the Mir station".) Kizim and Solovyov redocked with Mir shortly before midnight, Moscow Time, on the 26th. The Soyuz-T had barely 420 kg of propellant in its tanks, of which 250 kg was reserved for the de-orbit burn, and the rendezvous profile was specifically selected to be as economical as possible. This had actually begun more than a week earlier, with Mir about 2,000 km 'ahead' of Salyut 7. "The plane difference was 0.04 degrees," wrote Grahn. "Mir then lowered its orbit and Salyut 7 drifted to a distance of about 3,000 km behind Mir. To *make* Salyut 7 approach it, Mir raised its orbit on 25 June ... and Soyuz T-15 was then detached to move slightly ahead of Salyut 7 to complete the redocking to Mir. The raising of Mir's orbit also caused the orbital planes to become co-planar at the moment of redocking." At a distance of 100 m or so from Mir, Kizim once again switched off the Igla device, flew around to the front and docked; by now, the Progress 26 freighter had departed the aft port and been de-orbited. A few days later, on 2 July, Kizim surpassed the 361-day cumulative record for time in space, set by Valeri Ryumin in October 1980 ... and, on the 6th, he became the first human to have spent a full year in orbit, spread across three missions.

The Soyuz T-15 mission ended as it had begun: with live television coverage; yet another example, perhaps, of the impact of Mikhail Gorbachev's burgeoning *perestroika* effort as it smoothed a rocky road to *glasnost*. Just as Soviet television had broadcast 'live' coverage of Kizim and Solovyov's launch and parts of both of their EVAs at Salyut 7, it covered the landing in Kazakhstan at 3:34 pm Moscow Time on 16 July. This 125-day mission remains unique to this day and in the months that followed there was much talk in the West that the Salyut 7-Cosmos 1686 complex might be revisited – indeed, in August the old station was boosted into a high 'storage' orbit of 475 km – or even retrieved and returned to Earth aboard the

Soviet Shuttle, Buran. None of this came to pass, but until December 1989, when Cosmos 1686's orientation system finally failed, a glimmer of hope still existed that another mission might someday be despatched. In fact, the loss of orientation control was the beginning of the end. There was no longer any chance of a Soyuz T-13-style rescue. Speaking at around this time, Vladimir Dzhanibekov ruled out any possibility that another salvation mission *could* be attempted. As the Soviet Union entered its final months, with many of its former republics having already declared their independence and tax revenues to the central government in Moscow having long since dried up, the wherewithal to finance just *one* space station was intensely difficult. It was only Japanese, British, Austrian and German fee-paying visits to Mir that kept the programme alive. As the 1980s gave way to the 1990s and cosmonauts struggled aboard a space station which was plagued with financial and technical woes, hard-line Communists brought tanks onto the streets of Moscow in a failed attempt to depose Gorbachev and reverse his reforms.

If financing and running just *one* space station in this period of acute political and economic strife was difficult, then financing *two* stations was unthinkable. Hopes that the high storage orbit might keep Salyut 7-Cosmos 1686 aloft for ten years were dashed by higher than expected solar activity and, in early 1991, the complex was dangerously close to an uncontrolled re-entry. On 7 February, it plunged into the atmosphere in a blazing fireshow. Although it overshot its intended point of entry, which would have scattered the debris over an uninhabited stretch of the South Pacific, and instead came down close to the town of Capitán Bermúdez in eastern Argentina, there were no fatalities on the ground. The demise of the old station and its faithful companion marked the end of an era. Salyut 7 had secured its own share of the records – two ultra-long-duration missions, together with propaganda victories secured through the missions of Chrétien and Savitskaya and Sharma – but the 1990s would be the decade in which Mir loomed large, not just in terms of size and achievements and technological complexity, but in the imaginations and future dreams of East and West. In fact, Mir would truly take centre stage in the fortunes of the Russian *and* American human space programmes in the 1990s. It would be a difficult decade for post-Soviet Russia, and by the end of that decade the glory days would be over and Mir would be a shadow of what it once was, but it would have carved for itself an irreplaceable niche in the history of human space exploration.

"RATHER PAINFUL"

Since the beginning of this history, an unspoken fear has underpinned the relationship between the prime and backup members of a space crew; the former fears that it may lose its place on a mission, through illness, accident, death or some other unanticipated eventuality, whilst the latter must train intensively for a voyage which it will almost certainly never fly. There is, however, a second emotion; that of *euphoria*, felt by a backup team when they are suddenly catapulted onto the prime crew ... only *days* before launch. Had it not been for a quirk of fate, Colonel Yuri Viktorovich Romanenko could hardly have foreseen at the dawn of 1987 that by year's end he would be the world's unchallenged champion for the longest single space mission. Yet that is precisely what happened. A few days after Christmas 1987, he would touch down on Soviet soil, having spent almost *eleven months*, or 326 days, in orbit. Romanenko (whose son, Roman, would also grow up to become a cosmonaut) was a veteran parachutist with 39 jumps and a Soviet Air Force pilot with a wealth of experience in the Yak-18 and L-29 military trainers and the MiG-15, MiG-17 and MiG-21 jets ... even *before* he joined the hallowed ranks of the cosmonaut corps. Born on 1 August 1944 in the village of Koltubanovsky, more than a thousand kilometres south-east of Moscow, he was barely 33 years old and already a decorated lieutenant-colonel at the time of his first launch into space, paired with Georgi Grechko, in December 1977. A cursory glance at his family hints at a proud tradition in the armed forces: Romanenko's father had been a senior naval commander and his mother a combat medic. As a child, he enjoyed building model aircraft and ships and developed a love of ships, boxing, shooting and underwater fishing. On his second mission in September 1980, he flew with a Cuban cosmonaut named Arnaldo Tamayo-Méndez and several years later was invited to Cuba at the behest of Fidel Castro ... who not only organised a fishing tour, but participated in the event, freediving with Romanenko to a depth of several metres.

After completing his high school education in Kaliningrad in 1961, Romanenko enrolled in the Chernigov High Air Force School in Ukraine, graduating with honours. He rose to the rank of lieutenant and subsequently served as an instructor. He was selected as one of nine cosmonaut pilots in April 1970, from a pool of more than 400 finalists, passing not only the rigorous selection procedure, but also gaining clearance from the KGB and the Communist Party for admission into the country's spaceflying elite. Progression through the ranks of the Soviet Air Force was swift: Romanenko was promoted to captain in February 1971, major in February 1974, lieutenant colonel in December 1976 and finally, in March 1978, when he returned from his record-breaking mission with Grechko, he was decorated as a full colonel. When the Salyut 6 space station was launched into orbit, Romanenko watched from the ground as backup to Soyuz 25 commander Vladimir Kovalyonok, knowing that his turn would come soon. His expectation that he would fly a long-duration mission in 1978 with engineer Alexander Ivanchenkov changed rapidly when Soyuz 25 failed and he and Grechko found themselves propelled as prime crew onto the very next flight. With a radio callsign honouring one of the least known and most unexplored regions of Russia – Taymyr, whose mountainous and lake-studded peninsula is the most northerly point of the Eurasian landmass – Romanenko and Grechko explored new ground, as the Soviets strove to turn the strangest environment yet encountered by human beings into their long-term home.

Almost a decade later, as 1986 wore into 1987, Romanenko was serving as backup to Vladimir Titov for the first long-duration expedition to Mir. Titov, the reader will recall, had already established an unfortunate reputation for himself as one of the unluckiest cosmonauts ever to fly into space. His first mission, in April 1983, had failed to dock with Salyut 7, whilst his *second*, less than six months later, only left the ground with the aid of an escape tower which pulled Titov and crewmate Gennadi Strekalov to safety when their rocket exploded on the launch pad. By the end of

1986, he was back in training with Alexander Serebrov for an 11-month flight to Mir, during which they would fly the first manned version of the new Soyuz-TM spacecraft *and* host a visiting expedition with a Syrian guest cosmonaut. The soothsayers might have foreseen that ill-fortune would befall Titov yet again, but news of the crew exchange did not reach Western ears until the end of 1987. Only then did a press release from the Soviet news agency, Novosti, explain, almost in passing, that "Vladimir Titov was in training for his third mission, one year ago, and ... Serebrov was to be his partner ... " It became clear that the stroke of cruel luck had actually fallen on Serebrov, who apparently failed a medical check, a few weeks before launch, and the backup crew of Romanenko and civilian flight engineer Alexander Laveykin took their place.

Not only would their expedition mark the first dedicated mission to Mir *and* the first attempt to undertake a record long-duration stay aboard the new station, but Romanenko and Laveykin were also destined to become the first crew to fly the fourth-generation ferry: 'Soyuz-TM' ('Transport Modified'). As with Soyuz-T which preceded it, the new craft flew its maiden voyage in an unmanned capacity; in May 1986, the unnumbered 'Soyuz-TM' put the modifications through their paces, testing the new Kurs rendezvous system and perfectly evaluating a capability to fly around and dock without Mir having to reorient itself to face the incoming vehicle. Typically, Soyuz-TM would employ Kurs to make radar contact with the station from a distance of some 200 km, rather than the standard 30 km achievable with Igla. Soyuz-TM would then achieve a final radar 'lock' onto Mir at a distance of 20-30 km. The Kurs antenna was omni-directional, which offered additional flexibility, and the final approach could be completed by the spacecraft's on-board computer. Development of Soyuz-TM got underway with the signing of draft plans in April 1981. By the spring of the following year, most of the craft's engineering blueprints were complete, although the early launch of Mir and the unavailability of Soyuz-TM for its first manned flight until early 1987 prompted a gap of several months between the departure of Kizim and Solovyov and the arrival of Romanenko and Laveykin.

Physically, Soyuz-TM closely resembled its predecessor, with an independent lifetime in orbit of 14 days and an 'orbital storage' duration of about six months, but its systems were much improved. So too was its launch vehicle, which was equipped with lighter escape motor to enable an increased payload capacity in the region of 200–250 km into a 51.6-degree orbit. Rex Hall and Dave Shayler have remarked that, had Mir flown, as intended, into a 65-degree orbit, Soyuz-TM might have been unable to transport as many as three cosmonauts to the higher inclination. The changes made in 1984-85 effectively made Soyuz-TM "more flexible", Hall and Shayler wrote, "in executing its assigned mission, should the real-time situation require a deviation from the flight plan". Soyuz-TM's orbital module included an additional window and rendezvous controls, whilst the landing system of the descent module featured improved parachutes with new synthetic fibres woven into shroud lines and lighter canopies. The first crews to support Soyuz-TM missions to Mir were assembled in September 1985, with Titov and Serebrov scheduled to fly first, followed by Romanenko and the first ethnic Azerbaijani cosmonaut, Musa Manarov; however, the latter was replaced by Laveykin, due to a medical problem.

Six months later, a third crew of Alexander Volkov and Sergei Yemelyanov was assigned, although hopes of a first manned launch to Mir in late 1986 became untenable when it became clear that the first additional module – Kvant – would not be available until the late spring of 1987. Since Romanenko and Laveykin had trained in parallel with the prime crew, Serebrov's medical disqualification made it fairly straightforward for the backups to be substituted. With the removal of Serebrov, his place on Titov's crew was taken in March 1987 by the now-recovered Manarov. Both the Romanenko-Laveykin and Titov-Manarov teams would undertake missions in excess of 300 days, with great expectation in the West that a year-long expedition would be attempted. In the meantime, Yemelyanov was dropped from flight status in mid-1987; Hall and Shayler ascribed his removal to undisclosed "medical reasons", a supposition possibly supported by Yemelyanov's untimely death in December 1992 after a heart attack. *His* place alongside Volkov was initially taken by civilian engineer Alexander Kaleri, and ultimately – from March 1988 onwards – by Sergei Krikalev, a young rookie who would go on to carve out a particularly impressive six-flight career as a cosmonaut and today holds the world record for the most time spent cumulatively in space. The ability of Soyuz-TM to transport three-man crews to Mir in its 51.6-degree orbit also enabled the Soviets to manifest additional, short-duration crewmembers onto several missions: Buran pilot Anatoli Levchenko was one contender, together with a pair of Intercosmos pilots, one from Syria and a second from Bulgaria.

When a new Progress freighter was despatched towards Mir on 16 January 1987, there was much excitement in the West that the launch of a new long-duration crew was imminent. Surprisingly, Soyuz TM-2 did not fly until 12:38 am Moscow Time on 6 February, primarily due to heavy snowfall which had affected large areas of the Soviet Union and Eastern Europe in the opening weeks of the year. The names of Romanenko and Laveykin had been announced ten days before launch and the two cosmonauts completed a flawless two-day approach to Mir, performing a Kurs-assisted docking at the forward port, early on 8 February, as the station flew over China. After boarding the station and turning on the lights, they set to work 'demothballing' Mir, switching its temperature controls to 'Operational' mode, activating the water regeneration system, starting the automated experiments and beginning medical checks on themselves almost immediately. In what was perhaps a portent of what would come, Laveykin experienced – according to Moscow World Service – a "rather painful" adaptation to the weightless environment. (Certainly, after a rest day on 14 February, it was noted that he needed the free time badly to recharge his energy.) Alexander Ivanovich Laveykin was born in Moscow on 21 April 1951 and graduated from the Bauman Higher Technical School, before working on the engineering design of rockets. He was selected as one of ten cosmonaut candidates in December 1978 and completed his basic training without incident in October 1980. Today, Laveykin has earned a distinction as unenviable as that of Vladimir Vasyutin, having being obliged to terminate his mission, earlier than intended, due to an unforeseen medical issue. All that is known with certainty is that Laveykin suffered from "minor heart irregularities" during the course of his flight and that these were of sufficient severity to warrant a return to Earth, six months into the planned 11-month mission. Medical ethics, of course, make it doubtful that the full story will ever be divulged, but what is perhaps most remarkable is how the Soviets resiliently and flexibly dealt with the consequences: rather than bringing *both* men back to Earth, it was decided that Romanenko would *remain* in space to secure the long-duration record, whilst *another* cosmonaut, Alexander Alexandrov, already assigned to the Soyuz TM-3 visiting flight, would be hurriedly trained to take Laveykin's place for the remainder of the mission.

This eventuality was far from certain in the busy opening weeks of the flight, as Romanenko and Laveykin unpacked the Progress craft, which was undocked on 23 February and commanded to burn up in the atmosphere. Another Progress arrived in March and the cosmonauts seemed to be proceeding normally with their work, mildly grumbling that they were "impatiently" awaiting the launch of Mir's first additional module, Kvant, but conducting their own experiments in remote sensing and growing tulips aboard a special 'greenhouse'. Other Progress-delivered cargoes included geophysical cameras, an aerosol and hydrosol behaviour study and a semi-industrial crystal growth facility. When Kvant finally rose from Tyuratam, atop a Proton booster, at 3:06 am Moscow Time on 31 March, it was already known in the West that the module was dedicated to astrophysics research. Three months earlier, in December 1986, Flight Director and former cosmonaut Valeri Ryumin had revealed its purpose and announced that it would be launched before the scheduled July 1987 flight of the Syrian cosmonaut. What remained unclear was exactly *where* Kvant would be situated; the Soviets had spoken at length about their planned use of a manipulator system to swing the modules from the front port onto the radial ports, and this implied that the new arrival would reside at the front of Mir. In fact, Kvant would take up residency at the station's *rear* port ... for a very specific reason. Kvant's history is long and complex and it was originally designed for addition to Salyut 7, but its launch was repeatedly postponed due to technical difficulties and eventually assigned to Mir. However, this was much easier said than done. Salyut 7 was equipped with the Igla rendezvous device; so too was Kvant, which made it incompatible with the Kurs-based docking ports at the front of Mir. Rather than completely redesign Kvant to accept Kurs hardware, it was considered simpler and cheaper to install Igla on Mir's aft port, thereby enabling the new module to dock with the station. The *rear* port of Kvant, meanwhile, could then be outfitted with the Kurs hardware, which would allow future, Kurs-based visiting craft – Progress or Soyuz-TM – to dock.

Kvant's development began in September 1979, when orders were issued to start the development of a series of cylindrical pressurised modules, each 4.2 m in diameter, equipped with a single docking port and no integral propulsion system. Up to eight of these modules were planned: one for Salyut 7, four for Mir and a further three to be carried aboard Buran, either to operate in the craft's payload bay or to be added to a future space station. The module which later became Kvant was to be fitted with the 'Salyut-5B' digital computer and a gyrodyne flywheel orientation system, but when it was reassigned to fly to Mir it was realised that it was 800 kg too heavy for its Proton to boost it into a 65-degree orbit. The adjustment of Mir's orbit to 51.6 degrees alleviated some of this pressure, but as one problem seemed to vanish, another appeared. If Kvant, with its Igla-equipped forward port, was to dock at the rear of Mir, then the module would require to have a second docking port at the rear and be fitted with plumbing to enable Progress propellant deliveries to be routed into the station's tanks … and *this* posed another weight penalty. The absence of a propulsion system also meant that a disposable Functional Service Module (FSM) 'tug', based on the earlier TKS craft, would be needed to complete orbital adjustments and support the rendezvous with Mir. As a consequence, when Kvant headed for space in the small hours of 31 March 1987, at some 22,800 kg, it represented the heaviest payload yet launched atop a Proton.

Other issues of a technical and scientific nature had conspired to delay Kvant's launch, originally scheduled for late 1986. Difficulties with the completion and testing of its astrophysical instruments had forced engineers to tend to it for 14 months – *three times* longer than planned – in the integration hall before it could finally be transferred to Tyuratam for final checkout in November 1986. Software problems had also caused the Salyut-5B to be removed and the Soviets planned to launch the computer to the station aboard a subsequent module. In terms of physical appearance, Kvant measured 5.8 m long and 4.2 m wide and of its 18,500 kg weight, almost a quarter was devoted to its scientific instrumentation. It comprised a pair of pressurised compartments (one outfitted as a laboratory and the other as a living area, separated by a partition), girdled by an unpressurised segment at the rear, through which passed a transfer tunnel, and carried instruments to survey celestial sources at ultraviolet, X-ray and gamma ray wavelengths. The four primary astrophysical payloads were known collectively as 'the Roentgen Observatory', named in honour of the German physicist Wilhelm Conrad Röntgen, the discoverer of X-rays, and their international flavour was self-evident. They included a wide-angle camera and spectrometer, jointly constructed by scientists in Britain and the Netherlands, a high-energy X-ray experiment from West Germany, the European Space Agency's Sirene-2 gas scintillation proportional counter and the Soviet Union's Pulsar X-1 instrument. Also aboard Kvant were ultraviolet and magnetic spectrometers, the Glazar gamma ray instrument and the automated Svetlana electrophoresis machine. The nature of the astrophysical instruments, and the need for precise pointing accuracy, was aided by Kvant's six gyrodyne momentum flywheels, which could orient the entire Mir complex without propellant expenditure. The module also carried life-support apparatus, including Elektron (electrolysis) and Vika oxygen generators and Vozdukh ('Air') carbon dioxide scrubbers, as well as Mir's long-awaited third solar array for addition to the base block.

Docking with the station was expected on 5 April 1987 and only on the day of Kvant's launch did the Soviets reveal that it was directed towards the aft port of Mir and *not* one of the front ports. All seemed to be proceeding normally until the morning of docking, when Moscow Television Service reported that Romanenko and Laveykin were watching the approach profile, when – at a distance of just 200 m – "the rendezvous was suspended because of faults in the control system of Kvant". Phillip Clark explained that the thrusters aboard Kvant's tug failed to slow the module sufficiently and it flew past Mir. According to the US Senate's report on the incident, observations by the Kettering radio group pointed to a mean distance of …

just *ten metres*! At a press conference on the 6th, the Soviets explained with remarkable *glasnost* what had occurred: with senior official Alexander Dunayev making reference to "an irregular situation" which arose as Kvant proceeded "strictly according to the calculated flight path". According to Dunayev, the Igla rendezvous device lost its lock on Mir at a distance of 200 m, whilst Valeri Ryumin could only report that controllers would decide "in a few days" whether to make another docking attempt. In the meantime, Mir and Kvant drifted apart to a separation distance of some 400 km. Three days later, on 9 April, the second attempt *was* made, with a measure of success ... though it was not enough. At 4:36 am Moscow Time, the two spacecraft 'soft docked', but could not rigidise their embrace into a 'hard dock'. According to Clark, Kvant's probe had penetrated a matter of 36 cm into Mir's docking port, before getting inexplicably stuck. The docking collars of the two craft were thus separated by a matter of *centimetres*. Romanenko and Laveykin must have gazed with mounting distress through the station's windows, trying to identify *what* was causing the obstruction. It was impossible to see clearly from their vantage point and the only option seemed to be an EVA – the first ever to be performed from Mir – to inspect the situation and attempt a resolution. In the meantime, attitude control manoeuvres were suspended, lest Kvant pivot in the drogue of the base block's aft port, bang the two docking collars and cause further damage.

Late on the evening of 11 April, the two cosmonauts ventured outside Mir's forward transfer compartment and worked their way along the entire length of the base block to reach the interface with Kvant. When they reached the site of the trouble, it became clear that an "extraneous white object" was fouling the docking mechanism; it turned out to be a twisted fragment of cloth, probably a piece of trash or possibly a bag which had somehow freed itself during the undocking of the most recent Progress craft in late March. (Soviet officials rejected Western suggestions that it had been left attached to Kvant on the ground, citing photographic evidence of the module at Tyuratam.) Alexander Laveykin, whose first career spacewalk had begun with some alarm when his suit registered a minor drop in pressure, succeeded in removing and discarding it into space and the two cosmonauts watched with joy as flight controllers commanded Kvant's probe to retract and the two craft successfully completed a full hard docking. Romanenko and Laveykin returned inside their newly expanded home after three and a half hours and what a Soviet commentator described as a "heroic" EVA. April the twelfth – Cosmonautics Day, the anniversary of Yuri Gagarin's pioneering voyage – had dawned and the Soviet Union's latest pair of space heroes had saved Kvant and enabled its major international science mission to commence. The FSM tug was separated from the rear of the astrophysics module (by undocking from the port at the rear of the module), but its additional manoeuvres had cut into a reduced propellant supply, rendering it unable to perform a de-orbit burn. Instead, it boosted itself into a 'parking' orbit, 40 km 'above' Mir, from which it gradually decayed and made an uncontrolled re-entry in late August 1988.

Meanwhile, on 13 April, Romanenko and Laveykin entered Kvant for the first time and began unloading equipment into the base block. Less than two weeks later, another Progress completed a Kurs-guided docking with the aft port of the new module, creating, for the first time, "a truly modular space station ... all docked together". Unlike Cosmos 1443 and 1686, which each had only a single docking port and had to be physically detached in order to allow further visiting vehicles to arrive at Salyut 7, it was evident that Kvant was intended to remain attached to Mir's base block for a lengthy spell. In an impressive demonstration of the new station's complexity, no fewer than *four* spacecraft – the Soyuz-TM, the base block, Kvant and Progress – were now an integrated whole. "In this way," Tass crowed, "a sophisticated research complex ... had been built through successive dockings for the first time in the world." On the 30th the cosmonauts flexed the 165 kg of gyrodynes for the first time and pumped the first load of propellant through Kvant's transfer pipework and into the space station's tanks. During May, the Elektron and Vozdukh systems were activated and checked out. The Soviets admitted that the station was experiencing electrical shortages – Kvant's gyrodynes consumed around 90 watts of electricity and Mir's own materials processing furnaces took their own power toll – and a pair of EVAs had long been planned to remove the third solar array from the astrophysics module and install it onto the base block. However, Romanenko and Laveykin's workload was so immense that these EVAs were postponed from the beginning of May until the middle of June.

Preparations for the array installation began on 28 May, when the cosmonauts installed another storage battery in the base block and connected it to the electrical system. The first EVA, on 12 June, lasted for a little less than two hours and was quite different to the earlier work aboard Salyut 7, in which 'extensions' had been added to an already-extant and fully-wired panel: *this* time, Romanenko and Laveykin would be creating an entirely new array *and* its supporting structures *and* the associated circuitry to enable it to produce current. They installed an extendable 'hinged lattice girder' truss to the 'top' of Mir, then attached folded solar cells to both sides. During their work, they even operated *without* foot restraints, relying solely on tethers, which Laveykin said offered them "more freedom to manoeuvre ... but we had to cling to the ship with one hand!" The physical rigours associated with operating for several hours in the vacuum of space, clad in a bulky pressurised suit, required both men to undergo pre-EVA medicals in the days preceding the work and certainly Laveykin's data was giving flight surgeons a heightened sense of alarm; with the next manned visit scheduled for July, plans must already have been advanced to bring him home by this stage. Still, he and Romanenko headed outside again on 16 June for more than three hours, during which they placed an extendable truss atop the one deployed a few days earlier, affixed solar cells to either side, then linked up electrical buses and deployed the new array to its full height of 10.6 m. Twenty-four square metres of additional solar-energy-collecting area was provided in four additional arrays to Mir, expanding its power capabilities to somewhere in the region of 11.4 kilowatts. Over the next few days, working inside the base block, the two men hooked up the circuitry for the new arrays and wired them into the station's power system. It was a remarkable accomplishment. *Pravda* was not exaggerating when it lauded their work as having helped "to increase the effectiveness of the scientific research work on the Mir complex, substantially".

During their expedition, Yuri Romanenko and Alexander Laveykin performed a series of EVAs to install additional solar array cells and thus expand Mir's electrical capability.

As June wore into July, the focus was upon the imminent arrival of the Soyuz TM-3 visiting crew, an *expeditsya poseshchenya* (EP), which had long been anticipated to include a Syrian cosmonaut. In fact, as early as December 1986, the Soviets had announced – quite remarkably and a clear break from precedent – the *exact* date *and* time of the launch: 22 July 1987 at 6:30 am Moscow Time. The choice of a Syrian pilot is an interesting one. At the time, Syria, perched at the eastern end of the Mediterranean, was midway through the 30-year dictatorship of Hefez al-Assad, whose Ba'ath Party shared certain socialist traits with the Soviet Union and was ideologically opposed the 'imperialism' of the West. In a sense, therefore, it is hardly surprising that the Soviets should have approached yet another authoritarian regime with the offer to send one of their countrymen into space, but al-Assad was hardly a staunch Communist. In fact, since its formation by a pair of Syrian thinkers, more than five decades ago, the Ba'ath Party has pursued both Arab nationalist *and* socialist values, calling for a 'renaissance' and 'unification' of the entire Arab world into a single political state. When Hefez al-Assad, a Syrian army general and former defence secretary, seized power after a string of military coups in 1970, he brought dramatic change to his nation: guaranteeing women's equal rights, industrialising Syria and opening it up to foreign markets, investing in infrastructure, improving education, medicine and urban building, expanding the economy through oil exploration and increasing the spread of literacy among his people. At the same time, however, he increased the repression of the people through a tangled web of police informers and agents and although his accession was initially popular, having brought a measure of stability to Syria, the secular, anti-Western stance of al-Assad's government prompted him to crack down harshly on extreme religious groups and in the early 1980s he crushed opposition from the Muslim Brotherhood in a series of devastating artillery attacks on the city of Hama. To this day, the attacks represent one of the largest acts of violence against a leader's own people ever seen in the Arab world and, indeed, in 2011, the repressive regime of his son and successor, Bashar al-Assad, triggered the greatest popular uprising ever seen in this closeted, authoritarian state.

Muhammed Ahmed Faris, who became the second Arab spacefarer (after Sultan al-Saud, who flew a Shuttle mission in 1985) and the first Syrian cosmonaut, came from Aleppo – one of the oldest continuously-inhabited locations on the planet, with human occupation traced back to at least the sixth millennium BC – where he was born on 26 May 1951. As with so many of the Intercosmos pilots, little but the basic facts of his education and experience were revealed: he graduated from Military Pilot School in Aleppo in 1973 and rose to become a colonel and a pilot in the Syrian Air Force, specialising in navigation techniques. Faris and his backup, fellow air force pilot Munir Habib, were selected for cosmonaut training in September 1985, with original plans to fly to Salyut 7. In yet another example of openness on the part of the Soviets, the *entire* experiment load for the mission was revealed in December of the following year: the Syrian cosmonaut would conduct remote sensing studies as part of the al-Furat ('Euphrates') project, using Mir's KATE-140 topographical camera, grow crystals in the Kristallizator ('Crystalliser') furnace as part of the Palmyra experiment, observe Earth's ionosphere and purify samples of interferon and an influenza vaccine. Of these, al-Furat is worthy of note, since it included the observation of artificial reservoirs in Syria, and their drainage areas, together with the identification of new water resources, forests and sections of farmland.

As circumstances transpired, Soyuz TM-3 actually speared for the heavens almost an hour *ahead* of schedule, at 5:50 am Moscow Time on 22 July 1987. With Faris were veteran cosmonaut Alexander Alexandrov in the flight engineer's seat and a rookie commander, Lieutenant-Colonel Alexander Stepanovich Viktorenko. Born in Olginka, in northern Kazakhstan, on 29 March 1947, Viktorenko was selected as a cosmonaut in May 1978, alongside fellow Soviet Air Force pilot Nikolai Grekov, and the pair joined a larger detachment of civilian engineers and physicians in December of that year to commence training in earnest. However, whereas the remainder of the group graduated in October 1980, Viktorenko was delayed by a serious training accident and did not complete his final examinations until February 1982. Nevertheless, he quickly established a reputation for himself as a dedicated professional. Of the eight members of his 16-strong group who went on to actually fly in space, Viktorenko was fifth. In time, he would undertake four space missions, spend a cumulative 489 days in orbit and would be in command of Mir in March 1995 when the first American visitor, astronaut Norm Thagard, arrived. "A pleasant fellow, but also extremely competent", was how Thagard later described him. "Sasha Viktorenko was one of those who was probably more like you would expect to find in an American commander. Although we've got some American commanders who are fairly authoritarian ... I think if you looked at the scale of things, you'd find the Russians ... a little bit further down towards the less authoritarian end." After completing his high school education, Viktorenko entered the Advanced School of Military Aviation in Orenburg, graduated in 1969 and entered the Soviet Air Force, where he rose rapidly through the ranks. In 1985-86, he served as backup commander for both the Soyuz T-14 and T-15 missions. The flight engineer, Alexander Alexandrov, had already flown a long-duration mission to Salyut 7 and was thus ideally suited to change direction at short notice to replace the ailing Laveykin. In their history of the Soyuz programme, Rex Hall and Dave Shayler cast additional light on the situation: as well as noting "slight irregularities" in Laveykin's heartbeat, they explained that the young rookie "had trained hard for his first space mission and the stress of that training programme, the excitement of the first launch into space, the desire to perform well and the burden of the contingency EVA during the Kvant docking ... all contributed to the problem". Laveykin was in no immediate danger – and, certainly, would not have been despatched on a pair of complicated EVAs, as late as June, if he were unable to handle the rigours of spacewalking – but it was considered prudent for the experienced Alexandrov to take his place from July until the end of the mission in December.

Two days after launch from Tyuratam, Viktorenko guided Soyuz TM-3 to a perfect docking at the aft port of Kvant. Minor difficulties were experienced when unsealing the hatch, which required the cosmonauts to employ a lever, but the remainder of the visiting mission proceeded normally. Faris handled his experiments with vigour, observing his home country and utilising Mir's materials processing apparatus to 'grow' crystals of gallium antimonide and examine the effects of microgravity on the structure of eutectic alloys of aluminium and nickel. Yuri Romanenko lent a hand on the 25th, acting as an impromptu cameraman as the Syrian conducted the Palmyra experiment, using seven syringes to combine a pair of substances and examine the formation of crystals. Several days later, in the late evening of the 29th, a bitterly disappointed Laveykin took Alexandrov's seat in the Soyuz TM-2 craft and undocked from Mir with Viktorenko and Faris, touching down in Kazakhstan a few hours later, at 5:05 am Moscow Time on the 30th. The landing was not entirely uneventful. "Its parachute caught in high winds, the spacecraft missed buildings by just 2 km," reported *Flight International*.

Laveykin had completed 174 days in space, and was described as "pale" at the landing site, but was apparently in no mood for the intensive medical attention he received. At a press conference, specialists informed journalists that he was fine, that "no trouble of the cardiovascular system or other organs has been diagnosed" and that Laveykin had taken appropriate medicines during his mission, which had "ensured the necessary effect". However, psychologically, the cosmonaut's recovery was questioned. According to *Pravda*, he announced that he would eat dinner alone and Viktorenko and Faris were obliged to sit through a somewhat embarrassing press conference, *sans* Laveykin. When journalists *were* permitted to meet him, in his room, later in the evening, Laveykin tried to laugh it off: he simply did not understand, he told them, why there was so much medical interest in him – he was *not* ill – and, indeed, had it not been for the cardiograms from orbit, there was little cause for him to terminate his mission so early. In fact, the heart irregularities represented nothing more than a quirky symptom of his physiology and were not indicative of any serious cardiac problem. Clearly, Laveykin was simply venting frustration at having been yanked from what would have been a record-setting mission at the halfway point. In November 1987, Nurmukhamed Mukharlyamov, head of the clinical department of the Soviet Union's National Cardiological Centre, announced that supplementary examinations showed Laveykin to be "practically healthy", that there was "nothing wrong with the cosmonaut's health" and proclaimed him fit for further missions. However, like Vladimir Vasyutin before him, Laveykin would *never* fly into space again. He retired, "for physical reasons", from the cosmonaut corps in March 1994.

In the meantime, Romanenko and his new flight engineer, Alexandrov, plunged directly back to work for the remaining five months of the mission. On the very day that Soyuz TM-2 touched down, the resident cosmonauts undocked the TM-3 craft from Kvant and relocated it at the station's front port. Shortly thereafter, on 3 August, another Progress freighter was launched with equipment, supplies and mail. Operations with the astrophysical payload aboard Kvant got underway in spectacular fashion, when, on the 10th, it became the first observatory to detect X-rays from the newly-discovered Supernova 1987A on the outskirts of the Tarantula Nebula in the Large Magellanic Cloud. It represented the explosion of a blue supergiant, some 168,000 light years from Earth, and was first identified in February 1987 by ground-based astronomers in Chile and New Zealand. To date, it represents the closest observed supernova since 1604, when Johannes Kepler observed the death of a star in our own Milky Way. Although the brightness of 1987A peaked in May and steadily declined in the following months, its X-ray and radio emissions grew stronger, as the shock wave from the supernova crashed into a dense cloud of interstellar gas and dust. Romanenko and Laveykin had already begun observing it earlier in the year and, between June and September 1987, more than 115 sessions in Kvant would be devoted to its study. Closer to home, the cosmonauts continued their other work, performing a pre-planned emergency evacuation drill to Soyuz TM-3 at the end of August and undertaking plant growth studies of cedar seedlings. During the arrival of a Progress freighter in late September, it was noticed that Mir's new dorsal solar array seemed to be bending and shaking and ground controllers expressed concern that future dockings might dislodge it. It was decided that the *next* crew would undertake an EVA in the spring of 1988 to tend to it.

By the beginning of October, Yuri Romanenko had become the unquestioned space endurance record-holder, when he surpassed the achievement of Kizim, Solovyov and Atkov of 237 days in a single mission. The Soviets had already announced on a number of occasions that his expedition would continue "until the end of the year", marking a substantial 35 percent increase over the achievement of the Soyuz T-10B crew. (Previous long-duration missions had typically exceeded their predecessors by around 10 percent to satisfy the ruling of the Fédération

Aéronautique Internationale, needed to recognise a new world aviation record.) By now, however, Romanenko was beginning to grow weary and made his first complaint of fatigue to ground controllers. In one account, published by the journal *Izvestiya* on 2 October, he described his life aboard Mir: "I never used to notice noise from the fans at all," he said, "but now it sometimes wakes me up at night. Sometimes, I don't fall asleep right away in the evening, or I wake up in the middle of the night and them I'm not myself in the morning." His crewmate, Alexandrov, treated him "considerately" in the rare instances when Romanenko overslept. If the commander grew more tired than normal during his workday, he took "medicine at night" on the orders of the flight surgeons. Certainly, reports in the Western press confirmed that the cosmonauts' working day was reduced to around five and a half hours. By the beginning of November, counting 104 days' experience from two previous missions, Romanenko exceeded Leonid Kizim's cumulative record of 375 days in space.

According to Nurmukhamed Mukharlyamov, the psychological wellbeing of the cosmonauts was ensured through a balanced cycle of work and leisure, regular exercise and a healthy diet to help maintain vital elements such as potassium and calcium. The work continued. On 10 November, a novel propellant-consumption test was performed following the undocking of a Progress freighter; the craft moved away to a distance of a couple of kilometres, then re-rendezvoused with Mir and docked an hour later. Early in December, Romanenko passed the 300-day mark of his mission and by this time the crew's workload had been shortened still further to around four and a half hours per day. (Romanenko would typically sleep for nine hours and exercise for two and a half hours, per day.) In the meantime, on 26 November, another communications satellite, Cosmos 1897, was inserted into geostationary orbit, thereby enabling continuous voice and data coverage for Mir. The cosmonauts increased their daily exercise regime during this period, to prepare their bodies for the impending return to Earth, with Romanenko insistent upon running several kilometres daily on the station's treadmill. On 9 December the Soviets introduced the next crew to visit the station: Romanenko's backup, Vladimir Titov, would command Soyuz TM-4 on his second space mission, hoping to dock successfully this time, with Azerbaijan-born Musa Manarov as the flight engineer. These two men were expected to attempt a full year in orbit. The third seat aboard Soyuz TM-4 would be taken by Buran pilot Anatoli Levchenko, flying for eight days on an 'acclimatisation' mission into space, after which he would return to Earth with Romanenko and Alexandrov and perform a series of mock Shuttle landing approaches. This came as something of a surprise to many Western observers, who expected the crew to consist of Titov and Serebrov – logically enough, since they had been paired for Soyuz TM-2 – together with Oleg Atkov's backup, Valeri Polyakov. As circumstances transpired, Polyakov *would* spend time in orbit with Titov and Manarov, supervising their medical status during the latter phase of their long mission.

In the meantime, Titov, Manarov and Levchenko lifted off from Tyuratam at 2:18 pm Moscow Time on 21 December 1987 and successfully docked at the rear of Kvant a little over two days later. It seemed that Titov's long run of bad luck, at last, was behind him. Although the practice of 'handing over' station operations from an outgoing crew to an incoming crew had been first done in September 1985, the Romanenko-Titov exchange in December 1987 was quite different. Unlike the Soyuz T-13/T-14 handover, in which Viktor Savinykh remained aboard to 'bridge' both expeditions, the Soyuz TM-3/TM-4 handover involved a direct rotation of the *entire* crew. "No one already familiar with the operation would be left on-board," noted the 1987 Senate report. "This placed greater demands on Romanenko and Alexandrov during the one week they shared with the new crew on the space station, both in terms of acquainting Titov and Manarov with the station and its experiments *and* in packing their own things to go home." There were also experiments that still required completion before Soyuz TM-3 could depart, a situation which clearly irritated the exhausted Yuri Romanenko. At one stage, on 28 December, the day before undocking, he snapped at flight controllers that he and Alexandrov were "rushing around like squirrels in a wheel", trying to do all that was asked of them, and that the "superfluous" scientists, in charge of the experiments, should be removed from the control centre. Perhaps, the fuming Romanenko concluded, it might make better sense *not* to schedule time-critical experiments during a period of joint work with a new crew.

Finally, on the morning of the 29th, Soyuz TM-3 undocked from Mir, carrying Romanenko, Alexandrov and Levchenko back to a touchdown at 12:16 pm Moscow Time. The landing was a rough one. The temperature at the landing site, 80 km from the town of Arkalyk, was -12 degrees Celsius and meteorologists warned that icy conditions might impair the ability of search and rescue helicopters and aircraft from reaching the crew. Visibility in the thick fog extended to just 4 km. After touching down, strong winds blew the descent module onto its side and plans to erect a medical tent nearby had to be abandoned; the cosmonauts would instead be hurried into a medical helicopter. Serious concerns were expressed about Romanenko's health, although he was able to sit upright in the helicopter and, amazingly, he completed a run, unaided, for a *hundred metres* on the day after landing. (According to *Flight International*, he was able to walk unaided, just three hours after landing.) He was also spotted walking in a park with his wife and a physician and appeared to be moving "very confidently". In Romanenko's mind, another hurdle had been cleared in the effort to despatch a crew to the Red Planet. For the first time in history, *three* separate individuals from *three* separate missions had accrued *three* separate flight durations and were landing together in the same spacecraft. Interestingly, wrote Phillip Clark, the first time that Romanenko and Alexandrov even *met* Levchenko ... was in *orbit*, suggesting that the Buran pilots trained away from the main body of the cosmonaut corps.

Anatoli Semenovich Levchenko, a captain in the Soviet Air Force Reserves, came from Krasnokutsk in the Kharkiv Oblast of eastern Ukraine, where he was born on 5 May 1941. He graduated from Chernigov Higher Air Force School in 1964 and served for a time as a military pilot, then became a civilian test pilot for the Soviet Air Force Ministry. In July 1977, Levchenko was one of five test pilots selected by the Gromov Flight Research Institute, based in Zhukovsky, near Moscow, for the Soviet Shuttle. Of the five, only Levchenko and Igor Volk would actually fly Soyuz missions; the others – Oleg Kononenko, Rimantas Stankyavichus and Alexander Shchukin – were all killed an aircraft accidents before getting their chance. (Years later, it became clear that Volk would have commanded the first manned flight of Buran, with Levchenko as his backup.) It would appear that the plan for Levchenko, who was assigned to Soyuz TM-4 in March 1987, was similar to that of Volk on his mission: to fly the Tu-154 aircraft, whose instruments had been specially modified to mirror the displays aboard Buran, from Tyuratam to Akhtubinsk, then return to Tyuratam at the controls of a MiG-25 fighter. Little has ever come to light about the 'results' of Levchenko's flights – the Moscow Domestic Service noted only on 30 December that he had passed the test "with flying colours" – but what *is* known with certainty is that, tragically, he would not survive to see his old crewmates, Titov and Manarov, return to Earth. It appears that he underwent surgery on a brain tumour in the Nikolai Burdenko Neurosurgical Institute in Moscow and died there on 6 August 1988. *Complications from a brain tumour* is the short, clipped comment which appears as a footnote at the end of many biographies of Levchenko and he remains unique amongst cosmonauts in having died so soon after returning from a space mission.

Alexander Alexandrov completed 160 days in space, whilst Romanenko himself basked in the glory of a 326-day achievement as 1987 faded into history. The year had indeed been "rather painful" for Alexander Laveykin, in that his adaptation to microgravity conditions was troubled and then misinterpreted cardiac measurements erroneously suggested a heart problem. His disappointment at having been denied the opportunity to set a new record with his commander is entirely understandable. In January, it had also been rather painful for Alexander Serebrov and the unlucky Vladimir Titov, dropped as they were from the prime slot on Soyuz TM-2, shortly before launch, although, for Titov, at least, this was sweetened in December when he found himself in command of the first year-long mission. More success derived from the steady increase in Mir's capabilities. The arrival of Kvant in April, its near-failure and successful recovery, was balanced by the first murmurings of world-class science from a fully-fledged astrophysical observatory in Earth orbit – although by late 1987 the British and the Dutch were already experiencing difficulties with the performance of their instruments – and a continuous human presence in space was becoming a reality. As America struggled to prepare its Shuttle programme for a four-day mission to resuscitate the programme after the loss of Challenger, the Soviets were operating a permanently manned space station. Talk of the inferiority of Soviet technology, compared to that of the United States, may have been entirely accurate, but their technology *worked* and their missions were reliable and relatively frequent. The year 1988 would bring even greater success, with Titov and Manarov's marathon flight, visited by another pair of Intercosmos pilots from Bulgaria and, surprisingly, Afghanistan. A painful year, in so many ways, would give way to a euphoric one.

Nevertheless, at the start of the year, there was doubt in some minds that 'long' missions would continue to be attempted. Only days before the launch of Titov, Manarov and Levchenko, *Flight International* told its readers that the core crew was "likely to stay in space for just six months" and suggested that "the Soviets are having reservations about certain long-endurance flights". Yuri Romanenko's fatigue, homesickness and "significant, though expected, leg muscle wasting", from an extended period in microgravity, was also cited. Other plans for 1988 included the launch of two more modules for Mir and *Flight International* speculated that the Bulgarian and French cosmonauts might participate in EVAs to evaluate a new Soviet 'space bicycle', similar to NASA's Manned Manoeuvring Unit. As 1988 began and Vladimir Titov and Musa Manarov settled down to make Mir their home, it became increasingly clear that theirs would be a record-setting duration mission, with estimates ranging from a full year to as much as 400 days. One of their first acts was to create an inventory of the station's stock of consumables, with a view to placing orders for the next Progress freighter, which arrived late in January. Although both men had trained for long-duration flights, neither had any experience in this area: Titov had accumulated just two days in orbit on his Soyuz T-8 mission, whilst Musa Khiramanovich Manarov – the spelling of his first name has also appeared, variously, as 'Musakhi', 'Musat' or 'Musachi' – was making his first flight. Born on the southern shore of the Absheron Peninsula, in the industrial hub of Baku, it is hardly surprising that Manarov grew up with a keen engineering mind, a voracious interest in aviation and an ambitious outlook on the world around him; for Baku stands as the political, scientific, cultural and industrial centre of today's independent Azerbaijan. (At the time of writing, this impressive capital city was bidding to host the 2020 Summer Olympics.) Manarov was born here on 22 March 1951, at a time when Azerbaijan was a republic of the Soviet Union. He earned an engineering diploma from Moscow Aviation Institute in 1974, entered the Soviet Air Force – eventually rising to the rank of colonel in the reserves – and was selected as a cosmonaut candidate in December 1978. It would appear that Manarov was originally teamed with Romanenko on the Soyuz TM-2 backup crew in September 1985, but was temporarily removed from flight status due to an undisclosed medical issue and replaced by Laveykin. By March 1987, Manarov was apparently back on active status and, following the removal of Alexander Serebrov from the Soyuz TM-2 prime crew, he was paired with Vladimir Titov.

With Titov and Manarov aboard Mir for a long mission, the Soviets announced openly that there would be two visiting missions, involving international participation: a Bulgarian national, Alexandar Alexandrov (not to be confused with Romanenko's former crewmate), would join Soviet cosmonauts Anatoli Solovyov and Viktor Savinykh on the Soyuz TM-5 short-duration visit in June 1988. Then, in November, a Frenchman – either Michel Tognini or Soyuz T-6 veteran Jean-Loup Chrétien – would travel to Mir ... and become the first non-Soviet and non-American to perform an EVA. It was also hinted by Viktor Blagov that a physician – probably Valeri Polyakov – would visit Mir to monitor the health of Titov and Manarov during the final months of their mission. "Neither of the Soviet cosmonauts announced as members of the June mission are physicians," the US Senate report explained. "Whether the Soviets are leaving open the possibility of changing that crew or plan to send the doctor along on the November mission is unclear." *Flight International* also hinted in early January about "a possible all-Soviet visiting flight" at some stage and even that Manarov might be replaced aboard Mir (perhaps by Savinykh) after six months. What was not known, at least in early February 1988, was that a third visiting mission *would* indeed take place in late August; a mission which *would* bring physician Valeri Polyakov to the station to monitor Titov and Manarov as their epic expedition drew to a close. This mission, Soyuz TM-6, would earn renown for another reason, carrying into space perhaps the most unlikely of cosmonauts: a man whose very *homeland* had been under direct and devastating attack by the Soviet Union for almost a decade.

WITHDRAWAL

In mid-February 1989, Soviet tanks and armoured personnel carriers rumbled through the mountains of the Hindu Kush and the rugged Panjsher Valley and left behind a landlocked and desolate nation which they had spent years fruitlessly trying to subdue. In the long term, they had fared badly against an insurgency of native mujahideen fighters, financed, trained or directly supplied by the United States, Britain, China, Pakistan, Saudi Arabia and other sympathetic nations. As already explained, détente between the Soviet Union and the United States had veered sharply off course in the late 1970s, but this did not prevent Leonid Brezhnev and Jimmy Carter from adding their signatures to the agreements of a second round of Strategic Arms Limitation Talks in Vienna in June 1979. Six months later, though, in an event which would leave Carter "open-mouthed", Soviet forces swept across the border into Afghanistan and landed in the capital, Kabul, on Christmas Day. Quickly, they took government and media buildings and assaulted the Tajbeg Palace, where they killed the unpopular president, Hafizullah Amin. By February 1980, more than 100,000 Soviet troops had occupied Afghanistan, bringing with them thousands of tanks and amphibious armoured vehicles. Three dozen Islamic countries demanded the immediate withdrawal of Soviet troops and Carter prohibited American athletes from participating in the 1980 Summer Olympics in Moscow. His boycott was endorsed by other nations, including China, Japan and West Germany.

Mikhail Gorbachev's arrival as General Secretary in March 1985 had brought with it new ideas on foreign and domestic policy and it was under his rule that the first steps were taken to leave Afghanistan. Far from generating stability and imposing a socialist state on a population still ruled by tradition and religion, the mountainous country had become a quagmire and the conflict had turned into what one commentator described as "the Soviet Union's Vietnam War". In a two-step process from May 1988 until February 1989, the Soviets departed Afghanistan, but left the country in tatters. More than a million Afghan men, women and children were dead and an astonishing *half* of the world's refugees were Afghans. The Geneva Accords, signed in May 1988, left post-war Afghanistan in ruins, with no viable support for future governance. It was perhaps more than a little ironic that the first Afghan cosmonaut, Abdul Ahad Mohmand, rode Soyuz TM-6 to Mir in late August, but gained few of the honours that the other Intercosmos fliers received upon their return. Vicious inter-tribal warfare in the early 1990s eventually forced him to leave his homeland and he became another tragic statistic on the world's list of Afghan refugees. Today, he lives in Germany.

Abdul Ahad Mohmand was born on 1 January 1959 in Sardah, near the ancient city of Ghazni in east-central Afghanistan – a city presently recovering from the devastation laid down by the latest in a long line of military invasions; in fact, over the millennia, Ghazni has been occupied by Buddhists and Hindus and Muslims, becoming the capital of the Ghaznvid Empire in the tenth century and later being razed to the ground. In the 19th century, it was again destroyed by British-Indian forces and today it is being rebuilt *again* after the destruction of ten years' of the present conflict in the troubled country. Twenty-five percent of Ghazni's occupants speak Pashto, the native tongue of indigenous Afghans, and Mohmand himself was of Pashto ethnicity; in fact, the Mohmand clan is one of the most ubiquitous, with members living in Afghanistan and throughout urban and rural Pakistan. Mohmand graduated from the Polytechnical High School in Kabul in 1978, a year before the Soviet invasion, and entered the Afghan Air Force, trained in Russia as a fighter pilot and rose to the rank of captain. He completed the Gagarin Air Force Military Academy in Monino in 1987 and on 12 February of the following year was selected, together with a fellow Afghan Air Force pilot, Colonel Mohammed Dauran-Ghulam Masum, for a shortened cycle of training for a mission to Mir. When Mohmand was chosen, that summer, as Masum's backup for the Soviet-Afghan flight, he might have felt a pang of disappointment; for the backups were largely forgotten by history and the names of Intercosmos backups like Oldrich Pelèzák and Zenon Jankowski, Eberhard Köllner and Béla Magyari, Bùi Thanh Liêm and José López Falcón, Maidarzhavyn Ganzorig and Dumitru Dediu, Ravish Malhotra and Munir Habib are rarely remembered as anything more than footnotes at the end of reports about the exploits of their spacegoing partners. Little could Mohmand have possibly foreseen that he *would* be elevated onto the prime crew – and, in true Afghan fashion, with its racial emphasis on clans and tribal groupings, the decision to drop Masum boiled down to little more than mere ethnicity – but he could not have known that his flight would garner so little international reaction and, today, even among his fellow countrymen, the identity of Afghanistan's first cosmonaut is virtually forgotten.

INTERNATIONAL VISITORS

The long mission of Vladimir Titov and Musa Manarov was already well underway when, on 26 February 1988, the cosmonauts floated into open space for their first EVA, scheduled to last around four and a half hours. The primary task was to replace a section of the 'dorsal' solar array that had been installed the previous year by Romanenko and Laveykin, and this intricate task necessitated a certain amount of 'refresher' training. For several days before the excursion, the cosmonauts watched videotaped footage of their own practice sessions in the underwater hydrolab and spent time checking their suits and equipment in Mir's multiple docking adaptor. On the 26th, they finally moved outside and prepared to replace one of the array's four main segments. This required them to 'collapse' the lower extendable boom, in order to fold shut both array sections. Next, they fitted new segments, which boasted carbon-plastic composite, rather than metal, and improved solar cell 'leaves'. In fact, *six* of the new leaves produced as much power as *eight* conventional ones, meaning that the remaining two leaves could be instrumented and utilised for testing new solar cell materials. With this work completed, Titov and Manarov clambered aft along Mir and the Kvant module to inspect the Progress 34 freighter, whose rendezvous antenna had proven somewhat balky in deployment.

Back inside the station, the first half of the long-duration mission proceeded relatively quietly, with frequent Progress visitors and an emphasis upon scientific research. The cosmonauts studied fish behaviour, grew onions, installed and activated a biological crystal growth apparatus and performed astronomical observations with the Kvant hardware. In the case of the latter, long-exposure photography was acquired of galaxies and stellar groups in the ultraviolet part of the spectrum, although it became apparent that even small movements by the two men could 'shake' Mir and induce distortions. Motions were therefore limited during major observations. In total, during their mission, Titov and Manarov performed over 2,000 experiments, whose themes also included materials research, meteorology, technology, physiology and psychology and medical science. The lack of human company drew to a close at 5:03 pm Moscow Time on 7 June 1988, when Soyuz TM-5 roared into the early evening Tyuratam sky and docked at the rear port of Kvant two days later. (The launch had actually been *advanced* by two weeks, late in the planning stages, in order to improve lighting conditions for one of the mission's astronomical experiments.) Aboard Soyuz TM-5 were a pair of Soviet cosmonauts and a Bulgarian fighter pilot, the second of his countrymen to venture into space ... yet only the *first* ever to board a space station. To say that Bulgaria's manned space mission had been a long time coming was something of an understatement; more than nine years earlier, in April 1979, Georgi Ivanov had been aboard Soyuz 33, when an engine failure prevented a docking with Salyut 6. Shortly after this crushing disappointment, the Soviets offered Bulgaria a second opportunity to fly a man into space and the chance went to Ivanov's backup, Alexandar Panayotov Alexandrov.

For Alexandrov, who shared the same name and even the same *middle initial* as the Soviet cosmonaut who had flown Soyuz TM-3, it had been a long journey. He came from the town of Omurtag in the Targovishte Oblast of eastern Bulgaria, where he was born on 1 December 1951. At the time of his birth, the People's Republic of Bulgaria had been a staunch ally of the Soviet Union for more than five years, ever since the deposition of its monarchy by the Communist Fatherland Front. (In fact, a 1969 Soviet stamp later presented the friendship between the two nations as 'indestructible for eternity'.) By the mid-1950s, Todor Zhivkov had assumed absolute power and would control the country for more than three decades; under his regime, Bulgaria would remain a loyal vassal of the Soviet Union – even participating in the invasion of Czechoslovakia in August 1968 and distancing itself diplomatically from China – but he would pursue relatively moderate policies at home, permitting some freedom of expression, ending the persecution of the Church and renewing friendly relations with Yugoslavia and Greece. Despite operating a planned economy, he decreed that surplus produce could be freely sold and in 1965 Bulgaria was the first Communist nation to obtain a Coca Cola licence. Having said this, Zhivkov was intolerant of dissent and in September 1978 the Bulgarian émigré Georgi Markov was assassinated in London by Communist agents. His death – the notorious 'Umbrella Murder', in which a modified umbrella was used to inject a ricin-filled pellet into his leg – cast the People's Republic in a decidedly more unpleasant light and by the early 1980s, the Zhivkov regime was seen as corrupt, autocratic and increasingly erratic in its actions. In 1984, for example, an attempt was made to forcibly assimilate a Turkish minority by forbidding them to speak their own language and adopt Bulgarian names. In a sense, the People's Republic remained a socialist bastion right to the end and Zhivkov even (barely) survived Mikhail Gorbachev's reforms of the Soviet Union, being finally pushed aside on the basis of age and infirmity in November 1989 after mass reformist demonstrations throughout Bulgaria.

By the time Zhivkov was removed from power and the Communists had given up their stranglehold on the country, not one, but *two*, Bulgarian cosmonauts would have flown into space: Ivanov in April 1979 and Alexandrov in June 1988. It is interesting that the People's Republic of Bulgaria, as it was at the time, had been one of the first socialist states to propose undertaking a manned space mission with the Soviet Union. In August 1964, Russia's minister of defence organised a meeting in Moscow with the Bulgarian military attaché, Zakhari Zakhariev, to discuss such matters. Realistically, it was far too early in the Space Age for such plans to reach fruition, but a decade later, with the establishment of Soyuz and Salyut technology, the time seemed more conducive. When the call for candidates went out in late 1977, it was mandated that all applicants had to be Bulgarian People's Air Force fighter pilots and *had* to be graduates of the Georgi Benkovski Higher People's Air Force School and *must* have completed their studies at some point between 1964 and 1972. The rationale for this requirement was that in 1964 the school had begun to issue science degrees and therefore the first Bulgarian cosmonaut would be a suitably educated scientist. On top of this, the candidate needed at least three years of regular flying experience. An aeromedical commission evaluated several hundred candidates and the most suitable finalists were sent to Sofia for detailed medical examinations. In rank, they varied from squadron leader to the executive officer of an air regiment and eventually fifteen were winnowed down to just six: Ivanov and Alexandrov, together with Georgi Yovchev, Ivan Nakov, Clavdar Dzhourov and Kiril Radev. Of these, the latter pair resigned their places, reducing the number to four when they finally flew to Moscow for training. Yovchev turned out to have a heart problem, which eliminated him from consideration, and the Soviet doctors finally settled on Ivanov as the prime crewman for Soyuz 33, backed up by Alexandrov. By the time the mission took place, Alexandrov was already an accomplished pilot in the Bulgarian People's Air Force and had graduated from the academy in 1974.

With the failure of Georgi Ivanov's mission, speculation was rife in the Western press about what the Soviets planned for their next move. On 28 April 1979, *Flight International* speculated on the dilemma. "It is thought," the magazine explained, "that the Soviet Union has already begun preparations for another Soyuz launch, but the choice of crew poses a political dilemma. One possibility is the backup crew to Soyuz 33, Yuri Romanenko and Alexandar Alexandrov. However, *this* would mean that Bulgaria will be able to claim *two* men in space ... to the chagrin of other East European countries. This makes it more likely that the *original* crew, Rukavishnikov and Ivanov, will make a repeat attempt." The correct answer? *None of the above.* For reasons which remain unclear to this day, Ivanov and his Soviet commander, Nikolai Rukavishnikov, were never given another chance at the mission which had been frustrated by technical failure. One possibility is that Rukavishnikov was reassigned, later in 1979, as the backup flight engineer for the upcoming Soyuz T-3 mission.

By the summer of 1986, plans for a second Bulgarian flight had reached the level of senior politics. In August of that year, General Dobri Dzhourov, the minister of the People's Defence Army, visited the Soviet Union and signed an agreement in Moscow, authorising the mission. By way of financial recompense, Bulgaria would build and supply an array of scientific equipment, costing some $14 million, which would be transferred to the ownership of the Soviets ... although the *technology* needed to produce this hardware was *donated* by the Soviets! In the autumn of that year, 300 air force officers were screened and ten candidates were approved. Of this group, four finalists were accepted by Soviet physicians and two of these (Major Alexandar Alexandrov and Captain Krasimir Stoyanov) were selected in January 1987. The Bulgarian cosmonaut would operate almost 2,000 kg of equipment, delivered aboard Progress freighters, in support of 46 experiments.

In command of Soyuz TM-5 was a man who had waited for more than a decade for his first chance to venture into space; a man who would go on to record five missions and 651 days in orbit; a man who would narrowly, and by his own hand, miss out on commanding the first crew of the International Space Station; and a man who, even today, has spent more time spacewalking – over 82 hours in 16 excursions – than any other human being. That man was Anatoli Yakovlevich Solovyov, born on 16 January 1948 in Riga, today's capital of Latvia. He completed the Lenin Komsomol Chernigov Higher Military Aviation School in 1972 and served until 1976 as a senior pilot and group commander in the Far Eastern Military District. Selected as a cosmonaut candidate in August 1976, Solovyov commenced more than two years of evaluation and training and served as backup commander for the Soyuz TM-3 Syrian mission. His military experience and upbringing had imbued him with a staunch support for Communism – he had been a party member since 1971 – and, during the Shuttle-Mir effort with the United States in the 1990s, it has been remarked that he was "not especially friendly" towards the Americans and gave NASA reason to suspect that he was "a stern critic of the two countries' collaboration". Bryan Burrough described him as "the Chuck Yeager of the Russian programme ... There is a bit of the Old Soviet in Solovyov". If Alexander Viktorenko was one of the most 'Americanised' of the Soviet commanders, in terms of his personable nature, then Solovyov was the reverse. "He gave *orders*," wrote Burrough, "harsh orders, often shouted." Having said this, NASA astronaut Dave Wolf, who flew aboard Mir with him in 1997-98, praised Solovyov's stance and felt that a more 'patronising' tone would not have made him feel like an integrated

The Soyuz TM-4 descent module, carrying Anatoli Solovyov, Viktor Savinykh and Bulgaria's second man in space, Alexandar Alexandrov, descends under the canopy of its main parachute towards a touchdown in Kazakhstan on 17 June 1988.

member of the team. By *making* the demands that he did on his crew, Solovyov was actually demonstrating their immense *worth* in what was a team effort. Yet his credentials stand for themselves and do not just extend to EVA: at the time of writing, in December 2011, he holds fifth place on the list of most experienced spacefarers in history, is one of only half a dozen cosmonauts to have completed five missions and is one of only two cosmonauts to have commanded all five of his flights.

The docking with Mir on 9 June occurred an orbit later than planned, when faulty Kurs data caused the Soyuz to deviate from its planned final approach and required Solovyov to assume manual control. It was later determined that the rendezvous was actually proceeding normally, and the *readings* were at fault, although the gremlins in the craft's systems would return in September 1988, in what Rex Hall and Dave Shayler described as "one of the most dramatic end-of-mission scenarios in the history of manned space flight". If Anatoli Solovyov was beginning his space career with Soyuz TM-5, then his flight engineer, Viktor Savinykh, was embarking on his third and final mission. Savinykh had already spent eight and a half months of his life in orbit, spread across two previous voyages, and, but for the illness of Vladimir Vasyutin, might have established a world endurance record for himself in the spring of 1986. On 17 June, after a longer than normal international mission of almost ten days, Solovyov, Savinykh and Alexandrov undocked from Mir, aboard the old Soyuz TM-4 spacecraft, and touched down at 1:13 pm Moscow Time. In yet another example of the burgeoning *glasnost* effort, the mission had been accompanied by a blaze of publicity, with live television coverage offered to any countries interested in taking it. Interestingly, the cosmonauts came down in a *lake* ... albeit a dry one. "But for the extremely hot weather," *Flight International* reported on 2 July, "with temperatures reaching 42 degrees Celsius, the lake would have been full of water and re-entry would have resulted in an unscheduled splashdown."

In the wake of the visiting mission, Vladimir Titov and Musa Manarov resumed their duties aboard the station, undocking the new Soyuz TM-5 from Kvant's aft port and docking it at the front of Mir. A few weeks later, on 30 June, they ventured outside for the second EVA of their mission and spent a little over five hours repairing the joint British-Dutch X-ray instrument aboard Kvant, which had experienced difficulties soon after its launch. Although the telescope had not been designed for repairs, several tools were hurriedly fabricated by Dutch and Soviet engineers and were carried aloft by Anatoli Solovyov's crew. Titov and Manarov received instructions and familiarisation briefings from British scientists and Dutch researchers were on hand in the mission control centre as the spacewalk got underway. Firstly, the two cosmonauts cut away no fewer than 20 layers of thermal insulation to reach the 40 kg detector, but the lack of handholds was frustrating and forced them to take turns working, whilst the other acted as a human anchor. There were other difficulties, too. More clips held the detector in place than they had anticipated and a trio of screws, all locked in place by resin, put them behind on the timeline; they had to scrape one of them with a saw blade to make it budge. At length, with 70 percent of their work done, a special 'key', needed to remove a brass clamp, snapped. Mir drifted out of radio contact and by the time communications were restored, Titov and Manarov had been forced to give up their efforts and return to the station.

A second attempt to complete the repair was scheduled for 5 July, but was postponed to give the cosmonauts additional preparation time. In September, seven new tools arrived aboard a Progress craft, as well as upgraded Orlan-DMA suits, which boasted a lighter, more flexible and tougher structure than their predecessors, together with more durable motors in the life-support apparatus and improved gloves for hand mobility. The new suits each weighed 105 kg when fully charged and could support EVAs as long as seven hours, compared to a maximum of around five hours in the case of the older Orlan-DM. Significantly, the new ensembles included an autonomous radio and battery package, which eliminated the need for an umbilical link to Mir and made the cosmonauts more 'autonomous' when they were outside. In fact, these 'add-on' packages had been specifically designed with the Soviet 'space bicycle' in mind, which cosmonaut Alexander Serebrov would test for the first time in early 1990. However, for the first EVA in the new suits, on 20 October 1988, Titov and Manarov remained linked to Mir by the old Orlan-DM umbilicals. They spent five hours outside the station, lugging a new detector for the X-ray telescope along the entire length of Mir and only managing to slide it into place with some difficulty. Nonetheless, the task took an hour *less* than expected and the cosmonauts also fitted a foot restraint near the airlock, ready for use by Alexander Volkov and Frenchman Jean-Loup Chrétien on their spacewalk in December.

Titov and Manarov's second group of human visitors, who arrived on the last day of August, were particularly notable, not only because they included amongst their number an Afghan pilot, but also that there was *no* flight engineer; the commander, Vladimir Lyakhov, had been specially trained to fly Soyuz TM-6 in a solo capacity. In fact, Lyakhov was one of a handful of military cosmonauts – the others including Anatoli Berezovoi and Yuri Malyshev – who had been chosen in September 1985 to fly the Soyuz spacecraft alone, thereby providing a rescue capability for a two-man space station crew in the event of an emergency. The idea originated in the wake of the Salyut 7 problems earlier that year, but in the spring of 1988 the Soviets decided to put this capability to the test and teamed Lyakhov with the Afghan cosmonaut and a physician, Valeri Vladimirovich Polyakov.

For almost two decades, Polyakov has held the world record for the longest single space mission, during which he spent 437 days in orbit aboard Mir in 1994-95. (When added to the 241 days he spent aboard Mir in 1988-89, Polyakov currently ranks fourth on the list of the world's most experienced spacefarers, with a personal achievement of 678 days or more than 22 months.) Yet the name which today continues to grace the Guinness Book of Records is not even his birth name. He was actually born Valeri Ivanovich Korshunov on 27 April 1942 in the industrial city of Tula, a couple of hundred kilometres south of Moscow, but changed his middle patronymic and his family name to 'Vladimirovich' and 'Polyakov' at the age of 15, when he was adopted by his stepfather. Polyakov completed his secondary schooling in Tula in 1959 and entered the I.M. Sechenov First Moscow Medical Institute, from where he gained his doctorate. He then specialised in astronautical medicine at the Institute of Medical and Biological Problems at Moscow's Ministry of Public Health; this fascination with the behaviour of the human organism in the microgravity environment was inspired by the flight of the first medical specialist, Boris Yegorov, aboard Voskhod 1 in October 1964. A little over seven years later, in March 1972, Polyakov was selected, along with two other physicians, Georgi Machinsky and Lev Smirenny. However, due to his ongoing postgraduate research towards a candidate of medical sciences degree, Polyakov did not begin his cosmonaut training until October 1972. He received his degree in 1976 and joined another cosmonaut selection, two years later, to complete the remainder of his training. He was an early candidate for a dedicated medical research mission to Salyut 6 in late 1980 and served as physician Oleg Atkov's backup on the long-duration Soyuz T-10B flight in 1984. To American physician Norm Thagard, who spent time aboard Mir with Polyakov in early 1995, the cosmonaut reminded him of his colleague, Bill Thornton, in terms of expertise, physical size and nature. "He's this big guy," Thagard said of Polyakov, "an extroverted guy, but can be sort of like a bull in a china shop." As Thagard watched Polyakov and his crewmates, Alexander Viktorenko and Yelena Kondakova, depart

Mir in March 1995, the physician was extremely excited to be concluding his mission and returning home. "Rambunctious", in fact, was the adjective Thagard used ... to such an extent that Viktorenko, the commander, had to calm Polyakov down before re-entry.

Polyakov's first ride to orbit began at 7:23 am Moscow Time on 29 August 1988, with the Soyuz TM-6 launch televised live, and Afghan President Mohammad Najibullah announced that his country's armed forces would symbolically halt any further hostilities for the duration of the mission. (As circumstances transpired, as early as 1 September, more than four dozen rockets were fired at Kabul Airport, leaving five dead and two injured.) Lyakhov docked his craft at the rear port of Kvant and Mohmand spent the next week photographing and making other observations of Afghanistan from space in support of cartography, water-resource assessments and oil and gas exploration. The decision to fly Mohmand seems to have taken place quite late in the training process. According to Phillip Clark, the original prime crew was Lyakhov, Polyakov and 34-year-old Afghan Air Force Colonel Mohammed Dauran-Ghulam Masum, with Mohmand joined on the backup crew by Anatoli Berezovoi and physician Gherman Arzamazov. The reason for Masum's replacement by Mohmand at such short notice was for many years attributed to a bout of appendicitis. However, rumour has abounded that Masum's Tajik ethnicity rendered him 'unsuitable' for the mission and Mohmand – a member of a significant Afghan tribal group – was chosen instead. Ironically, Masum's additional name of 'Dauran-Ghulam' means *servant of luck*.

During his time aboard Mir, Mohmand also brewed Afghan tea for his fellow cosmonauts and even spoke, by audiovisual link, to his distraught mother, who was so worried about his mission that President Najibullah personally invited her to his office. It had already been revealed by the Soviets that Polyakov would remain aboard the station with Titov and Manarov for the remainder of their mission and would most likely stay until April 1989, joining the *next* long-duration crew, and late on 5 September Lyakhov and Mohmand boarded Soyuz TM-5 and undocked in the small hours of the following morning. Shortly thereafter, in accordance with the flight plan, the orbital module was jettisoned. Touchdown on Soviet soil was scheduled for around 5:15 am Moscow Time, but it subsequently became clear that the retrofire attempt failed. Apparently, barely 30 seconds before the de-orbit burn was due to take place, *both* the primary and backup horizon sensors were confused by solar glare as Soyuz TM-5 passed over the terminator and into sunlight. "Despite all the upgrades," wrote Rex Hall and Dave Shayler, "the Soyuz-TM still relied on the old flight requirement to pass into sunlight at least ten minutes prior to retrofire to allow the on-board sensors sufficient time to orientate and stabilise the spacecraft ... but because Soyuz TM-5 passed across the terminator at dawn, the system apparently became confused." The computer automatically cancelled the burn, but allowed the engine to remain armed. At length, the system apparently cleared its confusion and gave the engine a go-ahead for the burn ... but by this point the spacecraft was seven minutes past its original burn time and a further 3,500 km along its orbital path. After only six seconds, Lyakhov manually halted the burn, aware that it would have brought them down far off-target, somewhere to the north of

Manchuria. Two orbits later, a second attempt was made. On this occasion, the infrared sensors were disengaged and the spacecraft's inertial measurement unit was employed to control the burn. Here the reader will recall that Anatoli Solovyov had been obliged to cancel his automated rendezvous with Mir in June and it would appear that the parameters for *that* manoeuvre were still stored in Soyuz TM-5's memory. *Now*, three months later, the engine ignited and the autopilot ordered the backup computer to complete the manoeuvre which *should* have been executed in June. When a discrepancy was detected between the two flight programs, the burn was automatically stopped, just six seconds into the planned 230-second firing. Lyakhov attempted a manual restart, but after less than a minute the navigation system detected an incorrect orientation and shut the engine down. Several more manual attempts were also fruitless.

By mid-morning, more than five hours after the scheduled landing, the silence was finally broken by Radio Moscow: a problem had delayed the two cosmonauts by 24 hours. According to the Soviets, the spacecraft's systems had deviated from pre-set operational parameters and "the decision was taken *not* to take risks, analyse the situation thoroughly and compute another re-entry trajectory". This explanation belied the seriousness of the episode. Hall and Shayler have noted in their summary that the automatic termination of the engine burns on two occasions might have prevented Soyuz TM-5 from being inserted into a lower orbit, which might have caused atmospheric drag to begin an unplanned re-entry and perhaps triggering disaster. "Their orbit might not have been low enough to decay immediately," they wrote of Lyakhov and Mohmand's predicament, "and thus they could have re-entered over any part of their orbital ground track. Their re-entry angle might also have been incorrect, which could have subjected the capsule to heating and descent stresses beyond designed limits, thereby burning up the capsule *and* the two cosmonauts inside." Other reports, cited by Hall and Shayler, suggested that a timer was activated to separate the Soyuz instrument module at around this time and only the prompt action of Lyakhov to override it prevented the men from becoming stranded in orbit.

For now, however, they were safe, in a stable orbit, although conditions in the cramped descent module cannot have been pleasant, with no toilet or hygiene facilities; in fact, they were obliged to use the plastic sleeves of Mohmand's suit as makeshift toilet bags. Lyakhov also elected not to dip into their three-day emergency rations, meaning that they ate only cold food. During those fraught 24 hours, computer specialists on the ground discovered the error in Soyuz TM-5's old rendezvous program and for the first time it became clear that the emergency was the result of a *software* problem, rather than a failure within the engine. The absence of the orbital module and its docking apparatus ruled out any possibility of physically redocking with Mir and, although the craft *had* sufficient propellant to rendezvous with the station, getting Lyakhov and Mohmand aboard would require a difficult and dangerous contingency EVA transfer in their Sokol suits. For now, the men waited. As well as being cramped, uncomfortable and a little unhygienic, the Soyuz was quite chilly, at 10 degrees Celsius, and even within their suits they could *feel* the cold. Thankfully, a retrofire attempt at 3:00 am Moscow Time on 7 September was successful and Soyuz TM-5 touched down safely at 3:50 am. Mohmand and Lyakhov were rightly lauded for their efforts and received both medals and honours, but for Afghanistan the final withdrawal of Soviet troops in early 1989 left a nation in utter devastation, on the brink of civil war. Unable to reconcile the many ethnic factions and control the regional warlords within the country, Mohammad Najibullah resigned office in the spring of 1992, making way for the arrival of a new interim government. Yet Afghanistan's problems, both political and religious, ran deep and any kind of lasting settlement in this fragmented land was virtually impossible to achieve; indeed, a constant state of warfare, lawlessness and the total absence of peace and prosperity ultimately aided the hard-line Taliban in their sweeping rise to power in 1996. By this time, Abdul Ahad Mohmand had long since fled, along with thousands of others, to the West. Today, a quarter of a century after Mohmand pushed the frontiers of Afghan science and technology further than ever before, 42 percent of his kinsfolk still struggle to scrape a living on the equivalent of a dollar a day. A third of the population are unemployed and exist below the poverty line and, although the Afghan economy has grown steadily by around 10 percent per annum in the last decade, much of this has come about only through Western handouts and other donations. Far from igniting a spark of hope in the hearts of the Afghan people, Mohmand's mission was instead regarded as the cynical act of a superpower; a superpower responsible for the deaths of hundreds of thousands of Afghans. A token mission into space offered little recompense for a decade of warfare, oppression, military occupation, political surveillance, torture and execution. Not surprisingly, Mohmand's mission is now virtually forgotten and, as such, is one of the greatest tragedies in the story of humans in space. Perhaps, someday, another Afghan will follow in Mohmand's footsteps. Perhaps the flight of this second Afghan spacefarer will originate from a more settled and more prosperous land and will generate the levels of national pride that it truly deserves.

AN UNCERTAIN FUTURE

With the departure of Lyakhov and Mohmand in the first week of September, the Mir crew was expanded to three long-duration members for the first time, with Polyakov devoting his time to medical observations of Vladimir Titov and Musa Manarov. Yet the last few months of the year-long mission would not be quiet ones; the EVA on 20 October to bring one of Kvant's instruments back to full operating order has already been discussed and the end of 1988 would be marked by another remarkable accomplishment: a three-week flight by Frenchman Jean-Loup Chrétien, during which he became the first non-American and non-Soviet cosmonaut to perform an EVA. Since the beginning of the Space Age, France had nurtured an interest in a human space programme and in the wake of the missions of Chrétien in June 1982 and Patrick Baudry in June 1985, much enthusiasm had been demonstrated in undertaking further flights of an extended nature. According to Phillip Clark, France initially requested a two-month mission, "but they eventually accepted a flight lasting for about 38 days". In November 1986, two cosmonaut candidates were selected to travel to Star City for training: Chrétien and another French Air Force pilot, Michel Ange-Charles Tognini. By the late summer of 1988, Chrétien had been confirmed on the prime crew, joining veteran commander Alexander Volkov and rookie flight engineer Sergei Krikalev, with Tognini joining the backup crew alongside Alexander Viktorenko and Alexander Serebrov. The mission, called 'Aragatz', was provided free of charge by the Soviets, who received payment in kind through the scientific return from the French experiments; these included the Echograph electrocardiogram and equipment to measure neurosensory reactions, visual acuity, radiation effects and blood analysis.

Even in the final months before the launch of the Soviet-French mission, there was much discussion in the Western press about the precise configuration of the return of Titov and Manarov back to Earth at the end of their long flight. In mid-September, *Flight International* speculated (correctly, as it turned out) that they would return with the Frenchman, but added that "it is thought that Polyakov may need to monitor [them] during re-entry" and that "either Titov or Manarov could remain for a longer mission with Volkov and his flight engineer, perhaps to clock up 400 days". The latter would have produced an unusually short mission of just a couple of months for Volkov and Krikalev. A few weeks later, in October, the details had become clearer and the Soviets revealed that, far from pushing the year-long record still further, Soyuz TM-7 would commence a series of missions lasting around six months. Volkov and Krikalev would remain aboard Mir, together with Polyakov, until April-May 1989, whereupon they would be relieved by the Soyuz TM-8 crew of Viktorenko and Serebrov.

For all the excitement which accompanied Jean-Loup Chrétien's second mission, the 38-day agreement was trimmed first to around four weeks, and eventually three, by several factors, including a decision by French President François Mitterand to attend the launch at Tyuratam. As a result, the start of the mission was postponed by several days until 26 November 1988, enabling Mitterand to fly into the desolate Kazakh launch site aboard Concorde. Soyuz TM-7's rise into orbit occurred in darkness at 6:49 pm Moscow Time and marked the first flight into space of Sergei Konstantinovich Krikalev, a man who today has logged over 803 days in orbit, spread across six missions, and as such holds the unchallenged record for having spent more time off the planet than any other human being. At the time of Soyuz TM-7, Krikalev was barely 30 years old, making him one of the youngest cosmonauts ever launched. He came from Leningrad – today's St Petersburg – where he was born on 27 August 1958; his father, Konstantin, was an engineer and his mother, Nadezhda, a survivor of the Nazi siege of Leningrad. After high school, Krikalev specialised in chemistry and entered the Leningrad Mechanical Institute, graduating first in his class in 1981 with a degree in mechanical engineering. As part of his studies, he worked on the design and manufacturing of flight vehicles. Krikalev also served as an aircraft technician and learned to fly at a Leningrad aviation club. He joined Energia and was involved in the testing of space equipment and ground control operations; in fact, aged just 26, he worked on the Salyut 7 rescue team to develop contingency docking procedures with the inert station. (More than one source has commented that Krikalev's work assured him of Vladimir

Dzhanibekov's support in his application to become a cosmonaut.) Krikalev also excelled in athletics, swimming and aerobatics and was a member of the Soviet national flying teams, winning the accolade of Champion of Moscow in 1983 and Champion of the Soviet Union in 1986. By this time, he had entered cosmonaut training, having been selected in September 1985, and worked for a time on the Buran programme. He was officially appointed as a 'test-cosmonaut' in February 1987 and in March of the following year he was teamed with Volkov. It would appear that Krikalev replaced Alexander Kaleri, who had suffered cardiac problems. Significantly, Volkov and Krikalev's training encompassed preparations for the installation of two new modules for Mir ... and they would be bitterly disappointed when those modules fell ever further behind schedule, beyond their tour, and in April 1989 led to an interruption of the hoped-for permanent occupancy of the station.

Two days after leaving Earth, on 28 November, Alexander Volkov expertly guided Soyuz TM-7 to a docking at Kvant's aft port and, for the first time, *six men* were aboard Mir. Years later, Krikalev would remark that this was "a worst-case scenario", as far as overcrowding on the station was concerned, since it doubled the 'normal' population and the situation was not helped by the fact that Mir was packed with almost 600 kg of equipment and supplies for the French mission. Specifically, Chrétien would operate a series of medical and technological experiments in support of the French-led Hermes shuttle project ... and this imposed great strain on the space station's electrical supply. Also draining was a six-hour EVA on 9 December by Chrétien and Volkov, whose primary purpose was the installation of the 240 kg French-built ERA experiment, which took the form of a 'bundle' of metre-long interconnected carbon fibre tubes. It was designed to unfurl automatically in just a few seconds into a hexagonal structure, measuring 3.6×3.8 m, to demonstrate the construction and rigidity of large assemblies in the microgravity environment. Understanding of the dynamics of space structures was essential if effective European participation in the forthcoming Space Station Freedom effort was to be realised. Early plans for the spacewalk to be made on 12 December, but it was brought forward by several days to permit the cosmonauts to undertake a second excursion if ERA failed to properly deploy. Clad in the upgraded Orlan-DMA suit, Chrétien poked his helmeted head out of one of Mir's multiple docking ports, leaned outside and began to unfurl handrails recessed into the hull. Next, he used springs and hooks to affix a rack of space exposure samples and was joined by Volkov to assemble ERA. At length, Sergei Krikalev commanded the structure to unfurl ... but it remained stubbornly folded. The spacewalkers tried shaking it and Volkov's rather unorthodox suggestion – to give it a good *kick* – was rejected by ground controllers.

Shortly before Mir passed out of communications range, the cosmonauts were told to discard ERA if it failed to remotely open. For Volkov, whose previous mission had failed to achieve its objectives and who would soon become the commander of Mir, there was *no way* that he was going to mark his first spacewalk with an abject failure ... particularly in such an important experiment, supplied by a major international partner. Whilst out of radio contact, he gave ERA several kicks ... and it *opened*. The men announced the news when communications were restored, but said nothing of precisely *how* they had achieved success.

Volkov (left) and Chrétien practice their EVA in the 'hydrolab' at Star City.

Chrétien had flown before, but *this* mission incomparably surpassed the handful of days he had spent aboard Salyut 7. "I found it much more attractive this time," he told the NASA oral historian, "because you think about the *nice* side of space flight." It was possible to look for changes on Earth over a period of *weeks*, rather than days. Short missions were quick-paced, with little time to look around, but a longer flight offered the luxury of being able to appreciate *where* he was, to really enjoy the surroundings and really enjoy the mission. However, the *real* experience of the mission for Chrétien was the spacewalk; an ethereal experience, like no other. "You forget about your space suit very quickly," he recalled, "so you're really in the impression that you are *free floating* ... just *swimming*. It was fascinating." Orbital darkness did not bother him: the lights on his helmet and the natural albedo of Earth provided an eerie illumination of their own.

Other successes were being achieved. Vladimir Titov and Musa Manarov had long since exceeded the 326-day achievement of Yuri Romanenko and on 15 December they reached 359 days in orbit, satisfying the Fédération Aéronautique Internationale's ruling that a new record must surpass its predecessor by at least 10 percent. Four days later, the Soyuz TM-6 spacecraft was brought out of hibernation and its systems were activated in readiness for a return to Earth on the 21st. (Since 1988 was a leap year, Titov and Manarov would actually have logged almost 366 days aloft by the time they landed.) Concern was still evident about the performance of the vehicle's software and this actually overloaded the computer, causing the cancellation of the first de-orbit opportunity at 9:48 am Moscow Time. At length, the backup software program was employed and, in a change to procedure, the orbital module was retained *throughout* retrofire. Titov, Manarov and Chrétien landed, some 180 km south-east of Jezkazgan, at 12:57 pm Moscow Time. The Frenchman had spent 25 days in orbit, making him by far the most flight-experienced non-American and non-Soviet spacefarer, whilst Titov and Manarov had smashed yet another endurance milestone and had demonstrated the ability of humans to remain aloft for extreme periods in the most hostile environment known. The record-breakers – who had totalled 365 days, 22 hours and 38 minutes since their launch from Tyuratam on 21 December 1987 and who had circled Earth an astonishing 5,790 times – were welcomed back to the Home Planet under low clouds and thick fog and below-freezing temperatures on the desolate, windswept, snowy steppe of Kazakhstan. "*That* was a great landing," recalled Chrétien. "No one there. The helicopters ... could not find the Soyuz in the fog. I think it took them at least 30 minutes to find us!" It gave the three men a period of personal quiet time to acclimatise themselves to their new surroundings, before the hordes of physicians and journalists flocked to the scene.

If Titov and Manarov had been greeted by a terrestrial deep freeze, then Mir itself would fall victim to a freeze in human visitors in the following year, 1989. The Soviets had long planned to despatch a pair of 20,000 kg additional laboratories to the station – the first equipped with a dedicated EVA airlock and the second loaded with microgravity science equipment – but in February 1989 it was announced that the first of these modules, known as Kvant-2, would be delayed, primarily because the production of the other one, called 'Kristall' had slipped. (They would be positioned on opposing radial ports at the front of the station, and it was desirable that they be installed in rapid succession in order to preserve the balanced configuration.) Certainly, Volkov and Krikalev had trained to commission the new modules and perform a series of EVAs from Kvant-2's airlock to demonstrate the Soviet equivalent of America's Manned Manoeuvring Unit. Crew assignments indicate that the new modules were expected at around this time. Space analyst Jim Oberg had speculated in mid-1988 that Alexander Serebrov's lengthy 'disappearance' from either a prime or backup crew slot in the wake of Soyuz TM-2 was indicative of his being 'rested' in preparation for a dedicated mission involving one or more of the new modules. Instead of launching aboard Soyuz TM-8 in April 1989 to replace Volkov, Krikalev and Polyakov and continue the permanent occupancy of Mir, a four-month hiatus in manned operations was unavoidable. For the second time in its orbital life, Mir was to be left uncrewed until the early autumn ... but when Serebrov and Alexander Viktorenko finally boarded the station on 8 September 1989, they would switch on the station's lights, figuratively and literally, for almost a full decade of uninterrupted operations. Not until the dying months of the 20th century would the lights aboard Mir be dimmed and in those ten years, from September 1989 until August 1999, there would never be an occasion on which at least two humans circled Earth at *all* times. It is a remarkable accomplishment which cannot, and *should not*, be underestimated, for one should bear in mind that, even by the turn of the millennium, the pioneering flight of Yuri Gagarin remained very much in living memory.

In that decade, Mir would seesaw violently between triumph and near-disaster: it would expand, just as the Soviets originally intended, into an impressive 'dragonfly' (as Bryan Burrough called it) with modules, arrays, cranes and antennas sprouting in a myriad directions, with no fewer than *five* research laboratories, and would host dozens of crews from a dozen nations. On the other side of the coin, however, it would fall victim to the kind of funding issues which were an integral feature of post-Soviet Russia and from 1989 onwards the country's space budget established itself on a slippery slope which would see it plummet by 80 percent in the coming years. That lack of funding would delay modules, scrub missions and oblige cosmonauts to live aboard an outpost no longer fit for purpose. As early as April 1989, half of Mir's scientific equipment was effectively inoperable and the lack of additional modules made the interior cramped, as Krikalev had observed. Batteries would not properly hold charge, creating chronic power supply difficulties. These difficulties in space translated to more everyday issues on Earth; pressing and crucial issues, such as livelihoods. Across the vast expanse of the largest nation on Earth, from St Petersburg to Vladivostok, from Pevek and Anadyr in the east to Derbent and Baltiysk in the west, millions of citizens of the old Soviet Union – and, from 1992, the 'Commonwealth of Independent States' – would wait weeks, or even months, to receive even the most meagre paycheck. Lights in Star City would be dimmed to save electricity, flight controllers would moonlight as taxi drivers to make ends meet and the West and its allies would be increasingly courted: a Japanese television station would pay millions of dollars to fly one of its journalists to Mir and visitors would fork out $200 for a day exploring flight simulators and trying out space suits in places still shrouded in military secrecy only a few years earlier. If the mid-1980s offered a glimmer of hope for political reform, then the 1990s yielded a glimpse of the harsh reality of transitioning from single-party authoritarianism to multi-party pluralism; "the demons of democracy" as Burrough put it. What happened to the old Soviet Union during those times makes it easier to understand the attraction of Communism – in which the price of having a secure job and healthcare and a pension was paid by a lack of real freedom of expression. Svetlana Savitskaya's parents would have turned over in their graves to see what became of their once-proud nation.

With such severe austerity and profound economic strife on Earth, there was little to justify massive expenditure on a space station whose relevance to the average Russian in the street no longer carried the same sense of propagandist pride enjoyed in the days of Khrushchev or Brezhnev. Nor were the cosmonauts themselves any longer revered as the 'heroes' that they once were; rather, they were 'workers', earning the equivalent of a thousand dollars for an EVA, and Mir's commanders would worry about mission success for more than just the obvious reasons ... not only from a professional perspective or a sense of national pride, but with a wary eye focused on the bonuses (or lack thereof) that he could expect to receive in pay and perks upon his return to Earth. Such perks and professional accolades and a sense of national pride must certainly have been on Alexander Volkov's mind on the morning of 27 April 1989, when he separated Soyuz TM-7 from Mir and brought the capsule and his comrades, Krikalev and Polyakov, back to Earth at 5:59 am Moscow Time. Polyakov had experienced a unique mission: observing the physiological adaptation of two long-duration crews of cosmonauts, whilst undertaking a long-duration mission of his own. For Volkov and Krikalev, on the other hand, their *next* mission together, late in 1991, would come at one of the most divisive and tumultuous times in Russian history. Their country would have morphed from a vast Union, covering a sixth of the world's land area, spanning 10,000 km from east to west and 7,000 km from north to south, and a dozen time zones, into a loose arrangement known initially as the Commonwealth of Independent States. In some cases, the 'independence' of its former satellite republics was hotly contested and in the 1990s many of the old Soviet states were torn by political and ethnic conflict. Nowhere was 'The End' of the Soviet Union more clearly delineated than in space, when Sergei Krikalev launched to Mir in May 1991 as a 'Soviet' cosmonaut ... and returned to Earth, ten months later, as a citizen of an entirely new and very different nation. A new era had begun and its twists and turns would affect the future progress of our species in ways which could hardly be imagined.

5

New beginnings

MORALE BOOST

One day in June 1987, Brewster Shaw poked his head into Robert 'Hoot' Gibson's office with an idea. More than a year had elapsed since the destruction of Challenger and morale had hit rock-bottom. Several astronauts had already departed NASA or been reassigned elsewhere in the agency – Don Lind, Roy Bridges, Owen Garriott, Bob Stewart and Gordon Fullerton in 1986, alone – whilst the remainder were still mourning their fallen friends. Yet there was a glimmer of hope on the horizon. The Shuttle was gearing up to fly again, the first post-51L crew was named in January and modifications to the Solid Rocket Boosters were going well. A new class of astronaut candidates – the 12th – had been selected, but something more was needed to raise the corps' spirits. A 1950s-style Saturday night 'sock hop', with hoop skirts and jeans, was proposed and Shaw wanted to put together a four-man rock band for it. Both he and Gibson played guitar; so too did George 'Pinky' Nelson, who was quickly recruited. "Brewster played rhythm guitar, didn't want to play lead," Gibson remembered, years later. "Pinky could play bass or rhythm, didn't want to play lead. I could play rhythm and I didn't want to play lead, either, but my protest was the weakest, so I became lead guitarist!" Rookie astronaut Jim Wetherbee owned a drum kit and, despite having not played for 17 years, was invited to a rehearsal. When the big night finally arrived, the foursome played a handful of songs and closed with a medley of Chuck Berry classics. Little could they have known that their band – which Gibson named 'Max Q' to honour the maximum aerodynamic turbulence encountered by the Shuttle during ascent – would not only survive, but thrive and endure in a slightly different form to this very day.

It had been a difficult year since Challenger. NASA was rudely awakened to the reality that no more commercial primary payloads would fly aboard the Shuttle and the agency's early hopes that the reusable spacecraft would return to flight operations in the spring of 1987 had very quickly become untenable. Not surprisingly, the extensive redesign and recertification of the SRBs took centre

Though Mike Lounge dubbed them "political eyewash", partial-pressure suits would become a regular feature of each Shuttle crew from STS-26 onwards. Here, astronauts Dick Covey, Mike Lounge and Dave Hilmers are pictured in the simulator, clad in training versions of the suits.

stage. Morton Thiokol created a pair of full-scale, short-duration 'simulators' to evaluate the changes. The first was the joint environment simulator, which was test-fired on seven occasions between August 1986 and July 1988 and was designed to evaluate the field joint hardware, insulation and the performance of the O-ring seals. The second was the nozzle joint environment simulator, tested between February 1987 and August 1988, whose purpose was to test the integrity of the joints between the nozzle and the main booster casing. An engineering test motor was fired in May 1987 to evaluate the performance of SRB heaters, the additional O-ring and the ability of external graphite composite stiffener rings to prevent joint rotation. A pair of demonstration motors were fired in August and December to qualify the redesign features and in April 1988 the process of recertification for flight got underway with Qualification Motor Six.

In addition to the lengthy programme of modifications to the boosters, the enforced down time was spent attending to other critical areas on the orbiter itself: the troublesome main engines, the brakes and tyres, the development of partial-pressure suits to provide hypobaric protection and permit a measure of survivability for the crew and the implementation of an escape system. Newly-appointed Associate Administrator for Space Flight Dick Truly had admitted that the Shuttle was too mature in design to install any kind of 'escape pod' or ejection seats for the whole crew, but two concepts which *were* explored were a tractor rocket and a curved, telescoping pole to extract astronauts from the cabin in an emergency situation. Both methods, noted Dennis Jenkins, were useful only below speeds of about 350 km/h and below altitudes of some 6 km, whilst the vehicle was in controlled, gliding flight. Ultimately, the pole was selected. In an emergency, an astronaut would remove the pole from its mounting on the middeck ceiling, pyrotechnically jettison the side hatch, extend the pole, affix a lanyard hook on his or her suit and essentially *slide out*, parachuting to safety. The pole was meant to 'guide' the astronauts for 3-4 m, propelling them on a trajectory which would take them 'underneath' the Shuttle's port-side wing. Tested extensively in February and March 1988 by Navy jumpers from a Lockheed C-141 Starlifter aircraft, it was predicted that around 90 seconds would be required for a crew of eight to evacuate the orbiter. The development of the pole was closely watched by astronaut Steve Nagel. "The charter from the Rogers Commission," said Nagel, "was to provide a controlled, gliding flight escape system, and this does it. I think you're better off having a pressure suit and a parachute and some survival gear. Even if it's out of control, somebody might have a chance of climbing out the hatch. There's *plenty* of evidence in World War II of crew members getting out of bombers with wings off, if they happen to be close to an opening or a hatch." The pole modifications were completed on Discovery by April 1988. A return to pressure suits aided the astronauts' chances of survival ... and Dick Covey would claim some credit for having contributed to the choice of *colour*. "As they developed this idea of bailout," he said, "the first suits they got were *dark blue* and the life rafts were black of dark blue." The astronauts reasoned that, after bailing out of a crippled orbiter and floating in the ocean, hundreds of miles offshore, they would *never* be seen, so it was decided to use *orange* rafts instead. "Well, if we're going to do *that*," was Covey's line of thought, "why are we going to have *blue* suits? Why don't we have *orange* suits?" Many of the blue suits had already been developed for training purposes and there are several images of early post-51L crews wearing them.

Political eyewash was Mike Lounge's opinion of the suits. "And I really feel bad," he said in his oral history, "because it's an extra ... weight in the crew cabin that takes away from the payload-carrying capability of the Space Shuttle and it is just *no* value added. It's value *subtracted*. What little could you do in the event something went wrong, you could do *less* of it when you're burdened by these suits. I was *totally* against it and still am. They offer *no* value." Yet the suits were here to stay. Safety was the new byword for NASA. A new Office of Safety, Reliability and Quality Assurance had been established, reporting directly to NASA's new Administrator, Jim Fletcher, and chief astronaut John Young had proven highly critical of Shuttle management's appalling attitude towards safety. There were other concerns, too. President Ronald Reagan had already promised the nation that a permanent space station would be built ... and *that* needed four orbiters. In June 1986, plans to use an already extant set of Shuttle spares to build a new orbiter were opposed by the White House Chief of Staff, Donald Regan, and by powerful voices within Congress, who felt that $2.8 billion could be better spent on an entirely new vehicle. Young countered that President Reagan's space station plan would be unachievable *without* four operational vehicles and in September 1986 the construction of Orbiter Vehicle 105 – later named 'Endeavour' – was approved. It would be ready to make its maiden voyage in the spring of 1992. Until then, NASA would have to rely solely upon its three surviving Shuttles: Columbia, Discovery and Atlantis.

By the end of 1986, it was clear that the return to flight would take longer than expected and a provisional launch date for 'STS-26' – the first post-Challenger mission, to be flown by Discovery – in February 1988 was held in serious doubt. Dick Truly had already stated that the mission would be accorded all of the caution which had accompanied the maiden flight. Discovery's ascent trajectory was to be designed to minimise the risks of a Transoceanic Abort Landing, it would launch in warm weather and in the hours of daylight, it would land at Edwards Air Force Base in California ... and its crew would all be veteran astronauts. For a time, the most likely contender to command STS-26 was Bob Crippen, although in November 1986 he accepted a new position as deputy head of Shuttle Operations at NASA Headquarters. Yet, with Truly and George Abbey, Crippen *did* play a role in choosing the man who would command the flight and instinctively knew that it would receive much scrutiny. Rick Hauck seemed an obvious choice, having flown two Shuttle missions, including the dramatic Palapa and Westar retrieval. Much has been written about the level of 'fairness' or 'unfairness' associated with crew selections at this time, but Hauck's team actually made perfect sense, since for the most part it comprised the men who would have flown on the 61F Ulysses mission. Pilot Roy Bridges returned to the Air Force in May 1986 and was replaced by 51I veteran Dick Covey, whilst mission specialists Mike Lounge and Dave Hilmers remained. "Rick ... hinted that we might fly soon," said Lounge, "or we might be on the crew that flew the return." The name of the third mission specialist, Pinky Nelson, caused some consternation within the astronaut corps, for he had only recently flown on 61C and in the wake of Challenger had returned to academia. "Pinky was well-liked," Mike Mullane wrote in *Riding Rockets*, but admitted that his sabbatical to the University of Washington highlighted in many minds that he had not paid his dues to the recovery effort. "He had the additional plum," Mullane continued, "of flying back-to-back missions."

Nelson, for his part, was simply thrilled to be flying again ... although his wife and daughters burst into tears when they first heard the news. As for the politics of crew selections, his belief was that NASA assigned most of the original Ulysses crew to STS-26 and Nelson's previous EVA experience made him an obvious choice. Yet

Inside the cavernous Vehicle Assembly Building, Discovery is prepared for attachment to her External Tank and twin Solid Rocket Boosters for STS-26.

the negative sentiment in some quarters of the astronaut office continued for a time. It was sour grapes, perhaps, but others were equally vocal that the selection process was neither fair, nor rational, at this time. Hauck had been permitted to *select* Covey as his pilot, an extremely rare practice. "He figured out that he was going to fly the first flight," Covey told the NASA oral historian, "and I think at that point started *lobbying* to get me on ... as his pilot." For his part, Covey was tickled by the irony that he had been the *last* of his class to fly and was already in line for a second mission, but *his* selection as Hauck's pilot was unsurprising. Before Challenger, he was assigned as the ascent capcom for Hauck's 61F mission and had already worked extensively with the crew on abort scenarios. In July 1986, Covey was already working with Hauck, Lounge and Hilmers in the Shuttle Mission Simulator at JSC. He was part of the team.

Setting aside the enormity of the fact that STS-26 would be the first Shuttle voyage after a major disaster, the mission plan itself was relatively straightforward – a 'vanilla flight' – since it would be just four days long and feature the deployment of a Tracking and Data Relay Satellite (TDRS-C) atop an Inertial Upper Stage to replace the one lost aboard Challenger. Rick Hauck admitted that it would be "the *safest* flight we've flown", but this did not prevent the space agency's senior leadership from exaggerating the significance of the mission. STS-26 *was* a hugely important milestone, but for the jokers in the astronaut office, and particularly the crew of STS-27, it offered an opportunity for humour. One evening, Hauck's crew were present at a fund-raising event for a Challenger charity at the Wortham Center in Houston. At its conclusion, the master of ceremonies brought a young girl onto the stage to sing Lee Greenwood's *I'm Proud to be an American* and the crew of STS-26 was introduced. "At this cue," wrote Mike Mullane, "the orchestra pit platform began a slow rise. Artificial smoke swirled about it and the spotlights flashed through the vapour. And *there*, to the astonishment of every astronaut, were Rick Hauck and Dick Covey. They stood like carvings on Mount Rushmore: chins jutted out, chests puffed up, arms rigidly at their sides, steely eyes straight ahead." Shortly thereafter, the crew of STS-27 – commander Hoot Gibson, pilot Guy Gardner and mission specialists Mullane, Jerry Ross and Bill Shepherd – plotted their revenge. Two days later, at the astronaut office's Monday morning meeting, Gibson was asked if he had any STS-27 issues to discuss. At this stage, the plan went into action. Jerry Ross pressed a button on a boom box, triggering *I'm Proud to be an American*, whereupon Mullane and Shepherd set off a pair of CO_2 fire extinguishers to create a smoky effect and Gibson and Gardner, who had clipped ties onto their flight suits, slowly rose from their chairs in an outrageous parody of Hauck and Covey. Watching the proceedings was Kathy Sullivan. "They rise, *all* the way, until they're standing straight and tall," she said. "Then Mullane shuts off the boom box, the fire extinguisher goes out, they sit back down and Hoot says calmly, 'No, we don't have anything!'" The office exploded with laughter.

On Independence Day, 4 July 1988, at a darkened Kennedy Space Center, Discovery – mated to her External Tank and twin SRBs – left the Vehicle Assembly Building and made the slow roll to Pad 39B. Dave Hilmers was in attendance and spoke to the assembled crowd of NASA employees who had worked tirelessly for this day to come. "For over two years now," he said, "each one of us here, tonight, has had a dream, that one day a Shuttle would once again make its way to the launch pad to launch Americans into space." That launch had already slipped from February to June to August and, now, to September. Shortly before rollout, a tiny leak, deep within Discovery's left-hand Orbital Manoeuvring System (OMS) pod, was discovered, but was repaired on the pad. Then, on 10 August, after one false start, the orbiter's three main engines were test-fired for 22 seconds and on the 29th TDRS-C and its attached IUS booster arrived at Pad 39B and were installed into the payload bay. When STS-26 finally set off on 29 September, no fewer than *thirty-two* months would have elapsed since the loss of Challenger and her crew. The day itself was a calm, warm one, which Rick Hauck remembered lucidly, even many years later. Radiosonde balloon soundings had highlighted an upper-level wind shear which might pose a constraint to the launch and the astronauts left the Operations and Checkout Building for what they assumed would be a fruitless exercise. (Later, Hauck would jokingly thank Bob Crippen for convincing them that they *weren't* going to launch, thereby allowing them to enjoy the otherwise beautiful morning!) The winds *did* conspire to delay the launch by an hour and 38 minutes and technicians also had to attend to failed vent fan fuses in the cooling systems of Covey and Lounge's suits. At 11:28 am EST, the Launch Director polled his team for their final status. As Hauck's crew listened in, they expected Crippen to declare a "No Go" on the basis of the high-level winds ... but were surprised when he gave his consent for them to fly. The excitement began to build in the cockpit and at 11:37 am the marshy Florida landscape was rocked by the tremendous roar of three main engines and the golden flame of two boosters, carrying men into the heavens once again.

For Dick Covey, watching the main engines, it brought back memories of Challenger in more ways than one ... for he had been one of the capcoms, sitting in Mission Control on 28 January 1986, who spoke the last words to Dick Scobee. On that terrible day, Covey had been too engrossed in his procedures to glance over at a monitor and *see* the carnage, but fellow astronaut Fred Gregory, seated to his right as the lead capcom, saw it immediately and recognised it for what it was. As Gregory saw the video feed and Covey saw the data on his screen freeze, both men's jaws hit the floor. At first, they wondered if a contingency abort was in progress. Had the range safety destruct system accidentally triggered? Had the SRBs been prematurely jettisoned? Had a main engine exploded? Should Covey radio further instructions to the crew and, if so, *what* could he possibly say? For the longest time, no one could be certain. At length, it was Flight Director Jay Greene who spoke next. "Lock the doors," was his instruction, telling all controllers to secure their data for the investigation ... and, instantly, Gregory and Covey and *everyone* else in the room knew that they were beyond contingencies or abort scenarios and the possibility of recovery was gone. Their friends were either dead or in the process of dying.

Almost three years later, the *wait* for launch on STS-26 was exciting, but uncomfortable, and not just on account of the partial-pressure suits. "We were still using the urine collection devices," said Covey, "and *those* aren't particularly comfortable or easy to use. When you're on your back for four hours out there ... then it's a *long* time. It's uncomfortable." The launch itself resembled previous launches, although all five men were keenly aware of what had happened to their predecessors and the level of anxiety peaked as they neared the 73-second psychological barrier beyond which Challenger had been unable to pass. Before 51L, Rick Hauck felt that NASA had Shuttle launches 'wired', but now he was relieved when the SRBs were jettisoned, as planned, two minutes into the ascent. When the fateful *Go at throttle up* call had come from Mission Control, perhaps not wanting to mimic Dick Scobee's response, Hauck had replied simply, "Roger, Go." Covey remembered as the Mission Elapsed Time clock ticked past 88 seconds, "we're all kind of thinking about what happened the *last* time the Space Shuttle had gotten to that point". With six more minutes to go before Main Engine Cutoff, Hauck relayed the progress of the flight to Nelson on the middeck as Discovery passed through Mach 16 and onwards. At length, at 11:46 am, the sound of the engines was gone and the ghosts of Challenger, finally, were laid to rest.

Several months earlier, it had been Dave Hilmers who suggested that the crew should commemorate their lost friends in some way. "We shared a personal loss in the class," said Covey, "and personal loss of friends across those classes." Hilmers had written it and had given a copy to fellow astronaut Lacy Veach, one of their capcoms, before launch. On the third day of the flight, on Sunday 2 October, each crew member took turns to deliver their eulogy to the Challenger astronauts. "It was something that needed to be done," said Hauck. "It was a need that someone during the mission needed to say something that all of us could reflect on ... and I gathered from what was said later by people in the office ... that it captured the thoughts." Additionally, the crew felt the need to thank the ground teams who had invested so much sweat and tears in preparing them to fly. The Orbiter Processing Facility teams had labelled themselves 'Loud & Proud' and, on occasion, they would wear 'loud' Hawaiian shirts to work. When electrical power was provided to Discovery, the STS-26 crew was in attendance and were made honorary members of the 'Loud & Proud' crowd and presented with honorary Hawaiian shirts, which they took into space with them. "Once we had the weighty issues behind us," said Hauck, "we decided *now* is the time to break out the Hawaiian shirts." They were passing over Hawaii, incidentally, at the time, and downlinked some video of themselves, clowning around on the middeck in their shirts and sunglasses. *Life's a Beach* was Hauck's comment. For Hauck and Nelson, they had already begun thinking that STS-26 would be their final flight. "Rick and I had just a wonderful time during the mission," Nelson remembered. "We spent a fair amount of time, just the two of us, up on the flight deck, looking out the window and taking it all in."

Aside from the enormous responsibility of getting America's human space programme back on track, the deployment of TDRS-C occurred in a relatively straightforward fashion at 5:50 pm EST, some six hours and 13 minutes after launch. The IUS booster functioned perfectly, achieving geostationary orbit, and the satellite manoeuvred into its position above the Pacific Ocean at 171 degrees West longitude.

Four days in space came to a spectacular conclusion at 9:37 am PST (12:37 pm in Florida) on 3 October, when Hauck and Covey brought Discovery smoothly onto Runway 17 at a sweltering Edwards Air Force Base. "A great ending to a new beginning," was the congratulatory call from Blaine Hammond, the re-entry capcom, as the orbiter rolled perfectly to a stop. Vice President George Bush was in attendance, as was NASA Administrator Jim Fletcher, California Governor George Deukmeijian and famed aviator General Chuck Yeager. Hauck fluttered an American flag as he descended the steps from the orbiter, a move which *Time* magazine derided as a 'staged' example of a politician using the space programme for political gain. "I remember writing a letter to the editor of *Time*," said Hauck, "where this was opined and I stated the facts that ... we were *not* prompted by *anyone* to bring a flag. That was *our* idea and we were very proud to have the Vice President of the United States meet us at the bottom of the steps. It didn't matter whether he was Republican, Democrat or whatever."

SWINE FLIGHT

Before the Challenger accident, three main 'markets' had been identified for the Shuttle. Firstly, there was the commercial market, launching domestic and foreign payloads, followed by the NASA scientific market and finally the military market, dominated by the national security requirements of the Department of Defense. Former NASA Administrator Jim Beggs was unwavering in his conviction that removing commercial payloads from the Shuttle was a mistake – "The argument was that you *shouldn't* risk lives," he said, "but you're risking lives *anyway*, when you fly the Shuttle" – but in the wake of Challenger the space agency was left with, primarily, its own scientific missions, including Spacelab, and a lengthy backlog of DoD satellites, optimised to be launched *only* by the reusable orbiters. The reader will recall from Chapter 3 of this volume that the decision to partially declassify 51J, yet retain the top-secret classification over the *other* military Shuttle flights, is significant, for many of the payloads placed into orbit on eight missions between December 1988 and December 1992 were almost certainly reconnaissance or intelligence-gathering satellites of one sort or another. This is particularly true of the payload for the second post-Challenger flight, STS-27, which is almost universally accepted – though *not* officially acknowledged – to have been the first in a series of radar-imaging and all-weather surveillance platforms, known by the code name of 'Lacrosse'. (More recent satellites in this series have also been designated 'Onyx'.) Over the years, a handful of images of these payloads under construction have trickled into the public domain and it subsequently became clear that Lacrosse was one of the largest satellites ever to be deployed from the Shuttle.

Built by Lockheed Martin of Denver, Colorado, Lacrosse was financed by the National Reconnaissance Office and the Central Intelligence Agency. Under the directorship of George Bush, in 1976 the CIA initiated efforts to develop a network of radar imaging and all-weather surveillance satellites. A prototype imaging satellite was tested in 1982 and Lacrosse was approved the following year. Design features of the giant satellite included a large radar antenna and solar arrays to provide electrical power for its transmitter. Some reports have hinted that the solar arrays have a wing span of perhaps 50 m, leading to speculation that the power of the radar itself might have been in the region of 10-20 kilowatts, several times greater than previous orbital radars and possibly yielding a resolution of better than a *metre*. Orbiting at an altitude of around 450 km, inclined 57 degrees relative to the equator, each Lacrosse cost around $1 billion and rumour about its precise purpose has been rife for more than two decades. Some sources have suggested that B-2 Spirit bombers would have downlinked Lacrosse targeting data in real time to seek and destroy Soviet intercontinental ballistic missile launch silos. The website www.globalsecurity.org has noted that, although later Lacrosses were launched aboard expendable Titan boosters, the unit placed into orbit by STS-27 was secured in Atlantis' payload bay by means of standard trunnion attachment points and the crew themselves later received military citations – now declassified – which explicitly state that the satellite was deployed using the Canadian-built Remote Manipulator System (RMS).

The crew was formally announced in September 1987, with an anticipated launch date in the late summer or early autumn of the following year. Old 62A crewmates Guy Gardner, Mike Mullane and Jerry Ross would be aboard, commanded by Hoot Gibson and rounded out by a third mission specialist, William McMichael Shepherd, who became the first member of the 1984 class of astronauts to fly into orbit. Shepherd is perhaps best known today as having commanded the first crew aboard the International Space Station, but his pre-NASA experience as an underwater demolition expert and Navy SEAL brought with it an interesting anecdote: that he had *killed* a man with his bare hands, whilst on a covert military operation. Totally inaccurate, Shepherd later told an interviewer: "The story was *with knives*," he deadpanned. The tall tale started making its rounds in Houston soon after his selection in May 1984. "I heard about it," he said, "and I thought, this is just really *too good* to deny, so I just wouldn't comment on it for a long time, which made people *really* wonder!" Shepherd came from Oak Ridge, Tennessee, where he was born on 26 July 1949, the son of a naval aviator who had flown in combat during the Second World War. It is perhaps hardly surprising that Shepherd also followed a naval career, but by his own admission he had grown up with a natural love of boats and the water. After high school in Phoenix, Arizona, he entered the Naval Academy and received a degree in aerospace engineering in 1971. "I wanted to be a pilot," he told an interviewer, "but unfortunately could not pass my eye exam, so I became a diver." Shepherd subsequently commenced training as a member of Basic Underwater Demolition/SEAL (BUD/S) and later qualified as a Navy SEAL. In 1978, he obtained a master's degree in mechanical engineering and the degree of Ocean Engineer from MIT. Shepherd unsuccessfully tried for NASA in 1980 and was finally selected to join the next class in May 1984.

If Bill Shepherd was the first member of his class to fly into space, then Atlantis' pilot, Guy Spence Gardner, was one of the *last*; selected in May 1980, he had waited an unenviable eight and a half years before his first launch into orbit. He was born on 6 January 1948 in Altavista, Virginia, and attended George Washington High School in Alexandria, then entered the Air Force Academy. Gardner earned a degree in astronautics, mathematics and engineering sciences in 1969 and a master's credential in astronautics from Purdue the following year. Initial flight instruction at Craig Air Force Base in Alabama and subsequent training in the F-4 Phantom II led to his assignment to Thailand in 1972, where he flew 177 combat missions over Vietnam. Gardner returned to the United States as an F-4 instructor and operational pilot at Seymour Johnson Air Force Base in North Carolina. He completed test pilot school at Edwards Air Force Base in 1975 and served there as a test pilot and an instructor pilot. After unsuccessfully applying as a pilot candidate in the 1978 group of astronauts, Gardner was successful two years later. Had the Challenger accident not occurred, he would have piloted the first Shuttle mission into polar orbit from Vandenberg Air Force Base in July 1986.

Hoot Gibson and his crew quickly earned their mission the moniker of 'Swine Flight' from the office secretaries and were even given novelty pigs' snouts. (The nickname came about from Gibson's penchant for making animal-like snorting sounds whenever attractive women were in the vicinity.) As with Missions 51C and 51J before it, STS-27 was totally classified and virtually nothing – not even the exact launch time – would be revealed until 24 hours prior to liftoff. "All the software was classified," Mike Mullane related. "Everybody supporting it had to have clearances, but that was pretty transparent to us." Jerry Ross had worked on classified Shuttle payloads before his selection as an astronaut and he was aware of many of the protocols and procedures involved. "You have to work within secured facilities," he said, "facilities that are swept, so that you don't have any inadvertent electronic signals or voice going outside it, so it constrains you significantly on how and where you can do your business. When we travelled, we travelled on basically classified orders. I'd tell my wife I was leaving. I couldn't, in most cases, tell her when I was coming home." Not for five years *after* the flight were the astronauts permitted to reveal even the slimmest of facts: that they had used the RMS to deploy their payload. Many of the other DoD missions remained enshrouded in a similar cloak of secrecy, even decades later. A launch attempt on 1 December 1988 was scrubbed, due to unacceptable cloud cover and strong winds at the Kennedy Space Center, but all seemed ready for a mid-morning liftoff on the 2nd. The countdown clock ticked down to T-31 seconds, just prior to Auto Sequence Start, and was held whilst the weather at the Transoceanic Abort Landing site in Morocco was monitored. Guy Gardner had already started Atlantis' Auxiliary Power Units and a 'Go' or 'No-Go' decision had to be made quickly. At length, the good news came from the Launch Director: "Atlantis, the TAL weather is acceptable. We'll be picking up the count." Within moments, the clock started ticking again. At ten seconds, the Go was given for the ignition of the main engines, which promptly roared to life. Launch came at 9:30 am EST and within nine minutes the second post-51L crew was in orbit … though *not* safely.

The near-calamity which would engulf the mission could hardly have been further from anyone's mind in the hours after launch, as Mike Mullane uncradled the RMS from the port-side sill of the payload bay and grappled Lacrosse for deployment. "For once, the incredible beauty of the Earth passed unseen beneath me," he wrote. "I had eyes only for the payload, Atlantis and the robot arm … I steered the end of the arm over the payload grapple fixture and fired the snare, which rigidly latched the payload to the arm." Downstairs, in the airlock, Bill Shepherd watched through a porthole to ensure that the clearances between the satellite and the bay were acceptable. At length, he 'flew' Lacrosse on the arm to its release attitude and called to Gibson. Mission controllers gave them a Go for payload release and Lacrosse drifted away into the inky blackness. At this stage, one of the great unknown stories of STS-27 unfolds … the rumour of a 'secret' EVA.

Not surprisingly, Mullane made no reference to any such event in his memoir, but more than a decade later, in February 2001, astronauts Tom Jones and Bob Curbeam had just completed their third EVA to install the US-built laboratory module, Destiny, onto the International Space Station. As well as being a significant mission, this particular EVA was itself significant. It was the *hundredth* American spacewalk and NASA had trumpeted this fact beforehand. Then, just as Jones and Curbeam were about to make a comment, they were told on a private communications loop to keep quiet. In his autobiography, Jones noted that "somebody had done a recount" and the actual hundredth EVA had been a couple of days earlier, on their *second* spacewalk. "How could that happen?" asked Michael Cassutt in an August 2009 article for *Air & Space* magazine. "Had there been a secret spacewalk that never made it into the official tally?" The only hint came when one of the STS-27 astronauts made an offhand remark that they had experienced problems with Lacrosse after its deployment, which necessitated a rendezvous and some kind of 'repair'. Of course, a rendezvous with a troublesome payload does not necessarily imply that an EVA was needed to fix it – a balky antenna, for example, could have been easily unfurled with a nudge from the RMS – but the question of whether a 'secret' EVA occurred on this secret mission, even more than two decades later, has never been answered. In 1993, a high-ranking intelligence official travelled to Houston to present National Intelligence Achievement Awards to astronauts who had participated in classified Shuttle missions. For the first time, they could wear the medals in *public* and Hoot Gibson was permitted to disclose that he had "returned" to the STS-27 payload. "We separated from it and it had a problem," he later told an interviewer from the Smithsonian. "We re-rendezvoused with it and assisted with fixing it, separated again and left it." Those handful of words are amongst the most tantalising. In July 2009, Internet chatter on the website www.nasaspaceflight.com included speculation that Atlantis' crew manoeuvred close enough to Lacrosse to employ a low-gain antenna to command the activation of a failed high-gain dish. Still others remarked that EVA communications are conducted on unsecured UHF frequencies and that, if a spacewalk *did* occur, its progress *should* have been detectable on the ground. One observer remarked that 'concealing' an EVA would be hardly worth the effort, since it would have revealed nothing of substance about the payload. Until STS-27 is declassified, we will almost certainly never know the truth.

Early on the second day of the mission, 3 December, the astronauts were awakened with disturbing news. A review of the launch video had clearly shown something – probably a piece of ablative insulator – breaking away from the nose of the right-hand SRB about 85 seconds into the ascent ... and striking Atlantis' fragile thermal protection system. The crew had already seen a 'white' material on the forward windows. If the orbiter's heat shield had been breached, it could spell disaster during their fiery descent through the atmosphere at the end of the mission. Gibson's crew were instructed to use the cameras on the RMS to carry out a visual inspection. Mike Mullane started the procedure by gingerly manoeuvring the arm across the forward section of the payload bay and tilting its lower boom over the starboard side of Atlantis' nose. All looked good: a pristine 'checkerboard' of undamaged black tiles. Delicately, Mullane moved the arm further along the fuselage, towards the Shuttle's belly and the five men gaped in horror. White streaks were everywhere, clearly indicative of damage which had stripped off their outer black coating by a kinetic impact. "We could see that at least one tile had been completely blasted from the fuselage," Mullane wrote. "The white streaking grew thicker and faded aft beyond the view of the camera. It appeared that *hundreds* of tiles had been damaged and the scars extended outboard toward the carbon composite panels on the leading edge of the wing." Gibson's crew knew that a single missing tile was probably survivable, but the leading edges of the wings were subjected to the most intense heat during re-entry. Damage to *them* could mean only one thing: that the fliers of STS-27 were *dead men*. Some observers have offered this as additional evidence that Ross and Shepherd did *not* perform a secret EVA, arguing that they might otherwise have been directed to perform a visual inspection of several areas of damage.

The RMS did not have the capacity to reach far enough to inspect the leading edges. Mullane called Mission Control. "Houston," he said, with urgency clear in his voice, "we're seeing a *lot* of damage. It looks as if one tile is completely missing." He knew that flight controllers could see their downlinked video and was surprised when the capcom came back to them with an acknowledgement that the damage was not too severe. Gibson keyed his mike. "Houston, Mike is right," he emphasised. "We're seeing a *lot* of damage." Again, he was reassured that the damage did not represent a serious breach of the thermal protection system. The problem was made worse by the fact that the astronauts could not use their standard method of relaying imagery to the ground, due to the classified nature of the mission; they were forced to use a slow, encrypted transmission, which meant that the pictures received in Houston were of poor quality. Efforts to request higher resolution imagery were frustrated by the Department of Defense, which would not release it due to the level of classification. With only the encrypted images at their disposal, it seemed to flight controllers that the crew had been misled into seeing damage in what were actually poor lighting conditions and grainy images. Gibson was furious and convinced that Atlantis would *not* survive re-entry. To *him*, the pictures were clear. He knew *exactly* what he was seeing. However, there was little that could be done. He told his crew to enjoy the rest of the mission. There was no point in dying all tensed up.

Early on 6 December, the final steps were taken to prepare Atlantis for her return to Earth. All five astronauts donned their bulky orange suits and Gibson executed the firing of the twin OMS engines to begin the hour-long hypersonic descent through the atmosphere, towards a landing site at Edwards Air Force Base, on the opposite side of the planet. Mullane ought to have been strapped into his seat on the middeck, but had asked Gibson for permission to linger on the flight deck, behind the flight engineer, Jerry Ross, and shoot some video through the overhead windows. He would then get downstairs and into his seat as the G levels steadily increased. On autopilot, Atlantis descended across the sleeping Indian Ocean, over Australia and the expanse of the Pacific in barely 25 minutes. As they entered the 'sensible' atmosphere, compression against the orbiter's belly steadily heated the air molecules into a white-hot glow. "I wondered what was happening underneath us," Mullane wrote. "I had visions of molten aluminium being smeared backwards, like rain on a windshield. None of our instruments or computer displays showed Atlantis' skin temperature. Only Houston had that data." After an interminable period of waiting, punctuated only by the calm voice of Gibson, relaying altitude, velocity and G readings, Atlantis passed through the period of maximum atmospheric heating and wind noise could clearly be heard outside. Travelling at five times the speed of sound, Guy Gardner deployed a set of air data probes from the nose to provide airspeed and altitude data for guidance. Thirty kilometres above Earth, still travelling at almost four times the speed of sound, Gibson spotted the runway at Edwards. Travelling lower now, and passing below the speed of sound, the cockpit buzzed noticeably and observers at Edwards were startled by the Shuttle's trademark twin sonic booms, which echoed across the desert.

Throughout re-entry, in addition to his normal duties, Gibson had kept an eye on the gauge which showed the levels of deflection in the elevons at the rear of Atlantis' wings. "If we started to burn through," he told an interviewer, years later, "we would change the drag on that wing, which is *exactly* what happened to Columbia. I knew we would start developing a 'split' between the right and left wing elevon positions if we had excessive drag over on the right side. The automatic system would try to trim it out with the elevons. That is one of the things we always watched

The STS-27 astronauts and NASA Administrator Jim Fletcher (suited, second from the left) look over an area of damage on Atlantis after a miraculously safe touchdown.

on re-entry, because if you had half a *degree* of trim, something was *wrong*. Normally, you wouldn't see even a *quarter* of a degree of difference on that thing." Before entry interface, Gibson had privately concluded that if he *did* see an elevon split of beyond a quarter of a degree, he had about 60 seconds of life remaining in which to tell Mission Control *exactly* what he thought of their heat shield 'analysis' …

On this occasion, Atlantis was lucky, touching down on the lakebed Runway 17 at 3:36 pm PST (6:36 pm in Florida), after a mission lasting almost four and a half days. Although NASA acknowledged the damage in its post-mission report, published in February 1989, the agency officially described re-entry as "normal in all respects". Whilst this was, fortunately, the case, the experience of the crew and the condition of the vehicle were by no means normal. After wheelstop, the true horror of what the STS-27 crew had endured would become abundantly clear. "There was already a knot of engineers gathered at the right forward fuselage, shaking their heads in disbelief," wrote Mullane. "The damage was much worse than any of us had expected." No fewer than 707 tiles had been damaged, with one – in the area of the Shuttle's nose – missing completely. The latter had been located over a dense aluminium mounting plate for the L-band antenna, which may have prevented a burn-through at this point. All told, the damage extended from beyond Atlantis' nose cap, right along the belly and stopped just short of the leading edges of the wings. The lower surfaces of the elevons were untouched, but the OMS pods had suffered 14 impacts and the starboard rudder speed brake had also been hit repeatedly. The astronauts, engineers *and* NASA Administrator Jim Fletcher, who was in attendance, knew that Gibson's crew had cheated death by a hair's breadth. A Thermal Protection System Damage Review Team was immediately convened, chaired by John Thomas and Jay Honeycutt, and submitted its report in January 1989. They traced the most likely cause to a change in the manufacturing process, intended to improve the performance of the ablative materials needed to protect the boosters from aerodynamic heating. For Gibson and his men, it brought back memories of the single, primary directive of engineering: *Better* is the enemy of *good enough*. If it *worked*, there was no need to *fix* it. Thomas' team was present at the rollout review of the *next* Shuttle mission and scrutinised all External Tank and SRB manufacturing records to identify potential debris liberation sources. Years later, Hoot Gibson would remark that, had Atlantis been lost, the consequence would probably have been the end of the Shuttle programme.

As for STS-27's classified payload, the Lacrosse, it would remain in orbit for more than eight years, finally reaching the end of its operational life in the spring of 1997. In his interview for the Smithsonian, Gibson related that his crew showed their classified movie in a debriefing to the Joint Chiefs of Staff at the Pentagon. The movie itself was top-secret and had to be carried around by a courier, in a locked briefcase, *handcuffed* to his wrist! To this day, Lacrosse's precise purpose and achievements remain classified. Yet at one of their post-flight presentations in Houston, Gibson – demonstrating the sick humour of the fighter pilot – dropped one inaccurate clue. On 7 December, the day after Atlantis landed, a massive earthquake, reaching 6.9 on the Richter magnitude scale, had hit the Spitak region of Armenia, then part of the Soviet bloc. An aftershock, measuring 5.8, had followed within minutes. The entire city of Spitak was destroyed and at least 25,000 people were killed in the disaster, primarily due to poorly built apartments, combined with freezing winter temperatures, bad soil conditions and the nature of the notoriously earthquake-prone region, where the Arabian land mass met the Eurasian tectonic plate. Massive humanitarian aid was requested by Soviet General Secretary Mikhail Gorbachev. The United States, Belgium, Chile, China, France, Great Britain, Italy, Japan, Sweden, West Germany and others responded with rescue equipment, search teams and much-needed medical supplies. "I know many of you may have been very curious about our classified payload," Hoot Gibson told the gathered astronauts at the Monday morning meeting. The entire room hushed in anticipation. "While I can't go into its design features, I *can* say Armenia was its *first* target!" The military astronauts laughed, whilst the civilians cringed in disgust. As if to press the point further, with a twinkle in his eye, Gibson concluded: "And we only had the weapon set on *stun!*"

The response from the female astronauts: "Don't you guys *ever* grow up?"

EXCITEMENT AND DISAPPOINTMENT

When John Blaha was named as pilot on STS-29, the assignment brought with it both excitement and disappointment. There was excitement because, after waiting for more than eight years, he would finally get his chance to fly into space, tempered with disappointment that a crew member with whom he had previously trained – Anna Fisher – would *not* be aboard. In January 1985, they had both been assigned to Mission 61C, along with commander Mike Coats and fellow mission specialists Norm Thagard and Bob Springer. Within months, the situation had changed: in September, following the resignation of John Fabian, Thagard was moved onto the 61G Galileo mission, and Jim Buchli took his place. The five astronauts continued to train, joined by a pair of payload specialists – Royal Air Force test pilot Nigel Wood and Indonesian microbiologist Pratiwi Sudarmono – for Mission 61H in late June 1986. The loss of Challenger altered those plans markedly, but the 'core' crew remained together and performed a 56-hour Shuttle flight simulation at JSC in April 1987. Blaha was not alone in expecting them to stay together as a crew, but when the STS-29 announcement came in March 1988 Fisher was replaced by rookie astronaut Jim Bagian.

"I never understood that," recalled Blaha. "We had become good buddies and I really liked working with her." He was on a public relations trip in New York when he was named to STS-29, but Fisher would *not* be aboard. Blaha had no problem with Bagian, but by 1988 the old 61H crew had been together for three years. "I thought the crew that we had was working well together," Blaha continued. "We had four military academy people on the crew and Anna. I used to smile through the training briefings, because there was no question Anna was ten or 20 IQ points above the other three of us. I could tell by her eyes, when someone explained something to us, she *knew* within about ten seconds; she *had* it. We were all scratching our heads and, maybe 15 minutes later, we would catch up!" As the flight engineer, Fisher could catch problems even more quickly than Blaha, often telling him to throw a certain switch at a certain time. In Blaha's mind, Fisher united two sections of the crew: the 1978 astronauts, including herself, Coats and Buchli, with a pair of 1980 rookies. The pilots had long since formed a tight friendship and partnership, since Blaha had served as one of the capcoms on Coats' first flight, Mission 41D. Anna Fisher, meanwhile, fell pregnant at around the time of the STS-29 announcement and gave birth to her second daughter, Kara Lynne, in January 1989. She subsequently took a leave of absence from NASA to raise her young family. Although she would return to the space agency in 1996, and remains in management to this day, she never flew in space again.

John Elmer Blaha was jointly the senior-ranking member of the ninth group of astronauts, alongside Roy Bridges, when he was selected as an Air Force lieutenant-colonel in May 1980. He came from San Antonio, Texas, where he was born on 26 August 1942, the son of an Air Force pilot, although he received his high school education in Virginia. Blaha entered the Air Force Academy and received a degree in engineering science in 1965 and a master's in astronautical engineering from Purdue University in the following year. He earned his pilot's wings at Williams Air Force

Base in Arizona in 1967 and undertook 361 combat missions in Vietnam, flying the F-4 Phantom, the F-102 Delta Dagger, the F-106 Delta Dart and the A-37 Dragonfly. By the time he returned from Vietnam and was accepted into the famed test pilot school at Edwards Air Force Base, Blaha had his sights on someday becoming an astronaut. "I had read the biographies of some of the early astronauts," he explained, "and realised many flew different types of airplanes and attended the test pilot school." Apollo 11 veteran Edwin 'Buzz' Aldrin happened to be commandant of the school at this time and Blaha got to know him and shared his goal. Aldrin recommended that he stay at Edwards and teach in the NF-104 research aircraft, which achieved altitudes of over 100,000 feet. After test pilot school, Blaha followed Aldrin's advice and flew the NF-104 to an altitude of more than 104,400 feet – around *thirty kilometres*. "This was a research airplane that had a rocket on it," Blaha told the oral historian. "You flew the aircraft above 100,000 feet and wore a space suit. At that altitude, the sky was black and the Earth looked curved ... the Space Shuttle entry from 70,000 feet was identical to what we were doing in the NF-104." (Interestingly, Roy Bridges also flew the NF-104.) Blaha next served as an F-104 Starfighter instructor and taught his students to fly low lift-to-drag approaches, stability and control and spin flight test techniques. In 1973, he was assigned as an exchange pilot with the Royal Air Force at Boscombe Down in Wiltshire and flew a variety of aircraft, including the Jaguar, the Hawk, the Harrier and the Buccaneer. "Our time in England," he said, "was the best three years of our lives. We could eat breakfast together as a family and then I went to work at nine o'clock in the morning. By five o'clock in the evening, I was home." Graduation from the Air Command and Staff College at Maxwell Air Force Base in Alabama followed in 1976 and Blaha was assistant chief of staff for studies and analysis at the Pentagon when he answered NASA's first call for astronaut candidates. He was unsuccessful on his first attempt, but was selected two years later. In his book, *Dragonfly*, Bryan Burrough summarised many perceptions of Blaha from engineers, managers and fellow astronauts. In some ways, they parallel perceptions of John Young. Blaha often presented himself as "a dolt", who needed things repeated to him, in perhaps a similar fashion to how Andrew Chaikin described Young as exhibiting a misleading "aw, shucks" demeanour, which concealed a keen engineering mind. "But John's far from stupid," engineer Mark Bowman said of Blaha. "He just wants to make sure he *gets it*. He wants to start from scratch and go over and over and over."

Launch of STS-29 was postponed from the original date of 18 February 1989 to replace faulty liquid oxygen turbopumps on Discovery's three main engines. A new target of 11 March was also scrapped when a master events controller failed during checks and had to be replaced. The crew had already been told by Bob Crippen that their planned five-day flight might be reduced by 24 hours, to aid with Discovery's turnaround for her next mission. Privately, Blaha questioned the point of such a move: was *one day* really going to make a difference, he wondered? He wanted to *lengthen* his time in space, not *shorten* it. Finally, the five astronauts left for Pad 39B on the morning of the 13th and were strapped into their seats. Ground fog and high upper-level winds delayed the launch for almost *two hours* ... which proved particularly uncomfortable for the crew, since they were all lying on their backs with their legs elevated. "Mike Coats had a very bad backache," Blaha remembered, "so he finally decided he *had* to unstrap, because we were being delayed for such a long time. We ended up laying on our backs ... very close to the *five-hour* limit, before we launched. Mike unstrapped and actually ... was laying on his side and then he'd lay on the other side." At length, Discovery thundered into space at 9:57 am EST.

As well as being Blaha's first flight, STS-29 was also the maiden voyage of mission specialists Robert Clyde Springer, a Marine Corps colonel, and civilian physician James Philip Bagian. Interestingly, Springer – born in St Louis, Missouri, on 21 May 1942 – had actually been interviewed by NASA for both its 1978 *and* 1980 astronaut classes ... as a *pilot*, not as a mission specialist. In fact, his interview as a pilot candidate for the 1980 class occurred in April of that year, yet he was selected only a few weeks later as a mission specialist. Springer completed high school in Ohio and entered the Naval Academy, receiving his degree in naval science in 1964 and a commission into the Marine Corps. After finishing Basic School in Virginia, he reported for flight training in Pensacola, Florida, and Beeville, Texas, and was designated as a naval aviator in August 1966. Springer flew in Vietnam, completing 300 combat missions in the F-4 Phantom II and 250 combat missions in helicopters and served as an advisor to the South Korean Marine Corps in June 1968. Upon his return home, he completed a master's degree in operations research and systems analysis at the Naval Postgraduate School and in March 1971 was assigned to the Third Marine Aircraft Wing in El Toro, California, as a wing operations analysis officer. Springer later attended the Navy Fighter Weapons School – today's 'Top Gun' – and completed test pilot school at Patuxent River, Maryland, in 1975. Among his test pilot work were the first flights of the AHIT attack helicopter instrumented test programme and, after graduation from the Armed Forces Staff College in Norfolk, Virginia, in 1978, he assumed responsibility for joint operational planning for Marine forces in NATO and the Middle East. Springer was serving as *aide-de-camp* for the commanding general of the Fleet Marine Force-Atlantic when he was selected by NASA in May 1980.

Jim Bagian, of Armenian ancestry, had already been assigned to his *second* Shuttle flight before he launched on his *first*. He was named as a mission specialist for the first Spacelab Life Sciences flight in February 1984, alongside John Fabian and Rhea Seddon, although by the time of the 51L tragedy their flight had been repeatedly postponed from an original target of January 1986 until at least March 1987. Meanwhile, Bagian was also assigned, in pre-Challenger days, to the crew of Mission 61I, which would have retrieved the Long Duration Exposure Facility. Bagian was born in Philadelphia, Pennsylvania, on 22 February 1952 and completed high school in his home town, then entered Drexel University to study mechanical engineering. He received his degree in 1973 and a doctorate in medicine from Thomas Jefferson University in 1977. Whilst undertaking his doctorate, Bagian worked as a process engineer for the 3M Company and, later, as a mechanical engineer at the Naval Air Test Center at Patuxent River. By 1978, his involvement with NASA started with a position as a flight surgeon and research medical officer at JSC in Houston. He applied unsuccessfully for the 1978 astronaut class and was actually invited to Houston in the same interview group as Rhea Seddon, with whom he would later fly into space. Two years later, he was picked as an astronaut and his early years were demonstrative of his astonishing ability: he was a qualified free-fall parachutist and private pilot, had been active in the mountain rescue community since 1981, served as a member of the Denali Medical Research Project on Mount McKinley in Alaska, was a snow and ice rescue instructor on Mount Hood in northern Oregon, received his Professional Engineer Certification in 1986, was board-certified by the American College of Preventive Medicine in 1987 *and* reached the rank of Colonel in the Air Force Reserve. In the words of future crewmate Rhea Seddon, Bagian's credentials were "very outstanding".

The three rookies of STS-29 adapted well to their new environment. For his own part, John Blaha experienced none of the worrisome signs of space sickness. "From that first *millisecond* that we were in zero-G, I *never* felt bad," he told the oral historian. He would ultimately fly five times, including a long-duration mission to Mir, and proved vocal of his profound love of space travel. "After each of his flights," Bryan Burrough wrote, "he always said he felt like running back to the launch pad and climbing back onto the *next* rocket." Within hours of reaching orbit, at 4:10 pm EST, NASA's fourth Tracking and Data Relay Satellite – 'TDRS-D', soon to become TDRS-4 – was deployed perfectly. The next few days of the mission were spent focusing on middeck and other experiments: these included a pair of student investigations, involving chicken eggs and four live rats, a demonstration heat pipe for Space Station Freedom and protein crystal growth and plant growth studies. Despite Crippen's warning that the mission might be curtailed, Discovery *did* fly for almost five full days, touching down an orbit earlier than planned at Edwards Air Force Base at 6:35 am PST (9:35 am in Florida) on 18 March. For Blaha, his first re-entry and landing went by in the flicker of an eye. *Time compression* was the greatest surprise. An hour had elapsed between the OMS burn and the touchdown, yet to him it seemed as if only five minutes had gone by. Coats handed Blaha the stick for a few seconds and it compared favourably with his experience in the Shuttle Training Aircraft. Yet the problem of time compression remained and Blaha resolved to prepare himself more thoroughly for that final hour in readiness for his next mission.

For Mike Coats, STS-29 should have been the end of his astronaut career. He had already promised his wife that after commanding a flight, *that* would be the end. Unfortunately – or perhaps *fortunately* – for Diane Coats, a meeting with President George H.W. Bush, an invitation to the White House and a Shuttle-shaped doggie biscuit would change all that ...

MISSION TO VENUS

When the Centaur-G Prime was cancelled in the summer of 1986, it left a number of important payloads out in the cold. Two of these – the Ulysses solar probe and the Galileo Jupiter orbiter – have already been discussed, but the Department of Defense had a number of its own satellites which required the powerful liquid booster ... *and*

NASA had the Magellan mission, intended to be the first American flight to the planet Venus in more than a decade. Not since the Pioneer Venus Orbiter in 1978 had anything bearing the banner of the Stars and Stripes drawn close to the strange, cloud-enshrouded world, so close to Earth in terms of proximity and physical size, yet so astonishingly different in terms of temperature and atmospheric structure, composition and pressure. Since the early 1960s, it had been recognised that radar imaging could generate crude images of the Venusian surface and it was these which helped to peg the planet's 'year' at 243 Earth days – even longer than its year of 225 days – and ascertain the retrograde nature of its rotation. Late in the 1970s, efforts got underway to plan a new spacecraft, the Venus Orbiter Imaging Radar, to gaze beneath the haze of the planet's thick atmosphere and gain a clearer understanding of the surface. Unfortunately, this mission exceeded budget constraints and was cancelled in 1982. The venture itself retained its importance and a stripped-down version was recommended by the Solar System Exploration Committee and was accepted under the new name of 'Venus Radar Mapper' in 1983. Two years later, in September 1985, it was renamed 'Magellan', in honour of the 16th century Portuguese explorer Ferdinand Magellan, who had mapped and circumnavigated the globe, just as his namesake would do for Venus. In the months before Challenger, Magellan was to be launched atop a Centaur-G Prime on Mission 81I in April 1988, after which it would follow a so-called 'Type I Heliocentric Orbit' trajectory, carrying it 180 degrees around the Sun to rendezvous with Venus in just four months. The Pioneer Venus Orbiter mission and earlier Soviet Venera spacecraft had mapped a relatively small proportion of the planet's surface. Built by Marin Marietta, with mission management from the Jet Propulsion Laboratory, Magellan would create the most comprehensive map of Venus in history: a staggering 92 percent of the surface would be scrutinised under the stare of its powerful synthetic aperture radar.

Costing $295 million, Magellan was built from spare parts of earlier missions, including Voyager, Galileo and Ulysses. Physically, the spacecraft comprised a decahedral, three-axis-stabilised equipment 'bus', measuring 42.4 cm high and 2 m in diameter and weighing 3,453 kg. It was dominated by a large parabolic, dish-shaped antenna, composed of strong graphite-epoxy sheets mounted to an aluminium honeycomb structure. This device measured 3.7 m across and actually served two purposes: high-gain communications with Earth *and* as the synthetic aperture radar to map Venus' surface. A pair of square, paddle-like solar panels, each measuring 2.5 m long, were capable of supplying up to 1,200 watts of electrical power. Insertion into orbit around Venus was to be accomplished by a Star-48 solid-rocket motor. Although Magellan's high-gain antenna was ideal as a communications tool, it was uncommon as a synthetic aperture radar, which is normally much larger and can be directed without physically moving the spacecraft. Since the antenna was rigidly attached to Magellan, continuous attitude control was required during mapping cycles to image the planet's surface and transmit data back to Earth. Attitude control was by a propulsion module of 24 hydrazine-fed thrusters and a set of reaction wheels. It would turn to face its antenna at the planet whilst passing the low point in its highly elliptical orbit, taping the data, and then as it climbed to the high point of its orbit would turn to replay the tape to Earth. In fact, Magellan made around *three thousand* changes to its operating mode during each mapping period, many of which were repetitive and were stored in the memory of its twin computers.

Two hammer blows struck the project in the summer and autumn of 1986. The destruction of Challenger automatically delayed Magellan, but the cancellation of the Centaur-G Prime also meant that the mission required a new booster. The solid-fuelled Inertial Upper Stage was ultimately pressed into service, although it was far less powerful than the Centaur. With the resumption of Shuttle flights scheduled for the late summer of 1988, the next available launch 'window' to reach Venus under the most optimum conditions came in October 1989. That date quickly became untenable, for it was needed by Galileo, whose own trajectory to Jupiter involved a gravity-assisted boost from Venus. Consequently, Magellan's trajectory specialists settled on a four-week window of opportunity which extended from 28 April until 28 May 1989. The trajectory design was known as a 'Type IV Heliocentric Orbit', which required Magellan to complete one and a half circuits of the Sun, before arriving at its destination in August 1990 but offered the advantages of lower launch energy and Venus approach speed. It also meant that the spacecraft would approach the planet over its north pole and perform mapping swathes in a north-to-south direction, which was the reverse of that planned on the basis of an original 1988 launch. With this in mind, STS-30 was targeted to head into space on the opening day of the Venusian window. The crew had been assigned more than a year earlier, in March 1988, and for the most part consisted of the original team members assigned to the 61G Galileo mission: Dave Walker in command, Ron Grabe as pilot and Norm Thagard as the centre-seat flight engineer. James 'Ox' van Hoften had resigned from NASA and was replaced by 61B veteran Mary Cleave, whilst the final mission specialist place went to a former Air Force fighter pilot, Major Mark Charles Lee.

Lee was born in Viroqua, Wisconsin, on 14 August 1952 and after high school entered the Air Force Academy, from which he graduated with a degree in civil engineering in 1974. He trained as a pilot at Laughlin Air Force Base in Texas and later flew the F-4 Phantom II for several years out of Kadena Air Base in Okinawa, Japan, as part of a tactical fighter squadron. Lee returned to study in 1979, earning a master's degree in mechanical engineering from MIT, with a specialism in graphite-epoxy advanced composites. He was then assigned to Hanscom Air Force Base in Massachusetts, resolving mechanical and material deficiencies affecting the combat readiness of Airborne Warning and Control System (AWACS) aircraft. In 1982, Lee upgraded to the new F-16 Fighting Falcon jet and served as an executive officer and flight commander. Two years later, in May 1984, he was selected as a member of NASA's tenth group of astronaut candidates. In the summer of the following year, Lee became the first of his class to draw a mission assignment, when he was named to Loren Shriver's 61M crew.

In the wake of Challenger, Lee and Mary Cleave worked closely together on techniques for payload deployments from the Shuttle and when they were assigned to STS-30 they took primary responsibility for the release of Magellan and its IUS. "I never expected to get assigned *that* fast after the accident," admitted Cleave, who had flown 61B only a few weeks before Challenger, "so I was really surprised when

Atlantis reaches Pad 39B, completing another milestone in the weeks leading up to STS-30.

they assigned Mark Lee and I to do this Magellan deployment. It was the first time we deployed a spacecraft that was going to another planet from the Shuttle." One of the biggest issues was going from a lightweight flight garment on 61B to a bulky and uncomfortable partial-pressure suit, with limited cooling, on STS-30. Comparing the two missions, Cleave felt that her first flight was "really loose", with the astronauts clowning around and having a good time, but on STS-30 all that had changed. It was *safer*, she said, but *much* more serious. In their early training, Cleave asked Lee where he wanted to fly for ascent. Lee picked the flight deck and it was therefore Cleave's responsibility to take the seat in the darkened middeck, with the duties of handling the escape pole, if necessary, only a wall of lockers for company ... and an excuse to make as much *noise* as she liked! Whereas some astronauts would loathe the middeck seat during ascent, considering it to be claustrophobic, Cleave thoroughly enjoyed it. "I could *hoot*, I could *holler*," she said, "I could have a *marvellous* time ... and, man, that's a ride! It's really a great ride."

Cleave's great ride was a few days in coming. The countdown had proceeded normally towards a liftoff on the 28th, but the clock was stopped at T-31 seconds when a hydrogen recirculation pump – responsible for cooling Atlantis' three main engines – developed a short circuit and stalled. The problem was reviewed and another issue with the abnormal venting of a hydrogen circulation line was also rectified and launch was rescheduled for the start of a 64-minute 'window' on 4 May. The length of these windows extended slightly as April progressed into May. The window on 28 April lasted for barely 18 minutes, whilst those from the middle of May through to the close of the Venus window on 28 May increased to a maximum of 121 minutes. This was dictated by the need for daylight landing opportunities at the Transoceanic Abort Landing sites. The day of the second launch attempt dawned dreary and overcast, with strong crosswinds blowing across the Shuttle Landing Facility. Nonetheless, the five astronauts were strapped into their seats, although the countdown clock was held at T-5 minutes, just before the activation of the Auxiliary Power Units, to wait out the weather. It seemed as if another scrub was inevitable ... and then, *fifty-nine minutes* into the window, the winds dissipated, the clouds parted and Atlantis rocketed into orbit at 2:46:59 pm EST. Deployment of Magellan got underway almost immediately. In Cleave's mind, as soon as it left the payload bay, responsibility passed from JSC to the Jet Propulsion Laboratory, and the *longer* it remained aboard Atlantis, the more chances existed for problems to develop. The release occurred at 9:01 pm EST, a little more than six hours into the mission. Two burns from the IUS booster set Magellan on course for Venus and the spacecraft fired its own thrusters on three occasions over the next year to remain on track for the correct arrival time.

That arrival came on 10 August 1990, kicking off one of the most spectacular episodes of planetary exploration ever accomplished. For four years, Magellan acquired detailed radar data of craters, volcanoes of various kinds, flat plains, hills, ridges and other geological formations on Venus. Five 'cycles', each lasting approximately one Venusian 'year', or 225 Earth days, were completed, acquiring radar imagery and gathering gravitational data. Towards the end of its life, the spacecraft was used as part of 'aerobraking' tests to examine atmospheric densities at

Magellan departs Atlantis' payload bay, bound for Venus. The large synthetic aperture radar dish antenna and square solar arrays are clearly visible.

different altitudes. Magellan was finally commanded to burn up in the Venusian atmosphere in October 1994, ending a mission which far exceeded all scientific expectations. In total, the project had cost $680 million and had mapped almost the entire surface at a resolution ten times better than the Veneras.

All of this was in the future for the crew of STS-30, who spent the remainder of their four-day mission performing scientific experiments, including the Fluids Experiment Apparatus (FEA), under Cleave's supervision. On 7 May, the day before landing, one of the Shuttle's four General Purpose Computers failed and Cleave and Lee replaced it with a backup. The mission ended in spectacular fashion at 12:43 pm PST (3:43 pm in Florida) on the 8th, when Walker guided his ship to a smooth touchdown on the concrete Runway 22 at Edwards. Original plans to land on the dry lakebed, Runway 17, had earlier been called off, due to high crosswinds. In the days that followed, as Magellan commenced its 15-month voyage to Venus, an unfortunate incident would befall Dave Walker. A year hence, in the summer of 1990, the ramifications of that incident would cost him the command of his next Shuttle mission.

FIRST TASTE OF SPACE

Richard Noel Richards was five weeks from his first launch into space when the opportunity was cruelly snatched from him. In January 1985, he had been assigned as Jon McBride's pilot for the ASTRO-1 mission, scheduled for the following March to observe Halley's Comet. When Challenger lifted off a year later, McBride and Richards were in their seats in the JSC simulator, practicing launch aborts, and briefly stepped outside to watch their friends fly into orbit. Within 73 seconds, their mission, 61E, vanished in a heartbeat, the Shuttle programme collapsed to its knees and the astronaut office would never be the same again. In fact, the careers of Richards and 51L pilot Mike Smith were entwined in more ways than one. For starters, they were the only Navy pilots selected by NASA in May 1980 and they had *both* been assigned to their first missions at the same time. "I started getting all these simulation flights with Dick Scobee," Richards recalled, and he started to wonder if they were being primed for a crew assignment. Then, a few weeks later, Mike Smith started doing simulations with Scobee and Richards started simulations with McBride. Years later, Richards believed that the postponement of the first Vandenberg mission and its knock-on effect on other flights, further downstream, may have led to the decision. It would certainly be ironic to suppose that simple quirk of fate and timing might have kept Richards from flying as pilot on Challenger's final mission.

He was born in Key West, Florida, on 24 August 1946, but received much of his schooling in Missouri, then entered the University of Missouri to study chemical engineering on a scholarship from the Reserve Officers' Training Corps. With a father who had served as a submariner, it seemed inevitable that Richards would follow in his footsteps and join the Navy. "I'd already sort of committed that I was going to do four years in the Navy," he explained, "and at the time, that was the

Vietnam War going on, so the draft was in vogue, so any young male at that point knew he *had* to deal with it." Richards opted to tackle the issue head-on and get part of his college fees paid by the military. After receiving his degree in 1969, followed by a master's in aeronautical systems from the University of West Florida in 1970, he opted for naval aviation and began flight training in Pensacola, Florida, finally earning his aviator's wings from Corpus Christi, Texas. He never saw action in Vietnam and was instead based until 1973 with a 'shore' squadron, but had the bonus of flying the F-4 Phantom fighter.

Richards next deployed to the North Atlantic aboard the USS *America* and and the Mediterranean aboard the USS *Saratoga*, again flying the F-4. Despite the ongoing conflict in Vietnam, he reflected, "we had to maintain, because of our NATO commitments, two aircraft carriers in the Mediterranean all of that time, because of the perceived Russian threat over there." He returned to the United States to report to test pilot school at Patuxent River, Maryland, graduated in 1976 and worked for over three years as a project pilot for automatic carrier landing systems for the F-4. Shortly before his selection as an astronaut, he served as a carrier suitability project officer for the F-18 Hornet, performing its first shipboard catapults and arrested landing tests aboard the *America*. Richards had been drawn to the astronaut programme in 1977, when John Young visited Pax River to encourage the test pilots to apply. He was unsuccessful in his first attempt, but "got enough encouragement to try to reapply" and was duly selected two years later. Although he remained an active duty naval officer – eventually reaching the rank of captain – Richards admitted that he probably donned his military uniform on only a handful of occasions after 1980. "That was effectively the end of my Navy career," he said, "and the start of my NASA career."

The weeks and months after Challenger were devastating and one of his hardest jobs was supporting Mike Smith's widow, Jane, in her grief. Two years later, in February 1988, he finally received assignment as pilot on STS-28, a classified Department of Defense assignment. The crew that he would be joining had actually been assigned to Mission 61N in December 1985, although the pilot for *that* flight, Mike McCulley, was substituted for Richards. Commander Brewster Shaw would be joined by mission specialists Jim Adamson, Dave Leestma and Mark Brown. NASA had already decided that the return to flight process would see Discovery flying first, followed by Atlantis, and finally Columbia on STS-28. In total, more than 250 modifications were incorporated into the queen of the fleet during the post-51L down time. Among the most important of these were the addition of a reinforced carbon-carbon 'chin', just behind and beneath her nose, installation of the telescoping pole and improvements to her wiring, power distribution systems and thermal protection tiles and blankets. She also received upgraded General Purpose Computers, improved fuel cells and new controllers for her Auxiliary Power Units. Although these changes added 1,130 kg of weight to Columbia, some of this was saved by replacing 2,300 of her tiles with lighter thermal blankets. "There was an opportunity to look at the whole system," said Arnold Aldrich, who had been appointed head of the Shuttle programme in 1986, "and make it as good as it could be." After STS-1 and the rapid ramping-up of the flight rate, there had simply been no time for engineers and technicians to step back and tend to the Shuttle's flaws.

With the pressure on getting Discovery and Atlantis into space before the end of 1988, Columbia found herself last in the queue and her launch was delayed until July and eventually August in the following year. However, despite being her first post-Challenger mission, the curtain of secrecy surrounding STS-28 showed no sign of being drawn back. Not until many years later would a few details of exactly what Shaw's crew did in space finally begin to trickle out. For his part, Dave Leestma described preparations for the mission as very cloak-and-dagger in nature. "Sometimes you had to *disguise* where you were going," he said. "You'd file a flight plan in a T-38 and go somewhere else, to try to not leave a trail for where you were going or what you were doing, who was the sponsor of this payload or what its capabilities were or what it was going to do. You had to be careful, all the time, of what you were saying." STS-28 would transport the Department of Defense's fourth major Shuttle payload into orbit, but the United States intelligence community had begun to reduce its reliance on the reusable orbiter by reverting to expendable boosters. Only payloads which were too large, heavy or awkward to be reconfigured for an expendable launch remained on the Shuttle. "The DoD did not like dealing with NASA," said Leestma. "It was a constrained arrangement, but it worked very well and the DoD was happy with the product that they got in the end."

It had long since become standard practice in the build-up to such missions that the countdown was conducted in almost complete secrecy, with the public affairs commentary starting when Columbia emerged from the T-9 minute hold. Only after this point were the gathered spectators able to listen in to the clipped intercom exchanges between the crew and launch controllers. A software problem caused the clock to be held for longer than planned and a combination of haze and fog over the Shuttle Landing Facility meant that STS-28 set off 40 minutes late at 8:37 am EST on 8 August 1989. Watching from the VIP area was NASA Deputy Administrator J.R. Thompson, who declared "We're off to a good start on this mission." Considering that the flight was historic, as the space agency's flagship orbiter spread her wings once again, the official announcement from spokesman Brian Welch was flat and businesslike: two hours after launch, he said, Shaw's crew had been given a 'Go' for orbital operations. That was it. The 'primary payload' was deployed around seven and a half hours into the mission. At the time, John Pike, a space policy analyst for the Federation of American Scientists, speculated that this payload was a massive, 14,500 kg 'KH-12' satellite, one of the latest generation of 'Key Hole' photographic reconnaissance platforms, whose ancestry stretched back to the 1960s. Pike commented that the KH-12 was one of the Pentagon's most expensive payloads – with an estimated price tag in the region of $1 billion *per unit* – although other sources argued that STS-28's cargo might have been a much lighter Strategic Reconnaissance Satellite (SRS). Still others speculated that it was capable of manoeuvring itself to an orbital altitude of around 480 km, from which vantage point it could take photographs with a resolution as fine as a single metre.

More recently, it came to light that the payload was probably a member of the second-generation Satellite Data System (SDS-B), a family of Air Force telecommunications platforms. In fact, doubts over whether it was a KH-12 were raised within weeks of the launch, when ground-based civilian observers noted that the satellite 'flashed', as sunlight reflected from its solar panels, at regular intervals. This phenomenon, they concluded, was not normally consistent with a reconnaissance system. Certainly, it was not deployed by the RMS mechanical arm, which was not carried on STS-28, and the first photographs of an SDS-B entered the public arena in the spring of 1998, when the National Reconnaissance Office released images and videotapes of a pair of military satellites. One was identified as an SDS-B and Hughes was acknowledged as its prime contractor. Physically, it was not dissimilar to the Syncom/Leasat payloads already deployed on earlier Shuttle flights, but somewhat longer. In a 2009 article for *Air & Space* magazine, Michael Cassutt quoted an Air Force officer who was familiar with the SDS-B project. "It's strange," he told Cassutt, "to work on a *secret* project for ten years, then see it on network television!"

The Air Force began to develop the first-generation SDS in 1973 to provide America's intelligence community with a network of orbiting relays, capable of transmitting real-time data and images from reconnaissance satellites which were out of range of ground stations. Another of their responsibilities was to support voice and data communications for covert military activities. The second-generation SDS-B – which first flew on STS-28 – operated in high-apogee and low-perigee orbits, ranging from as close as 480 km and as far as 38,000 km, at steep inclinations which achieved their highest point over the northern hemisphere. This enabled them to cover two-thirds of the globe, relay spy satellite data of the entire Soviet land mass and cover the entire north polar region in support of Air Force communications. Such wide coverage was not possible to geostationary-orbiting satellites. The SDS-B featured a pair of 4.5 m dish antennas and a third, smaller dish for Ku-band downlink. Overall, the satellite measured 4 m long and 3 m wide, with a launch mass estimated at close to 3,000 kg. In total, three cylindrical SDS-Bs were deployed by the Shuttle, on STS-28, STS-38 in November 1990 and STS-53 in December 1992. Although it is unclear as to *how* they were deployed, some observers have assumed that they were released in a similar fashion to the Syncoms, in a 'frisbee' fashion. Others have noted that the solid rocket booster used for the SDS-B was an Orbus-21, physically identical to the motor later fitted to Intelsat 6-3 by spacewalking astronauts during STS-49 in May 1992. This has prompted alternative suggestions that the SDS-B was deployed 'vertically' from a special cradle in the payload bay.

In whatever manner that SDS-B left Columbia, it is certain that the deployment was completed on the first day of the mission, because Shaw and Richards performed a separation manoeuvre at 4:58 pm EST on the 8th. A second payload, weighing just 125 kg, was also deployed and has been rumoured to have been a kind of 'ferret' satellite for radio and radar signals intelligence. The remainder of the mission went like clockwork and the astronauts tended a number of military experiments in the middeck and a pair of Getaway Special canisters in the payload bay. Together with Richards, there were two other rookies amongst the crew: Lieutenant-Colonel James Craig Adamson of the Army and Lieutenant-Colonel Mark Neil Brown of the Air Force. Adamson came from Warsaw, New York, where he was born on 3 March 1946. He received a degree in engineering from the Military Academy in 1969 and was commissioned into the Army as a second lieutenant. During his career, he undertook pilot and paratrooper training, Arctic water and mountain survival training, nuclear weapons training and graduated from the Navy's test pilot school at Patuxent River, Maryland. Adamson flew as a scout pilot and air mission commander over Cambodia during the Vietnam conflict and later earned a master's degree in aerospace engineering at Princeton University. He subsequently taught aerodynamics – including fluid mechanics and aircraft performance – as an assistant professor at West Point. He arrived at JSC in 1981 as a research pilot and aerodynamicist in Mission Control and was selected as an astronaut three years later. His classmate, Mark Brown, had also worked for NASA in a different capacity before becoming an astronaut. Brown was born in Valparaiso, Indiana, on 18 November 1951 and received his degree in aeronautical and astronautical engineering from Purdue. Upon graduation in 1973, he entered the Air Force, received his wings and flew the T-33 Shooting Star and the F-106 Delta Dart as a fighter-interceptor pilot at K.I. Sawyer Air Force Base in Michigan. He was selected to attend the Air Force Institute of Technology in 1979 and earned a master's degree in astronautical engineering the following year. Shortly afterwards, Brown was transferred by the Air Force to JSC as a flight dynamics engineer, participating in the development of contingency procedures for the Shuttle. Admission into NASA's tenth group of astronauts followed in May 1984.

Richards and Leestma were Navy, Shaw and Brown were Air Force and Adamson was the only Army member of the crew, but their common string was that they all shared a military background. "We were all cut from the same cloth," Richards remembered. "There wasn't too much that we needed to talk about. We all understood each other and it was comfortable for us." Maintaining the requirements of the classification was a pain, although there were a handful of non-classified experiments ... included an instrumented female *skull*, donated to research the penetration of radiation into the human cranium in space. Hundreds of thermoluminescent dosimeters were mounted in the skull to record radiation levels and it was flown twice more – in February and April 1990 – to measure the effects of different orbital altitudes and inclinations.

Five days after launch, at 6:37:08 am PST (9:37:08 am in Florida) on 13 August, Brewster Shaw guided Columbia smoothly onto the dry lakebed Runway 17 at Edwards Air Force Base. "Super team and great machine," radioed Capcom Frank Culbertson as the vehicle rolled to a halt. For Dick Richards, one anecdote stood out from the STS-28 re-entry. As Columbia passed through Mach 10 and super-heated air streamed across the vehicle, creating and depositing pockets and blobs of white-hot plasma, something splattered across one of his cockpit windows ... and *stayed there*. After touchdown, he mentioned it to Don Puddy, the head of Flight Crew Operations and asked him to get one of the technicians to take a look at it. The substance, still in liquid form, was scooped up into a plastic coffee cup and taken away for analysis. That *analysis* was not quite what either man expected. A few days later, Richards caught up with Puddy.

"What did they say about what that material was?"

"You're not gonna *believe* this," Puddy replied with a grin. The technician had taken the coffee cup, filled with *something* from the upper reaches of the atmosphere and placed it onto a counter. Without realising, *another technician had come along, grabbed the cup, poured coffee into it and knocked it straight back. "That,"* he said in disgust, "is the *worst* tasting coffee I've *ever* had!" From that day to this, Richards never discovered what had deposited itself on his window at ten times the speed of sound ... but, in a new and somewhat dubious space 'first', someone had at least *tasted* it.

THE ROMANCE OF ADVENTURE

When the Galileo spacecraft drifted out of Atlantis' payload bay on the evening of 18 October 1989, on the first leg of its six-year voyage to Jupiter, the sight was a moving one for Shannon Lucid. As STS-34's lead mission specialist, she was primarily responsible for the deployment of one of the most important payloads ever launched by NASA. For almost a dozen years, Lucid had lived and worked with the reality that her job was an overwhelmingly technical one, drawing from its roots in engineering and pure science ... but as Galileo and its IUS booster floated silently into the inky void, she beheld a new reality: the *romance* of adventure. Emblazoned across the base of the spacecraft which would one day circle Jupiter and deposit an instrumented probe into its atmosphere were two names: 'Galileo' in script and 'NASA' in worm-like block capitals. To Lucid, those two words symbolised exactly what the mission stood for: the script represented the romance of adventure and exploration, whilst the worm was indicative of the outstanding engineering and scientific talent which had brought this awesome project from the drawing board to fruition. Yet Galileo's journey to the launch pad had been a long and tortured one and its voyage to Jupiter would be longer and harder still.

The mission traced its genesis back to the mid-1970s and was actually one of the projects for which Joe Allen had helped to gain approval in his three-year stint as NASA's head of legislative affairs in Washington, DC. Named in honour of the Italian scientist, Galileo Galilei, whose endeavours in the early 17th century included the discovery of Jupiter's four large moons, Ganymede, Callisto, Europa and Io. Originally known as 'Jupiter Orbiter and Probe' (JOP), the name 'Galileo' seemed an obvious one and the project received Congressional approval on the first day of October 1977, with a planned launch four years later. However, delays to the first flight of the Shuttle and the limited capability of Boeing's Inertial Upper Stage to boost Galileo on its way to Jupiter raised concerns. In 1979, *Washington Post* journalist Thomas O'Toole highlighted that problems with certifying the Shuttle's main engines to operate at the 109-percent performance level needed to lift Galileo posed additional obstacles. By now, the launch had slipped until 1982 at the earliest. O'Toole noted that if the 109-percent-capable engines were not ready for this date, Galileo could slip even further. Timing was critical, since a 1982 launch depended upon a Mars gravity assist and if it was delayed much further, the potential existed to halve the scientific mission at Jupiter, from 11 to only *five* orbits of the giant planet. At length, in late 1980, under pressure from Representative Edward Boland, a Democrat from Massachusetts, NASA was obliged to abandon the IUS plan and initiate planning for a launch on General Dynamics' liquid-propelled Centaur-G Prime, which Administrator Robert Frosch had earlier opposed. The situation for Galileo's future dimmed substantially for much of 1981, with Congressional mutterings of closing down the California Institute of Technology's Jet Propulsion Laboratory, which managed many of NASA's planetary projects. A massive letter-writing campaign to George Keyworth, head of the White House's Office of Science and Technology Policy, was spearheaded by Galileo investigator and famed physicist James van Allen. In a speech to the National Academy of Sciences, van Allen identified Galileo as one of the most exciting missions of exploration ever undertaken and that its cancellation would prove devastating. Thankfully, in December 1981 the Office of Management and Budget relented, reinstated Galileo and it was rescheduled for 1983. There was a caveat, however: Galileo would *not* use the powerful Centaur-G Prime. In January 1982, NASA rescoped the mission, returned to the less powerful IUS fitted with a third, 'injection stage' to provide increased propulsion. As a consequence, Galileo's launch was rescheduled for August 1985, but the absence of the powerful Centaur meant that it would take *five years*, instead of two, and the spacecraft would be injected into a two-year-long elliptical solar orbit, would require a gravity assisted boost from Earth in June 1987 and would finally reach Jupiter in January 1990.

By the summer of 1982, some members of Congress – led by New Mexico Senator Harrison 'Jack' Schmitt, a former Moonwalker and chairman of the Senate Space Subcommittee of the Science, Commerce and Transportation Committee – were pushing vigorously for a *return* to the Centaur and a reduced journey time. Despite worries about additional expense in changing boosters *again*, coupled with concerns about further delays to the mission, in July President Reagan approved the move and NASA was forced to replan. The Centaur *would* be used to boost Galileo, but launch would be unavoidably postponed until May 1986, with a two-year flight time to the giant planet. At this stage, the mission truly entered the phase of equipment testing. In the early summer of 1983, the parachute for the instrumented probe, which would descend into Jupiter's atmosphere, successfully passed full-scale tests, and by September of that year the main spacecraft and probe were integrated. A model of the Centaur passed its own tests in September 1984 and the actual flight model was rolled out of General Dynamics' plant in San Diego in August of the following year. By this time, NASA Administrator Jim Beggs had endorsed other possible tasks for Galileo, most notably a flyby of the asteroid Amphitrite, which it was hoped might unlock secrets of the primordial solar nebula from which the Sun and planets formed. An Amphitrite flyby would delay the Jupiter arrival from August to December 1988, however, and it was decided to make a final decision after launch. In December 1985, only weeks before the loss of Challenger, Galileo was transported, cross-country by truck, escorted by police, state troopers and other guards, and arrived safely at the Kennedy Space Center for launch the following May.

When Challenger exploded in the skies above Florida, Galileo was undergoing final checkout and preparation for attachment to its Centaur-G Prime. In the weeks after the accident, NASA Acting Administrator William Graham spoke of the possibility of a return to flight in the spring of 1987, which kept alive the option to launch Galileo in the *next* Jovian 'window' in June of that year. Eventually, the modifications to the SRBs and the orbiters themselves inevitably pushed the return to flight further to the right. On 19 June 1986, newly-reappointed NASA Administrator Jim Fletcher formally cancelled Centaur-G Prime and new options needed to be found. One of these was an 'enlargement' of the IUS, possibly coupled with an additional booster, such as a Special Payload Assist Module (PAM-S). However, as already noted, the IUS was insufficient to send Galileo directly to Jupiter and alternate trajectories, involving planetary gravity assists, were explored. Even before NASA settled on October-November 1989 as the most appropriate 'window' for Jupiter, Galileo's planners were already working towards this date, creating a complex flight profile, known as the Venus-Earth-Earth Gravity Assist (VEEGA), in which the spacecraft would perform a flyby of Venus in February 1990, return to Earth in December and be placed into a two-year elliptical solar orbit. Returning a second time to Earth in December 1992, it would pick up sufficient energy to reach Jupiter in December 1995. The VEEGA technique was highly conservative of Galileo's on-board propellant, with predictions indicating that up to 80 kg would remain, even after the arrival at Jupiter and completion of its primary mission. The trajectory also permitted possible rendezvous with up to three asteroids – Ausonia, Gaspra and Ida – and eventually the latter two were selected. However, since the spacecraft would fly much closer to the Sun than had been planned, additional thermal shielding was added in the three-year down time after Challenger. It is interesting that Galileo also 'leapfrogged' Ulysses in the launch pecking order. "NASA based its decision on optimising data return from the two missions," wrote Michael Meltzer in *Mission to Jupiter*. "Launching Ulysses first would have resulted in too long a wait before Galileo reached Jupiter and began transmitting prime data from the Jovian system."

As launch neared, with an opening of the Jupiter window at 1:29 pm EST on 12 October 1989, there were still last-minute concerns about Galileo ... although *these* were not focused upon its mission, but upon its power system. Since the spacecraft would be travelling more than half a billion kilometres further from the Sun than Earth, the use of solar cells for electrical provision was impractical. (In fact, more than 65 *square metres* of cells would have been necessary to harness sufficient solar energy!) Therefore, General Electric supplied a pair of Radioisotope Thermoelectric Generators (RTGs), fuelled by fracture-resistant 'pellets' of plutonium-238, whose decay produced heat which was in turn converted into electricity. To keep them at a safe distance from the sensitive scientific instruments, the RTGs were mounted on a boom, which extended them 5 m away from the main body of the spacecraft. Both power plants produced 570 watts of electricity at launch, which steadily decreased by around half a watt per month and reached around 493 watts by the time Galileo reached Jupiter. Atlantis also required modification to incorporate an RTG coolant line and purging system in her payload bay. In the late 1980s, of course, 'nuclear' was a dirty word; a word which conjured images of military superpowers, the faceless Department of Defense and greedy power corporations. Peace marches were undertaken and representatives of several anti-nuclear groups gathered at the gates of the Kennedy

Space Center to express their fear that a Challenger-like explosion could spread radioactive plutonium across much of the United States' eastern seaboard. The allegation that NASA was playing "ecological roulette" with the lives of Floridians was not groundless. Memories of the 'messy' crash of the Soviet Union's nuclear-fuelled Cosmos 954 satellite in Canada, a decade earlier, were fresh in many minds, and even the noted physicist Carl Sagan remarked that "there is nothing absurd about *either* side of *this* argument". Final approval to proceed with the Galileo launch came from President George H.W. Bush himself in September 1989. Three days before the scheduled launch, outraged protestors staged a mock 'death scene' at the Cape and even threatened to sit on Pad 39B itself to prevent Atlantis from launching into orbit. Franklin Chang-Díaz, one of the STS-34 mission specialists, was astonished by the controversy surrounding a mission which was a *scientific* odyssey. "It was striking to drive through the gates ... and *see* all these demonstrators, trying to stop the launch," he told a Smithsonian interviewer, years later. "The topic of nuclear power is going to come up over and over again as we move into space. It's a key issue we are going to have to resolve, because the survival of people in space, far away from Earth, will *totally* depend on the use of nuclear power."

The launch window for Jupiter would close on 21 November, after which the *next* opportunity would not arise until 1991, so there existed a very real risk that the mission might be cancelled. Security was increased at the Kennedy Space Center, as guards armed with M-16 assault rifles and 9 mm semi-automatic pistols patrolled the perimeter of the launch site. A faulty main engine controller put paid to the 12 October attempt and launch was rescheduled for the 17th, then the 18th when rain showers drifted within 30 km of the Shuttle Landing Facility. During these few days, final efforts to stop the launch were rejected by the Circuit Court of Appeals in Washington, DC. In her summary, Chief Justice Patricia Wald declared that she could find no evidence that NASA had improperly compiled its environmental assessment reports for Galileo and on 16 October a number of activists were arrested at the Cape for trespassing. Launch on the 18th was postponed by about three and a half minutes in order to update Atlantis' computers for a change in the Transoceanic Abort Landing site, which had moved from Ben Guerir in Morocco to Zaragoza in Spain, due to heavy rain at the former. At 12:53 pm EST, Atlantis and her crew of five roared into the clear Florida skies and reached orbit perfectly, eight and a half minutes later.

For Don Williams, named to command STS-34 in November 1988, the assignment was something quite distinct from his previous mission, 51D. "There's some amount of loneliness at the top," he told the NASA oral historian, "and having that authority and with it comes the responsibility for accomplishing the mission. With those first two comes the most important one, in my mind, which I learned early on as a midshipman at Purdue ... is with the authority and responsibility comes the *accountability* and if something goes wrong, it's the person *in command's* fault! The same thing is true when you command a mission. You're accountable for the performance of the crew, for the accomplishment of the mission, for getting the objectives completed successfully and for getting the spacecraft back so somebody else can use it again. *That's* the name of the game." Command was important to Williams. In fact, by his own admission, it had been his primary goal: to command the Shuttle. It had taken *longer* than he hoped – 11 years – but had finally arrived. "Okay, this is what you came here for," he told himself. "Let's go do it."

Six hours into the mission, at 7:15 pm, under the watchful eye of Shannon Lucid, Galileo and the IUS were tilted to their deployment position and set free. "Galileo is on its way to *another world*," exulted Williams. "It's in the hands of the best flight controllers in the world. Fly safely!" Chang-Díaz felt a very personal affinity with Galileo. To him, it was a memorable occasion, because it represented his childhood desire to leave Earth and travel to other planets. Shortly thereafter, Williams and his pilot, Michael James McCulley, manoeuvred Atlantis to a safe separation distance and the IUS fired to boost Galileo onto a course for Venus, which it would reach in a little over three months' time.

McCulley came from San Diego, California, where he was born on 4 August 1943. He attended high school in Tennessee and enlisted in the Navy, serving aboard one diesel-powered and two nuclear-powered submarines. He entered Purdue University in 1965 to study metallurgical engineering and in January 1970 received his naval officer's commission and *both* bachelor's *and* master's degrees. During his naval career, he flew the A-4 Skyhawk and A-6 Intruder, attended the Empire Test Pilot School in England and the Naval Test Pilot School at Patuxent River in Maryland and undertook sea duties aboard the USS *Saratoga* and the USS *Nimitz*. Selected as an astronaut in May 1984, McCulley was the first pilot in his class to draw a flight assignment. "And I think I know why that was," he told the Smithsonian interviewer. One night in the autumn of 1985, he was at the Kennedy Space Center, and was summoned at short notice to step in for another pilot and perform practice approaches in the Shuttle Training Aircraft. On *this* occasion, though, he was joined in the cockpit by none other than chief astronaut John Young and Congressman Don Fuqua, chair of the House Science Committee. "My performance in the training aircraft," said McCulley, "was usually at least average, although I doubt I was at the top of the heap in terms of my ability to always land it in the right spot at exactly the right time. But *that* was one of those nights when I could do no wrong. In fact, about halfway through, I was really hoping the airplane would break, so I could quit while I was ahead, because every single approach I was making was absolutely, totally and completely *dead-on*." After landing, Fuqua, Young and McCulley's instructor were happy. A few weeks later, in December 1985, McCulley was assigned as pilot of Mission 61N. For McCulley, a Shuttle launch was unlike anything even the *simulator* could have prepared him for. "The vehicle *explodes* ... literally explodes ... off the pad," he recalled. "The simulator shakes you a little bit, but the actual liftoff shakes your entire body and soul!"

His crewmate, Ellen Louise Baker, surely shared similar feelings. She was born in Fayetteville, North Carolina, on 27 April 1953, but was raised in New York City, the daughter of Dr Mel Shulman and politician Claire Shulman. She completed high school in Queens in 1970 and earned a degree in geology from the University of Buffalo at the State University of New York in 1974 and a doctorate in medicine, four years later. She trained in internal medicine at the University of Texas Health Science Center in San Antonio and was board-certified in 1981. Her career with NASA commenced that same year, as a medical officer at JSC, and she later served as a physician in the Flight Medicine Clinic, before selection into the astronaut corps in May 1984.

An hour after deployment, the IUS fired to commence Galileo's six-year odyssey, kicking off Shannon Lucid's romantic adventure to explore a distant planet. It is not the purpose of this volume to discuss the mission itself, but the spacecraft – which weighed 2,564 kg and stood *seven metres* tall – proved itself to be a remarkable example of triumph over adversity. A little more than a year into its cruise, and several months after its first flyby of Earth, the high-gain antenna only partially unfurled, threatening to ruin the mission. 'Workaround' techniques were devised to use the low-gain antenna in its stead and the spacecraft returned remarkable images from the asteroids Gaspra (in October 1991) and Ida (in August 1993) and, far from conducting *two* years of scientific exploration at Jupiter, Galileo spent almost *eight* years in operation. During that time, it measured the chemical composition of the giant planet's atmosphere, directly observed its ammonia clouds and mysterious Great Red Spot, analysed the causes and effects of volcanism on Io and yielded tantalising clues for liquid oceans beneath the frozen surfaces of Europa and Ganymede and the extent of Jupiter's gigantic magnetosphere was mapped and modelled for the first time. On its way to the planet, in July 1994, Galileo also observed the impact of Comet Shoemaker-Levy 9 into the Jovian clouds. Not until the end of 2003 was the mission finally terminated by having the spacecraft dive into the planet's atmosphere.

Having set Galileo on its way, for all intents and purposes, the primary mission of STS-34 was over. A problem with one of Atlantis' Auxiliary Power Units triggered an alarm which woke the crew on 22 October, and other minor glitches centred on the flash evaporator system and cryogenic oxygen manifolds. Predicted high winds at Edwards Air Force Base on the 23rd prompted a decision to bring the Shuttle home two orbits earlier than planned and Williams and McCulley brought their ship to a smooth touchdown at 9:33 am PST (12:33 pm in Florida), just 20 minutes short of five full days after launch. Despite the excitement of an unplanned EVA on his first mission, Williams regarded STS-34 in a somewhat different light. He and his crew had done something remarkable for science. "We knew that Galileo was going to be a *lasting* programme," he said, "as opposed to the first flight, where we deployed the two satellites. The Galileo mission, we knew, if it was successful, the spacecraft was going to end up in orbit around Jupiter several years later and then there were going to be several years of data and images sent back. It was going to be a living, ongoing programme and *we* got to be a part of it."

A GOLD STAR FOR AN ADMIRAL

The delays in preparing Columbia for her first post-Challenger launch and the need to despatch Galileo to Jupiter during a critical 'window' of opportunity in October 1989 had already pushed Fred Gregory's STS-33 mission from August until the middle of November. By that time, his flight – a classified Department of Defense assignment, scheduled for four days – had changed in other ways, too. Together with his crew of pilot Dave Griggs and mission specialists Manley 'Sonny' Carter, Story Musgrave and Kathy Thornton, he had been working on the flight since November 1988. Then, just six and a half months later, an unexpected and shocking tragedy occurred. On 17 June,

Griggs was preparing for his role in a weekend air show, flying alone in a single-engine, 1940s-era aircraft, just south of the town of Earle, Arkansas. Shortly after nine o'clock that morning, according to eyewitnesses, Griggs was performing aileron rolls, when one wing accidentally touched the ground and the aircraft crashed into a wheat field. Dave Griggs – test pilot, Vietnam veteran, accomplished astronaut and rear admiral in the Naval Reserve – was killed instantly. The accident had occurred whilst Griggs was off-duty and not in a NASA aircraft, but it sent shockwaves through the astronaut corps. After the funeral, Mike Mullane remembered the wake at the Outpost tavern in Houston and saw Griggs' former crewmate, Kathy Thornton, her cheeks soaked with tears, walk in and place one of the wreaths of flowers onto the bar.

Kathryn Ryan Cordell Thornton secured a minor record of her own on STS-33, becoming the first and only woman to fly aboard a classified Department of Defense Shuttle mission. Many have speculated that her assignment came about because she had served as a research physicist at the Army's Foreign Science and Technology Center, before becoming an astronaut, and had worked on a NATO-awarded post-doctoral fellowship at the Max Planck Institute for Nuclear Physics in Heidelberg, West Germany. She was born on 17 August 1952 in Montgomery, Alabama, and after high school entered Auburn University to study physics. She received her degree in 1974 and a PhD from the University of Virginia, five years later. After completing her post-doctoral research in West Germany, she was employed by the Army in Charlottesville, Virginia, in 1980, and entered NASA's astronaut corps in May 1984, alongside her future STS-33 crewmate, Manley Lanier Carter Jr. Nicknamed 'Sonny', Carter came from Macon, Georgia, where he was born on 15 August 1947 and studied chemistry at Emory University in Atlanta. Whilst an undergraduate, he played collegiate football (and later captained the team), ran as a member of Emory's track team *and* was an intramural wrestling champion. Carter earned his degree in 1969 and entered medical school, graduating in 1973. Yet a sporting career was never too far away. Even as he laboured away at his doctorate in medicine, he played professional football for three seasons with the Atlantic Chiefs of the North American Soccer League. He completed his internship in internal medicine at Grady Memorial Hospital in Atlanta, then entered the Navy in 1974 to train as a flight surgeon. Flight training followed and Carter received his aviator's wings in 1978, flew F-4 Phantom jets and completed a nine-month deployment to the Mediterranean, aboard the USS *Forrestal*. He graduated from test pilot school in 1984, the same year that he was selected by NASA.

Aside from the tragedy of Dave Griggs' death, the reality was that launch of STS-33 was only five months away and a new pilot had to be appointed. A first-timer was not advisable at such short notice and many of the experienced pilots had already been assigned to other missions. At length, in discussion with Don Puddy, it was Fred Gregory who requested that Blaha be reassigned to his crew. Gregory felt that Blaha's experience on Discovery's most recent flight would be beneficial to STS-33. Puddy agreed. Blaha had already been assigned, back in April, as pilot of STS-40, the first Spacelab Life Sciences mission, tentatively scheduled for June 1990 ... but *that* flight was far enough into the future for a rookie pilot to be named in his stead. Within days, at the end of June 1989, Blaha was formally announced to join Gregory's crew. When

The "beauty and power" of the Shuttle is amply illustrated in this view of Discovery spearing for the heavens in the first post-51L night launch. The plumes from the three main engines, with their glittering 'Mach diamonds', are particularly evident.

the crew produced their official mission patch, a small gold star was placed near Blaha's name. It commemorates Rear Admiral Stanley David Griggs, the original pilot, who flew STS-33 in spirit.

The last three years had been tough on all of them, including Fred Gregory himself. Not only had he lost his pilot in a horrifying accident, but he had been sitting next to Dick Covey in Mission Control on the day that Challenger exploded *and* had worked closely with the two teacher candidates. "I had spent a lot of time with Christa and Barbara," he told the NASA oral historian, "because I had teachers in *my* family." Gregory spent many hours talking through the planned lessons and the importance of the Teacher in Space project. His assignment to STS-33 had come in a late-evening call from Don Puddy, after he returned from a flying assignment. "I don't believe that I told my *wife* immediately," he said. "I wanted to really *milk it* a little bit ... and I *did*." Years later, he would remember the crew as the best ever assembled, with the entire team performing like a flawless symphony. "We would get in a simulator," he reflected, "and our training crew, I'm sure, was attempting to kill us, and *we* were trying to make absolute *fools* of them."

Launch was originally set for 20 November 1989, but was delayed due to problems with the integrated electronics which controlled the ignition and separation of Discovery's twin SRBs. NASA rescheduled the attempt for the small hours of the 22nd and Gregory's crew remained at the Cape. Story Musgrave had already flown two Shuttle missions and his completion of a master's degree in literature at the University of Houston in 1987 had imbued him with skills as a poet and wordsmith. On the day of launch, at three in the morning, he took Carter and Thornton to Pad 39B, surprising the guards and technicians, who greeted them with enthusiasm. It was "a primitive, primal experience to never be forgotten," Musgrave wrote on his website, "welcomed by everybody. That machine looming in the lights, Jupiter overhead, a crescent Moon on the ocean horizon, the fog moving in and out over Discovery ... What exuberant exhilaration! What beauty and power." That power was unleashed at 7:23 pm EST that same evening, when STS-33 speared for the heavens in the first nocturnal launch of the post-Challenger era and only the third in the Shuttle's eight-year history.

Deployment of their classified payload followed on Discovery's seventh orbit, approximately ten and a half hours after launch. The Air Force would later admit that the satellite was boosted into orbit atop an Inertial Upper Stage and it is today generally believed that it was an electronic intelligence platform, possibly codenamed 'Magnum' or 'Orion'. It was a successor to the earlier Rhyolite-Aquacade spacecraft and represented the latest in an era of satellites developed under the auspices of the National Reconnaissance Office, the Central Intelligence Agency and the National Security Agency. With a mass of around 2,600 kg, the STS-33 satellite is thought to have featured a large, gold-coloured mesh antenna, measuring perhaps 78 m in diameter when fully unfurled, and was physically similar to the payload deployed on Mission 51C in January 1985. This antenna 'farm' was so large that it was presumably attached to a gimbal mechanism for steering, which permitted it to monitor specific points of interest, such as ballistic missile flight test telemetry and observers have speculated that the rear of the Magnum-Orion comprised a pair of solar arrays and a downlink communications antenna. Other objectives included electronic, radio communications and radar emissions intelligence and the satellite was apparently boosted to geostationary orbit by its IUS. Since the Department of Defense had openly admitted that only the payloads which could *not* be reconfigured for an expendable launch were kept aboard the Shuttle, the physical size of the satellite was immense ... as was its cost, with some estimates placing Magnum-Orions at around $750 million each. Touchdown of Discovery occurred in the late afternoon gloom at Edwards Air Force, a day late due to strong winds, at 4:30 pm PST (7:30 pm in Florida) on 27 November.

PUSHING THE ENVELOPE

When Challenger reached orbit in April 1984, at the beginning of Mission 41C, the main purpose of her flight was to retrieve and repair NASA's crippled Solar Max. This had been accomplished in a pair of spectacular EVAs and some deft handling of the RMS mechanical arm. However, in order for the repair to go ahead, *another* satellite had first to be deployed: a 12-sided, bus-sized structure called the Long Duration Exposure Facility (LDEF). As its name implies, it was intended to accommodate experiments which required long-term exposure to the harsh environment of low-Earth orbit. No one could possibly have known, at the time of its launch, exactly *how* long-term that would be. Original plans called for Brewster Shaw's crew to retrieve it on Mission 51D in February 1985, but then a team under the command of Dan Brandenstein spent several months preparing for the task and, eventually, at the time of the 51L accident, it had been postponed to September 1986. Three years *after* Challenger, LDEF was in a precarious state. Trajectory specialists estimated that by early March 1990, at the very latest, it would be unable to maintain itself in orbit and would tumble back to Earth, burning up in the dense layers of the atmosphere. Since many of its experiments promised to yield valuable scientific data, particularly as NASA devised new materials for Space Station Freedom, it was imperative for a Shuttle mission to return it home.

That was the task of Brandenstein's STS-32 crew, announced in November 1988 for launch in November of the following year. The assignment of Brandenstein is perhaps unsurprising, since he had already trained for the LDEF retrieval and – at the beginning of March 1985 – had been within three weeks of actually carrying it out. Having said this, many of the pilots in the astronaut office regarded STS-32, with its rendezvous commitment, as one of the 'plum' mission assignments and felt that Brandenstein, who had replaced John Young as chief of the corps in April 1987, was picking the best flight for himself. Whatever the reality, one other crew member had also trained extensively for the LDEF retrieval. Mission specialist Bonnie Dunbar would be at the controls of Columbia's RMS to grapple the giant satellite and manoeuvre it into a berth in the payload bay. The other members of the STS-32 crew were all rookies, but all would contribute hugely to the space programme in the following years. Pilot Jim Wetherbee would go on to become the *only* American astronaut to command five missions and flight engineer Marsha Ivins would enjoy a 36-year career with the space agency and install the Destiny laboratory module onto the International Space Station, whilst David Low – son of the former NASA

Deputy Administrator George Low – would prove instrumental in drawing the Russians to the negotiating table as partners in the ISS project.

George David Low was one of the youngest astronauts ever chosen by NASA when he joined the tenth class in May 1984, aged only 28. He was born in Cleveland, Ohio, on 19 February 1956, and grew up with the space programme all around him, since his father had worked for both NASA and its predecessor, the National Advisory Committee for Aeronautics. The elder Low had been intimately involved in the planning of Projects Mercury, Gemini and Apollo and later headed the Apollo Spacecraft Program Office in Houston, forming part of the team which committed Apollo 8 to the audacious goal of orbiting the Moon. He later served as Deputy Administrator and Acting Administrator of the agency in 1969–76 and saw his son admitted into the astronaut corps, only to die just two months later, in July 1984. (Sadly, father and son would both succumb to cancer.) As a child, David Low was fascinated by science and declared his intent to become an astronaut, aged only nine years old. He entered Washington & Lee University to study physics and engineering and graduated in 1978. He undertook a master's degree in physics and engineering at Cornell University in 1980 and a *second* master's, this time in aeronautics and astronautics, from Stanford in 1983. During this period, he worked in the Spacecraft Systems Engineering Section of the Jet Propulsion Laboratory, working on the systems engineering design of the Galileo spacecraft and the Mars Geoscience/Climatology Orbiter (later the Mars Observer). One of his contemporaries, Frank Culbertson, labelled him as "more academic than the rest of us", but admitted that Low was a good operator and a skilled mechanic, who worked on cars, but understood the physics behind them and communicated this understanding well. Described by United Press International as "an intense young astronaut" and "a man not given to frivolity", Low admitted that the influence of his father represented a yardstick by which he measured his own life and how he treated others.

One of Low's key roles on STS-32 was the deployment of the Navy's Syncom 4-5, the fifth and final satellite in a series which traced its genesis back to 1978. Four of the drum-shaped Hughes spacecraft had been placed into orbit on a series of pre-Challenger Shuttle flights and, although one suffered a catastrophic failure of its UHF electronics, another was triumphantly retrieved, 'hot-wired' and returned to operations. The Syncom to be released under Low's auspices, therefore, would be essential in completing a 'minimum' constellation of four satellites needed by the Navy.

Columbia required several tries before finally making it into space for STS-32. Launch was originally planned for 18 December 1989, which – judging from the ten-day duration – would have made Brandenstein's crew the first team of Shuttle astronauts to remain in orbit over Christmas. This fact evidently played so much on their minds that they privately organised an impromptu crew portrait to be taken, in which they posed in Santa suits, hats and dark glasses. Fortunately, their NASA name tags at least made them identifiable. (Unfortunately, problems with getting Pad 39A ready for its first launch in almost four years resulted in a delay until 8 January 1990, so the Santa joke fell flat.)

Since the return to flight of STS-26, most missions had lasted around five days, but STS-32 was to break this cycle by approaching – or even exceeding – the duration record set by the Spacelab-1 crew in December 1983. (The press kit reported that the flight was to last nine days and 21 hours.) Although the deployment of Syncom and the retrieval of LDEF would consume only the first three days and did not specifically require a lengthy mission, NASA wanted to exercise the opportunity to demonstrate the Shuttle's capabilities, because it planned to modify Columbia for flights lasting up to a month. Processing of the orbiter involved modifications to support the longer mission. A fifth set of cryogenic oxygen and hydrogen tanks were installed underneath the orbiter's payload bay floor and by the end of November 1989 the Shuttle had been rolled out to Pad 39A, marking the first use of this launch complex since Mission 61C. After a delay until 8 January to finish work on the pad, the weather became the next issue. "Our main concern," said Air Force meteorologist Ed Priselac on the 6th, "is that low-level cloudiness will not clear out of here very quickly." The threat also included rain showers and high-altitude clouds, which reduced the prospects of acceptable weather on the 8th to just 40 percent. The odds of successfully launching that day were reduced yet further by the relatively short, 54-minute 'window', which had been precisely timed to allow Columbia to rendezvous with LDEF on the third day of the mission. NASA engineers also expressed concerns that pad hardware used to load cryogenic propellants into the External Tank might leak, although these concerns proved unfounded. Otherwise, the attempt on 8 January proceeded smoothly: the crew were strapped into their seats by mid-morning, although the clock was held at T-9 minutes by unsatisfactory conditions at the Shuttle Landing Facility. In an effort to keep the option of launching open, the clock was restarted and counted down to T-5 minutes – the point at which Wetherbee would start the Auxiliary Power Units – but was held again. Just when it seemed that the weather might just co-operate, a faulty electronics component signalled a possible glitch with Pad 39A's sound suppression water system. A team of engineers were hurriedly despatched to check the system and were satisfied that everything was normal, but then the weather closed in once again and prompted a scrub.

Brandenstein's crew had more luck the next day and STS-32 thundered into space precisely on the opening of the hour-long launch window at 7:35 am EST. A picture-perfect ascent established them on an orbital 'racetrack' to catch up with LDEF and retrieve it on 12 January. In the meantime, the astronauts spent their first day in space by concentrating on two major objectives: checking out the RMS arm, which Bonnie Dunbar called "a beautiful piece of hardware", and preparing to deploy Syncom. At 8:18:39 am on the 10th, a little under 25 hours after launch, the satellite was released as Columbia flew above Africa. Low radioed to Mission Control that the deployment looked good. A few minutes later, Brandenstein and Wetherbee performed a separation manoeuvre to create a safe distance before the first engine burn. Syncom's manufacturer, Hughes, was exceptionally pleased with the performance of their product. "It was as good as you can get," said spokesman Tom Bracken. "Everything looks great." A series of manoeuvres by the satellite's own propulsion system were required to achieve its 'slot' in geostationary orbit. The first, at 8:53 am, involved Syncom firing its solid-fuelled motor to boost itself into an elliptical transfer orbit. This was later circularised and the perigee raised to geostationary altitude. During this time, Dunbar uncradled the RMS and used one of its cameras to photograph the first Syncom burn. Several additional manoeuvres were made by the satellite to achieve its final orbit, which it had accomplished by the 13th. Following a month-long period of checks, it was declared operational and joined its siblings. Later in 1990 and 1991, it was used to support military communications during Operations Desert Storm and Desert Shield in Iraq.

With the successful deployment behind them, the crew turned their attention to the LDEF retrieval. At the time of their launch, they trailed the satellite by about 2,730 km and, being lower, were closing at about 60 km per orbit. Three flawless manoeuvres were performed by the pilots on the 9th and 10th to reduce this distance and Flight Director Bill Reeves exulted to journalists that everything was proceeding smoothly towards the rendezvous. On the morning of the 12th, the astronauts were awakened to music from Mission Control: *Bring it Home*, set to the melody of *Let it Snow*. Due to the extreme sensitivity of LDEF's experiments, the final approach was as unobtrusive as possible to minimise the risk of contamination. From a distance of 40 km down to just 1.6 km, Columbia's radar and star tracker locked onto their target and permitted Brandenstein to move closer. He then took manual control of his ship, passing 'below' and 'ahead' of the satellite, then pitching the Shuttle's nose 'upwards' to achieve a position directly 'above' LDEF. Jim Wetherbee related that this was essential to keep the closure rates slow and smooth when approaching such a massive object.

"I decided when I was about ten years old that this is what I wanted to do for a job," Wetherbee told an interviewer before one of his later Shuttle flights. "I don't really know why." Perhaps it was one of the dramatic Gemini missions, with their rendezvous and spacewalking activities, and he remembered as a child sneaking into a classroom to listen to a small, nine-volt transistor radio. Instead of scolding Wetherbee, the teacher had allowed him to sit at the back of the class and plot the Gemini's orbital progress on a map of the world to show the other children. To James Donald Wetherbee, it was the challenge of *exploration* which most excited him. Born in Flushing, New York, on 27 November 1952, he grew up *tall* in more ways than one ... in fact, even today, at 1.93 m – six feet and four inches – he remains one of the tallest spacefarers ever launched. He attended high schools in New York and earned a degree in aerospace engineering from the University of Notre Dame in 1974, then entered the Navy, hoping to become a pilot and land on aircraft carriers. Certainly, aviation was important to him, since his father had served as an Army Air Corps aviator in the Second World War and later became an American Airlines captain and chief pilot. In his youth, becoming an astronaut was a dream, but Wetherbee recognised that the chances of bringing that dream to fruition were very low. He received his aviator's wings in December 1976 and trained initially in the A-7E Corsair II, flying for three years aboard the USS *John F. Kennedy* and accumulating more than a hundred carrier landings. Wetherbee's next step was test pilot school, from which he graduated in 1981, and he was serving as a project officer for the weapons delivery systems and avionics integration of the new F-18 Hornet fighter. He later flew the Hornet as an operational pilot. It was his wife, Robin, who spotted NASA's call for astronauts and encouraged Wetherbee to apply. Selected in May 1984, with the rank of a lieutenant, Wetherbee was the youngest and most junior of the pilots ... yet he would fly more often than any of them. He had drummed for the marching band at Notre Dame as an undergraduate and had put his kit away when he entered the Navy – "I *couldn't* take the drums on an aircraft carrier!" – but little did he know that in the summer of 1987 he would be recruited by Hoot Gibson and Brewster Shaw to join the Max Q rock band. A little more than a year later, he would be recruited again … only this time to his first Shuttle crew.

Now, at the start of a new decade, Wetherbee was astonished as the gigantic LDEF 'hung' just a few tens of metres above Columbia's payload bay. By this point, the astronauts – who had watched computer-generated views in simulators for more than a year – saw the real thing, waiting to be captured. It had travelled almost 1.3 billion km, or roughly the distance between Earth and Saturn, and had completed 32,000 orbits. As they looked at it, the crew noticed that LDEF had suffered some damage during its six years aloft: a small solar cell had apparently dislodged itself and was flying in formation with the satellite, whilst a number of holes, apparently from micrometeoroid impacts, were evident. After the retrieval, the Interim Operational Contamination Monitor in the payload bay revealed that it had also sustained a fair amount of particulate debris. Dunbar focused the RMS wrist camera on LDEF's starboard side and prepared to grapple it. Brandenstein then performed a yaw manoeuvre to align the wrist camera with the grapple fixture. By this point, the satellite was above the cabin, as the pilots maintained formation in an inverted orientation. When she saw the grapple fixture in the wrist camera monitor, Dunbar went in for the kill. She rotated the camera by 180 degrees to the correct retrieval position and at 10:16:05 am EST, whilst over the Atlantic Ocean, near Brazil, she grasped it.

Brandenstein keyed his mike. "Houston, Columbia," he said, "we have LDEF!"

"You've made many scientists very happy that their LDEF experiments are finally coming home," replied Capcom Tammy Jernigan in Mission Control. In the background, the crew could hear the sound of applause. In fact, it was the first of many accolades for the astronauts that day. Lead Flight Director Al Pennington called it "the culmination of a lot of work by a lot of people" and NASA Administrator Dick Truly expressed his admiration as he "watched America's space programme at its best". According to LDEF Chief Scientist William Kinard, the excitement of investigators from around the world, from the United States to Europe and even as far afield as Australia, was audible. Immediately after the retrieval, Columbia's General Purpose Computers commanded the RMS to align LDEF with the payload bay's berthing guides and Dunbar lowered the satellite into position at 3:49 pm. "It looks like LDEF is going to join us for the ride home," Bill Reeves remarked.

Marsha Sue Ivins, meanwhile, had spent the past four and a half hours painstakingly photographing each and every surface of the satellite for the benefit of engineers. To assist, Dunbar had rotated LDEF slowly on the end of the mechanical arm. Due to the sheer size of the satellite, there was a real risk that the crew might not be able to latch it properly into the payload bay and, in a worst-case scenario, might have to leave it behind it orbit. "So we wanted to at least get the pictorial data," Dunbar recalled. "We also put video down to the ground. *That* took several hours, but we were able to do that and then got it latched and brought it back." Four and a half hours of focused inspection and photography work, though, was nothing, compared to the *ten years* that Ivins had waited to *become* an astronaut. Like so many others of her generation, Ivins had grown up with dreams of flying into space. She was born in Baltimore, Maryland, on 15 April 1951. "When I was ten years old," she told an interviewer, shortly before her final Shuttle flight, "Alan Shepard made the first flight in the American space programme." Ivins remembered that the event captured her imagination and she refused to allow the fact that all of the astronauts were *male* and *military pilots* to get in her way. She realised that they were also engineers and, "for no other reason", went to the University of Colorado at Boulder to study aerospace engineering. Ivins received her degree in 1973 and began working at the Johnson Space Center in July of the following year. Her initial role was as an engineer for the displays and controls of the Shuttle, which was then in its earliest stages, and in 1980 she became a flight engineer in the Shuttle Training Aircraft. She applied for the astronaut programme on *three* occasions, was unsuccessful in 1978 and 1980, and made the cut in May 1984.

By the time that Ivins' photography was done and the RMS stowed, the crew had been awake for almost 17 hours, but, according to Brandenstein, "*all* the faces up here are smiling and happy". Clearly, the triumph had raised their spirits; to such an extent, in fact, that they transmitted cartoon picture to Mission Control the following morning, showing LDEF literally imprisoned within a web of overgrown tomato seeds. "We saw something ... *strange*," Brandenstein grinned, "so we got it on the video recorder and thought we would show it to you." The tongue-in-cheek cartoon was a quip at the 12 million tomato seeds flown by students aboard LDEF, which had seemingly overgrown after their long stay in orbit!

With both primary objectives of their mission – the deployment of the Syncom and the successful recovery of LDEF – now behind them, the crew settled down to a week of scientific and medical experiments. "This is the second longest Shuttle mission we've had so far," said David Low on 13 January, "so we can do some good science experiments up here and get some very good medical data." His words would prove ironic, if not a little prophetic, for Columbia would break her own endurance record, set during STS-9 in December 1983, and fly for almost 11 days. One avenue of study for the astronauts was materials processing in microgravity and Dunbar spent a great deal of her time tending the Fluids Experiment Apparatus (FEA) in the middeck. This device was capable of heating, cooling, mixing, stirring or imposing centrifugal force on gases, liquids or solids and had been carefully designed to meet industrial requirements. Dunbar supervised the processing of seven samples of indium – a well-characterised material with a low melting point – to assess the effect of disturbances induced by the Shuttle's thruster firings or the movements of crew members. It was anticipated that results could lead to more advanced versions for Space Station Freedom. The FEA was activated a few hours after STS-32 reached orbit and ran successfully for almost a full week, until a sensor indication showed that it had exceeded its touch-temperature limit. The unit shut itself down, as programmed, but the astronauts reported that it did not seem to be hot. Nevertheless, after a week of operations, it had achieved more than three-quarters of its objectives. Meanwhile, Ivins and Low tended to a series of protein crystal growth investigations. Ivins was also responsible for the American Flight Echocardiograph, an off-the-shelf ultrasound device, specially modified for carriage aboard the Shuttle. It had the potential to non-invasively generate three-dimensional, cross-sectional imagery of the heart or soft tissues and display it on a monitor. During the mission, the echocardiograph was used in conjunction with a lower-body negative pressure instrument, which resembled a collapsible set of 'trousers' which drew fluids into the legs as a countermeasure for the punishing effects of a return to terrestrial gravity. Several commanders had objected to the medical experiments, but since becoming chief of the astronaut office, Brandenstein had seen them as part of the job. "I was very demanding on the experiments that they do have *real* merit," he told the NASA oral historian, "and be well-organised and have a test plan, not a willy-nilly type experiment." His enthusiasm was perhaps pushed a little far by physician-astronaut Sonny Carter, who talked the crew into "the granddaddy of all experiments" – a *muscle biopsy*. Surprisingly, Brandenstein's crew were game and wilfully volunteered to have hunks of muscle pulled out of their legs both *before* and *after* the mission.

Despite its impressive success, the flight was not entirely smooth sailing. On 11 January, several litres of water oozed from a leaking dehumidifier onto the middeck. After switching it off and activating a backup, Brandenstein joked that his crew had won the Plumber of the Year award ... but *not* Housekeeper of the Year. A more serious problem arose on the 14th, when he was awakened by Mission Control, following an indication of trouble with one of Columbia's Inertial Measurement Units, a critical part of the navigational hardware. Although he reset the unit and returned to sleep, Flight Director Al Pennington worried that any other problems could result in an early end to the mission. Another key issue factoring into the chance of an early landing was the weather outlook at Edwards Air Force Base, where skies were forecast to be overcast, with a chance of snow flurries, on 17 and 18 January. Thankfully, a shortened mission was not necessary and revised weather estimates predicted dry conditions, scattered clouds and light winds – coupled with frigid temperatures – for the 19th. With LDEF in her payload bay, Columbia would tip the scales at a mammoth 103,400 kg, which meant that the dry lakebed Runway 17 would be too soft and perhaps cause controllability problems. NASA therefore decided to land on the concrete runway, but even that posed its own challenges. The presence of LDEF shifted the orbiter's centre of gravity 'forward', meaning that without deft handling of the vehicle, the nose gear might 'slap' down too hard onto the runway. In a press conference, fielded whilst in orbit, Brandenstein told journalists that he needed to maintain sufficient speed after main gear touchdown in order to gently rotate the nose down.

Despite hopes that the weather would co-operate for a landing at 1:59 am EST (4:59 am in Florida) on 19 January, a wave-off seemed likely because a dusting of snow at Edwards the previous day had left water on the runways, and there was a chance of fog. According to spokesman Kyle Herring on the evening of the 18th, the weather remained marginal. Notwithstanding this potential obstacle, the crew marched through their pre-landing checks of Columbia's flight surfaces and controls ... but to no avail. As feared, the fog prompted a 24-hour delay. "We're looking and watching the weather," said Bill Reeves on the evening of the 19th. "Edwards is improving for tomorrow." In fact, no fewer than *four* landing opportunities existed on 20 January, followed by three more on the 21st and the Shuttle could conceivably

Spectacular view of Tropical Storm Sam, captured from Columbia.

remain aloft until the 22nd if needed. The mission *could* have landed in Florida or at White Sands in New Mexico, but NASA elected to hold out for Edwards, aware that its wide runway provided a more forgiving environment with the heavy LDEF aboard. As they waited for the weather to improve, the STS-32 crew enjoyed a light day on the 19th, quietly surpassing the STS-9 record in the mid-afternoon. Brandenstein also became the most experienced Shuttle astronaut, having notched up more than 570 hours on three missions. In doing so, he surpassed Bob Crippen, but admitted that each and every hour was enjoyable. On the 17th, he had celebrated his 47th birthday in orbit and an inflatable plastic cake had been smuggled aboard by his crewmates. He also received a chorus of pre-recorded greetings from the rest of the astronaut office and a message from basketball star Larry Bird, who congratulated him on the "slam dunk with LDEF". When questioned by Mission Control about his age, Brandenstein, alluding to Einstein's theory of relativity, replied that he had hoped that flying at Mach 25 would have slowed down the aging process ...

The return to Earth on 20 January was successful, although a switch failure in one of the General Purpose Computers during de-orbit preparations led Mission Control to wave off the first landing opportunity of the day. Ironically, the particular computer which failed carried the backup flight software to be used if the four primaries failed. To play things safe, the backup software was loaded into one of the primaries and the failed unit was shut down for the rest of the flight. It was a nail-biting time. Brandenstein and Wetherbee were minutes away from firing the OMS engines to begin the irreversible de-orbit burn, with touchdown scheduled for midnight PST (3:00 am in Florida), when the computer glitch arose. The engines were finally fired for five minutes at 12:30 am PST (3:30 am in Florida), slowing Columbia and dropping her into the upper reaches of the atmosphere. No radio 'blackout' was experienced, because constant communications were possible through the Tracking and Data Relay Satellites.

Soaring through the darkness, Columbia touched down on concrete Runway 22 at 1:35:35 am PST (4:35:35 am in Florida). "Welcome home. Outstanding job," radioed Capcom Mike Baker from Mission Control, as Brandenstein brought the vehicle smoothly along the centreline. "You showed the Shuttle at its best, deploying and retrieving satellites." All six wheels stopped a minute later, setting a new Shuttle duration record of ten days, 21 hours, one minute and 39 seconds ... a record which would endure for another two and a half years. "Records are there to be broken," admitted Bonnie Dunbar, "but we were just glad to get another day in space." For Dunbar, the breaking of records would characterise the rest of her career. On her next mission, also aboard Columbia, in the summer of 1992, she would exceed her own record from STS-32, by spending 14 days aloft. *That* would be the first in a series of long-haul flights, featuring a new system known as the Extended Duration Orbiter (EDO). Columbia herself returned to Florida on 26 January and LDEF was removed from her payload bay. Excited scientists *had* been permitted to photograph their precious payload from the aft flight deck windows at Edwards, but it was not until the orbiter was back on the East Coast that they could get their hands on the satellite properly. For the next two months, radiation levels were monitored, infrared video surveys were made, contamination was recorded and thousands of photographs were taken. LDEF was intact, but certainly weathered after six years in space. Clear evidence existed of 'pitting' as a result of micrometeoroids punching into her outer surfaces, and some erosion to a Kevlar thermal cover on her space-facing end. "I think the conclusion that we all came away with," said William Kinard, "is that you have to be cautious in designing a spacecraft." Organic materials like Kapton, Mylar, paint binders and bare composites exhibited severe erosion from atomic oxygen. 'Coated' composites generally survived and maintained their mechanical properties, but due to the extended nature of the mission a few of the thin polymeric films and blankets had been completely destroyed; they had deposited their debris on adjacent areas of the spacecraft. A low-density particulate 'cloud' was also spotted in LDEF's wake and many of the satellite's surfaces showed brownish discolouration. Yet the six years endured by LDEF in the harshest environment known to mankind had actually provided a tremendous service to the engineers and technicians, who were at that time working on the kinds of materials best suited for Space Station Freedom.

'PLAY MISTY FOR ME'

Something strange happened in March 1990. Ground-based observers were busy tracking the orbital progress of a classified Department of Defense payload, recently deployed by Atlantis' crew on STS-36, when they spotted something unexpected. The massive satellite, which reportedly weighed around 17,000 kg, had proven to be extremely bright and an easy object to follow in the night sky, but on the 16th, barely two weeks after its deployment, the Soviet Novosti news agency reported that it appeared to have broken up into several large 'pieces'. Had America's latest national security sentinel malfunctioned and *exploded*, they gloatingly wondered? The Pentagon quickly rebutted such claims, insisting that "hardware elements ... would decay over the next six weeks". In total, five pieces of debris (designated '1990-019 C-G') were monitored and speculation was rife over whether they represented a catastrophic loss of the satellite or were little more than jettisoned payload shrouds or instrument covers. One magazine published images of the STS-36 crew, led by John Creighton, in their quarters before launch and cynically asked if they would have done better to stay at the breakfast table! *Had* the satellite – later identified as 'Air Force Program-731' (AFP-731) or 'Misty' – exploded or broken up ... or was something else afoot? As with so much in the 'deep black' world of Department of Defense space operations, all was *not* what it seemed.

The visual brightness of AFP-731 reached a magnitude of -1 under favourable conditions, similar to the very large KH-9 Hexagon and KH-11 Kennan imaging reconnaissance satellites and it is thought today that the STS-36 payload was probably around the same size, shape and weight as the Hubble Space Telescope. Indeed, the website www.globalsecurity.org has noted that Misty weighed in the region of 16,640 kg at launch, with half a dozen propellant tanks, a short, offset telescope with a large, black-coloured photo shutter window to permit wider fields of view. Electrical power came from a set of "curved, body-hugging solar arrays", the website explained, consisting of "three segments ... attached to a deploying boom mechanism that allows the panels to be rotated in one plane to track the Sun". These arrays were composed of "battle-hardened" gallium arsenide. Misty had been deployed into space on 1 March, a day after Atlantis roared into the highest-inclination orbit ever achieved by the Shuttle. Mike Mullane, one of the mission specialists, described this inclination – tilted 62 degrees to the equator, which offered the astronauts a broader view of Earth than any other crew in history, even reaching the Arctic and Antarctic Circles – as the only declassified component of their flight. Launching *into* this orbit had been a long time coming. When the crew was assigned in February 1989, they expected to fly in February of the following year. That much, at least, came true, but what the astronauts could not have anticipated was that they would endure no fewer than *five* postponements, before finally blasting off on their *sixth* attempt. Originally scheduled for 22 February, their launch was to occur in the early hours of the morning, which obliged them to adjust their sleep cycles accordingly in the Kennedy Space Center crew quarters. "We were going to bed at 11:00 am and waking at 7:00 pm," Mullane wrote. "Breakfast was at 8:00 pm, lunch at midnight and supper at 6:00 am. A *vampire* kept better hours!" The reason for the first 24-hour delay was Creighton himself; for several days, he had been bothered by a steadily worsening cough and, although he tried to avoid the flight surgeon, he could not conceal it for very long and was diagnosed with an upper respiratory infection. He had not improved by the 23rd or the 24th and the weather also took a turn for the worse, resulting in two more postponements. Creighton was moved out of the crew quarters, to avoid infecting the others, and placed in an old space suit room, painted brilliant white and illuminated by a full ceiling of fluorescent lights. His crewmates wished him a speedy recovery, but the opportunity for humour was not far away. "We placed his food tray on the floor," wrote Mullane, "and used a long-handled push broom to shove it close to his table and then immediately retreated from the room." Creighton managed to croak a laugh as they greeted him with plastic bags over their heads. Years later, he remembered that a combination of the awful sleep-shifting, sheer exhaustion and possibly catching a bug had most likely conspired against him.

Despite the precautions, the others began to show the signs of sickness, too. First among them was Atlantis' pilot, John Howard Casper, an Air Force colonel, who had unsuccessfully applied for the 1978 astronaut class, before being selected in May 1984. Casper was born in Greenville, South Carolina, on 9 July 1943 and after high school entered the Air Force Academy to study engineering science. He earned his degree in 1966 and completed a master's qualification in astronautics from Purdue University, early the following year. Following initial flight instruction, Casper received his wings at Reese Air Force Base in Texas, trained on the F-100 Super Sabre and was despatched to Vietnam, where he flew more than 200 combat missions. Upon his return to the United States, he continued to fly the F-100, as well as the F-4 Phantom II, and was assigned as an exchange pilot to a tactical fighter wing at the Royal Air Force's Lakenheath base in Suffolk. Casper then attended test pilot school at Edwards Air Force Base, graduated in 1974 and headed the F-4 Test Team, performing weapons separation and avionics testing. He later worked at the Pentagon as deputy chief of the Special Projects Office, developing Air Force positions on requirements, operational concepts, policy and force structure for tactical and strategic programmes. By the spring of 1990, six years after his selection and only days away from his first launch into orbit, Casper was showing signs of sickness and was placed on medication. So too was mission specialist Dave Hilmers. The only astronauts left 'healthy' were Mullane and the third mission specialist, Pierre Joseph Thuot, nicknamed 'Pepe'. Thuot was born in Groton, Connecticut, on 19 May 1955. After finishing high school in Virginia, he entered the Naval Academy to study physics. He graduated in 1977 and entered naval flight training in July of the same year. After receiving his wings in August 1978, Thuot worked as a radar intercept officer in the rear seat of the F-14 Tomcat fighter and was later deployed to the Mediterranean Sea aboard the USS *John F. Kennedy* and the Caribbean Sea aboard the USS *Independence*. He later attended the Navy Fighter Weapons School

– 'Top Gun' – and Naval Test Pilot School at Patuxent River, Maryland, graduating in 1983 as a test engineer. By the time of his selection into NASA's astronaut corps in June 1985, Thuot had gained a master's degree in systems management from the University of Southern California and had accrued considerable experience as a flight instructor at Pax River. On STS-36, he became the first of his class to be assigned a mission and the first to actually fly. Mike Mullane remembered Thuot as a fast mover and a fast thinker. "Pepe was a 24-volt guy in a 12-volt world," he wrote. "He reminded me of a hummingbird in the way he darted at whatever he was doing, whether he was turning the page of a checklist, punching in a phone number or flipping cockpit switches." From a personal perspective, when I was a teenager, I contacted Thuot to ask him about his career. One comment in particular stood out and proved illustrative of his work ethic. "Whatever you do in life," he told me, "always make sure that you enjoy what you're doing and *aim high*. Pick challenging goals and work hard to achieve them."

With the launch of STS-36 now rescheduled for 25 February, it seemed highly unlikely, considering Creighton's illness, that Atlantis would be able to go. As circumstances transpired, a malfunction was detected in a range safety backup computer at T-1 minute and 55 seconds. The clock continued counting down to 31 seconds and was held whilst engineers tackled the issue. During the hold, the prolonged liquid oxygen drainback resulted in the lower inlet temperature limits on Atlantis' three main engines being exceeded, violating Launch Commit criteria. Unacceptable weather for a Return to Launch Site abort caused a fifth attempt on the 26th to be scrubbed and a 48-hour delay was enforced to give the ground crews some rest. Finally, on 28 February, the astronauts lay uncomfortably on their backs for several hours, waiting for a break in the weather. Rain showers lashed the Cape, whilst the Transoceanic Abort Landing sites in Spain were also coded as 'No Go'. At length, fellow astronaut Mike Coats, flying the Shuttle Training Aircraft on weather reconnaissance, gave them the all-clear and confirmed with the pilots in Spain that the launch could go ahead. At 2:41 am EST, the countdown came out of an extended hold at T-9 minutes. John Casper brought up the Auxiliary Power Units shortly thereafter and at 2:48 am the crew were instructed to close their helmet visors. For Mullane, seated downstairs on the middeck for launch, the next few minutes were a blur. Ascent was unusual, since the normal maximum inclination was around 57 degrees. To achieve a 62-degree orbital tilt, Atlantis performed a 'dog-leg' exercise – flying downrange on a normal flight azimuth, then manoeuvring to a higher azimuth whilst above the Atlantic – and, although this created a penalty in terms of performance, it was the only possible means of reaching the high-inclination orbit.

"Normally, the highest inclination you'll ever get is 57 degrees," said Creighton, "which keeps you just off the East Coast, so in case anything bad happened, where you *blew up*, you're not going to rain debris down on a major city in the United States. This particular flight was the one exception – the *only* time in US manned space flight – where we've ever gone beyond 57 degrees. It's kind of hard to hide that fact after you launch, when you're up there and the Russians are tracking you, so that was declassified after we launched." (Prior to the loss of Challenger, the plan was for a mission requiring such a high inclination for its payload to launch from

Astonishing view of Atlantis punching through the clouds during her ascent trajectory on 28 February 1990. The Vehicle Assembly Building and Launch Control Center are clearly visible in the foreground.

Vandenberg Air Force Base on a north-south trajectory.) For STS-36, that payload was deemed of such importance to the national security that 'normal' flight rules, which prohibited overflights of land, were suspended and Atlantis passed near Cape Hatteras, Cape Cod and parts of Canada during her climb to orbit.

Having said that, their altitude was one of the Shuttle programme's lowest, often reaching only a little higher than 200 km. For Mike Mullane, this presented stunning views of Earth, which made the Home Planet seem "hugely close". The wind-rippled waves of the oceans were visible in astonishing clarity and, flying over the Caribbean, the humps and valleys of the sea floor stood out in stark relief. Elsewhere, supertankers in the Persian Gulf could be seen, with their V-shaped wakes glinting in the Sun, whilst further south, ribbons of plankton stretched many kilometres out to sea on the fringes of the Antarctic. On one occasion, Mullane saw a flotilla of icebergs in the Southern Ocean and used gyroscopically stabilised binoculars to take a look at the distant land mass of Antarctica. "The pole was nearly 1,800 miles distant," he wrote, "so I had no view of it. Instead, I focused on the rugged coastal mountain chains. The occasional black of a windswept cliff was the only colour in an otherwise sheet-white topography." Floating horizontally, he found that he could roll himself into a ball and plant his face against Atlantis' forward flight deck windows, which, with the cockpit lights switched off, made it seem as if he were snorkelling in the Aegean Sea, watching the iridescence of sea life through a face mask. Over the Pacific, as trade winds of the northern and southern hemispheres met and mixed in the heat and humidity of the equator, ominous clouds flickered with electricity – "sputtering fluorescent light bulbs," Mullane called them – as thunderstorms rippled across the region. He had already decided that STS-36 would be his final mission and was determined to spend this quiet time at the windows on his last full evening in space. Every hour and a half, Atlantis brushed the Arctic and Antarctic Circles and the astonishing diversity of Earth was displayed, map-like, beneath him: the dense Siberian taiga, the rolling dunes of the African deserts, punctuated in their north-eastern corner by the green of the Nile Delta, the snow-capped mountaintops of the Himalayas and the Andes ... and, finally, the place he called 'home' and from where he had received much of his schooling: the sprawling city of Albuquerque, New Mexico. Bordered by the Sandia Mountains to the east and with the mighty Rio Grande flowing through it, from north to south, it was the place to which Mullane and his wife, Donna, intended to return after this flight.

More than two decades later, Atlantis' flight remains cloaked in secrecy, although it is believed that AFP-731 was deployed on the second day in orbit. "The payload had been expected to be deployed at 27 hours into the mission," wrote observer Ted Molczan, based in the Canadian capital, Toronto. However, observations made some 28-31 hours into the mission revealed only the orbiter and, somewhere between 34-35 hours, Atlantis' orbit changed slightly, providing clear evidence of a separation manoeuvre from the payload. "Therefore, deployment probably occurred between 31.5 and 35.3 hours," concluded Molczan. "It is possible that it occurred earlier and that the spacecraft were too close together to separate it with binoculars." Certainly, the "separation burn" took place on the afternoon of 1 March and the payload was sighted, in an orbit of 248 x 260 km, some 57 seconds in time 'behind' Atlantis, on the morning of the 2nd. John Creighton and his crew returned to Earth on the 4th, touching down at Edwards Air Force Base at 10:08 am PST (1:08 pm in Florida). Sketchy details trickled out in the following weeks: that AFP-731 was some sort of electro-optical reconnaissance platform, possibly with a signals intelligence component, and that it utilised a mechanism known as the Stabilised Payload Deployment System (SPDS), which 'rolled' it over the payload bay wall and released it in an offset, near-vertical angle of between 65-80 degrees. The 82 kg SPDS, described in a 1989 paper by JSC engineers Guy King and Ted Tsai, was an electromechanical structure, capable of rotating the payload out of the bay at a pre-determined angular position and separating it on command. Mounted on the port side of the bay, it took the position normally occupied by the RMS. "After the payload is stabilised," wrote King and Tsai, "it is released through a double swivel toggle release mechanism, located within the release head." By the time Atlantis landed, observers had already noticed that the satellite had increased its altitude from 254 km to 271 km and on 7 March it executed a much larger manoeuvre. After the sighting of the five mysterious objects in late March, nothing more was seen until mid-October 1990, when a team of European observers tracked something at an altitude of 811 km, inclined 65 degrees to the equator. Although the United Nations had received no notification of anything operating in that region, analysis of AFP-731's track suggested that it was the closest possible contender and it was suggested that the 'debris' seen in March was actually associated with the transfer of the satellite to its operational orbit. Certainly, this orbit was adjusted again in early

November – possibly to monitor the deteriorating situation in the Persian Gulf; Saddam Hussein having invaded Kuwait the previous August – although attempts to 'find' it in the weeks which followed were unsuccessful. It has also been suggested that another of AFP-731's targets were the Soviet Union's military assets in the Arctic, particularly the strategic archipelago of Novaya Zemlya. Over the years, the satellite's 'vanishing' trick was assumed to represent an example of a new-generational spacecraft, capable of demonstrating optical or radar stealth to prevent adversaries from monitoring it or predicting exactly when it would overfly their territories. Not until 1996-97 was AFP-731 seen again by civilian observers, purely by chance, in binoculars, and it has been speculated that it possessed an inflatable, conical 'shield', composed of very thin polymers and coated with highly reflective gold or aluminium. This shield, it is theorised, was designed to suppress the satellite's optical or radar signature. Today, it is often known by the code name of 'Misty' and is thought to have been built by Lockheed Martin, specifically as a 'low-observable' spacecraft at a unit cost of around $360 million. Its original purpose, according to www.globalsecurity.org, was to permit the Reagan administration to catch the Soviets cheating on arms control agreements. Two more Misty satellites have since been inserted into orbit and it is generally agreed that the first-generation payload, deployed on STS-36, was probably de-orbited sometime after 1997.

Whilst the two military communications satellites deployed on Mission 51J were unveiled a little more than a decade after launch, it seems highly unlikely that any concrete facts will be revealed about intelligence platforms like Misty or Lacrosse for many years to come. John Creighton's crew had *all* been subjected to intensive background checks by the intelligence community during the astronaut application process, but flying on a Department of Defense mission required a new level of security clearance. "There was additional background information," he remembered, "because you were going to be cleared Top Secret for a clearance. Most of the military guys had already had a Top Secret clearance when they were in the military, but it's only good for specific purposes. I'm sure they did a background check on us before we were announced as a crew, unbeknownst to any of us." Creighton's recollection is that only about two dozen other souls knew *exactly* what happened for the entirety of STS-36, including NASA Administrator Dick Truly and the mission's flight directors. "Most of the people in Mission Control didn't know specifically what we did," he added. Yet for five military officers, the sense of pride at completing a mission for the national security interest was pervasive. "Even though I can't talk about them," Mike Mullane said, "I feel very, very proud about those DoD missions. I felt like that was something that had a significant impact on America's security ... and *I* was part of it."

WINDOWS ON THE FUTURE

"If there were *ever* two missions that were completely opposite in terms of the public attention that was given to them," astronaut Loren Shriver told NASA's oral historian, "it would be my first and second missions." It was no understatement. His first Shuttle flight, in January 1985, had been totally cloaked in secrecy, since it deployed a classified Department of Defense payload into orbit, whereas his *second*, five years later in April 1990, launched NASA's scientific showpiece: the $1.5 billion Hubble Space Telescope. "It seems like, sometimes, everybody in the *world* was interested in that and what it would be and what it could do. There was a *lot* of publicity surrounding the mission. Years later, Shriver would remember much goodwill and good feelings amongst his crew about being part of a momentous event as the telescope – a joint US-European project – finally rose from the drawing board to creation to actual operation, some 550 km above Earth. "It's a *big* vehicle," said Kathy Sullivan, "with a *lot* of cross-section. You need to get it into a very low-density region, very high. Its control systems ... are *wimpy* in a sense. Magnetic torquers and control moment gyros are not high-impulse things, so you want to get it pretty high, so that the pointing systems can keep it very still for long observations." Today, Hubble has earned itself a well-deserved reputation as one of the most successful space-based observatories ever launched. In two decades of operations, its instruments have peered deeper into the heavens than ever before. It has acquired images of distant galaxies, created breakthroughs in physics and cosmology by accurately determining the Universe's rate of expansion, detected planets around far-off stars, witnessed the impact of a comet into Jupiter, tracked cloud movements in the atmospheres of Uranus and Neptune and created the best achievable 'map' of the surface of Pluto. In fact, the quality of that map is not expected to be surpassed until New Horizons completes its rendezvous with the tiny world in 2015. "There was no doubt in my mind," said Charlie Bolden, "from the moment I was assigned to the Hubble deployment mission about the historical significance of what we were doing. *That* was one *monster* flight!"

With the advent of the Space Age, it was hardly surprising that plans for a space-based telescope would be an important step forward and an attractive option for the fields of astronomy and astrophysics. Yet the ideas far pre-dated even the launch of Sputnik. Shortly after the Second World War, physicist Lyman Spitzer of Yale University had argued that an orbiting telescope would offer enormous advantages over ground-based instruments, its abilities unimpaired by the distorting effect of Earth's atmosphere and its sensors able to detect high-energy emissions, including X-rays, from distant celestial sources. Following the creation of NASA, the first real efforts to develop a space telescope got underway and in 1975 the agency tried to sell the project to the politicians. Funding was initially denied by the House Appropriations Subcommittee, who reasoned that it was too ambitious, too expensive at around $400 million and lacked the required support from the National Academy of Sciences. This prompted large-scale lobbying from NASA and leading astronomers *and* a supportive report from the National Academy of Sciences. International co-operation was directed by Congress and the newly-formed European Space Agency (ESA) was invited to participate, with its role encompassing the creation of inexpensive solar panels for the telescope. The size of the mirror was reduced from 3 m to 2.4 m and together these measures halved the cost from $400 million to $200 million. There were other reasons for the reduction in mirror size. "The Shuttle could not lift a 3 m telescope to the required orbit," wrote Andrew

The "monster flight" of STS-31 reached its height with the release of the Hubble Space Telescope on 25 April 1990. Here, the telescope – which almost completely filled Discovery's payload bay – is lifted into its pre-deployment position by Steve Hawley, deftly operating the RMS arm. The telescope's closed aperture door is visible at the top of the image and one of its folded solar arrays can be seen at centre-left.

Dunar and Stephen Waring in their book *Power to Explore*. "In addition, changing to a 2.4 m mirror would lessen fabrication costs by using manufacturing technologies developed for military spy satellites. The smaller mirror would also abbreviate polishing time from 3.5 years to 2.5 years."

In 1977 Congress granted approval for what was then known as the 'Large Space Telescope'. The primary candidates for the fabrication of the observatory's mirror were Perkin-Elmer Corporation, whose bid ran to $64.2 million, and Eastman

Kodak, teamed with the defence contractor Itek, at almost $99.8 million. Despite being significantly higher, the Kodak-Itek joint bid included *two* independent tests of the grinding and polishing quality of the finished optics ... a 'double-checking' provision which Perkin-Elmer did not offer and which would not go unnoticed when investigators dug into the cause of the telescope's spherical aberration, more than a decade later. Perkin-Elmer received approval from NASA to proceed with their bid in 1979. Meanwhile, Lockheed would build the spacecraft itself and the Europeans would make the solar arrays. In anticipation of the research bonanza, a Space Telescope Science Institute (STScI) was established at the Johns Hopkins University in Maryland in 1983 and the telescope itself was scheduled for launch by the Shuttle in 1985. By this time, it had been named in honour of the American astronomer Edwin P. Hubble, who, in the earlier part of the century, had not only conducted extensive research into the structure of stars and galaxies, but also made the surprising discovery that the Universe was expanding. The mirror was one of the most complex headaches of the project – both *before* and *after* launch. Optically, Hubble was a Cassegrain reflector of Ritchey-Chrétien design and its two hyperbolic mirrors offered good imaging performance across a wide field of view ... whilst also having shapes which were difficult to fabricate and test. Perkin-Elmer employed custom-made polishing machines to precisely grind the mirror and, in case problems were encountered, NASA required the company to subcontract to Kodak to build a backup mirror using traditional polishing techniques. (The Kodak mirror is today on permanent display in the Smithsonian.) In 1979, the construction of the Perkin-Elmer mirror began and was completed two years later, washed in hot, deionised water and coated with aluminium and protective magnesium fluoride. NASA remained sceptical about Perkin-Elmer's ability to competently fabricate the mirror and the delays ultimately pushed Hubble's launch back from April 1985 to first the summer and then the autumn of 1986. By this time, the total cost of the project had risen to a little more than $1 billion. At the time of its completion, Hubble housed five instruments: the Wide Field Planetary Camera, the Goddard High Resolution Spectrograph, the High Speed Photometer, the Faint Object Camera and the Faint Object Spectrograph. These devices gave the telescope a range which encompassed not only the visible area of the electromagnetic spectrum, but also the ultraviolet. Physically, Hubble was a cylindrical spacecraft, measuring over 13 m in length and weighing nearly 11,000 kg, which meant that it virtually filled the payload bay. It had been designed to be serviced by future Shuttle crews and, as such, was fitted with EVA-friendly hand holds, and would be deployed using the RMS mechanical arm.

On STS-31, that arm would be under the primary direction of astronaut Steve Hawley. Nicknamed 'the Attack Astronomer', he had proven himself to be one of NASA's most capable spacefarers, having chalked up two previous missions and had been appointed deputy chief astronaut in April 1987. He had worked on Hubble issues for some time, having been assigned in September 1985 to Mission 61J, the flight scheduled to deploy the telescope in August of the following year. Hawley has quipped that he was selected for the mission because he was such a good RMS operator, but he was convinced that the need for an *astronomer* on this most *astronomical* of flights was critical, "for the simple reason that we want to make sure

... that the needs and requirements of the customer are understood and dealt with appropriately". Obviously, as an astronomer, Hawley would not actually be *using* the telescope, nor would there be any 'real' astronomy for him to perform, but he believed that it helped Hubble's science team by having someone who knew what they wanted to accomplish, knew what the constraints were and, in a nutshell, *cared* about the payload. By the time that Challenger was lost, further delays had pushed Hubble's launch until October 1986. The problems faced by Perkin-Elmer have already been mentioned, but the manufacturer of the telescope's bodywork, Lockheed, had also suffered its own difficulties. By the end of 1985, it was over-budget by 30 percent and three months behind schedule, bringing Hubble dangerously close to breaking the 'ceiling' which Congress had imposed on the budget. If Challenger smashed the dreams of so many within America's space programme, it also provided additional breathing room for the hurdles to be overcome. Hubble came through a major thermal vacuum test with flying colours in June 1986 and the enforced down time was used to add more powerful solar arrays, enhance redundancy capabilities, improve software and install better connectors. Nickel-cadmium batteries, which were prone to failure, were replaced with nickel-hydrogen ones, and by the early spring of 1990 the Hubble team felt supremely confident that their observatory would herald a new dawn in the study of astronomy.

Elsewhere, at British Aerospace's plant in Bristol, the development of the twin solar arrays was monitored closely by the crew. In her oral history, Kathy Sullivan remembered one trip to England. The crew's approach had always been to take actual EVA tools to the actual flight hardware and after an exhausting flight to Heathrow, she finally arrived in Bristol, with Bruce McCandless and the rest of the crew, for the tests. She expected to don a clean room suit and get straight to work. The British Aerospace managers had other ideas. "We found ourselves in these rather more formal *Welcome the flight crew* events," she said, "which I hadn't expected." A brief walkthrough of the solar array, suspended in a rig above a water table, followed, after which the astronauts expected to dig into their EVA procedures. Right? *Wrong*. It was lunch time and the astronauts were ushered into a management dining room, "with senior company officials and linen tablecloths", where they beheld to their horror a number of engineers and technicians drinking pints of ale or glasses of wine! Sullivan and McCandless wondered who in their *right minds* would consume alcohol, minutes before handing delicate flight hardware? Answer: The *English*, obviously. Sullivan enjoyed her time in Bristol, although subsequent conversations with Kathy Thornton – who flew the first Hubble servicing mission in 1993 – demonstrated that in addition to visiting British Aerospace, *her* crew also got an extended tour of English historic landmarks, including Stonehenge. "Maybe that's what you get if you've successfully *fixed* their solar array," Sullivan said. "We didn't get that. We just went over there, worked and came home." Aside from the humour of the episode, British Aerospace's treatment of the STS-31 crew was entirely understandable; for in addition to the technical role, Hubble had major European involvement and the arrival of the Shuttle astronauts who would deploy it was afforded the right level of significance.

As long ago as 1983, NASA Administrator Jim Beggs had encouraged his

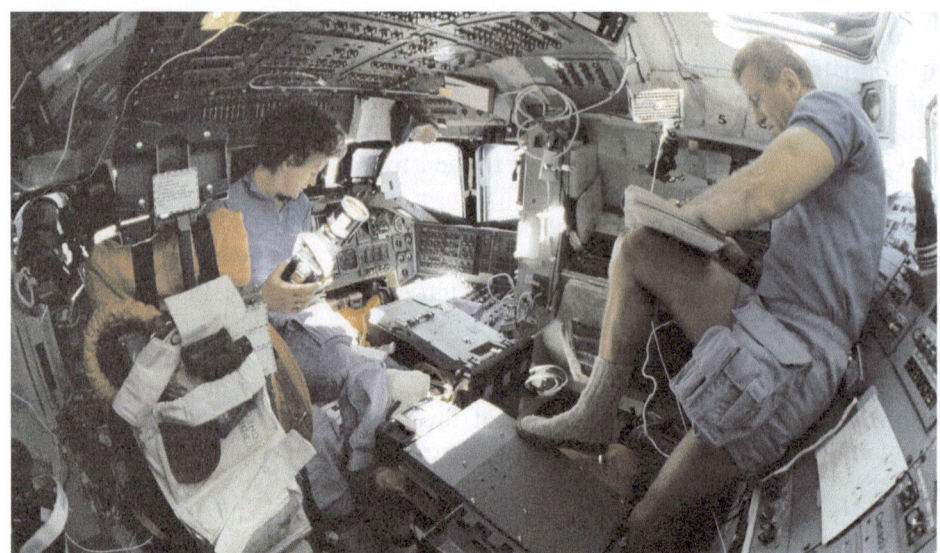

This view amply illustrates the smallness of the Shuttle's flight deck. Kathy Sullivan (left) perches at the edge of the commander's seat, whilst Loren Shriver abuts the aft control panel. The wrap-around windows of the forward cockpit are clearly visible, as are the overhead windows above Shriver and the aft windows overlooking the payload bay. The black joystick at far right, directly behind Shriver's back, is the primary RMS controller which Steve Hawley had earlier used to deploy Hubble.

subordinates to treat Hubble in terms of significance on a par with the Shuttle itself and had even labelled it "the eighth wonder of the world". With this in mind, it is unsurprising that the man selected to command the deployment was chief astronaut John Young. He had barely started training with his crew when Challenger was lost and in April 1987 – for reasons probed in depth elsewhere – he was removed from his role at the helm of the astronaut office, taken off flight status and became the special assistant to JSC Director Aaron Cohen for engineering, operations and safety. Some have seen Young's reassignment as evidence that he had angered NASA's top brass with his much-publicised and aggressive safety memos, issued in the wake of 51L, whilst others, including Mike Mullane, have questioned his management style. Still more have argued that it was respect for his commitment to operational flight safety which led to the new post. Whatever the reason, when the STS-31 crew was named in March 1988, with launch set for June of the following year, Young was notably absent from the list. In his place was Loren Shriver, veteran of 51C. The rest of the original crew remained the same: Charlie Bolden would serve as the pilot, with mission specialists McCandless, Hawley and Sullivan. "It was originally just me and Bruce McCandless, assigned to dig in on the EVA and the servicing aspects," Sullivan told the oral historian. "He'd been the office representative and advisor to Hubble on the whole philosophy and design of the EVA servicing for quite some time. Hubble was supposed to be serviceable on orbit for 15 years. We were assigned ahead of the full crew, so that we could spend extra time being sure that all the EVA tools and procedures were in place to accomplish those servicing aims." Using the engineering diagrams of the telescope, Sullivan and McCandless began working on the design of tools which would permit future crews to capture and service Hubble. "It's a *major* national scientific investment," Sullivan added, "and we know we're committed to servicing this one. We can't consider it acceptable . . . to have little stuff that bites you again. You just can't get up there and discover, *Oops, sorry, wrong wrench, can't do this!* That's just *not* acceptable."

As the Shuttle returned to normal operations after STS-26, McCandless, Sullivan and their crewmates were keenly aware that, despite the scientific importance of their mission, they were not an 'infrastructure-critical' flight and would have to take their place in the line behind the Tracking and Data Relay Satellites, the Department of Defense assignments and the Magellan and Galileo planetary missions that were tied to specific launch windows. Eventually, they found themselves targeted for the March-April 1990 timeframe. "The year 1990 was close to a solar maximum year," said Sullivan, "so the envelope of the atmosphere is physically larger." This had implications for the precise altitude of Hubble's orbit. "There was a long time watching the solar activity and doing all the calculations to determine when we will actually be at a solar max," she continued, "because if we're on the down side of a solar max cycle, then you could go a little lower, because the atmosphere would be deflating. This would give you more performance margin for the deployment flight." As the launch slipped from 1986, in a time of reduced solar activity, to 1990, closer to the maximum, the Hubble deployment altitude was raised to a little more than 610 km. This high altitude meant that a long-duration OMS firing of more than *five minutes* was needed for orbital insertion and the effect upon the Shuttle's performance was that 50 percent of the available propellant was consumed by the time Discovery achieved orbit! On top of this was the need for sufficient margin to re-rendezvous with Hubble, if necessary, after deployment, and perhaps repair it, then re-release it, perform another separation manoeuvre and still have sufficient propellant stores for the de-orbit and return to Earth. (The de-orbit burn from Hubble's altitude was expected to require an OMS firing of almost five minutes, some 60 percent longer than other Shuttle missions.) In the weeks before launch, it was evident that STS-31 would have much lower reserves of propellant at the start of its mission than had been typically on other flights. As a result, a significant amount of training time was devoted to how the crew responded to propellant leak alarms; on an 'ordinary' mission, the first prudent step would have been to verify if the alert was a false one, but on STS-31 the assumption *had* to be taken that it *was* a leak and preparations to either substantially lower their altitude or de-orbit had to be made quickly. All of those steps had to be performed in parallel.

Yet the launch was eagerly anticipated by the astronomical community. "After 45 years of dreaming," wrote John Noble Wilford of the *New York Times* on 9 April 1990, "and almost 20 years of planning, development and delays, the Hubble is ready to be taken into orbit." The first attempt to get Discovery and Hubble into space, on 10 April 1990, was actually eight days earlier than had been published on previous planning manifests. The crew was strapped into the vehicle and the countdown proceeded to T-4 minutes – after the activation of the Auxiliary Power

Units – when, all at once, a scrub was called due to abnormal pressure and turbine speeds in APU No. 1. The device was replaced and retested and the launch rescheduled for the 24th. This time, the countdown was held at T-31 seconds when computer software failed to shut down the liquid oxygen outboard fill and drain valve on ground support equipment. When this had been completed satisfactorily, the clock resumed and Discovery headed aloft – a few minutes later than planned – at 8:33:51 am EST. "Our window on the Universe," was the launch announcer's commentary as the tenth post-Challenger Shuttle mission got underway. Shortly after achieving orbit, the payload bay doors were opened to expose Hubble to the space environment for the first time. "It's a pretty huge device," remembered Loren Shriver, "in terms of filling up almost the entire payload bay and it was covered with nice, shiny insulation and it was just a pretty impressive piece of equipment." In fact, it blazed brightly under the glare of the unfiltered Sun, and after his return to Earth Hawley would advise planners of the Hubble servicing flights to schedule the rendezvous and capture of the telescope during periods of orbital *darkness*. The payload bay lights would be more than sufficient to illuminate Hubble, he said, and avoid momentarily blinding the crew or the RMS cameras.

Deployment occurred on the second day of the mission, 25 April, when Hawley gingerly grappled the massive payload and positioned it above the bay, with its large aperture door facing forward. The twin high-gain antennas were released, springing 'downward' and 'outward', and the next step was to open British Aerospace's solar arrays, which would unfurl like kitchen blinds on rollers. The first array came out perfectly and Shriver reflected that – although he had *seen* this in the factory in Bristol – it was an incredible sight to behold, from such an immense altitude above the Home Planet. Deployment of the *second* array was more problematic. It began to unroll for a few centimetres, then stopped. Everyone's heart skipped a beat. Time was of the essence. After cutting orbiter umbilical power to the telescope, they had less than two hours on internal batteries to get the arrays open to enable Hubble to draw on its own electricity supply. If that did *not* happen, the batteries would die and so would the telescope. The worry went up in notches with the realisation that Hubble might be in severe difficulty. An EVA repair loomed as the only possible solution. Having reduced Discovery's cabin pressure the previous night, Bruce McCandless and Kathy Sullivan were already downstairs, clad in their space suits, in the airlock, with tools, ready to venture outside. McCandless was convinced that *he* knew the source of the problem: a glitch with a tension monitoring module. This was a unit of software to detect any excessive strain on the array and prevent it from tearing or binding. It would stop the array-deployment process, McCandless explained, and the spacewalkers would then have to go outside to fix it manually. To give them more time, they had partially completed their pre-EVA procedures *before* Hubble was grappled. "What's left," said Sullivan, "is to button up the suit, breathe 40 minutes of pure oxygen, close the hatch, depress, get outside."

It was a double-edged sword, for despite having the opportunity of a spacewalk, it also seemed likely that Sullivan's other task of being the primary photographer for the deployment would be missed. Before launch, she had resolved to literally "wallpaper the telescope with photos" … in fact, she *wanted* a cover shot on *Aviation*

Week & Space Technology. Now, after working for five years on Hubble, she was locked up in the airlock and might not even see the moment of its release. With only minutes remaining before she and McCandless would have been directed to fully depressurise the airlock, Mission Control told them that a young engineer had assured them that the problem was an erroneous software indication ... from the *tension monitoring module*. He requested permission to command the module to 'No Op' ('No Operation', effectively taking it 'out of the loop') and was certain that this would enable the stubborn second array to unfurl. Flight Director Bill Reeves concurred and Capcom Story Musgrave radioed up the news: Hawley was to orient Hubble in a ready-to-deploy attitude and if the attempt to 'No Op' worked, the array would probably start unfurling immediately and they would have to release the telescope as quickly as possible. *It worked.* Charlie Bolden was astounded, but McCandless offered a wizened grin. He had worked on Hubble for so long that he knew, instinctively, what had caused the problem.

Steve Hawley's primary concern during the deployment, aside from the array, was the very remote chance that the RMS could fail. The joints of the arm were intentionally limited in terms of their speed, thereby offering him some margin to respond to contingencies, and that meant that the motion he could command was restricted. As he lifted Hubble, it 'wobbled' – a lot more than the simulator had taught him that it should – and not until after landing did he realise that the signal 'noise' in the RMS joints contributed to random imparted motions. In his post-mission debriefings, he recommended that future simulations should take RMS joint noise into account ... a recommendation which would prove hugely beneficial for Swiss astronaut Claude Nicollier, who would grapple Hubble on the *first* servicing mission and for Hawley himself, who would fly the *second*. Charlie Bolden remembered the irony that the crew had actually practiced a solar array deployment failure during their last integrated simulation, shortly before launch. In that instance, McCandless and Sullivan had manually wound out the array, although they knew that doing such an action for real had the potential to severely damage the telescope. "Once you did that," he said, "it took it out of its automatic mode and it would no longer be able to take care of itself ... sort of like taking a baby from the womb, putting it on a respirator and putting it in a position where the rest of its life it would need something. *That* was what that would have meant for Hubble ... until you send another crew up and put on another set of solar arrays and reset the clock."

Disappointment at being unable to perform an EVA was stretched a little further when McCandless and Sullivan were unable to see the moment of deployment at 3:37:51 pm EST. Bolden certainly felt for them, still cooped up in the airlock. "So we deploy Hubble, coming off the Pacific Ocean, across the west coast of South America," he told the NASA oral historian, "and it's just the most beautiful thing you can imagine. It comes off the end of the arm and *down*. We're looking at the Andes Mountains and it goes right across the coast between Bolivia and Venezuela." Shortly afterwards, he and Shriver pulsed Discovery's RCS thrusters on two occasions to raise their orbit slightly, causing them to fall steadily 'behind' the telescope.

As the telescope drifted away from them, the three men on the flight deck gaped in awe at what they were seeing. Then, all of a sudden, and in unison, they all barked: *"Camera! Somebody get a camera!"*

Bolden helped to save the day in terms of the photography. An IMAX large-format camera was already filming the deployment from the rear of Discovery's payload bay and Bolden also acquired some spectacular footage from the interior of the cabin, with the hand-held IMAX device. Having sealed McCandless and Sullivan in the airlock, he shot a length of film as he floated upstairs to the flight deck, focusing firstly on Shriver at the orbiter's controls, then on Hawley at the RMS controls, and *then* through the windows to reveal Hubble in all its enormity and grandeur. It was not like a 'normal' camera, where it was possible to see what was going through the lens; instead, Bolden had nothing to gauge whether he was doing the right thing. After the flight, his biggest surprise was not only that the camera worked, but that it also stayed in focus. Later, IMAX would include the STS-31 imagery as part of its 'Blue Planet' and 'Destiny in Space' documentaries.

Four days later, at 6:49 pm PST (9:49 pm in Florida), Loren Shriver and Charlie Bolden brought Discovery onto the concrete runway at Edwards Air Force Base, capping a mission which promised to open a new set of eyes on the Universe. The winds at the Californian site were described as very high – in fact, Shriver's brother and uncle were staying there in a camper van and had been unable to sleep, so strong was the buffeting – but the touchdown was smooth and precise. Before he became an astronaut, Shriver had worked on the development of upgraded carbon brakes on the F-15 Eagle and he had no doubts about the capabilities of Discovery. The capabilities of Hubble, on the other hand, were something quite different. In the first few weeks, the problems seemed reasonably benign: a few communications glitches, drifting star trackers and snagged coaxial cables were part and parcel in the process of wringing out a new spacecraft. More serious concerns arose when temperature changes bent materials in the solar arrays' booms, the effect of which was magnified by the orientation mechanism in such a way that it 'bounced' the whole telescope. The result was a 'jittering' in Hubble's images and, since the booms only stabilised in the final few minutes of orbital daylight, the pointing system was only able to meet its design specifications for a fraction of its orbit. Engineers at the Marshall Space Flight Center worked with their counterparts at Lockheed to change the control program in the spacecraft's computer and successfully counteracted the vibrations. On 21 May, Hubble returned its first images of a double star in the Carina system and these were lauded as being much clearer than were achievable with ground-based instruments.

Four weeks later, calamity befell the mission in a manner which could hardly have been anticipated: on 24 June Hubble failed a *focusing* test. Its secondary mirror had been adjusted to focus the incoming light from a celestial source, but a fuzzy ring – like a halo – encircled even its best images, creating a blur. Additional tests revealed that the telescope was suffering from a 'spherical aberration' in its primary mirror; in essence, Perkin-Elmer had ground it to the *wrong* specification, removing too much glass and polishing it *too flat* ... by a mere *fiftieth* of the width of a human hair. The consequence was that Hubble was unable to acquire sharp images. With mounting horror, NASA realised that its attempts to sell its scientific showpiece on the basis of its ability to see further into the cosmos than ever before, with unprecedented clarity, now became very hollow indeed. The promised white knight of astronomy was turning instead into a white elephant. Even Hubble's chief scientist, Ed Weiler, admitted that it was comparable only to "a very good ground telescope on a very good night". Marshall staff were astounded and Senator Barbara Mikulski, a Democrat from Maryland, exploded that Hubble wasted taxpayers' money and was little more than "a techno-turkey". Meanwhile, Senator Al Gore – then a Democrat for Tennessee and later Vice President during the Clinton administration – observed that, for the *second* time in less than *half* a decade, NASA's quality control shortcomings had been publicly exposed. The media had a field day. On 28 July 1990, the *New York Times* pointed out that – had Kodak-Itek's bid been accepted – the mirror would have been subjected to *two* independent checks of its grinding and polishing accuracy, which certainly would have caught the error and enabled engineers to rectify it before launch. NASA responded to critics by asserting that, with 20-20 hindsight, it would have cost in excess of $100 million to incorporate additional testing and independent checking of the telescope optics into Perkin-Elmer's contract, but the effect on the general public was the same. The once-proud space agency was rendered a laughing-stock on late-night TV talk shows. David Letterman compiled a pejorative list of Top Ten Hubble Excuses, whilst others criticised the Marshall Space Flight Center for having been in charge of *both* the Hubble development *and* the Shuttle's SRBs. Several analysts noted that NASA's attitude had changed from the 1960s, in which problems were anticipated and incorporated into planning, to the late 1970s and 1980s when there was little effort to prepare for unforeseen obstacles. In the words of John Logsdon, "the agency was not being honest with itself or with anyone else".

In early July, NASA established an investigating committee, chaired by Lew Allen, head of the Jet Propulsion Laboratory. His report – published in November – harshly criticised the incorrect assembly of the reflective null corrector, an optical device used to determine the figure of Hubble's mirror. The location of a lens in the device was improperly measured and the null corrector guided the polisher to shape a perfectly smooth mirror ... with the *wrong* curvature. Analysis revealed that the curvature flaw in the primary mirror exactly matched the flaw in the null corrector. A *second* null corrector, made only with lenses, was also built to measure the vertex radius of the finished mirror. It, too, clearly identified an error in the primary mirror. However, neither of these warning signs were heeded and Allen's report noted that "*both* indicators of error were discounted at the time as being *themselves* flawed". During the fabrication process, technicians had simply *assumed* the perfection of the mirror and of the reflective null corrector and had rejected information from other independent tests, convincing themselves that no problems existed. These errors were ultimately traced back to 1981-82, when Perkin-Elmer and the Marshall Space Flight Center had been distracted by serious cost and schedule difficulties. Allen's report was particularly critical of Perkin-Elmer's quality control and communications failures, as well as Marshall's own failure to correct them. In orbit, the spherical aberration was particularly obvious in its effect on Hubble's Wide Field Planetary Camera and Faint Object Camera, both of which suffered in terms of spatial resolution and their ability to acquire images of distant sources. Having said this, the aberration was well characterised and stable and, over time, enabled astronomers to optimise the results obtained by Hubble with sophisticated techniques, such as 'deconvolution', whereby software algorithms and microwave image processing methods were employed in an effort to remove many of the blurring effects of optical distortion. Spectroscopy was less severely affected, because the instruments required less focused light, and by increasing exposure times it became possible to still gather valuable images. However, the jittering of the solar arrays rendered the High Speed Photometer virtually useless. Nevertheless, by the end of 1991, Hubble had made almost two thousand quality observations of hundreds of astronomical objects, including storms on Saturn and images of Pluto's moon, Charon.

At the start of the following year, a quarter of all the papers presented before the American Astronomical Society's meeting drew on Hubble data. A repair was critical in order to restore the telescope to its pre-flight billing and, although the primary mirror itself could not be replaced, a new device – the Corrective Optics Space Telescope Axial Replacement (COSTAR) – was developed to revitalise its vision. COSTAR was manifested onto the first Shuttle servicing mission to Hubble. For the STS-31 astronauts, there would be involvement in these efforts, both directly and peripherally. Kathy Sullivan had already been assigned to another mission and Steve Hawley had gone to work for NASA's Ames Research Center in Moffett Field, California – although he would later return to flight status – so they watched from afar, but Bruce McCandless dug in with the development of the corrective optics and Charlie Bolden, for a time, hoped to command the repair mission. "Deep down," he said later, "I was kind of keeping my fingers crossed that *that* would be my next mission." Eventually, that command went to another astronaut, although Bolden led two significant Shuttle flights and is today NASA's Administrator, guiding the agency's human space effort beyond Earth orbit for the first time in half a century.

In its first decade of operations, the Shuttle had scored some remarkable successes and taken immense strides forward in both science and technology. Many engineers and managers have spoken eloquently that the orbiters, with their intricate patchwork of thermal tiles, their ability to knife hypersonically through the atmosphere, their reusability and their sheer muscle, proved a far greater technical challenge than landing on the Moon. Yet, all too often, human space exploration falls foul of mistakes made by *humans*. In some cases, such as Challenger, those mistakes resulted in the loss of human life, whilst in others, such as Hubble, they led to humiliation and ridicule in an era where space budgets were increasing frowned upon and seemed to continuously shrink with each new year. NASA was planning to build a permanent orbital outpost, Space Station Freedom, starting in the mid-1990s, and needed the politicians and the public firmly on its side. The spectacular repair of Hubble by the crew of STS-61 in December 1993 would be one of the agency's most defining moments and – although many senior administrators harboured little faith that it would succeed – it actually contributed greatly to changing attitudes toward the Shuttle. As for Hubble itself, hardly anyone remembers the spherical aberration today; Wikipedia gushes in far more detail about its astounding scientific discoveries, its contributions to physics, astronomy and cosmology and its importance to humanity as a whole, with the mirror problem now little more than a footnote. The trauma of Challenger would never be forgotten, but Hubble, its triumphant repair and the resurgence of the Shuttle would turn NASA's attention away from the errors of the past and focus them on new visions for the future.

As the spring burned into the summer of 1990, that future offered faces which were both bleak and promising. A year earlier, President George H.W. Bush had proposed a return to the Moon and plans for a manned expedition to Mars, although his so-called 'Space Exploration Initiative' would breathe its last with all the excitement and drama of a damp squib. Carrying somewhat greater promise was the construction of Freedom, a co-operative venture with the European Space Agency, Canada and Japan. This was not the skin-deep 'co-operation' of the Soviet Union's Intercosmos and Glavcosmos programmes in which guests flew into space for political point-scoring; the Europeans were building their own laboratory, as were the Japanese, whilst the Canadians had committed to supply a robotic servicer, which later evolved into today's Canadarm2. Freedom would require more EVAs during its construction phase than the United States had accomplished in three decades and their complexity would be significantly higher than anything previously attempted. A difficult path lay ahead. President Ronald Reagan's 1984 directive to build the station "within a decade", perhaps emulating the words of John Kennedy, seemed almost impossible in the wake of Challenger, and steadily dwindling budgets and ever-increasing costs gradually turned it, like Hubble, from a white knight into a white elephant. Twelve months after Hubble entered orbit, the station project came to within *one* Congressional vote of cancellation. It is often said that great endeavours sprout from disappointment, tragedy and failure and in some cases hit absolute rock-bottom before the process of rebuilding can commence. With the transition from the 1980s into the 1990s, both the Soviet and American human programmes had hit absolute rock bottom. The Soviets had scored triumph after triumph with their Salyut 7 and Mir space stations, but the breakup of the Union left a once-proud space programme in tatters. On the American side, Challenger left a long shadow in the country's space fortunes. As with the depression which accompanied the rebuilding of the Shuttle programme, the station would reach its own nadir in the early 1990s, before it slowly reached maturity, gained a surprising new partner and, at the dawn of the new millennium, finally established a permanent *international* human presence in space. Only then could the hope for the future steadily begin to brighten.

Bibliography

'NASA to Recruit Space Shuttle Astronauts.' NASA Lyndon B. Johnson Space Center, Houston, Texas, 8 July 1976

'Spacelab Simulation Crew Undergoes Medical Tests'. NASA Lyndon B. Johnson Space Center, Houston, Texas, 1 December 1976

'NASA Gives Rockwell Go for Third Orbiter Start.' NASA Lyndon B. Johnson Space Center, Houston, Texas, 20 June 1977

'Over 8,000 Apply for Shuttle Astronaut Program at JSC.' NASA Lyndon B. Johnson Space Center, Houston, Texas, 15 July 1977

'NASA Selects 35 Astronaut Candidates.' NASA Lyndon B. Johnson Space Center, Houston, Texas, 16 January 1978

'New Astronaut Candidates Arrive at JSC for Training.' NASA Lyndon B. Johnson Space Center, Houston, Texas, 29 June 1978

'Joe Allen Returns to Johnson Space Center Astronaut Office.' NASA Lyndon B. Johnson Space Center, Houston, Texas, 26 July 1978

'Mission Specialists for Spacelab-1 Named at JSC.' NASA Lyndon B. Johnson Space Center, Houston, Texas, 1 August 1978

'Shuttle Orbiters Named After Sea Vessels.' NASA Lyndon B. Johnson Space Center, Houston, Texas, 1 February 1979

'Space Shuttle Orbiter Procurement Contract Signed.' NASA Lyndon B. Johnson Space Center, Houston, Texas, 5 February 1979

'RMS Contract Award.' NASA Lyndon B. Johnson Space Center, Houston, Texas, 25 May 1979

'England Returns to Astronaut Program.' NASA Lyndon B. Johnson Space Center, Houston, Texas, 7 June 1979

'NASA to Recruit Space Shuttle Astronauts.' NASA Lyndon B. Johnson Space Center, Houston, Texas, 1 August 1979

'NASA Tests New Space Manoeuvring Backpack.' NASA Lyndon B. Johnson Space Center, Houston, Texas, 10 September 1979

'Hispanics are Encouraged to Apply for Astronaut Program.' NASA Lyndon B. Johnson Space Center, Houston, Texas, 12 September 1979

'Astronauts May Repair Orbiter Heatshield in Flight.' NASA Lyndon B. Johnson Space Center, Houston, Texas, 20 September 1979

'NASA to Develop Manned Manoeuvring Unit.' NASA Lyndon B. Johnson Space Center, Houston, Texas, 2 October 1979

'Heads-Up Display for Orbiter.' NASA Lyndon B. Johnson Space Center, Houston, Texas, 16 October 1979

'Martin Receives TPS Repair Contract.' NASA Lyndon B. Johnson Space Center, Houston, Texas, 22 January 1980

'NASA Signs Martin Marietta to Build Manned Manoeuvring Unit.' NASA Lyndon B. Johnson Space Center, Houston, Texas, 29 February 1980

'NASA Signs Canadians to Build Shuttle Robot Arm.' NASA Lyndon B. Johnson Space Center, Houston, Texas, 14 April 1980

'NASA Selects 19 Astronaut Candidates.' NASA Lyndon B. Johnson Space Center, Houston, Texas, 29 May 1980

'Investigators File Report on Cause of Space Suit Backpack Fire.' NASA Lyndon B. Johnson Space Center, Houston, Texas, 10 June 1980

'France names cosmonauts for Russian spaceflight.' *Flight International*, 21 June 1980

'Two Europeans Accepted for Space Shuttle Mission Specialist Training.' NASA Lyndon B. Johnson Space Center, Houston, Texas, 7 July 1980

'First Spacelab-1 Hardware Shipped.' NASA Lyndon B. Johnson Space Center, Houston, Texas, 9 October 1981

'NASA Names Crews for Three Missions.' NASA Lyndon B. Johnson Space Center, Houston, Texas, 22 February 1982

'Three Shuttle Crews Selected.' NASA Lyndon B. Johnson Space Center, Houston, Texas, 19 April 1982

'Expectant Astronauts.' NASA Lyndon B. Johnson Space Center, Houston, Texas, 28 April 1982

'Europe creates a workplace in space.' *Flight International*, 15 May 1982

'Hughes signs for down-under comsat.' *Flight International*, 29 May 1982

'Salyut 7 comes alive.' *Flight International*, 29 May 1982

'Franco-Soviet crews fly to Baikonur.' *Flight International*, 19 June 1982

'France to put man in space.' *Flight International*, 26 June 1982

'India to join in USSR spaceflight.' *Flight International*, 4 September 1982

'Second woman in space.' *Flight International*, 4 September 1982

'Svetlana returns from space.' *Flight International*, 11 September 1982

'Miniature TV Camera to be Used on Fifth Shuttle Mission.' NASA Lyndon B. Johnson Space Center, Houston, Texas, 14 September 1982

'Spacelab fliers named.' *Flight International*, 2 October 1982

'NASA Names STS-10 Astronaut Crew.' NASA Lyndon B. Johnson Space Center, Houston, Texas, 20 October 1982

'STS Flight Assignment Baseline: November 1982.' NASA Headquarters, Washington, DC, November 1982

'STS-5 Space Suit Inquiry.' NASA Lyndon B. Johnson Space Center, Houston, Texas, 19 November 1982

'How to repair a satellite.' *Flight International*, 20 November 1982

'Team Reports on Space Suit Failure.' NASA Lyndon B. Johnson Space Center, Houston, Texas, 2 December 1982

'Cosmonauts 'hand-launch' Iskra-3.' *Flight International*, 11 December 1982

'Tenth Shuttle crew named.' *Flight International*, 11 December 1982

'Fifth Crewmember Named for STS-7 and STS-8.' NASA Lyndon B. Johnson Space Center, Houston, Texas, 21 December 1982

'Space Suit Failure.' NASA Lyndon B. Johnson Space Center, Houston, Texas, 28 January 1983

'Inquiry Team Reports on Space Suit Failures.' NASA Lyndon B. Johnson Space Center, Houston, Texas, 1 February 1983

'STS-11 and STS-12 Crews Named.' NASA Lyndon B. Johnson Space Center, Houston, Texas, 4 February 1983

'Crew Members Named for STS-13, Spacelab-2 and 3.' NASA Lyndon B. Johnson Space Center, Houston, Texas, 18 February 1983

'All change on Challenger engine.' *Flight International*, 26 February 1983

'Salt spray delays Shuttle.' *Flight International*, 26 March 1983

'IUS Investigation Board Members Named'. NASA Lyndon B. Johnson Space Center, Houston, Texas, 7 April 1983

'Astronaut Recruitment.' NASA Lyndon B. Johnson Space Center, Houston, Texas, 16 May 1983

'Flight Data File Crew Activity Plan STS-12.' Mission Operations Directorate, NASA Lyndon B. Johnson Space Center, Houston, Texas, 23 May 1983

'Second Relay Satellite Deleted from Shuttle Flight 8 Manifest.' NASA Lyndon B. Johnson Space Center, Houston, Texas, 27 May 1983

'Payload Specialist Named.' NASA Lyndon B. Johnson Space Center, Houston, Texas, 29 June 1983

'Insat-1C has launch slot.' *Flight International*, 9 July 1983

'Gap in Shuttle schedule.' *Flight International*, 16 July 1983

'MDC has an astronaut.' *Flight International*, 16 July 1983

'TDRS-A completes a high-jump.' *Flight International*, 16 July 1983

'Satcom-K is signed up.' *Flight International*, 23 July 1983

'All change at Salyut 7.' *Flight International*, 27 August 1983

'Costly Insat.' *Flight International*, 10 September 1983

'Insat is power-hungry.' *Flight International*, 17 September 1983

'Nocturnal Shuttle did well.' *Flight International*, 17 September 1983

'Shuttle Crew Announcements.' NASA Lyndon B. Johnson Space Center, Houston, Texas, 21 September 1983

'Shuttle boosters delay Spacelab-1.' *Flight International*, 22 October 1983

'NASA changes SRB nozzles.' *Flight International*, 29 October 1983

'Progress 18 buys Salyut time.' *Flight International*, 29 October 1983

'Space Transportation System Space Shuttle Payload Flight Assignments: November 1983 Baseline.' NASA Headquarters, Washington, DC, November 1983

'Flight Data File Crew Activity Plan STS-16.' Mission Operations Directorate, NASA Lyndon B. Johnson Space Center, Houston, Texas, 4 November 1983

'When to fly Spacelab-1?' *Flight International*, 5 November 1983

'STS Flight Assignments.' NASA Lyndon B. Johnson Space Center, Houston, Texas, 17 November 1983

'Spacelab-1 tries again.' *Flight International*, 26 November 1983

'Spacewalks add power to Salyut 7.' *Flight International*, 26 November 1983

'Fireproof Columbia.' *Flight International*, 31 December 1983

'Space Transportation System Space Shuttle Payload Flight Assignments: January 1984 Baseline.' NASA Headquarters, Washington, DC, January 1984

'Skynet 4: passport to a UK astronaut?' *Flight International*, 7 January 1984

'EVAs to dominate STS-11 mission.' *Flight International*, 28 January 1984

'51D, 61D Crew Announcements.' NASA Lyndon B. Johnson Space Center, Houston, Texas, 2 February 1984

'Palapa fares better than Westar.' *Flight International*, 11 February 1984

'51K Crew Announcement.' NASA Lyndon B. Johnson Space Center, Houston, Texas, 14 February 1984

'Why Westar and Palapa failed.' *Flight International*, 18 February 1984

'Canadian to fly Shuttle soon.' *Flight International*, 25 February 1984

'STS-41B National Space Transportation System Mission Report.' NASA Lyndon B. Johnson Space Center, Houston, Texas, March 1984

'Shuttle hopefuls named.' *Flight International*, 24 March 1984

'Repairs in space.' *Flight International*, 7 April 1984

'The launch and forget satellite.' *Flight International*, 7 April 1984

'India joins the cosmonaut club.' *Flight International*, 14 April 1984

'Born-again Solar Max brings Shuttle cheer.' *Flight International*, 21 April 1984

'Flight Data File Crew Activity Plan STS-41F.' Mission Operations Directorate, NASA Lyndon B. Johnson Space Center, Houston, Texas, 23 April 1984

'Space Transportation System Space Shuttle Payload Flight Assignments: May 1984 Baseline.' NASA Headquarters, Washington, DC, May 1984

'STS-41C National Space Transportation System Mission Report.' NASA Lyndon B. Johnson Space Center, Houston, Texas, May 1984

'Astronaut T.J. Hart to Leave NASA.' NASA Lyndon B. Johnson Space Center, Houston, Texas, 10 May 1984

'Florida countdown.' *Flight International*, 19 May 1984

'NASA Selects 17 Astronaut Candidates.' NASA Lyndon B. Johnson Space Center, Houston, Texas, 23 May 1984

'NASA Announces Crew Members for Future Space Shuttle Flights.' NASA Lyndon B. Johnson Space Center, Houston, Texas, 7 June 1984

'Discovery is poised to fly.' *Flight International*, 16 June 1984

'Mattingly to Leave Astronaut Corps for Navy Post.' NASA Lyndon B. Johnson Space Center, Houston, Texas, 16 July 1984

'Space Transportation System Space Shuttle Payload Flight Assignments: August 1984 Baseline.' NASA Headquarters, Washington, DC, August 1984

'NASA Announces Updated Flight Crew Assignments.' NASA Lyndon B. Johnson Space Center, Houston, Texas, 3 August 1984

'STS-41D National Space Transportation System Mission Report.' NASA Lyndon B. Johnson Space Center, Houston, Texas, October 1984

'NASA Announces Flight Assignments and Changes.' NASA Lyndon B. Johnson Space Center, Houston, Texas, 22 October 1984

'STS-41G National Space Transportation System Mission Report.' NASA Lyndon B. Johnson Space Center, Houston, Texas, November 1984

'STS-51A National Space Transportation System Mission Report.' NASA Lyndon B. Johnson Space Center, Houston, Texas, December 1984

'Shuttle test for tiles and upper stage.' *Flight International*, 19 January 1985

'NASA Names Crew to Deploy Satellites in Year-End Flights.' NASA Lyndon B. Johnson Space Center, Houston, Texas, 29 January 1985

'Toys Given Role in Space for Education.' NASA Lyndon B. Johnson Space Center, Houston, Texas, 11 February 1985

'Crews for First Vandenberg Mission, DoD Flight Named.' NASA Lyndon B. Johnson Space Center, Houston, Texas, 15 February 1985

'Shuttle shortcut for Senator.' *Flight International*, 16 February 1985

'Skynet 4 EHF delivered.' *Flight International*, 16 February 1985

'Challenger delayed again.' *Flight International*, 23 February 1985

'Palapa could refly in November.' *Flight International*, 23 February 1985

'STS-51C National Space Transportation System Mission Report.' NASA Lyndon B. Johnson Space Center, Houston, Texas, March 1985

'Boost for PAM-D2.' *Flight International*, 2 March 1985

'NASA aims for Shuttle record.' *Flight International*, 2 March 1985

'NASA scraps Shuttle schedule.' *Flight International*, 9 March 1985

'Shuttle chaos again.' *Flight International*, 16 March 1985

'NASA shuffles astronauts.' *Flight International*, 6 April 1985

'NASA Changes 51B Landing Site to Edwards Air Force Base.' NASA Lyndon B. Johnson Space Center, Houston, Texas, 24 April 1985

'Shuttle leaves Leasat adrift.' *Flight International*, 27 April 1985

'STS-51D National Space Transportation System Mission Report.' NASA Lyndon B. Johnson Space Center, Houston, Texas, May 1985

'Saudi Arabian Payload Specialist.' NASA Lyndon B. Johnson Space Center, Houston, Texas, 2 May 1985

'Is Leasat 3 lost for good?' *Flight International*, 4 May 1985

'RAF wins space race.' *Flight International*, 4 May 1985

'Spacelab-3 problems resolved?' *Flight International*, 11 May 1985

'Spacelab-3 success claimed.' *Flight International*, 18 May 1985

'NASA Names Astronaut Crews for Ulysses, Galileo Missions.' NASA Lyndon B. Johnson Space Center, Houston, Texas, 31 May 1985

'STS-51B National Space Transportation System Mission Report.' NASA Lyndon B. Johnson Space Center, Houston, Texas, June 1985

'NASA Selects 13 Astronaut Candidates.' NASA Lyndon B. Johnson Space Center, Houston, Texas, 4 June 1985

'Star Wars experiment on next Shuttle.' *Flight International*, 8 June 1985

'NASA Names Astronaut Crew for Space Shuttle Mission 61I.' NASA Lyndon B. Johnson Space Center, Houston, Texas, 17 June 1985

'AT&T stops at three.' *Flight International*, 29 June 1985

'Salyut revisited.' *Flight International*, 29 June 1985

'STS-51G National Space Transportation System Mission Report.' NASA Lyndon B. Johnson Space Center, Houston, Texas, July 1985

'UK astronaut experiments approved.' *Flight International*, 6 July 1985

'Shuttle mission a success.' *Flight International*, 6 July 1985

'RCA PS named.' *Flight International*, 20 July 1985

'New Soviet craft docks with Salyut.' *New York Times*, 23 July 1985

'P&W improves RL-10.' *Flight International*, 27 July 1985

'Shuttle teacher named.' *Flight International*, 3 August 1985

'Spacelab-2 hits trouble.' *Flight International*, 10 August 1985

'Salyut was frozen.' *Flight International*, 17 August 1985

'Satcom Ku ready.' *Flight International*, 17 August 1985

'Spacelab-2 makes good.' *Flight International*, 17 August 1985

'Shuttle set for Leasat rescue.' *Flight International*, 24 August 1985

'Soyuz T-13 achieves successes.' *Flight International*, 31 August 1985

'STS-51F National Space Transportation System Mission Report.' NASA Lyndon B. Johnson Space Center, Houston, Texas, September 1985

'NASA Names Crews for Upcoming Space Shuttle Flights.' NASA Lyndon B. Johnson Space Center, Houston, Texas, 19 September 1985

'GD prepares Centaur.' *Flight International*, 21 September 1985

'Indonesia chooses Shuttle candidates.' *Flight International*, 21 September 1985

'Leasat 3 looks good.' *Flight International*, 21 September 1985

'Pointing system performance praised.' *Flight International*, 21 September 1985

'Insurance losses top $752 million.' *Flight International*, 28 September 1985

'New crew for Salyut.' *Flight International*, 28 September 1985

'Skynet 4 progresses.' *Flight International*, 28 September 1985

'STS-51I National Space Transportation System Mission Report.' NASA Lyndon B. Johnson Space Center, Houston, Texas, October 1985

'More politicians to fly Shuttle.' *Flight International*, 5 October 1985

'Phase II SSME ready.' *Flight International*, 5 October 1985

'Venus mapper is Magellan.' *Flight International*, 5 October 1985

'Atlantis takes flight.' *Flight International*, 12 October 1985

'Spacelab-2 reflight studied.' *Flight International*, 26 October 1985

'Spacelab-D1 ready to go.' *Flight International*, 26 October 1985

'Space Transportation System Space Shuttle Payload Flight Assignments: November 1985 Baseline.' NASA Headquarters, Washington, DC, November 1985

'Australia plans free-fliers.' *Flight International*, 9 November 1985

'Journalist to fly on Shuttle.' *Flight International*, 9 November 1985

'So far so good for Leasat.' *Flight International*, 16 November 1985

'Teal Ruby revealed.' *Flight International*, 16 November 1985

'Flight Data File Crew Activity Plan STS-61E.' Mission Operations Directorate, NASA Lyndon B. Johnson Space Center, Houston, Texas, 22 November 1985

'Spacelab-D1 controlled from Europe.' *Flight International*, 23 November 1985

'Space Station rehearsal planned for Shuttle mission.' *Flight International*, 23 November 1985

'Space-made drug named by MDC.' *Flight International*, 30 November 1985

'STS-51J National Space Transportation System Mission Report.' NASA Lyndon
     B. Johnson Space Center, Houston, Texas, December 1985
'Shuttle begins 23rd mission.' *Flight International*, 7 December 1985
'Vandenberg launch schedule date slips.' *Flight International*, 21-28 December 1985
'Flight Data File Crew Activity Plan STS-61F.' Mission Operations Directorate,
     NASA Lyndon B. Johnson Space Center, Houston, Texas, 14 January 1986
'Major changes in 1986 plan.' *Flight International*, 25 January 1986
'Teacher flies next Shuttle.' *Flight International*, 25 January 1986
'Booster burnthrough caused Shuttle explosion.' *Flight International*, 8 February 1986
'Booster failure: the evidence grows.' *Flight International*, 15 February 1986
'STS-61A National Space Transportation System Mission Report.' NASA Lyndon
     B. Johnson Space Center, Houston, Texas, March 1986
'NASA halts Shuttle sales drive.' *Flight International*, 22 March 1986
'Shuttle off until spring '87.' *Flight International*, 22 March 1986
'Military favoured in new Shuttle schedule.' *Flight International*, 29 March 1986
'Soviets develop new manned spacecraft.' *Flight International*, 29 March 1986
'Astronaut Bridges Reassigned.' NASA Lyndon B. Johnson Space Center, Houston,
     Texas, 11 April 1986
'Conservative rules for first Shuttle.' *Flight International*, 12 April 1986
'More business for Ariane.' *Flight International*, 26 April 1986
'STS-61B National Space Transportation System Mission Report.' NASA Lyndon
     B. Johnson Space Center, Houston, Texas, May 1986
'NASA hands Shuttle data to commission.' *Flight International*, 3 May 1986
'Cosmonauts move to Salyut.' *Flight International*, 17 May 1986
'Fletcher takes over at NASA.' *Flight International*, 24 May 1986
'Regan resists new orbiter.' *Flight International*, 7 June 1986
'Shuttle to lose Centaur-G Prime?' *Flight International*, 7 June 1986
'Salyut missions continue.' *Flight International*, 14 June 1986
'Shuttle accident inevitable, report suggests.' *Flight International*, 21 June 1986
'No Shuttle before 1988.' *Flight International*, 19 July 1986
'NASA follows Shuttle report recommendations.' *Flight International*, 26 July 1986
'Challenger crew unaware, tape confirms.' *Flight International*, 9 August 1986
'Shuttle restrictions urged.' *Flight International*, 9 August 1986
'Science threatened by orbiter funding.' *Flight International*, 6 September 1986
'Ariane and Shuttle plans revealed.' *Flight International*, 18 October 1986
'NASA schedules Shuttle flights.' *Flight International*, 18 October 1986
'Shuttle booster redesign approved.' *Flight International*, 25 October 1986
'Astronaut given key Shuttle role.' *Flight International*, 22 November 1986
'Rocketdyne improves Shuttle engines.' *Flight International*, 22 November 1986
'Flight Crew Announced for Next Shuttle Mission.' NASA Lyndon B. Johnson
     Space Center, Houston, Texas, 9 January 1987
'Chief Astronaut John Young Appointed to New Post.' NASA Lyndon B. Johnson
     Space Center, Houston, Texas, 15 April 1987
'JSC Schedules Long-Duration Simulation.' NASA Lyndon B. Johnson Space
     Center, Houston, Texas, 21 April 1987

'NASA Selects 15 New Astronaut Candidates'. NASA Lyndon B. Johnson Space Center, Houston, Texas, 5 June 1987

'STS-61C National Space Transportation System Mission Report.' NASA Lyndon B. Johnson Space Center, Houston, Texas, August 1987

'Soviets switch Mir crew.' *Flight International*, 1 August 1987

'Soviets launch large platform.' *Flight International*, 8 August 1987

'STS-27 Crew Named.' NASA Lyndon B. Johnson Space Center, Houston, Texas, 15 September 1987

'Mir crew period halved.' *Flight International*, 12 December 1987

'Marathon space mission ends.' *Flight International*, 2-9 January 1988

'Soviets open 1988 space account.' *Flight International*, 23 January 1988

'NASA Announces Shuttle Crew.' NASA Lyndon B. Johnson Space Center, Houston, Texas, 4 February 1988

'Payload Flight Assignments: NASA Mixed Fleet.' NASA Headquarters, Washington, DC, March 1988

'NASA Announces Three New Shuttle Crews.' NASA Lyndon B. Johnson Space Center, Houston, Texas, 17 March 1988

'Telescoping Pole Chosen for Crew Escape System.' NASA Lyndon B. Johnson Space Center, Houston, Texas, 7 April 1988

'Cosmonauts splash down in dry lake.' *Flight International*, 2 July 1988

'US Air Force grounds astronauts.' *Flight International*, 16 July 1988

'Permanently Manned Space Station to Carry the Name of Freedom.' NASA Lyndon B. Johnson Space Center, Houston, Texas, 18 July 1988

'Afghan visits Mir.' *Flight International*, 10 September 1988

'Soviets plan shorter Mir mission.' *Flight International*, 15 October 1988

'Soviets set space endurance records.' *Flight International*, 19 November 1988

'Four New Shuttle Crews Named.' NASA Lyndon B. Johnson Space Center, Houston, Texas, 30 November 1988

'STS-27R OV-104 Orbiter TPS Damage Review Team Summary Report.' NASA Marshall Space Flight Center, Huntsville, Alabama, February 1989

'Magellan: The Unveiling of Venus.' NASA Jet Propulsion Laboratory, California Institute of Technology, Pasadena, California, March 1989

'Atlantis set for April 28 launch.' *Flight International*, 15 April 1989

'Astronaut S. David Griggs Killed in Air Crash.' NASA Lyndon B. Johnson Space Center, Houston, Texas, 17 June 1989

'Partial Shuttle Crew Assignments Announced.' NASA Lyndon B. Johnson Space Center, Houston, Texas, 29 June 1989

'Telescope is set to peer at space and time.' *New York Times*, 9 April 1990

'Losing bid offered 2 tests on Hubble.' *New York Times*, 28 July 1990

'The Hubble Space Telescope Optical Systems Failure Report.' NASA-Caltech Jet Propulsion Laboratory, Pasadena, California, November 1990

'The down-to-earth Rakesh Sharma.' *The Hindu*, 4 April 2010

'What a hoot: A chat with Astronaut Hoot Gibson.' Uncredited interview, www.gibson.com, 28 July 2005

'Astronaut's Brother Recalls a Man Who Dreamed Big.' Uncredited article, www.npr.org, 28 January 2011

Broache, Anne, 'Footloose.' *Smithsonian Magazine*, August 2005

Burrough, Bryan (1998), *Dragonfly: NASA and the Crisis Aboard Mir*. London: Fourth Estate

Cassutt, Michael, 'The Manned Spaceflight Engineer Program.' *Spaceflight*, January 1989

Cassutt, Michael, 'Max Q Live: In space no one can hear you sing.' *Air & Space Magazine*, 1 March 2009

Cassutt, Michael, 'Secret Space Shuttles.' *Air & Space Magazine*, 1 August 2009

Clark, Phillip (1988), *The Soviet Manned Space Programme*. London: Salamander

Cooper, Henry S.F., Jr. (1987) *Before Lift-Off: The Making of a Space Shuttle Crew*. Baltimore, Maryland: The Johns Hopkins University Press

Dawson, Virginia P. and Bowles, Mark D. (2004), *Taming Liquid Hydrogen: The Centaur Upper Stage Rocket, 1958-2002*. Office of External Relations, NASA Headquarters, Washington, DC

Day, Dwayne A., 'A Lighter Shade of Black: The (Non) Mystery of STS-51J.' *The Space Review*, 4 January 2010

Dunar, Andrew J. and Waring, Stephen P. (1999), *Power to Explore: The History of Marshall Space Flight Center, 1960-1990*. NASA George C. Marshall Space Flight Center, Huntsville, Alabama

Ervin, Elaine, 'Bob Stewart: Soldier, Astronaut and Compelling Apologist.' Reasons to Believe, www.reasons.org, October 2000

Evans, Ben (2005), *Space Shuttle Columbia: Her Missions and Crews*. Chichester: Praxis

Evans, Ben (2006), *Space Shuttle Challenger: Ten Journeys into the Unknown*. Chichester: Praxis

Evans, Ben (2009), *Escaping the Bonds of Earth*. Chichester: Praxis

Evans, Ben (2010), *Foothold in the Heavens*. Chichester: Praxis

Evans, Ben (2011), *At Home in Space*. Chichester: Praxis

Froelich, Walter (1984), *Spacelab: An International Short-Stay Orbiting Laboratory*. NASA Headquarters, Washington, DC

Germani, Clara, 'Space heroine as hardline hopeful.' *Baltimore Sun*, 14 November 1995

Hall, Rex D. and Shayler, David J. (2003), *Soyuz: A Universal Spacecraft*. Chichester: Praxis

Heppenheimer, T.A. (1999), *The Space Shuttle Decision*. NASA Office of Policy and Plans, NASA Headquarters, Washington, DC

Jenkins, Dennis R. (2000), *Hypersonics Before the Shuttle*. NASA Office of Policy and Plans, NASA Headquarters, Washington, DC

Jenkins, Dennis R. (2001), *Space Shuttle: The History of the National Space Transportation System The First 100 Missions*. Hinckley: Midland Publishing

Jenkins, Dennis R., Lander, Tony, and Miller, Jay (2003), *American X-Vehicles*. NASA Office of External Relations, NASA Headquarters, Washington, DC

Jones, Tom (2006), *Sky Walking: An Astronaut's Memoir*. New York: HarperCollins

King, Guy L. and Tsai, Ted, 'The Design and Analysis of a Double Swivel Toggle Release Mechanism for the Orbiter Stabilised Payload Deployment System.' Mechanical Design and Analysis Branch, Structures and Mechanics Division, NASA Lyndon B. Johnson Space Center, Houston, Texas, 1989

Lamar, Jacob V., Hannifin, Jerry and Leavenworth, Geoffrey, 'Jake Skywalker: A Senator boards the Shuttle.' *Time*, 22 April 1985

Lawton, John and Moody, Patricia, 'A Prince in Space.' *Saudi Aramco World*, January/February 1986

Leavitt, Melvin, 'Mission Specialist One.' *ERA*, April 1985

Lenda, J.A. (1978), *Manned Manoeuvring Unit*. Martin Marietta Corporation, Denver, Colorado

Linenger, Jerry M. (2000), *Off the Planet: Surviving Five Perilous Months Aboard the Space Station Mir*. New York: McGraw-Hill

Lord, Douglas R. (1987), *Spacelab: An International Success Story*. Scientific and Technical Information Division, NASA Headquarters, Washington, DC

Meltzer, Michael (2007), *Mission to Jupiter: A History of the Galileo Project*. NASA History Office, Washington, DC

Mihelich, Peggy, 'Legendary astronaut still finds herself star-struck.' CNN, 26 January 2007

Millbrooke, Anne, 'More Favored than the Birds: The Manned Manoeuvring Unit in Space.' In *From Engineering Science to Big Science* (Pamela E. Mack, ed.), NASA Office of Policy and Plans, NASA History Office, Washington DC, 1998

McMahan, Tracy and Neal, Valerie (1984), *Repairing Solar Max: The Solar Maximum Repair Mission*. NASA Goddard Space Flight Center, Greenbelt, Maryland

Molczan, T.J. (1990), 'Observations of STS-36 and its payload.' Accessed from www.fas.org, 13 November 2011

Mullane, Mike (2006), *Riding Rockets: The Outrageous Tales of a Space Shuttle Astronaut*. New York: Scribner

Neal, Valerie (1986), *Renewing Solar Science: The Solar Maximum Repair Mission*. NASA Goddard Space Flight Center, Greenbelt, Maryland

Pike, Christopher Anson, 'Canyon, Rhyolite and Aquacade.' *Spaceflight*, Vol. 37, November 1995

Portree, David S.F. and Treviño, Robert C. (1997), *Walking to Olympus: An EVA Chronology*. NASA History Office, NASA Headquarters, Washington, DC

Reichhardt, Tony (ed.), *Space Shuttle: The First 20 Years*. London: DK Publishing, Inc. 2002

Shayler, David J. and Burgess, Colin (2007), *NASA's Scientist-Astronauts*. Chichester: Praxis

'Soviet Space Programs: 1981-87: Piloted Space Activities, Launch Vehicles, Launch Sites and Tracking Support'. Prepared at the Request of Hon. Ernest F. Hollings, Chairman, Committee on Commerce, Science and Transportation, United States Senate, February 1988

White, Robin, 'The Man Who's Flown Everything.' *Air & Space Magazine*, May 2009

Young, Carolynn (ed.), *The Magellan Venus Explorer's Guide*. NASA Jet Propulsion Laboratory, California Institute of Technology, Pasadena, California, August 1990

Index